T0204019

Carbons for Electrochemical Energy Storage and Conversion Systems

Advanced Materials and Technologies Series

Series Editor

Yury Gogotsi

Drexel University
Philadelphia, Pennsylvania, U.S.A.

Carbons for Electrochemical Energy Storage and Conversion Systems

Edited by
François Béguin
Elzbieta Frackowiak

CRC Press
Taylor & Francis Group
Boca Raton London New York

CRC Press is an imprint of the
Taylor & Francis Group, an **informa** business

CRC Press
Taylor & Francis Group
6000 Broken Sound Parkway NW, Suite 300
Boca Raton, FL 33487-2742

First issued in paperback 2019

© 2010 by Taylor and Francis Group, LLC
CRC Press is an imprint of Taylor & Francis Group, an Informa business

No claim to original U.S. Government works

ISBN-13: 978-1-4200-5307-4 (hbk)
ISBN-13: 978-0-367-38486-9 (pbk)

This book contains information obtained from authentic and highly regarded sources. Reasonable efforts have been made to publish reliable data and information, but the author and publisher cannot assume responsibility for the validity of all materials or the consequences of their use. The authors and publishers have attempted to trace the copyright holders of all material reproduced in this publication and apologize to copyright holders if permission to publish in this form has not been obtained. If any copyright material has not been acknowledged please write and let us know so we may rectify in any future reprint.

Except as permitted under U.S. Copyright Law, no part of this book may be reprinted, reproduced, transmitted, or utilized in any form by any electronic, mechanical, or other means, now known or hereafter invented, including photocopying, microfilming, and recording, or in any information storage or retrieval system, without written permission from the publishers.

For permission to photocopy or use material electronically from this work, please access www.copyright.com (http://www.copyright.com/) or contact the Copyright Clearance Center, Inc. (CCC), 222 Rosewood Drive, Danvers, MA 01923, 978-750-8400. CCC is a not-for-profit organization that provides licenses and registration for a variety of users. For organizations that have been granted a photocopy license by the CCC, a separate system of payment has been arranged.

Trademark Notice: Product or corporate names may be trademarks or registered trademarks, and are used only for identification and explanation without intent to infringe.

Library of Congress Cataloging-in-Publication Data

Carbons for electrochemical energy storage and conversion systems / editors, François Béguin and Elzbieta Frackowiak.
       p. cm. -- (Advanced materials and technologies series)
    Includes bibliographical references and index.
    ISBN 978-1-4200-5307-4 (hard back : alk. paper)
     1. Electric batteries--Materials. 2. Power electronics--Materials. 3. Energy storage--Materials. 4. Carbon compounds--Electric properties. I. Béguin, François. II. Frackowiak, Elzbieta. III. Title. IV. Series.

TK2901.C37 2010
621.31'242--dc22   2009019805

Visit the Taylor & Francis Web site at
http://www.taylorandfrancis.com

and the CRC Press Web site at
http://www.crcpress.com

Contents

Preface

Since the beginning of this millennium, our world has changed dramatically. Many of the most dire predictions about fuel supply and environmental impact forecasted a generation ago are beginning to come true much more quickly than we thought possible. While many of us grew up with the misguided belief that oil and coal are an endless feedstock, we now know that if the supply is not dwindling, it will soon become more expensive to exploit. And while we debated over the validity of greenhouse concerns, the problem has now become far too large to ignore.

Climatic perturbations such as hurricanes, tsunamis, melting icebergs, and incredibly hot summers have become common in most regions of the world. At the same time, the price of oil has started to increase due to industrialization in emergent countries. These two developments have led politicians to realize the importance of limiting oil consumption. In most developed countries (e.g., Europe, Japan, the United States, and Canada), several programs have been launched for better energy consumption such as limiting the energy spoiling, improving the energy efficiency of systems (e.g., domestic heating/cooling, powering of transportation systems, and industrial uses) and changing citizens' behavior with regard to energy consumption and environment protection.

There is certainly a general agreement that electricity and hydrogen represent cleaner energy carriers than oil or coal, provided that the latter are not used for their production. Nuclear and renewable energies should progressively assume a larger role for electricity production. However, because of the uncontrolled nature of their production, renewable energies are far from ideal and they require storage systems to regulate their production and consumption. Similarly, although seemingly attractive, powering an electric vehicle is not an easy task, as many criteria such as cost, volume, and weight of systems have to be taken into account. Powering an electric vehicle requires onboard delivery either from a rechargeable battery or from a fuel cell fed by a hydrogen reservoir. High-performance energy storage and conversion systems must be developed to meet the growing industrial and societal demands. In this domain, materials and particularly carbons play a major role. Research and industrial efforts to develop high-performance carbons and carbon-based devices have assumed much importance in the last few years.

Carbon is a very interesting and unique element in the periodic table. It exists under few allotropic forms, but what is most interesting is that in its sp^2 hybridization state it presents an unlimited number of nanotextural arrangements giving rise to extremely varying physical properties (mechanical properties, electrical conductivity, porosity, etc.). Carbon as a material is constituted not only of carbon, but also of heteroatoms such as oxygen, nitrogen, and hydrogen, which are generally present as surface functional groups at the edges of graphene layers. The properties of carbons, in particular the electrochemical ones, are influenced by the surface functionality. Carbons can be found in various forms or morphologies (e.g., powders, fibers, spherical particles, monoliths, and nanotubes), which allow them to be adapted for a large number of industrial applications. This high versatility of carbon is not matched by any other known material. Last but not least, most carbon materials can be produced from available precursors through simple processes. In summary, carbon is a very suitable material for electrochemical applications because of its low cost, high electrical conductivity, versatile nanotexture, and surface functionality.

Although carbons are used widely in energy storage and conversion systems, there are no updated books that describe their physical, chemical, and electrochemical properties. The same is the case with scientific conferences, which are generally highly specialized. Given this situation, we came up with the idea to organize a conference on "Carbon for Energy Storage and Environment Protection" (CESEP) in October 2005 at Orléans, France. This proved to be a success and is now a regular feature, organized every 2 years. Therefore, when we were contacted by CRC Press at the end of 2006, we enthusiastically accepted the proposal to write a book devoted to "carbons for electrochemical energy storage and conversion systems."

As we wanted to write a book for researchers, engineers, and students, we decided to collaborate with prominent scientists in the field of carbon science and its energy-related applications, to get the largest coverage of the topics. This book consists of 12 chapters: one is devoted to electrochemistry, five address the general properties of carbons, and the remaining six discuss their applications for energy storage/conversion. Each chapter aims at giving the most detailed information using familiar terms.

As most scientists involved in these applications are materials scientists rather than electro-chemists, Chapter 1 introduces the basic principles in electrochemistry and the methods. Chapter 2 describes the different forms of traditional sp^2 carbons and Chapter 3 introduces novel techniques and processes for preparing advanced carbons.

Chapters 4 through 6 describe the main physicochemical properties (porous texture, surface functionality, and electronic structures) that control the electrochemical behavior of carbons.

Chapters 7 through 12 discuss three main systems—lithium-ion batteries, supercapacitors, and fuel cells—from the point of view of both research and industrial applications. They represent, undoubtedly, the most original source of information in this book.

We would like to thank all those who have been helpful in the realization of this book and all colleagues who kindly devoted their time to contribute chapters; colleagues, students, and postdocs in Orléans and in Poznan who helped us; and the staff at Taylor & Francis for giving us this oppor-tunity and also for being patient. We dedicate this book to our parents.

François Béguin and Elzbieta Frackowiak

Editors

François Béguin is a professor of materials chemistry in Orléans University, France. His research activities are devoted to chemical and electrochemical applications of carbon materials, with a special focus on the development of nanocarbons with controlled porosity and surface functionality for applications to energy conversion/storage and environment protection. He is the leader of the "Energy–Environment" group at Orléans University. The main topics researched by this group are lithium batteries, supercapacitors, electrochemical hydrogen storage, and reversible electrosorption of pollutants. Béguin owns several patents related to the synthesis of nanostructured carbon materials (nanotubes, carbons from seaweeds, etc.) and their use for electrochemical systems. He has published over 230 publications in high-ranking international journals and his works are cited in more than 3000 papers. He is also involved in the writing of several books dealing with carbon materials and energy storage. He is a member of the International Advisory Board of the Carbon Conferences, the chairman of the Advisory Board for the International Symposium on Intercalation Compounds, and has launched the first international conference on Carbon for Energy Storage and Environment Protection. He is a member of the editorial board of the journal *Carbon*. He is also the director of two national programs in the French Agency for Research (ANR): one on energy storage (Stock-E) and the other on hydrogen and fuel cells (H-PAC).

Elzbieta Frackowiak is a professor at the Poznan University of Technology, Poland. She is an electrochemist, and her research interest is focused on energy storage/conversion. Her main scientific interests are electrode materials for primary and secondary cells, production of carbon nanotubes by catalytic and template methods, investigation of intercalation/insertion processes in lithium-ion batteries, and investigation of the processes in fuel cells such as hydrogen storage in carbon materials and oxidation of methanol in acidic medium using nanotubular supports for catalyst particles. Her research interests are especially devoted to the application of different activated and template carbon materials for hydrogen storage and supercapacitors, use of composite electrodes from conducting polymers, doped carbons and transition metal oxides for supercapacitors, and application of ionic liquids as new "green" electrolytes.

Frackowiak has more than 150 publications and several chapters and patents to her name. The number of her citations exceeds 2500. She is a member of the advisory board of the International Conferences on Intercalation Compounds and was the organizer of ISIC12 in Poznan in 2003. She is a member of the advisory board of the journal *Energy and Environmental Science* as well as the coordinator of many international projects. She is Chair Elect of Division 3 "Electrochemical Energy Conversion and Storage" of the International Society of Electrochemistry.

Contributors

François Béguin
Centre de Recherche sur la Matière Divisée
Centre National de la Recherche Scientifique
 and Orléans University
Orléans, France

Ralph J. Brodd
Broddarp of Nevada, Inc.
Henderson, Nevada

Diego Cazorla-Amorós
Departamento de Química Inorgánica
Grupo Materiales Carbonosos y
 Medio Ambiente
Universidad de Alicante
Alicante, Spain

John Chmiola
Department of Materials Science and
 Engineering
A.J. Drexel Nanotechnology Institute
Drexel University
Philadelphia, Pennsylvania

and

Environmental Energy Technologies Division
Lawrence Berkeley National Laboratory
Berkeley, California

Morinobu Endo
Department of Electric and Electronic
 Engineering
Faculty of Engineering

and

Institute of Carbon Science and Technology
Shinshu University
Nagano, Japan

Toshiaki Enoki
Department of Chemistry
Tokyo Institute of Technology
Tokyo, Japan

Jean-François Fauvarque
Laboratoire d'Electrochimie Industrielle
Conservatoire National des Arts et Métiers
Paris, France

Elzbieta Frackowiak
Institute of Chemistry and Technical
 Electrochemistry
Poznan University of Technology
Poznan, Poland

Roland Gallay
Maxwell Technologies
Rossens, Switzerland

Dietrich Goers
TIMCAL Ltd.
Bodio, Switzerland

Yury Gogotsi
Department of Materials Science and
 Engineering
A.J. Drexel Nanotechnology Institute
Drexel University
Philadelphia, Pennsylvania

Hamid Gualous
Institut-Franche-Comté Electronique
 Mécanique Thermique et Optique-Sciences
 et Technologies
Centre National de la Recherche Scientifique
Unité Mixte de Recherche
Université de Franche-Comté
Besançon, France

Michio Inagaki (retired)
Aichi Institute of Technology
Toyota, Japan

Yong Jung Kim
Institute of Carbon Science and Technology
Shinshu University
Nagano, Japan

Takashi Kyotani
Institute of Multidisciplinary Research for
    Advanced Materials
Tohoku University
Sendai, Japan

Claude Lamy
Electrocatalysis Laboratory
Centre National de la Recherche Scientifique
Université de Poitiers
Poitiers, France

Ángel Linares-Solano
Departamento de Química Inorgánica
Grupo Materiales Carbonosos y
    Medio Ambiente
Universidad de Alicante
Alicante, Spain

Dolores Lozano-Castelló
Departamento de Química Inorgánica
Grupo Materiales Carbonosos y
    Medio Ambiente
Universidad de Alicante
Alicante, Spain

Petr Novák
Department of General Energy
Electrochemistry Laboratory
Paul Scherrer Institute
Villigen, Switzerland

Ki Chul Park
Department of Electric and Electronic
    Engineering
Faculty of Engineering
Shinshu University
Nagano, Japan

Ljubisa R. Radovic
Department of Energy and Mineral
    Engineering
The Pennsylvania State University
University Park, Pennsylvania

and

Department of Chemical Engineering
University of Concepción
Concepción, Chile

Encarnación Raymundo-Piñero
Centre de Recherche sur la Matière Divisée
Centre National de la Recherche Scientifique
    and Orléans University
Orléans, France

Patrice Simon
Inter-University Research and Engineering
    Centre on Materials
Université Paul Sabatier
Toulouse, France

Michael E. Spahr
TIMCAL Ltd.
Bodio, Switzerland

Fabián Suárez-García
Departamento de Química Inorgánica
Grupo Materiales Carbonosos y
    Medio Ambiente
Universidad de Alicante
Alicante, Spain

1 Principles of Electrochemistry and Electrochemical Methods

Jean-François Fauvarque and Patrice Simon

CONTENTS

1.1 INTRODUCTION

1.1.1 ENERGY STORAGE

Plants, the most successful energy systems on earth, have been using solar energy for millions of years to convert carbon dioxide and water into oxygen and carbohydrates, that are subsequently converted into oil and coal. While fuel is usually considered as the main source of energy, it would be useless without the presence of oxygen; accordingly, oxygen (in air) can be considered as an important energy reservoir. Solar energy is also at the origin of most of the renewable energies: hydroelectric, wind mills, solar cells, biomass, etc. Only nuclear plants can deliver large amount of energy not originating from sun.

Fossil fuels, nuclear heat, renewable energies all can be converted to electrical energy. However, this conversion is performed with large energy losses (only about 30% of the nuclear heat is converted to electrical energy). Moreover, since the electrical energy is almost impossible to store, energy production must timely fit with the demand: electrical energy storage is thus an important issue to be addressed. Electrical energy storage is also an important problem for portable devices (e.g., mobile phones) and for autonomous systems (e.g., electric vehicles). Since limited amount of electrical energy can be stored in magnetic fields (superconductive coils) or electric fields (capacitors), the development of reliable storage technologies is critical, and current storage technologies require the conversion of electrical energy into mechanical or chemical energy.

Transformation of electrical energy into mechanical energy is mainly achieved by pumping water into hydraulic dams. The energy efficiency is good (about 80%), and the storage in the dam is achieved for several months, allowing for interseasonal energy storage; however, appropriate locations must be found. Another way to convert electrical energy into mechanical one is to use fly-wheels, with the issues linked with the mobile parts: mechanical failures and fast "self-discharge."

Secondary batteries or electrolysis cells can be used to transform electrical energy into chemical energy. Water electrolysis produces hydrogen (with energy efficiencies of about 70%) which can be stored for further usage in fuel cells. Fuel cells can be viewed as components of an energy storage/conversion system provided that the reactants have been produced by some energy consuming process; the same applies for redox flow systems. However, fuel cells and redox flow systems have not yet reached a large commercial state of development and, today, the most convenient way for electrical energy storage is to use rechargeable batteries.

1.1.2 Electrochemical Energy Storage and Conversion Devices

Detailed information about batteries and supercapacitors can be found in specialized textbooks [1–5]. Some useful concepts are reviewed below.

An electrochemical cell (the elementary block from which all batteries and supercapacitors are constructed) is composed of two active masses, one positive and one negative, two current collectors (leads), and a separator placed in between. The current collectors are electronically conducting and ensure electrons transfer from and to the electrodes (active masses). The separator must be ionically conducting—thus allowing the transfer of electrical current—and electronically isolating to avoid self-discharge. Active masses are subject to electrochemical reactions; they must have both electronic and ionic conductivity; they are ionically connected to the separator and electronically connected to one current collector. Both the separator and the active masses require an ionically conducting medium, such as a liquid electrolyte (most common) or even gelified, polymeric or solid electrolyte. The entire unit is generally housed in a sealed container. From a thermodynamic point of view, these sealed devices may be considered isolated systems, only exchanging energy, either electrical or heat, with the external environment. The ideal sealed electrochemical device will not exchange mass with the external environment, in the form of neither reactant nor product. Such a system is characterized by its weight, volume, maximum energy content, and a nominal voltage. Usually, electrochemical cells do not have any mobile parts and can deliver the stored energy within milliseconds and adjust the power delivered with time constants in the range of tens of milliseconds. A second class of electrochemical devices is the so-called "open systems," including fuel cells and fluid redox systems (flow batteries). The maximum energy content depends on the kind and the amount of active materials. The power is controlled by the rate at which the reactants are delivered and the products extracted. Metal–air batteries are intermediate systems where the maximum energy content is related to the mass of active metal and the power related to the air flow rate.

Electrochemical storage systems (ESS) like batteries are generally assembled with more than one cell. Except special cases, identical cells connected in series or in parallel are used, most preferably with the same state of charge and the same state of performance reliability.

The topic of this book is focused on active masses containing carbon, either as an active mass (e.g., negative mass of lithium-ion battery or electrical double layer capacitors), as an electronically conducting additive, or as an electronically conductive support for catalysts. In some cases, carbon can also be used as a current collector (e.g., Leclanché cell). This chapter presents the basic electrochemical characterization methods, as applicable to carbon-based active materials used in energy storage and laboratory scale devices.

1.2 ELECTROCHEMICAL CELL CHARACTERISTICS

1.2.1 Thermodynamics

Each active mass contains a redox couple. For example, the negative mass (Reaction 1.1) of a Ni/MH battery contains a metallic hydride MH:

$$M + H_2O + e^- \leftrightarrow MH + OH^- \tag{1.1}$$

M is the oxidized form of redox (Reaction 1.1), MH being the reduced one.

Under discharge, the oxidation of MH is associated with the reduction of NiOOH into $Ni(OH)_2$ in the positive active mass. When the external circuit is open (no current), the active mass (Reaction 1.1) confers to the negative current collector, a potential versus the standard hydrogen electrode (SHE) according to the Nernst law:

$$E_1 = E_1^\circ + (RT/nF) \ln(Ox_1)/(Red_1) \tag{1.2}$$

where

E_1° is standard redox potential of redox couple (Reaction 1.1) versus SHE

$R = 8.314\,J/mol/K$

T is the absolute temperature in Kelvin

n is the number of exchanged electrons

$F = 96,500\,C$

Ox_1, chemical activity of the oxidized part of redox (Reaction 1.1), regularly expressed in mole per cubic meter, usually in mole per liter

Red_1, chemical activity of the reduced part of redox (Reaction 1.1), the activities of pure solids being equal to one.

In this example, M and MH are pure solid and distinct phases, electronically conductive, and their activities are the same. M and MH are in equilibrium with di-hydrogen at pressure P_1 (depending only on temperature T):

$$2MH \leftrightarrow 2M + H_2 \tag{1.3}$$

Thus, potential E_1 of the redox couple (1) is the same as that of a reversible hydrogen electrode at pressure P_1, temperature T, and pH $= 14 - \log_{10}(OH^-)$, as far as M and MH are simultaneously present in the active mass.

At $T = 298\,K$,

$$E_1 = -0.06\,pH - 0.03\,\log_{10}(P_1) \tag{1.4}$$

In this example, the positive mass (Reaction 1.2) contains oxidized nickel hydroxide:

$$NiOOH + H_2O + e^- \leftrightarrow Ni(OH)_2 + OH^- \tag{1.5}$$

Actually, NiOOH and $Ni(OH)_2$ together form a single-phase compound, since $Ni(OH)_2$ (bivalent Ni^{II}) may be considered as the result of an electrochemical intercalation of a proton between the sheets of nickel oxyhydroxide (trivalent Ni^{III}).

$$E_2 = E_2^\circ + RT/F \ \ln\left[\left(Ni^{III}\right)\middle/\left(Ni^{II}\right)\right] - RT/F \ \ln(OH^-) \tag{1.6}$$

NiOOH is electronically conductive, $Ni(OH)_2$ is an isolator; therefore, cobalt oxyhydroxide or carbon (graphite) are often added to the positive mass for improving the electronic conductivity. Both positive and negative masses contain concentrated (7 M) aqueous potassium hydroxide for improving the ionic conductivity.

The open cell voltage (OCV) U is the difference between E_2 and E_1

$$U = E_2 - E_1 = E_2^\circ + RT/F \ \ln\left[\left(Ni^{III}\right)\middle/\left(Ni^{II}\right)\right] - RT/2F \ \ln(P_1), \tag{1.7}$$

$$E_2^\circ = 1.35\,V \text{ at } 25°C$$

(the positive mass is seen to be thermodynamically unstable versus water oxidation)

U is related to the free enthalpy variation ΔG of Reaction 1.8:

$$NiOOH + MH \leftrightarrow Ni(OH)_2 + M \tag{1.8}$$

$$U = -\Delta G/nF \tag{1.9}$$

where $n = 1$ in this example.

When the cell is at the charged state, ΔG is negative; the system is not at equilibrium, but no reaction occurs if the electrons and ions are not transferred from the negative to the positive mass. Once the electron transfer is achieved by closing the external circuit, the whole reaction occurs spontaneously and the current flows. The potential differences occur only at the interfaces between the electrolyte and the electronic conducting materials. It can be seen that U does not depend on (OH^-), as long as (OH^-) is the same in the negative and positive masses (This is the case if diffusion of KOH is possible between the two masses or if the separator is selectively OH^- conducting). U slightly depends on the state of charge (through $RT/F \ln(Ni^{III})/(Ni^{II})$). However, U is not a sensitive state-of-charge indicator).

In standard electrochemical notation, the above cell would be referred to as MH/M/aqueous KOH/Ni(OH)$_2$/NiOOH. However, the current denomination is Ni/MH (nickel/metal hydride cell).

Commercial cells using the above chemistry are chemically unbalanced; the negative active mass is generally oversized and contains M/MH in excess. The capacity of the cell is thus limited by the capacity of the positive mass. When the cell is overcharged (not excessively), di-oxygen gassing is observed at the positive mass and reduced at the negative mass (in sealed cells, with a separator properly designed). During (slight) overdischarge, di-hydrogen may be produced at the positive mass and recombined back at the negative mass to form MH. However, overdischarge may be harmful, causing oxidative corrosion of M. In rest periods, di-hydrogen gas is slowly oxidized at the positive current collector, causing slow self-discharge. NiOOH is thermally unstable above 50°C and decomposes into di-oxygen, causing fast self-discharge at this temperature or even thermal runaway. Cyclability of the positive mass is usually good, better than that of the negative mass, subject to slow corrosion of M. When the capacity of the negative mass becomes lower than that of the positive mass, the cell performance degrades very fast. Unbalancing the capacities of active masses is usually achieved when assembling the batteries for cyclability or safety reasons. For example, lithium ion cell uses overcapacitive negative carbon electrode to avoid lithium metal deposition during charge.

Current collectors in Ni/MH batteries are usually made of nickel foam, passivated by a NiOOH layer at the positive side. Separators are usually microporous plastic films impregnated with liquid electrolyte. It is, therefore, necessary that the active masses be insoluble, or at least sparingly soluble, in order to avoid mixing of components via diffusion through the separator, thus leading to self-discharge.

1.2.2 KINETICS

Electrochemical transformations are related to the amount of charge Q passed through the circuit (units: Ah or As). Accordingly, electrochemical kinetics is measured by deriving the charge versus time. The rate of the chemical transformations is thus linked to the current I (Faraday's law).

1.2.2.1 Polarization

When the external leads of a battery are connected to an energy consuming device (a load), electrical current flows: electrons move from the negative electrode to the positive electrode via the leads and ions move inside the electrolyte. During this polarization, the positive mass potential V_2 is decreased below E_2 (rest potential of the positive electrode, see Equation 1.4), an electrochemical reaction occurs transforming Ox$_2$ to Red$_2$. The positive mass is negatively polarized and is called a cathode (attracting cations). Simultaneously, the negative mass reaches a potential $V_1 > E_1$ (see Equation 1.6) and an electrochemical reaction occurs transforming Red$_2$ to Ox$_2$. The negative mass is polarized positively and it acts as an anode (attracting anions). Ionic migration through the separator keeps the active masses neutral. An electric field is established inside the cell, related to the local conductivity σ and current density j, $j = \sigma \times \vec{E}$. An ohmic drop is established between the two current collectors, strongly dependent on the ionic conductivity of the electrolyte and on the electronic conductivity of the active masses and current collectors. This explains why cells are constructed as thin as possible, with a small distance between the positive and negative current collectors, to limit ohmic drops.

During the reverse process, on charge, the positive mass becomes an anode and the negative mass a cathode. It is important to note that negative masses are usually and improperly denoted as anodes and positive mass cathodes, but this is true only during discharge. On discharge, voltage V between the two electrodes is lower than U_{OCV}. The difference $U_{OCV} - V$ is called the polarization of the battery. Also note that during charge, the applied voltage will always be greater than U_{OCV}.

The power delivered by the battery is defined as the product $P = I \times V$. At the open circuit, both the current and the delivered power are zero. Discharging the battery through lower impedance increases the current. If the external impedance is zero, the battery delivers the short-circuit current I_{max}, with $V = 0$. The power delivered is again zero. Thus, for some value of the current, the delivered power is at the maximum.

1.2.2.2 Ohmic Drop

If the battery is assumed to have an internal resistance $\rho°$ independent of I and delivers a current I to an external load resistance R, the power delivered is

$$P = (U - \rho° \times I) \times I \tag{1.10}$$

and

$$I = (U - \rho° \times I)/R \tag{1.11}$$

$\rho° \times I$ being an ohmic potential drop.

The maximum power is obtained for $I = 0.5 \times U/\rho°$, with $V = 0.5 \times U$ and $R = \rho°$, and then

$$P_{max} = U^2/(4 \times \rho°) \tag{1.12}$$

Note that U and $\rho°$ depend on the state of charge of the battery and, thus, P_{max} decreases when the state of charge decreases. The change of U versus the state of charge is called the thermodynamic discharge curve; however, more interesting for users are the discharge plots obtained at a defined constant current (Figure 1.1). Any attempt to increase the current I beyond P_{max} will decrease the delivered power. Accordingly, all the batteries should be used at V between U and $U/2$. At P_{max}, half of the chemical energy is lost as heat in the generator. Economic management of an ESS requires cell voltages of $0.75U$ or higher, the corresponding current being $I < (U/4 \times \rho°)$ for an optimal use.

1.2.2.3 Electron Transfer Resistance

At very low discharge rates, the internal resistance of the generator is controlled by the electron charge transfer resistance at the active masses (see the Butler–Volmer's law). At high rates, this charge transfer resistance changes as $1/I$ and becomes negligible when I is sufficiently high, in comparison with the ohmic resistance: the polarization curve, V versus I, becomes linear. The standard operation range for an electrochemical device is in the linear portion of the polarization curve at

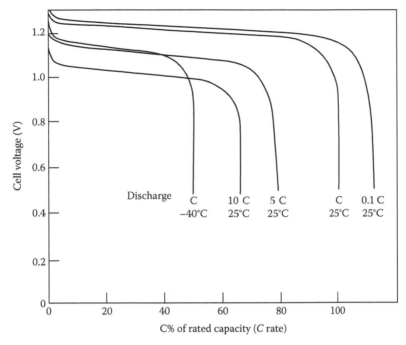

FIGURE 1.1 Typical discharge curves of a Cd/Ni cell. (From Linden, D. and Reddy, T.B. in *Handbook of Batteries*, 3rd edn., Mc Graw Hill, New York, 2002. With permission.)

$V > 0.75U$. (For a 12 V nominal battery, discharge is not allowed to proceed under 9 V.) Conversely, if the current I is kept below $0.1 \times U/\rho^\circ$ during the application, it means that the generator is oversized for the application.

1.2.2.4 Diffusion Limitations

At a very high discharge rate, the current may be limited by diffusion effects that limit the accessibility of the reactants to the electrochemical reaction sites. These diffusion effects are sometimes related to the transport numbers of ions in the electrolyte. For example, in aqueous KOH, the transport number of K^+ is 0.3 and the transport number of OH^- is 0.7. When one Faraday (96,485 C/mol) is passed from the anode to the cathode, 0.7 mol of OH^- is transferred to the anode, where 1 mol is consumed and 0.3 mol of K^+ has been taken out of the anode, that is to say, the KOH amount has been lowered at 0.3 mol. At the same time, 0.3 mol of supplementary KOH has appeared at the cathode that leads to a concentration polarization, usually low. This concentration gradient is normally counterbalanced by the opposite diffusion of KOH from the cathode to the anode. This diffusion process, as well as the associated diffusion current, are limited: there is a maximum stationary current corresponding to a zero concentration of KOH at the anode and a maximum concentration at the cathode. This is clearly visible on the polarization curve, where it corresponds to the abrupt decrease of voltage V at high current. However, it should be reminded that pulsed currents may be much higher than the diffusion limiting current, especially, if rest periods are allowed for the balancing KOH concentration in the active masses. Similarly, diffusion limitations may occur in SLI (starting, lightning, and ignition) lead acid batteries during engine starting. It can be solved by achieving rest periods of tens of seconds between two attempts of starting if the first is unsuccessful.

Batteries are given a nominal capacity C, which is the capacity delivered at constant current (C rate) during a predetermined time (1 h for instance) of discharge. Discharge at lower rate (e.g., 0.1 times the discharge current of C rate) may yield greater capacity and a discharge curve closer to the

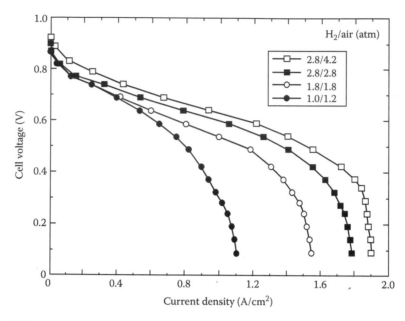

FIGURE 1.2 Polarization curves of a hydrogen–air fuel cell. (From Kolde, J.A., et al., in *Proton Conducting Membranes Fuel Cells* 1. *The Electrochemical Society Meetings*, Electrochemical Society, Pennington, NJ, Vol. 95–23, 1995, 193–201. With permission.)

thermodynamic one. At higher rates, the discharge plots exhibit a voltage decrease mainly linked with ohmic drops, as well as a capacity decrease linked mainly to an incomplete use of the limiting active mass. A marked slope of V versus state-of-charge plot may help in estimating the state of charge of the battery. However, battery users prefer a constant voltage during the discharge with a specific voltage drop at the end of the discharge, signaling the need for recharging.

At lower temperature, ionic conductivity is lowered, ohmic polarization is increased, and the battery capacity decrease is observed mainly due to diffusion limitations.

A typical polarization curve (Figure 1.2) shows three different domains, depending on the current density. At low current density, the voltage decrease (V) with the current density (j) follows a logarithm equation (see Butler–Volmer); when the current density is increased, the V versus j plot is almost linear (ohmic). At high current density, diffusion limitations occur. In the present example, the limitation is related to the diffusion of oxygen through a layer of nitrogen and water vapor.

1.2.3 Reversibility/Irreversibility

A reversible electrochemical reaction is characterized by exchange current j^0 between the electrode and the electrochemically active materials.

Redox systems in solution may behave reversibly, and polarization η of a polarizable electrode in contact with the solution usually follows the Butler–Volmer law:

$$J = J^\circ \times \left\{ \exp\left[(1-\alpha) \times \eta \times nF/RT \right] - \exp\left[-\alpha \times \eta \times nF/RT \right] \right\} \qquad (1.13)$$

The "irreversibility" can be described as the energy losses between charge and discharge (for a given active mass) that strongly depend on exchange current j^0. This leads to the definition of electron transfer resistance R_t, usually calculated for $(\eta, J) \to 0$:

$$R_t = R \times T/(nF \times J^\circ) \qquad (1.14)$$

(with $R = 8.314\,\text{J/mol/K}$, T = absolute temperature in Kelvin)

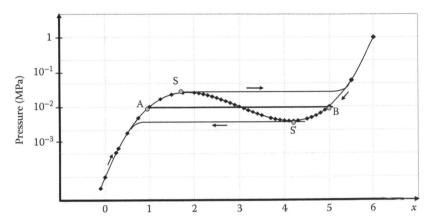

FIGURE 1.3 Conceptual spinodal transformation of $LaNi_{4.5}H_x$ ($\Delta G = 0$ for $P_{theoretical} = 10^{-2}$ MPa).

At low currents, near equilibrium, the power losses are close to $R_t \times J^2$. At high currents, the electron transfer resistance becomes $R \times T/(nF\alpha J)$, inversely proportional to the current density. It is generally negligible in comparison with the ohmic resistance.

The situation may be different for solid redox couples that often show an intrinsic irreversibility. For example, let us consider the log $P = f(x)$ plot for the $LaNi_5H_x$ system (Figure 1.3). The flat horizontal line AB is the stable thermodynamic domain where both M and MH coexist. Starting from M to form MH hydride, P increases and E decreases but the MH phase does not appear at point A, but at point S where the phase M is unstable versus any composition fluctuations (a phenomenon analogous to overfusion). Similarly, starting from MH, E increases, but M does not appear at point B, but at point S'. When discharge at low current is achieved at lower potential than recharge at low current (hysteresis), the active mass cannot reach the thermodynamic equilibrium state. The difference $Q \times \Delta E$ is lost as heat in addition to any electron transfer, ohmic or diffusion losses.

The potential of the MH electrode at 298 K changes with the hydrogen pressure according to

$$E = -0.06 \, \text{pH} - 0.03 \log_{10} p(H_2) \tag{1.15}$$

Figure 1.3 represents a tentative equation for the pressure:

$$P = 10^{-2} \times [1 + (x-1)(x-3)(x-5)] \tag{1.16}$$

that is a degree 3 polynomial equation similar to those used to describe the behavior of a condensable real gas.

Figure 1.4 shows the plots of $LaNi_{4.5}Mn_{0.5}$ hydriding and dehydriding [14]. A hysteresis between hydriding (charge) and dehydriding (discharge) is clearly apparent. To the hysteresis in pressure ΔP corresponds a hysteresis in voltage ΔE, and an irreversible conversion of electrical energy into heat. Of course, appropriate additives may reduce the ΔE gap for this spinodal transformation. The Butler–Volmer law becomes approximately valid when the gap ΔE is sufficiently low. Electrochemical kinetics (electrolyte conductivity, electron transfer, and phase transformations) are, of course, strongly dependent on the temperature.

1.2.4 CYCLABILITY

Although cyclability varies under regular use with cell chemistry and design, the most common factors leading to failure are given next.

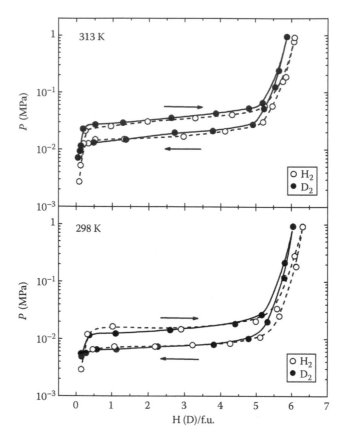

FIGURE 1.4 Plots for LaNi$_{4.5}$Mn$_{0.5}$ hydriding and dehydriding. (From Leardini, F., et al., *J. Electrochem. Soc.*, 154, A507, 2007. With permission.)

1.2.4.1 Active Masses Breathing

This is the change in the active masses volume (active mass breathing). Oxidized and reduced forms do not usually have the same volumetric density. This may cause a defect in electronic percolation and loss of active mass. The lower the difference in density, the higher the cyclability. Usually, proton and lithium intercalation compounds have low volumetric variation on cycling.

1.2.4.2 Dissolution of an Active Mass

Partial dissolution of an active mass often leads to poor cyclability by compaction and loss of the double percolation, electronic and ionic conductivity, necessary for a successful operation. This is especially true for PbSO$_4$, in diluted H$_2$SO$_4$ (discharged battery). Metallic deposition in the separator of a dissolved metal may cause short circuits, accelerated self-discharge, and may prevent subsequent charging.

1.2.4.3 Corrosion

Corrosion of the current collector or of the active mass itself can lead to a reduction in the cyclability, due to the increase in the contact resistance, formation of a passive insulating film oxide, etc. Corrosion may also involve the electrolyte and cause extra ohmic losses.

1.2.5 ROBUSTNESS

Robustness refers to the resistance versus deterioration of the performances when the active masses are run under abnormal conditions: overcharge, overdischarge, extreme temperature, high charge/discharge rate, aging, etc. The most robust commercial batteries are Ni/Cd batteries. Lead acid batteries can sustain overcharge but cannot be kept for a longer time at the discharged state. Lithium-ion batteries are very sensitive to overcharge.

1.3 EXPERIMENTAL TESTING OF MATERIALS AS ACTIVE MASSES

The following sections are devoted to the laboratory methods traditionally used for determining the characteristics (1) of individual active electrodes and (2) of laboratory test cells. Detailed information is found in usual electrochemistry textbooks [6–13].

1.3.1 EXPERIMENTAL

Active materials testing is usually conducted in a three-electrode cell configuration with the working electrode (materials to be tested), a reference electrode supposed to be operational in the electrolyte used, and a counter electrode designed not to interfere with the working electrode. In some cases, successful testing may also be achieved using a two-compartment cell. The three-electrode configuration ensures that the observed behavior is characteristic of the chosen half-cell reaction. While testing of prototype cells is critical to the development of technology, the use of a three-electrode configuration is needed to understand the basic chemistry.

1.3.1.1 Working Electrode

To get relevant information about active materials, the working electrode is made as similar as possible to the electrode of an operational device. However, current collectors are usually made with corrosion resistant materials, with good electronic conductivity, and no concern is taken about its relative mass. Materials such as gold, platinum, and vitreous carbon are commonly used. The active mass is usually tested in small amounts, mixed with electronically conducting materials, such as acetylene black, and a binder, such as polyvinylidene fluoride PVDF or polytetrafluoroethylene. The working electrode may be flat, with a $1\,cm^2$ surface, for example, a rotating disk electrode (RDE), or a microcavity electrode, or any geometrical convenient electrode.

1.3.1.2 Reference Electrode

Selection of a suitable reference electrode is critical to the success of any experimental program. The appropriate reference electrode is dependent on the medium used.

Hg/HgO is often the electrode of choice in alkaline aqueous medium, silver/silver acetate in many nonaqueous medium such as acetic acid, etc. When the experiment requires large currents to flow between the working and the counterelectrodes, a particular attention must be paid to place the reference electrode at an equipotential line close to the working electrode or to make an appropriate ohmic drop correction.

1.3.1.3 Counter Electrode

Care is not always sufficiently brought to this electrode. An insoluble redox system, much more capacitive than the working electrode is often a good choice, or possibly a large double layer capacitor electrode, which does not pollute the electrolyte.

1.3.1.4 Electrolyte Conductivity

A high electrolyte conductivity is essential. It is correctly measured by impedance spectroscopy at high frequency.

1.3.1.5 Electrolyte: Domain of Electrochemical Stability

The potential window over which the electrolyte is electrochemically stable may be estimated using a polarizable (blocking) electrode, such as platinum. The current is monitored as a function of the electrode potential, and the zero-current region (or nearly zero-current) defines the domain of electrochemical stability of the electrolyte. Of course, this potential range will depend mainly on the nature and the surface of the electrode (roughness).

1.3.1.6 Methods

The most common techniques for testing electrodes are sweep voltammetry, galvanostatic potentiometry, rotating disk electrochemistry, and impedance spectroscopy. Detailed information about these techniques may be found in most classical electrochemical textbooks [6–13], and we will present here the basics of these techniques.

1.3.2 Potentiostatic Methods

1.3.2.1 Voltammetry at a Stationary Electrode

This method is one of the most used to characterize active masses. It quickly provides useful information about potential range of activity, capacity, cyclability, and kinetics. The result is a current versus potential (or versus time). Sweep voltammetry is easily conducted with commercially available potentiostat–galvanostat. Common sweep rates are in the range of 0.001–100 mV/s and common current densities from 0.01 to 10 mA/cm². Cyclic voltammetry is usually applied for estimating the reversibility of the electrochemical reaction.

Charging the double layer capacity yields a current proportional to the electrode surface—negligible if the electrode is not designed as a supercapacitor one—and proportional to the sweep rate. To this current is added the faradic current of the active mass linked with the redox reactions, large only in the potential range of electroactivity of the active mass, $E° \pm 200$ mV. Diffusional effects may be negligible in the measurement of the capacity of the active mass if the potential sweep begins at a completely charged state (discharged) and ends at a completely discharged state (charged). Reversing the sweep should give a mirror image of the forward sweep curve for reversible surface processes. Integration of the current versus time gives the capacity in coulomb (or mAh) of the active mass. If the electrochemical process is sufficiently fast compared to the sweep rate, the integration of the current versus time gives an accurate value of the capacity, a characteristic independent of time. This may be obtained at sufficiently low sweep rate. If this is the case, reversing the sweep rate gives the capacity of the reverse redox reaction. Comparison of the two values gives the faradic reversibility and coulombic efficiency. This technique is widely used for the estimation of the specific capacity and potential domain of active masses, as presented in the following examples.

Figure 1.5 shows the voltammogram of nano-sized $LiFePO_4$ particles (10–30 nm diameter). $LiFePO_4$ is mixed with 5% of PVDF (binder) and 15% acetylene black (electronically conducting in amount sufficient for good percolation) in N-methylpyrrolidone (NMP). The slurry is cast onto an aluminum foil (current collector) and dried in an oven at 120°C for 4 h. Lithium foil is used as an anode and Celgard™ 2400 as the separator. The electrolyte is a mixture of 1 M $LiPF_6$-ethylene carbonate (EC)/dimethylcarbonate (DMC) (1/1,v/v). The cells were assembled in an argon-filled glove box. Cyclic voltammetry (CV) curves were conducted from 2.5 to 4.1 V at a scanning rate of 0.1 mV/s. Oxidation of $LiFePO_4$ to $FePO_4$ gives two distinct solid phases, and at this sweep rate Li^+ concentration in solution can be viewed as constant. The electrochemical transformation should occur at a fixed potential. Oxidation of $LiFePO_4$ begins at 3.45 V with a maximum current at 3.55 V. Oxidation current becomes negligible at 3.8 V. Reversing the sweep rate, reduction begins at 3.45 V with a maximum current at 3.3 V and becomes negligible at 3 V. The curve is almost symmetric from the point $I = 0$, $V = 3.45$ indicating a good faradic reversibility. The widths of the peaks are

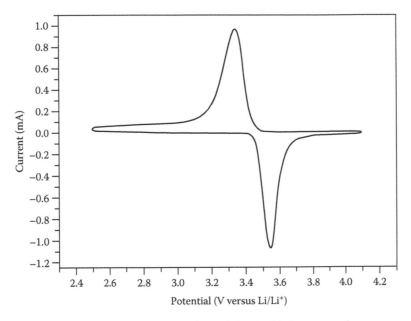

FIGURE 1.5 Voltammogram of nano-sized $LiFePO_4$ particles. In this figure, negative currents are for oxidation and positive for reduction (corresponding to the discharge of a battery). (From Wang, L., et al., *Electrochim. Acta*, 52, 778, 2007.)

indicative of some delay in the solid phases conversion, and may be related to the low diffusion of lithium ions in the solid phase [15]. In this example, there is only one peak of oxidation and one peak of reduction, indicating the presence of only one redox couple.

Cyclic voltammetry may also be used to identify the different redox couples and different phases occurring during the charge and the discharge of an electroactive material. An example can be found with the nano-carbon coated V_2O_5 electrodes described in [16] and shown in Figure 1.6. At least eight peaks can be seen in the discharge voltammogram between 4 and 2 V versus Li/Li^+.

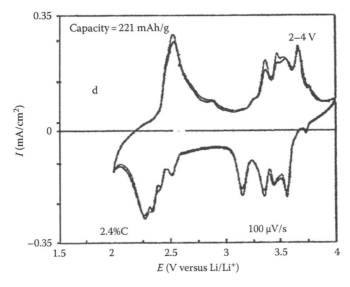

FIGURE 1.6 First and second CVs of composite C-V_2O_5 electrodes. (From Koltypin, M., et al., *J. Electrochem. Soc.*, 154, A605, 2007. With permission.)

Increasing the sweep rate increases the sensitivity and the precision of the capacity measurement, but time-dependent phenomena may interfere with the measurement at higher sweep rate, especially resistive (in relation with the current intensity), electron transfer kinetics, and diffusion phenomena. Capacity measurements and redox potentials determination should be independent of the sweep rate.

If the electrochemical reaction is controlled by diffusion, the peak height is proportional to the square root of the sweep rate. Integrating the current versus time may still give good approximations of the specific capacity if the current at the end of the sweep is sufficiently low. Forward and reverse current peaks are observed at different potentials, the difference depending on electron transfer kinetics.

1.3.2.2 Potentiostatic Intermittent Titration Technique

Diffusion limitations may occur if one reactant is soluble and in dilute concentration in the vicinity of the electrode, a situation uncommon in electrochemical generators, except at high current densities (e.g., Li^+ cation starvation in lithium-ion polymer battery or intercalation materials with an electrolyte with low transfer number for lithium cation). More often, diffusion limitations occur in solid phases where some species have to migrate to the interface. Examples are lithium migration in lithium intercalation materials or hydrogen migration in metallic hydrides. In these cases, diffusion coefficients are better estimated using a potentiodynamic amperometry. The potential of the electrode is stepped from an initial value at which no electrochemical reaction occurs, (e.g., fully oxidized) to an active potential (e.g., more negative than E_2). The current versus time is recorded; it decreases as the inverse of the square root of time, until the diffusion layer thickness is in the range of the grain size. Chronocoulometry versus the square root of time may give information about the diffusion constants and specific capacity. It must be kept in mind that 10^{-10} cm^2/s is a large diffusion coefficient in solid phase, which means about 30 s for the creation of a diffusion layer of 1 µm thickness. In voltamperometry, this is the minimum time for a voltage sweep of 100 mV in order to obtain a correct value of the specific capacity of the solid active mass and even a correct value of the redox potential (potential sweep rate lower than 3 mV/s).

A milder version of the potentiostatic intermittent titration technique (PITT) consists in modifying the initial potential by small steps, may be 1 mV, recording the current after a constant delay, and waiting until it has decreased to almost zero before modifying again the potential (staircase voltammetry). This version gives more information (activity potential and capacity) about the different redox couples present in the electrodes which are successively electroactive, but needs more sensitive equipment.

Huggins and colleagues proposed the PITT in 1979 [17]. This technique also aims to determine the diffusion coefficient of diffusing species from Fick's equations. The solid particles are supposed to have initially a homogeneous concentration [18].

A potential step E_s is applied to a solid electrode and the resulting current (I) versus time (t) is measured (see Figure 1.7)

The diffusion constant can be obtained from [17,18]:

$$I(t) = zFs(c_s - c_0)\left[\frac{D_{PITT}}{\pi t}\right]^{1/2}, \quad \text{for } t \ll \frac{r^2}{D_{PITT}} \tag{1.17}$$

where $(c_s - c_0)$ is the concentration at the surface of the particle at time t and $t = 0$ (Cottrel equation). $(c_s - c_0)$ can be estimated from the total charge passed through the electrode according to Equation 1.18:

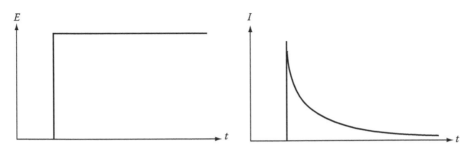

FIGURE 1.7 Schematic representation of the time dependence of current after a constant potential pulse. (From Wen, C.J., et al., *J. Electrochem. Soc.*, 126, 2258, 1979.)

$$\Delta Q = \int_0^\infty I(t)\,\mathrm{d}t = zFnV_M\left(c_s - c_0\right) \tag{1.18}$$

where

   n is the number of gram atoms
   V_M is the molar volume of the host material for the Li ion intercalation for instance

From Equations 1.17 and 1.18,

$$D_{PITT} = \pi r^2 \left[\frac{\left(I\sqrt{t}\right)}{\Delta Q}\right]^2 \tag{1.19}$$

If the current decay does not follow the Cottrel equation, $I\sqrt{t}$ is not constant and it is replaced with $\max(I\sqrt{t})$. ΔQ is the charge passed through the electrode when the current has decreased to less than 1% of its initial value [18].

This method is accurate for determining the activity potential and checking for the presence of different redox couples or variation of crystallographic phases. Reversing the sweep gives good information about thermodynamic reversibility. But, the measurement is time-consuming and needs a specially designed apparatus, currents may be low, and their integration is inaccurate.

1.3.2.3 Ohmic Drop Evaluation

If the reference electrode is correctly placed in the same equipotential as the working electrode, ohmic drop in the electrolyte may be negligible, but this may not be the situation in the bulk of the electrode. Quantitative determination of the electrode resistance is difficult when using voltammetry only. It is much easier using galvanostatic cycling.

The influence of the resistance can be easily seen on the cycling behavior of a supercapacitor electrode. For an ideal supercapacitor electrode (absence of any ohmic resistance), the cyclic voltammogram should be a rectangle, with a constant current equal to the capacitance of the electrode multiplied by the sweep rate. In real systems, with a resistance R and on sweep inversion, the current varies under an exponential law of time constant RC. Since the capacitance C is large, even a low series resistance R may induce a time constant of several seconds, as can be seen in Figure 1.8. The presence of a very large time constant suggests that something is wrong in the experimental setup.

Electron transfer kinetics may be more difficult to determine using stationary electrodes sweep voltammetry. Residual ohmic drop may interfere strongly with the determination of kinetics constants. Impedancemetry and RDEs provide useful alternatives and will be discussed in some detail.

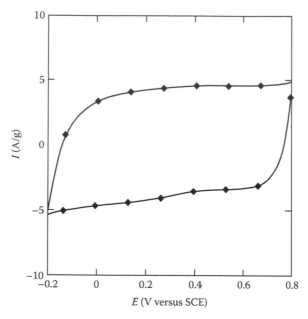

FIGURE 1.8   Voltammogram of a carbon supercapacitor electrode. (Adapted from Toupin, M., et al., *J. Power Sources*, 140, 203, 2005.)

1.3.3   GALVANOSTATIC METHODS

1.3.3.1   Galvanostatic Charge and Discharge

This evaluation is critical to the analysis and prediction of the active materials performance under practical operating conditions. The working electrode is submitted to a constant current *I* (charge or discharge), and voltage versus time is recorded between minimal and maximal values. The current *I* is chosen as a fraction (*C*/2, *C*/5, etc.) or a multiple (*C*, 2*C*, 5*C*, etc.) of the rated discharge capacity of the electrode, *C*, where *C* is the charge delivered in 1 h. The primary knowledge derived from these experiments is, of course, the capacity of the electrode as a function of the rate, the change of the potential as a function of the state of charge, the thermodynamic reversibility, the cyclability, and the estimation of the ohmic drop. Attention must be paid to avoid any interference with the products of the counterelectrode (separate compartments and oversized counterelectrode) and, of course, with the reference electrode (careful choice of the reference electrode location and stability). Testing of air-sensitive materials is best conducted under argon atmosphere in a glove box or in special sealed cells. Multichannel galvanostatic cycling devices are commercially available, however still expensive. Cycling can be performed in thermostatic boxes for investigating the role of temperature. Electrodes for supercapacitors are usually tested under high current densities, and the apparatus must be able to record the voltage at a very high sampling rate (a few milliseconds). Figure 1.9 shows the galvanostatic behavior of a supercapacitor cell. The inset in Figure 1.9 shows that the ohmic drop is more easily obtained from galvanostatic experiments than from voltammetry.

An example of galvanostatic cycling of a lithium battery cathode materials can be found in [21]; an electrode slurry was prepared by mixing the $LiNi_{0.5}Mn_{0.5}O_2$ powder with 10 wt% Super P carbon (Timcal) and 10 wt% PVDF in NMP solution. The slurry was cast on an Al foil and dried in a vacuum oven at 120°C for 12 h. The electrode disks (1.77 cm²) were punched out and dried in a vacuum oven at 70°C for 6 h before storing them in an argon-filled glove box. A total of 2016 coin cells (diameter 20 mm, thickness 1.6 mm) were assembled in the glove box, which consisted of the $LiNi_{0.5}Mn_{0.5}O_2$ and lithium foil electrodes (used as reference electrode and as overcapacitive counterelectrode) and two pieces of Celgard 2500 separator, 1 M $LiPF_6$ dissolved in 1:1 EC:DMC solvent (Merk) was used as the electrolyte. The coin cells were cycled in the galvanostatic mode

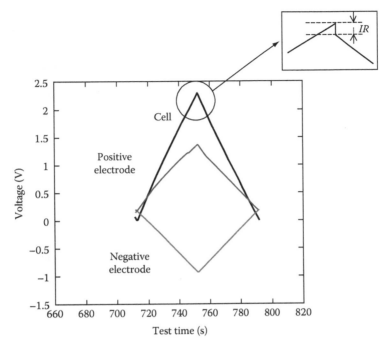

FIGURE 1.9 Galvanostatic cycling of the cell assembled with the PICA A-activated carbon in tetraethyl ammonium methane sulfonate 1.7 M in acetonitrile. (From Taberna, P.L., et al., *J. Electrochem. Soc.*, 150 A292, 2003.)

between desired voltage limits (5.3 and 2.0 V, and 4.5 and 2.0 V) at $C/50$ (current density $= 5.6\,\text{mA/g}$ and $0.027\,\text{mA/cm}^2$) using Solartron 1470 battery tester.

Examination of the solid line curve in Figure 1.10a shows that the charge begins at 3.7 V and reaches up to 4.5 V, discharge begins at 4.5 V, with a very small ohmic drop (about 50 mV) and ends roughly at the same 3.7 V potential, but with less than 90% faradaic efficiency. Figure 1.10b shows a charge curve with intermittent OCV steps of 5 h duration after an increment of charge

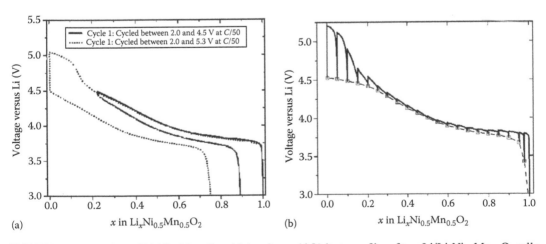

FIGURE 1.10 Cycling of $Li_xNi_{0.5}Mn_{0.5}O_2$ at high voltage. (a) Voltage profiles of two $Li/Li_xNi_{0.5}Mn_{0.5}O_2$ cells in the first cycle from 4.5 to 2.0 V and 5.3–2.0 V at 5.6 mA/g. (b) Voltage profile on the first charge of a $Li/Li_xNi_{0.5}Mn_{0.5}O_2$ cell with intermittent OCV steps of 5 h duration. The cell was cycled between 5.3 and 2.0 V at 14 mA/g. (From Yabuuchi, N., et al., *J. Electrochem. Soc.*, 154, A566, 2007.)

of $dx = 0.05$. This allows for the determination of the thermodynamic voltage (dashed line), the differential capacity plot dQ/dV, or the slope of steady state, open circuit voltage (OCV) versus lithium content. Figure 2 of Ref. [21] shows the discharge capacity versus the discharge rate. A Ragone plot is given showing the specific power versus specific energy, but there is no indication about the related weight (does it include the whole coin cell or is it only the weight of the active mass?). Also, the voltage used is probably the voltage versus the lithium electrode, higher than the voltage of an actual lithium-ion battery. The specific energy is thus overestimated, in that work, by comparison of what could be achieved in an actual battery.

1.3.3.2 The Galvanostatic Intermittent Titration Technique

The galvanostatic intermittent titration technique (GITT) has been first proposed by Weppner and Huggins in 1977 [22]. This method is of particular interest for the measurement of ion transport properties in solid intercalation electrodes, used in lithium-ion batteries, for instance [18]. The determination of the diffusion constants relies on Fick's law. The GITT method records the transient potential response of a system to a perturbation signal: a current step (I_s) is applied for a set time τ_s, and the change of the potential (E) versus time (t) is recorded (Figure 1.11) [18,22].

As described elsewhere [18,22], the diffusion constant D_{GITT} of the diffusing species can be calculated from Equation 1.20:

$$D_{GITT} = \frac{4}{\pi}\left(\frac{1}{SFz_A}\right)^2 \left[\frac{I_s\left(dE/dc\right)}{dE/d\sqrt{t}}\right]^2, \quad \text{with } t \ll r^2/D_{GITT} \tag{1.20}$$

where
  z_A is the charge number of the diffusing ion
  F is the Faraday constant
  S is the particle surface area
  r is the particle length.

This equation is valid as long as the time (t) is short enough, i.e., assuming a constant diffusivity of the species in a semi-infinite solid [17].

When both the step time (τ_s) is short and the current step is small, it becomes

$$\frac{dE}{d\sqrt{t}} \approx \frac{\Delta E_t}{\sqrt{t}} \quad \text{and} \quad \frac{dE}{dc} \approx \frac{\Delta E_s}{\Delta c} \tag{1.21}$$

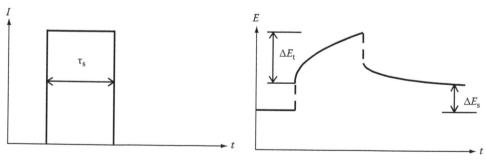

FIGURE 1.11 Schematic representation of the time dependence of potential after a constant current pulse. (From Weppner, W. and Huggins, R.A., *J. Electrochem. Soc.*, 124, 1569, 1977.)

where ΔE_s and ΔE_t are, respectively, the change in the equilibrium potential after the current pulse and the total transient potential change during the current pulse (after eliminating the ohmic drop).

In these conditions, Equation 1.20 then transforms into Equation 1.22:

$$D_{GITT} = \frac{4r^2}{\pi\tau_s}\left(\frac{\Delta E_s}{\Delta E_t}\right)^2 \tag{1.22}$$

The measurement of E_s and E_t during a constant current step of time τ_s leads to the determination of the diffusion constant D_{GITT} of an ion in a solid electrode with a particle length of r.

During galvanostatic cycling of $Li_xNi_{0.5}Mn_{0.5}O_2$ [21], GITT measurements were performed according to the following: a constant current at $C/20$ during 1 h ($dx = 0.05$) followed by an open circuit stand for 5 h. During the OCV duration, the evolution of the potential versus the square root of the time was analyzed, yielding a slope dE/dt. This allowed for the determination of the chemical diffusion coefficient of lithium ions around 10^{-13} cm^2/s. Of course, such an evaluation is time-consuming.

1.3.3.3 Evaluation of Supercapacitor Electrodes

Galvanostatic techniques are well adapted to the characterization of supercapacitor electrodes. Figure 1.9 shows a typical potential versus time plot under constant current charge/discharge between fixed potential limits. A pure capacitive behavior leads to the same charge and discharge times; if not, this is an indication that an irreversible faradaic reaction (often reduction or oxidation of the solvent) occurs. The capacitance C is given by the slope of the discharge line. When reversing the current I, there is a sudden drop in the potential equal to $2 \times R_s \times I$, allowing for the evaluation of the series resistance R_s at that potential. The linear variation of potential versus time is then obtained after a delay of about $2R_s \times C$.

Galvanostatic evaluation is often conducted at current densities of about 10–50 mA/cm^2 and capacitances in the range of 1 F/cm^2 can be easily achieved. The discharge time is approximately 100–200 s for a 2 V range. A good series resistance is below 1 Ω cm^2 in organic electrolytes (a time constant of about 1 s). An important component of the series resistance may be the interfacial resistance between the current collector and the active mass.

1.3.3.4 Cyclability

Long-term galvanostatic cycling is often used for testing cyclability and faradaic efficiency over long periods of time. Approximate values may have been obtained using cyclic voltammetry. Galvanostatic measurements are often considered more reliable for the measurement of capacity during charge and during discharge, because the current may be kept rigorously constant and the time may be determined with accuracy. Capacity and capacitance can be determined with accuracy better than 1%. Multichannel battery testing devices offer the possibility to simultaneously record several cells. The main difficulty with individual electrodes is the lack of tightness of individual electrode assemblies versus atmospheric moisture or oxygen over a very long period of time. Although cyclability is an actual property of a single electrode, batteries and supercapacitors are usually tested in complete sealed cells. Testing separately both electrodes and the complete cell gives more information. A limitation is the long time frame required for successful completion of reliable experiments.

1.3.3.5 Self-Discharge

Self-discharge measurement is a special case of galvanostatic measurement, recording the evolution of potential versus time at zero current, or measuring the residual capacity after some delay after charging. Self-discharge is a property of an electrode in a given medium. But self-discharge may strongly depend on the cell design. Accordingly, the self-discharge of an individual electrode is only a rough indication of what it can be in an actual cell.

For similar reasons, aging and robustness versus anomalous conditions of cycling are usually conducted on actual sealed cells.

1.3.4 ROTATING DISK ELECTRODE VOLTAMMETRY

This method is well suited for slightly soluble electroactive materials. It has been widely used for the study of oxygen reduction reaction and hydrogen oxidation reaction, the two main reactions occurring in fuel cells.

The following shows the results found in the laboratory of Fauvarque and colleagues [23], according to a method described by Savinell and colleagues [24]. The aim of the study was the demonstration of the importance of oxygen solubility in Nafion™. Figure 1.12 shows a typical reduction curve of oxygen at platinum RDE, below −150 mV versus SCE, where the current I is limited by the oxygen diffusion. Above −150 mV, the current depends on both diffusion and electron transfer. At a fixed potential (e.g., −80 mV), a plot of $1/I = 1/I_{max} + 1/I_k$ (Koutecky–Levich) against the square root of the inverse of the rotation speed yields the value of the kinetic current for the oxygen reduction at that potential. A complete description can be found in usual textbooks [5–8].

The polarization curve is obtained step by step, at every potential until obtention of a steady-state value. The polarization curve must be identical during forward or backward potential scan. If not, either the steady state has not been obtained, or, more frequently, the surface of the electrode has been modified by the electrochemical reaction. Covering the platinum electrode by a Nafion film reduces the limiting current I_1, by the addition of a supplementary diffusion resistance, depending on the thickness of the Nafion film (Figures 1.13 and 1.14).

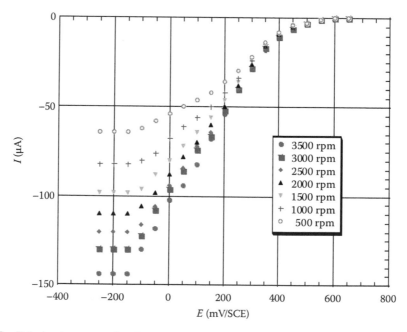

FIGURE 1.12 Polarization curves for the reduction of oxygen on a bare platinum disk in oxygen-saturated 0.5 M H_2SO_4 at different rotation speeds. (From Ayad, A., et al., *J. Power Sources*, 149, 66, 2005.)

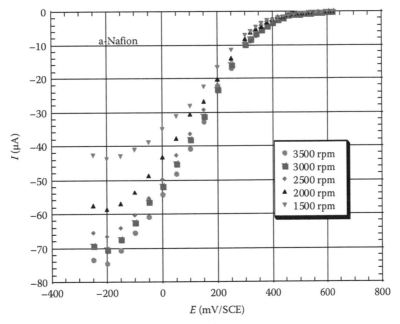

FIGURE 1.13 Polarization curves for the reduction of oxygen on a platinum disk covered with a film of a-Nafion in oxygen-saturated 0.5 M H_2SO_4 at different rotation speeds. (From Ayad, A., et al., *J. Power Sources*, 149, 66, 2005.)

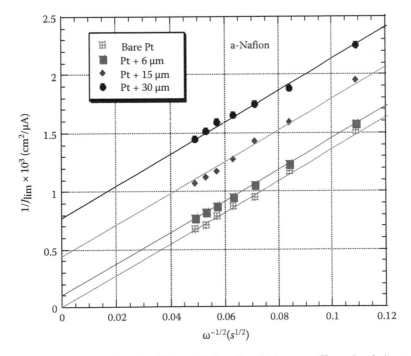

FIGURE 1.14 Koutecky–Levich lines for different Nafion film thicknesses. (From Ayad, A., et al., *J. Power Sources*, 149, 66, 2005.)

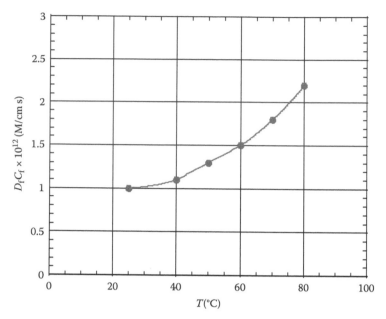

FIGURE 1.15 Variation of the $D_f C_f$ product with temperature. (From Ayad, A., et al., *J. Power Sources*, 149, 66, 2005.)

From these results, the permeability of oxygen in a Nafion film can be obtained as the product of its concentration by its diffusion coefficient in the film $D_f C_f$. Figure 1.15 shows the variation of the oxygen permeability with the temperature. Like many perfluorinated materials, the Nafion membrane is especially a good solvent of oxygen.

It is also possible to evaluate the electrochemical performance of electrodes similar to those in service in Proton Exchange Membrane (PEM) fuel cells. The current collector of the RDE is then glassy carbon. The active layer is prepared by mixing 30% platinized carbon (the platinum is deposited on Vulcan carbon black XC72 (from E-TEK) with the polymer. To prepare the mixture of polymer and platinized carbon, a solvent in which the polymer is soluble is used. The ink obtained is sonicated for an hour to obtain good homogeneity and then deposited by means of a micropipette on the glassy carbon disk (diameter 5 mm). In the case of Nafion, the alcohol used evaporates rapidly at ambient temperature leaving in all cases the active matter glued to the glassy carbon. Once the active layer has been deposited, the rotating electrode is immersed in an oxygen-saturated 0.5 M H_2SO_4 solution. The polarization curves for the reduction of oxygen are recorded at ambient temperature; during the entire experiment, the electrolyte remains in an atmosphere of oxygen. The results obtained with 30% platinized carbon, E-TEK, follow the Koutecky–Levich law and show that the amount of polymer used as glue to prepare the active layer does not alter the oxygen transport process. The calculated polymer film thickness is less than 1 μ, as is suggested by the low values of the intercepts in Figure 1.16.

The different deposits contain the same amount of platinum and different percentages of the polymer. The results on the reduction of oxygen show that the ratio of the currents, I/I_{max}, presents a maximum when varying the proportion of polymer: 25% for Nafion (Figure 1.17). The presence of an optimum value for the polymer proportion can be easily explained: increasing the amount of polymer increases the ionic percolation and the number of the active sites, but increasing the amount of polymer causes excessive masking of platinum that slows down the diffusion of oxygen.

1.3.5 Electrochemical Impedance Spectroscopy

This section aims to present the basics of electrochemical impedance spectroscopy (EIS) and to give some example of application in the electrochemical energy storage area. For more detailed information about this technique, the reader can refer to more specific works [10–13].

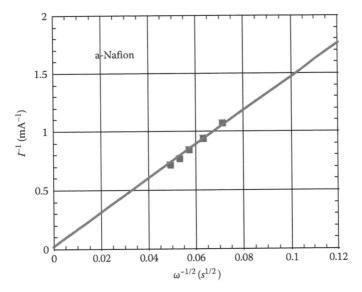

FIGURE 1.16 Evolution of $1/i$ on 30% platinized carbon plus polymer versus the reciprocal of the square root of the rotation speed (sample with 25% Nafion). (From Ayad, A., et al., *J. Power Sources*, 149, 66, 2005.)

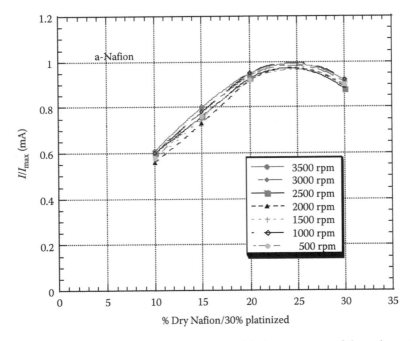

FIGURE 1.17 Evolution of I/I_{max} at ambient temperature with the percentage of dry polymer in the active layer. (From Ayad, A., et al., *J. Power Sources*, 149, 66, 2005.)

1.3.5.1 General Presentation

All the previous techniques described in this chapter used large perturbations of the system for recording the transient response of the system. It is the case, for instance, with potential sweeps (CV) and potential or current step (PITT and GITT). Another way to characterize an electrochemical system is to perturb the system initially at the steady state by the use of an alternative signal of small amplitude: this method is used in EIS.

When polarized under constant DC potential, the current flowing in an electrochemical cell is the sum of various contributions from all the reactions occurring at the electrode/electrolyte interface; accordingly, it is almost difficult to identify the different contributions of the current. The EIS technique is a powerful tool that can often discriminate all these different contributions on the basis of their respective time constant. As explained above, in the EIS technique, an alternating signal of small amplitude is injected in an electrochemical system at the steady state. As a consequence, all the different processes responsible for the current flowing through the cell go back to the stationary state through their own time constant:

1. High-rate electrochemical processes are active at high frequencies, such as charge transfer reactions
2. Low-rate processes are active at low frequencies, such as mass transfer (diffusion)

The measurement is done by applying a perturbation signal $\delta E(\omega)$ to the constant potential E_{ST} where the system was initially at the steady state, with the pulsation $\omega = 2\pi f$. The whole potential of the electrode is then

$$E = E_{ST} + \delta E(\omega) \tag{1.23}$$

with

$$\delta E(\omega) = |\delta E(\omega)| \exp(j\omega t) \tag{1.24}$$

In response to the perturbation, the current flowing through the electrode can be written as

$$I = I_{ST} + \delta I(\omega) \tag{1.25}$$

with

$$\delta I(\omega) = |\delta I(\omega)| \exp(j\omega t + \psi) \tag{1.26}$$

with ψ the phase angle between the current response and the potential.

The electrochemical impedance is then defined as the ratio between $\delta E(\omega)$ and $\delta I(\omega)$:

$$Z(\omega) = \delta E(\omega)/\delta I(\omega) = Z'(\omega) + jZ''(\omega) \tag{1.27}$$

Two main plots are used to show the change of electrochemical impedance with frequency. The first one is the logarithm of the impedance Z versus the logarithm of the frequency, $Log|Z|$ versus $Log(f)$. However, if this representation is useful for electricians, electrochemists prefer to use the representation in the Nyquist plane, i.e., the opposite of the imaginary part of the impedance $-Z''$ versus the real part of the impedance Z', together with the change of the phase angle ψ with the frequency. Both plots are useful. In addition to qualitative examination of Nyquist plots, the electrochemical systems are often modeled using electric equivalent circuits. The use of such circuits can be useful to model the electrode/solution interface to explain the electrochemical behavior of the systems. However, these representations must be used carefully since one can always find a theoretical circuit fitting the experimental data, without any evident scientific correlation.

At the electrode/solution interface, two different currents can be observed:

1. A capacitive current I_{DL}, which is used to charge the double layer present at the solid/liquid interface [12]. It corresponds to the charge of a capacitor (electrons in the electrode side, ions from the electrolytes on the electrolyte side) with a capacitance of about $15\,\mu F/cm^2$.
2. A faradic I_F current linked with the faradic (charge transfer) reactions.

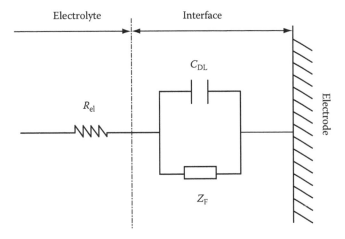

FIGURE 1.18 Equivalent circuit of a solid electrode/electrolyte interface (in the absence of any film on the electrode surface).

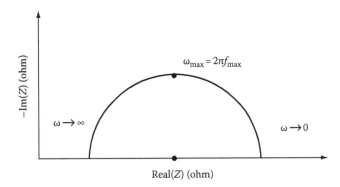

FIGURE 1.19 Complex-plane impedance plot (Nyquist plane) for an electrochemical reaction under kinetic control.

The electrode/solution interface, in the simpler case (no passive layer on the electrode surface) can be modeled using the following equivalent circuit (Figure 1.18), where R_{el} stands for the electrolyte resistance, C_{DL} for the double layer capacitance, and Z_F the faradic impedance.

The impedance of the equivalent circuit (Figure 1.19) is given by

$$Z(\omega) = R_{el} + Z_F / (1 + jZ_F C_{DL}\omega) \tag{1.28}$$

1.3.5.2 Electrode Reactions under Kinetics (Charge Transfer) Control

When the rate determining step of the electrode reaction is the charge transfer process (kinetic control), the faradic impedance Z_F in Figure 1.18 can be described as R_{CT}, the charge transfer resistance [7,8]. The impedance plot in the Nyquist plane describes a semicircle, as shown in Figure 1.19.

The impedance of the cell can be written as

$$Z(\omega) = R_{el} + \left[\frac{R_{CT}}{1 + R_{CT}^2 \times C_{DL}^2 \times \omega^2} \right] - j \left[\frac{R_{CT}^2 \times C_{DL} \times \omega}{1 + R_{CT}^2 \times C_{DL}^2 \times \omega^2} \right] \tag{1.29}$$

leading to

$$Z(\omega) = Z'(\omega) + jZ''(\omega) \qquad (1.30)$$

with

$$Z'(\omega) = R_{el} + \left[\frac{R_{CT}}{1 + R_{CT}^2 \times C_{DL}^2 \times \omega^2} \right] \qquad (1.31)$$

$$Z''(\omega) = -j \left[\frac{R_{CT}^2 \times C_{DL} \times \omega}{1 + R_{CT}^2 \times C_{DL}^2 \times \omega^2} \right] \qquad (1.32)$$

These equations fit with the equation of a semicircle according to

$$\left[Z'(\omega) - \left(R_{el} + \frac{R_{CT}}{2} \right) \right]^2 + Z''(\omega)^2 = \frac{R_{CT}^2}{4} \qquad (1.33)$$

The center of the circle is on the real axis, at $Z' = R_{el} + R_{CT}/2$. The frequency where the imaginary part of the impedance is maximum allows the determination of C_{DL} according to [13]:

$$\omega_{max} R_{CT} C_{DL} = 1 \qquad (1.34)$$

with $\omega_{max} = 2\pi f_{max}$, f_{max} being the frequency at the top of the semicircle.

1.3.5.3   Mixed Kinetic and Diffusion Control

We will consider here the case where the mass transfer limitation is due to the diffusion of the species. When the mass transfer becomes predominant in the low-frequency range, experimental plots obtained in the Nyquist plane shift from the semicircular shape. In that case, indeed, the impedance Z_F can no longer be described as only a charge transfer resistance, but as a combination of R_{CT} with the impedance of diffusion Z_D. Z_D changes with the frequency; it takes into account the relaxation processes inside the diffusion layer. Different cases can be described depending on the diffusion layer thickness.

(a) *Infinite thickness of the diffusion layer*
    If the diffusion layer has an infinite length, the impedance is given by Equation 1.35 [11]

$$Z_F(\omega) = R_{CT} \left[1 + \frac{\lambda}{(j\omega)^{1/2}} \right] \qquad (1.35)$$

with $\lambda = K_f / D_{ox}^{1/2} + K_b / D_{red}^{1/2}$
where K_b and K_f are the reaction kinetics rates of the electrochemical reactions

$$ox + ne \xrightarrow{\ K_f\ } red$$

$$red \xrightarrow{\ K_b\ } ox + ne$$

and D_{red} and D_{ox} are the diffusion coefficients of, respectively, the reduced and oxidized species. $Z_F = R_{CT}\lambda/(j\omega)^{1/2}$ is called the Warburg impedance [5,10].

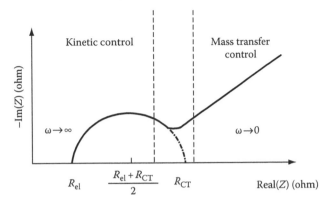

FIGURE 1.20 Complex-plane impedance plot (Nyquist plane) for an electrochemical system, with the mass transfer and kinetics (charge transfer) control regions, for an infinite diffusion layer thickness.

In the case of the infinite diffusion layer, the whole impedance of the cell is given by Equation 1.36

$$Z_F(\omega) = \left[R_{el} + R_{CT}\left(1 + \frac{1}{(2\omega)^{1/2}}\right) - R_{CT}^2 \lambda^2 C \right] - j\left[\frac{R_{CT}\lambda}{(2\omega)^{1/2}} \right] \tag{1.36}$$

The impedance plot in the Nyquist plane as presented in Figure 1.20 shows two different parts: a loop at high frequency and a line at low frequency, also named "Warburg line," at 45° angle with the real axis [10].

(b) Finite diffusion layer
If concentration relaxation occurs in a diffusion layer with a thickness δ_N (Nernst diffusion layer), the impedance Z_F is then [5,10]

$$Z_F(\omega) = R_{CT}\left(1 + \frac{K_f \, \text{th}\left(\delta_N\sqrt{\frac{j\omega}{D_{ox}}}\right)}{\sqrt{D_{ox}}\sqrt{j\omega}} + \frac{K_b \, \text{th}\left(\delta_N\sqrt{\frac{j\omega}{D_{red}}}\right)}{\sqrt{D_{red}}\sqrt{j\omega}} \right) \tag{1.37}$$

The Warburg impedance has been substituted by Z_D, the diffusion impedance:

$$Z_D(\omega) = R_{CT}\frac{K_f \, \text{th}\left(\delta_N\sqrt{\frac{j\omega}{D_{ox}}}\right)}{\sqrt{D_{ox}}\sqrt{j\omega}} + \frac{K_b \, \text{th}\left(\delta_N\sqrt{\frac{j\omega}{D_{red}}}\right)}{\sqrt{D_{red}}\sqrt{j\omega}} \tag{1.38}$$

When $\omega \to 0$, $Z_D(\omega)$ tends to a defined value R_D according to Equation 1.39

$$Z_D(\omega) = R_{CT}\left(\frac{K\delta_N}{D}\right), \quad \text{see Figure 1.21} \tag{1.39}$$

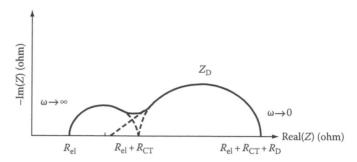

FIGURE 1.21 An example of a complex-plane impedance plot (Nyquist plane) for an electrochemical system under mixed kinetic/diffusion control, with the mass transfer and kinetics (charge transfer) control regions, for a finite thickness δ_N of the diffusion layer. Assumption was made that $K_f \ll K_b$ at the bias potential of the measurement, and $D_{ox} = D_{red} = D$, leading to $R_D = R_{CT}\left(K_b \delta_N / D\right)$.

When $\omega \to \infty$, th$\left(\delta_N \sqrt{j\omega/Di}\ \right)\big/\sqrt{j\omega/Di} \approx 1\big/\sqrt{j\omega/Di}$ and Z_D is equal to the Warburg impedance (Equation 1.40)

$$Z_D = \frac{R_{CT}\lambda}{(j\omega)^{1/2}} \tag{1.40}$$

Different forms of the impedance plots can be obtained for an electrochemical system described by a mixed kinetic/diffusion control process, depending on the parameters of diffusion and charge transfer. An example of a Nyquist plot is presented in Figure 1.21.

As previously explained, we have presented in this part of the chapter the basics of EIS, and we have limited our study to simple cases. More particularly, we did not consider in our approach the presence of a solid film on the electrode surface or the adsorption–desorption reactions at the electrode surface. Either will significantly modify the electrochemical response to a sinusoidal perturbation of small amplitude.

1.3.6 CHARACTERIZATION BY ELECTROCHEMICAL IMPEDANCE SPECTROSCOPY OF POROUS ELECTRODES

Electrochemical generators often have porous electrodes that exhibit a special behavior by impedance spectroscopy. This is especially true for supercapacitors that present negligible charge transfer reactions under normal conditions of use. In this section, we will focus on the impedance behavior of the electrochemical double layer capacitors (EDLCs) also called as "supercapacitors" or "ultracapacitors," i.e., energy storage systems using porous carbon electrodes. Once polarized, EDLCs store the energy through the reversible adsorption/desorption of ions from an electrolyte onto high-surface area carbons. In such electrochemical systems, no charge transfer reactions (i.e., redox reactions) occur since the energy is stored under purely electrostatic form by charging what is called the double layer capacitance (for complete description of the basics of EDLCs operation, please see Chapter 8). Many of the following results can be easily extended to porous electrodes of batteries.

1.3.6.1 Limitations of the Basic Equivalent Circuits

A basic electric equivalent circuit to describe an EDLC is presented in Figures 1.22a and b, which shows the Nyquist (Figure 1.22b) plot of an ideal capacitor C, in series with a resistance

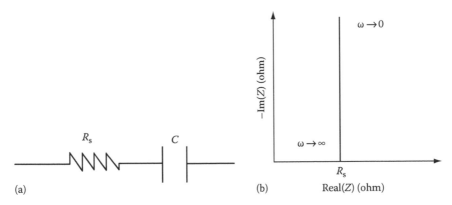

FIGURE 1.22 (a) Capacitor C with a series resistance R_s. (b) Complex-plane impedance plot for the equivalent circuit presented in (a).

R_s (Figure 1.22a). The double layer capacitance is represented by the capacitance C, and R_s is the series resistance of the EDLC, also named the equivalent series resistance (ESR). This series resistance shows the nonideal behavior of the system. This resistance is the sum of various ohmic contributions that can be found in the system, such as the electrolyte resistance (ionic contribution), the contact resistance (between the carbon particles, at the current collector/carbon film interface), and the intrinsic resistance of the components (current collectors and carbon). Since the resistivity of the current collectors is low when Al foils or grids are used, it is generally admitted that the main important contribution to the ESR is the electrolyte resistance (in the bulk and in the porosity of the electrode) and to a smaller extent the current collector/active film contact impedance [25,26]. The Nyquist plot related to this simple RC circuit presented in Figure 1.22b shows a vertical line parallel to the imaginary axis.

Figure 1.23 shows the Nyquist plot of a 2.3 V/10 F supercapacitor laboratory cell. When compared to Figure 1.22b, it appears that the basic equivalent circuit proposed is unable to describe the real behavior of an EDLC.

Why such a difference between the real EDLC and the model proposed here? In other words, what are the specific features the equivalent circuit does not take into account? The oversimplified equivalent circuit presented in Figure 1.22 considers two planar electrodes face to face, with a constant thickness of dielectric material between them. The reality is much more complex since EDLCs use three-dimensional porous electrodes, and the porous electrodes are responsible for the particular shape of the Nyquist plot presented in Figure 1.23.

1.3.6.2 The Transmission Line Model from De Levie

De Levie was the first to describe the electrochemical behavior of porous electrodes [27]. We will not present here the details of the mathematical treatment developed by De Levie, and we ask the reader to refer to his papers [27,30] for more information. However, we will present a summary of the results obtained by De Levie that have been of great importance for the understanding of the electrochemistry at porous electrodes as a whole, not just for EDLC applications.

Basically, the impedance behavior of a porous electrode cannot be described by using only one RC circuit, corresponding to a single time constant RC. In fact, a porous electrode can be described as a succession of series/parallel RC components, when starting from the outer interface in contact with the bulk electrolyte solution, toward the inner distribution of pore channels and pore surfaces [4]. This series of RC components leads to different time constant RC that can be seen as the electrical response of the double layer charging in the depth of the electrode. Armed with this evidence, De Levie [27] proposed in 1963 a (simplified) schematic model of a porous electrode (Figure 1.24a) and its related equivalent circuit deduced from the model (Figure 1.24b).

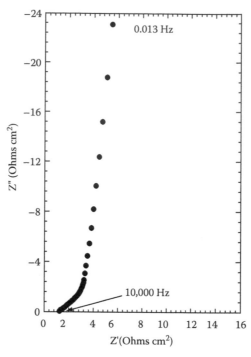

FIGURE 1.23 Complex-plane impedance plot for a 2.3 V/10 F supercapacitor laboratory cell using acetonitrile-based electrolyte. Bias voltage = 0 V; room temperature.

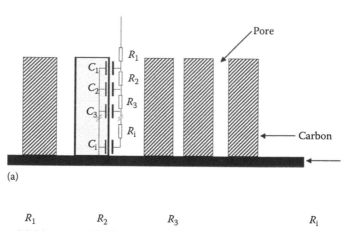

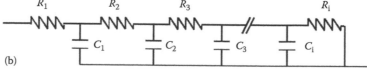

FIGURE 1.24 (a) Schematic representation of a porous solid electrode (here a carbon electrode) assuming parallel pores, identical in size and shape. (b) Equivalent circuit for the porous electrode presented in Figure 1.24a (TLM). (From Conway, B.E., *Electrochemical Supercapacitors*, Kluwer Academic/Plenum Publishers, New York, 1999; De Levie, R., *Electrochim. Acta*, 8, 751, 1963 (in 3 parts).)

Figure 1.24a shows a porous electrode whose pores are uniform in size and shape. The opposite electrode needed to close the circuit loop is not represented but should be located above the presented electrode. When a potential difference is applied between the two electrodes, the double layer is charged. If we assume that the ions involved in the charging of the electrode shown in

Figure 1.24a (cations, for instance, if the electrode is negatively polarized) come from the bulk electrolyte, the transportation of these ions from the electrolyte to the electrode outer surface induces a resistance R_s due to the ionic resistivity of the solution. The first surface available for double layer charging is located at the top of the carbon porous electrode, and the first capacitance C_1 can be charged. When going down into the depth of the electrode, the transportation of the ions in the column of electrolyte in the porous electrodes leads to another ohmic resistance, R_1. Then, a second site for double layer charging along the pore of the carbon can be found, with a related capacitance noted as C_2. This operation can be repeated again and again along the pore of the carbon, until the bottom of the electrode (the current collector) is reached. The equivalent circuit issued from this representation is then a series of capacitances and resistances in series/parallel as shown in Figure 1.24b.

The first resistance R_s is the resistance of the electrolyte outside the pores; the R_i elements are the electrolyte resistances inside the pores of the electrode; and C_i are the double layer capacitances along the pores. This model is called the Transmission Line Model (TLM) by De Levie. A careful selection of a set of R_i, C_i values allows to calculate back the experimental plot such as the one presented in Figure 1.23 [28]. It can be noted that constant phase element (CPE) can be used to replace the capacitance C for better fitting, the CPE impedance Z_{CPE} being $Z_{CPE} = 1/j(C\omega)^\alpha$.

In the light of the results obtained by De Levie [27], the ac behavior of an EDLC presented in Figure 1.23 can now be separated into three different parts. At a very high frequency, only the outer surface of the electrode can be accessible to the ions. The EDLC impedance is then purely ohmic, and the value of the impedance is R_s: the EDLC behaves like a pure resistance. In the second part, when the frequency is decreased, the influence of the electrode porosity and thickness on the ions migration rate from the electrolyte inside the electrode can be seen. According to the equivalent circuit presented in Figure 1.24, more and more RC elements are active since their impedance can no more be neglected: the Nyquist plot shows an increase in both the real and imaginary parts of the impedance, until a constant value of Re(Z) is reached. Schematically, ions can now penetrate deeper into the porous electrode; both the capacitance and the resistance of the EDLC change. In the last part of the Nyquist plot, at low frequency, both the capacitance and the resistance have reached their maximum values, and the impedance plot tends to reach a vertical line: the crossing of the low-frequency vertical line with the mid-frequency line defines what is often called "the knee frequency" [29]. Below this frequency, the whole capacitance is reached, and for a higher value, the capacitance strongly depends on the frequency.

This model proposed by De Levie [27,30] has been updated several times during the last decades, taking into account, for instance, the pore shape [31] or the distribution of pores with nonuniform size [32]; however, today, it is of great interest since the set of equations and the model proposed by De Levie allow a rapid characterization of the frequency behavior of the EDLCs.

1.3.6.3 The Complex Capacitance Model

As described previously, the De Levie model is useful to describe the electrochemical ac (or dc) behavior of a porous electrode. This TLM describes the porous electrode as an interpenetrated network of RC elements, whose contribution depends on the frequency of the ac signal. In other words, the capacitance and the resistance of the EDLC change with the frequency. However, it is difficult to get from these equations the change of the capacitance with the frequency useful to characterize an EDLC electrode. This is why other approaches have been developed and focused more onto the EDLC problematic, with the aim to quantify the change of the porous electrode capacitance (or the EDLC device capacitance) with the frequency of the ac signal.

An alternative approach between the basic R_sC series circuit and the De Levie model (TLM) is to consider a porous electrode as a whole capacitance by simply using the impedance data [20]:

$$Z(\omega) = 1/[j\omega\, C(\omega)] \tag{1.41}$$

$$Z(\omega) = Z'(\omega) = jZ''(\omega) \tag{1.42}$$

The combination of Equations 1.41 and 1.42 leads to

$$C'(\omega) = -Z''(\omega)/[\omega |Z(\omega)|^2] \tag{1.43}$$

$$C''(\omega) = Z'(\omega)/[\omega |Z(\omega)|^2] \tag{1.44}$$

where
  $|Z(\omega)|^2$ is the impedance modulus
  $C'(\omega)$ and $C''(\omega)$ are, respectively, the real part and the imaginary part of the capacitance $C(\omega)$

$C'(\omega)$ is the capacitance of the electrode (or the cell); the low frequency value of $C'(\omega)$ is the capacitance obtained during constant current discharge measurements.

$C''(\omega)$ is the imaginary part of the capacitance, corresponding to an energy dissipation by an irreversible process. For instance, it is because of the dielectric losses in water occurring during the movement of the molecules that food and drinks get heated in a microwave oven [20].

The set of equations (Equations 1.43 and 1.44) now gives the opportunity to quantify the capacitance change of a porous electrode (or an EDLC device) with the frequency of the ac signal during impedance spectroscopy measurements [20]. An example is given in Figures 1.25a and 1.26b that show, respectively, the variation of the real part of the capacitance $C'(\omega)$ and the imaginary part of the capacitance $C''(\omega)$ versus frequency, for a 2.3 V/2 F laboratory cell (see the Nyquist plot in Figure 1.23).

As expected, the capacitance of the cell increases when the frequency is decreased (Figure 1.25a); below the "knee frequency," the capacitance tends to be less dependent on the frequency and should be constant at lower frequencies. This knee frequency is an important parameter of the EDLC; it depends on the type of the porous carbon, the electrolyte as well as the technology used (electrode thickness, stack, etc.) [20]. The imaginary part of the capacitance (Figure 1.25b) goes through a maximum at a given frequency noted as f_0 that defines a time constant $\tau_0 = 1/f_0$. This time constant was described earlier by Cole and Cole [33] as the dielectric relaxation time of the system, whereas

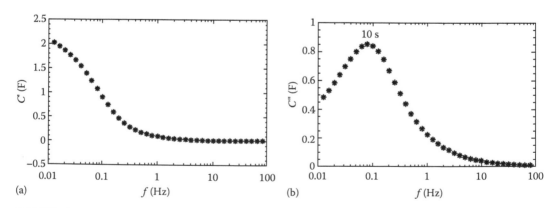

FIGURE 1.25 (a) Change of the real part of the capacitance (C') versus frequency for a 2.3 V/2 F laboratory cell, obtained from Equation 1.43. (b) Change of the imaginary part of capacitance (C'') versus frequency for a 2.3 V/2 F laboratory cell, obtained from Equation 1.44. (From Taberna, P.L., et al., *J. Electrochem. Soc.*, 150, A292, 2003.)

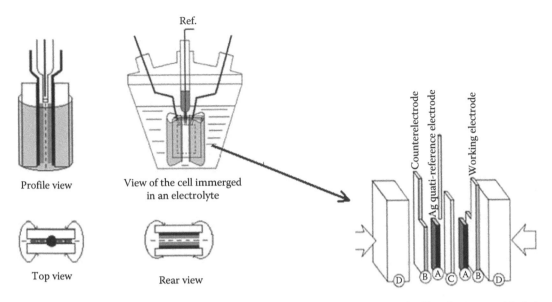

FIGURE 1.26 Three-electrode laboratory cell. (A) Porous carbon electrode film (carbon + binder); (B) treated Al current collectors; (C) porous separator; (D) Teflon plates (for stack pressure); and (E) stainless steel clamps (not shown, see arrows). (From Gamby, J., et al., *J. Power Sources*, 101, 109, 2001.)

Miller [34] mentioned this time constant as the factor of merit of the capacitor. At this frequency, half of the low-frequency capacitance C_{LF} is reached. For $f < f_0$, the capacitance C is always greater than C_{LF}, and for higher frequencies, $C < C_{LF}$.

This complex capacitance model, even if simplified, gives precious quantitative information about the change of the capacitance of an EDLC device versus the frequency. The knowledge of the ac behavior is indeed important since EDLC, as power devices, are often used in ac modes. Finally, it must be precise that improvements of such approach have been recently developed in a series of papers [35,36], where the De Levie TLM is associated with the complex capacitance model to estimate the porous structure of the carbon electrodes using discrete Fourier transformation.

1.4 TESTING LABORATORY CELLS

1.4.1 Coin Cells

The most common cells used in laboratories are coin cells. They are built easily with standard devices and can be sealed under controlled atmosphere in a glove box (see Equation 1.18). They operate with limited amounts of electrolyte as in most commercial batteries. Caution must be taken to minimize the contact resistance between the active masses and the cell case parts, working as current collectors. Cyclability can be easily studied using galvanostatic charge and discharge. They are the best devices for testing the effect of the separators on the performances. They can be easily assessed by EIS, more specifically for examination of the effect of cycling or aging on internal resistance or electron transfer kinetics. They have only one drawback: no reference electrode as in the three-electrode cells, making it difficult to assess the role of each electrode.

1.4.2 Three-Electrode Cells

In order to know the performances of individual electrodes working in the conditions of actual cells, it is necessary to add a reference electrode. Laboratory cells have been designed for that

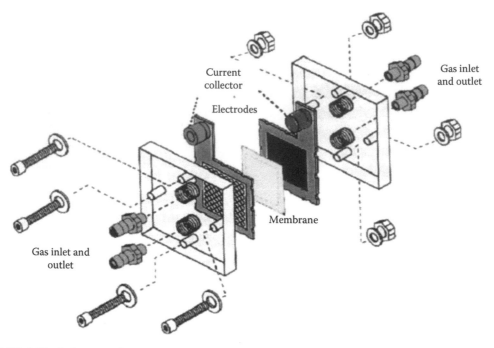

FIGURE 1.27 Laboratory fuel cell.

purpose, including a reference electrode in or near the separator, often a silver wire coated with some insoluble silver salt.

Many cells have been described in the literature. The cell shown in Figure 1.26 is easily mounted and tested, preferably in a glove box, if the materials are air-sensitive.

1.4.3 TESTING FUEL CELLS

Materials for fuel cells may be assessed in experimental simple cells such as those designed by H-tec (Figure 1.27) (4 × 4 cm electrode surface atmost). But more sophisticated cells must be used for testing larger electrodes or assemblies. Many examples can be found in the literature. Tests are often conducted above room temperature.

1.4.4 TESTING COMMERCIALLY AVAILABLE CELLS AND BATTERIES

Testing commercially available batteries should be done using appropriate commercial testing devices (from Bitrode, Arbin, Biologic, etc.). Small cells can be tested using laboratory potentiostats–galvanostats. Prior to testing, it may be wise to check the weight of the battery, its OCV (using appropriate rest time after charging), and, if possible, short-circuit current and to compare the values with manufacturer specifications.

1.5 CONCLUSIONS

Electrochemical generators are the most common way for the storage of electrical energy. The energetic yield is greater than 50%, usually around 80%. Electrochemical methods are used for the assessment of the batteries as a whole or of their components: active masses (electrodes), electrolytes, separators, and current collectors. The most important are, of course, electrodes, and their behavior can be analyzed using different electrochemical techniques. The most common techniques are

Cyclic voltammetry for a quick determination of the domain of electroactivity, the number of active redox couples (phases), the electrochemical reversibility, and, to some extent, the capacity and the cyclability.

Galvanostatic charge and discharge that are preferably used for the determination of the capacity (capacitance for supercapacitors), the faradic reversibility, the ohmic drop, and the cyclability.

Step and pulse techniques, appropriate for determining diffusion coefficients in solids.

Rotating disk electrode, well adapted for the estimation of diffusion limitations in solution and, sometimes, for the determination of kinetics parameters.

EIS, especially useful for the estimation of the cell resistance, the kinetics parameters, the double layer capacitance, and for the assessment of porous electrodes.

It must be kept in mind that nonelectrochemical techniques are also very useful for assessing electrode materials and more specially optical spectroscopy, x-ray absorption and diffusion, electron microscopy, calorimetry, and quartz crystal microbalance (see, for example, Ref. [6] for a quick insight about these techniques). They are used as a complement to the electrochemical ones.

REFERENCES

Selected Books on Batteries and Supercapacitors

1. D. Linden and T. B. Reddy, *Handbook of Batteries*, 3rd edition, Mc Graw Hill, New York (2002).
2. J. A. Kolde, B. Bahar, M. S. Wilson, T. A. Zawodzinski, and S. Gottesfeld, in *Proton Conducting Membranes Fuel Cells* 1. The Electrochemical Society Meetings, Electrochemical society, Pennington, NJ, Vol. 95-23, pp. 193–201 (1995).
3. C. A. Vincent and B. Scrosati, *Modern Batteries: An Introduction to Electrochemical Power Sources*, 2nd edition, John Wiley & Sons, New York (1997).
4. R. M. Dell and D. A. J. Rand, *Understanding Batteries*, The Royal Society of Chemistry, Cambridge (2001).
5. B. E. Conway, *Electrochemical Supercapacitors*, Kluwer Academic/Plenum Publishers, New York (1999).

Selected Textbooks on Electrochemical Methods

6. Southampton Electrochemistry Group, *Instrumental Methods in Electrochemistry*, Ellis Horwood Limited, John Wiley & Sons, Chichester (1985).
7. C. M. A. Brett and A. M. Oliveira Brett, *Electrochemistry: Principles, Methods and Applications*, 5th edition, Oxford University Press, Oxford (2002).
8. A. J. Bard and L. R. Faulkner, *Electrochemical Methods: Fundamental and Applications*, 2nd edition, Wiley, New York (2000).
9. H. H. Giraud, *Physical and Analytical Electrochemistry*, Presses Polytechniques et Universitaires Romandes, Lausanne (2007).
10. C. Gabrielli, *Identification of Electrochemical Processes by Frequency Response Analysis*, Solartron Instrumentation Group, Solartron Schlumberger, Farnborough, U.K. (1980); see also http://www.solartronanalytical.com/downloads/technotes/technote04.pdf.
11. J. R. Scully, D. C. Silverman, and M. W. Kendig (eds), *Electrochemical Impedance: Analysis and Interpretation*, ASTM, Philadelphia (1993).
12. J. O'M. Bockris and A. K. N. Reddy, *Modern Electrochemistry*, Plenum Press, New York (1970).
13. D. D. Macdonald, *Transient Techniques in Electrochemistry*, Plenum Press, New York (1977).
14. F. Leardini, J. F. Fernandez, and F. Cuevas, *J. Electrochem. Soc.*, 154 (2007) A507–A514.
15. L. Wang, Y. Huang, R. Jiang, and D. Jia, *Electrochim. Acta*, 52 (2007) 778–783.
16. M. Koltypin, V. Pol, A. Gedanken, and D. Aurbach, *J. Electrochem. Soc.*, 154 (2007) A605–A613.
17. C. J. Wen, B. A. Boukamp, and R. A. Huggins, *J. Electrochem. Soc.*, 126 (1979) 2258.
18. B. C. Han, A. Van Der Ven, D. Morgan, and G. Ceder, *Electrochim. Acta*, 49 (2004) 4691–4699.
19. M. Toupin, D. Bélanger, I. R.Hill, and D. Quinn, *J. Power Sources*, 140 (2005) 203–210.
20. P. L. Taberna, P. Simon, and J. F. Fauvarque, *J. Electrochem. Soc.*, 150 (2003) A292–A300.
21. N. Yabuuchi, S. Kumar, H. H. Li, Y. Kim, and Y. Shao-Horn, *J. Electrochem. Soc.*, 154 (2007) A566–A578.

22. W. Weppner and R. A. Huggins, *J. Electrochem. Soc.*, 124 (1977) 1569–1571.
23. A. Ayad, J. Bouet, and J. F. Fauvarque, *J. Power Sources*, 149 (2005) 66–71.
24. S. K. Zecevic, J. S. Wainright, M. H. Litt, S. L. Gojkovic, and R. F. Savinell, *J. Electrochem. Soc.*, 144 (1997) 2973–2982.
25. J. Zheng, *J. Power Sources*, 137 (2004) 158–162.
26. C. Portet, P. L. Taberna, P. Simon, and E. Flahaut, *J. Electrochem. Soc.*, 153 (2006) A649–A653.
27. R. De Levie, *Electrochim. Acta*, 8 (1963) 751 (in 3 parts).
28. J. R. Miller, In *Proceedings of the Second International Symposium on EDLC and Similar Energy Storage Sources*, S. P. Wolsky and N. Marincic (ed.), Florida Educational Seminars, Boca Raton, FL (1992).
29. M. Keddam and H. Takanouti, In *Electrochemical Capacitors*, I. F. M. Delvick, D. Ingerssol, X. Andrieu, and K. Naoi (ed.), Electrochemical Society Series, Pennington, NJ, PV 96-25 (1996), p. 220.
30. R. De Levie, *Electrochim. Acta*, 9 (1964) 1231.
31. H. Keiser, K. D. Beccu, and M. A. Gutjhar, *Electrochim. Acta*, 21 (1976) 539.
32. H. K. Song, Y. H. Yung, K. H. Lee, and L. H. Dao, *Electrochim. Acta*, 44 (1999) 3513.
33. K. S. Cole and R. H. Cole, *J. Chem. Phys. C*, 341 (1941).
34. J. R. Miller, In *Proceedings of the 8th International Symposium on EDLC and Similar Energy Storage Sources*, S. P. Wolsky and N. Marincic (eds), Florida Educational Seminars, Boca Raton, FL (1998).
35. J. H. Jang and S. M. Oh, *J. Electrochem. Soc.*, 151 (2004) A571–A577.
36. J. H. Jang, S. Han, T. Hyeon, and S. M. Oh, *J. Power Sources*, 123 (2003) 79–85.
37. J. Gamby, P. L. Taberna, P. Simon, J. F. Fauvarque, and M. Chesneau, *J. Power Sources*, 101 (2001) 109–116.

2 Structure and Texture of Carbon Materials

Michio Inagaki

CONTENTS

2.1   INTRODUCTION

Carbon materials have always played important roles for human beings, charcoal as heat source and adsorbent since prehistorical era, flaky natural graphite powder as pencil lead, soots in black ink for the development of communication techniques, graphite electrodes for steel production, carbon blacks for reinforcing the tires in order to develop motorization, membrane switches for making computers and control panels thinner and lighter, etc. In electrochemistry, electric conductive carbon rods and carbon blacks support the development of primary batteries; a compound of graphite with fluorine, graphite fluoride, improved the performance of primary batteries; and the reaction of lithium intercalation/deintercalation into the galleries of graphite was greatly promoting the development of lithium-ion rechargeable batteries. Carbon nanotubes and fullerenes developed recently are promoting the development of nanotechnology in various fields of science and engineering.

Carbon materials are predominantly composed of carbon atoms, only one kind of element, but they have largely diverse structures and properties. Diamond has a three-dimensional structure and graphite has a two-dimensional nature, whereas carbon nanotubes are one-dimensional and buckminsterfullerene C_{60} is zero-dimensional. Fullerenes behave as molecules, although other carbon materials do not. Graphite is an electric conductor and its conductivity is strongly enhanced by AsF_5 intercalation, becoming almost comparable to metallic copper, whereas diamond is completely insulating. Diamond which is the hardest material is used for cutting tools, and graphite is so soft that it can be used as a lubricant.

In this chapter, an overview on structures and textures of carbon materials is presented, through introducing a concept of carbon families, before going into the chapters concerning the detailed applications of carbon materials for electrochemical energy storage.

2.2   CARBON FAMILIES

2.2.1   CARBON–CARBON BONDS

The carbon atom has four electrons in its outermost 2s and 2p orbitals, namely $2s^2 2p^2$. When it forms a chemical bond with other atoms, it can have three different hybridizations, sp, sp^2, and sp^3. The hybridized sp orbital is composed of two σ electrons, associated with two π electrons, the sp^2 of three σ electrons with one π electron, and the sp^3 of four σ electrons. It is well-known that various combinations of these hybrid orbitals produce a variety of organic hydrocarbons, from aliphatic to aromatic ones. This situation is illustrated in Figure 2.1, by showing typical hydrocarbons. The carbon–carbon bonds (C–C bonds) using sp^2 hybrid orbitals give a series of aromatic hydrocarbons (benzene, anthracene, phenanthrene, etc.). The C–C bonds using sp^3 hybrid orbitals give various aliphatic hydrocarbons, such as methane, ethane, propane, adamantane, etc. Mixing the different types of hybridizations, sp, sp^2, and sp^3, in a molecule expands the variety of hydrocarbons.

Figure 2.1 also shows how the inorganic carbon materials, diamond, graphite, fullerenes, and carbynes, result from the extension of these organic molecules to giant ones [1,2]. A simple repetition of C–C bonds with sp^2 hybrid orbital gives planar hexagons of carbon atoms, as benzene, anthracene, ovalene, etc. The extension of this series of aromatic hydrocarbons reaches graphite, in which the σ bonds using the sp^2 hybrid orbitals are reinforced by the delocalized π electrons. The giant planes tend to stack on each other due to the interactions between the π-electron clouds of the stacked layers, giving rise to weak van der Waals-like interactions. The somewhat curved corannulene molecule is formed by associating a pentagon of sp^2 hybridized carbon atoms with five hexagons. By the polymerization of these corannulene molecules to form a closed shell, various sizes of carbon clusters are formed and the resultant carbon materials are called fullerenes,

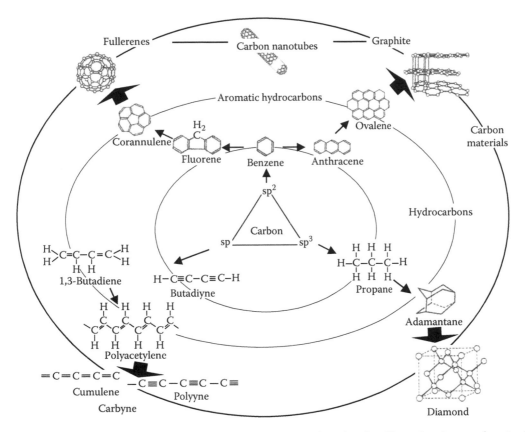

FIGURE 2.1 Construction of organic hydrocarbons and inorganic carbon families, using the sp, sp², and sp³ hybrid orbitals.

where carbon hexagons have to be located between the pentagons. The smallest closed shell is the buckminsterfullerene C_{60}, consisting of 12 pentagons and 20 hexagons of carbon atoms. As shown schematically in Figure 2.1, the infinite repetition of C–C bonds using sp³ hybrid orbitals results in a three-dimensional network of carbon atoms, which is known to be diamond. The carbon atoms in the organic compound named as adamantane and in diamond are exactly in the same atomic positions. The infinite repetition of C–C bonds with sp hybrid orbital gives the carbon materials called carbynes, in which the carbon atoms form linear chains either with double bonds (one σ-bond with one π-bond) or with the alternative repetition of single (one σ-bond) and triple bonds (one σ-bond with two π-bonds), the former being called cumulene type and the latter polyyne type.

Based on the extension by the repetition of three kinds of C–C bonds to infinite molecules, we may define carbon families, consisting of diamond, graphite, fullerenes, and carbynes [1]. In each family, the structure shows characteristic diversities; representative structures are listed in Figure 2.2. The detailed diversity in structure will be explained in the following sections for each family. Most structures in each family are thermodynamically metastable, and the family is represented by the allotrope name. In Figure 2.2, some possibilities to accept foreign species are also illustrated.

2.2.2 GRAPHITE FAMILY

The carbon family having a sp² bonding type is represented by graphite, in which the layers of carbon hexagons are stacked in parallel. The ABAB... sequence belongs to the hexagonal crystallographic system (hexagonal graphite) [3]. The ABCABC... stacking regularity is also possible,

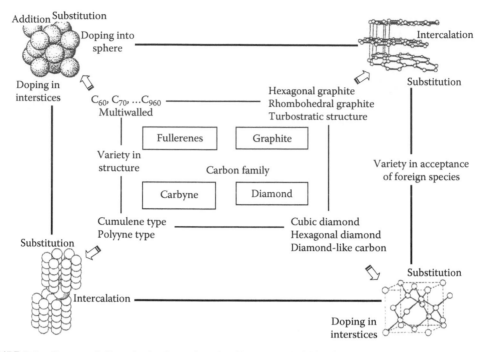

FIGURE 2.2 Structural diversity in the carbon family and possibilities for accepting foreign species.

and belongs to the rhombohedral system (rhombohedral graphite) [4]. The unit cells of these two allotropes are shown in Figure 2.3, together with their crystallographic data, space group, equivalent points, and lattice parameters. The rhombohedral and hexagonal cells are shown together in the figure presenting rhombohedral graphite.

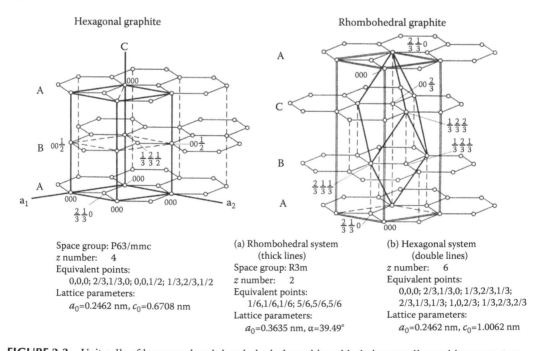

FIGURE 2.3 Unit cells of hexagonal and rhombohedral graphite with their crystallographic parameters.

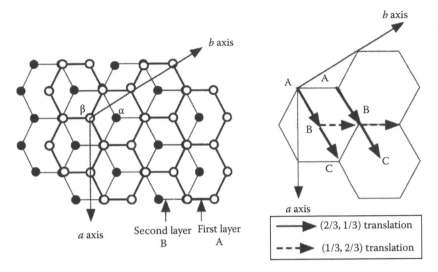

FIGURE 2.4 Translation of adjacent layers in hexagonal graphite to construct a regular stacking sequence.

The structural relation between hexagonal and rhombohedral graphite is explained in Figure 2.4. In hexagonal graphite, the second layer (B layer) is translated by (2/3, 1/3) along the a and b axes from the first layer (A layer). With a further translation by (1/3, 2/3) for the third layer, the total translation becomes unity in the two directions, which is exactly the same as for the A layer. Therefore, it is said to be the ABA stacking. However, a second translation again by (2/3, 1/3) is possible, and in this case the third layer is not at the same position as either the A layer or the B layer, and is thus denoted as C layer. By repeating the same translation (2/3, 1/3), the fourth layer coincides with the A layer, giving the ABCA stacking.

The rhombohedral structure of graphite was shown to form by applying shearing forces, for example, by grinding [5,6]. With increasing grinding time, additional lines indexed as 101 and 102 based on the rhombohedral structure appear around 44° in 2θ in the diffraction pattern of hexagonal graphite (Figure 2.5), because of the introduction of ABC stacking regularities (stacking faults in the ABA stacking regularity) [7]. The rhombohedral phase tends to saturate at around 33%, which seems to be reasonable because the rhombohedral stacking is formed mainly due to the introduction of stacking faults in the crystallites having the hexagonal stacking regularity.

In addition to these regular stacking modes, the parallel stacking of the layers without any regularity occurs mostly in the carbon materials prepared at low temperature, below 1300°C. In this case, the layers of hexagons are usually small in size, and only few layers are stacked in parallel. This random stacking of layers is called turbostratic structure [8]. In this random stacking, two kinds of displacement are possible, translation and rotation, as shown schematically in Figure 2.6. Although they are difficult to be differentiated in practical carbon materials by conventional techniques, their presence was recently shown through a detailed analysis of scanning tunneling microscopy (STM) images [9].

Figure 2.5 (top) shows the x-ray powder diffraction pattern of natural graphite with a high crystallinity. The diffraction lines of graphite have to be classified into three groups, with 00l, $hk0$, and hkl indices, mainly because of the strong anisotropy in structure. The lines with 00l indices are due to the reflection from the basal planes (hexagonal carbon layers and graphite basal planes), where only the l even indices are allowed because of the extinction rule due to the ABAB stacking sequence of the layers. The lines with $hk0$ indices are due to the reflection from the crystallographic planes perpendicular to the basal planes, and the lines with hkl indices come from the planes declined to the basal planes, where the three-dimensional structure has to be established with graphitic stacking of layers. Therefore, the hkl lines are called three-dimensional lines.

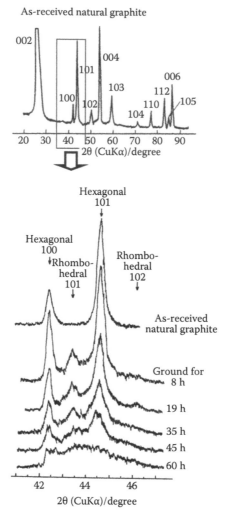

FIGURE 2.5   X-ray diffraction pattern of natural graphite and its change with grinding.

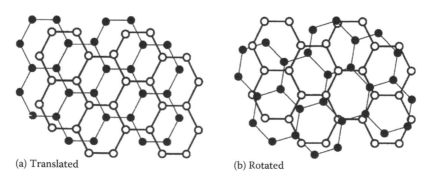

(a) Translated   (b) Rotated

FIGURE 2.6   Translation and rotation of neighboring hexagonal carbon layers to construct the turbostratic structure.

On the other hand, the powder pattern that is given by low-temperature-treated carbons is quite different from natural graphite, because it consists of random stackings of small layers, as shown on a petroleum coke heat-treated up to 1000°C in Figure 2.7. The diffraction pattern is characterized by very broad 00l lines due to the small number of stacked layers, and by unsymmetrical hk lines and not hkl lines due to the random turbostratic stacking of layers. The diffraction lines due to the planes perpendicular to the hexagonal carbon layers have a missing l index because of the absence of regularity in the direction along the normal to the layers; therefore, they are expressed as 10 and 11 in Figure 2.7.

When graphite is observed by scanning tunneling microscopy (STM), the hexagonal carbon layer looks like a triangular arrangement of spots (Figure 2.8a). Because of the interaction with the lower-lying atoms in the B layer, mainly the carbon atoms designated by α in Figure 2.4a are detected under STM. In the turbostratic structure, however, the interactions with the lower-lying atoms do not exist, because of the irregular stacking, and consequently, the spots form a hexagonal arrangement (Figure 2.8b) [9].

The grinding of natural graphite for a long time results in broadening all the diffraction lines, including the 100 line for hexagonal graphite and the 101 line for rhombohedral graphite (Figure 2.5).

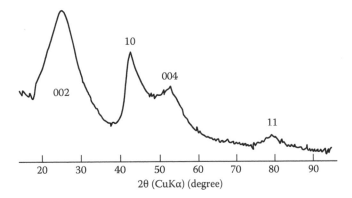

FIGURE 2.7   X-ray powder diffraction pattern of a petroleum coke heat-treated at 1000°C having a typical turbostratic structure.

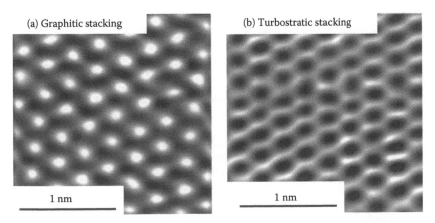

FIGURE 2.8   STM images for the graphitic and turbostratic stackings. (Courtesy of Prof. K. Oshida, Nagano National College of Technology, Nagano, Japan. With permission.)

After 60 h of grinding, the 100 line becomes broad and unsymmetrical, indicating that introducing too many stacking faults leads to a turbostratic structure [7].

Since this turbostratic structure can be partly transformed into a regular stacking of layers with a heat treatment at high temperatures, a wide range of structural diversity may result in the graphite family [10,11]. The graphitization degree is defined as the probability of occurrence of the regular graphitic AB stacking in turbostratic stackings. In different scales, the graphite family can have various textures, mainly because the basic structural units (BSU), i.e., the stacked hexagonal carbon layers, are highly anisotropic. The texture at the nanometric scale (nanotexture) will be separately discussed on the carbon materials classified in the graphite family.

2.2.3 FULLERENE FAMILY

Although based on sp²-type hybridization of carbon atoms, the bonding nature in fullerenes is different from graphite in the fact that the molecule is curved by the introduction of pentagons, as seen in corannulene (Figure 2.1). The repetition of this curved molecule can result in various closed shells, from the smallest closed shell of C_{60} to giant fullerenes [12,13]. The introduction of additional hexagons while keeping a closed shell morphology leads to giant fullerenes, as shown on the left-hand side of Figure 2.9. In this fullerene family, the variety of structures is mainly due to the number of carbon atoms building the closed shell.

Fullerenes behave as molecules [14], i.e., most of them can be dissolved into an organic solvent, giving a characteristic color, which is exceptional in the carbon families because all other materials are not dissolved in any organic solvent.

As shown in the center of Figure 2.9, another way to increase the number of hexagons is to make two groups of six pentagons apart from each other, which results in a single-wall carbon nanotube. To some extent, the carbon nanotube seems to locate in between the graphite and fullerene families, because it has some of the graphite characteristics together with those of fullerenes. The closed ends of a nanotube consist of carbon pentagons with hexagons, which is a characteristic of the fullerene family, and the tube wall consists of carbon hexagons and can be considered to be rolled from an hexagonal carbon layer, which is the fundamental structural unit composing the graphite family. A single-wall carbon nanotube is characterized by its diameter and chiral vector [14,15]. The illustration on the right-hand side of Figure 2.9 presents an example.

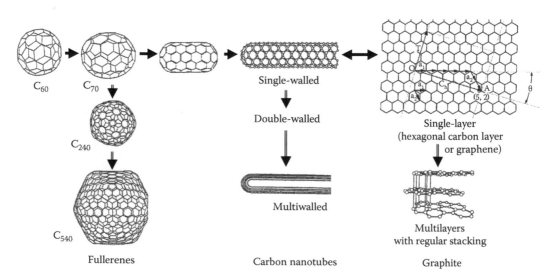

FIGURE 2.9 Structural relationships between fullerenes, carbon nanotubes, and graphite.

2.2.4 Diamond Family

Diamond consists of sp³-hybridized carbons, in which the purely covalent chemical bonds extend three-dimensionally. Diamond is very hard and electrically insulating. To construct a diamond crystal, a periodic and regular repetition of the C–C bond is required in a long range. For a couple of carbon atoms indicated as A and B in Figure 2.10a, the carbon atom A has to be connected with four carbon atoms, including B, to make a tetrahedron because of directional sp³ bonds, and the atom B has also to be surrounded by four carbon atoms, including A. Looking down these two tetrahedra centered on the A and B atoms along their connecting line, there are two possibilities in mutual relation between the two basement triangles consisting of three carbon atoms, which give two crystal structures. If these two basement triangles are rotated with respect to each other by 60°, as shown in Figure 2.10b, the resultant diamond crystal belongs to the cubic system (cubic diamond, Figure 2.10b'). If there is no rotation between these two basement triangles (Figure 2.10c), the structure belongs to the hexagonal system (hexagonal diamond, Figure 2.10c') [16]. Most of the diamond crystals, either occurring in nature or synthesized, are cubic.

In the case where a long-range periodicity is not attained between interconnected tetrahedra (Figure 2.10a), in other words, in the case of a random rotation between the two basement triangles of tetrahedra, an amorphous structure is formed [17]. Because of a random rotation between tetrahedra, some carbon atoms cannot form chemical bonds with the neighboring carbon atoms, most of which are supposed to be connected with hydrogen atoms for stabilization. This is usually obtained as a thin film, mainly because a random repetition of tetrahedra is difficult to keep on a long distance, and is called diamond-like carbon (DLC) [17]. DLC is as hard as diamond crystal, because it is also based on sp³-hybridized atoms, and it contains a relatively large amount of hydrogen.

A metastable crystalline phase of carbon atoms, consisting of a face-centered cubic unit cell with a parameter of 0.3563 nm, was found in either thin films or nanoparticles [18].

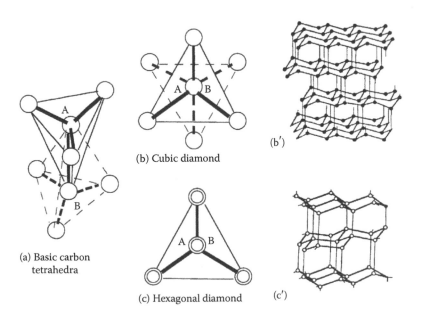

(a) Basic carbon tetrahedra

(b) Cubic diamond (b')

(c) Hexagonal diamond (c')

FIGURE 2.10 Construction of the cubic and hexagonal diamond structures based on tetrahedra of carbon atoms.

2.2.5 CARBYNE FAMILY

Carbynes are supposed to be formed of sp-hybridized carbon atoms bound linearly, where two π electrons have to be involved, giving two possibilities, i.e., an alternative repetition of single and triple bonds (polyyne) and a simple repetition of double bonds (cumulene) (Figure 2.1) [19]. The detailed structure of carbynes is not yet clarified, but some structural models have been proposed [19–22]. A structural model is illustrated in Figure 2.11, where some numbers of sp-hybridized carbon atoms form chains that associate together by van der Waals interaction between π-electron clouds to make layers, and then the layers are stacked. Foreign atoms are intercalated between the layers that are supposed to stabilize the carbyne structure. In the carbyne family, the variety of structures seems to be mainly due to the number of carbon atoms forming a linear chain, in other words, to the layer thickness, and to the density of chains in a layer.

2.2.6 MODES OF ACCEPTANCE OF FOREIGN SPECIES

Each carbon family also shows various characteristic possibilities for accepting foreign species. Diversities in foreign atom acceptance are illustrated for each family in Figure 2.2.

The substitution of either boron or nitrogen atoms for carbon atoms is possible for all carbon families, even though the amount substituted is very limited [23,24]. Since this substitution is expected to modify the properties of carbon materials without noticeable structural changes, various modifications have been investigated on different types of carbon materials. Nitrogen-substituted diamond crystals occur rarely in nature. Boron substitution has extensively been studied to improve the functions of carbon materials in the graphite family. Its effects on accelerating the development of graphitic stacking [25,26] and on improving the electrode performance in lithium-ion rechargeable batteries have been reported [27]. Recently, nitrogen-containing carbon materials were reported to have a high capacitance in electrochemical capacitors [28], though the state of the nitrogen atoms in the carbon matrix was not yet well understood. Nitrogen substitution in buckminsterfullerene, $C_{59}N$ (heterofullerene), was also reported [29].

Intercalation between the layers of graphite and carbynes is effective and useful for the modification of their properties. Only the intercalation of either iron or potassium is reported to stabilize

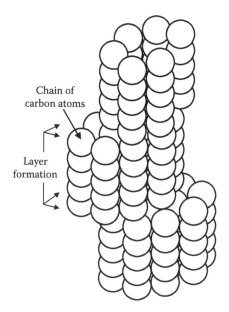

FIGURE 2.11   Structural model for a carbyne.

the structure of carbynes [19]. By contrast, a wide variety of foreign species, atoms, ions, and even molecules, have been intercalated into the carbon materials belonging to the graphite family, and interesting results have been obtained [30,31]. Either donors (such as lithium, potassium, etc.) or acceptors (such as sulfuric acid, bromine, ferric chloride, etc.) can be intercalated in the galleries between neighboring hexagonal carbon layers. The intercalation of sulfuric acid is widely used for the production of exfoliated graphite, which is converted to a graphite sheet for shielding and packing [32]. Intercalation/deintercalation of lithium ions is the fundamental electrochemical reaction for the discharge/charge of lithium-ion rechargeable batteries [33]. AsF_5 and SbF_5 graphite intercalation compounds were reported to have high electrical conductivity [34,35], higher than metallic copper, but they were never used in practice mainly because of instability of the compounds in air. The buckminsterfullerene C_{60} crystal can be doped with alkali metals (AMs) (e.g., potassium), which occupy all the tetrahedral and octahedral interstices between the molecules, giving rise to a superconductive material $C_{60}(AM)_3$ [36].

In the fullerene family, in addition to the substitution of carbon atoms in the shell and doping in the interstices between molecules, there are two additional possibilities to accept foreign species, the insertion into the inner space of the closed clusters [37] and the covalent bonding of organic radicals [38]. Rare earth elements can be inserted into fullerene cages (endohedral metallofullerenes). The Gd-metallofullerenol $Gd@C_{82}(OH)_n$ has been reported to be useful as a contrast agent for magnetic resonance imaging [39].

In the carbyne family, doping by intercalation into the space between the carbon chains in a layer is theoretically possible, but to date experimental results have not been reported.

2.3 FORMATION OF CARBON MATERIALS

2.3.1 CARBONIZATION PROCESS

When carbon materials, in which the carbon atom is the principal constituent, are produced by heat treatment of some organic polymers, i.e., carbon precursors, up to a temperature around 2000°C in an inert atmosphere, the process is called carbonization. The general scheme of the carbonization process is shown in Figure 2.12 for a solid carbon precursor, such as phenolic resin or cellulose, by indicating the main outgassed components and changes in residues, although it strongly depends on the carbon precursor used.

At the beginning of precursor pyrolysis, aliphatic and then aromatic molecules with low molecular weight are released as gases, mainly because some of the C–C bonds are weaker than C–H bonds. Associated with this hydrocarbon release, cyclization and aromatization proceed in the

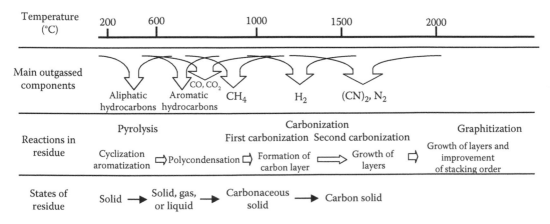

FIGURE 2.12 Carbonization process of a solid precursor.

residues and then polycondensation of aromatic molecules occurs. From around 600°C, mainly foreign atoms such as oxygen and hydrogen are released as CO_2, CO, and CH_4. In this stage, the residues are in either liquid or solid phase depending on the starting precursors. Above 800°C, the main gas evolved is H_2 because of polycondensation of aromatics, and the residues may be called carbonaceous solids that still contain hydrogen. The residual solids carbonized above 800°C are often called carbon materials, but still contain hydrogen and a small amount of foreign atoms, such as O, N, and S. To eliminate the latter, a heat treatment up to 2000°C is required, strongly depending on the chemical state of these foreign atoms in the carbonaceous solid. The pyrolysis of precursors and carbonization of the pyrolysis products, i.e., carbonaceous solids, often overlap with each other. The carbonization process may be divided into two steps, the first carbonization being associated with the evolution of hydrocarbon gases and the second carbonization with the evolution of light gases. Usually, the whole process from the precursor to the carbon material is called "carbonization." This process depends strongly on the carbon precursors and also on the heat treatment conditions, mainly temperature, heating rate, pressure, and concentration (particularly in the case of gaseous precursors).

As an example, the evolution of gases during the carbonization of a polyimide film (commercially available Kapton film with 25 μm thickness) together with the changes in weight and size of the film up to 1000°C is shown in Figure 2.13 [40]. The carbonization proceeds in three steps: (i) the first step occurs in a rather narrow temperature range of 500°C–600°C, showing an abrupt weight loss associated with the evolution of a large amount of carbon monoxide and a pronounced shrinkage along the film (pyrolysis); (ii) the second step occurs in the temperature range from 600°C to 1000°C (first carbonization) with a small weight loss, the evolution of small amounts of methane and hydrogen, and a little shrinkage; (iii) the third step is accompanied by the evolution of nitrogen, a negligibly small shrinkage, and a small weight loss (second carbonization). The release of nitrogen in the third step was found to continue up to temperatures higher than 2000°C, leaving a large number of pores in the film if it was heated continuously up to 2400°C [41]. In the case of polyimides, the "carbonization" process is rather simple, as shown above, and it is known that the starting molecular structure of the polyimide used governs the structure and nanotexture of

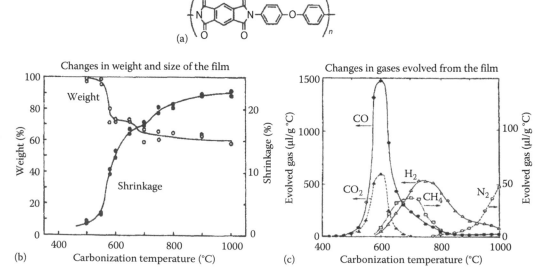

FIGURE 2.13 Changes in weight, film size, and amount of evolved gases with carbonization temperature for a polyimide (Kapton) film.

TABLE 2.1

Carbonization Processes for the Production of Various Carbon Materials

Carbonization Process	Precursors	Carbon Materials	Characteristics
Gas-phase carbonization	Hydrocarbon gases, decomposition in space	Carbon blacks	Fine particles, so-called "structure"
	Hydrocarbon gases, deposition on substrate	Pyrolytic carbons	Various textures, preferred orientation
	Hydrocarbon gases, with metal catalysts	Vapor-grown carbon fiber (VGCFs), carbon nanofibers	Fibrous morphology, various nanotextures
	Hydrocarbon gases, without any catalyst	DLC	Thin film, sp^3 bond, amorphous structure
	Carbon vapor	Carbon nanotubes	Tubular, single-wall, and multiwalled
	Carbon vapor	Fullerenes	Spherical, molecular nature
Solid-phase carbonization	Plants, coal, and pitches	Activated carbons	Highly porous adsorptivity
	Furfuryl alcohol, phenol resin, cellulose, etc.	Glass-like carbon	Amorphous structure, gas impermeability, conchoidal fracture surface
	Poly(acrylonitrile), pitch, cellulose, and phenol resin with stabilization	Carbon fibers	Fibrous morphology, high mechanical properties
	Polyimide films	Carbon films	Film wide range of graphitizability
Liquid-phase carbonization	Pitch, coal tar	Cokes Mesocarbon microbeads (MCMB)	Spherical particles
	Cokes with binder pitches	Polycrystalline graphite blocks	Various densities, various degrees of orientation

the resultant carbons. However, the carbon precursors used for industrial carbonization processes, such as pitches, have a much more complicated molecular composition so that their carbonization processes are much more complicated.

In Table 2.1, the carbonization processes are classified on the basis of the intermediate phases of the precursor used [42], and the representative carbon materials formed are listed with their characteristics.

As it will be explained below, the nanotexture of carbon materials is principally established during the carbonization process. Moreover, nanotexture has determining effects on the structural changes at high temperatures and also on the properties of carbons. Therefore, carbonization is the most important process for producing carbon materials.

2.3.2 GAS-PHASE CARBONIZATION

When carbon materials are produced through the decomposition of hydrocarbon molecules in gaseous phase, the process is called gas-phase carbonization. A large variety of carbon materials over the whole carbon family was produced through gas-phase carbonization (Table 2.1). For a high concentration of hydrocarbon gases, carbon blacks are obtained. When a solid substrate is placed in the carbonization system, pyrolytic carbons are produced. If some fine metallic particles, such as Fe or Ni, are added to the carbonization system, various carbon materials, carbon fibers (VGCFs) and carbon nanofibers with various nanotextures and morphologies are produced. From the carbon vapor produced by electric-arc discharge or laser ablation, carbon nanotubes and fullerenes are formed.

2.3.2.1 Carbon Blacks

Carbon blacks formed through incomplete combustion of either gaseous or mist-like hydrocarbons are very important industrial products [43]. They have been used as colorants in inks since the third century AD and are now applied in large amount for rubber reinforcement. They are characterized by spherical primary particles, which are in different sizes and more or less coalesced into aggregates, what is called "structure" in the field of rubber reinforcement (Figure 2.14). On the basis of the reaction process and raw materials, carbon blacks are classified into furnace blacks, channel blacks, lamp blacks, thermal blacks, and acetylene blacks. Ketjenblack, which is produced as a by-product of heavy oil gasification and has a size of primary particles of 30–40 nm and a high surface area (ca. 1300 m²/g) [44], and also acetylene black, have been frequently used as conductive additives in the electrodes of primary and secondary batteries and also for electrochemical capacitors.

2.3.2.2 Pyrolytic Carbons

Carbon deposition occurs on the surface of a substrate inserted into the carbonization system using hydrocarbon gases, such as methane and propane [45]. This process is a kind of chemical vapor deposition (CVD) and the products are called pyrolytic carbons. In order to control the structure, the deposition conditions have to be controlled. The deposition can occur on either static or dynamic substrates. In the former, the substrate is placed in a furnace, which is heated either by direct passing of electric currents or from the surroundings. In the latter, small substrate particles are fluidized

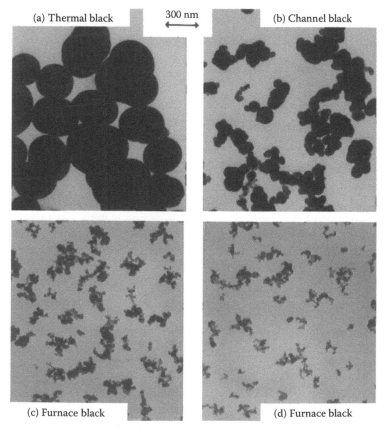

FIGURE 2.14 Transmission electron micrographs of carbon blacks.

in the furnace. The factors controlling the structure and properties of pyrolytic carbons are (i) the kind of precursor hydrocarbon gas, (ii) the gas concentration, (iii) the deposition temperature, (iv) the contact time between the gas molecule and the substrate heated at a high temperature, which is governed by the gas flow rate and the size of the furnace, and finally (v) the geometrical arrangement of the furnace, particularly the ratio of the substrate surface area to the volume occupied by the gas [46]. In order to prepare pyrolytic carbon with high density at 1400°C–2000°C, it is necessary to use a low methane pressure as low as 1.7×10^{-2} mbar. High methane pressure of 4 mbar gives low density pyrolytic carbon in this temperature range, which is known to be mainly due to the formation of carbon black particles incorporated into the pyrolytic carbon [45].

2.3.2.3 Vapor-Grown Carbon Fibers and Carbon Nanofibers

VGCFs are formed on a substrate from benzene vapor in a flow of high-purity hydrogen using fine iron particles as catalyst [47,48]. The transmission electron microscopy (TEM) analysis using various modes (bright-field images, selected area electron diffraction, dark-field images with 002 and 10 diffractions, and 002 and 10 lattice fringes) shows that the resultant VGCFs are constituted of two parts, as shown in Figure 2.15: a hollow tube with few straight layers (carbon nanotube) (Figure 2.15b and c) and oriented small layers deposited on the wall of the hollow tube (Figure 2.15d) [49]. The long and straight layers of the central hollow tube are aligned exactly

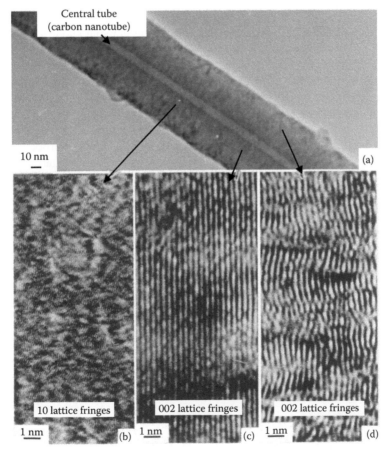

FIGURE 2.15 Transmission electron micrograph and lattice fringe images for VGCF. (Courtesy of Prof. M. Endo, Shinshu University, Nagano, Japan. With permission.)

parallel to the fiber axis, and the small layers deposited on the wall of the tube are preferentially and concentrically oriented along the fiber axis. The formation of similar VGCFs was reported by using different precursor hydrocarbon gases, carrier gases, catalyst metals, and deposition conditions [50–52]. To get a high yield of VGCFs, the floating catalyst method was developed [53], in which nanometric catalyst metal particles are formed *in situ* by the decomposition of its precursor, instead of the method where the catalyst is seeded on the surface of a substrate plate (seeding catalyst method).

Since most VGCFs can be changed to well-graphitized carbon fibers and can be prepared with high purity, without the coexistence of other carbon forms like carbon blacks, several works were performed to prepare fibrous carbons through the decomposition of various gases, not only hydrocarbons but also CO, under different conditions [54–58]. Similar works reported recently are aiming to prepare highly pure carbon nanotubes with a high yield [59,60]. Now, these fibrous carbons are called nanofibers (or nanofilaments), and are differentiated from carbon nanotubes and also from VGCFs [61]. By selecting the decomposition conditions, carbon helical microcoils with a controlled pitch were prepared [62–64]. Fibrous carbons with a stacking of cup-like carbon layers along the fiber axis, cup-stacked nanofibers, were also prepared in a similar process [65,66]. Carbon nanohorns, showing a high storage capacity for methane, were also synthesized by a similar technique [67]. The addition of a small amount of VGCF into the electrode of lead-acid rechargeable batteries was reported to improve the performance [68].

2.3.2.4   Carbon Nanotubes

Carbon nanotubes were found in the carbon deposits on the graphite anode after arc discharge in He atmosphere [69]. The temperature at the graphite electrode was estimated to reach up to 2500°C. By selecting the arc discharge conditions between the graphite electrodes, relatively high yields of carbon nanotubes were reported. The hollow tubes thus obtained show a wide range of diameters from 1 to 50 nm, and their walls consist of different numbers of carbon layers. Most of them are closed at the end with the help of pentagons, the smallest diameter being the same as the size of the smallest fullerene C_{60}. The carbon nanotubes consisting of a single carbon layer were also found later [70,71]. Figure 2.16 shows the TEM image of a carbon nanotube. Vertically aligned single-wall carbon nanotubes were synthesized from an alcohol with a high formation rate, their length reaching up to 5 μm [72]. The decomposition of a SiC single crystal wafer under reduced pressure (1×10^{-4} mbar) at 1700°C was found to give carbon nanotubes, which are well aligned perpendicular to the precursor wafer [73,74].

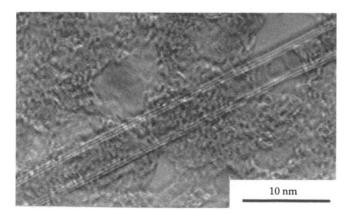

10 nm

FIGURE 2.16   Transmission electron micrograph of a carbon nanotube. (Courtesy of Prof. M. Endo, Shinshu University, Nagano, Japan.)

2.3.2.5 Fullerenes

The carbon cluster C_{60} was first found in the soot formed by laser ablation of graphite in He atmosphere and was named as buckminsterfullerene [12,13]. C_{60} was also found in the carbon deposits formed by arc discharge between graphite electrodes [75]. This carbon cluster C_{60} could be extracted using solvents, such as toluene, carbon disulfide, etc., and could also be separated from the deposits by vaporization at around 400°C in vacuum [76]. Carbon clusters with different sizes, such as C_{70}, C_{82}, ..., C_{960}, were identified and this series of clusters was called fullerenes. Among the different gas atmospheres, which were examined for the formation of fullerenes, He atmosphere gives relatively high yield of fullerenes in most cases.

2.3.3 SOLID-PHASE CARBONIZATION

In contrast to gas-phase carbonization, most thermosetting resins, such as phenol-formaldehyde and furfuryl alcohol, and also cellulose can be converted to carbon materials by solid-phase carbonization. When the carbonization of most of these precursors proceeds rapidly, the resultant carbon materials become porous. If the carbonization is performed so slowly that the resultant carbonaceous solids can shrink completely, the so-called glass-like carbons are produced, which contain a large number of closed pores.

2.3.3.1 Activated Carbons

In activated carbons, a large number of open pores are formed, which are usually evaluated by surface area and pore size distribution [77–79]. They have been used as adsorbents since prehistoric age, and are now used widely not only in our daily life but also in various industries for production processes and for the treatment of the waste. For the predominant applications of activated carbons as adsorbents, the pore texture is the most important property to be controlled. In order to develop the small pores, micropores with a width less than 2 nm, a slight oxidation of carbonized materials is usually carried out, which is called "activation." Various carbon precursors, not only thermosetting resins, including natural biomass as plants, but also some thermoplastic resins, including pitches and coals, are used. In Figure 2.17a, the scanning electron microscopy (SEM) micrograph of a carbon prepared from coconut shell, which is used, for example, as a filter in cigarettes, is shown. The pores observed by SEM are macropores, which originated from the cell structure of the precursor coconut shell.

These macropores are not effective for adsorption of various molecules, but their presence before activation is preferable for creating micropores in the walls. The pore texture of most activated carbons is illustrated in Figure 2.17b, where macropores (>50 nm width) and mesopores (2–50 nm

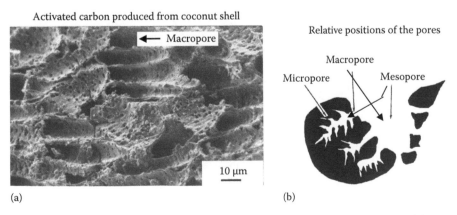

Activated carbon produced from coconut shell

Relative positions of the pores

(a) (b)

FIGURE 2.17 Scanning electron micrograph of an activated carbon and pore structure model.

width) work as paths for oxidizing agents to create micropores during the activation process and also for adsorbate molecules to reach the micropores during adsorption application. Figure 2.18 shows a conventional process for the production of activated carbon from thermosetting and thermoplastic precursors. For thermoplastic precursors, the so-called stabilization, which is a slight oxidation, is required prior to carbonization. For activation, different processes are employed: oxidation using diluted oxygen gas, air, steam, CO_2, etc., and chemical oxidation using H_3PO_4, $ZnCl_2$, KOH, etc. [77–79]. During gas activation, the pores may grow from micropores to mesopores and finally to macropores. Figure 2.19 shows the changes in pore size during air activation of glass-like carbon spheres [80]. At the very beginning of oxidation, ultramicropores (<0.7 nm) are formed, presumably by the opening of closed pores existing in the glass-like carbon matrix, and then the ultramicropores

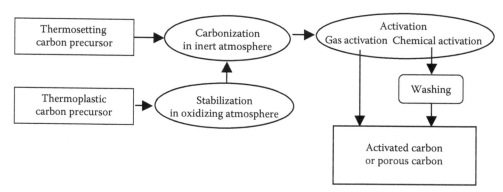

FIGURE 2.18   Preparation process of activated carbon from thermosetting and thermoplastic precursors.

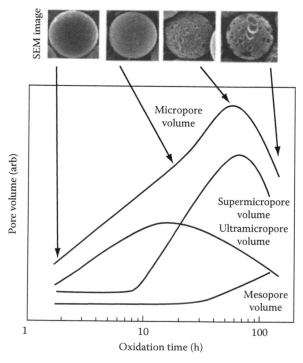

FIGURE 2.19   Progressive change from ultramicropores through supermicropores to mesopores during air activation of glass-like carbon spheres.

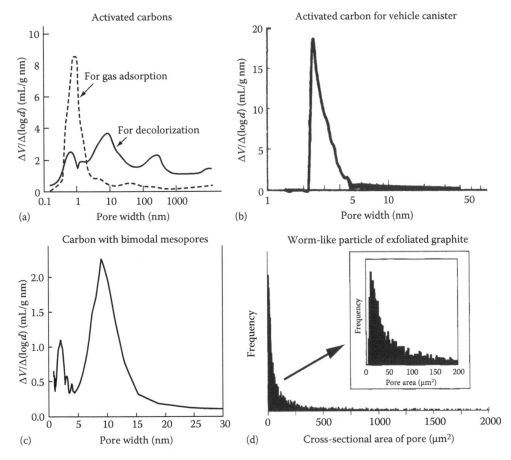

FIGURE 2.20 Pore size distributions in various porous carbons.

grow to supermicropores (0.7–2 nm). For chemical activation, the carbonization and activation processes are often combined in one heat treatment. Recently, a great success to get a high surface area
reaching 3000 m²/g was obtained using KOH for activation [81].

The pore size distributions of activated carbons used for the adsorption of small gas molecules
and decolorization of water are shown in Figure 2.20a. By the control of the micropore size, the
activated carbons may be successfully used as molecular sieves for various gases. A proper combination of micropores and mesopores is pointed out to be important for the electrodes of electric
double layer capacitors to get a high performance [82,83]. Mesopores are effective for the adsorption
of gasoline vapor composed from large hydrocarbon molecules (Figure 2.20b) [84]. Figure 2.20c
shows a bimodal distribution of mesopores in the carbon prepared from Mg citrate and gluconate
mixture, which has an excellent rate performance in the electric double layer capacitors [83]. Figure
2.20d shows a distribution of cross-sectional area of macropores formed in worm-like particles of
exfoliated graphite, which play an important role in the recovery of spilled heavy oils [85,86] and
macromolecular biomedical fluids [87].

2.3.3.2 Glass-Like Carbons

Glass-like carbons (glassy carbons) are produced by the pyrolysis of different precursors, such
as phenol–formaldehyde resin, poly(furfuryl alcohol), cellulose, etc., through an exact control
of the heating process [88,89]. They are characterized by an amorphous structure and also by

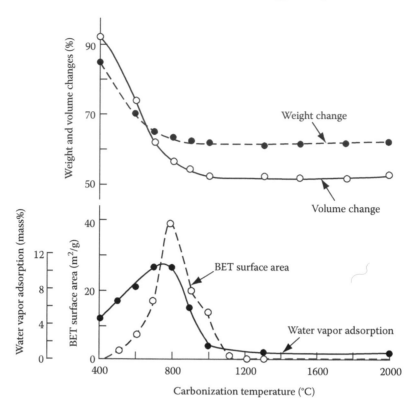

FIGURE 2.21 Changes in weight, volume, BET surface area, and water vapor adsorption during the carbonization of poly(furfuryl alcohol) to form glass-like carbon. (Courtesy of Dr. S. Yamada, Tokai Carbon Co. Ltd., Shizuoka, Japan. With permission.)

properties very similar to inorganic glasses, such as high hardness, brittle conchoidal fracture, and gas impermeability. The most important parameter for their production is a very slow heating slower than the rate of shrinkage to compensate the formation of open pores due to the evolution of decomposition gases from the precursor block. In Figure 2.21, the changes in weight, volume, BET specific surface area (surface area determined by the Brunauer, Emmett, Teller method), and adsorption of water vapor with carbonization temperature are shown for poly(furfuryl alcohol) condensates [88]. The rapid decreases in weight and volume up to 700°C–800°C are due to the precursor pyrolysis and correspondingly the Brunauer, Emmett and Teller (BET) surface area and adsorption of water vapor increase, for which micropores are known to be responsible. Above 800°C, however, the BET surface area and water vapor adsorption decrease quickly, suggesting that most of the pores are closed. The commercially available glass-like carbons have a bulk density of 1.3–1.5 g/cm³, and show a very low gas permeability of 10⁻¹² cm²/s, almost gas-impermeable. It suggests that a large number of pores exist in glass-like carbons, but all of them are closed. The pores formed initially during the pyrolysis and carbonization of the precursor below 800°C are closed above 800°C, while the bulk of the product shrinks, so that the heating rate for carbonization is crucial for the production of glass-like carbons. The formation of closed pores with a rather homogeneous size of about 3 nm was reported on a glass-like carbon, the volume of which reached around one-third of the bulk [90].

2.3.3.3 Carbon Fibers

In order to produce carbon fibers from poly(acrylonitrile) (PAN) and various pitches, stabilization is essential after the spinning, which consists of a chemical reaction using different oxidizing gases, such as air, oxygen, chlorine, hydrochloric acid vapor, etc. [91]. The stabilized fibers are then

TABLE 2.2
Classification of Carbon Fibers Based on Their Precursors and Their Characteristics

Precursor	Carbon Fibers	Characteristics
PAN	PAN-based carbon fibers	Different grades and types
Isotropic pitch	Isotropic-pitch-based carbon fibers	General-purpose grade
		Random cross-sectional nanotexture
Mesophase pitch	Mesophase-pitch-based carbon fibers	High modulus types
		Various cross-sectional nanotextures
Cellulose	Cellulose-based carbon fibers	General-purpose grade
		Random nanotexture
Phenol	Phenol-based carbon fibers	General-purpose grade
		Random nanotexture
Hydrocarbon gases	VGCFs	High graphitizability
		Ring texture in cross section

subjected to solid-state carbonization (extrinsic solid-phase carbonization) in inert atmosphere. The resultant carbon fibers are called PAN- and pitch-based carbon fibers, the latter being classified into two series, isotropic-pitch- and mesophase-pitch-based, depending on the pitch precursors subjected to spinning. When cellulose and phenol fibers are used as precursors, the stabilization process is not necessary and the spinned fibers are subjected directly to carbonization (intrinsic solid-state carbonization). In Table 2.2, carbon fibers are classified on the basis of their precursor, and their characteristics are summarized. Based on their mechanical properties, carbon fibers are classified into two grades, general-purpose and high-performance grades. The latter is further divided into two types, high modulus and high strength types [91].

2.3.3.4 Carbon Films

Polyimides have been developed as thermoresistant polymers and used in different fields, especially in the field of electronics. Polyimide films with different molecular structures have been developed because of their practical and promising applications. The heat treatment of these films at high temperatures leads to a wide variety of carbon structures, from highly crystalline graphite to amorphous glass-like carbon films [92]. This is a typical case in which the molecular structure of the organic precursors and the texture of the films govern the structure and texture of the resultant carbon films, i.e., crystallinity and preferred orientation of hexagonal carbon layers. As shown on a polyimide Kapton film in Figure 2.13b [40], the yield after carbonization is relatively high, about 50 wt%, and the linear shrinkage along the film is slightly larger than 20%. Even if these changes in weight and size are not so small, it is noteworthy to mention that the film can keep its original form, although a little care to avoid mechanical shock is required. Figure 2.22 shows a crane made by paper folding, the form of which is kept after carbonization up to 1300°C and even after graphitization at about 3000°C.

2.3.4 Liquid-Phase Carbonization

Liquid-phase carbonization occurs for some precursors, such as pitches, which become viscous fluids before carbonization. This process has been used to produce various polycrystalline graphite blocks for steel refining and electrical discharge machining, jigs for the growth of semiconductor crystals, structural components of nuclear reactor, etc.

(a) Polyimide film   (b) After carbonization   (c) After graphitization

FIGURE 2.22   A crane made from a polyimide film by the paper folding technique and after its carbonization and graphitization. (Courtesy of Dr. H. Hatori, AIST, Tsukuba, Japan. With permission.)

2.3.4.1   Polycrystalline Graphite Blocks

To produce graphite blocks, coke is used as a filler and a pitch as a binder [93]. For the filler, cokes derived from petroleum and coal tar pitches are usually used, but natural graphite, carbon blacks, and also recycled graphite particles are sometimes included. In advance, the cokes are carbonized to avoid a large number of volatiles that might introduce some shape distortion and cracks in the product and also might reduce the density. For the binder, petroleum and coal tar pitches are used in most cases, because they have relatively high carbon yield as about 60 wt%, and also because they give a carbon similar to the filler cokes after carbonization. However, thermosetting resins, such as phenol and epoxy resins, are sometimes employed. The particle size of the filler and the mixing ratio of the filler to the binder have to be controlled, in accordance with the requirements of the applications. The filler and the binder are mixed at a temperature higher than the softening point of the binder. The mixture thus prepared, which is usually called carbon past, is formed after warming up at a temperature around 150°C by either extrusion, molding, or cold isostatic pressing (CIP or rubber pressing). The formed blocks are carbonized at a temperature of 700°C–1000°C (sometimes called calcination) and then graphitized at a high temperature above 2500°C. The carbon blocks heat-treated at a high temperature are often called "polycrystalline graphite" or "synthetic graphite." For producing graphite blocks, liquid-phase carbonization is involved in two steps, (i) in the process of filler coke preparation and (ii) in the carbonization of the binder pitch.

Forming is an important process for the fabrication of polycrystalline graphite blocks, because it governs the preferred orientation of the filler particles. The extrusion gives a preferred orientation of either flaky or needle-like filler particles along the direction of extrusion. Electrodes for metal processing with large diameter, carbon rods with different sizes, and also leads for automatic pencil with a diameter as thin as 0.3 mm are fabricated by this forming process. In the molding process, which is applied for the fabrication of carbon brushes for electric motors and electric contacts, the filler particles are statistically aligned perpendicular to the compressing direction. By isostatic pressing, the compressive force is applied hydrostatically; in this case, the filler particles are randomly oriented, which leads to producing isotropic high-density graphite blocks.

2.3.4.2   Cokes and Mesocarbon Microbeads

Cokes are also the product of liquid-phase carbonization of pitches, as mentioned above. The nanotexture of the resultant cokes can be changed by applying a shear stress during liquid-phase carbonization, giving the so-called needle-like cokes, which are now important raw materials in the production of large-sized graphite electrodes for metal refining [93].

In the course of liquid-phase carbonization of pitches, optically anisotropic spheres are formed first, which are called mesophase spheres. By further heating, the spheres grow and coalesce with

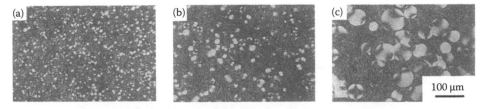

FIGURE 2.23   (a) Formation, (b) growth and (c) coalescence of mesophase spheres in a pitch.

FIGURE 2.24   Scanning electron micrograph of MCMB. (Courtesy of Kawasaki Steel Co. Ltd., Tokyo, Japan. With permission.)

each other to give solids (coke) with different textures composed of optically anisotropic units, called bulk mesophase [94,95]. A series of polarized-light microscope images (Figure 2.23) shows the sequence from the formation of small mesophase spheres to their growth and partial coalescence. The anisotropic mesophase spheres can be separated from the isotropic pitch matrix by using a solvent [96]. The separated spheres, shown in Figure 2.24, are called mesocarbon microbeads (MCMB) and have been used as the anode material of lithium-ion rechargeable batteries [97]. Mesophase-pitch-based carbon fibers are produced by spinning from a pitch containing a predominant amount of bulk mesophase [98].

2.3.5   NOVEL CARBONIZATION PROCESSES

The recent developments in science and technology require a more exact control of structure/ nanotexture and properties of various materials, including carbon materials. In order to meet the requirements for carbon materials, various novel carbonization processes have been proposed. In relation to electrochemistry, the following processes have to be mentioned: template method, polymer blend method, defluorination of fluorinated hydrocarbons, and carbonization of organic aerogels [99].

2.3.5.1   Template Method

The template carbonization technique was first developed for the preparation of thin, oriented graphite films using two-dimensional spaces in clay minerals with a layered structure [100,101]. A series of works applying this technique reveals a high possibility to control the dimensionality of morphology by selecting suitable template materials: one-dimensional carbon nanofibers were synthesized using anodic aluminum oxide films, two-dimensional graphite layers using layered compounds, and three-dimensional microporous carbons using zeolite. Carbon nanofibers were

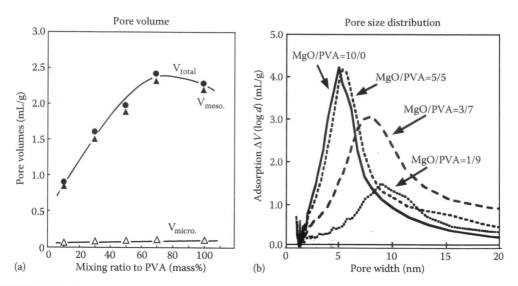

FIGURE 2.25 Pore volume and pore size distribution of mesoporous carbons obtained from the mixture of Mg citrate with poly(vinyl alcohol).

prepared by carbon deposition from propylene gas at 800°C on the inner walls of nanosized channels in an anodic aluminum oxide film [102]. Aluminum oxide is subsequently dissolved either by HF at room temperature or NaOH aqueous solution at 150°C in an autoclave. Porous carbons with a high BET surface area, higher than 2000 m²/g, were prepared using the three-dimensional channels of zeolite [103] and mesoporous silica [104]. The size and shape of the pores in the resultant carbon depend strongly on the template used. The materials obtained by this technique are extensively presented in Chapter 3.

MgO was also used as a template for obtaining mesoporous carbons. A thermoplastic carbon precursor [such as poly(vinyl alcohol), poly(ethylene terephthalate), and pitch] is mixed with a MgO precursor (MgO, Mg acetate, Mg citrate, and Mg gluconate) and heat-treated at 900°C in inert atmosphere. Then, MgO formed through the pyrolysis of the precursor is dissolved by a diluted acid solution, liberating the mesoporous carbon [105–107]. Since the mesopore size of the resultant carbon is governed by the size of MgO thus formed, the process may be classified as an autogenous template method. A large number of mesopores with a size of about 5 nm is easily obtained using Mg citrate [108]. Figure 2.25 shows the pore volume and pore size distribution as functions of the mixing ratio between Mg citrate and poly(vinyl alcohol) (PVA). From the mixture of Mg citrate and gluconate, a carbon with bimodal mesopore size was obtained (Figure 2.20c).

The template carbons demonstrate interesting performance in electric double layer capacitors [109], especially the mesoporous carbons prepared using MgO as the template [105,110]. Asymmetric electric double layer capacitors constructed from these mesoporous carbons coupled with a microporous activated carbon display high capacitance and high rate performance [111].

2.3.5.2 Polymer Blend Method

In order to control the pore texture in carbon materials, blending of two kinds of carbon precursors, the one giving a relatively high carbonization yield and the other having a very low yield, was proposed and called polymer blend method [112]. This idea gave certain success to prepare macroporous carbons from poly(urethane–imide) films prepared by blending poly(amide acid) and phenol-terminated polyurethane prepolymers [113]. By coupling this polymer blend method with

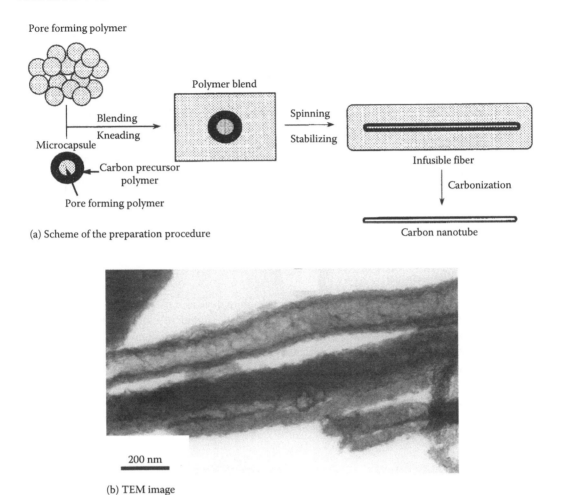

(a) Scheme of the preparation procedure

(b) TEM image

FIGURE 2.26 Scheme of the polymer blend method to prepare carbon nanotubes and their TEM image. (Courtesy of Prof. A. Oya, Gunma University, Kiryu, Japan. With permission.)

spinning, carbon nanofibers were successfully prepared [114]. The preparation procedure is schematically shown in Figure 2.26a, and a TEM image of the resultant carbon nanofibers is shown in Figure 2.26b.

2.3.5.3 Defluorination

Porous carbons with a high surface area are obtained through the defluorination of poly (tetrafluoroethylene) (PTFE) with alkali metals (AMs) [115]. The carbons derived from PTFE were found to have a large number of mesopores and to give a high electric double layer capacitance [116,117].

2.3.5.4 Carbon Aerogels

Mesoporous carbon aerogels are prepared by the pyrolysis of resorcinol–formaldehyde organic aerogels [118,119]. The effects of the type of organic precursor aerogel and of the preparation conditions, such as the drying process, on the pore texture of carbon aerogels have been extensively studied [120–122]. Also, the doping by some metals into the organic aerogels was reported to be effective for the development of micropores [123].

2.4   NANOTEXTURE IN THE GRAPHITE FAMILY

2.4.1   CLASSIFICATION OF NANOTEXTURES

In the graphite family, the fundamental structural unit is a stack of layers of carbon hexagons, which has a strong anisotropy because the bonds in the layers are covalent and those between the layers are of van der Waals type. The way the structural units tend to agglomerate is determined by the schemes and degrees of preferred orientation, and results in a variety of textures in carbon materials, which govern the graphitization behavior at high temperatures.

A classification has been proposed on the basis of the scheme of preferred orientation of the anisotropic structural units (Figure 2.27) [1,124]. Since these are textures constructed by fundamental structural units at a nanosized scale, they are called nanotextures. Firstly, random and oriented nanotextures are differentiated and, secondly, the latter is divided according to the scheme of orientation, in parallel to a reference plane (planar orientation), along a reference axis (axial orientation), and around a reference point (point orientation).

2.4.2   PLANAR ORIENTATION

The extreme case of planar orientation is a single graphite crystal, i.e., a perfect orientation with large-sized hexagonal carbon layers. Single crystals can be found in either naturally occurring graphite (natural graphite) or in precipitated graphite flakes from molten iron (kish graphite). However, finding large-sized crystals (more than 10 mm) is extremely difficult. The plates of the so-called highly oriented pyrolytic graphite (HOPG) [125] have a very high degree of planar orientation of the hexagonal carbon layers, but the size of the layers is not so large. In other words, the structure is close to a perfect orientation in the direction perpendicular to the plates, but it is polycrystalline in the parallel direction [126]. A SEM image on a fractured cross section of an

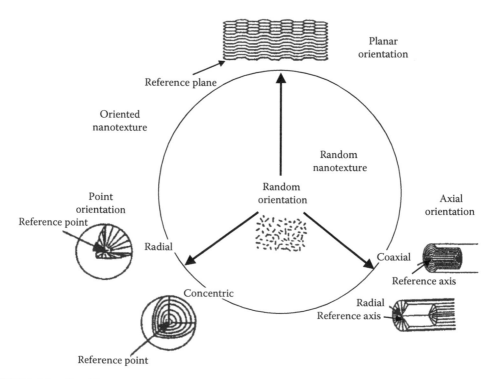

FIGURE 2.27   Classification of nanotexture for carbon materials in graphite family.

HOPG and a channeling contrast image of an HOPG plate surface are shown in Figure 2.28a and b, respectively. A well-ordered stacking of the hexagonal carbon layers is seen in Figure 2.28a, whereas Figure 2.28b demonstrates the polycrystalline nature, consisting of small grains of graphite basal planes in different orientations.

Various intermediate nanotextures between the perfectly planar and random orientations are found in pyrolytic carbons and coke particles, depending on the preparation and heat treatment conditions. In pitch-derived cokes, the planar orientation of the layers is improved with increasing heat-treatment temperature (HTT), as demonstrated through HRTEM [127]. As shown on a coke particle in Figure 2.29, the size of the 002 lattice fringes increases and their parallel stacking is improved with increasing HTT.

Needle-like coke has been widely used for the industrial production of graphite electrodes for steel refining. The changes in transverse magnetoresistance ($\Delta\rho/\rho_0$) with the rotation angles φ and θ of the magnetic field along and perpendicular to the axis of a needle-like particle (TL and T

| (a) SEM image of the edge surface | (b) Electron channeling image of the basal surface |

FIGURE 2.28 Scanning electron micrograph of a cross section of an HOPG and electron channeling image of an HOPG plate basal surface. (Courtesy of Prof. A. Yoshida, Musashi Institute of Technology, Tokyo, Japan. With permission.)

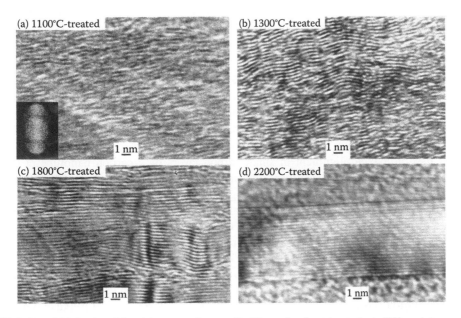

FIGURE 2.29 002 lattice fringe images of a needle-like coke heat-treated at different temperatures. (Courtesy of Mme. A. Oberlin, Tokai Science University, Tokyo, Japan. With permission.)

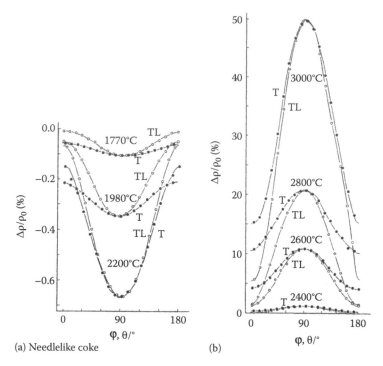

(a) Needlelike coke   (b)

FIGURE 2.30 Dependence of the magnetoresistance ($\Delta\rho/\rho_0$) with the rotation angles φ and θ for a needle-like coke particle. (Courtesy of Prof. Y. Hishiyama, Musashi Institute of Technology, Tokyo, Japan. With permission.)

rotation), respectively, are shown in Figure 2.30 for different HTTs [128]. Below 2200°C, the $\Delta\rho/\rho_0$ value at $\varphi = \theta = 90°$ is negative and decreases with increasing HTT. Above 2200°C, it turns to a positive value and increases to very large values above 2400°C, indicating the growth of carbon hexagonal layers and the improvement in stacking order (graphitization). The $\Delta\rho/\rho_0$ value at $\varphi = 0°$ and 180° does not coincide with that at $\theta = 0°$ and 180°, which suggests that this needle-like coke particle has an intermediate nanotexture between planar and axial orientations.

2.4.3 Axial Orientation

An axial orientation of the layers is found in various fibrous carbon materials, in other words, a fibrous morphology is possible because of this axial orientation scheme. In carbon fibers, the coaxial and radial alignments of the layers along the reference axis (i.e., fiber axis) are possible (Figure 2.27). In Figure 2.31, the variety in the nanotexture of different carbon fibers is illustrated [129] and examples are shown in Figure 2.32.

VGCFs have a coaxial alignment of small carbon layers on a carbon nanotube at the center (Figure 2.15). After the 3000°C treatment, the small layers coalesce with each other forming long and smooth layers. The degree of concentric alignment is greatly improved and the cross section of the fibers is polygonized, mainly due to crystallite growth (Figure 2.32a) [130]. For mesophase-pitch-based carbon fibers, different nanotextures of the cross section, from radial and concentric to random are possible (Figure 2.31) [131]. The difference in the cross-sectional nanotexture becomes more pronounced after high-temperature treatment (Figure 2.32b and c).

A concentric alignment of the layers is realized in multiwalled carbon nanotubes and various carbon nanofibers, most of which are grown through CVD using fine catalyst particles. However, different orientation schemes can be found along their axes, from parallel (tubular type) through perpendicular (platelet type) to herringbone type (Figure 2.33) or cup-stacked type. Examples of herringbone type and tubular type nanotextures are shown in Figure 2.34 for both as-prepared

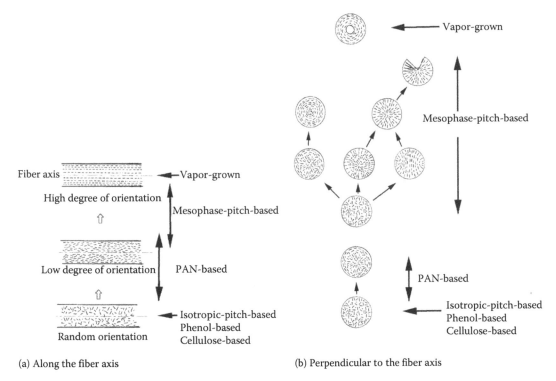

(a) Along the fiber axis

(b) Perpendicular to the fiber axis

FIGURE 2.31 Nanotextures in the sections along and perpendicular to the fibre axis of various carbon fibers.

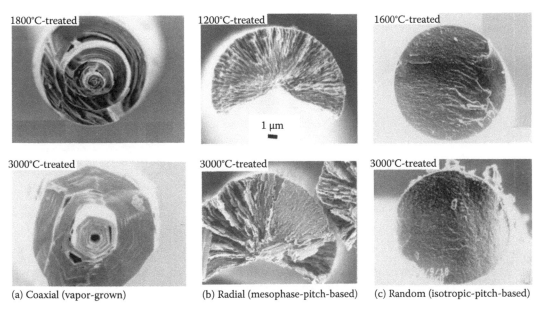

(a) Coaxial (vapor-grown) (b) Radial (mesophase-pitch-based) (c) Random (isotropic-pitch-based)

FIGURE 2.32 Cross section of as-prepared and high-temperature-treated carbon fibers showing different nanotextures.

and high-temperature-treated carbon nanofibers [59]. After the high-temperature treatment, the nanotexture is recognized more clearly, mainly because of the layer growth. Impregnation of molten poly(vinyl alcohol) into the channels of anodic aluminum oxide films gives nanofibers with a platelet nanotexture [132].

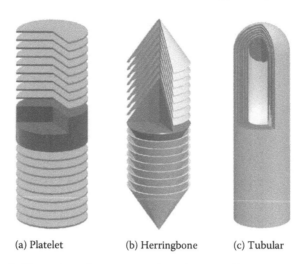

(a) Platelet   (b) Herringbone   (c) Tubular

FIGURE 2.33 Schematic illustration of nanotextures in carbon nanofibers.

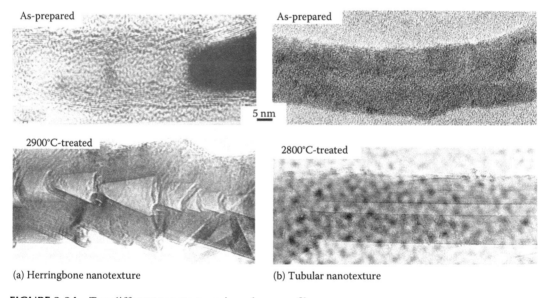

(a) Herringbone nanotexture   (b) Tubular nanotexture

FIGURE 2.34 Two different nanotextures in carbon nanofibers.

In single-wall carbon nanotubes, various chiralities are realized from arm-chair type through zigzag type to chiral type (Figure 2.35), which are known to govern their properties [15].

2.4.4 POINT ORIENTATION

In the point orientation, concentric and radial alignments also have to be differentiated (Figure 2.27). The extreme case of concentric point orientation is found in the fullerene family. This orientation is also found in the spherical particles of carbon blacks, the diameter of which is from few tens to few hundreds of nanometers, minute hexagonal carbon layers being preferentially oriented along the surface (Figures 2.36a and 2.37a) [43]. The concentric orientation of the carbon layers is more

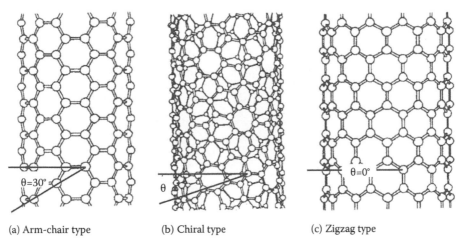

(a) Arm-chair type   (b) Chiral type   (c) Zigzag type

FIGURE 2.35 Schematic illustration of the various structures of single-wall carbon nanotube.

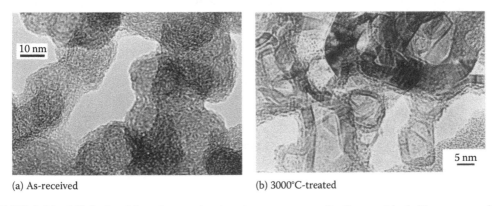

(a) As-received   (b) 3000°C-treated

FIGURE 2.36 002 lattice fringe image showing the nanotexture of a furnace black (a) as-prepared and (b) high-temperature-treated. (Courtesy of Tokai Carbon Co. Ltd., Shizuoka, Japan. With permission.)

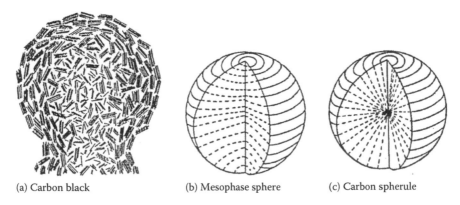

(a) Carbon black   (b) Mesophase sphere   (c) Carbon spherule

FIGURE 2.37 Models of nanotexture for small-sized carbon black, mesophase sphere, and carbon spherule.

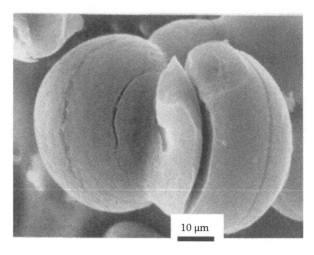

FIGURE 2.38   Scanning electron micrograph of MCMB after heat treatment at 2800°C.

marked near the surface, whereas the layers are randomly oriented at the center. When large-sized carbon black particles are heat-treated at high temperatures, the particles are polygonized and a hollow core appears (Figure 2.36b), because of the crystallite growth.

A radial alignment of the layers to form a sphere is found in the carbon spherules, which are formed by pressure carbonization of a mixture of polyethylene and poly(vinyl chloride) (Figure 2.37c) [133–135]. The radial point orientation scheme appears near the surface of most mesophase spheres (MCMB), but at their center the orientation of the layers is not radial (Figure 2.37b) [95,136]. As a consequence of the radial orientation near the surface, tangential cracks are formed after heat treatment at high temperatures (Figure 2.38) [137]. These cracks might explain the good performance of MCMB as anode materials for lithium-ion rechargeable batteries; they are supposed to accommodate the expansion due to lithium intercalation.

2.4.5   RANDOM ORIENTATION

A randomly oriented nanotexture occurs in glass-like carbons and in carbon materials obtained by carbonization of polymer precursors (e.g., phenol resin and sugar). The fundamental structural units comprising most glass-like carbons are so small that they are difficult to be observed by TEM. Therefore, their nanotexture was often discussed on the basis of TEM observations on high-temperature-treated materials, where the layers become somewhat larger. A 002 lattice fringe image of sugar coke heat-treated at a high temperature is shown in Figure 2.39a [138], and a model of nanotexture is proposed for the carbons derived from various precursors, such as phenolic resin, sugar, etc., including glass-like carbons (Figure 2.39b) [139]. In this model, closed pores are formed by concentric hexagonal layers. By taking into consideration the large number of closed micropores in most glass-like carbons and their gas impermeability, this model is believed to be realistic.

2.4.6   IMPORTANCE OF NANOTEXTURE

Various treatments, such as high-temperature treatment, intercalation of different foreign species, and doping by boron and nitrogen, are frequently applied to carbons in order to modify and improve their nanotexture and properties.

Heat treatment at high temperatures, in most cases up to 3000°C, in inert atmosphere is an essential process for the production of graphite blocks. The principal purpose of this treatment is the improvement in structure, i.e., the increase in the graphitization degree. The development of the

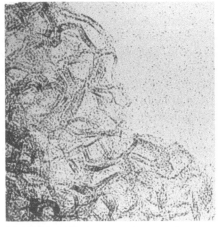

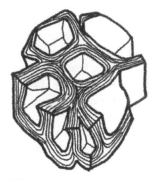

(a) 002 lattice fringe image of sugar coke   (b) Model proposed

FIGURE 2.39 (a) 002 lattice fringe image of sugar coke and (b) nanotexture model for glass-like carbon. (Courtesy of Mme A. Oberlin and Prof. M. Shiraishi, Tokai Science University, Tokyo, Japan. With permission.)

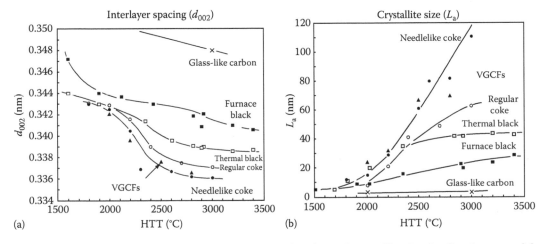

FIGURE 2.40 Influence of HTT on the interlayer spacing d_{002} and crystallite size L_a of carbon materials with different nanotextures.

graphite structure is evaluated by different parameters. It is exactly evaluated by the graphitization degree P_1, which is the probability to have the AB stacking sequence for neighboring hexagonal carbon layers. Conventionally, however, the average interlayer spacing between neighboring layers, d_{002}, and the layer size, L_a, which are determined by x-ray diffraction analysis, are often used as parameters to estimate the graphitization degree. The d_{002} values of highly crystallized graphite and amorphous carbon are 0.3354 nm and more than 0.344 nm, respectively. Figure 2.40a and b show the variation of d_{002} and L_a versus HTT for various carbon materials with different nanotextures [140]. The needle-like coke mainly with a planar-orientation nanotexture and VGCFs with a coaxial orientation scheme show a rapid decrease in d_{002} and a rapid increase in L_a with increasing HTT, d_{002} approaching 0.3354 nm and L_a being more than 100 nm. Carbon blacks, thermal and furnace blacks, which have a coaxial point orientation, give relatively high d_{002} values of 0.339–0.341 nm and small L_a of 20–50 nm, even after heat treatment up to 3400°C. Glass-like carbon with a random orientation gives much larger d_{002} and very small L_a even after 3000°C treatment.

Since the behavior with the treatment at high temperatures is quite different from carbon to carbon, it was proposed to classify the carbon materials into two classes before this treatment, e.g., the one that can change to graphite and the other that cannot. The former was called either graphitizing, graphitizable, or soft carbon, and the latter either nongraphitizing, nongraphitizable, or hard carbon [141–143]. However, such a classification is not used now, because intermediate behaviors are very often found, as clearly seen in Figure 2.40. The difference in graphitization behavior is now known to be mainly due to the nanotexture. In order to change glass-like carbon with random nanotexture to graphite, heat treatment under high pressure is required, accompanied by the disruption of the original nanotexture [144].

As already illustrated in Figure 2.31, mesophase-pitch-based carbon fibers can have different cross-sectional nanotextures. Figure 2.41 shows that the cross-sectional nanotexture also governs the graphitization behavior of the fibers [131]. Fibers with a straight radial nanotexture in their cross section have a relatively high graphitization degree, expressed by low d_{002} and high L_a values, but those presenting a zigzag radial nanotexture have a poor graphitization degree.

In the case of intercalation reactions, the nanotexture of the host carbon materials has a strong effect. The intercalation of sulfuric acid into natural graphite can proceed at room temperature in concentrated sulfuric acid with a small amount of oxidant, such as nitric acid. The resulting intercalation compound is commonly used in industry for the preparation of exfoliated graphite [32]. However, in order to intercalate sulfuric acid into carbon fibers, electrolysis is needed [145,146]. Potassium as a vapor, on the other hand, can be easily intercalated in various carbon materials, even in low-temperature-treated carbon fibers [147].

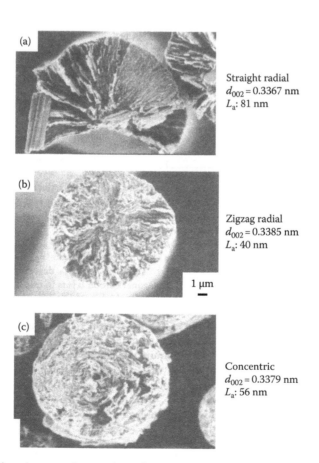

(a) Straight radial
$d_{002} = 0.3367$ nm
L_a: 81 nm

(b) Zigzag radial
$d_{002} = 0.3385$ nm
L_a: 40 nm

1 μm

(c) Concentric
$d_{002} = 0.3379$ nm
L_a: 56 nm

FIGURE 2.41 Scanning electron micrographs and structural parameters of mesophase-pitch-based carbon fibers with different cross-sectional nanotextures.

2.5 MICROTEXTURES IN CARBON MATERIALS

Most particles with planar and axial orientation, such as cokes, carbon fibers, and carbon nanotubes, for example, are also anisotropic, and as a consequence their agglomeration can create a further variety in texture. Therefore, in addition to the nanotexture and also graphitization degree of each particle, it is necessary to take into consideration the texture formed by the preferred orientation of these anisotropic particles in order to understand the properties of different carbon materials. The texture due to the preferred orientation of anisotropic particles may be called microtexture, because the particles are often of micrometer or millimeter size. The microtexture is usually formed during the forming process of bulky carbon materials. In large-sized graphite electrodes for metal refining, for example, the particles of needle-like coke tend to be oriented along the rod axis during their forming process through extrusion with pitch binder. To prepare carbon-fiber-reinforced plastics (composites), different microtextures based on the carbon fibers' orientation, some examples of which are shown in Figure 2.42, have been applied to get high strength and high modulus of the composites.

Two methods have been employed for realizing the isotropy of carbon materials that fundamentally consist of anisotropic structural units or crystallites: (i) the random aggregation of micrometer- or millimeter-sized particles, even though these particles are anisotropic and (ii) the random agglomeration of nanometer-sized crystallites in the bulk. The former is realized in the so-called isotropic high-density graphite blocks, where small-sized coke particles are formed using isostatic pressing [93]. The latter, i.e., aggregation of nanosized carbon layers, was realized in glass-like carbons, which are isotropic and have a nongraphitizing nature [148].

The presence of pores in the formed carbons influences the properties of the material. In such a case, the microtexture, including the shape and size of the pores, has to be taken into account in addition to that due to the orientation of anisotropic particles. Optical microscopy images on polished sections of isotropic high-density graphite blocks having different bulk densities from 1.735 to 1.848 g/cm³ are shown in Figure 2.43 [149]. Although the difference in bulk density looks rather small, a marked difference is easily seen in the micrographs, showing different shapes, sizes, and distributions of the pores in the cross section. Different pore parameters of these carbon materials, such as density, average cross-sectional area, roundness, fractal dimension, etc., were determined with the help of image analysis [149]. Figure 2.44 shows the plots of mechanical properties (elastic modulus,

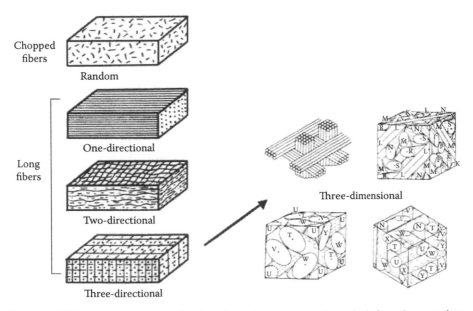

FIGURE 2.42 Different arrangements of carbon fibers for the preparation of reinforced composites.

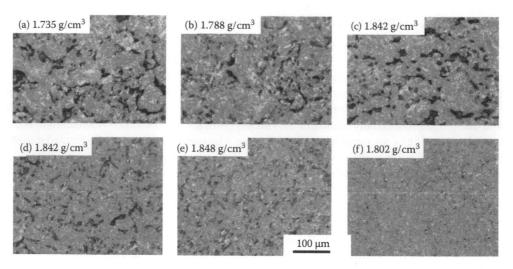

FIGURE 2.43 Optical micrographs of polished cross section of high-density isotropic carbons. (Courtesy of Prof. K. Oshida, Nagano National College of Technology, Nagano, Japan. With permission.)

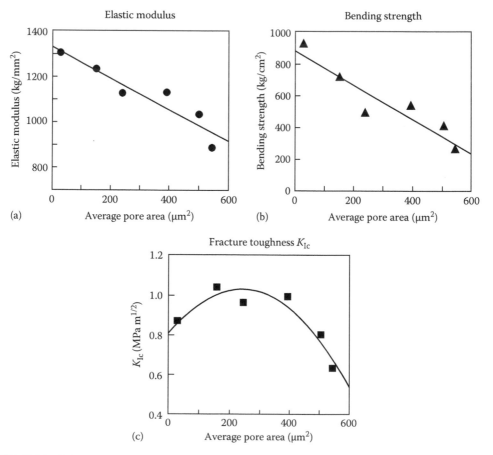

FIGURE 2.44 Relationship between the mechanical properties and average pore area of high-density isotropic carbons.

bending strength, and fracture toughness K_{Ic}) versus the average cross-sectional area of the pores. A dependence of the mechanical properties of these carbons on pore size is clearly shown.

Macropores in exfoliated graphite, which can sorb up to 80 g of heavy oil per 1 g of exfoliated graphite [86], were analyzed with the aid of image analysis [150]. An example of a histogram for the cross-sectional area of macropores formed in worm-like particles is shown in Figure 2.20d.

2.6 CONCLUSION

In this chapter, the structures and textures of carbons at different scales are explained. The carbon materials are classified into four families, diamond, graphite, fullerene, and carbyne on the basis of hybridized sp^3, flat sp^2, curved sp^2, and sp orbitals used, respectively. Each family has its own characteristic diversity in structure and also in the possibility of accepting foreign species. The formation of these carbon materials from organic precursors (carbonization) is shortly described by dividing the process into three phases (gas, solid, and liquid), based on the intermediate phases formed during carbonization. The importance of nanotexture, mainly due to the preferred orientation of the anisotropic BSU in the graphite family, i.e., planar, axial, point, and random orientation schemes, is particularly emphasized.

Before closing this chapter, it has to be emphasized that carbon materials have a wide range of structures and textures, which strongly depend on the preparation conditions. When they are applied for electrochemistry, their detailed structure and texture must be exactly understood. The following chapters will present the practical applications of various carbons in various electrochemical devices, such as lithium-ion rechargeable batteries, electric double layer capacitors, fuel cells, and primary batteries.

REFERENCES

1. Inagaki, M. *New Carbons—Control of Structure and Functions*, London: Elsevier, 2000.
2. Inagaki, M. and Kang, F. Carbon families. In *Carbon Materials Science and Engineering*, Beijing, China: Tsinghua University Press, 2006; 23.
3. Bernal, J.D. *Proc R Soc A* 1924; 106: 749.
4. Lipson, H. and Stokes, A.R. *Nature* 1942; 149: 328.
5. Bacon, G.E. *Acta Crystallogr* 1950; 3: 137.
6. Boehm, H.P. and Hofman, U.Z. *Z Anorg Chem* 1955; 278: 58.
7. Inagaki, M., Mugishima, H., and Hosokawa, K. *TANSO* 1973; 1973(74): 76.
8. Warren, B.E. *J Chem Phys* 1934; 2: 551.
9. Endo, M., Oshida, K., Kobori, K., Takeuchi, K., Takahashi, K., and Dresselhaus, M.S. *J Mater Res* 1995; 10: 1461.
10. Franklin, R.E. *Acta Crystallogr* 1951; 4: 253.
11. Inagaki, M. and Kang, F. Structure and texture of carbon materials. In *Carbon Materials Science and Engineering*, Beijing, China: Tsinghua University Press, 2006; 32.
12. Kroto, R.F., Hearth, J.R., O'Brien, S.C.O., Curl, R.F., and Smalley, R.E. *Nature* 1985; 318: 162.
13. Curl, R.F. and Smalley, R.E. *Sci Am* 1991; 1991(October): 32.
14. Dresselhaus, M.S., Dresselhaus, G., and Eklund, P.C. *Science of Fullerenes and Carbon Nanotubes*, San Diego: Academic Press, 1996.
15. Saito, R., Dresselhaus, G., and Dresselhaus, M.S. *Physical Properties of Carbon Nanotubes*, London, UK: Imperial College Press, 1998.
16. Bundy, F.P. and Kasper, J.S. *J Chem Phys* 1967; 46: 3437.
17. Aisenberg, S. and Chabot, R. *J Appl Phys* 1971; 42: 2953.
18. Jarkov, S.M., Tctarenko, Y.N., and Churilov, G.N. *Carbon* 1998; 36: 595.
19. Kudryavtsev, Y.P., Evsyukov, S., Guseva, M., Babaev, V., and Khvostov, V. Carbyne—A linear chainlike carbon allotrope. In *Chemistry and Physics of Carbon*, Vol. 25, New York: Marcel Dekker, 1996; 1–69.
20. Kavan, L. Electrochemical carbonization of fluoropolymers. In *Chemistry and Physics of Carbon*, Vol. 23, New York: Marcel Dekker, 1991; 69.
21. Whittaker, A.G. and Kintner, P.L. *Science* 1978; 200: 763.

22. Yamada, Y. and Inagaki, M. *TANSO* 1997; 1997 (178): 122.
23. Lowell, C.E. *J Am Ceram Soc* 1967; 50: 241.
24. Konno, H., Nakahashi, T., and Inagaki, M. *Carbon* 1997; 35: 669.
25. Hishiyama, Y., Irumano, H., and Kaburagi, Y. *Phys Rev B* 2001; 63: 245406.
26. Endo, M., Kim, C., Karaki, T., Tamaki, T., Nishimura, Y., Matthewa, S., Brown, S.D.M., and Dresselhaus, M.S. *Phys Rev B* 1998; 58: 8991.
27. Dahn, J.R., Reimers, J.N., Sleigh, A.K., and Tiedje, T. *Phys Rev B* 1992; 45: 3773.
28. Kodama, M., Yamashita, J., Soneda, Y., Hatori, H., Nishimura, S., and Kamegawa, K. *Mater Sci Eng B* 2004; 108: 156.
29. Hummelen, J.C., Knight, B., Pavlovich, J., Gonzalez, R., and Wudl, F. *Science* 1995; 269: 1554.
30. Inagaki, M. *J Mater Res* 1989; 4: 1560.
31. Enoki, T., Suzuki, M., and Endo, M. *Graphite Intercalation Compounds and Applications*, Oxford U.K.: Oxford University Press, 2003.
32. Inagaki, M., Kang, F., and Toyoda, M. Exfoliation of graphite via intercalation. In *Chemistry and Physics of Carbon*, Vol. 29, New York: Marcel Dekker, 2004; 1.
33. Dahn, J.R., Sleigh, A.K., Shi, H., Way, B.M., Weydanz, W.J., Reimers, J.N., Zhong, Q., and von Sacken, U. Carbons and graphites as substitutes for the lithium anode. In *Lithium Batteries, Industrial Chemistry Library*, Vol. 5, London, U.K.: Elsevier, 1994; 1.
34. Vogel, F.L. *J Mater Sci* 1977; 12: 982.
35. Matsubara, H., Yamaguchi, Y., Shioya, J., and Murakami, S. *Synth Met* 1987; 18: 503.
36. Tanigaki, K. *J Phys Chem Solids* 1993; 54: 1645.
37. Shinohara, H. *Rep Prog Phys* 2000; 63: 843.
38. Zhou, S., Burger, C., Chu, B., Sawamura, M., Nagahama, N., Toganoh, M., Hakler, U.E., Isobe, H., and Nakamura, E. *Science* 2001; 291: 1944.
39. Kato, H., Kanazawa, Y., Okumura, M., Taninaka, A., Yokawa, T., and Shinohara, H. *J Am Chem Soc* 2003; 125: 4391.
40. Inagaki, M., Harada, S., Sato, T., Nakajima, T., Horino, Y., and Morita, K. *Carbon* 1989; 27: 253.
41. Hishiyama, Y., Yoshida, A., Kaburagi, Y., and Inagaki, M. *Carbon* 1992; 30: 517.
42. Inagaki, M. and Kang, F. Nanotexture development in carbon materials. In *Carbon Materials Science and Engineering*, Beijing, China: Tsinghua University Press, 2006; 47.
43. Donnet, J.B. and Voet, A. *Carbon Black*, New York: Marcel Dekker, 1976.
44. Maeno, S. *TANSO* 2006; 222: 140.
45. Diefendorf, R.J. *J Chim Phys* 1960; 17: 127.
46. Bokros, J.C., LaGrange, L.D., and Schoen, F.J. Control of structure of carbon for use in bioengineering. In *Chemistry and Physics of Carbon*, Vol. 9, New York: Marcel Dekker, 1972; 103.
47. Endo, M., Koyama, T., and Hishiyama, Y. *Jpn J Appl Phys* 1974; 13: 1933.
48. Endo, M., Koyama, T., and Hishiyama, Y. *Carbon* 1976; 14: 133.
49. Oberlin, A., Endo, M., and Koyama, T. *J Cryst Growth* 1976; 32: 335.
50. Katsuki, H., Matsunaga, K., Egashira, M., and Kawasumi, S. *Carbon* 1981; 9: 148.
51. Tibbetts, G.G. *J Cryst Growth* 1985; 73: 431.
52. Ishioka, M., Okada, T., Matsubara, K., and Endo, M. *Carbon* 1992; 11: 674.
53. Endo, M. and Shikata, M. *Ouyou Butsuri* 1985; 54: 507.
54. Boehm, H.P. *Carbon* 1973; 11: 583.
55. Baird, T., Fryer, J.R., and Grant, B. *Carbon* 1974; 12: 591.
56. Audier, M., Coulon, M., and Oberlin, A. *Carbon* 1980; 18: 73.
57. Rodriguez, N.M. *J Mater Res* 1993; 8: 3233.
58. Soneda, Y. and Makino, M. *Carbon* 2000; 38: 478.
59. Hamwi, A., Alvergnat, H., Bonnamy, S., and Béguin, F. *Carbon* 1997; 35: 723.
60. Martin-Gullon, I., Vera, J., Conesa, J.A., Gonzalez, J.L., and Merino, C. *Carbon* 2006; 44: 1572.
61. Inagaki, M. and Kang, F. Fibrous carbon materials. In *Carbon Materials Science and Engineering*, Beijing, China: Tsinghua University Press, 2006; 359.
62. Motojima, S., Kawaguchi, M., Nozaki, K., and Iwanaga, H. *Appl Phys Lett* 1990; 56: 321.
63. Motojima, S., Hasegawa, I., Asakura, S., Ando, K., and Iwanaga, H. *Carbon* 1995; 33: 1167.
64. Chen, X., In-Hwang, W., Shimada, S., Fujii, M., Iwanaga, H., and Motojima, S. *J Mater Res* 2000; 15: 808.
65. Endo, M., Kim, Y.A., Hayashi, T., Fukai, T., Oshida, K., Yanagisawa, T., Higaki, S., and Dresselhaus, M.S. *Appl Phys Lett*, 2002; 80: 1267.

66. Endo, M., Kim, Y.A., Hayashi, T., Yanagisawa, T., Muramatsu, H., Ezaka, M., Terrones, H., Terrones, M., and Dresselhaus, M.S. *Carbon* 2003; 41: 1941.
67. Iijima, S., Yudasaka, M., Yamada, R., Bandow, S., Suenaga, K., Kokai, F., and Takahashi, K. *Chem Phys Lett* 1999; 309: 165.
68. Endo, M., Kim, Y.A., Hayashi, T., Nishimura, K., Matusita, T., Miyashita, K., and Dresselhaus, M.S. *Carbon* 2001; 39: 1287.
69. Iijima, S. *Nature* 1991; 354: 56.
70. Iijima, S. and Ichihashi, T. *Nature* 1993; 363: 603.
71. Bethune, D.S., Kiang, C.H., deVries, M.S., Gorman, G., Savoy, R., Vazquez, J., and Beyers, R. *Nature* 1993; 363: 605.
72. Murayama, S., Kojima, R., Miyauchi, Y., Chiashi, S., and Kohno, M. *Chem Phys Lett* 2002; 360: 229.
73. Kusunoki, M., Rokkaku, M., and Suzuki, T. *Appl Phys Lett* 1997; 18: 2620.
74. Kusunoki, M., Shibata, J., Rokkaku, M., and Hirayama, T. *Jpn J Appl Phys* 1998; 37: 1605.
75. Krätschmer, W., Fostiropoulos, K., and Huffman, D.R. *Chem Phys Lett* 1990; 170: 167.
76. Krätschmer, W., Lowell, D., Lamb, K., Fostiropoulos, K., and Hoffman, D.R. *Nature* 1990; 347: 354.
77. Derbyshire, F., Jagtoyen, M., and Thwaites, M. Activated carbons—production and applications. In *Porosity in Carbons*, London, U.K.: Edward Arnold, 1995; 227.
78. Rodriguez-Reinoso, F. and Linares-Solano, A. Microporous structure of activated carbons as revealed by adsorption methods. In *Chemistry and Physics of Carbon*, Vol. 21, New York: Marcel Dekker, 1995; 1.
79. Marsh, H. and Rodriguez-Reinoso, F. *Activated Carbon*, London, U.K.: Elsevier, 2006.
80. Inagaki, M., Nishikawa, T., Oshida, K., Fukuyama, K., Hatakeyama, Y., and Nishikawa, K. *Ads Sci Tech* 2006; 24: 55.
81. Marsh, H., Yan, D.S., O'Grady, T.M., and Wennerberg, A. *Carbon* 1984; 22: 603.
82. Frackowiak, E. and Béguin, F. *Carbon* 2001; 39: 937.
83. Morishita, T., Ishihara, K., Kato, M., and Inagaki, M. *Carbon* 2007; 45: 209.
84. Johnson, P.J., Setsuda, D.J., and Williams, R.S. Activated carbon for automotive application. In *Carbon Materials for Advanced Technologies*, Amsterdam, The Netherlands: Pergamon, 1999; 235.
85. Toyoda, M., Aizawa, J., and Inagaki, M. *Desalination* 1998; 115: 199.
86. Toyoda, M., Iwashita, N., and Inagaki, M. Sorption of heavy oils into carbon materials. In *Chemistry and Physics of Carbon*, Boca Raton, FL: Taylor & Francis, 2007; 30: 177.
87. Kang, F., Zheng, Y., Zhao, H., Wang, H., Wang, L., Shen, W., and Inagaki, M. *New Carbon Mater* 2003; 18: 161.
88. Fitzer, E., Schaffer, W., and Yamada, S. *Carbon* 1969; 7: 643.
89. Noda, T., Inagaki, M., and Yamada, S. *J Non-Cryst Solids* 1969; 1: 285.
90. Nishikawa, K., Fukuyama, K., and Nishizawa, T. *Jpn J Appl Phys* 1998; 37: 6486.
91. Donnet, J.B., Wang, T.K., Rebouillat, S., and Peng, J.C.M. *Carbon Fibers*. 3rd edn., New York: Marcel Dekker, 1998.
92. Inagaki, M., Takeichi, T., Hishiyama, Y., and Oberlin, A. High quality graphite films from aromatic polyimides. In *Chemistry and Physics of Carbon*, Vol. 26, New York: Marcel Dekker, 1999; 245.
93. Inagaki, M. and Kang, F. Polycrystalline graphite blocks. In *Carbon Materials Science and Engineering*, Beijing, China: Tsinghua University Press, 2006; 269.
94. Brooks, J.D. and Taylor, G.H. The formation of some graphitizing carbons. In *Chemistry and Physics of Carbon*, Vol. 4, New York: Marcel Dekker, 1968; 243.
95. Honda, H. *TANSO* 1983; 1983 (113): 60.
96. Yamada, Y., Imamura, K., Kakiyama, H., Honda, H., Oi, S., and Fukuda, K. *Carbon* 1974; 12: 307.
97. Tatsumi, K., Iwashita, N., Sakaebe, H., Shioyama, H., Higuchi, S., Mabuchi, A., and Fujimoto, H. *J Electrochem Soc* 1995; 142: 716.
98. Fujimaki, H. and Otani, S. *Ceramics* 1976; 11: 612.
99. Inagaki, M., Kaneko, K., and Nishizawa, T. *Carbon* 2004; 42: 1401.
100. Kyotani, T., Sonobe, N., and Tomita, A. *Nature* 1988; 331: 331.
101. Sonobe, N., Kyotani, T., Hishiyama, Y., Shiraishi, M., and Tomita, A. *J Phys Chem* 1988; 92: 7029.
102. Kyotani, T., Tsai, L., and Tomita, A. *Chem Mater* 1995; 7: 1427.
103. Kyotani, T., Nagai, T., Inoue, S., and Tomita, A. *Chem Mater* 1997; 9: 609.
104. Sakamoto, Y., Kameda, M., Terasaki, O., Zhao, D.Y., Kim, J.M., Stucky, G., Shin, H.J., and Ryoo, R. *Nature* 2000; 408: 44953.
105. Morishita, T., Soneda, Y., Tsumura, T., and Inagaki, M. *Carbon* 2006; 44: 2360.

106. Morishita, T., Suzuki, T., Nishikawa, T., Tsumura, T., and Inagaki, M. *TANSO* 2006; 223: 220.
107. Inagaki, M., Kato, M., Morishita, T., Morita, K., and Mizuuchi, K. *Carbon* 2007; 45: 1121.
108. Morishita, T., Ishihara, K., Kato, M., Tsumura, T., and Inagaki, M. *TANSO* 2007; 2007 (226): 19.
109. Sevilla, M., Alvarez, S., Centeno, T.A., Fuertes, A.B., and Stoeckli, F. *Electrochim Acta* 2007; 52: 3207.
110. Fernández, J.A., Morishita, T., Toyoda, M., Inagaki, M., Stoeckli, F., and Centeno, T.A. *J Power Sources* 2008; 175: 675.
111. Wang, L., Morishita, T., Toyoda, M., and Inagaki, M. *Electrochim Acta* 2007; 53: 882.
112. Ozaki, J., Endo, N., Ohizumi, W., Igarashi, K., Nakahara, M., Oya, A., Yoshida, S., and Iizuka, T. *Carbon* 1997; 35: 1031.
113. Takeichi, T., Zuo, M., and Ito, A. *High Perform Polym* 1999; 11: 1.
114. Hulicova, D., Sato, F., Okabe, K., Koishi, M., and Oya, A., *Carbon* 2001; 39: 1438.
115. Jansta, J., Dousek, F.P., and Patzelova, V. *Carbon* 1975; 13: 377.
116. Shiraishi, S., Kurihara, H., Tsubota, H., Oya, A., Soneda, Y., and Yamada, Y., *Electrochem Solid State Lett* 2001; 4: A5.
117. Tanaike, O., Yoshizawa, N., Hatori, H., Yamada, Y., Shiraishi, S., and Oya, A. *Carbon* 2002; 40: 457.
118. Pekala, R.W. and Kong, F.M. *Polym Properties* 1989; 30: 221.
119. Hanzawa, Y., Kaneko, K., Yoshizawa, N., Pekala, R.W., and Dresselhaus, M.S. *Adsorption* 1998; 4: 187.
120. Tamon, H., Ishizaki, H., Mikami, M., and Okazaki, M. *Carbon* 1997; 35: 791.
121. Tamon, H., Ishizaki, H., Yamamoto, T., and Suzuki, T. *Carbon* 1999; 37: 2049.
122. Yamamoto, T., Nishimura, T., Suzuki, T., and Tamon, H. *Carbon* 2001; 39: 2374.
123. Bekyarova, E. and Kaneko, K. *Adv Mater* 2000; 12: 1625.
124. Inagaki, M. *TANSO* 1985; 1985 (122): 114.
125. Bokros, J.C. Deposition, structure, and properties of pyrolytic carbon. In *Chemistry and Physics of Carbon*, Vol. 5, New York: Marcel Dekker, 1969; 1.
126. Yoshida, A. and Hishiyama, Y. *J Mater Res* 1992; 7: 1400.
127. Oberlin, A. and Terriere, G. *J Microsc* 1972; 14: 1.
128. Hishiyama, Y. and Kaburagi, Y. *TANSO* 1979; 1979 (98): 89.
129. Inagaki, M. and Kang, F. Fibrous carbons. In *Carbon Materials Science and Engineering*, Beijing, China: Tsinghua University Press, 2006; 359.
130. Yoshida, A., Hishiyama, Y., Ishioka, M., and Inagaki, M. *TANSO* 1995; 1995 (168): 169.
131. Inagaki, M., Iwashita, N., Hishiyama, Y., Kaburagi, Y., Yoshida, A., Oberlin, A., Lafdi, K., Bonnamy, S., and Yamada, Y. *TANSO* 1991; 1991 (147): 57.
132. Konno, H., Sato, S., Habazaki, H., and Inagaki, M. *Carbon* 2004; 42: 2756.
133. Inagaki, M., Ishihara, M., and Naka, S. *High Temp High Press* 1976; 8: 270.
134. Inagaki, M., Kuroda, K., and Sakai, M. *Carbon* 1983; 21: 231.
135. Washiyama, M., Sakai, M., and Inagaki, M. *Carbon* 1988; 26: 303.
136. Augie, D., Oberlin, M., Oberlin, A., and Hyvernat, P. *Carbon* 1980; 18: 337.
137. Inagaki, M., Hayashi, S., Kamiya, K., and Naka, S. *High Temp High Press* 1971; 3: 355.
138. Oberlin, A. High-resolution TEM studies of carbonization and graphitization. In *Chemistry and Physics of Carbon*, Vol. 22, New York: Marcel Dekker, 1989; 1.
139. Shiraishi, M. *Introduction to Carbon Materials*, Tokyo, Japan: Carbon Society of Japan, 1984; 29.
140. Inagaki, M. and Kang, F. Structural development in carbon materials (graphitization). In *Carbon Materials Science and Engineering*, Beijing, China: Tsinghua University Press, 2006; 112.
141. Franklin, R.E. *Proc R Soc A* 1951; 209: 196.
142. International Committee for Characterization and Terminology of Carbon. *Carbon* 1982; 20: 445.
143. Mrozowski, S. Mechanical strength, thermal expansion and structure of cokes and carbons. In *Proceedings of the Conference on Carbon*, New York, University of Buffalo, 1956; 31.
144. Inagaki, M. and Meyer, R.A. Stress graphitization. In *Chemistry and Physics of Carbon*, Vol. 26, New York: Marcel Dekker, 1999; 149.
145. Toyoda, M., Shimizu, A., Iwata, H., and Inagaki, M. *Carbon* 2001; 39: 1697.
146. Toyoda, M., Kaburagi, Y., Yoshida, A., and Inagaki, M. *Carbon* 2004; 42: 2567.
147. Endo, M., Ueno, H., and Inagaki, M. *Trans Inst Elect Eng Jpn* 1985; 105: 329.
148. Inagaki, M. and Kang, F. Non-graphitizing and glass-like carbons. In *Carbon Materials Science and Engineering*, Beijing, China: Tsinghua University Press, 2006; 343.
149. Oshida, K., Ekinaga, N., Endo, M., and Inagaki, M. *TANSO* 1996; 1996 (173): 142.
150. Inagaki, M., Toyoda, M., Kang, F., Zheng, Y.P., and Shen, W. *New Carbon Mater* 2003; 18: 241.

3 Carbide-Derived Carbons and Templated Carbons

Takashi Kyotani, John Chmiola, and Yury Gogotsi

CONTENTS

3.1　INTRODUCTION

Carbon in all its forms (diamond, graphite, and amorphous) is historically a technologically important material being used in everything from graphite electrodes for metal refining, activated carbon for water purification, diamond-cutting tools, jewelry and IR windows, and graphite for lubrication, for example [1]. Because of its electrochemical inertness and good conductivity, graphitic carbon also finds use in many ubiquitous electrochemical applications, such as battery and fuel cell electrodes, and in the production of many chemicals [2–4]. In recent years, more advanced applications have arisen that utilize the wide range of carbon's desirable properties, such as the carbon–carbon composites in the space shuttle's wing edges, and structural materials for next generation sport equipment, automobiles, and aircraft, for example [5]. With the recent discovery of carbon nanotubes (CNTs) and fullerenes, the promise of tailoring properties precisely by controlling the nanoscale structure of carbon has opened up new avenues for exploration.

Porous (activated) carbon (charcoal) has been known for at least 3000 years and is most widely used as a low-tech material for such things as wastewater remediation, gas purification, and treating poisoning by ingestion [6]. Many different porous carbon materials [7–10], made from different precursors, ranging from coconut shells [11] to phenolic resins [12], or CNTs [13,14] are used, for instance, in the construction of electrochemical capacitors. The conventional synthesis of porous carbon materials involves treating carbonaceous precursors (coal, charcoal, carbon fibers, or carbonized organic materials) in an oxidizing environment (water, CO_2, O_2, KOH, NaOH, etc.) at elevated temperature [6]. Different temperatures, treatment times, oxidant concentrations, etc. lead to different carbon structures and changes in the surface chemistry and nanotexture of the resulting carbon that are difficult if not impossible to precisely and independently control. Unfortunately, the pore size and structure of carbon prepared from polymer and natural precursors are difficult to control because of a random orientation of polymer chains of the former and structural inhomogeneities of the latter.

The traditional understanding of how to increase the capacitance involves maximizing the surface area that is accessible to the electrolyte ions. Over the past decade, the focus was to create a carbon material with the highest surface area of pores above 2 nm (mesopores) [15]. This approach was mostly born of the fact that capacitance stops being a linear function of surface area above surface areas of ~1200 m^2/g [9,16–20]. More recent work has shown the possibility that the smallest pores (<1 nm) are responsible for the lion's share of capacitance [21,22]. It was also suggested that different orientations of graphite planes can influence the capacitance [18]. Studies on highly oriented pyrolytic graphite (HOPG) oriented with the basal planes parallel and perpendicular to the surface have shown an order of magnitude increase in surface capacitance for the latter [23]. To date, correlations between structural parameters and properties in traditional activated carbons are poorly understood because of difficulties in preparing a series of samples with controllable nanotexture and porosity. Better understanding of processing structure–property relations in porous carbons may lead to many new electrochemical applications.

Novel, inexpensive synthesis routes for producing materials with precisely controlled nanotexture must be developed to improve the performance of batteries and electrochemical capacitors, as well as to enable new electrochemical applications of carbons. Two alternatives, carbide-derived carbon (CDC) and templated carbon, have shown a promise to offer the requisite control necessary to push device performance to the next level and will be explored in this chapter.

3.2 CARBIDE-DERIVED CARBON

3.2.1 SYNTHESIS

In general, CDC is synthesized by high-temperature extraction of metals and metalloids from carbide precursors. This can be accomplished using halogens [24,25], vacuum decomposition [26], or etching in supercritical water [27]. High-temperature chlorination (Equation 3.1) and vacuum decomposition (Equation 3.2) are the most widely used methods for CDC production:

$$MC + x/2\,Cl_2 \rightarrow MCl_x + C \tag{3.1}$$

$$MC(s) \xrightarrow[\text{vacuum}]{} M(g) + C(s) \tag{3.2}$$

Processing of CDC was recently reviewed by Nikitin and Gogotsi [28]. Carbon derived from selective etching of metals by halogens has shown great promise as the active material in electrochemical capacitors [15,29–31] because of the structure and properties that can be fine-tuned [21,32] and control of surface functional groups [33].

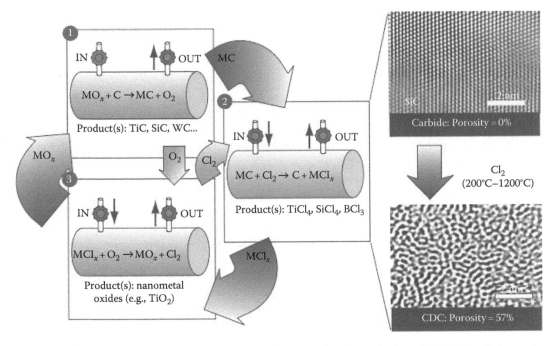

FIGURE 3.1 Schematic of hypothetical closed-loop process for the production of CDC. Step 1 shows the transformation of carbide-producing metal oxides into metal carbides. Step 2 shows the conversion of metal carbides into CDC and metal chlorides. Step 3 shows the oxidation of metal chlorides into nanoscale metal oxides and chlorine. Also shown are high-resolution TEM (HRTEM) micrographs of SiC and SiC-CDC.

Many different carbide precursors were chlorinated to produce CDC, ranging from complex ternary carbides [32,34–36] to more common carbides such as B_4C [37], Fe_3C [38], ZrC [39], SiC [24–26], and Mo_2C. The chlorination process following Equation 3.1 was originally used to produce silicon tetrachloride from silicon carbide [40], with the carbon being burned, until pure silicon became less expensive than silicon carbide due to the semiconductor industry. This process is still used to produce a variety of metal chlorides for semiconductor industry and metal-organic synthesis, however. As the process for CDC formation is the same as that used in industry for the synthesis of various metal chlorides and processes that exist for the production of metal carbides from metal chlorides, it can be envisioned as a closed cycle process for the production of both CDC and technologically important metal chlorides (Figure 3.1). Metal carbides are usually produced by reacting carbide-forming metal oxides and carbon at high temperature (Figure 3.1-1). The metal carbides can then produce CDC and metal chlorides (Figure 3.1-2) by Equation 3.1. The metal chlorides produced in step 2 can then either be further transformed into widely used nanoscale metal oxides, such as silica or titania (Figure 3.1-3), with sequestration of the chlorine to use in step 2 or sold as a chemical ($SiCl_4$ or $TiCl_4$). The metal oxides generated in this process can also be used in step 1 for carbide production, thus recycling the metal and theoretically making the only input carbon, which is transformed to CDC in this manufacturing process.

3.2.2 POROSITY

As first shown by Boehm in 1975 [41] on CDC synthesized from TaC and SiC, and later for most carbide precursors [33], the resulting carbons have type I isotherms in the Brunauer classification, which are indicative of microporous carbon having pore sizes less than 2 nm and relatively high surface areas up to $2000\,m^2/g$ [37,39,42]. The pore size of CDC can be tailored by the selection of carbide precursors with different spatial distributions of carbon atoms in the initial carbide lattice, changing the nanotextural ordering in the CDC by varying the synthesis temperature, and posttreatment in a variety of oxidizing atmospheres (activation) [43,44]. Therefore, both microporous and mesoporous carbons with different pore shapes can be produced by this method.

3.2.3 MICROSTRUCTURE

Linear reaction kinetics for chlorination of SiC powders, whiskers, and sintered ceramics to carbon coating thicknesses above 100 μm was demonstrated [26,45], showing that there is virtually no thickness limitation to CDC coatings. Moreover, as the coating is conformal, intricate shapes can be prefabricated using the hard carbide precursor, and complex morphologies can be realized. Equally important for applications is the wide range of microstructures that CDC possesses depending on the precursor carbide and synthesis conditions. Curved nanostructures ranging from onions [46,47] to nanotubes [48] have also been produced from carbide precursors. In fact, tubular nanostructures produced from carbide precursors may have preceded Iijima's discovery of CNTs [49].

Raman spectroscopy shows a narrowing of the full width at half-maximum of the G-band of graphite (~1580 cm^{-1}) with increasing synthesis temperature indicating increasing ordering for all CDCs studied. Different carbide precursors lead to different levels of graphitization for a given synthesis temperature [37,39,42]; however, in general, samples produced below 1000°C are x-ray amorphous, and only at synthesis temperatures of 1200°C and above is there significant graphitization. Structure collapse leading to specific surface areas (SSAs) below 100 m^2/g and formation of graphite were observed in samples produced at 1300°C–1800°C [50]. The increase in graphitization leads to requisite increasing electrical conductivity with increasing synthesis temperature [21,51]. It was recently shown by Dimovski et al. [38] and Leis et al. [52] that iron in the precursor carbide can also cause significant graphitization.

In addition to porous carbons and graphite, graphene and CNTs, which possess unique electrical, mechanical, and optical properties, can be produced by vacuum decomposition of SiC. Kusunoki et al. [53–55] observed the formation of CNTs growing normally to the carbon-terminated (000-1) C-face of hexagonal SiC with primarily zigzag chirality [56] and graphite growing on the Si-terminated (0001) Si-face. Nagano et al. claimed CNT formation on both carbon and silicon faces of HF-etched hexagonal 6H-SiC [57] and cubic 3C-SiC [58], with tubes growing normal to the surface. CNT formation on 6H, 3C [59], and 4H [60] SiC single crystals, powders [53,61], as well as amorphous [62] and 3C [57] films has been reported in literature. Vacuum decomposition of SiC has also become a common method for the synthesis of graphene. Forbeaux et al. [63] reported that graphite grows epitaxially on the Si-face of SiC. Our recent work on vacuum decomposition of β-SiC whiskers showed well-ordered graphene layers but no nanotube formation [26]. The most recent studies [60,64] suggested the growth of CNTs at 1400°C–1700°C on both C- and Si-faces of SiC single crystals, but with different growth rates, following the reaction:

$$SiC + \tfrac{1}{2}O_2(g) \rightarrow SiO(g) + C \qquad (3.3)$$

Equally intriguing and interesting for tribological applications is the possibility to create nanodiamonds during the synthesis of CDC [65]. However, similar to graphene and nanotubes, these are outside the scope of this chapter. Because of the conformal nature of the carbide-to-carbon conversion and lower theoretical density of CDC than graphite, the main feature of CDC is highly developed and well-controlled porosity [66]. It will be described in more detail in the following sections.

3.2.4 STUDIES ON CARBIDE-DERIVED CARBON

Porosity is important for many applications of CDC. Using CDC from SiC, TiC, and Mo$_2$C, it has been shown that pore size is important for Li diffusion and capacity [67,68]. The importance of pore size was also shown using CDC for cryogenic hydrogen storage [42,66]. It was shown that tuning the pore size to have a maximum volume of pores smaller than 1 nm can increase storage to ~3 wt% at 77 K and 1 atm, as well as methane uptake at room temperature. Using CDC for cytokine adsorption [69] showed that adsorption scales with the volume of pores larger than the size of the cytokine in question (5–10 nm). Porous CDC coatings have also shown good tribological properties [45,70]. Therefore, the effect of various process parameters on the porosity of CDC will be described in detail in the following sections.

3.2.4.1 Effect of Starting Carbide

Using different carbide precursors for CDC synthesis essentially amounts to have a different spatial distribution of carbon atoms in the precursor and affords a degree of freedom beyond typical porous carbon synthesis techniques for tailoring properties. For example, Figure 3.2a and b shows the layered distribution of carbon in Ti_3SiC_2 and uniform distribution of carbon in 3C-SiC [33], respectively. Upon removal of the titanium and silicon atoms, it can be seen that the different carbon distributions in the initial carbide lattice would lead to different pore sizes between Ti_3SiC_2-CDC (Figure 3.2a) and SiC-CDC (Figure 3.2b). This is reflected in experimental pore size distributions calculated from nonlocal density functional theory treatment of argon adsorption isotherms taken at 77 K (Figure 3.2c and d). At the same synthesis temperature, 1200°C, the pore size distribution of SiC-CDC is much narrower and shifted to lower average pore size than that of Ti_3SiC_2-CDC.

A wide range of carbide precursors leads to a wide range of theoretical pore volumes (Figure 3.3). Without further posttreatment, pore volumes from ~55% to ~85% are possible just by selecting different starting carbides. Alternatively, by altering the stoichiometry of the starting carbide, there is another degree of freedom for pore size control. It was shown using TiC that the average pore size was ~0.7 nm while $TiC_{0.5}$ precursors gave 2.8 nm pores [71]. A summary of data for several carbides is shown in Table 3.1.

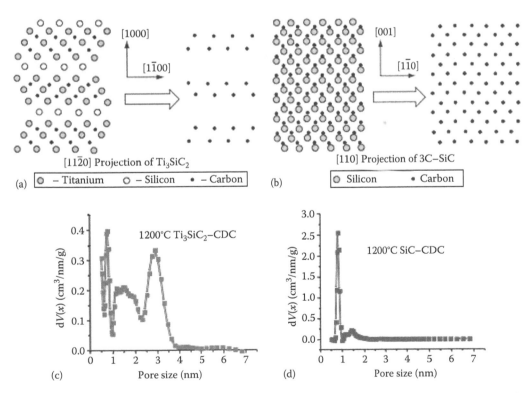

(a) [11$\bar{2}$0] Projection of Ti_3SiC_2

○ – Titanium ◎ – Silicon • – Carbon

(b) [110] Projection of 3C–SiC

◎ Silicon • Carbon

(c) 1200°C Ti_3SiC_2–CDC

(d) 1200°C SiC–CDC

FIGURE 3.2 Depiction of differences in pore size between SiC-CDC and Ti_3SiC_2-CDC after conversion. (a) A [11$\bar{2}$0] projection of Ti_3SiC_2 shows that upon removal of Ti and Si atoms, the resulting carbon distribution is layered. (b) A [110] projection of 3C-SiC shows that upon removal of Si, the resulting carbon distribution is uniform, and the spacing between carbon atoms is closer than that of (a). The resulting pore size distributions for (c) 1200°C Ti_3SiC_2-CDC and (d) 1200°C SiC-CDC show larger pores for the former.

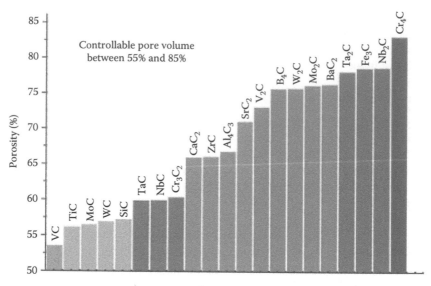

FIGURE 3.3 Calculated pore volumes for CDC from various metal carbides assuming conformal transformation. (From Dash, R.K., Nanoporous carbons derived from binary carbides and their optimization for hydrogen storage, PhD Thesis, Drexel University, Philadelphia, PA, 2006.)

TABLE 3.1
Pore Structure of CDCs Produced from a Variety of Carbides

Synthesis T (°C)	Pore Width (nm)	BET SSA (m²/g)	Total Pore Volume (cm³/g)	Mesopore Volume (cm³/g)	Micropore Volume (cm³/g)	Refs.
B₄C-CDC						
600	1.25	1150	—	—	—	[66,73]
750	—	1750	—	—	—	[66,73]
800	1.15	2000	—	—	—	[66,73]
850	—	2050	—	—	—	[66,73]
900	—	1975	—	—	—	[66,73]
1000	1.2	1850	—	—	—	[66,73]
1100	—	1750	—	—	—	[66,73]
1200	1.9	1525	—	—	—	[66,73]
SiC-CDC						
600	—	675	—	—	—	[66]
800	—	1425	—	—	—	[66]
900	—	1275	—	—	—	[66]
1000	—	1400	—	—	—	[66]
1100	—	1450	—	—	—	[66]
1200	—	1300	—	—	—	[66]
TiC-CDC						
500	0.678	1039.5	—	—	—	[21]
600	0.738	1269	—	—	—	[21]
700	0.764	1401	—	—	—	[21]
800	0.806	1595	—	—	—	[21]
1000	1.09	1625	—	—	—	[21]

TABLE 3.1 (continued)
Pore Structure of CDCs Produced from a Variety of Carbides

Synthesis T (°C)	Pore Width (nm)	BET SSA (m²/g)	Total Pore Volume (cm³/g)	Mesopore Volume (cm³/g)	Micropore Volume (cm³/g)	Refs.
Ti₂AlC-CDC						
400	—	400	—	—	—	[74]
600	1.4	650	0.8	0.35	0.45	[73,74,165]
800	1.75	1050	1.9	1.52	0.38	[73,74,165]
1000	2.6	1125	—	—	—	[73,74]
1100	—	1550	—	—	—	[74]
1200	2.8	1475	1.95	1.33	0.62	[73,74,165]
Ti₃SiC₂-CDC						
300	0.535	—	—	—	—	[32]
400	0.575	—	—	—	—	[32]
600	0.68	—	0.75	0.3	0.45	[32,165]
700	0.65	—	—	—	—	[32]
800	0.75	—	1.68	1.32	0.36	[32,165]
1000	0.65	—	—	—	—	[32]
1200	0.75	—	1.25	0.82	0.43	[32,165]
VC-CDC						
500	0.6	1163	0.51	0.04	0.47	[166]
600	0.6	1266	0.57	0.039	0.53	[166]
700	0.6	1277	0.62	0.059	0.56	[166]
800	0.6	1286	0.65	0.06	0.59	[166]
900	0.7	1305	0.66	0.03	0.63	[166]
1000	0.75	864	0.5	0.1	0.4	[166]
1100	1.1	236	0.18	0.08	0.1	[166]
WC-CDC						
700	1.27	474	—	—	0.26	[167]
1000	0.83	2036	—	—	1	[167]
TaC-CDC						
700	0.82	1963	—	—	0.83	[167]
1000	0.83	2275	—	—	1.01	[167]
NbC-CDC						
700	0.79	1864	—	—	0.78	[167]
1000	0.84	1721	—	—	0.78	[167]
HfC-CDC						
700	0.79	1590	—	—	0.69	[167]
1000	0.86	1833	—	—	0.82	[167]
ZrC-CDC						
350	0.76	500	0.225	0.225	0	[39]
400	0.74	500	0.225	0.224	0.001	[39]
600	0.8	875	0.475	0.45	0.025	[39]
800	0.85	1350	0.6	0.525	0.075	[39]
1000	1.21	1500	0.71	0.45	0.26	[39]
1200	1.41	1875	0.9	0.45	0.45	[39]

3.2.4.2 Effect of Synthesis Conditions

Increasing pore size with increasing synthesis temperature is a global trend for all CDCs: TiC [21,42,72], B_4C [37,73], ZrC [39,73], and Ti_2AlC [73,74], for example. In general, the starting carbide templates the initial pore size, and then any range of average pore sizes from less than 1 nm to greater than 10 nm can be achieved by changing the synthesis temperature. It was shown using Ti_3SiC_2-CDC synthesized in the 300°C–1100°C range that the pore texture changes considerably with synthesis temperature [32]. The isotherms of low temperature (600°C synthesis temperatures and below) were all of type I [75] in Brunauer classification indicating the presence of only pores smaller than 2 nm. At synthesis temperatures above 600°C, the isotherms were of type IV, which shows the presence of pores larger than 2 nm. Considering the pore size distributions calculated using the Horvath–Kawazoe method, it can be seen that with moderate temperature control, the pore size can be tuned with sub-Angstrom accuracy (Figure 3.4). Also, the pore size distributions are narrower than that of even single-walled CNTs. This pore widening with increasing synthesis temperature comes about because of increasing carbon mobility with increasing temperature [76].

Small angle x-ray scattering (SAXS) was used to corroborate porosity trends unveiled using gas adsorption techniques (Figure 3.5) [32,43]. SAXS detects density variations within the samples and can probe pores much smaller than the 0.4 nm lower limit of gas adsorption techniques. CDC showed a monodisperse region of micropores and a lower Q region that depended systematically on the particular CDC under consideration. For rock-salt structures, TiC and ZrC, and sphalerite SiC, there is a first neighbor separation of ~0.4 nm, and the SAXS profiles show narrow pore size distributions, whereas for rhombohedral B_4C with a larger and nonuniform nearest neighbor separation, the pore size distribution is much wider.

3.2.4.3 Effect of Posttreatment

Posttreating CDC using physicochemical methods has been pursued in order to modify the pore texture and surface chemistry beyond what is achievable with the chlorination alone. As CDC

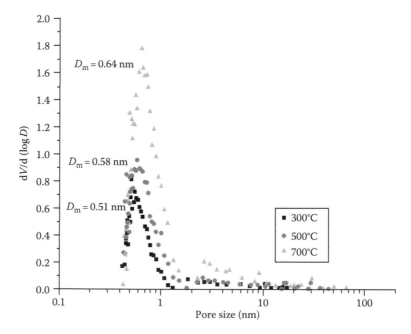

FIGURE 3.4 Pore size distributions for Ti_3SiC_2-CDC synthesized between 300°C and 700°C showing gradual increase in pore size with increasing synthesis temperature. (From Laudisio, G., et al., *Langmuir*, 22, 8945, 2006. With permission.)

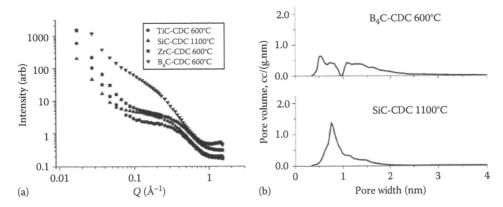

FIGURE 3.5 (a) Small angle x-ray scattering curves for various CDCs showing the variation in scattering with carbide. (b) Pore size distributions calculated using NLDFT on argon isotherms taken at 77 K for 600°C B$_4$C-CDC and 1100°C SiC-CDC show that despite the higher synthesis temperature of SiC-CDC, because of the closer and very uniform spatial distribution of carbon atoms in SiC over B$_4$C, the pore size is more narrow and shifted to smaller values. (From Laudisio, G., et al., *Langmuir*, 22, 8945, 2006. With permission.)

synthesis is a closed cycle process, it is easy to introduce posttreatment gases to change porosity and surface chemistry. Considerable efforts have been given to increasing carbon pore size without compromising surface area for supercapacitor applications [77]. In general, for activated carbons, widening pores are accomplished by increasing the amount of material burned during activation. As discussed above, by choosing different starting carbides and synthesis conditions, CDC can be made with very precisely tuned porosity. As the cost of some metal carbides is high, it is desirable to have some method to control the pore size of CDCs prepared from inexpensive precursors such as SiC. This is done by known methods such as exposing the carbon to an oxidizing environment at elevated temperature (activation).

A method was reported whereby CDC was produced from TiC and SiC and then subsequently treated in a water/argon mixture at 900°C [78,79]. This activation was found to increase the pore size, surface area, and capacitance. Even though there is more complexity to the synthesis, this was a desirable method to increase the capacitance due to the fact that TiC and SiC are the least expensive carbides. Another method of pore widening involves infiltrating the pores with water and rapidly heating the infiltrated carbon above the temperature where water vapor etches carbon [80]. The objective of this method was to selectively widen only the smallest pores.

Arulepp et al. [81] reported an *in situ* technique of CDC activation whereby TiO$_2$ was mixed with TiC and chlorinated. During synthesis, titanium in the TiO$_2$ was etched, leaving controlled amounts of oxygen to etch the carbon. It was shown that the oxidation treatment improved the volumetric capacitance but not the gravimetric capacitance. This was perhaps because of unreacted TiO$_2$ in the sample. The carbon was ideally polarizable at $E < 2.8$ V and showed a power density of 2.5 kW/kg.

Hydrogen treatment has been explored as a method to improve performance in a number of applications. It was shown that a treatment at 800°C in 5% H$_2$/Ar increased performance of CDC coatings for tribological applications [82]. It was postulated that the mechanism responsible was due to passivation of surface with C–H bonds and elimination of high-energy dangling bonds and reactive groups. Hydrogen annealing was also shown to be beneficial for hydrogen storage. H$_2$ annealing at 600°C increases surface area and pore volume by removing chlorine trapped in pores after synthesis without greatly changing the pore size distribution [42]. This increased hydrogen storage above 3 wt% at 77 K and 1 atm. It was also shown that H$_2$ treatment increased hydrogen storage by up to 75% for some other CDC materials [83]. This was believed to occur because hydrogen effectively cleaned small pores, increasing both the volume of the smallest pores and SSA. This can be appreciated by considering that up to 40 wt% chlorine has been observed in the as-produced Ti_3SiC_2-CDC samples [84].

3.2.4.4 CDC in Electrochemical Capacitors

A number of studies utilizing CDC from various precursors as the active material in electrochemical capacitors have been undertaken over the past decade, a few of which were briefly highlighted above. A study of the electrochemical response of TiC-CDC in tetraethylammonium tetrafluoroborate showed that CDC was a worthwhile material for supercapacitor applications [85]. Earlier patents [71,86] utilizing CDC blanks produced by first decomposing hydrocarbons in the interparticle porosity of a blank formed from carbide particles and then chlorinating the blank were filed, but performance was low. Later patents [79] outlined using different carbide precursors showed that the capacitance is dependent on the CDC structure, which can be controlled by precursor selection. For instance, in going from Al_3C_4-CDC to SiC-CDC in a KOH electrolyte, the capacitance decreases from ~260 to ~200 F/g. Likewise, in going from TiC-CDC to B_4C-CDC in 1.5 M tetraethylammonium tetrafluoroborate in acetonitrile solution, the capacitance drops from 100 to 75 F/g. A study on TiC, α-SiC, Mo_2C, Al_4C_3, and B_4C CDCs [87] showed that the electrochemical response rate depends fairly appreciably on starting carbide. This was justified by showing that the total pore volume and micropore volume depend on the carbide precursor. Cyclic voltammograms taken at a low scan rate of 5 mV/s, showed the highest current density on both the anodic and cathodic sweeps for SiC-CDC, while B_4C-CDC had the lowest. At a scan rate of 50 mV/s, the voltammogram of SiC-CDC became distorted and showed low current density at negative polarizations corresponding to adsorption of the larger $(CH_3CH_2)_3CH_3N^+$ cation. The highest current density for positive polarizations corresponding to adsorption of the smaller BF_4^- anion was also observed for SiC-CDC. This widely different behavior for the adsorption of the anion and cation in SiC-CDC showed the potential for high selectivity for electrolyte ions of different sizes with highly nanoporous CDCs.

Burke [3] showed that TiC-CDC has higher gravimetric capacitance (220 and 120 F/g in KOH and organic electrolyte, respectively) than SiC-CDC (175 and 100 F/g in KOH and organic electrolyte, respectively) and SiC-CDC has higher volumetric capacitance (126 and 72 F/cm³ in KOH and organic electrolyte, respectively) than TiC-CDC (110 and 60 F/cm³ in KOH and organic electrolyte, respectively). TiC-CDC was also studied in H_2SO_4 along with ZrC-CDC [88]. Gravimetric capacitance of 190 and 150 F/g was found for ZrC-CDC and TiC-CDC, respectively. High volumetric capacitance of 140 and 110 F/cm³ was achieved because of the relatively low total pore volume in comparison with the available surface area of the CDC samples studied. This study also pointed to a departure from the widely held theory that pores significantly smaller than the solvated ion do not contribute to double-layer capacitance [89–92]. It is important to note that different groups working on the same material produced different results due to differences in synthesis conditions, especially synthesis temperature which affects pore size.

Systematic studies were conducted on TiC and ZrC CDCs [88] as well as B_4C and Ti_2AlC CDCs [73] in 1 M H_2SO_4 to determine the optimum synthesis conditions. The CDCs were prepared in the temperature range 600°C–1200°C, which produced samples having tailored porosity between ~0.7 and ~3 nm and microstructures ranging from completely amorphous to turbostratic with isolated graphite ribbons. This systematic study showed that decreasing the surface area of pores smaller than 2 nm has a much larger effect on capacitance than the subsequent increase in surface area of pores larger than 2 nm. These results corroborate those obtained by Vix-Guterl et al. [22] who said that pores smaller than 0.7 nm contribute most to specific capacitance. The results on CDC also showed that the characteristic time constant [93] decreases with increasing synthesis temperature showing that CDC can be tailored for either high energy applications or high power applications.

It was shown by Permann et al. [94] that the specific capacitance of TiC-CDC in an organic electrolyte consisting of $(CH_3CH_2)_3CH_3NBF_4$ is in the range of 70–90 F/cm³ or 100–130 F/g depending

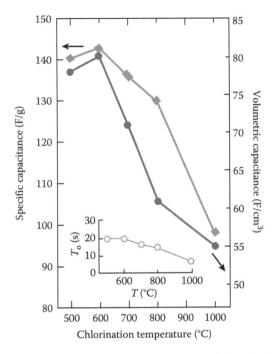

FIGURE 3.6 Evolution of gravimetric and volumetric capacitance with synthesis temperature for TiC-CDC in an electrolyte consisting of $(CH_3CH_2)_4NBF_4$ salt in acetonitrile. As synthesis temperature is increased, both volumetric and gravimetric capacitance decreases. The characteristic time constant (inset), which is a function of both resistance and capacitance, decreases with increasing synthesis temperature. (From Chmiola, J., et al., *Science*, 313, 1760, 2006. With permission.)

on the synthesis conditions, which were chlorinated in the range of 600°C–1000°C and annealed in hydrogen at 800°C. These numbers were comparable to the results achieved by Chmiola et al. [21] on TiC-CDC using $(CH_3CH_2)_4NBF_4$. Cells made from TiC-CDC at all synthesis temperatures were stable between 0 and 2 V in acetonitrile and propylene carbonate (PC). As a general rule, it can be said that capacitance decreases with increasing synthesis temperature though surface area and pore size generally increase [21] (Figure 3.6).

3.2.4.5 CDC as "Designer" Carbon Material

Because of the ease of property tunability in CDCs, they have been used to determine global trends applicable to other porous carbons. A series of tetraalkylammonium tetrafluoroborate salts in PC were used with TiC-CDC to determine the effect of electrolyte ion size on electrochemical performance [95]. At negative polarizations, it was found that the relaxation time constants were practically independent of cation size while at positive polarizations it increased in the order of electrolyte salts $MeEt_3NBF_4 < Me_2Et_2NBF_4 < Et_4NBF_4$ with increasing electrolyte viscosity and decreasing molar conductivity. Also, the dependence of the phase angle vs. frequency plots shows that the relaxation time constant is independent of the cation and the low-frequency behavior is mainly determined by the solvent characteristics. A similar study was conducted that also involved Li^+ cations and ClO_4^- anions that arrived at similar conclusions [96].

The effect of solvent properties on double-layer capacitance of TiC-CDC was explored using triethylmethylammonium tetrafluoroborate in PC, DMK (dimethyl ketone), γ-butyrolactone, and acetonitrile [97,98]. The capacitance was shown to decrease in the order acetonitrile > γ-butyrolactone > DMK > PC and was dependent on polarization. The same trend was found in relating solvent to cycling efficiency. Interestingly, CDC was shown to have a lower time constant than an advanced activated carbon used in commercial supercapacitors, even though the capacitance of the former was higher showing ease of ion migration in CDC.

CDC has been able to study double-layer intricacies difficult to achieve with commercial activated carbons. The diffusion of ions in pores smaller than the size of solvated ions has been demonstrated in both aqueous [99–101] and organic electrolytes [102]. Surprisingly, despite faster kinetics in supercapacitors, they were found to have similar diffusion coefficients to solid-state lithium diffusion in graphite as well as hydrogen diffusion in metal hydrides. It was shown that the current passed and diffusion coefficients at positive and negative polarization were highly dependant on the potential [99,100]. At positive polarizations [100] below 300 mV vs. Hg|HgO, there is a current peak that decreases with cycling. Cycling in this potential range also leads to decreased diffusion coefficients which, with the current peak, is consistent with functionalization by a single oxygen-containing species such as a phenol or quinone group. At potentials between 300 and 500 mV vs. Hg|HgO, there is an even more dramatic current peak and reduction in the diffusion coefficient and distortion in the cyclic voltammogram consistent with functionalization by a more highly oxygenated species. Neither of these oxidation events were found to be severe enough to change the microstructure of carbon as evidenced by Raman spectroscopy. At negative polarizations [99], the effect of hydrogen in the nanoporous texture was shown. Diffusion was ~10 times faster at negative polarizations. At potentials greater than −500 mV vs. Hg|HgO, the currents were purely capacitive, but between −500 and −1000 mV vs. Hg|HgO, water is dissociated and hydrogen is adsorbed at defined sites in the nanoporous texture of CDC. At longer times, there was a recombination of atomic hydrogen to H_2 as evidenced by decaying current peaks after holding the cell at open circuit. At negative potentials of −1000 mV, there is significant hydrogen evolution. Unlike the carbon oxidation processes at positive potentials in alkaline solutions, negative polarization does not seem to affect ion transport due to the absence of bulky functional groups introduced on the carbon surface. The transport of the organic salt $(CH_3CH_2)_4NBF_4$ in nanoporous SiC-CDC was also studied using microelectrodes [102]. This study hinted at problems with the traditional understanding of ion motion in supercapacitors. First, the cations were shown to be released at defined potentials, indicating some sort of activation energy is required, unlike the purely electrostatic interactions that are assumed. Also, the Cottrell model and its applicability to supercapacitors were called into question. Diffusion in nanoporous CDC was also compared to diffusion in activated carbon with a wide pore size distribution [103] and shown to have slightly lower kinetics of ion insertion and withdrawal than traditional carbons.

The above work using CDC as a model carbon system showed that the traditional understanding of double-layer capacitance is short-sighted, and other factors besides carbon structure need to be understood to maximize performance. To begin to understand the effect of constraining ions, a series of samples from TiC-CDC was synthesized with all pores tuned to be smaller than the size of the electrolyte ion and a single layer of associated solvent molecules [21]. It was shown that both the specific (gravimetric) and volumetric capacitances decreased with increasing synthesis temperature, even though the surface area and pore size both increased. Even for the sample with the smallest pore size (500°C TiC-CDC), there was only a minimal decrease in specific capacitance when the current density was increased to 100 mA/cm², which illustrates that subnanometer-sized pores are still readily accessible even at very high discharge rates. When the specific capacitance was normalized by SSA, the effect of pore size, irrespective of surface area, was ascertained (Figure 3.7). For TiC-CDC, increasing the pore size appeared to have a detrimental effect on the normalized capacitance. Choosing initial carbides that yielded larger pore sizes (Ti_2AlC and B_4C) and data from two studies in literature on the same electrolyte system [11,22], there appears to be a decrease in capacitance with decreasing pore size for pores larger than 1 nm. However, at a pore size less than 1 nm, the trend reversed and there was a sharp increase in capacitance with decreasing pore size. Two carbons with narrow pores that looked like outliers in the study of Gamby et al. actually follow the trend reported here when put in this context. It is also important

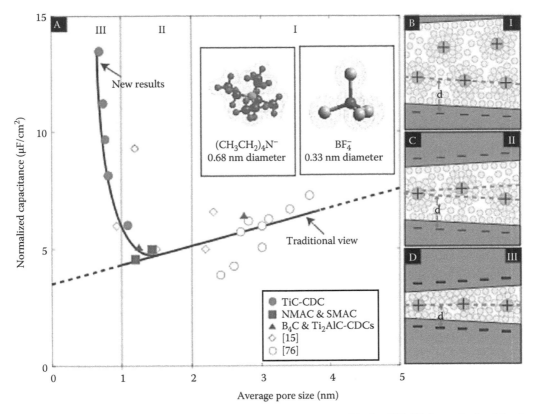

FIGURE 3.7 A plot of specific capacitance normalized by BET SSA for TiC-CDC synthesized in the 500°C–1000°C range as well as B_4C-CDC and Ti_2AlC CDC and data from two other studies [11,22] on typical activated carbons in an electrolyte consisting of $(CH_3CH_2)_4NBF_4$ in acetonitrile. The normalized capacitance decreased with decreasing pore size until a critical value was reached, unlike the traditional view which assumed that capacitance continually decreased.

to account for to the energy stored in the removal of the solvation shell surrounding each ion in solution and the subsequent release upon resolution [104]. A recent study on an ionic liquid with no solvent added has shown a correlation between the ion size and pore size in CDC [105]. The maximum capacitance was observed for CDC with the average pore size closely matching the maximum ion dimension.

3.3 TEMPLATED CARBON

3.3.1 INTRODUCTION

Templated carbons have also shown the promise of providing precise control of carbon porosity. The template carbonization method consists of the carbonization of an organic compound in nanospace of a template inorganic substance and the liberation of the resulting carbon from the template. So far, various types of unique carbon materials have been synthesized using this method. For example, ultrathin graphite film has been prepared from the carbonization of organic polymer in the two-dimensional opening between the lamellae of layered clay such as montmorillonite and taeniolite [106–108]. This research revealed that even a typical nongraphitizable carbon precursor like polyfurfuryl alcohol (PFA) can be graphitized very well by the template method using the layered clays [108]. This finding was beyond the bounds of the conventional common knowledge of carbon science, where it was said that the final structure of a carbon material strongly depends on the nature of the original precursor rather than its nurture (the conditions of carbonization process). Besides this two-dimensional approach, the template technique allows one to prepare one- and three-dimensional carbons such as CNTs and nanoporous carbons. The present section demonstrates how effectively the carbon nanostructure can be controlled by the template carbonization technique and introduces the electrochemical characteristics of these nanostructured carbons.

3.3.2 TEMPLATE SYNTHESIS OF UNIFORM CNTs

3.3.2.1 Synthesis and Structure

Using uniform and straight nanochannels of an anodic aluminum oxide (AAO) film as a template, CNTs can be prepared by pyrolytic carbon deposition on the AAO film [109–116]. Briefly, the AAO film was subjected to carbon deposition from the pyrolytic decomposition of propylene at 800°C, which resulted in a uniform pyrolytic carbon coating on the inner wall of the template nanochannels. Then, the AAO template was removed with HF washing, and only carbon was left as an insoluble fraction. The formation process of carbon tubes using this chemical vapor deposition (CVD) technique is illustrated in Figure 3.8.

Figure 3.9 shows scanning electron microscope (SEM) photographs of the carbon samples prepared using two types of AAO films with different channel diameters (30 and 230 nm). These photographs reveal that in both cases the samples consist of only cylindrical tubes and their outer diameter is the same as the channel diameter of the corresponding AAO film. No other form of carbon was found in the microscopic observation. In the SEM photographs with low magnification (Figure 3.9a and c), many bundles of tubes can be observed and the length of all the tubes in a bundle corresponds to the thickness of the parent template film. Carbon tubes with such uniform diameters and lengths cannot be synthesized by conventional arc-evaporation and catalyst CVD techniques, which generally produce tubes of different sizes together with many types of impurities including metal particles. The presence of many bundles in Figure 3.9 implies that most of the tubes are connected at both ends of each tube, because the carbon deposition also took place on the external flat surface of the AAO film. From a high-resolution transmission electron microscopy (HRTEM) image for the carbon tubes thus prepared, it was shown that the size of most graphene layers in carbon walls is less than 10 nm and they wrinkle to a great extent. This structure is far from graphite, but all the layers are oriented toward the direction of the carbon tube axis, and these can easily be graphitized by further thermal treatment at temperatures above 2400°C [115,117]. Another important feature of this template-synthesized carbon tube is

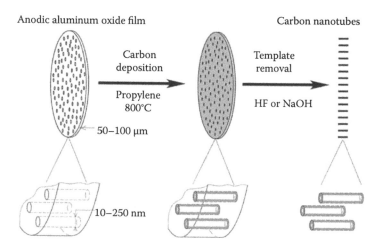

FIGURE 3.8 The formation process of CNT by the template method using an AAO film. (From Chmiola, J., et al., *Science*, 313, 1760, 2006. With permission.)

FIGURE 3.9 SEM photographs of the CNT prepared using AAO films with a channel diameter of (a, b) 30 nm and (c, d) 230 nm.

that the tubes are not capped at both ends, unlike conventional CNTs synthesized by the arc-discharge and the catalytic CVD methods. In conclusion, the template technique allows one to prepare multi-walled carbon tubes of uniform diameter and length without any metal catalyst.

3.3.2.2 Selective Fluorination onto Nanotube Inner Surface

Chemical modification of CNTs changes or improves their chemical and electrical properties, thereby expanding their application fields. All of the efforts for the chemical modification have been directed toward the outer surface of CNTs. No one has, however, attempted to differentiate between the outer and inner surfaces or to modify only the inner one while leaving the outer one as it is. One of the reasons for this is that both ends are generally closed for most CNTs, but even if they were open, such differentiation would be essentially impossible; any chemical treatment to the inner surface always affects the outer one. Only the template technique enables such selective chemical modification of the inner surface of nanotubes. With this technique, CNTs with outer and inner surfaces that have different properties can be prepared, and unique adsorption behaviors and electrical properties can be expected from such CNTs with heteroproperties.

If the carbon-deposited AAO film (see Figure 3.8), i.e., the nanotubes embedded in the AAO film, is chemically treated, only the inner wall surface could be modified, because the inner surface is exposed to the atmosphere but the outer surface is completely covered with the template in the stage of carbon-deposited film. Based on this concept, an attempt was made to fluorinate only the inner surface of the CNTs [118]. It is well known that fluorination is a quite effective way to introduce strong hydrophobicity to carbonaceous materials and it perturbs the carbon π electron system. Consequently, by the selective fluorination of the nanotube inner surface, it would be possible to produce CNTs whose inner surface is highly hydrophobic and electrically insulating, whereas its outer surface is conductive.

After propylene CVD at 800°C over an AAO film with a channel diameter of 30 nm, its carbon-coated film surface was fluorinated by direct reaction of the film with elemental fluorine.

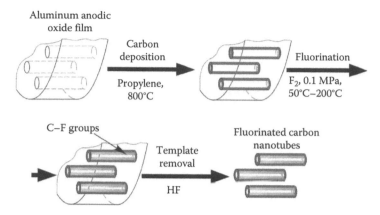

FIGURE 3.10   Fluorination process to the inner surface of CNT.

The film was placed in a nickel reactor and allowed to react with 0.1 MPa of dry fluorine gas (purity 99.7%) for 5 days at a predetermined temperature in the range of 50°C–200°C. The fluorinated CNTs were then liberated by dissolving the AAO template with HF at room temperature. A schematic drawing of the fluorination process is illustrated in Figure 3.10, where it should be noted that the carbon deposit on the external flat surface of the AAO film is essentially the same as the deposit on the inner surface of the AAO nanochannels. The information on the inner surface of the CNTs can therefore be obtained from the analysis of the external surface of the parent coated film.

In order to examine the chemical state on the carbon surface, x-ray photoelectron spectroscopy (XPS) analysis was performed for both the fluorinated carbon-coated AAO film and the liberated nanotubes from this fluorinated film. In the resulting XPS C1s spectrum of the fluorinated film (Figure 3.11a), the most intense component is the peak assigned to the covalent C–F bond (about 290 eV),

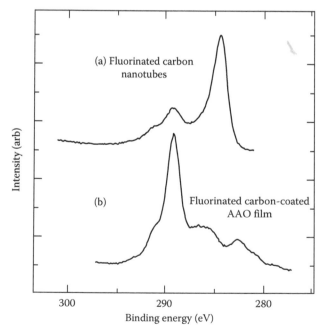

FIGURE 3.11   XPS C1s spectra of the (a) CNT prepared from the films and (b) fluorinated carbon-coated AAO films.

whereas the spectrum of the nanotubes shows a clearly resolved peak at 284.4 eV (Figure 3.11b), which can be attributed to sp^2 carbon atoms of the carbon skeleton. The XPS spectrum in Figure 3.11a does not directly reflect the CNT surface, but the carbon on the external flat surface of the film. However, the chemical form of the inner surface of the CNTs can be expected to be the same as that of the external flat surface of the film, as described in the previous paragraph. It is safe to say that the inner surface of the nanotubes was fluorinated by the treatment. The information on the outer surface of the CNTs cannot be obtained from such XPS measurements as in Figure 3.11a, but, after the removal of the template, the outer surface is exposed so that the XPS spectrum in Figure 3.11b can be regarded as the information from the tube's outer surface. These XPS spectra therefore provide evidence for the fact that fluorine atoms were introduced only to the inner surface of the tubes but the outer surface remained almost unchanged even upon the treatment. From TEM observation, no apparent difference was found between the fluorinated and pristine tubes, thus indicating that the fluorinated treatment under the present conditions does not cause any damage to the tubular shape.

3.3.2.3 Effect of Fluorination on Adsorption and Electrochemical Properties

The fluorinated carbon-coated AAO film has an interesting adsorption characteristic that has not been reported so far. Figure 3.12 shows N$_2$ adsorption/desorption isotherms at −196°C for the pristine carbon-coated AAO film and the films fluorinated at different temperatures [119]. The isotherm of the pristine film is characterized by the presence of a sharp rise and a hysteresis in a high relative pressure range. Such a steep increase can be ascribed to the capillary condensation of N$_2$ gas into the nanochannels of the AAO films, that is, the inner space of the nanotubes embedded in the AAO films. The amount of N$_2$ adsorbed by the condensation into the fluorinated channels is lower than that of the pristine one. Moreover, the amount drastically decreases with an increase in the severity of fluorination. Since TEM observation revealed that the inner structure of the fluorinated CNTs was not different from that of the pristine nanotubes, the reason why the N$_2$ isotherm was so changed as in Figure 3.12 cannot be attributed to the alteration of the pore texture upon the

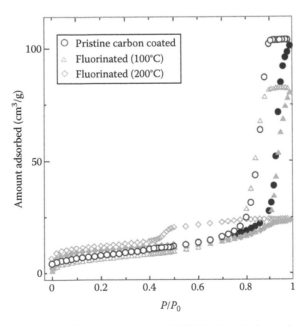

FIGURE 3.12 N$_2$ adsorption/desorption isotherms at −196°C for the pristine carbon-coated AAO film and the films fluorinated at different temperatures (100°C and 200°C).

fluorination treatment. One of the possibilities for this is poor interaction between liquid N_2 and the fluorinated carbon layer in the nanochannels.

The group of Martin and coworkers [110,114] examined the electrochemical properties of template-synthesized CNTs and demonstrated that the membrane of such nanotubes can reversibly intercalate lithium ions, although there was some irreversibility. In this context, it is interesting to study the effect of fluorination on the lithium ion intercalation. Electrochemical insertion of lithium into the pristine carbon-coated AAO film and the carbon-coated film fluorinated at 50°C was studied by cyclic voltammetry (CV) and discharge–charge experiments using Li/1 M $LiClO_4$ – (EC + DEC)/WE cells, where EC, DEC, and WE, respectively, correspond to ethylene carbonate, diethyl carbonate, and working electrode (the pristine and fluorinated carbon-coated films) [119]. In the test cells, lithium can be inserted into the CNTs in the cathodic sweep (discharge) while extracted in the anodic sweep (charge). Figure 3.13 shows the CVs of the pristine and fluorinated films. The open circuit voltage (OCV) of the pristine film is 2.98 V, whereas the fluorination of the inner surfaces leads to a sharp increase in the OCV to 3.64 V. The increase of OCV is due to the formation of C–F bonds. It can be seen from Figure 3.13a that the CV of the pristine film is characterized by two peaks (A and B). Peak A (at 1.4 V) and peak B (at 1.26 V) can be attributed to the reduction of surface oxygen-containing functional groups and the electrolyte $LiClO_4$, respectively. The discharge scan is completed with a sharp current declination from 0.25 V due to lithium ion intercalation between the graphene layers parallel to the tube axis. On the reverse sweep (first charge), a current plateau, starting at 0.4 V and extended to around 1.5 V is observed. This plateau is attributed to different extraction processes of lithium from various insertion sites of the CNTs embedded in the AAO film, occurring at different oxidation potentials. In the case of the fluorinated carbon-coated AAO film (Figure 3.13b), the first discharge sweep starts with the reduction of C–F bonds (peak C at 3.0 V). The peaks A and B observed in the pristine nanotubes become negligibly small in Figure 3.13b. Note that the scale of the ordinate in this plot is smaller than that in Figure 3.13a. The replacement of oxygen-containing functional groups by elemental fluorine leads to such observation.

The galvanostatic charge–discharge curves of the carbon-coated AAO film are shown in Figure 3.14a. A large irreversible capacity in the reduction processes was observed for the first cycle. This is attributable to a relatively high BET surface area of 28 m^2/g and the presence of oxygen-containing functional groups on the carbon surface. The reversibility of the insertion and extraction process was improved considerably in the second and consecutive cycles. For example, the first cycle coulombic efficiency of 22% was enhanced to about 80% with a reversible capacity of 460 mAh/g in

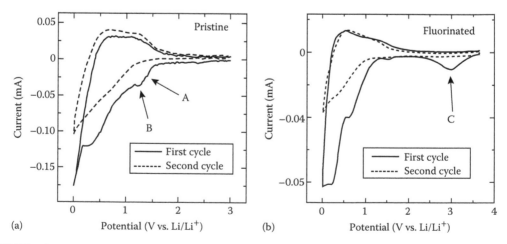

FIGURE 3.13 Cyclic voltammograms for (a) the pristine carbon-coated AAO film and (b) the carbon-coated film fluorinated at 50°C.

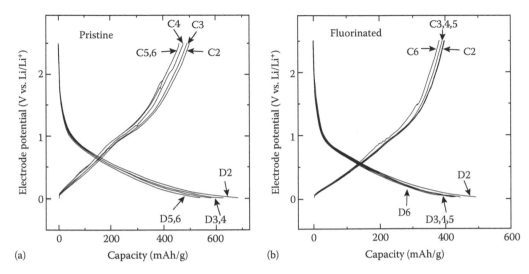

FIGURE 3.14 Discharge–charge curves of lithium insertion into (a) the pristine carbon-coated AAO film and (b) the fluorinated film under a current density of 50 mA/g. D2–D6 correspond to 2nd–6th discharge curves and C2–C6 to 2nd–6th charge ones.

the sixth cycle. The effect of fluorination on the charge–discharge curve can be seen from Figure 3.14b. Although the irreversible capacity at the first cycle was found to be almost the same for both pristine and fluorinated films, there is a substantial increase in coulombic efficiency from 23% to about 90% with a reversible capacity of about 400 mAh/g after the second cycle. The changes in the inner surface properties caused by the fluorination reduced the formation of Solid Electrolyte Interphase (SEI), which resulted in such increase of coulombic efficiency of more than 10% after the second cycle.

3.3.3 Template Synthesis of Microporous Carbons

Porous carbons that possess a well-tailored micropore structure are extremely attractive and now in great demand for such applications as storage media for methane or hydrogen gas and electrode of electric double-layer (EDL) capacitor (supercapacitor). Many novel approaches to control pore structure have, thus, been proposed [120]. Among them, great attention has been paid to the template carbonization method, as well as the carbon formation from carbide precursors as described in Section 3.2. So far, many researchers have prepared novel porous carbons with the template technique using a variety of inorganic porous templates [121–126]. In 1999, two Korean research groups independently obtained mesoporous carbon with a regular structure using a silica mesoporous molecular sieve as a template and demonstrated that their method is quite effective for the precise control of carbon mesoporosity [127,128]. However, such mesoporous silica templates cannot be used for the synthesis of microporous carbon, and consequently, the control of microporosity.

The group of Kyotani et al. [129] has been investigating the template synthesis of porous carbon using zeolite Y, which has an open microporous structure and possesses regular three-dimensional channels. In 2000, they prepared long-range ordered microporous carbons with the structural regularity of zeolite Y [130]. Such regularity was actually observed using TEM. Furthermore, it was found that, under a certain synthesis condition, the carbon possessed almost no mesoporosity and its surface area and micropore volume reached 3600 m^2/g and 1.5 cm^3/g, respectively [131–133]. Then, they further improved the preparation method and, as a result, they have succeeded in synthesizing ordered microporous carbon with a surface area of more than 4000 m^2/g and micropore volume of 1.8 cm^3/g [134]. Several other zeolite materials such as EMT and BEA zeolites have also been successfully used for carbon replication [131,135]. The common feature of these zeolite materials is

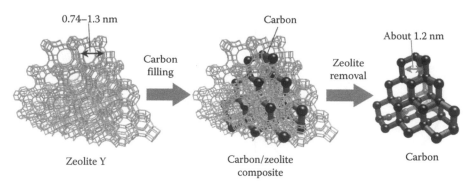

FIGURE 3.15 Scheme of the template carbonization using zeolite Y.

that they possess relatively large micropores composed of 12-membered pore aperture. The whole scheme and concept of the template synthesis of the ordered microporous carbon can be illustrated as in Figure 3.15.

This kind of carbon with such unique pore structure, extremely large surface area, and micropore volume has never been reported. Sections 3.3.3.1 and 3.3.3.2 will introduce the details of the production method of the ordered microporous carbons and compare the extraordinary pore structure of this carbon with commercial microporous carbons with a large surface area. In Section 3.3.3.3, it will be demonstrated that a similar ordered microporous carbon containing N atoms can be prepared by the template technique, and the N-doping influences the adsorption behavior of H_2O molecules [136]. In Section 3.3.3.4, the use of this unique carbon as an electrode for supercapacitor will be introduced [137].

3.3.3.1 Synthesis and Structure

A two-step method was used to fill carbon into the zeolite nanochannels. In the first step, dry zeolite Y powder (Na-form, $SiO_2/Al_2O_3 = 5.6$, Toso Inc., HSZ-320NAA) was impregnated with FA. The mixture of FA and zeolite powder was stirred for 8 h and then filtered and washed with mesitylene to remove any FA on the external surface of the zeolite powder. The polymerization of FA in zeolite was carried out by heating the powder at 80°C for 24 h and then at 150°C for 8 h in N_2. The resultant zeolite/PFA composite was heated under a N_2 flow at 5°C/min up to 700°C. When the temperature reached 700°C, propylene gas (2% in N_2) was passed through the reactor for 4 h. This second step resulted in pyrolytic carbon deposition into the remaining openings of the composite. After the propylene CVD, some of the zeolite/carbon composite was further heat-treated at 900°C for 3 h under a N_2 flow. In addition to this two-step method, the template carbon was prepared from a single carbon precursor using either FA impregnation or propylene CVD, using a carbonization temperature and holding period for PFA or propylene of 700°C and 4 h, respectively.

All the zeolite/carbon composites prepared by these methods were treated with an excess amount of 46% aqueous HF solution at room temperature for 3 h and subsequently refluxed in concentrated HCl solution at 60°C for 3 h to dissolve the zeolite framework. The solution was filtered and the insoluble fraction was extensively washed with water. Finally, the resultant insoluble carbon was air-dried at 120°C overnight.

For convenience, the preparation step is defined here using the following abbreviations: impregnation of FA and subsequent carbonization of PFA at 700°C as PFA7; propylene CVD at 700°C as P7; PFA carbonization followed by propylene CVD at 700°C (the two-step method) as PFA-P7; and heat treatment at 900°C (the two-step method) as H.

TABLE 3.2
Carbon Fraction in the Composites and Results of Elemental Analysis of the Carbons

Method	Carbon Content in the Composite (g/g of Zeolite)	Elemental Analysis of the Carbon (wt%)			
		C	H	N	O (diff.)
PFA7	0.12	82	3	0	15
P7	0.24	89	2	0	9
PFA7-P7	0.30	88	3	0	9
PFA-P7-H	0.29	90	2	0	8
PFA-AN8-H	0.25	88	2	6	4

The carbon contents of the composites were calculated from the results of elemental analysis, and the results are summarized in Table 3.2. The carbon content in the PFA7 composite was 0.12 g/g and much lower than those prepared by the propylene CVD, which ranged from 0.24 to 0.30 g/g. Thus, the CVD process is quite effective for high carbon loading in zeolite. The carbon content significantly increased when the CVD temperature was increased from 700°C to 800°C, but this increase was at least partly due to the deposition of extra carbon on the external surfaces of the zeolite particles. The results of the elemental analysis of the resulting carbons are also given in Table 3.2, which shows that the carbons contain relatively large amounts of oxygen. Since the carbon precursor of P7 carbon contains no oxygen, some oxygen was probably incorporated into the carbon during the acid washing process.

Figure 3.16 shows XRD patterns of the original zeolite Y and the carbons prepared using the zeolite as template. The zeolite pattern (Figure 3.16a) is characterized by the appearance of many

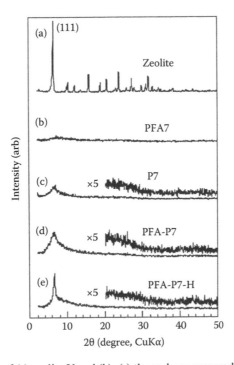

FIGURE 3.16 XRD patterns of (a) zeolite Y and (b)–(e) the carbons prepared using zeolite Y as a template.

sharp peaks due to the framework topology of zeolite Y. The strongest peak appears at 6.19°, and it can be indexed as the diffraction from the (111) planes. No apparent peak was observed for the carbon obtained from the FA-impregnation method (Figure 3.16b). There was a weak and broad peak at around 6° for the carbon obtained from the propylene-CVD method (Figure 3.16c). The presence of this XRD peak implies that this carbon had a structural regularity with a periodicity of about 1.4 nm, although the ordering was poor judging from the weak intensity and the broadness of the peak. In contrast, PFA-P7 carbon prepared by the two-step (FA-impregnation and then CVD) method showed a strong but broad peak at around 6° (Figure 3.16d). Interestingly, this peak became sharper after heat treatment at 900°C (PFA-P7-H carbon in Figure 3.16e).

Figure 3.17 shows SEM images of zeolite Y and PFA-P7-H carbon. Zeolite Y has clear crystal faces on each particle. The size and morphology of the carbon particles are very similar to those of the original zeolite particles. Figure 3.18 presents an HRTEM image of the PFA-P7-H carbon, which showed a strong and sharp XRD peak at 6° (Figure 3.16e). A periodically ordered array can be seen from the HRTEM image, and the arrangement and periodicity (about 1.3 nm) of this array structure correspond to those of supercages on the {111} planes of zeolite Y. The electron diffraction pattern (inset in Figure 3.18) is characterized by two pairs of spots whose arrangement corresponds

FIGURE 3.17   SEM images of zeolite Y and PFA-P7-H carbon.

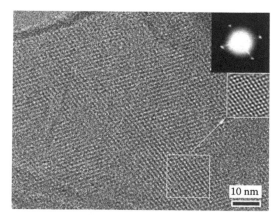

FIGURE 3.18   HRTEM image of PFA-P7-H carbons. An inset is a corresponding electron diffraction pattern. The image just below the electron diffraction pattern is a noise-filtered image taken from the area indicated by a square.

to the pattern of four spots from the {111} reflection of fcc zeolite Y with <110> zone axis. In addition to the four spots, sometimes we could observe another pair of very weak spots at the positions corresponding to the {220} reflection of zeolite Y. However, all these spots could be observed only in the diffraction with <110> zone axis and no diffraction spot appeared in the case of other zone axes. This finding suggests that the carbon inherited its regular ordering only from the zeolite crystal planes parallel to the <110> direction, mostly from the ordering of the {111} planes, which directly correspond to a row of supercages in the zeolite Y crystal.

As shown in Table 3.2, the carbon content in the composite prepared by the FA-impregnation method was much lower than that of the other composites, but this composite still had a relatively high BET surface area from the N_2 adsorption results (350 m²/g). These results indicate incomplete filling of carbon in the zeolite channels. Due to the low occupancy, the carbon could not retain the three-dimensional regular structure of zeolite Y after the liberation from the zeolite channels. As a result, the obtained carbon had no ordered structure reflecting that of the parent zeolite. The carbon content of the composite obtained by the propylene-CVD method was much higher than that of the composite obtained by the FA-impregnation method (Table 3.2). Due to such high carbon loading into the zeolite channels, the carbon could retain the zeolite regular structure to some extent even after the framework was completely removed by the acid washing. This explains the appearance of a small and broad peak at 6° in the XRD pattern of the P7 carbon (Figure 3.16c). Subsequent heating of the PFA-P7 composite at 900°C to form the PFA-P7-H carbon resulted in a very sharp peak around 6° in the XRD pattern (Figure 3.16e). This finding implies that the carbon inside zeolite channels may be more carbonized during the heat treatment, and consequently, the structure becomes more stable and robust. The resulting carbon has therefore a long-range ordering after liberation from the template.

3.3.3.2 Unique Pore Structure

As described in Section 3.3.3.1, the PFA-P7-H carbon better inherits the structural regularity from the parent zeolite than the other carbons. Moreover, the pore structure of this carbon is unique and it has never been found in any type of carbon. Its detail will be explained in this section in comparison with three commercial activated carbons: MSC-30 (Kansai Coke and Chemicals), M-30 (Osaka Gas), and ACF-20 (Osaka Gas). The former two carbons were prepared from petroleum coke and mesocarbon microbeads, respectively, and both were activated with KOH. The last sample was activated with carbon fibers (ACFs). All are characterized by a large BET SSA.

Nitrogen adsorption–desorption isotherms of all carbons are of type I, indicating the development of microporosity. All the carbons examined here possess a large BET SSA (Table 3.3). In particular, the value of the PFA-P7-H carbon reaches more than 4000 m²/g. For obtaining more reasonable surface area, Kaneko and Ishii [138] recommended drawing a high-resolution α_s-plot using the N_2 isotherm data (Table 3.3). Among the carbon samples examined here, the PFA-P7-H carbon has the largest surface area and, to the best of our knowledge, this value (3730 m²/g) is larger than any other value reported so far.

Dubinin–Radushkevich (DR) plots for all the carbons are shown in Figure 3.19. Whereas the commercial carbons show significant upward deviations from linearity at high pressure $[\ln(P_0/P)^2 < 10]$, the PFA-P7-H carbon maintains linearity even in such a high pressure range. This suggests that the commercial carbons have mesopores together with micropores, whereas the PFA-P7-H carbon has a narrow pore size distribution. The micropore volume of PFA-P7-H (1.8 cm³/g) is much larger than for the other three carbons (Table 3.1); this value may be the largest one reported in literature. Although the micropore volume of PFA-P7-H is the largest, its mesopore volume is the smallest. Only micropores are developed in the case of the PFA-P7-H carbon.

For obtaining more detailed information on the pore structure, pore size distribution (PSD) curves were determined by the application of the density functional theory (DFT) method to the N_2 adsorption isotherms (Figure 3.20). The PSD curve of PFA-P7-H is very sharp and most of the

TABLE 3.3

SSA and Pore Volume for the Five Different Porous Carbons

Carbon	Specific Surface Area (m²/g)		Pore Volume (cm³/g)	
	BET[a]	α_s[b]	V_{micro}[c]	V_{meso}[d]
PFA-P7-H	4080	3730	1.8	0.2
MSC-30	2770	2780	1.1	0.4
M-30	2410	2480	1.0	0.8
ACF-20	1930	1850	0.7	0.5
PFA-AN8-H	3310	n.d.	1.3	0.3

[a] Determined from the BET equation using the data at $P/P_0 = 0.01$–0.05.
[b] Determined from α_s plot.
[c] From DR equation using N_2 isotherm.
[d] By subtracting the micropore volume (obtained from the N_2 isotherm) from the volume of N_2 adsorbed at $P/P_0 = 0.95$.

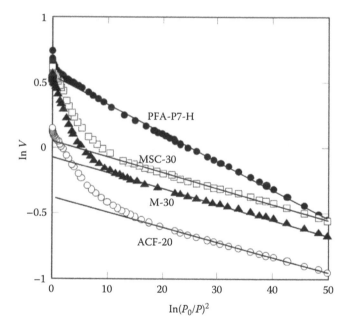

FIGURE 3.19 DR plots for the four porous carbons. (From Matsuoka, K., et al., *Carbon*, 43, 876, 2005. With permission.)

pore sizes fall within the range 1.0–1.5 nm, whereas the curves of the other carbons are broad and the pore size ranges from 0.5 to 3 nm or above. The value of the PSD maximum of PFA-P7-H at about 1.2 nm is comparable to the periodicity of this ordered carbon (1.4 nm). Although the value of pore size (1.2 nm) cannot directly be linked to the periodicity (1.4 nm), it is reasonable to assume that the pore structure of PFA-P7-H is derived from its periodically ordered array structure (this assumption can be deduced from the synthesis process in Figure 3.15). On the other hand, the conventional activation methods using oxidizing gas or KOH develop not only microporosity but also mesoporosity when excess activation is performed to obtain a large surface area. The precise

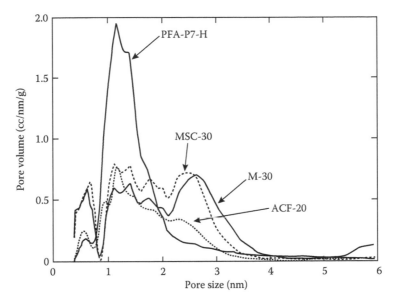

FIGURE 3.20 Pore size distribution curves determined by applying the DFT method to the N_2 adsorption isotherms of the four carbons.

control of microporosity in large surface area carbons as demonstrated in Section 3.3.3 is impossible for conventional activation methods.

Owing to the unique structure of the ordered microporous carbons synthesized by the template method, they exhibit interesting behavior when they are hot-pressed at 300°C up to 147 MPa [139]. In this experiment, besides the PFA-P7-H carbon, another type of zeolite templated carbon was employed, which was prepared only by acetylene CVD (600°C, 4 h) over zeolite Y and subsequent heat-treatment at 900°C (referred to as Ac6-H carbon). Upon hot-pressing, the powdery carbons were easily pelletized without any binder and the bulk density was significantly increased from 0.2 to 0.7–0.9 g/cm³. As a result, both the surface area and micropore volume per unit volume were increased. Surprisingly, it was found that the hot-pressing treatment reduced the average micropore size, and it was tunable by simply adjusting pressure during the treatment (Figure 3.21). In contrast, such changes in the density and pore texture were not observed for the commercial KOH-activated carbons (MSC-30 and M-30) when they were hot-pressed under the same conditions.

3.3.3.3 Nitrogen Doping

Many researches have focused their attention to N-containing porous carbons, because the introduction of N atoms endows carbons of polar nature. Their physicochemical properties would thus be different from those of N-free porous carbons and are more desirable for application to the electrodes of EDL capacitors [140,141]. Porous carbons containing N atoms can be obtained using different methods: (1) reaction of porous carbons with N-containing gases [142–144], (2) cocarbonization of N-free and N-containing precursors [145–147], and (3) carbonization of raw materials containing N atoms [148]. However, due to the complexity of nanotextured carbons, it is very difficult to tailor their porous texture, especially microporosity. Thus, the template method was employed to synthesize N-containing microporous carbons [136].

The two-step method described before was applied in the preparation of N-containing carbons. In the first step, FA was polymerized in the nanochannels of zeolite Y. The resulting PFA/zeolite composite was heated up to 800°C and then subjected to CVD of acetonitrile over the composite

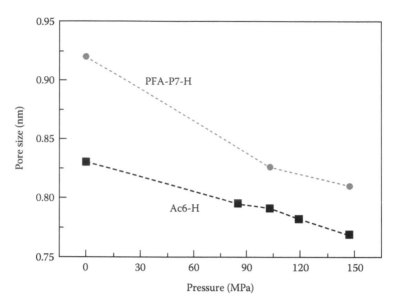

FIGURE 3.21 Plots of average micropore size against pressure applied during hot-pressing at 300°C. The average micropore diameter was calculated from the slope and intercept of an α_s plot of each sample. (From Hou, P.X., et al., *Carbon*, 45, 2011, 2007. With permission.)

for 2 h, followed by heat treatment at 900°C under a N_2 flow. Finally, the carbon part was liberated from the zeolite framework by HF washing. For convenience, this carbon is referred to as PFA-AN8-H.

The structural regularity of PFA-AN8-H was confirmed by TEM and XRD. The elemental analysis of this carbon evidences the presence of nitrogen (the last row of Table 3.2), and XPS reveals that quaternary N is the main N-functionality. In order to investigate the effect of N-doping, the PFA-AN8-H carbon was compared with the N-free (PFA-P7-H) carbon having a similar type of microporous structure. PFA-P7-H does not contain any N, but its O content is twice as large as that of the N-doped one (Table 3.2). PFA-AN8-H shows a sharp XRD peak derived from the regularity of zeolite Y, and the intensity and sharpness of this peak are almost the same as those of the PFA-P7-H carbon. It is therefore likely that the degree of ordering of PFA-AN8-H is similar to that of PFA-P7-H. The BET SSA, micropore, and mesopore volumes of the N-doped carbon were determined from its N_2 isotherm, and they are summarized in the last column of Table 3.3. The SSA and micropore volume of PFA-AN8-H are smaller than those of the PFA-P7-H carbon. Despite some differences in microporosity between the two carbons, it was found that their PSD curves determined by the DFT theory are similar; both carbons have a surprisingly sharp PSD peak, and most of the pore sizes fall within the range 1.0–1.5 nm. Hence, the two carbons possess a very similar ordered microporous structure with a very narrow PSD.

The H_2O adsorption–desorption isotherms of the above two carbons are plotted in Figure 3.22. Their isotherms are of type-V, and the shape is characterized by a sharp adsorption uptake accompanied by a clear hysteresis occurring over a medium relative pressure (P/P_0) range. Such characteristics have often been observed in H_2O isotherms of microporous carbons such as ACF [149,150]. Mowla et al. found that the width of the hysteresis loop in H_2O isotherms for microporous carbons depends on their pore size; no hysteresis is observed for carbons with a pore size of less than 0.8 nm, but a wide loop exists for carbons having a larger pore size [151]. The latter is indeed the case for the present carbon samples.

It is noteworthy that the pressure where rapid H_2O adsorption takes place on the PFA-AN8-H carbon is lower than that of PFA-P7-H. In other words, the N-containing porous carbon has a stronger affinity to H_2O than the N-free carbon. Such a lower shift of the uptake pressure due to

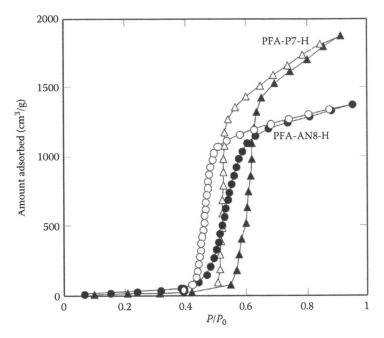

FIGURE 3.22 H_2O adsorption–desorption isotherms at 25°C for the undoped and N-doped carbons.

N-doping has already been reported for ACF and activated carbon [150,152]. It is well known that the uptake pressure and the shape of the H_2O isotherm are functions of both micropore size and surface chemical properties. In this case, however, the influence of micropore size can almost be excluded and the observed difference in the uptake pressure be attributed solely to carbon surface chemistry. It is therefore reasonable to conclude that the inner pore surface of the N-doped carbon is more hydrophilic than that of the undoped one. Since the O content of the former carbon is lower than that of the latter, the above results indicate that in this case the presence of N groups is more effective for H_2O adsorption.

3.3.3.4 Use as Electrode of Supercapacitor

As already described in the beginning of this chapter, large surface area carbons are promising as electrodes of supercapacitors, but the attainment of extremely large surface area inevitably leads to the broadness in pore size distribution, especially the increase in pore size. CDCs in which the pore size is precisely controlled at a nano- or subnanolevel show promise to optimizing the electrosorption of ions. Because of their surprisingly uniform micropore size and controlled pore structure, the ordered microporous carbons prepared using the zeolite template can give some hints for understanding how the pore size and structure influence the final performance of supercapacitors. The capacitance of the PFA-P7-H carbon was measured in nonaqueous electrolyte [137], and a primary result of this experiment is introduced in this section. Moreover, the effect of N-doping is examined using PFA-AN8-H.

To prepare a working electrode, carbon powder (PFA-P7-H or PFA-AN8-H) was mixed with polytetrafluoroethylene (PTFE) (5 wt%) to form a pellet and it was sandwiched in Ni mesh as a current collector. The EDL capacitance properties were measured by CV and galvanostatic charge/discharge in a three-electrode cell vs. Ag/AgCl reference electrode. $1 M$-$(C_2H_5)_4NBF_4$ in PC was used as a nonaqueous electrolyte.

The box-like shape of the CV curves of PFA-AN8-H is characteristic of charging the EDL with quick charge propagation (Figure 3.23). Figure 3.24 shows the current density dependence of the

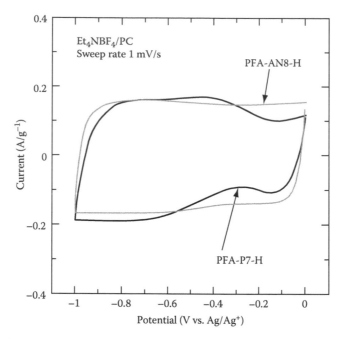

FIGURE 3.23 Cyclic voltammograms for the undoped and N-doped carbons in organic electrolyte.

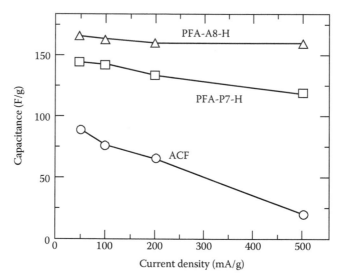

FIGURE 3.24 Current density dependence of capacitance for PFA-A8-H, PFA-P7-H, and ACF (AD'ALL A20 from Unitika, Ltd., Japan).

capacitance for both carbons together with that of commercially available ACF (Unitika, Ltd., BET SSA; 1670 m²/g). The two template carbons exhibit much larger capacitance than the ACF, and the current density dependence of the former is less pronounced. These findings demonstrate that a pore size as small as 1.2 nm is large enough for the diffusion of $(C_2H_5)_4N^+$. Considering the size of solvated $(C_2H_5)_4N^+$ [153], it confirms that the ions are desolvated and penetrate into the pores. The observed high rate capability may be explained by the easy diffusion of the electrolyte ions in the

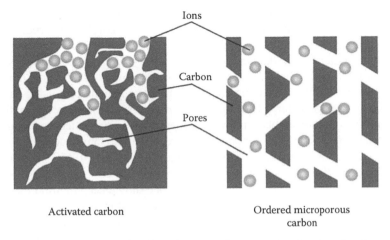

FIGURE 3.25 Diffusion of electrolyte ions into the pores of a conventional activated carbon and the ordered microporous carbon prepared using the zeolite template.

uniform, ordered; and open pores of the resulting carbons in comparison with the nonuniform and complicated pore structure in the conventional activated carbons, as illustrated in Figure 3.25. Very recently, Ania et al. [154] have demonstrated that the N-doped zeolite templated carbon displays a large gravimetric capacitance in aqueous media because of the combined electrochemical activity of the heteroatoms and of the accessible porosity.

The specific capacitance per unit surface area of PFA-P7-H and PFA-AN8-H at a low current density of $50\,mA/g$ are 0.035 and $0.050\,F/m^2$, respectively, whereas the corresponding value of the ACF is $0.051\,F/m^2$. Since the ordered microporous carbons possess only uniform and ordered micropores with a diameter of about $1.2\,nm$, all the pores are expected to be accessible for the electrolyte ions, which should have resulted in higher specific capacitance (even per unit surface area) than that for the ACF in which there is a broader pore size distribution and the fraction of nonaccessible pores may be larger. One of the possibilities for the unexpected small capacitance of the ordered microporous carbons might be their weak mechanical strength. When the physical mixture of the carbon powder and PTFE is put under shear stress in a mortar to prepare an electrode pellet, the pore structure of the ordered microporous carbons might be broken to some extent. Another possibility is that some pores were blocked by the binder. Actually, the SSA of each resulting pellet was found to be 15%–20% less than the expected surface area which is calculated from the ratio of the carbon part in the pellet. In other words, the ordered microporous carbons still have potential for obtaining larger capacitances than the values in Figure 3.24.

3.4 CONCLUSIONS

Developing novel methods for property control of porous carbons is paramount to both maximizing device performance and understanding the fundamental issues governing operation. CDC, because of its unique property tunability, is well positioned to become widely used in many applications. In many cases, it has already shown superior performance to traditional activated carbons. As the active material in supercapacitors, in particular, CDC has shown gravimetric and volumetric capacitances in technologically relevant electrolytes on par or in excess of all other carbon materials. Perhaps more important, using CDC has allowed systematic and fundamental studies of the pore texture and properties of carbon and pointed to a contradiction in the presently held understanding of pore size influence on capacitance that, if properly understood, could lead to large performance gains.

It was also shown that the template technique is a very powerful method for the control of a carbon nanostructure and the production of various types of novel nanocarbon materials. This method produces many types of CNTs, some of which were introduced in the Section 3.3. In addition to the fluorinated nanotubes, double coaxial nanotubes [155,156], water-dispersible nanotubes [157], nanotubes filled with magnetic metal in controlled manner [158], and highly crystallized $Ni(OH)_2$-filled nanotube [159,160] can be prepared (the latter may be promising as $Ni(OH)_2$ electrode for an alkaline storage battery). None of them have been synthesized by conventional methods. Moreover, the use of a zeolite template allows one to precisely control micropores in porous carbon and, as a result, a large surface area carbon with an ordered micropore structure has been obtained for the first time. High gravimetric hydrogen storage capacity (at −196°C) of zeolite-templated carbons prepared using β-zeolite has been recently reported [161].

Although the template technique is quite attractive, one should keep in mind that this technique requires both the use of an expensive template and its removal by a severe treatment such as HF washing, which hampers the practical use of the template technique. To use more inexpensive templates such as natural zeolites and clays would be one of the solutions. Furthermore, if the template were water-soluble, the process would be much simpler and could be performed at low cost. Instead of the inorganic templates, if one could use an organic template, the removal process would not be needed anymore, because the organic templates could be decomposed during the carbonization process. However, such an organic material is usually decomposed before the carbonization of a carbon precursor is finished; very recently, several groups have succeeded in the synthesis of ordered mesoporous carbon using an organic copolymer as a template [162–164].

ACKNOWLEDGMENTS

J. Chmiola and Y. Gogotsi would like to thank the U.S. Department of Energy, Office of Basic Energy Sciences, Arkema France, and DARPA for funding. They would also like to thank Dr. R.K. Dash, Y–Carbon; Dr. G. Yushin, Georgia Institute of Technology; Dr. C. Portet, Drexel University Prof. P. Simon, University of Paul Sabatier; and Prof. J.E. Fischer, University of Pennsylvania for helpful discussions. J. Chmiola was supported by a NSF Graduate Research Fellowship and NSF IGERT Fellowship. T. Kyotani would like to thank the Ministry of Education, Science, Sports and Culture in Japan for funding. He would also like to express his appreciation to Prof. H. Touhara, Shinshu University, Japan, for his electrochemical experiments using the templated carbons.

REFERENCES

1. Peirson, H.O. *Handbook of Carbon, Graphite, Diamond.* Park Ridge, NJ: Noyes Publications, 1994.
2. Flandrois, S. Battery Carbons. In *Encyclopedia of Materials: Science and Technology,* pp. 484–488. Amsterdam, the Netherlands: Elsevier, 2001.
3. Burke, A. Ultracapacitors: Why, how, and where is the technology. *J. Power Sources* 91, 2000: 37–50.
4. Kinoshita, K. *Carbon: Electrochemical and Physicochemical Properties.* New York: Wiley-Interscience, 1988.
5. Gogotsi, Y. *Carbon Nanomaterials.* Boca Raton, FL: CRC Press, 2006.
6. Marsh, H. and Reinoso, F.R. *Activated Carbon.* Amsterdam, the Netherlands: Elsevier, 2005.
7. Endo, M., Kim, Y.J., Takeda, T., Maeda, T., Hayashi, T., Koshiba, K., Hara, H., and Dresselhaus, M.S. Polyvinylidene chloride-based carbon as an electrode material for high power capacitors with an aqueous electrolyte. *J. Electrochem. Soc.* 148, 2001: A1135–A1140.
8. Endo, M., Takeda, T., Kim, Y.J., Koshiba, K., and Ishii, K. High power electric double layer capacitor (EDLC's): From operating principle to pore size control in advanced activated carbons. *Carbon Sci.* 1, 2001: 117–128.
9. Yang, K.L., Yiacoumi, S., and Tsouris, C. Electrosorption capacitance of nanostructured carbon aerogel obtained by cyclic voltammetry. *J. Electroanal. Chem.* 540, 2003: 159–167.

10. Saliger, R., Fischer, U., Herta, C., and Fricke, J. High surface area carbon aerogels for supercapacitors. *J. Non-Cryst. Sol.* 225, 1998: 81–85.

11. Gamby, J., Taberna, P.L., Simon, P., Fauvarque, J.F., and Chesneau, M. Studies and characterization of various activated carbons used for carbon/carbon supercapacitors. *J. Power Sources* 101, 2001: 109–116.

12. Endo, M., Maeda, T., Kim, Y.J., Dresselhaus, M.S., Ohta, H., Inoue, T., Hayashi, T., and Nishimura, Y. Morphology and organic EDLC applications of chemically activated Ar-resin-based carbons. *Carbon* 40, 2002: 2613–2626.

13. Chen, J.H., Li, W.Z., Wang, D.Z., Yang, S.X., Wen, J.G., and Ren, Z.F. Electrochemical characterization of carbon nanotubes as electrode in electrochemical double-layer capacitors. *Carbon* 40, 2002: 1193–1197.

14. Emmenegger, C., Mauran, P., Sudan, P., Wenger, P., Hermann, V., Gallay, R., and Zuttel, A. Investigation of electrochemical double-layer (EDLC) capacitors electrodes based on carbon nanotubes and activated carbon materials. *J. Power Sources* 124, 2003: 321–329.

15. Chmiola, J. and Gogotsi, Y. Supercapacitors as advanced energy storage devices. *Nanotechnol. Law Bus.* 4, 2007: 577–584.

16. Gryglewicz, G., Machnikowski, J., Lorenc-Grabowska, E., Lota, G., and Frackowiak, E. Effect of pore size distribution of coal-based activated carbons on double layer capacitance. *Electrochim. Acta.* 50, 2005: 1197–1206.

17. Hahn, M., Baertschi, M., Barbieri, O., Sauter, J.-C., Kotz, R., and Gallay, R. Interfacial capacitance and electronic conductance of activated carbon double-layer electrodes. *Electrochem. Solid State Lett.* 7, 2004: A33–A36.

18. Qu, D. Studies of the activated carbons used in double-layer supercapacitors. *J. Power Sources* 109, 2002: 403–411.

19. Kim, Y.J., Masutzawa, Y., Ozaki, S., Endo, M., and Dresselhaus, M.S. PVDC-based carbon material by chemical activation and its application to nonaqueous EDLC. *J. Electrochem. Soc.* 151, 2004: E199–E205.

20. Frackowiak, E., Lota, G., Machnikowski, J., Vix-Guterl, C., and Béguin, F. Optimization of supercapacitors using carbons with controlled nanotexture and nitrogen content. *Electrochim. Acta.* 51, 2006: 2209–2214.

21. Chmiola, J., Yushin, G., Gogotsi, Y., Portet, C., Simon, P., and Taberna, P.L. Anomalous increase in carbon capacitance at pore sizes less than 1 nanometer. *Science* 313, 2006: 1760–1763.

22. Vix-Guterl, C., Frackowiak, E., Jurewicz, K., Friebe, M., Parmentier, J., and Béguin, F. Electrochemical energy storage in ordered porous carbon materials. *Carbon* 43, 2005: 1293–1302.

23. Kim, T., Lim, S., Kwon, K., Hong, S.-H., Qiao, W., Rhee, C.K., Yoon, S.-H., and Mochida, I. Electrochemical capacities of well-defined carbon surfaces. *Langmuir* 22, 2006: 9086–9088.

24. Ersoy, D.A., McNallan, M.J., and Gogotsi, Y. Carbon coatings produced by high temperature chlorination of silicon carbide ceramics. *Mater. Res. Innov.* 5, 2001: 55–62.

25. Gogotsi, Y.G., Jeon, I.-D., and McNallan, M.J. Carbon coatings on silicon carbide by reaction with chlorine containing gases. *J. Mater. Chem.* 7, 1997: 1841–1848.

26. Cambaz, Z.G., Yushin, G.N., Vyshnyakova, K.L., Pereselentseva, L.N., and Gogotsi, Y. Conservation of shape during formation of carbide-derived carbon on silicon carbide nano-whiskers. *J. Am. Ceram. Soc.* 89, 2006: 509–514.

27. Jacobson, N.S., Gogotsi, Y.G., and Yoshimura, M. Thermodynamic and experimental study of carbon formation on carbides under hydrothermal conditions. *J. Appl. Chem.* 5, 1995: 595–601.

28. Nikitin, A. and Gogotsi, Y. Nanostructured carbide derived carbon. In *Encyclopedia of Nanoscience and Nanotechnology*, edited by H.S. Nalwa, pp. 553–574. Stevenson Ranch, CA: American Scientific Publishers, 2003.

29. Conway, B.E. *Electrochemical Supercapacitors: Scientific Fundamentals and Technological Applications*. New York: Kluwer, 1999.

30. Becker, H.I. Low voltage electrolytic capacitor. US Patent #2800616, 1957.

31. Béguin, F. and Frackowiak, E. Nanotextured carbons for electrochemical energy storage. In *Nanomaterials Handbook*, edited by Y. Gogotsi, pp. 713–737. Boca Raton, FL: CRC Press, 2006.

32. Gogotsi, Y., Nikitin, A., Ye, H., Zhou, W., Fischer, J.E., Yi, B., Foley, H.C., and Barsoum, M.W. Nanoporous carbide-derived carbon with tunable pore size. *Nat. Mater.* 2, 2003: 591–594.

33. Yushin, G., Nikitin, A., and Gogotsi, Y. Carbide derived carbon. In *Nanomaterials Handbook*, edited by Y. Gogotsi, pp. 237–280. Boca Raton, FL: CRC Press, 2006.

34. Yushin, G., Hoffman, E., Nikitin, A., Ye, H., Barsoum, M.W., and Gogotsi, Y. Synthesis of nanoporous carbide-derived carbon by chlorination of titanium silicon carbide. *Carbon* 43, 2005: 2075–2082.

35. Hoffman, E.N., Carbide-derived carbon from MAX phases and their separation applications. PhD Thesis, Drexel University, Philadelphia, PA, 2006.

36. Chmiola, J., Yushin, G., Dash, R.K., Hoffman, E.N., Fischer, J.E., Barsoum, M.W., and Gogotsi, Y. Double-layer capacitance of carbide derived carbons in sulfuric acid. *Electrochem. Solid State Lett.* 8, 2005: A357–A360.

37. Dash, R.K., Nikitin, A., and Gogotsi, Y. Microporous carbon derived from boron carbide. *Micropor. Mesopor. Mat.* 72, 2004: 203–208.

38. Dimovski, S., Nikitin, A., Ye, H., and Gogotsi, Y. Synthesis of graphite by chlorination of iron carbide at moderate temperatures. *J. Mater. Chem.* 14, 2004: 238–243.

39. Dash, R.K., Yushin, G., and Gogotsi, Y. Structure and porosity analysis of microporous and mesoporous carbon derived from zirconium carbide. *Micropor. Mesopor. Mat.* 86, 2005: 50–57.

40. Hutchins, O. Method for the production of silicon tetrachloride. US Patent #1271713, 1918.

41. Boehm, H.P. and Warnecke, H.H. Structural and molecular sieve properties of carbons prepared from metal carbides. In *Proceedings of the 12th Biennial Conference on Carbon*, pp. 149–150. Oxford: Pergamon, 1975.

42. Dash, R., Chmiola, J., Yushin, G., Gogotsi, Y., Laudisio, G., Singer, J., Fischer, J., and Kucheyev, S. Titanium carbide derived nanoporous carbon for energy-related applications. *Carbon* 44, 2006: 2489–2497.

43. Laudisio, G., Dash, R.K., Singer, J.P., Yushin, G., Gogotsi, Y., and Fischer, J.E. Carbide-derived carbons: A comparative study of porosity based on small-angle scattering and adsorption isotherms. *Langmuir* 22, 2006: 8945–8950.

44. Lin, R., Taberna, P.L., Chmiola, J., Guay, D., Gogotsi, Y., Simon, P. Microelectrode study of pore size, ion size and solvent effects on the charge/discharge behavior of microporous carbons for electrical double layer capacitors. *J. Electrochem. Soc.* 156(1), 2009: A7–A12.

45. Ersoy, D., McNallan, M.J., and Gogotsi, Y.G. Carbon coatings produced by high temperature chlorination of silicon carbide ceramics. *Mater. Res. Innov.* 5, 2001: 55–62.

46. Zhang, J., Ekstrom, T.C., Gordeev, S.K., and Jacob, M. Carbon with an onion-like structure obtained by chlorinating aluminum carbide. *J. Mater. Chem.* 10, 2000: 1039–1041.

47. Jacob, M., Palmqvist, U., Alberius, P.C.A., Ekstrom, T., Nygren, M., and Lidin, S. Synthesis of structurally controlled nanocarbons—in particular the nanobarrel carbon. *Solid State Sci.* 5, 2003: 133–137.

48. Weltz, S., McNallan, M., and Gogotsi, Y. Carbon structures in silicon carbide derived carbon. *J. Mater. Process. Tech.* 179, 2006: 11–22.

49. Boehm, H.P. The first observation of carbon nanotubes. *Carbon* 35, 1997: 581–584.

50. Kravchik, A.E., Kukushikina, J.A., Sokolov, V.V., and Tereshchenko, G.F. Structure of nanoporous carbon produced from boron carbide. *Carbon* 44, 2006: 3263–3268.

51. Kyutt, R.N., Smorgonskaya, E.A., Danishevskii, A.M., Gordeev, S.K., and Grechinskaya, A.V. Structural studies of nanoporous carbon produced from silicon carbide. *Phys. Solid State* 41, 1999: 808–810.

52. Leis, J., Perkson, A., Arulepp, M., Nigu, P., and Svensson, G. Catalytic effect of metals of the iron subgroup on the chlorination of titanium carbide to form nanostructural carbon. *Carbon* 40, 2002: 1559–1564.

53. Kusunoki, M., Rokkaku, M., and Suzuki, T. Epitaxial carbon nanotube film self-organized by decomposition of silicon carbide. *Appl. Phys. Lett.* 71, 1997: 2620–2622.

54. Kusunoki, M., Suzuki, T., Hirayama, T., Shibata, J., and Kaneko, K. A formation mechanism of carbon nanotube films on SiC(000-1). *Appl. Phys. Lett.* 77, 2000: 531–533.

55. Kusunoki, M., Suzuki, T., Kaneko, K., and Ito, M. Formation of self-aligned carbon nanotube films by surface decomposition of silicon carbide. *Phil. Mag. Lett.* 79, 1999: 153–161.

56. Kusunoki, M., Suzuki, T., Honjo, C., Hirayama, T., and Shibata, N. Selective synthesis of zigzag-type aligned carbon nanotubes on SiC (000-1) wafers. *Chem. Phys. Lett.* 366, 2002: 458–462.

57. Nagano, T., Ishikawa, Y., and Shibata, N. Effects of surface oxides of SiC on carbon nanotube formation by surface decomposition. *Jpn. J. Appl. Phys.* 42, 2003: 1380–1385.

58. Nagano, T., Ishikawa, Y., and Shibata, N. Preparation of silicon-on-insulator substrate on large free-standing carbon nanotube film formation by surface decomposition of SiC film. *Jpn. J. Appl. Phys.* 42, 2003: 1717–1721.

59. Yamauchi, T., Tokunaga, T., Naitoh, M., Nishigaki, S., Toyama, N., Shoji, F., and Kusunoki, M. Influence of surface structure modifications on the growth of carbon nanotubes on the SiC (0001) surfaces. *Surf. Sci.* 600, 2006: 4077–4080.

60. Harrison, J., Sambandam, S.N., Boeckl, J.J., Mitchel, W.C., Collins, W.E., and Lu, W. Evaluation of metal-free carbon nanotubes formed by SiC thermal decomposition. *J. Appl. Phys.* 101, 2007: 104311.

61. Wang, S., Humphreys, E.S., Chung, S.-Y., Delduci, D.F., Lustig, S.R., Wang, H., Parker, K.N., Rizzo, N.W., Subramoney, S., Chiang, Y.-M., and Jagota, A. Peptides with selective affinity for carbon nanotubes. *Nat. Mater.* 2, 2003: 196–299.

62. Botti, S., Asilyan, L.S., Ciardi, R., Fabbri, F., Lorety, S., Santoni, A., and Orlanducci, S. Catalyst-free growth of carbon nanotubes by laser annealing of amorphous SiC films. *Chem. Phys. Lett.* 396, 2004: 1–5.

63. Forbeaux, I., Themlin, J.-M., and Debever, J.-M. High-temperature graphitization of the 6h-SiC (0001) face. *Surf. Sci.* 442, 1999: 9–18.

64. Cambaz, Z.G., Yushin, G., Osswald, S., Mochalin, V., and Gogotsi, Y. Noncatalytic synthesis of carbon nanotubes, graphene and graphite on SiC. *Carbon* 46(6), 2008: 841–849. DOI: 10.1016/j.carbon.2008.02.013.

65. Gogotsi, Y., Weltz, S., Ersoy, D.A., and McNallan, M.J. Conversion of silicon carbide to crystalline diamond-structured carbon at ambient pressure. *Nature* 411, 2001: 283–287.

66. Gogotsi, Y., Dash, R.K., Yushin, G., Yildirim, T., Laudisio, G., and Fischer, J.E. Tailoring of nanoscale porosity in carbide-derived carbons for hydrogen storage. *J. Am. Chem. Soc.* 127, 2005: 16006–16007.

67. Kotina, I.M., Lebedev, V.M., Ilves, A.G., Patsekina, G.V., Tuhkonen, L.M., Gordeev, S.K., Yagovkina, M.A., and Ekström, T. Study of the lithium diffusion in nanoporous carbon materials produced from carbides. *J. Non Cryst. Sol.* 299–302, 2002: 815–819.

68. Kotina, I.M., Lebedev, V.M., Ilves, A.G., Patsekina, G.V., Tuhkonen, L.M., Gordeev, S.K., Yagovkina, M.A., and Ekstrom, T. The phase composition of the lithiated samples of nanoporous carbon materials produced from carbides. *J. Non Cryst. Sol.* 299–302, 2002: 820–823.

69. Yushin, G., Hoffman, E.N., Barsoum, M.W., Gogotsi, Y., Howell, C.A., Sandeman, S.R., Phillips, G.J., Lloyd, A.W., and Mikhalovsky, S.V. Mesoporous carbide-derived carbon with porosity tuned for efficient adsorption of cytokines. *Biomaterials* 27, 2006: 5755–5762.

70. Carroll, B., Gogotsi, Y., Kovalchenko, A., Erdemir, A., and McNallan, M. Effect of humidity on the tribological properties of carbide-derived carbon (CDC) films on silicon carbide. *Tribo. Lett.* 15, 2003: 41–44.

71. Avarbz, R.G., Vartanova, A.V., Gordeev, S.K., Zjukov, S.G., Zelenov, B.A., Kravtjik, A.E., Kuznetsof, V.P., Kukusjkina, J.A., Mazaeva, T.V., Pankina, O.S., and Sokolov, V.V. Process of manufacturing a porous carbon material and a capacitor having the same. US Patent 5,876,787, 1996.

72. Chmiola, J., Yushin, G., Gogotsi, Y., Portet, C., Simon, P., and Taberna, P.-L. Effect of pore size on electrochemical behavior of carbide derived carbon for supercapacitor applications. In *231st ACS Spring Meeting*. Atlanta, GA, 2006.

73. Chmiola, J., Yushin, G., Dash, R.K., Hoffman, E.N., Fischer, J.E., Barsoum, M., and Gogotsi, Y. Double-layer capacitance of carbide-derived carbons in sulfuric acid. *Electrochem. Solid State Lett.* 8, 2005: A357–A360.

74. Hoffman, E.N., Yushin, G., Barsoum, M.W., and Gogotsi, Y. Synthesis of carbide-derived carbon by chlorination of Ti₂AlC. *Chem. Mater.* 17, 2005: 2317–2322.

75. Rodriguez-Reinoso, F. and Sepulveda-Escribano, A. Porous carbons in adsorption and catalysis. In *Handbook of Surfaces and Interfaces of Materials*, edited by H.S. Nalwa, pp. 309–351. San Diego, CA: Academic Press, 2001.

76. Gordeev, S.K., Kukushkin, S.A., Osipov, A.V., and Pavlov, Y.V. Self-organization in the formation of a nanoporous carbon material. *Phys. Solid State* 42, 2000: 2314–1317.

77. Endo, M., Maeda, T., Takeda, T., Kim, Y.J., Koshiba, K., Hara, H., and Dresselhaus, M.S. Capacitance and pore-size distribution in aqueous and nonaqueous electrolytes using various activated carbon electrodes. *J. Electrochem. Soc.* 148, 2001: A910–A914.

78. Maletin, Y.A., Strizhakova, N.G., Izotov, V.G., Mironova, A.A., Kozachkov, S.G., Danilin, V.G., Podmogilny, S.N., Arulepp, M., Kukusjkina, J.A., Kravtjik, A.E., Sokolov, V.V., Perkson, A., Leis, J., Zheng, J., Gordeev, S.K., Kolotilova, J.Y., Cederstrom, J., and Wallace C.L. A supercapacitor and a method of manufacturing such a supercapacitor. European Patent WO 02/39468, 2002.

79. Maletin, Y., Strizhakova, N., Kozachkov, S., Mironova, A., Podmogilny, S., Danilin, V., Kolotilova, J., Izotov, V., Konstantinovich, G.S., Aleksandrovna, J.K., Vasilevitj, V.S., Efimovitj, A.K., Perkson, A., Arulepp, M., Leis, J., Wallace, C.L., and Zheng, J. Supercapacitor and a method of manufacturing such a supercapacitor. US Patent 6,697,249, 2004.

80. Leis, J., Arulepp, M., and Perkson, A. Method to modify the pore characteristics of porous carbon and porous carbon materials produced by this method. European Patent WO/2004/094307, 2004.

81. Arulepp, M., Leis, J., Latt, M., Miller, F., Rumma, K., Lust, E., and Burke, A.F. The advanced carbide-derived carbon based supercapacitor. *J. Power Sources* 162, 2006: 1460–1466.

82. Erdemir, A., Kovalchenko, A., McNallan, M.J., Weltz, S., Lee, A., and Gogotsi, Y. Effects of high-temperature hydrogenation treatment on sliding friction and wear behavior of carbide-derived carbon films. *Surf. Coat. Technol.* 188–189, 2004: 588–593.

83. Yushin, G., Dash, R., Jagiello, J., Fischer, J.E., and Gogotsi, Y. Carbide-derived carbons: Effect of pore size on hydrogen uptake and heat of adsorption. *Adv. Funct. Mat.* 16, 2006: 2288–2293.

84. Yushin, G.N., Hoffman, E.N., Nikitin, A., Ye, H.H., Barsoum, M.W., and Gogotsi, Y. Synthesis of nanoporous carbide-derived carbon by chlorination of titanium silicon carbide. *Carbon* 43, 2005: 2075–2082.

85. Lust, E., Nurk, G., Janes, A., Arulepp, M., Permann, L., Nigu, P., and Moller, P. Electrochemical properties of nanoporous carbon electrodes. *Condens. Matter Phys.* 5, 2002: 307–327.

86. Avarbz, R.G., Vartanova, A.V., Gordeev, S.K., Zjukov, S.G., Zelenov, B.A., Kravtjik, A.E., Kuznetsof, V.P., Kukusjkina, J.A., Mazaeva, T.V., Pankina, O.S., and Sokolov, V.V. Double layer capacitor with porous carbon electrodes and method for manufacturing these electrodes. European Patent WO 97/20333, 1996.

87. Janes, A., Permann, L., Arulepp, M., and Lust, E. Electrochemical characteristics of nanoporous carbide-derived carbon materials in non-aqueous electrolyte solutions. *Electrochem. Commun.* 6, 2004: 313–318.

88. Chmiola, J., Yushin, G., Dash, R., and Gogotsi, Y. Effect of pore size and surface area of carbide-derived carbons on specific capacitance. *J. Power Sources* 158, 2006: 765–772.

89. Zhou, H., Zhu, S., Hibino, M., and Honma, I. Electrochemical capacitance of self-ordered mesoporous carbon. *J. Power Sources* 122, 2003: 219–223.

90. Yoon, S., Lee, Y., Hyeon, T., and Oh, S.M. Electric double-layer capacitor performance of a new mesoporous carbon. *J. Electrochem. Soc.* 147, 2000: 2501–2512.

91. Liu, H.-Y., Wang, K.-P., and Teng, H. A simplified preparation of mesoporous carbon and the examination of the carbon accessibility for electric double layer formation. *Carbon* 43, 2005: 559–566.

92. Shiraishi, S., Kurihara, H., Shi, L., Nakayama, T., and Oya, A. Electric double-layer capacitance of meso/macroporous activated carbon fibers prepared by the blending method. *J. Electrochem. Soc.* 149, 2002: A855–A861.

93. Taberna, P.L., Simon, P., and Fauvarque, J.F. Electrochemical characteristics and impedance spectroscopy studies of carbon-carbon supercapacitors. *J. Electrochem. Soc.* 150, 2003: 292–300.

94. Permann, L., Latt, M., Leis, J., and Arulepp, M. Electrical double layer characteristics of nanoporous carbon derived from titanium carbide. *Electrochim. Acta.* 51, 2006: 1274–1281.

95. Lust, E., Jänes, A., and Arulepp, M. Influence of electrolyte characteristics on the electrochemical parameters of electrical double layer capacitors. *J. Solid State Electr.*, 8, 2004: 488–496.

96. Lust, E., Nurk, G., Jänes, A., Arulepp, M., Nigu, P., Möller, P., Kallip, S., and Sammelselg, V. Electrochemical properties of nanoporous carbon electrodes in various nonaqueous electrolytes. *J. Solid State Electr.* 7, 2004: 91–105.

97. Janes, A., Permann, L., Nigu, P., and Lust, E. Influence of solvent nature on the electrochemical characteristics of nanoporous carbon—1 M $(C_2H_5)_3CH_3NBF_4$ electrolyte solution interface. *Surf. Sci.* 560, 2004: 145–157.

98. Lust, E., Janes, A., and Arulepp, M. Influence of solvent nature on the electrochemical parameters of double layer capacitors. *J. Electroanal. Chem.* 562, 2003: 33–42.

99. Zuleta, M., Bjornbom, P., and Lundblad, A. Characterization of the electrochemical and ion-transport properties of a nanoporous carbon at negative polarization by the single-particle method. *J. Electrochem. Soc.* 153, 2006: A48–A57.

100. Zuleta, M., Bjornbom, P., and Lundblad, A. Effects of pore surface oxidation on electrochemical and mass-transport properties of nanoporous carbon. *J. Electrochem. Soc.* 152, 2005: A270–A276.

101. Zuleta, M., Bursell, M., Bjornbom, P., and Lundblad, A. Determination of the effective diffusion coefficient of nanoporous carbon by means of a single particle microelectrode technique. *J. Electroanal. Chem.* 549, 2003: 101–108.

102. Zuleta, M., Bjornbom, P., Lundblad, A., Nurk, G., Kasuk, H., and Lust, E. Determination of diffusion coefficients of BF_4^- inside carbon nanopores using the single particle microelectrode technique. *J. Electroanal. Chem.* 586, 2006: 247–259.

103. Malmberg, H., Zuleta, M., Lundblad, A., and Bjornbom, P. Ionic transport in pores in activated carbons for EDLCs. *J. Electrochem. Soc.* 153, 2006: A1914–A21.

104. Chmiola, J., Largeot, C., Taberna, P.-L., Simon, P., and Gogotsi, Y. Desolvation of ions in subnanometer pores, its effect on capacitance and double-layer theory. *Angewandte Chemie* 47(18), 2008: 3392–3395. DOI: 10.1002/anie.200704894.

105. Largeot, C., Portet, C., Chmiola, J., Taberna, P.-L., Gogotsi, Y., and Simon, P. Relation between the ion size and pore size for an electric double-layer capacitor. *J. Am. Chem. Soc.* 130, 2008: 2730–2731.

106. Kyotani, T., Sonobe, N., and Tomita, A. Formation of highly orientated graphite from polyacrylonitrile by using a two-dimensional space between montmorillonite lamellae. *Nature* 331, 1988: 331–333.

107. Kyotani, T., Mori, T., and Tomita, A. Formation of a flexible graphite films from poly(acrylonitrile) using a layered clay film as template. *Chem. Mater.* 6, 1994: 2138–2142.

108. Sonobe, N., Kyotani, T., and Tomita, A. Formation of graphite thin film from polyfurfuryl alcohol and polyvinyl acetate carbons prepared between the lamellae of montmorillonite. *Carbon* 29, 1991: 61–67.

109. Lee, J.S., Gu, G.H., Kim, H., Jeong, K.S., Bae, J., and Suh, J.S. Growth of carbon nanotubes on anodic aluminum oxide template: Fabrication of a tube-in-tube and linearly joined tube. *Chem. Mater.* 13, 2001: 2387–2391.

110. Che, G., Lakshmi, B.B., Martin, C.R., and Fisher, E.R. Metal-nanocluster-filled carbon nanotubes: Catalytic properties and possible applications in electrochemical energy storage and production. *Langmuir* 15, 1999: 750–758.

111. Li, J., Papadopoulos, C., and Xu, J. Growing Y-junction carbon nanotubes. *Nature* 402, 1990: 253–254.

112. Li, J., Moskovits, M., and Haslett, T.L. Nanoscale electroless metal deposition in aligned carbon nanotubes. *Chem. Mater.* 10, 1998: 1963–1967.

113. Che, G., Lakshmi, B.B., Martin, C.R., and Fisher, E.R. Chemical vapor deposition based synthesis of carbon nanotubes and nanofibers using a template method. *Chem. Mater.* 10, 1998: 260–267.

114. Che, G., Lakshmi, B.B., Fisher, E.R., and Martin, C.R. Carbon nanotubule membranes for electrochemical energy storage and production. *Nature* 393, 1998: 346–349.

115. Kyotani, T., Tsai, L.F., and Tomita, A. Preparation of ultrafine carbon tubes in nanochannels of an anodic aluminum oxide film. *Chem. Mater.* 8, 1996: 2109–2113.

116. Kyotani, T., Tsai, L.F., and Tomita, A. Formation of ultrafine carbon tubes by using an anodic aluminum oxide film as a template. *Chem. Mater.* 7, 1995: 1427–1428.

117. Delpeux-Ouldriane, S., Szostak, K., Frackowiak, E., and Béguin, F. Annealing of template nanotubes to well-graphitized multi-walled carbon nanotubes. *Carbon* 44, 2006: 799–823.

118. Hattori, Y., Watanabe, Y., Kawasaki, S., Okino, F., Pradhan, B.K., Kyotani, T., Tomita, A., and Touhara, H. Carbon-alloying of the rear surfaces of nanotubes by direct fluorination. *Carbon* 37, 1999: 1033–1038.

119. Touhara, H., Inahara, J., Mizuno, T., Yokoyama, Y., Okanao, S., Yanagiuch, K., Mukopadhyay, I., Kawasaki, S., Okino, F., Shirai, H., Xu, W.H., Kyotani, T., and Tomita, A. Property control of new forms of carbon materials by fluorination. *J. Fluor. Chem.* 114, 2002: 181–188.

120. Kyotani, T. Control of pore structure in carbon. *Carbon* 38, 2000: 269–286.

121. Meyers, C.J., Shah, S.D., Patel, S.C., Sneeringer, R.M., Bessel, C.A., Dollahon, N.R., Leising, R.A., and Takeuchi, E.S. Templated synthesis of carbon materials from zeolites (Y, Beta, and Zsm-5) and a montmorillonite clay (K10): Physical and electrochemical characterization. *J. Phys. Chem. B* 105, 2001: 2143–2152.

122. Johnson, S.A., Brigham, E.S., Ollivier, P.J., and Mallouk, T.E. Effect of micropore topology on the structure and properties of zeolite polymer replicas. *Chem. Mater.* 9, 1997: 2448–2458.

123. Kawashima, D., Aihara, T., Kobayashi, Y., Kyotani, T., and Tomita, A. Preparation of mesoporous carbon from organic polymer/silica nanocomposite. *Chem. Mater.* 12, 2000: 3397–3401.

124. Bandosz, T.J., Jagiello, J., Putyera, K., and Schwarz, J.A. Pore structure of carbon-mineral nanocomposites and derived carbons obtained by template carbonization. *Chem. Mater.* 8, 1996: 2023–2029.

125. Zakhidov, A.A., Baughman, R., Iqbal, Z., Cui, C., Khayrullin, I., Dantas, S.O., Marti, J., and Ralchenko, V.G. Carbon structures with three-dimensional periodicity at optical wavelengths. *Science* 282, 1998: 897–901.

126. Knox, J.H. and Kaur, B. Structure and performance of porous graphitic carbon in liquid chromatography. *J. Chromatogr.* 352, 1986: 3–25.

127. Lee, J., Yoon, S., Hyeon, T., Oh, S.M., and Kim, K.B. Synthesis of a new mesoporous carbon and its application to electrochemical double-layer capacitors. *Chem. Commun.* 1999: 2177–2178.

128. Ryoo, R., Joo, S.H., and Jun, S. Synthesis of highly ordered carbon molecular sieves via template-mediated structural transformation. *J. Phys. Chem. B* 103, 1999: 7743–7746.

129. Kyotani, T., Nagai, T., Inoue, S., and Tomita, A. Formation of new type of porous carbon by carbonization in zeolite nanochannels. *Chem. Mater.* 9, 1997: 609–615.

130. Ma, Z.X., Kyotani, T., and Tomita, A. Preparation of a high surface area microporous carbon having the structural regularity of Y zeolite. *Chem. Commun.* 2000: 2365–2366.

131. Kyotani, T., Ma, Z.X., and Tomita, A. Template synthesis of novel porous carbons using various types of zeolites. *Carbon* 41, 2003: 1451–1459.

132. Ma, Z.X., Kyotani, T., and Tomita, A. Synthesis methods for preparing microporous carbons with a structural regularity of zeolite Y. *Carbon* 40, 2002: 2367–2374.

133. Ma, Z.X., Kyotani, T., Liu, Z., Terasaki, O., and Tomita, A. Very high surface area microporous carbon with a three-dimensional nano-array structure: Synthesis and its molecular structure. *Chem. Mater.* 13, 2001: 4413–4415.

134. Matsuoka, K., Yamagishi, Y., Yamazaki, T., Setoyama, N., Tomita, A., and Kyotani, T. Extremely high microporosity and sharp pore size distribution of a large surface area carbon prepared in the nanochannels of zeolite Y. *Carbon* 43, 2005: 876–879.

135. Gaslain, F.O.M., Parmentier, J., P. Valtchev, V.P., and Patarin J. First zeolite replica with a well resolved X-ray diffraction pattern. *Chem. Commun.* 2006: 991–993.

136. Hou, P.X., Orikasa, H., Yamazaki, T., Matsuoka, K., Tomita, A., Setoyama, N., Fukushima, Y., and Kyotani, T. Synthesis of a nitrogen-containing microporous carbon with a highly ordered structure and effect of nitrogen-doping on H_2O adsorption. *Chem. Mater.* 17, 2005: 5187–5193.

137. Kyotani, T. Template synthesis of ordered microporous carbons and their electric double layer capacitor performance. *Prep. Pap.-Am. Chem. Soc., Div. Fuel Chem.* 51, 2006: 33–35.

138. Kaneko, K. and Ishii, C. Superhigh surface area determination of microporous solids. *Colloid. Surf.* 67, 1992: 203–212.

139. Hou, P.X., Orikasa, H., Itoi, H., Nishihara, H., and Kyotani, T. Densification of ordered microporous carbons and controlling their micropore size by hot-pressing. *Carbon* 45, 2007: 2011–2016.

140. Hulicova, D., Yamashita, J., Soneda, Y., Hatori, H., and Kodama, M. Supercapacitors prepared from melamine-based carbon. *Chem. Mater.* 17, 2005: 1241–1247.

141. Kodama, M., Yamashita, J., Soneda, Y., Hatori, H., Nishimura, S., and Kamegawa, K. Structural characterization and electric double layer capacitance of template carbons. *Mater. Sci. Eng. B* 108, 2004: 151–161.

142. Yang, C.M., El-Merraoui, M., Seki, H., and Kaneko, K. Characterization of nitrogen-alloyed activated carbon fiber. *Langmuir* 17, 2001: 675–680.

143. Jansen, R.J.J. and van Bekkum, H. Amination and ammoxidation of activated carbons. *Carbon* 32, 1994: 1507–1516.

144. Stöhr, B., Boehm, H.P., and Schlögl, R. Enhancement of the catalytic activity of activated carbons in oxidation reactions by thermal treatment with ammonia or hydrogen cyanide and observation of a superoxide species as a possible intermediate. *Carbon* 29, 1991: 707–720.

145. Machnikowski, J., Grzyb, B., Weber, J.V., Frackowiak, E., Rouzaud, J.N., and Béguin, F. Structural and electrochemical characterization of nitrogen enriched carbons produced by the co-pyrolysis of coal-tar pitch with polyacrylonitrile. *Electrochim. Acta* 49, 2004: 423–432.

146. Raymundo-Piñero, E., Cazorla-Amorós, D., Linares-Solano, A., Find, J., Wild, U., and Schlögl, R. Structural characterization of N-containing activated carbon fibers prepared from a low softening point petroleum pitch and a melamine resin. *Carbon* 40, 2002: 597–608.

147. Singoredjo, L., Kapteijn, F., Moulijn, J.A., Martin-Martinez, J.M., and Boehm, H.P. Modified activated carbons for the selective catalytic reduction of NO with NH_3. *Carbon* 31, 1993: 213–222.

148. Lahaye, J., Nansé, G., Bagreev, A., and Strelko, V. Porous structure and surface chemistry of nitrogen containing carbons from polymers. *Carbon* 37, 1999: 585–590.

149. Alcañiz-Monge, J., Linares-Solano, A., and Rand, B. Mechanism of adsorption of water in carbon micropores as revealed by a study of activated carbon fibers. *J. Phys. Chem. B* 106, 2002: 3209–3216.

150. Yang, C.M. and Kaneko, K. Adsorption properties of nitrogen-alloyed activated carbon fiber. *Carbon* 39, 2001: 1075–1082.

151. Mowla, D., Do, D.D., and Kaneko, K. Adsorption of water vapor on activated carbon: A brief overview. In *Chemistry and Physics of Carbon*, edited by L.R. Radovic, pp. 229–262. New York: Marcel Dekker, 2003.

152. Cossarutto, L., Zimny, T., Kaczmarczyk, J., Siemieniewska, T., Bimer, J., and Weber, J.V. Transport and sorption of water vapour in activated carbons. *Carbon* 39, 2001: 2339–2346.

153. Frackowiak, E. and Béguin, F. Carbon materials for the electrochemical storage of energy in capacitors. *Carbon* 39, 2001: 937–950.

154. Ania, C.O., Khomenko, V., Raymundo-Piñero, E., Parra, J.B., and Béguin, F. The large electrochemical performance of microporous doped carbon obtained by using a zeolite template. *Adv. Funct. Mat.* 17, 2007: 1828–1836.

155. Yang, Q.H., Xu, W.II., Tomita, A., and Kyotani, T. The template synthesis of double coaxial carbon nanotubes with nitrogen-doped and boron-doped multiwalls. *J. Am. Chem. Soc.* 127, 2005: 8956–8957.

156. Yang, Q.H., Hou, P.X., Unno, M., Yamauchi, S., Saito, R., and Kyotani, T. Dual Raman features of double coaxial carbon nanotubes with N-doped and B-doped multiwalls. *Nano Lett.* 5, 2005: 2465–2469.

157. Orikasa, H., Inokuma, N., Okubo, S., Kitakami, O., and Kyotani, T. Template synthesis of water-dispersible carbon nano "test tubes" without any post treatment. *Chem. Mater.* 18, 2006: 1036–1040.

158. Wang, X.H., Orikasa, H., Inokuma, N., Yang, Q.H., Hou, P.X., Oshima, H., Itoh, K., and Kyotani, T. Controlled filling of permalloy into one-end-opened carbon nanotubes. *J. Mater. Chem.* 17, 2007: 986–991.

159. Orikasa, H., Karoji, J., Matsui, K., and Kyotani, T. Crystal formation and growth during the hydrothermal synthesis of B-Ni(OH)$_2$ in one-dimensional nano space. *Dalton Trans.* 34, 2007: 3757–3762.

160. Matsui, K., Kyotani, T., and Tomita, A. Hydrothermal synthesis of single crystal Ni(OH)$_2$ nanorods in a carbon-coated anodic alumina film. *Adv. Mater.* 14, 2002: 1216–1218.

161. Yang, Z.X., Xia, Y.D., and Mokaya, R. Enhanced hydrogen storage capacity of high surface area zeolite-like carbon materials. *J. Am. Chem. Soc.* 129, 2007: 1673–1679.

162. Meng, Y., Gu, D., Zhang, F.Q., Shi, Y.F., Cheng, L., Feng, D., Wu, Z.X., Chen, Z.X., Wan, Y., Stein, A., and Zhao, D.Y. A family of highly ordered mesoporous polymer resin and carbon structures from organic-organic self-assembly. *Chem. Mater.* 18, 2006: 4447–4464.

163. Tanaka, S., Nishiyama, N., Egashira, Y., and Ueyama, K. Synthesis of ordered mesoporous carbons with channel structure from an organic-organic nanocomposite. *Chem. Commun.* 2005: 2125–2127.

164. Liang, C.D., Hong, K.L., Guiochon, G.A., Mays, J.W., and Dai, S. Synthesis of a large-scale highly ordered porous carbon film by self-assembly of block copolymers. *Angew. Chem. Int. Ed.* 43, 2004: 5785–5789.

165. Hoffman, E.N., Yushin, G., El-Raghy, T., Gogotsi, Y., and Barsoum, M.W. Micro and mesoporosity of carbon derived from ternary and binary metal carbides. *Micropor. Mesopor. Mat.*, 112, 2008: 526–532. DOI: 10.1016/j.micromeso.2007.10.033.

166. Janes, A. Synthesis and characterization of nanoporous carbide-derived carbon by chlorination of vanadium carbide. *Carbon* 45, 2007: 2717–2722.

167. Urbonaite, S., Juarez-Galan, J.M., Leis, J., Rodriguez-Reinoso, F., and Svensson, G. Porosity development along the synthesis of carbons from metal carbides. *Micropor. Mesopor. Mat.*, 113(1–3), 2008: 14–21. DOI: 10.1016/j.micromeso.2007.10.046.

168. Dash, R.K. Nanoporous carbons derived from binary carbides and their optimization for hydrogen storage. PhD Thesis, Drexel University, Philadelphia, PA, 2006.

4 Porous Texture of Carbons

Dolores Lozano-Castelló, Fabián Suárez-García,
Diego Cazorla-Amorós, and Ángel Linares-Solano

CONTENTS

4.1 INTRODUCTION

Porous carbon materials mostly consist of carbon and exhibit appreciable apparent surface area and micropore volume (MPV) [1–3]. They are solids with a wide variety of pore size distributions (PSDs), which can be prepared in different forms, such as powders, granules, pellets, fibers, cloths, and others.

The properties of porous carbons are important in a wide variety of industrial applications, including both liquid and gas phases [1,2]. Access of liquids and gases to the carbon pores and the development of porosity during activation are critical in the selection and production of these materials.

Since the porosity of carbons is responsible for their adsorption properties, the analysis of the different types of pores (size and shape), as well as the PSD, is very important to foresee the behavior of these porous solids in final applications. We can state that the complete characterization of the porous carbons is complex and needs a combination of techniques, due to the heterogeneity in the chemistry and structure of these materials. There exist several techniques for the analysis of the porous texture, from which we can underline the physical adsorption of gases, mercury porosimetry, small angle scattering (SAS) (either neutrons—SANS or x-rays—SAXS), transmission and scanning electron microscopy (TEM and SEM), scanning tunnel microscopy, immersion calorimetry, etc.

Physical adsorption of gases is, undoubtedly, the most widely used technique [4]. Due to the considerable sensitivity of nitrogen adsorption isotherms to the pore texture in both microporous and mesoporous ranges and to its relative experimental simplicity, measurements of subcritical nitrogen adsorption at 77 K are the most used. However, this technique has some limitations, and other complementary techniques are needed for the characterization of microporous solids.

In this chapter, we present in some detail gas adsorption techniques, by reviewing the adsorption theory and the analysis methods, and present examples of assessment of PSDs with different methods. Some examples will show the limitations of this technique. Moreover, we also focus on the use of SAXS technique for the characterization of porous solids, including examples of SAXS and microbeam small-angle x-ray scattering (μSAXS) applications to the characterization of activated carbon fibers (ACFs). We remark the importance of combining different techniques to get a complete characterization, especially when "not accessible" porosity exists.

4.2 PHYSICAL ADSORPTION OF GASES

4.2.1 FUNDAMENTALS OF ADSORPTION

When an outgassed solid (the adsorbent) is confined to a closed space and exposed to a gas or vapor (the adsorptive) at a given pressure and temperature, an adsorption process takes place. The adsorptive molecules are transferred to, and accumulate in, the interfacial layer, as a consequence of an attractive force between the surface of the solid and the adsorptive (the adsorptive actually adsorbed by the adsorbent is named adsorbate). After some time, the pressure becomes constant and the thermodynamic equilibrium of adsorption is achieved.

In general, two types of adsorption are distinguished, physical adsorption and chemisorption, which depend on the type of interaction established between the adsorbent and the adsorptive. In a chemisorption process, specific chemical interactions between the adsorbent and the adsorptive occur, and the process is not reversible. On the other hand, physical adsorption includes attractive dispersion forces and, at very short distances, repulsive forces, as well as contribution from polarization and electrostatic forces between permanent electrical moments and the electrical field of the solid, if the adsorptive or the adsorbent has a polar nature. In this case, the process is fully reversible (or almost reversible). Thus, the overall interaction energy $\phi(z)$ of a molecule of adsorptive at a distance z from the surface of the adsorbent is given by the general expression

$$\phi(z) = \phi_D + \phi_P + \phi_F + \phi_{FQ} + \phi_R \qquad (4.1)$$

where
   ϕ_D is the attractive energy due to dispersion forces
   ϕ_P is the energy due to the polar nature of the adsorbent or the adsorptive
   ϕ_F is the energy due to permanent dipole moments of the adsorptive
   ϕ_{FQ} is the energy due to permanent quadrupole moments of the adsorptive
   ϕ_R is the energy due to repulsion forces

The terms ϕ_D and ϕ_R are the nonspecific contributions, which are always present on a physisorption process, whereas the other terms provide the specific contributions, which may be important for the overall adsorption energy according to the polar nature of the adsorptive and the adsorbent.

The physical adsorption allows obtaining important information about the solid under study (pore volume, surface area, adsorbent–adsorbate interaction energy, etc.). Different procedures have been developed to measure the amount of gas adsorbed on an appropriately outgassed adsorbent. In practice, the mostly used experimental systems are volumetric and gravimetric [5,6]. In the first one, the amount of adsorptive adsorbed by the solid, in a system of a given volume, is indirectly determined as the difference between the amount of adsorptive remaining in the gas phase after the adsorption equilibrium is achieved and the amount of adsorptive present in the system at the beginning of the adsorption by applying an appropriate equation of state for the gas. In the gravimetric systems, the amount of adsorptive adsorbed by the solid is directly determined by the increase in weight of the adsorbent. Actually, the excess amount of gas in the adsorbed layer or Gibbs adsorption is measured experimentally, but if the equilibrium pressure is sufficiently low and the adsorption is not too weak, the excess and the total adsorption are practically the same and the general term of the amount adsorbed is applicable to both quantities.

The amount of gas adsorbed (n) on the surface of the solid is proportional to the mass of the adsorbent, and it depends on the pressure (P), the temperature (T), and the nature of both the adsorbent and the adsorptive. Therefore,

$$n = f\left(P, T, \text{gas, solid}\right) \tag{4.2}$$

The amount of adsorbate (n) may be expressed in different units (moles, grams, and cubic centimeters at STP or liquid volume); however, recommendations have been made to express it in terms of moles per gram of the outgassed solid [7].

For given solid and gas maintained at a fixed temperature, the amount of adsorbed gas is a function of the pressure, and Equation 4.2 gives

$$n = f\left(P\right)_{T,\text{gas,solid}} \tag{4.3}$$

The representation of the amount adsorbed versus equilibrium absolute pressure (P) or relative pressure (P/P^0) gives the adsorption isotherm. If the temperature is below the critical temperature of the gas, P^0 is the liquid vapor pressure at the adsorption temperature.

The different types of adsorbents (carbon materials, clays, zeolites, etc.) and the large family of possible adsorptives give physical adsorption isotherms with different shapes. The IUPAC has classified the physical adsorption isotherms into six principal groups [8]. However, because different types of pores are usually present in the adsorbent, intermediate shapes are observed. These adsorption isotherms are schematically represented in Figure 4.1.

1. Type I isotherms are characteristic of microporous solids having relatively small external surface area (activated carbons, molecular sieve zeolites, metal organic frameworks, etc.). They are usually obtained by most gases and vapors on activated carbons.
2. Type II isotherms are the normal forms observed for macroporous and nonporous adsorbents (i.e., N_2 at 77 K on graphite, nanotubes, and carbon blacks). The type II isotherm represents unrestricted monolayer–multilayer adsorption. The point B is indicative of the stage in which the monolayer is complete and the multilayer adsorption begins.
3. Type III isotherms are not common and are characteristic of very weak adsorbate–adsorbent interaction and are typical of cooperative adsorption (i.e., adsorption of water vapor on graphitized carbon blacks).

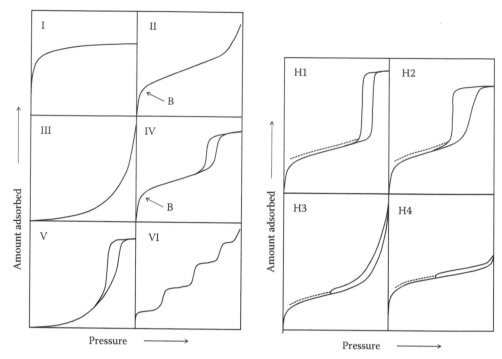

FIGURE 4.1   Types of physisorption isotherms (left) and hysteresis loops (right).

4. Type IV isotherms are characterized by the presence of a hysteresis loop (i.e., adsorption and desorption branches are not coincident) due to the capillary condensation on the mesopores. They are characteristic of adsorbents that have a wide proportion of mesopores (i.e., compacted carbon blacks under pressure, nanostructured carbons prepared using mesoporous silica as templates, etc.).
5. Type V isotherms are related to the type III isotherms and are also characteristic of weak adsorbate–adsorbent interactions (i.e., water vapor adsorption on charcoal).
6. Type VI isotherms are typical of adsorbents having a very uniform nonporous surface. Each step represents an adsorbed monolayer (i.e., noble gas adsorption on graphitized carbon blacks).

The IUPAC has also classified the hysteresis loops as a function of their shapes into four groups [8], namely, H1, H2, H3, and H4. Figure 4.1 (right) represents the different types of hysteresis loops. The H1 and H4 hysteresis loops are extreme cases. In type H1, both the branches (adsorption and desorption) are almost vertical and nearly parallel over an appreciable range of gas uptake (see Figure 4.1), whereas in H4, the adsorption and the desorption branches are horizontal and parallel over a wide range of relative pressures. The shapes of the hysteresis loops are also related to the pore structure of the solids, and interesting information on the porous texture (pore structure refers to the characteristics of the pores, i.e., pore shape, and texture includes all the parameters that define the porosity of the material: pore structure, PSD, pore volume, surface area, etc.) of the adsorbent can be deduced. Thus, PSD of the adsorbent can be obtained by applying different equations (mainly based on the Kelvin theory) to the desorption branches [4,7].

Regarding the adsorptives used for the characterization of solids by physical adsorption, in principle, it is possible to use any gas or vapor (N_2, CO_2, Ar, He, H_2O, CH_4, benzene, nonane, etc.), but in practice only a limited number of adsorptives are employed.

The selected adsorptive must fulfill the following general characteristics:

1. Be chemically inert, in order to avoid specific interactions with the adsorbent.
2. Have a saturation pressure (P^0) relatively high at the adsorption temperature, in order to cover the whole range of porosity.
3. Have the most spherical shape as possible, in order to minimize the error in the calculation of its molecular transversal section. This is important for the estimation of the surface area of the adsorbent (see below).

Among all adsorptives, N_2 adsorption at 77 K is the most used and, usually, has a special status of recommended adsorptive. Nitrogen almost fulfills the previous characteristics, the analysis temperature (77 K) is easily achieved and, at this temperature, N_2 adsorption covers relative pressures from 10^{-8} to 1, which results in adsorption in the whole range of porosity. One problem is that the saturation pressure at 77 K (P^0 = 1 bar) is quite low and, therefore, it is necessary to have suitable systems for achieving and measuring very low pressures (turbomolecular pumps, high-sensitivity pressure transducers, etc.). But, the main disadvantages of N_2 adsorption at 77 K are that when used for the characterization of microporous solids, diffusion problems of the molecules inside the narrow micropores (size <0.7 nm) occur; the kinetics of adsorption is very slow; and the equilibrium time for the adsorption may be extremely long. This may be a source of error for the evaluation of microporosity and, in most of the cases, this adsorptive cannot be used for the characterization of materials with narrow micropores (i.e., carbon molecular sieves [CMSs], charcoals, etc. [9–13]). To overcome this problem, the use of other adsorptives has been proposed.

He adsorption at 4.2 K has been proposed [14–16] as a promising method for the accurate determination of microporosity. The He atom is the smallest one; it has a spherical shape and interacts weakly with any solid surface [14]. He adsorption requires lower equilibrium times, and the amount adsorbed is higher than in the case of N_2 at 77 K. From this research, the authors concluded that the micropore analysis by N_2 adsorption at 77 K is insufficient and may give misleading conclusions [14]. In spite of the interesting results obtained with He, the experimental conditions used (adsorption at 4.2 K) make this technique unavailable for routine characterization of microporous solids.

CO_2 adsorption, either at 273 or 298 K [9–11,17–21], is another useful alternative for the assessment of narrow microporosity. In such case, though the critical dimension of the CO_2 molecule is similar to that of N_2, the higher temperature of adsorption used for CO_2 results in a larger kinetic energy of the molecules, which are able to enter the narrow porosity. In this way, CO_2 has been demonstrated to be an appropriate complementary adsorptive for the analysis of microporosity [11–13,22,23]. The main disadvantage of CO_2 at both temperatures is that its P^0 is quite high (34.7 and 64.1 bar at 273 and 298 K, respectively), and it is necessary to carry out the adsorption up to high pressures in order to cover the whole range of porosity.

4.2.2 Adsorption Theories and Analysis Methods

In this section, several theories and equations for the analysis of the adsorption isotherm data will be presented. The starting assumptions used in each equation will be discussed for a proper understanding of its limitations and correct use in the characterization of porous carbons.

4.2.2.1 Langmuir Theory

By means of molecular-kinetic arguments, Irving Langmuir [24,25] developed an adsorption equation, which describes very well the shape of type I isotherms (see Figure 4.1), even though this equation was initially developed for an open surface (nonporous solid). The starting point is the dynamic concept of the adsorption equilibrium in which the rates of adsorption and desorption are equal. Other assumptions were made:

1. Only one molecule per site can be adsorbed on the free surface of the adsorbent.
2. The surface of a solid is formed by energetically homogeneous sites.
3. The saturation is reached on completion of the monolayer (the adsorption is restricted to a monolayer).
4. The forces (repulsive or attractive) between the adsorbed molecules are negligible, when compared to the adsorbent–adsorptive interactions.

With these assumptions, the final equation can be expressed as

$$\frac{P}{n} = \frac{1}{Bn_{\mathrm{m}}} + \frac{P}{n_{\mathrm{m}}}$$

(4.4)

where

n is the amount adsorbed per gram of solid at equilibrium pressure P
n_{m} is the amount adsorbed per gram of solid to complete a monolayer
B is an adsorption coefficient of the adsorbate–adsorbent system

When the experimental data follows the Langmuir equation, a linear plot of P/n versus P is obtained, and from the slope and the interception with the P/n axis, n_{m} and B are calculated. The surface area of the adsorbent can be obtained from n_{m} (see below).

4.2.2.2 BET Theory (Including the Specific Surface Area of Solids)

The BET theory, developed by Brunauer, Emmett, and Teller [26], is based on the kinetic model of adsorption proposed by Langmuir [24,25] and was extended to describe the multilayer adsorption by the introduction of some assumptions listed below:

1. The first adsorbed layer is similar to the Langmuir model.
2. In all the layers, except the first one, the heat of adsorption is equal to the molar heat of liquefaction (q_{L}) of the adsorptive at the adsorption temperature.
3. In all the layers, except the first one, the evaporation–condensation conditions are identical.
4. When $P = P^0$, the adsorptive condenses to a bulk liquid on the surface of the adsorbent (the number of layers becomes infinite).

With these assumptions, the well-known BET equation is obtained [4,7,26], here written in the most convenient form for plotting the experimental data:

$$\frac{P}{n\left(P^0 - P\right)} = \frac{1}{n_{\mathrm{m}}C} + \frac{C-1}{n_{\mathrm{m}}C}\frac{P}{P^0}$$

(4.5)

where n is the amount adsorbed at equilibrium pressure P; n_{m} is the amount adsorbed per gram of solid to complete a monolayer, and C is a constant related exponentially with the heat of adsorption of the first layer, and in practice is taken as

$$C = \mathrm{e}^{(q_1 - q_{\mathrm{L}})/RT}$$

(4.6)

where q_1 and q_{L} are the molar heat of adsorption of the first layer and of the second and subsequent layers (equal to the molar heat of liquefaction of the adsorptive), respectively.

Therefore, the difference $(q_1 - q_L)$ is the net heat of adsorption.

Equation 4.5 gives a linear relation between $P/n(P^0 - P)$ and P/P^0, with the slope equal to $(C-1)/n_m C$ and with $1/n_m C$ as the interception with the Y-axis. From this, the values of both n_m and C are obtained. The value of C is affected by the shape of the isotherm and increases when the point B (see type II isotherm in Figure 4.1 and the corresponding explanation in the text) becomes sharper, that is, the interactions between the adsorbate and the adsorbent are stronger. The value of C is also affected by the selected relative pressure range, and it increases in the low relative pressure range. High values of C are usually found when Equation 4.5 is applied to type I isotherms.

Equation 4.5 gives an adequate description of type II isotherms over a limited P/P^0 range only, usually, within the range 0.05–0.35, due to the limitations of the BET theory. At relative pressures below $P/P^0 = 0.05$, the assumption that the surface is energetically homogeneous does not apply to the great majority of adsorbents. The lack of success at higher relative pressures $P/P^0 > 0.35$ is due to the contribution of lateral interaction between neighboring adsorbed molecules, which is not included in the model.

The specific surface area of the adsorbent can be obtained from the adsorption data if the number of molecules in the monolayer and the area effectively occupied by an adsorbed molecule in the monolayer, A_m (i.e., its cross-sectional area), are known. The amount adsorbed per gram of solid to complete a monolayer, n_m, can be obtained by applying Equations 4.4 and 4.5 to the adsorption isotherm. The surface area of the adsorbent (S) expressed in m²/g, is

$$S = n_m A_m N_A \cdot 10^{-18} \qquad (4.7)$$

where N_A is the Avogadro constant. The monolayer capacity (n_m) and the cross-sectional area of the adsorbed molecule (A_m) are expressed in mol/g and nm², respectively.

The cross-sectional surface area of the N_2 molecule at 77 K, assuming that it is packed like a liquid on the surface of the adsorbent, is $A_m = 0.167$ nm². But the cross-sectional surface area of a given adsorptive may not be constant, because it depends somewhat on the nature of the adsorbent, and the conventional picture of an A_m value for a monolayer completely filled with adsorbate molecules in a liquid-like packing does not correspond to the physical reality. This is evident since anomalous results (significantly different values) have been obtained when the surface area of a given solid was obtained from the adsorption isotherms of different adsorbates [4,7].

In general, in spite of its limitations and because of its simplicity, Equation 4.5 is still used, even to determine the surface area in the microporous adsorbents. Its use for these adsorbents is only recommended as a measurement of the monolayer equivalent area (area which would result if the amount of adsorptive required to fill the micropores would be spread as a close-packed monolayer of molecules).

4.2.2.3 Concept and Use of Standard Isotherms (the *t*-Plot Method and the α_S-Method)

Considering the nature of the forces involved in the physical adsorption process (see Section 4.2.1), it is evident that the adsorption isotherm of a given adsorptive on a particular solid at a given temperature depends on the nature of both the gas and the solid, and therefore, each adsorbate–adsorbent system has a unique isotherm. In spite of this, a number of attempts have been made to express the adsorption isotherm data in a normalized form. It was seen that, for a large number of nonporous solids (type II isotherms), the plot of n/n_m versus P/P^0 can be represented by a single curve, called the "standard isotherm." Among these related attempts, the *t*- and α_S-methods are the most widely used.

The *t*-plot method of Lippens and de Boer [27] assumes that a multilayer of adsorbed nitrogen is formed freely on the solid surface, and the thickness of adsorbed nitrogen increases with, and depends essentially on, the equilibrium relative pressure by the equation

$$t = 0.354\left(\frac{n}{n_{\mathrm{m}}}\right) = f\left(\frac{P}{P^0}\right) \qquad (4.8)$$

where the 0.354 nm value corresponds to the thickness of nitrogen at 77 K, assuming hexagonal close-packing of the molecules in the adsorbed film.

The experimental isotherm is transformed into a t-plot by plotting the amount adsorbed, n, versus t (the standard thickness of the reference nonporous material at the corresponding P/P^0).

By using this method, valuable information on the porous texture of the adsorbent can be obtained by comparing the nitrogen adsorption isotherm of a given porous solid, with the t-curve of a standard nonporous solid. Any difference in shape between the experimental isotherm of the solid under study and that for the reference one is revealed as a nonlinear region of the t-plot and a finite intercept of the Y-axis. In the case of an unhindered adsorption, the t-plot is a straight line passing through the origin, and its slope is a direct measurement of the surface area of the adsorbent.

An upward deviation is an indication that capillary condensation takes place in mesopores at high relative pressures; so, the adsorption increases more rapidly than that of the standard nonporous solid. A downward deviation indicates the presence of micropores in the solid. Because of its simplicity, the t-method has been widely used to analyze microporous carbons. The MPV, the external surface area, and the total surface area are the quantities often calculated by this method for activated carbons.

A different way of comparing an experimental isotherm to a standard one was proposed by Sing [28]. By a simple modification of the t-method, it is possible to avoid the necessity of knowing the thickness of the adsorbed gas. Instead of the thickness of the adsorbed layer, n_{m} is replaced by n_S as normalization factor, where n_S is the amount adsorbed at a preselected relative pressure. In practice, $P/P^0 = 0.4$ is chosen in the case of N_2 isotherms at 77 K, and it was justified, since monolayer coverage and micropore filling, if present, is complete, and capillary condensation has not yet begun [28]. Therefore, the corresponding reduced adsorption is $n/n_{0.4}$, the so-called α_S, where $n_{0.4}$ is the amount adsorbed by a nonporous reference solid at $P/P^0 = 0.4$. The reduced isotherm for the nonporous solid or α_S-curve is the graphic representation of α_S versus P/P^0. The α_S-plot for an adsorbent under study is constructed in a manner analogous to the t-plot. If there is neither capillary condensation nor micropore filling on the test sample, the α_S-plot is a straight line through the origin. Assuming that the surface of the standard is known (usually, but not necessary, from the BET nitrogen isotherm), the specific surface (S_α) of the test sample can be obtained by direct comparison to the reference isotherm:

$$S_\alpha(\text{sample}) = \frac{b_\alpha(\text{sample})}{b_\alpha(\text{reference})} \cdot S(\text{reference}) \qquad (4.9)$$

where b_α are the corresponding slopes of the α_S-curve (reference) and the α_S-plot (sample).

As in the case of t-plot, the deviations from the linearity can be used for the determination of different textural parameters of the sample (mainly, MPV, external surface area, and total surface area). In principle, the α_S-method is not restricted to nitrogen adsorption, but can be applied to any gas–solid system and can be used to identify the individual mechanisms of adsorption.

As an example of α_S-method application, Figure 4.2 shows the α_S-plot obtained from the high-resolution N_2 isotherms at 77 K of one ACF prepared from Nomex® (E.I. DuPont de Nemours & Company, Inc., Wilmington, DE) [poly-(m-phenylene isophthalamide)] by physical activation with CO_2 at a burn-off degree of 72% and using Spheron 6 carbon black as reference (BET surface area = 109 m^2/g) [29]. From the slope of the straight line at $\alpha_S > 1$, the external surface area can be obtained, and from its interception with the Y-axis, one can calculate the total MPV. From the slope of the straight line, which fits values of $0.5 < \alpha_S < 0.7$ and goes through the origin, one can calculate the total surface area [29].

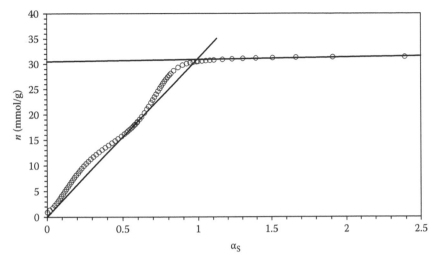

FIGURE 4.2 α_S-plot obtained from the high-resolution N_2 isotherm at 77 K of an ACF prepared from Nomex® (E.I. DuPont de Nemours & Company, Inc., Wilmington, DE) by physical activation with CO_2 (72% burn-off degree). Spheron 6 carbon black is used as reference.

Two deviations from linearity at $\alpha_S < 1$ can be observed (at $\alpha_S < 0.5$ and at $\alpha_S > 0.7$, respectively) on the α_S-plots. Both of them are due to an enhancement in the surface–molecule interactions as a function of the pore size. The first one is associated with the so-called "primary micropore filling" [4,7,30], and empirically suggests the presence of narrow micropores (pore width <0.7 nm), where the interaction between the pseudographitic surface and the adsorbate molecule is strongly enhanced by the overlap with the potential from the opposite pore wall (see below). The second one is associated with the so-called "secondary micropore filling" [4,7,30] and is attributed to the presence of pores with widths >1.0 nm, where adsorption in the empty space between the monolayer adsorbed on the pore walls is, presumably, accelerated regarding multilayer adsorption on the flat surface. The α_S-plots allow identifying the individual mechanisms of adsorption on the carbon materials. A detailed study of the micropore size evolution during CO_2 activation of ACF using the α_S method can be found in the work of Suárez-García et al. [29].

The use of standard isotherms as a tool to characterize the porosity of solids (even for microporous solids as it was shown) by means of one of the two methods described above may be appropriate. However, it is necessary to keep in mind that all of them are subject to the same limitation, that is, the difficulty of an appropriate choice of a nonporous reference material. The α_S-method chooses the standard isotherm according to the chemical nature of the sample to be studied, whereas the t-method chooses it independent of the nature of the nonporous adsorbent.

4.2.2.4 Peculiarities of Adsorption in Microporous Carbons (the Polanyi Potential Theory; Dubinin and Related)

The adsorption on microporous solids is not so well understood as in nonporous or mesoporous solids. When the pore size is similar to the size of the adsorbate molecule and the adsorption temperature is below the critical temperature, the adsorption does not take place by a progressive completion of a monolayer followed by multilayer adsorption, but by the filling of the MPV with the adsorbate in a liquid-like condition. There are a number of problems associated with adsorption in micropores including the following important points [23]:

1. BET surface areas are unrealistically high compared with the calculated area for 1 g of carbon in the form of an extended graphite layer plane, counting both sides (2630 m²/g).
2. The interaction energy between a free surface of a solid and an adsorbate molecule is rather lower than in a micropore, as a consequence of the overlap of the adsorption potential from the neighboring walls. This overlap leads to a strong adsorption of the gas by the micropore and, then, to an enhancement of the heat of adsorption.
3. The adsorption process in the micropores occurs by a volume-filling mechanism rather than by a surface-coverage mechanism. Thus, n_m values obtained either by BET or Langmuir equation are similar to the limited uptake of the type I isotherms. Therefore, the amount adsorbed for different adsorptives (expressed as volume of liquid) at a relative pressure near unity is very similar, which implies that micropore filling occurs rather than the surface area coverage of the adsorbent.
4. The equilibration time for the adsorption in some microporous materials, like CMS and carbonized chars, may be extremely long that may be a source of error for the evaluation of microporosity. For example, this occurs for N_2 at 77 K in samples with narrow microporosity (size below 0.7 nm), where the size of the adsorbate molecule is similar to the size of the pore entrance. In this case, contrary to the exothermic nature of the adsorption process, an increase in the temperature of adsorption leads to an increase in the amount adsorbed. In this so-called activated diffusion process, the molecules will have insufficient kinetic energy, and the number of molecules entering the pores during the adsorption equilibrium time will increase with temperature [9,23].

All the aforementioned points about peculiarities of adsorption in micropores show that special attention is needed when microporous solids (i.e., activated carbons, ACFs, nanotubes, CMSs, charcoals, etc.) are characterized by the physical adsorption methods.

A major development in understanding the adsorption of gases and vapors on microporous carbons was provided by the potential theory of adsorption by Polanyi. This theory assumes that, on the adsorbent surface, the gas molecules are compressed by attractive forces acting between the surface and the molecules, and these forces decrease with the increasing distance from the surface. The force of attraction at any given point near the surface is measured by the adsorption potential (A), which can be defined as the work done to transfer a molecule from the gas phase to a given point above the surface.

Polanyi described the adsorption space as a series of equipotential surfaces, each one with a given adsorption potential (A_i) and each one enclosing a volume (W_i). As one moves away from the surface, the adsorption potential decreases until it goes to zero and the adsorption space increases up to a limiting value, W_0 (where the potential becomes zero). On the surface, $W = 0$ and $A_i = A_{max}$. The building of the volume enclosed within the adsorption space is described by a function of the type $A = f(W)$.

Considering that dispersion and electrostatic forces are independent of temperature and assuming that the adsorption potential at a constant volume filling is also temperature-independent,

$$\left(\frac{\partial A_i}{\partial T}\right)_W = 0 \tag{4.10}$$

Equation 4.10 means that the curve $A = f(W)$ should be the same for a given gas–solid system at all temperatures. This relationship between (A) and (W) is called the "characteristic curve."

Polanyi expressed the adsorption potential (A) for a volume filling (W) as the work necessary to compress the adsorptive from its equilibrium pressure, P_1, to the compressed adsorbate pressure, P_2:

$$A = \int_{P_1}^{P_2} \frac{RT}{P} dP = RT \ln \frac{P_2}{P_1} \tag{4.11}$$

where $A = -\Delta G$, and ΔG is the equivalent free energy change.

The state of the compressed adsorbate in the adsorption space depends on the temperature. Three situations can be found: (i) when the adsorption temperature is well below the critical temperature (T_c), the adsorbed vapor may be considered as liquid-like; (ii) when the temperature is just below T_c, most of the adsorbate will be liquid-like, and also, partially, as a compressed gas; and (iii) when the temperature is above T_c, the adsorbate will be as a compressed gas.

The first case is, by far, the most used one. Therefore, the adsorption potential will take the form

$$A = RT \ln \left(\frac{P^0}{P} \right) \tag{4.12}$$

Equation 4.12 represents the work needed to compress the adsorptive from its equilibrium pressure in the gas phase to its saturation pressure P^0. From this equation, since it is assumed that the liquefied adsorbate is incompressible, and knowing the normal density of the liquid at the given adsorption temperature, it is possible to obtain the volume of the filled adsorption space by

$$W = \frac{nM}{\rho} = nv_m \tag{4.13}$$

where
n is the amount adsorbed, expressed in moles
M is the molecular weight of the adsorptive
ρ and v_m are the liquid density and the molar volume of the adsorptive at the adsorption temperature, respectively

The temperature invariance of the adsorption potential (fundamental postulate of Polanyi's theory) has been widely proved, especially, by the extensive work led by Dubinin [31–34].

Dubinin and coworkers, during the course of their extensive studies on activated carbons, have developed the so-called theory of volume filling of micropores. Based on numerous experimental data, Dubinin and collaborators have added a second postulate to the Polanyi theory, which complements it. For an identical degree of filling of the volume of adsorption space, the ratio of adsorption potentials for any two vapors is constant:

$$\frac{A}{A_0} = \beta \tag{4.14}$$

where β is called the affinity (also similarity) coefficient and should be regarded as the relative differential molar work of adsorption of a given vapor (A) relative to the differential molar work of adsorption of the vapor chosen as standard (A_0); for benzene (chosen as standard adsorptive by Dubinin) $\beta = 1$. The affinity coefficient is closely related to the ratio of the molar volume or of the parachors for the liquid-like phases of the vapor and benzene.

According to experimental data, and assuming that PSD is Gaussian, Dubinin and Radushkevich obtained an equation, which relates the degree of micropore filling (θ) with the differential molar work of adsorption:

$$\theta = \frac{W}{W_0} = \exp \left[-K \left(\frac{A}{\beta} \right)^2 \right] \tag{4.15}$$

where

    W_0 is the total MPV

    K is a constant dependent on the pore structure and is related to the characteristic energy (E_0)

By combining Equations 4.12 and 4.15, and for plotting purposes, the well-known Dubinin–Radushkevich equation (DR) is obtained:

$$\log W = \log W_0 - D \log^2 \left(\frac{P^0}{P} \right) \tag{4.16}$$

where $D = 2.303(RT/E_0\beta)^2$; a plot of log W against $\log^2(P^0/P)$ will be a straight line having an intercept equal to the total MPV W_0. From its slope, the value of D is obtained.

The application of Equation 4.16 is extensively reported in the literature. Thus, different adsorptives at different adsorption temperatures fit on straight lines over a very wide range of P/P^0, and all intersect the ordinate axis at one point. This observation indicates the constancy of the limiting volume of adsorption space, W_0. By introducing β into the potential equation, all the characteristic curves of the different adsorptives, on a given adsorbent, should coincide. Figure 4.3 plots the characteristic curves corresponding to the adsorption of N_2 at 77 K (P/P^0 from 10^{-7} to 1) and CO_2 at 273 K (both subatmospheric and high-pressure data, covering relative fugacities f/f^0 from 10^{-4} to 0.9) on a carbon fiber activated with CO_2 at 50% burn-off degree (data reported by Cazorla-Amorós et al. [17]).

As it can be seen in Figure 4.3, the characteristic curve for N_2 superimposes over that for CO_2 up to a value of $(A/\beta)^2$ lower than about 400 (kJ/mol)^2, confirming what was previously explained. From this value, a large downward deviation is observed in the N_2 characteristic curve. This behavior has been observed with other microporous carbon materials and, in fact, it is a common problematic feature of the characteristic curves for N_2 adsorption in microporous carbons and hence, a limitation of its use in such low relative pressures (i.e., from 10^{-3} to 10^{-7}) [10,11,17]. This downward deviation

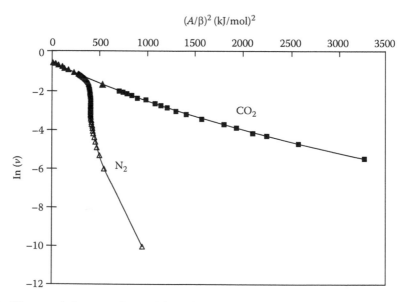

FIGURE 4.3 Characteristic curves for an ACF activated with CO_2 (50% burn-off degree): (■) CO_2 adsorption at subatmospheric pressures; (▲) CO_2 at high pressures; and (△) N_2 at 77 K. (From Cazorla-Amorós, D., et al., *Langmuir*, 12, 2820, 1996. With permission.)

shows that N_2 adsorption at 77 K in the narrow microporosity is influenced by diffusional limitations [17], and for that, N_2 adsorption isotherms at 77 K are not suitable for the narrow microporosity characterization. It must be noted that discrepancies about this large downward deviation exist; for example, it has also been related by other authors to a change in the adsorption mechanism at very low relative pressures [35].

Actually, very often, especially for highly microporous carbons, we observe that Equation 4.16 does not fit a straight line or the linear region is restricted to a very limited range of P/P^0 [7,36]. It is evident that when deviations occur, the extrapolation to the ordinate to obtain W_0 becomes uncertain. To overcome this problem, a number of more general equations have been proposed. For example, Dubinin and Astakhov [37] put forward a more general form:

$$\log W = \log W_0 - D' \log^N \left(\frac{P^0}{P} \right) \tag{4.17}$$

where N is an empirical constant. Values of N between 2 and 6 have been reported. However, in view of the empirical nature of N, it is not surprising to find that usually the best values are not integers. To overcome this, Stoeckli [38] suggested that the DR equation only holds for carbons with a narrow micro-PSD. According to this, the overall isotherm on a heterogeneous microporous carbon can be construed by summing the contributions of the different groups of micropores. Thus,

$$W = \sum_j W_{0,j} \exp\left[-K_j \left(RT/\beta \right)^2 \log^2 \left(\frac{P^0}{P} \right) \right] \tag{4.18}$$

where W_j represents the pore volume of the jth group. The previous equation can also be expressed for a continuous distribution:

$$W = \int_0^\infty f(K) \exp\left[-K \left(RT/\beta \right)^2 \log^2 \left(\frac{P^0}{P} \right) \right] \tag{4.19}$$

where $f(K)$ is a function of the micro-PSD. By assuming a Gaussian PSD, Stoeckli simplified Equation 4.19 and obtained an equation which is a function of $f(K)$ and which can be applied for the determination of the micro-PSD [39–43].

4.2.3 Assessment of Pore Size Distributions

Although textural parameters such as the monolayer capacities, surface area, pore volumes, etc. are important, the complete characterization of the adsorbent needs the knowledge of PSD. CMS is a clear example to show the importance of its PSD, which controls its application [12,21,44]. Thus, PSD governs to a great extent the adsorption properties and hence its applications. There are many examples in the literature where carbon materials with similar surface area and pore volumes have very different performances in a particular application. This can be found, for example, in hydrogen storage at room temperature or in the development of supercapacitors. In the first case, the activated carbons with a narrow PSD and a mean pore size of around 0.6 nm show higher hydrogen adsorption capacities [45,46]; in the second application, PSD exerts a strong influence either in the accessibility of the ions in solution to the porosity or in the optimization of the ion adsorption in the double layer [47,48].

PSD in the mesopore (2–50 nm) and macropore (>50 nm) regions can be reasonably obtained by different methods based on the Kelvin equation (like BJH, see below), since this equation describes rather well the equilibrium adsorption in these pores. However, there are no definitive methods for determining the PSD in the micropore region, due to the special features of the interaction between the adsorbate molecule and the adsorbent in these pore sizes. As we have shown in the previous section, overlapping of the potentials exerted by the two opposite walls makes the adsorbed gas behave very differently from that on open surfaces or in the bulk.

In the following sections, different methods and equations for determining the PSD of adsorbents, including their assumptions and limitations, will be presented. The main problem of all computational methods used for PSD determination is that they assume that the pores are rigid, not interconnected, and with a well-defined shape (perfect cylinders or parallel slits), which is not true in almost all the real microporous solids.

4.2.3.1 PSD Based on Kelvin Equation

The classical methods presented in the literature [4–7] for determining the PSD are only applicable to mesoporous solids, and they are based on the Kelvin equation.

The adsorption in mesopores is characterized by two sequential processes: (i) monolayer formation on the pore surface and (ii) filling of the remaining pore space when a critical filling pressure is reached. The first process can be modeled, for example, with the BET equation, and the second one with the Kelvin and related equations.

The equilibrium between two fluids at different pressures (for example, liquid and its vapor, at P^L and P^g, respectively) in a spherical interface of curvature radius, r, and surface tension, γ, is given by the Laplace equation

$$P^g - P^L = \frac{2\gamma}{r} \tag{4.20}$$

If instead of a spherical surface, any interface having two principal radii of curvature r_1 and r_2 is considered, then

$$P^g - P^L = \frac{2\gamma}{r_m} \tag{4.21}$$

where r_m is the mean curvature radius defined by $1/r_m = (1/2)(1/r_1 + 1/r_2)$.

Starting from the condition of thermodynamic equilibrium ($\mu^L = \mu^g$) and displacing it to another equilibrium state at constant temperature, Equation 4.21 becomes

$$dP^g - dP^L = d\left(\frac{2\gamma}{r_m}\right) \tag{4.22}$$

by assuming that (i) the molar volume of the liquid (V_L) is negligible compared with the molar volume of the gas (V_g), (ii) the adsorptive is in ideal gas state, and (iii) the liquid is incompressible. By integrating between the limits (r_m, P) and (∞, P^0), we obtain the Kelvin equation

$$\ln\left(\frac{P}{P^0}\right) = \frac{2\gamma V_L}{RT r_m} \tag{4.23}$$

The conversion of the mean curvature radius (r_m) to pore size involves assuming a model of pore shape and also knowing the contact angle (θ) between the capillary condensed phase and the adsorbed film on the walls. In the case of a cylindrical shape, the relation between r_m and the pore radius r_p is

$$r_p = r_m \cos\theta + t \qquad (4.24)$$

where t is the thickness of the adsorbed film. In the case of slit-shaped pores, the relation with the pore width (H) is

$$H = 2(r_m + t) \qquad (4.25)$$

Many different mathematical procedures based on the Kelvin equation have been proposed for the calculation of the meso-PSD from nitrogen adsorption isotherms. The most popular method was proposed by Barrett et al. [49] (known as BJH method), but others like Cranston and Inkley [50], Dollimore and Heal [51], and Robert [52] methods are also currently used.

All of these methods assume the following [4]: (i) Kelvin equation is applicable over the complete mesopore range; (ii) the meniscus curvature is controlled by the pore size and shape, and θ (contact angle) is 0; (iii) the pores are rigid and of well-defined shape; (iv) the distribution is confined to the mesopore range; (v) the filling (or emptying) of each pore does not depend on its location, and (vi) the adsorption on pore walls follows the same mechanism as on the open surface.

Although Kelvin equation is based on well-defined thermodynamic principles, its applicability to the mesopore size range, especially the lower ones, is questionable due to the uncertainties between the meniscus curvature and the pore size and shape. More extensive discussion about Kelvin-based meso-PSD methods and their applicability can be found in the literature [4,5,7].

4.2.3.2 Horvath–Kawazoe Method

Horvath and Kawazoe [53] working on nitrogen adsorption on CMSs proposed a novel method for determining the micro-PSD, assuming a slit-shaped geometry for the pores. This method is based on the idea that the P/P^0 required for filling the micropores of a given size and shape is a function of the adsorbate–adsorbent interaction energy. They made the assumption that the entropy contribution to the adsorption energy is small compared to the large change of the internal energy. If the micropores of the carbon material with slit-shaped geometry are described as two graphitic walls separated by a distance H (distance between the centers of the carbon atoms), then the interaction energy $\phi(z)_{pore}$ of an adsorbate molecule with two opposite walls can be written as

$$\phi(z)_{pore} = \phi(z) + \phi(H - z) \qquad (4.26)$$

Combining Equation 4.26 with a 10-4 potential function, Horvath and Kawazoe arrived at the following equation for the nitrogen adsorption at 77 K in CMSs:

$$\ln\left(\frac{P}{P^0}\right) = \frac{62.38}{H - 0.64}\left[\frac{1.895\times10^{-3}}{(H - 0.32)^3} - \frac{2.709\times10^{-7}}{(H - 0.32)^9} - 0.05014\right] \qquad (4.27)$$

where H is in nm. From this equation, a relationship between the pore size and the pressure is obtained and, therefore, the PSD can be obtained. Examples of application of this method in different microporous carbons can be found elsewhere [54–57].

4.2.3.3 PSD Based on Dubinin Equation

As described earlier, one of the first methods used to obtain PSD from the Dubinin equation is the so-called Dubinin–Stoeckli method [38–43]. For strongly activated carbons with a heterogeneous collection of micropores, the overall adsorption isotherm is considered as a convolution of contributions from individual pore groups. Integrating the summation and assuming a normal Gaussian equation for the distribution of MPV with respect to the K parameter (Equation 4.19), Stoeckli obtained an equation useful to estimate the micro-PSD.

A similar approach, in this case, based on the DR equation (Equation 4.16), was proposed by Cazorla-Amorós et al. [10] for the determination of the micro-PSD. Similarly, compared to other methods of determining PSDs (i.e., density functional theory [DFT], see below), this method assumes that the experimental isotherms can be described by the general adsorption isotherm (GAI) [58]:

$$n(P) = \int_0^\infty v(H,P) f(H) \, dH \tag{4.28}$$

where

$n(P)$ is the experimental isotherm
$v(H,P)$ is the local adsorption isotherm of a pore with mean width (H) (distance between the carbon layer, but not between the centers of carbon atoms)
$f(H)$ is the unknown PSD

In this method, the local adsorption isotherms are obtained from the DR equation (Equation 4.16), and the relation between the characteristic energy (E_0) and the mean pore width (H) is obtained from the equation proposed by Stoeckli et al. [39] when E_0 is between 42 and 20 kJ/mol

$$H(\text{nm}) = \frac{10.8}{E_0 - 11.4} \tag{4.29}$$

and from the Dubinin equation [34] for lower values of E_0:

$$H(\text{nm}) = \frac{24}{E_0} \tag{4.30}$$

Equation 4.29 corresponds to pore sizes between 0.35 and 1.3 nm, and Equation 4.30 to higher pore sizes. PSDs were determined in the form of histograms (following the approach developed by Sosin and Quinn for high-pressure methane isotherms [59]). In each pore size range, the amount of gas adsorbed per total MPV was calculated as

$$d_P^i = \frac{\int_{H1}^{H2} v(H,P) \, dH}{H_2 - H_1} \tag{4.31}$$

where

d_P^i for an ith pore size range is a function, which depends on P
H_1 and H_2 are the lower and upper limits of the ith pore size range

With this approach, Equation 4.28 is transformed into a discrete form

$$n(P) = \sum_{i=1}^n v^i d_P^i \tag{4.32}$$

where v^i are the unknown pore volumes of the ith pore size range (i.e., PSD).

The authors have applied this approach to low- and high-pressure CO_2 adsorption isotherms (using the fugacity instead of pressure) for determining the PSD of several carbon materials, including CMSs [10]. The obtained PSD confirmed the characteristics of these CMSs. Other examples of application of this method are included in Section 4.2.3.5.

4.2.3.4 PSD from DFT and GCMC

DFT and Monte Carlo (MC) simulations are modern methods for extracting information on PSDs. Both methods are used for obtaining the local isotherms as functions of pore width and geometry. Knowing the set of local isotherms, the PSDs are obtained by an inverse process solving the GAI equation (Equation 4.28).

Common to these methods is the choice of the potential energies: (1) intermolecular, (2) intramolecular, and (3) fluid–solid potential energy. The first one is the fluid–fluid potential and, for example, can be calculated from the 12-6 Lennard-Jones potential

$$u_{ff} = 4\varepsilon_{ff}\left[\left(\frac{\sigma_{ff}}{r}\right)^{12} - \left(\frac{\sigma_{ff}}{r}\right)^{6}\right] \tag{4.33}$$

where
ε_{ff} is the depth of the potential well
σ_{ff} is the collision diameter

The intramolecular potential energy is usually not considered for simple molecules, but it should be considered for molecules like CO_2 because of possible bond stretching and bending [60]. The third one depends on the solid nature and on the pore shape. In the case of carbon materials with slit-shaped pores, a Steele 10-4-3 potential can be used for solid–fluid interaction:

$$u_{sf}(z) = 4\Pi\varepsilon_{sf}\rho_s\sigma_{sf}^2\Delta\left[\frac{1}{5}\left(\frac{\sigma_{sf}}{z}\right)^{10} - \frac{1}{2}\left(\frac{\sigma_{sf}}{z}\right)^{4} - \frac{\sigma_{sf}^4}{6\Delta(z+0.61\Delta)^3}\right] \tag{4.34}$$

where
ρ_s is the density of the carbon center
Δ is the interlayer graphite spacing
z is the distance to one pore wall
σ_{sf} and ε_{sf} are fluid–solid molecular parameters, which can be obtained by matching the adsorption data on nonporous carbon black using the Henry constant in the fitting

The solid–fluid potential energy of one molecule and a slit-shaped pore of width H is

$$u_{sf}(z) + u_{sf}(H-z) \tag{4.35}$$

4.2.3.4.1 DFT Method
The pioneering work applying DFT for determining the PSD of carbon materials was reported by Seaton et al. [61], who used the local DFT. More recently, the nonlocal DFT (NLDFT), developed by Tarazona et al. [62–64] is used for determining the PSD. The density function approach [65] to describe inhomogeneous fluids at a solid interface consists of constructing the grand potential functional $\Omega[\rho(r)]$ of the average density $\rho(r)$ and minimizing it with respect to $\rho(r)$, to obtain the

equilibrium density profile and the thermodynamic properties. The grand potential functional for a one-component fluid at a fixed temperature is

$$\Omega\big[\rho(r)\big] = F[\rho] + \int \big(V(r) - \mu\big)\rho(r)\,dr \tag{4.36}$$

where

    $V(r)$ is an external potential
    μ is the chemical potential
    $F[\rho]$ is the intrinsic Helmholtz free energy functional, which contains the ideal gas term and
        contributions from fluid–fluid interactions:

$$F[\rho] = F_{HS}[\rho] + \frac{1}{2}\int U(r,r')\rho(r)\,dr \tag{4.37}$$

where $U(r,r')$ is the summation of the attractive part of the fluid–fluid potential of a molecule at r with all other fluid molecules in the system and which, for example, can be calculated from the 12-6 Lennard-Jones potential equation (Equation 4.33). $F_{HS}[\rho]$ are the repulsive forces between the molecules, which are modeled by hard spheres, and with the nonlocal approach of hard sphere proposed by Tarazona et al. [62–64] can be written as

$$F_{HS}[\rho] = \int f_{id}\big[\rho(r)\big]\,dr + \int \Delta\Psi_{HS}\big(\bar{\rho}(r)\big)\rho(r)\,dr \tag{4.38}$$

where the first term is the ideal gas contribution and the second one is the configurational part of the free energy per molecule, which depends on the smoothed density profile $\bar{\rho}(r)$:

$$\bar{\rho}(r) = \int w(r-r')\rho(r')\,dr' \tag{4.39}$$

where $w(r-r')$ is a normalized weight function, which may also be density-dependent.

Substituting Equations 4.37 and 4.38 into Equation 4.36, the grand potential functional can be written as

$$
\begin{aligned}
\Omega = kT \int \Big(\ln\big(\Lambda^3\rho(r)\big) - 1 \Big)\rho(r)\,dr + \int \Delta\Psi\big(\bar{\rho}\big)\rho(r)\,dr \\
+ \frac{1}{2}\int U(r,r')\rho(r)\,dr + \int \big(V(r) - \mu\big)\rho(r)\,dr
\end{aligned}
\tag{4.40}
$$

where $\mu = kT \ln(\Lambda^3\rho_0)$, ρ_0 is the bulk gas density, and Λ is the de Broglie wavelength. Since the chemical potential μ is fixed by the pressure of the gas phase, the equilibrium density profile $\rho(r)$ as a function of the equilibrium pressure can be obtained by minimizing the grand potential functional equation (Equation 4.40). More information about the solution procedure can be found in [60,62,66].

The amount adsorbed at each pressure can be obtained by integrating the equilibrium density profile $\rho(r)$ and subtracting the quantity of adsorptive which would be present in the absence of wall forces $\rho_0(r)$. The integration limits depend on the pore geometry, for example, in the case of slit-shape pores, the densities are only functions of the normal distance (z) from the wall; thus,

$$Q_{ad} = \int_0^H \big(\rho(z) - \rho_0(z)\big)\,dz \tag{4.41}$$

where $\rho_0(z) = 0$ for $z < \sigma_{sf}$ and $\rho_0(z) = \rho_b$ for $z > \sigma_{sf}$. Here, σ_{sf} is the collision diameter used in the 12-6 Lennard-Jones potential equation and is defined as the distance at which the resulting interaction potential energy is zero. ρ_b is the bulk density of the adsorptive at the same pressure and temperature. The upper limit of the integration (H) for a free surface is chosen large enough, so that Q_{ad} is essentially constant upon further extension of the limit, and it is chosen as the pore center for a slit pore. Extending to different pressures, the local adsorption isotherm, $v(H,P)$, in a pore of a given size is obtained.

4.2.3.4.2 GCMC Method

MC numerical simulation in grand canonical ensemble (GCMC) is used for calculating the adsorption in pores, where the chemical potential, temperature, and pore volume are specified. The number of particles (hence the density) is obtained from the simulation of a given chemical potential by minimizing the system energy through the displacement, the insertion, and the removal of molecules in the simulation box. The probabilities of the different movements are mentioned in the following paragraphs.

The displacement of particles, where a particle is selected at random and given a new conformation, has the following probability:

$$P = \min\left\{1, \exp\left[-\left(U_N - U_{OLD}\right)/kT\right]\right\} \tag{4.42}$$

where U_N and U_{OLD} are the energies of the new and the old configurations, respectively. The movement is accepted if the energy of the new configuration is lower than that of the old configuration. In the case that the energy of the new configuration is higher than the old one, the movement is also accepted if the probability obtained from Equation 4.42 is higher than a random number between 0 and 1. Otherwise, the movement is rejected.

The probability of the insertion of a particle at a random position in the simulation box, which already contains N particles is

$$P = \min\left\{1, \frac{V}{\Lambda^3(N+1)}\exp\left\{\left[\mu - U(N+1) + U(N)\right]/kT\right\}\right\} \tag{4.43}$$

And in the case of removal of a random particle from the simulation box, which contains N particles, the probability is

$$P = \min\left\{1, \frac{\Lambda^3 N}{V}\exp\left\{-\left[\mu + U(N-1) - U(N)\right]/kT\right\}\right\} \tag{4.44}$$

where
    Λ is the de Broglie wavelength
    μ is the chemical potential
    U is the configuration energy

The adsorption isotherm is obtained by relating the chemical potential with the gas bulk density or pressure by using an equation of state. Therefore, the set of local isotherms is obtained by defining the simulation boxes, which describe pores of various widths.

4.2.3.4.3 Determination of the PSD

The NLDFT and GCMC methods can be carried out to obtain a set of local isotherms for pores of different widths. The PSD is obtained by solving the GAI equation (Equation 4.28) in its discrete form:

$$n(P) = \sum_i v(H_i, P) f(H_i) \qquad (4.45)$$

It must be remarked that, as in the case of other methods for determining the PSD, the PSD obtained with NLDFT and GCMC must be regarded not as a definitive one but as an effective PSD. This is such because (1) the pore shape is assumed to be a perfect slit or cylinder (which is hardly present in real solids); (2) defects on the graphene surface are not accounted in the models; (3) interference of functional groups in the graphene layers are not taken into account; and (4) pore connectivity is not included in the models. Of course, NLDFT and GCMC, as any other PSD method, must not be applied under nonequilibrium adsorption data (as it usually happens with N_2 adsorption isotherms at 77 K with carbon materials having narrow microporosity, which causes kinetic problems of this adsorbate at these conditions, as we have indicated previously). Unfortunately, this lack of adsorption equilibrium is usually not considered and hence, the usefulness of these PSDs can occur.

Some examples where the authors apply the GCMC method for determining the PSD on different carbon materials can be found in the literature [67–75]. However, the DFT method is more used than the GCMC one, because the DFT method is included in the software of commercial adsorption apparatuses.

In Section 4.2.3.5, some examples of the application of different methods for determining the PSD of carbon materials will be shown and their advantages and disadvantages will be indicated.

4.2.3.5 Some Examples

Figure 4.4 shows the high-resolution (from 10^{-6} to 1 P/P^0) adsorption–desorption isotherms of N_2 at 77 K in (a) normal and (b) semilogarithmic scales, on three porous carbon materials. AC1 and AC2 are the two activated carbons prepared from Spanish anthracite by chemical activation with KOH [21], and ACF1 is an ACF prepared from Nomex by physical activation with CO_2 [29]. As it can be seen, all isotherms are of type I, according to the IUPAC classification [8], typical of essentially microporous samples. ACF1 and AC2 isotherms show a very sharp "knee," indicating that pore filling takes place at very low relative pressures, which reveals the presence of narrow micropores (micropores less than 0.7 nm). On the other hand, the AC1 isotherm shows a broad "knee" indicating the presence of wide micropores. This is also evident from the shape of the isotherms in the semilogarithmic scale (Figure 4.4b), where we can see more clearly the pressure of filling in each sample. Thus, the porosity of ACF1 is practically filled at a relative pressure of 0.0001–0.001; in the case of AC2, there is an important N_2 uptake at $P/P^0 < 0.001$, and the filling takes place at 0.01–0.1. Finally, the pore filling in AC1 occurs at P/P^0 higher than 0.1. From these differences in the N_2 isotherms, one should expect different PSDs for these samples and that the pore size and the width of the distribution increase in the order ACF1 < AC2 < AC1.

Similar conclusions can be obtained from the analysis of the CO_2 adsorption isotherms. In Figure 4.5, the subatmospheric and high-pressure adsorption isotherms of CO_2 at 273 K on the previous samples and on a CMS1 [21] are shown. We must mention that the N_2 adsorption isotherm corresponding to this CMS1 was not included in Figure 4.4, because there is no adsorption under the usual experimental conditions used. It is such because this sample presents a very narrow microporosity, where N_2 at 77 K has diffusional problems and the adsorption process is kinetically controlled and, therefore, extremely long time is necessary to reach the equilibrium at each point of the isotherm. On the contrary (see Figure 4.5a and b), this is not the case with CO_2 adsorption at 273 K, where the adsorption equilibrium is fast (due to the higher temperature of analysis) allowing the isotherm to be obtained easily. One can observe in Figure 4.5a that there is a good continuation between the subatmospheric measurements (from about 0.001 to 0.1 MPa) carried out in a volumetric setup and the high-pressure measurements (up to 3 MPa) carried out in a gravimetric setup. Like in the N_2 adsorption, all CO_2 isotherms are of type I, according to the IUPAC classification [8],

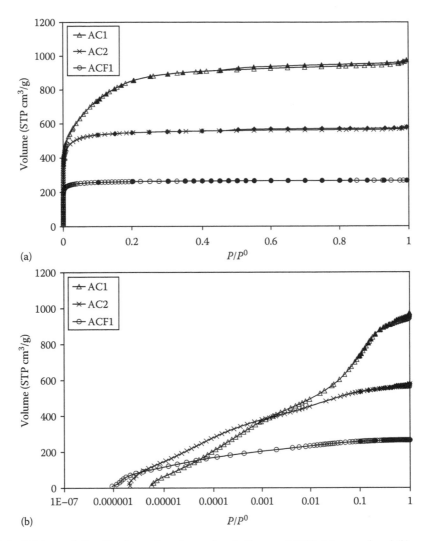

FIGURE 4.4 High-resolution N_2 adsorption–desorption isotherms at 77 K: (a) normal and (b) semilogarithmic scale, for three porous carbon materials. Open symbols: adsorption; full symbols: desorption. ACF1: ACF prepared from Nomex aramid fiber by physical activation with CO_2 [29]; AC1 and AC2: activated carbons prepared from Spanish Anthracite by chemical activation with KOH [21].

typical of essentially microporous samples. The CMS1, ACF1, and AC2 samples exhibit CO_2 isotherms with a very sharp "knee" (see Figure 4.5a), indicating the presence of narrow micropores, whereas the isotherm of the AC1 sample has a wider "knee" (see Figure 4.5a). The CMS1 and ACF1 samples attain near saturation at very low pressures (see Figure 4.5b), whereas the AC1 sample shows a low CO_2 adsorption at subatmospheric pressures, which increases practically linearly with the pressure, indicating the absence of narrow micropores. From the analysis of the CO_2 adsorption isotherms, one can conclude that the pore size and the width of the distribution must increase in the order CMS1 < ACF1 < AC2 < AC1.

In the following, the N_2 (Figure 4.4) and CO_2 (Figure 4.5) isotherms of these four samples, with different PSDs, will be used to assess and compare the methods analyzed previously.

Figure 4.6 shows the PSDs obtained from the high-resolution N_2 adsorption isotherms at 77 K (Figure 4.5) by applying the Horvath–Kawazoe method (Figure 4.6a), Dubinin–Astakhov method

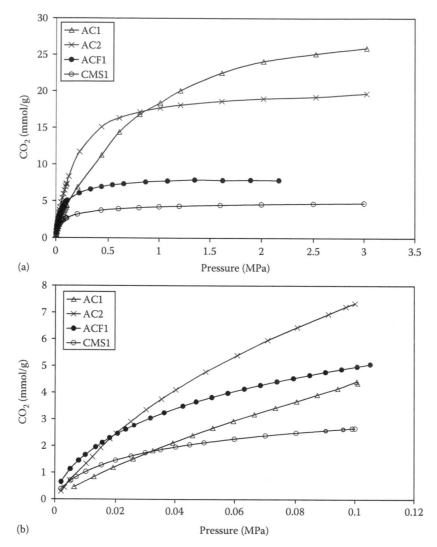

FIGURE 4.5 CO_2 adsorption isotherms at 273 K for four porous carbon materials: (a) subatmospheric and high-pressure isotherms; (b) subatmospheric isotherms. The samples AC1, AC2, and ACF1 are described in Figure 4.4. Sample CMS1: CMS prepared as reported in Ref. [21].

(Figure 4.6b), and NLDFT method (Figure 4.6c). All these PSDs were obtained using the software provided by the manufacturer of the commercial volumetric analyzer, where the N_2 adsorption isotherms were measured. Note that CMS1, for the reason commented before (lack of N_2 adsorption at 77 K due to diffusional problems), cannot be analyzed. The three methods qualitatively show the same evolution of the PSD, that is, the expected one deduced from the N_2 adsorption isotherm analysis. The micropore size and the width of the distribution increase in the order ACF1 < AC2 < AC1.

However, it must be emphasized that the PSDs are very different depending on the method used. The Horvath–Kawazoe method predicts that the porosity of these samples is mainly restricted to the narrow microporosity (pore size <0.7 nm). The Dubinin–Astakhov method shows Gaussian distributions with maxima at 1.3, 1.5, and 1.6 nm for ACF1, AC2, and AC1, respectively. Finally, the DFT method (Figure 4.6c) for these samples gives a more complex PSD having maxima in the narrow micropore region. ACF1 shows a narrow PSD restricted mainly to the pore size <1 nm. AC2 presents the maxima of the PSD at about 0.6–0.8 nm, with a small contribution of micropores with

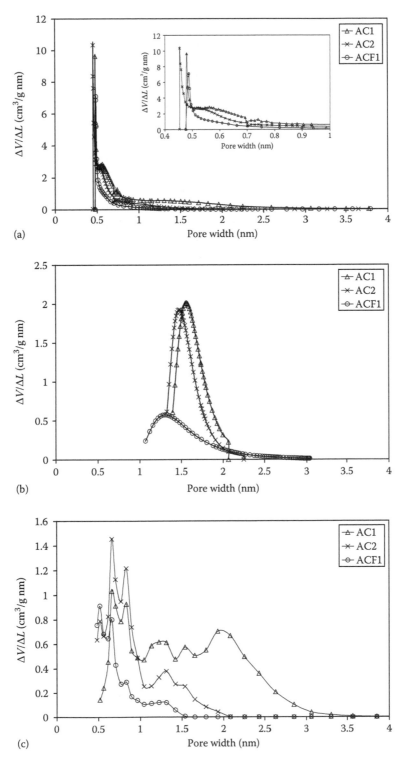

FIGURE 4.6 PSDs obtained from the high-resolution N_2 adsorption isotherms at 77 K (Figure 4.4) by applying different theories: (a) Horvath–Kawazoe; (b) Dubinin–Astakhov, and (c) DFT method.

size between 1 and 2 nm. Finally, AC1 has also narrow micropores but its PSD is wide showing an important contribution of wider micropores and small mesopores.

Comparing the PSDs obtained from the different methods used, we could conclude that DFT seems, at a first glance, to report the most coherent PSD. However, this favorable PSD for DFT merits recalling that the DFT method has several problems derived from the model assumption which are reflected on the PSD shape. For example, PSDs obtained from DFT and also from GCMC, systematically show minima at pore sizes of about 0.6 and 1.0 nm [76] (see Figure 4.6), which correspond to two and three molecular diameters of the adsorbate. These minima are due to packing effect since two or three molecules of adsorbate could be perfectly packed in pores with these sizes. These minima affect the PSD and give erroneous results.

PSD of the CMS1 obtained from the CO_2 adsorption isotherm at 273 K is included in Figure 4.7. The dashed curve (named "CO_2") was obtained from both subatmospheric and high-pressure data by applying the Dubinin-based method developed by Cazorla-Amorós et al. [10], and the curve named as "DFT" corresponds to the PSD obtained by applying the DFT method to the subatmospheric data. Figure 4.7 also includes the PSD obtained from the high-pressure isotherm of methane at 298 K by applying the method developed by Sosin and Quinn [59]. Considering the nature of CMS1, the PSD results presented in Figure 4.7 are interesting because they are quite reliable. First, in spite of the different experimental conditions used (equipments, pressure range, temperatures, adsorbates, and methods of analysis), a similar PSD was obtained from these three methods in the region of pore sizes below 0.8 nm. Second, the results confirm that this CMS has a very narrow PSD with a reasonable mean pore size of about 0.5 nm. Only the PSD from DFT gives an important contribution of pores of size larger than about 0.8 nm, which is not reasonable for this material which does not adsorb N_2 at 77 K and has molecular sieving properties [21,77]. In fact, this sample has shown good separation capabilities for CH_4 and CO_2 [21], which means that the pore size should be between 0.3 and 0.4 nm. This sample was also characterized in an earlier study by a completely different technique: image analysis of High Resolution Transmission Electron Microscopy (HRTEM) micrographs [78]. In that work, it was shown that HRTEM gives complementary results to gas adsorption techniques and can be used to investigate porous carbons, especially CMS, which are difficult to characterize with the well-established N_2 adsorption at 77 K. In that study [78], the PSD

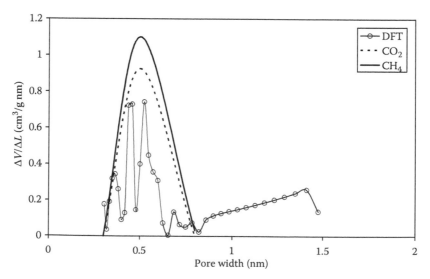

FIGURE 4.7 PSDs obtained by different methods on the CMS1 sample. DFT and CO_2 are the PSDs obtained by applying the DFT method and the Dubinin-based method proposed by Cazorla-Amorós et al. [10] to the CO_2 adsorption isotherm at 273 K (Figure 4.5), respectively. CH_4 is the PSD obtained from the high-pressure methane isotherm at 298 K by applying the method developed by Sosin and Quinn [59].

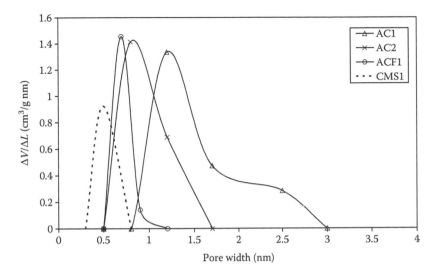

FIGURE 4.8 PSDs obtained from the subatmospheric and high-pressure CO_2 isotherms at 273 K (Figure 4.5a) by applying the Dubinin-based method proposed by Cazorla-Amorós et al. [10].

obtained by HRTEM for the sample CMS1 was wider than that obtained by CO_2 adsorption, suggesting that HRTEM was detecting the closed porosity existing in the sample, which was not accessible to gas adsorption.

PSDs for the other samples were also obtained from the CO_2 adsorption isotherms reported in Figure 4.5a by applying the method proposed by Cazorla-Amorós et al. [10]. These PSDs shown in Figure 4.8 follow the expected evolution. The mean pore size shifts to higher values, and the PSD widens in the order CMS1 < ACF1 < AC2 < AC1. The mean pore sizes obtained were 0.5, 0.7, 0.85, and 1.25 nm for CMS1, ACF1, AC2, and AC1, respectively.

As we have previously indicated, the PSDs obtained by any other method must be considered an "effective" PSD for this method and, in principle, it has only to be considered from a qualitative point of view. This is clearly shown in Figure 4.9, where the PSDs obtained from different methods are plotted on each sample. Clearly, each method gives a different PSD. Horvath–Kawazoe and Dubinin–Astakhov methods give the extreme results: the first one gives the smallest mean pore size, whereas the second one gives the highest. On the other hand, in spite of using different adsorbates and methods, DFT and DR-CO_2 provide closer results for all the samples, which suggest that the "effective" PSD could be in between these two methods.

The discrepancies shown in Figure 4.9 should not be forgotten when comparing the results from different authors. In principle, a given method could be used, never for a comparative purpose between published data, but, for example, for analyzing the development of the porosity of carbon materials under a given study. An example of such use can be seen in Figure 4.10, where the DFT method has been used to follow the PSDs of three ACFs with similar apparent surface area, prepared using different activation methods: physical activation (CO_2) and chemical activation (KOH and NaOH). It can be observed that the narrowest PSD is obtained with KOH, whereas the PSD becomes wider in the case of NaOH activation and much wider in the case of physical activation with CO_2. Thus, the widening of the PSD during activation follows the order KOH < NaOH < CO_2.

As stated earlier, one has to be very careful when comparing the PSDs reported from different authors, not only because they can use different methods, as shown before, but because when using the same method the software of the equipment used can give different results. This is clearly shown in the examples plotted in Figures 4.11 and 4.12. Figure 4.11 shows the cumulative pore volume obtained by applying the DFT method to the same N_2 adsorption isotherms (Figure 4.4) for samples ACF1, AC1, and AC2 using the software provided by two different commercial volumetric

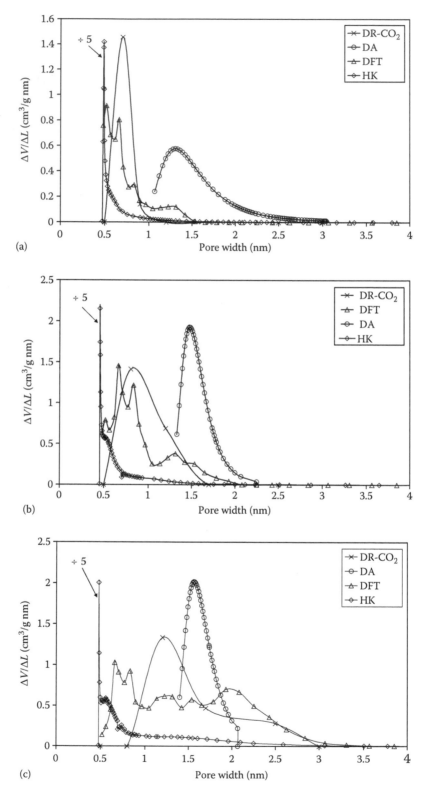

FIGURE 4.9 Comparison of the PSD obtained for different samples by applying different methods: (a) Sample ACF1, (b) sample AC2, and (c) sample AC1. DR-CO_2 is the PSD obtained by applying the Dubinin-based method proposed by Cazorla-Amorós et al. [10] to CO_2 at 273 K. HK, DFT, and DA are the PSDs obtained by applying Horvath–Kawazoe, DFT, and Dubinin–Astakhov methods to the N_2 adsorption isotherm at 77 K, respectively.

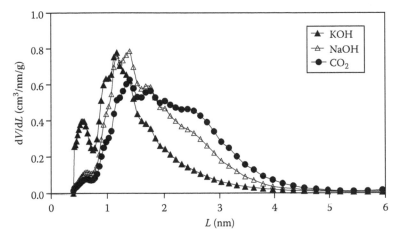

FIGURE 4.10 PSD by DFT method corresponding to ACF with similar apparent BET surface areas prepared with different activating agents.

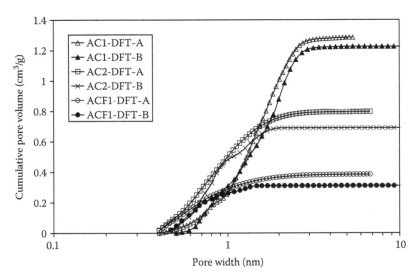

FIGURE 4.11 Cumulative pore volume corresponding to the three samples presented in Figure 4.4. These cumulative pore volumes have been obtained by applying the DFT methods to each N_2 adsorption isotherms using the software provided by two different commercial volumetric adsorption analyzers.

adsorption analyzers. The cumulative pore volumes of each sample obtained from both software (DFT-A and DFT-B) are different not only in the maximum uptake but also in shape, even though both software were applied to the same experimental adsorption isotherms. Therefore, PSDs obtained with each software are different, as shown in Figure 4.12.

4.3 SMALL-ANGLE X-RAY SCATTERING

SAS techniques represent an alternative to gas adsorption methods. The two most common subatomic scattering beams used in SAS are x-rays (SAXS) and neutrons (SANS). Both SAXS and SANS can provide structural details of porous materials on a scale covering a range from 1 to about 1000 nm. This scale ranges from lengths slightly larger than the diameters of single atoms to distances almost large enough to be resolved in an ordinary optical microscope. Since much

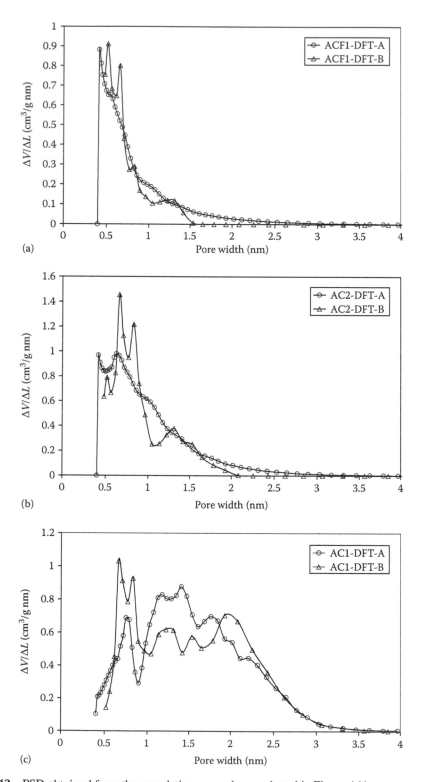

FIGURE 4.12 PSD obtained from the cumulative pore volumes plotted in Figure 4.11.

of the texture in porous solids lies in this interval, SAS can play an important role in the study of porosity.

SAXS techniques offer a number of advantages for the characterization of porous materials [79–81]: (1) they are sensitive to both closed and open porosity, (2) SAXS intensity profiles are sensitive to shape and orientation of the scattering objects, (3) they can be used to investigate samples that are saturated with liquids, and (4) they can be used to investigate the pore texture of materials under operating conditions. However, the equipment required for SAXS experiments is not as available as other adsorption equipment.

In this section, after a brief review of the SAS theory, some examples of SAXS (mainly centered on ACF) will be discussed to show its application for the characterization of porous solids.

4.3.1 SMALL-ANGLE SCATTERING THEORY

Scattering theory is covered in detail in textbooks [82,83]; hence, only a brief introduction to some of the basic principles will be given here.

When x-ray or neutrons from a source with a wavelength, λ, are incident on a sample, a small fraction of the beam is scattered (i.e., reemitted) in directions different from that of the incident beam. The radiation scattered into a solid angle element $\Delta\Omega$ is recorded by a detector. Figure 4.13 shows a schematic diagram of a SAS system. 2θ is the angle between the wave vectors (s_0) and (s) of the incident and scattered beams. The scattering vector, q, is the change in the wave vector (s_0) due to scattering; $q = s - s_0$, and its magnitude is $|q| = q = 2|s_0|\sin\theta$ for elastic scattering, where $|s_0| = |s|$. Since $|s_0| \equiv 2\pi/\lambda$, then

$$q = |q| = |s - s_0| = \frac{4\pi}{\lambda}\sin\theta \qquad (4.46)$$

Scattering angles $2\theta < 5°$ correspond to $dq < 1$, where d is the characteristic size of the scattering object. With $q = 2\pi/d$, one obtains Bragg's law $2d\sin\theta = \lambda$. The magnitude of the scattering vector is inversely proportional to the size of an object.

The flux of the radiation scattered, $I(\theta, \lambda)$, into a solid angle element $\Delta\Omega$ may be expressed in general terms in the following way:

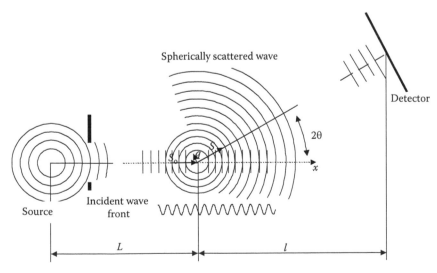

FIGURE 4.13 Two-dimensional projection of the scattering of a plane wave. The wave vectors of the incident and scattered waves are s_0 and s, and the scattering vector is $q = s - s_0$. The scattering angle is 2θ.

$$I(\theta,\lambda) = I_0(\lambda)\Delta\Omega\eta(\lambda)TV\frac{d\sigma(q)}{d\Omega} \tag{4.47}$$

where

I_0 is the incident flux

η is the detector efficiency (sometimes called response)

T is the sample transmission

$(d\sigma(q)/d\Omega)$ is a function known as the *(microscopic) differential cross section*

The objective of SAS experiments is to determine the differential cross section, since it contains all the information on the shape, size, and interactions of the scattering bodies in the sample. The differential cross section is given by

$$\frac{d\sigma(q)}{d\Omega} = N_p V_p^2 (\Delta\rho)^2 P(q)S(q) + B_{inc} \tag{4.48}$$

where

N_p is the concentration number of scattering bodies (given the subscript "p" for "particles")

V_p is the volume of one scattering body

$(\Delta\rho)^2$ is the *contrast* term, which is described later

$P(q)$ is a function known as the form or shape factor

$S(q)$ is commonly called the interparticle structure factor

B_{inc} is the incoherent background signal

$(d\sigma(q)/d\Omega)$ has dimensions of (length)$^{-1}$ and is normally expressed in units of cm^{-1}. In some textbooks, the microscopic differential cross section is replaced by the product $N_p \cdot (d\sigma(q)/d\Omega)$, a quantity known as the *macroscopic differential cross section* $(d\Sigma/d\Omega)$ (cm^{-2}).

In a scattering measurement, the scattered intensity $I(q)$ is recorded for several values of q. The intensity of the scattered beam and the way in which the intensity varies with the scattering angle are determined by the structure of the sample. One can, therefore, obtain information on the sample structure by the analysis of the scattered intensity at different scattering angles.

SAXS is observed only if in a specimen differences exist in the x-ray coherent scattering length density, ρ. The average x-ray coherent scattering length density for a given material is the number of x-ray coherent scattering lengths per unit volume [79]:

$$\rho = \sum_i \frac{b_i}{\Delta V} = \sum_i b_i \frac{DN_A}{M_w} \tag{4.49}$$

for atoms of type $i = 1, 2, ..., k$, in the volume element ΔV. The x-ray scattering length of one atom, b, is proportional to the number of electrons in the shell, that is, the x-ray scattering length increases with the atomic number. Here, M_w is the molecular weight, D is the mass density, and $N_A = 6.022 \times 10^{23}$ mol^{-1}.

A scattering object is *homogeneous* with respect to x-ray scattering if the scattering length densities, ρ, are equal everywhere in the object at a length scale larger than the interatomic distances. A porous material can be considered as a *monodisperse* system, which is a two-phase system consisting of a dispersion medium and dispersed pores, which are similar in size, shape, and internal structure.

The *contrast* for scattering by dispersion of pores (phase 1) in a medium (phase 2) is $(\Delta\rho)^2 = (\rho_g - \rho_s)^2$, where ρ_g and ρ_s are the x-ray coherent scattering length densities for the pores and solid, respectively. The scattered intensity $I(q)$ is proportional to the contrast.

Contrast-matching SAXS (CM-SAXS) consists of performing SAXS on samples, where the accessible porosity has been filled with a liquid which has the same x-ray scattering length density as that of the solid matrix, so that scattering is eliminated. Therefore, for systems with "closed" (not accessible) and "open" (accessible) porosity, scattering from open porosity is eliminated enabling any "closed" porosity to be investigated. If no "closed" porosity is present, then scattering is completely eliminated.

4.3.1.1 Approximations

4.3.1.1.1 Scattering at Low q
A simplification which describes scattering at low q can be made for the case of a two-phase system composed of particles having a homogeneous scattering length density distributed in a second phase. If we assume that the particles are monosized, then the intensity can be expressed as [83]

$$I(q) = I(0)\exp\left(\frac{-R_g^2 q^2}{3}\right) \tag{4.50}$$

where R_g is the radius of gyration, or *Guinier radius*, of a scattering object. This equation can be used when $q \cdot R_g < 1$. R_g is obtained directly from the slope m ($m < 0$) of the linear region [$1 \leq q^2$ (nm^{-2}) ≤ 3] in a graph of $\ln I$ versus q^2 [83], using Equation 4.51:

$$R_g = \sqrt{-3m} \tag{4.51}$$

In the case of spherical particles, the radius of a particle can be obtained from the Guinier radius according to

$$r = R_g\sqrt{\frac{5}{3}} \tag{4.52}$$

4.3.1.1.2 The High q Limit
Porod [84] demonstrated that in the limit of high q, the scattering is dependent on the interfacial area, S, between the two phases of the system:

$$I(q) = 2\pi S\left(\rho_s - \rho_g\right)^2 q^{-4} \tag{4.53}$$

This asymptotic decrease in intensity in the high q tail is described in the *Porod law* region and arises when $qR \geq 4$, where R refers to the dimension of the scattering heterogeneity.

A useful parameter for the analysis of porous materials, which can be derived from the application of the Porod law, is the Porod invariant (PI). PI is defined as [82]

$$PI = \int q^2 I(q)\,dq \tag{4.54}$$

PI is related to the void fraction (ϕ) of the material under investigation, as indicated in Equation 4.55, where V is the sample volume:

$$\frac{1}{V}PI = 2\pi(\Delta\rho)^2 \phi(1-\phi) \tag{4.55}$$

4.3.2   SAXS FOR THE POROUS TEXTURE CHARACTERIZATION IN AC

As previously discussed in this chapter, gas adsorption techniques are the most common approaches to the characterization of the pore texture, nitrogen adsorption at 77 K being the most popular technique. However, as mentioned above, the SAXS technique represents an alternative to gas adsorption methods.

Thus, the objective of this section is to make a comparison between the results obtained by SAXS technique and by gas adsorption (nitrogen at 77 K and CO_2 at 273 K). These results include pore size and porosity development. For this study, ACFs, which have been prepared using CO_2 and steam as activating agents up to different burn-off degrees, have been used.

ACFs were prepared from commercial carbon fibers according to the procedure described in the literature [85]. Two series of ACFs obtained from CO_2 (series CFC) and steam (series CFS) activation have been used in this study. The burn off of the fibers is between 11% and 54%. The nomenclature of each sample includes the burn-off degree.

Table 4.1 contains the MPVs obtained by applying the DR equation [86] to the N_2 and CO_2 isotherms adsorption at 77 and 273 K, respectively. For comparison purposes, the table also includes the BET surface area [4]. The volume of narrow microporosity (pore size smaller than 0.7 nm) has been assessed from CO_2 adsorption at 273 K and subatmospheric pressures (V_{CO_2}) [9,10,17]. From N_2 adsorption, the total MPV (V_{N_2}) (pore size lower than 2 nm) has been calculated. The densities used for liquid N_2 at 77 K and adsorbed CO_2 at 273 K were 0.808 and 1.023 g/cm³, respectively [9,10,17]. In Table 4.1, it can be observed that the original fiber does not adsorb N_2 at 77 K and does adsorb CO_2 at 273 K. Similar to an example presented in Section 4.2.3.5, this sample has a very narrow microporosity, where N_2 at 77 K has diffusion problems but adsorption of CO_2 at 273 K is fast. In Table 4.1 it can also be seen that the higher the burn-off degree, the higher is the porosity development, independent of the activating agent used. This table also shows that all the samples, except for the original fiber and the ACF with the lowest burn off (CFC11), present a V_{N_2} equal (CFC40, CFS27) to or higher (CFC50, CFS54) than the V_{CO_2}. Thus, samples CFC50 and CFS54 have the widest micro-PSD and contain a significant contribution of supermicroporosity (0.7 < pore size < 2 nm), estimated as $V_{N_2} - V_{CO_2}$ [85]. The contrary occurs in the original fiber and the sample CFC11 ($V_{CO_2} > V_{N_2}$). It indicates the existence of narrow microporosity, where N_2 adsorption at 77 K has diffusional limitations [9,10,17].

The DR plots can be used to estimate the mean pore size of the samples by using Equation 4.29 (for characteristic energy between 42 and 20 kJ/mol [39]) and Equation 4.30 (for lower E_0 [34]). The β values 0.33 and 0.35 have been used for N_2 and CO_2, respectively [10,17].

Table 4.2 contains the results obtained for the different samples. It can be observed that the mean pore size increases with burn off for both N_2 (L_{N_2}) and CO_2 (L_{CO_2}) adsorption. For the samples CFC11 (lowest burn off) and the original fiber, the mean pore size is given by L_{CO_2}, because these

TABLE 4.1

Characterization of the Porous Texture of Fibers by Physical Adsorption of N_2 at 77 K and CO_2 at 273 K

Sample	BET (m²/g)	V_{N_2} (cm³/g)	V_{CO_2} (cm³/g)
Original fiber	—	—	0.16
CFC11	—	0.13	0.23
CFC40	846	0.37	0.36
CFC50	1770	0.78	0.35
CFS27	644	0.29	0.29
CFS54	1500	0.67	0.30

TABLE 4.2

Characteristic Energies and Average Pore Width Calculated from the Adsorption Data of N_2 at 77 K and CO_2 at 273 K

Sample	E_0 (N_2) kJ/mol	L_{N_2} (nm)	E_0 (CO_2) kJ/mol	L_{CO_2} (nm)
Original fiber	—	—	35.4	0.45
CFC11	—	—	33.5	0.48
CFC40	22.8	0.95	27.4	0.68
CFC50	15.1	1.59	25.7	0.76
CFS27	23.1	0.92	29.0	0.61
CFS54	13.4	1.79	24.8	0.81

two samples have diffusional problems for N_2 adsorption at 77 K. For the rest of the samples, the mean pore sizes calculated from N_2 adsorption are higher than those deduced from CO_2 adsorption. This is because the samples contain wider micropores, which are not completely measured by CO_2 adsorption at subatmospheric pressures [10,17]. Then, for the samples where $V_{N_2} > V_{CO_2}$, L_{N_2} must be considered as the mean pore size in order to take into account all the microporosities.

In addition to the adsorption characterization, these samples have been characterized by SAXS using the CuKα (1.54 Å) radiation. The scans have been made between $0.1 \leq \theta \leq 3°$ with steps of $0.05°$ and counting times of 60 s per point. For these measurements, around 0.2 g of fiber was packed into a cell.

Figure 4.14 shows the SAXS plots for the ACFs prepared by CO_2 and steam activation. All the scattering curves have the same trend, and the main difference between them is that the intensity increases with the burn-off degree due to the development of porosity. In order to estimate the mean pore size from the SAXS data, the general approach based on Guinier equation has been used (see Equation 4.50). Table 4.3 presents the Guinier radii for the ACF and the original fiber. These values are quite similar to those obtained in a previous study done with ACF [87]. Table 4.3 also contains the pore width calculated for spherical particles (see Equation 4.51).

It can be observed that the pore size increases with burn-off, which is the same trend as that observed from the adsorption data (Table 4.2). For all the samples, the mean pore size calculated by SAXS is larger than that calculated by adsorption. This may be due to the fact that for SAXS and adsorption different distances are measured for the same pore by the nature of their interactions or to the different approximations assumed in both cases.

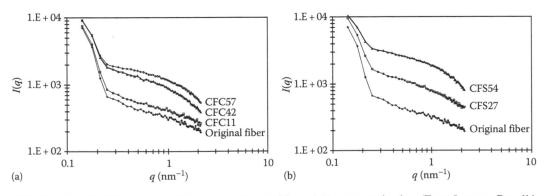

FIGURE 4.14 SAXS plots for ACFs prepared by (a) CO_2 and (b) steam activation. (From Lozano-Castelló, D., et al., *Studies in Surface Science and Catalysis, Characterisation of Porous Solids V*, vol. 128, Elsevier Science, the Netherlands, 523–532, 2000. With permission.)

TABLE 4.3

Guinier Radius and Mean Pore Width Obtained by SAXS

Sample	Guinier Radius (R_g) (nm)	Pore Diameter (nm)
Original fiber	0.67	1.73
CFC11	0.69	1.78
CFC40	0.73	1.88
CFC50	0.75	1.94
CFS27	0.73	1.88
CFS54	0.79	2.04

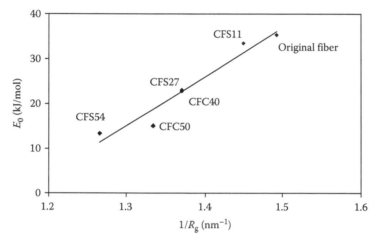

FIGURE 4.15 Relationship between the characteristic energy (E_0) and Guinier radius. (From Lozano-Castelló, D., et al., *Studies in Surface Science and Catalysis, Characterisation of Porous Solids V*, vol. 128, Elsevier Science, the Netherlands, 523–532, 2000. With permission.)

Figure 4.15 compares the information obtained from gas adsorption and SAXS to check the coherence of both approaches. Considering that E_0 is inversely proportional to pore size, the plot shows $1/E_0$ versus R_g. The figure reflects that, although there is some difference in the absolute pore size obtained (see Tables 4.2 and 4.3), both techniques are highly concordant. In addition, with the SAXS technique, which is completely different from gas adsorption, the usefulness of CO_2 adsorption for the characterization of samples with very narrow porosity, not accessible to N_2 at 77 K (i.e., non-ACF), has been confirmed.

Another example of SAXS application will be discussed next. In spite of the good tendency obtained with the two series of ACF (CO_2 and steam), our previous study using positron annihilation lifetime spectroscopy [88], and in our previous measurements of tensile strength and fiber diameter [85], it was shown that the ACF had a different evolution of porous texture with burn-off depending on the activating gas used (steam and CO_2). The confirmation of the structural changes produced in the activation process using both activation agents could be carried out analyzing the porosity development across the fiber diameter, instead of using a bundle of fibers packed in a cell. This could be done by SAXS, if a beam size much smaller than the fiber diameter and with high intensity would be available. The availability at European Synchrotron Radiation Facility (ESRF) (Grenoble, France) of x-ray microbeams with sizes down to 0.5 μm (Microfocus Beamline, ID13), together with a position-resolved x-ray scattering method makes this technique suitable for analyzing single fibers. In the next section, a description of the μSAXS technique is presented, together with some examples of application of this technique for the characterization of single ACF.

4.3.3 MICROBEAM SMALL ANGLE X-RAY SCATTERING

As previously mentioned, the availability of the x-ray microbeams, together with a position-resolved x-ray scattering method makes this technique suitable for analyzing single fibers. Experiments demonstrating the successful use of this technique in fibrous materials were carried out, e.g., on cellulose [89–92] and carbon fibers [93,94]. In the work carried out with carbon fibers [93,94], the internal structure of single carbon fibers from different precursors (PAN-based fiber and mesophase-pitch-based fibers) was investigated. However, in the literature, there was no report on the use of this technique for the characterization of ACFs, until some further works were published [77,95–97]. In these works, the usefulness of the Microfocus Beamline (ID13) in the characterization of single ACF was shown. Some examples of the application of this technique for the characterization of ACF prepared from different precursors (isotropic and anisotropic carbon fibers) by physical activation (CO$_2$ and steam) and chemical activation (KOH and NaOH) will be presented in this chapter.

The most remarkable characteristics of the Microfocus Beamline (ID13) in the ESRF in Grenoble, France are included below:

1. Monochromatic beams of 0.5, 2, 5, or 10 µm in diameter can be used for SAS experiments.
2. An area detector (MAR-CCD) with an active diameter of 130 mm is available. The use of an area detector has the advantage of having a better angular resolution than a quadrant detector, where the measurements have to be carried out by rotating the sample in the plane normal to the direction of the incident beam. Thus, two-dimensional scattering patterns can be obtained.
3. It has the so-called "scanning setup," with a Huber goniometer and a high-resolution microscope located at a calibrated distance from the beam position, so that a specific point can be selected on the sample and automatically transferred into the beam.

A scheme of the experimental setup is shown in Figure 4.16. In the examples presented in this chapter, the beams selected were of 0.5 and 2 µm diameter, and they were produced by Kirkpatrick–Baez

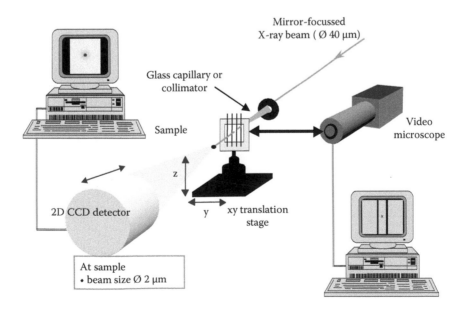

FIGURE 4.16 Scheme of the experimental setup in the µSAXS system. (From Lozano-Castelló, D., et al., *Studies in Surface Science and Catalysis, Characterisation of Porous Solids VI,* vol. 144, Elsevier Science, the Netherlands, 51–58, 2002. With permission.)

mirrors (wavelength $\lambda = 0.948$ Å). The fibers were examined with their axes perpendicular to the x-ray beam. The carbon fibers were mounted on an aluminum frame. Several fibers of a given sample were glued in each frame to repeat the measurements and to check the reproducibility. All measured data were corrected for background. Data evaluation was done using the software package FIT2D.

4.3.3.1 Characterization of Pore Distribution in ACF Prepared from Isotropic Pitch-Based Carbon Fiber

4.3.3.1.1 Carbon Fiber Activation with CO_2 and Steam

The materials used for these examples include a commercial petroleum pitch-based carbon fiber (Kureha Chemical Industry Co.), which is named as CF, and three ACFs, prepared from CF by activation in CO_2 up to 29% and 50% burn off (CFC29 and CFC50) and in steam up to 48% burn-off (CFS48). Table 4.4 contains the MPVs obtained by applying the DR equation to the N_2 and CO_2 isotherms at 77 and 273 K, respectively. The fiber diameter for each sample is also included in Table 4.4.

To show the suitability of microbeam µSAXS technique to characterize single carbon fibers, Figure 4.17 presents the two-dimensional scattering patterns corresponding to the center of each sample, after background and sample volume correction. These two-dimensional scattering patterns

TABLE 4.4

Fiber Diameter and Porous Texture Characterization by Physical Adsorption of N_2 at 77 K and CO_2 at 273 K

Sample	V_{DR} (N_2) (cm³/g)	V_{DR} (CO_2) (cm³/g)	Fiber Diameter (µm)
CF	—	0.18	28
CFC29	0.44	0.42	28
CFC50	0.72	0.33	28
CFS48	0.59	0.32	26

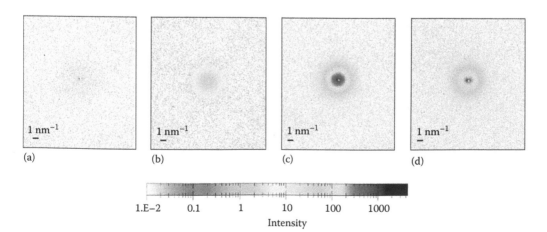

FIGURE 4.17 Two-dimensional scattering patterns after sample volume and background corrections corresponding to (a) CF, (b) CFC29, (c) CFC50, and (d) CFS48. (From Lozano-Castelló, D., et al., *Carbon*, 40, 2727, 2002. With permission.)

have been plotted for a maximum value of scattering vector (q) of $10\,nm^{-1}$. These figures show that in all the cases (carbon fiber and ACFs prepared using different activating agents) the scattering is the same in all the directions. These results indicate that the porosity is isotropically created along the fiber axis, which confirms the previous experiments [87]. As shown in Section 4.3.3.2, this behavior depends on the raw material and preparation process of the ACF.

In Figure 4.17a, it can also be observed that the original carbon fiber (CF) has a low scattering, as it is expected according to its low MPV (assessed by CO_2 adsorption at 273 K). The comparison of the scattering pattern of the original carbon fiber (Figure 4.17a) and the ACF prepared with CO_2 (Figure 4.17b and c) shows that the scattering intensity increases with the burn-off degree. The scattering corresponding to the fiber activated with steam (Figure 4.17) is intermediate between the two samples prepared with CO_2, which is reasonable considering the porous texture characterization results (see Table 4.4). These conclusions agree with those obtained from SAXS experiments carried out using large amount of fibers packed in a cell (see Section 4.3.2.).

To understand the effect of the percentage of activation in the microporosity, it is necessary to study not only the increase in pore volume, but also the widening of the porosity. In doing that, CM experiments have been carried out. As explained in Section 4.3.1, CM-µSAXS consists of performing SAXS on samples, where the accessible porosity has been filled with a liquid, which has the same x-ray scattering length density as that of the solid matrix. Therefore, for systems with "closed" (not accessible) and "open" (accessible) porosity, scattering from open porosity is eliminated enabling any "closed" porosity to be investigated. If no "closed" porosity is present, then scattering is completely eliminated after CM.

Dibromomethane is a solvent with an x-ray scattering length density ($18.95 \times 10^{10}\,cm^{-2}$) very similar to that for graphite ($19.18 \times 10^{10}\,cm^{-2}$). In addition, the size of the molecule of dibromomethane, calculated from the liquid density, is around 0.61 nm. Thus, this solvent will be able to access only the micropores with size larger than that value. These two characteristics make this solvent suitable for this type of experiment.

Figure 4.18 shows the two-dimensional scattering patterns for the CM experiments corresponding to the original carbon fiber and the two ACFs prepared by CO_2 activation. In Figure 4.18a, corresponding to the original carbon fiber, scattering is still observed after filling the capillary with

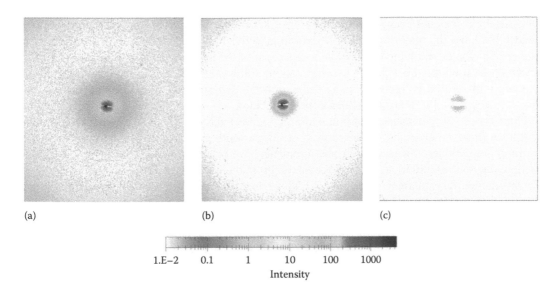

(a)   (b)   (c)

1.E−2   0.1   1   10   100   1000

Intensity

FIGURE 4.18 Two-dimensional scattering patterns for the CM experiments corresponding to (a) CF, (b) CFC29, and (c) CFC50. (From Lozano-Castelló, D., et al., *Chem. Eng. Technol.*, 26, 852, 2003. With permission.)

solvent. This indicates that this sample presents a very narrow microporosity, which is not accessible for dibromomethane. This result agrees with CO_2 adsorption results, which demonstrated the presence of narrow microporosity and reinforces the idea that N_2 at 77 K has several limitations to study the narrow microporosity (remember that there is no adsorption of N_2 at 77 K). The existence of this "not accessible" porosity in a sample with similar properties to those of the original carbon fiber (a CMS) was also confirmed in a previous work done using a different technique, HRTEM image analysis [78]. It should be mentioned that the scattering pattern corresponding to the original fiber presented in Figure 4.18a has a higher intensity than that presented in Figure 4.17a. The reason for this different intensity is that in CM experiments, a higher volume of sample was used for the analysis.

In Figures 4.18b and c, it can be observed that in the ACF, the use of dibromomethane eliminates the scattering intensity. In the case of the sample with intermediate burn off (CFC29), the scattering is only partially reduced. This must be most probably due to the existence of some microporosity, which is still very narrow for being accessible to the solvent. On the other hand, in the case of the highest burn-off degree (sample CFC50), the scattering is completely eliminated, which indicates that all the porosity of this sample is accessible to the solvent used for these experiments.

These results clearly show that the intensity observed in the two-dimensional scattering patterns comes from the porosity of the samples, which can be filled partially or totally with dibromomethane. According to the CM results, the very narrow porosity existing in the original carbon fiber becomes wider during the activation process. It has been seen that the higher the burn off, the higher would be the CM effect, due to the existence of a wider mean pore size.

These examples show that μSAXS experiments are an easy way to analyze the isotropy of the porosity in ACFs and to observe the development of porosity with the activation process.

Related to the porosity characterization in ACF, another important information would be to get an insight about the activation across the fiber. This can be done by obtaining a map of pore distribution across the fiber diameter, that is, by analyzing different regions of the same fiber. This type of characterization would be useful to understand better the activation process and to see the differences observed when the activation is carried out with CO_2 or with steam. As mentioned in Section 4.3.2, our previous results suggested that the textural changes produced by the activation process are different for both the activating agents (CO_2 essentially develops narrow microporosity and the fiber diameter does not change significantly after activation, whereas steam activation results in a wider PSD and the fiber diameter decreases after activation) [85,87,88,98]. The scanning setup (Huber goniometer and a high-resolution microscope) of the μSAXS technique, available at ID13 (ESRF), was used for this type of characterization. Some examples, corresponding to the results obtained with two ACFs prepared using different activating agents (CO_2 and steam) and similar burn-off (samples CFC50 and CFS48), are included below.

These types of experiments consist in scanning the fiber across its diameter with an accuracy of 0.5 μm. In these experiments, the scans were of 40 μm width and were divided into 10 steps with a duration of 20 s/step. The scattering measurements have been corrected considering the volume fraction, due to the cylindrical shape of the fibers. From the corrected scattering results, the PI has been estimated. As explained in Section 4.3.1, PI is related to the void fraction of the material under investigation, and it gives a useful comparison on how the void fraction of materials changes after treatment. Figure 4.19 includes the estimated relative PI values (PI values relative to the maximum PI value for each sample) versus the beam position. On the X-axis of this figure, the beam position equal to zero corresponds to the center of the fiber. Hence, this figure can be considered a map of porosity distribution across the fiber diameter.

For both samples, maximum scattering is observed in the external zone of the fibers, indicating a higher concentration of pores in this outer zone, as the activation process proceeds. In addition, it can be observed that the PI profiles are different for the CO_2 and steam-activated materials. In the case of steam, the scattering from all the internal zones is very similar and much lower than

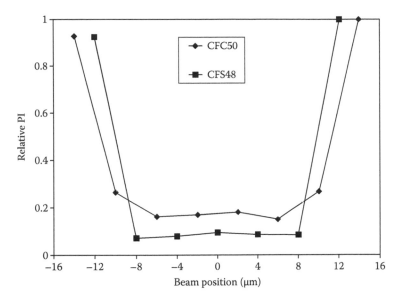

FIGURE 4.19 Relative PI values (calculated relatively to the maximum PI value for each sample) corresponding to the measurement carried out across the fiber diameter for the samples CFC50 and CFS48. (From Lozano-Castelló, D., et al., *Carbon*, 40, 2727, 2002. With permission.)

that from the external part of the fiber. On the other hand, in the case of CO_2, the porosity is more developed in the inner regions of the fibers compared to the steam ACFs.

All these results indicate that the activation process with CO_2 produces a more extensive porosity development across the fiber diameter than steam. However, steam activation focuses on the external zone of the fibers, which agrees with the decrease in fiber diameter for this process compared with CO_2 (see samples CFC50 and CFS48 in Table 4.4). This means that the CO_2 molecules penetrate more easily than H_2O into the carbon matrix. These results not only confirm the conclusions obtained in previous works using other techniques (i.e., gas adsorption and measurement of mechanical properties) [85,98], but also provide the first direct proof of the different behaviors of CO_2 and steam as activating agents, and show the suitability of these novel experiments for characterizing different regions of ACFs.

4.3.3.1.2   Carbon Fiber Activation with KOH and NaOH
The type of analysis done in Section 4.3.3.1.1 for "physically" activated carbon fibers (activated with CO_2 and steam) has been extended to ACF prepared by chemical activation with KOH and NaOH, using the same raw material (isotropic carbon fibers [Kureha Chemical Industry Co.]).

The petroleum pitch-based carbon fibers (sample CF) were chemically activated using KOH and NaOH [hydroxide/carbon ratio (wt/wt) of 4/1]. Chemical activations were carried out up to 750°C (holding time of 1 h), followed by a washing stage. More details of the activation process can be found elsewhere [99]. The characterization of the porosity was done using physical adsorption of N_2 at 77 K and CO_2 at 273 K. Table 4.5 contains the preparation conditions and pore texture characterization results. Scattering experiments were carried out at the Microfocus Beamline (ID13; ESRF; Grenoble). The beam selected for the experiments of these examples was a 0.5 μm beam produced by Kirkpatrick–Baez mirrors (wavelength $\lambda = 0.948$ Å). The domain of q values investigated with this setup is between 0.1 nm^{-1} (small angle resolution of 63 nm, approximately) and 10 nm^{-1}. These μSAXS measurements were carried out by scanning the fiber across its diameter with a step size of 1 μm and with accuracy better than 0.1 μm. More details of these measurements have been included elsewhere [97].

TABLE 4.5

Preparation Conditions and Porous Texture Characterization Corresponding to the Raw and Activated Isotropic Carbon Fibers

Sample	Chemical Activation	BET Surface Area (m²/g)	V_{DR} (N$_2$) (cm³/g)	V_{DR} (CO$_2$) (cm³/g)
CF (Kureha)	—	—	—	0.18
CFNa65	NaOH, 750°C, 1 h, 4/1	1085	0.42	0.34
CFK60	KOH, 750°C, 1 h, 4/1	980	0.40	0.34

Table 4.5 shows that the raw pitch-based carbon fiber has quite a high volume of micropores, reaching values of 0.40 and 0.42 cm³/g, after chemical activation with KOH or NaOH (samples CFK60 and CFNa65, respectively). To analyze in more detail the way this microporosity develops, two-dimensional scattering patterns were obtained at the center of the fiber, before and after chemical activation with KOH and NaOH (scattering patterns not shown here). An important increase of scattering intensity after activation of the original fiber, as expected due to the increase of pore volume in the ACF, is observed. Similar to the results obtained by μSAXS for "physically" activated carbon fibers (see Section 4.3.3.1.1), the development of porosity in "chemically" activated carbon fibers prepared from the same raw material (petroleum pitch-based carbon fiber) is isotropic, that is, the porosity development does not present any preferential orientation along the fiber axis. It should be mentioned that all the two-dimensional scattering patterns obtained at different regions of the chemically activated carbon fibers, across their diameters, also gave the same isotropic scattering. A detailed analysis of each of those two-dimensional scattering patterns across the fiber diameter has been done, and the scattering measurements have been normalized by the total sample volume analyzed by the beam. From the corrected scattering results, PI has been obtained and plotted against the beam position in Figure 4.20, for the ACF prepared by NaOH and KOH activation

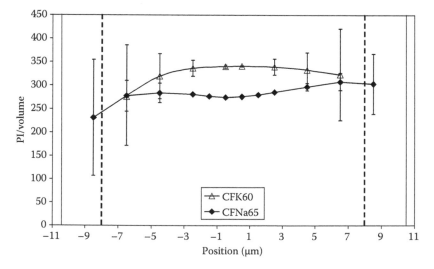

FIGURE 4.20 Volume corrected invariant versus beam position for the chemically activated carbon fiber prepared from pitch-based carbon fiber (samples CFK60 and CFNa65). The vertical lines indicate the fiber diameter (CFK60—dashed lines and CFK65—full lines). (From Lozano-Castelló, D., et al., *Carbon*, 44, 1121, 2006. With permission.)

(samples CFNa65 and CFK60, respectively). It is seen that the PI values obtained for each sample across the fiber diameter are similar, indicating a high concentration of pores even in the inner regions of the fiber. Additionally, the scattering profiles, as a function of the position of the fibers, are similar for the NaOH- and KOH-activated materials.

This trend observed for chemically activated carbon fibers is very different from that obtained for the ACF prepared by "physical" activation of the same raw carbon fiber with CO_2 and steam, where the maximum scattering appears at the external zone of the fiber (see Figure 4.19). In the case of NaOH and KOH activation, the results indicate that the activating agents penetrate the fiber, which is a new interesting observation, which points out that alkaline hydroxides penetrate much better than CO_2 and steam. The different evolution of porosity obtained for the chemically activated carbon fibers compared to the physically activated carbon fibers could be explained considering the important differences between the mechanisms of both the activation methods.

4.3.3.2 Characterization of Pore Distribution in ACF Prepared by Chemical Activation of Anisotropic PAN-Based Carbon Fiber

As previously mentioned, small angle x-ray (and neutron) scattering is one of the most suitable techniques to study the isotropy of the materials, because it is sensitive to the orientation and shape of the scattering objects. As explained in Section 4.3.3, a remarkable characteristic of the ID13 beamline (ESRF) is the availability of an area detector (MAR-CCD). With this detector, we can obtain two-dimensional scattering patterns, which allow to check if the porosity created during the activation process is preferentially oriented (i.e., to analyze the degree of isotropy). In this section, examples corresponding to the characterization of ACF prepared by chemical activation (KOH and NaOH) of an anisotropic PAN-based carbon fiber (Hexcel) (fiber diameter between 5 and 7 μm) are included. These ACFs were prepared and characterized (gas adsorption technique and scattering experiments) using the conditions explained in Section 4.3.3.1.2 for isotropic carbon fibers. Table 4.6 contains the preparation conditions and pore texture characterization results.

The optical micrographs, taken in polarized light, point out the existence of two different regions in the PAN-based carbon fiber [100]: a central isotropic region and an outer circumferential layer. The existence of these two regions was also observed in a previous study by TEM [97], where it was observed that the raw PAN-based carbon fiber consists of an external 2 μm ring and a core of 3 μm diameter.

Figure 4.21a contains the two-dimensional scattering patterns corresponding to the PAN-based carbon fiber (sample Hx) in the center of the fiber, presenting very different characteristics from that presented in Figure 4.17 for the isotropic carbon fiber. It can be seen that in the case of the raw PAN-based carbon fiber, anisotropic scattering develops into a fan shape perpendicular to the fiber axis. As demonstrated in the literature [101], a fan-like scattering along the equator is produced by a dilute system of microvoids with a preferred orientation along the fiber axis. This type of anisotropic scattering is observed for all the two-dimensional patterns obtained by scanning across the fiber diameter. The two-dimensional scattering patterns corresponding to the samples HxNa39 and HxK28,

TABLE 4.6
Preparation Conditions and Porous Texture Characterization Corresponding to the Raw and Activated Anisotropic Carbon Fibers

Sample	Chem. Activation	BET Surface Area (m²/g)	V_{DR} (N₂) (cm³/g)	V_{DR} (CO₂) (cm³/g)
Hx (Hexcel)	—	—	—	0.09
HxNa39	NaOH, 750°C, 2 h, 4/1	717	0.30	0.30
HxK28	KOH, 750°C, 2 h, 4/1	727	0.31	0.29

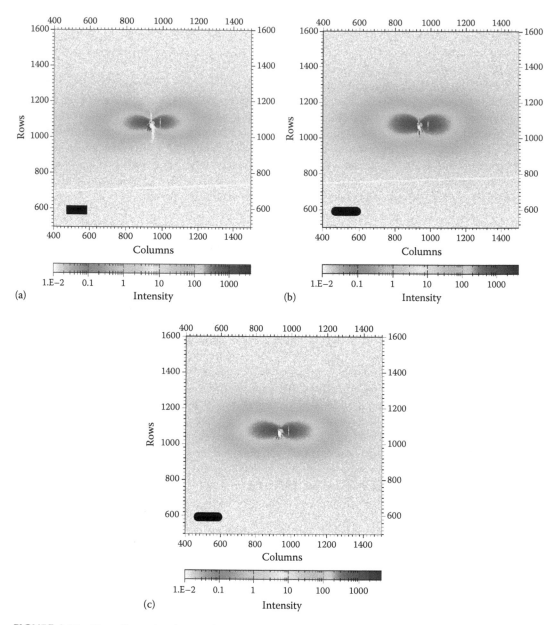

FIGURE 4.21 Two-dimensional scattering patterns obtained at the center of the anisotropic fiber (scale bar 1 nm^{-1}): (a) raw PAN-based carbon fiber (Hx), (b) HxNa39, and (c) HxK28.

which have been prepared under the same activation conditions but using different activating agents (NaOH and KOH, respectively), are presented in Figure 4.21b and c. The increase in the intensity of the two-dimensional scattering patterns, with respect to the original fiber (Hx), indicates an increase in porosity after activation. This fact is also deduced from the gas adsorption characterization results presented in Table 4.6, where it is seen that this raw material (Hx) develops an important MPV after activation with KOH and NaOH (around 0.3 cm^3/g). It can be seen that, after activation, the equatorial fan-shaped scattering remains in both samples, indicating that pores created during activation follow the preferential orientation of the microvoids along the fiber axis existing in the raw PAN-based carbon fiber. The orientation of the graphene layers along the fiber axis and the fact that the porosity generation occurs via removal of these layers, allow the anisotropy of the porosity generated to be understood.

Intensity has been measured by this technique as a function of the azimuthal angle (φ) at constant q, in different regions across the diameter of the original PAN fiber. These measurements confirmed the existence of a more anisotropic outer region and a higher misorientation of the pores at the center of the raw PAN fiber [97]. Moreover, the use of a very narrow microbeam (0.5 μm) is made possible to scan the very thin PAN-based carbon fiber (7 μm in diameter) and also the chemically activated carbon fiber across its diameter, analyzing the evolution of porosity during activation in each region. The analysis of the two-dimensional scattering patterns showed that the two regions remain after chemical activation with KOH or NaOH and that porosity is developed in both regions, indicating that the activating agents (KOH and NaOH) are reaching the center of the fibers. The porosity developed in the external anisotropic region of the fiber is very similar for both activating agents. However, the development of porosity in the fiber core with NaOH is much higher than that with KOH. These results indicate that, although both activating agents reach the fiber core, the most disordered region of the fiber is more easily activated by NaOH than KOH, which agrees with a previous work carried out with carbon nanotubes [102]. The detailed analysis of these results obtained with KOH- and NaOH-activated PAN-based carbon fibers has been published elsewhere [97].

4.4   CONCLUDING REMARKS

Considering the variety of fields where nanoporous carbons are used, it is extremely important to have a suitable characterization of their porous texture. This will allow the effect of the porous texture on the behavior of the carbon material in a given application to be understood. This knowledge is also essential to optimize its performance and to develop new uses of porous carbons.

Physical adsorption of gases and vapors is a powerful tool for characterizing the porosity of carbon materials. Each system (adsorbate–adsorbent temperature) gives one unique isotherm, which reflects the porous texture of the adsorbent. Many different theories have been developed for obtaining information about the solid under study (pore volume, surface area, adsorbent–adsorbate interaction energy, PSD, etc.) from the adsorption isotherms. When these theories and methods are applied, it is necessary to know their fundamentals, assumptions, and applicability range in order to obtain the correct information. For example, the BET method was developed for type II isotherms; therefore, if the BET equation is applied to other types of isotherms, it will not report the surface area but the apparent surface area.

Similarly, all the methods for determining the PSDs give the "effective" PSD and are useful for observing the trends and for qualitative or semiquantitative purposes. Although these methods are based on well-established thermodynamic principles, assumptions like pore size modeling (i.e., relation between pore size and meniscus curvature in Kelvin-based methods or pore shape modeling in DFT method—perfect parallel infinite graphene layers in the slit-shape model) reveal that the PSDs obtained from different methods cannot be compared to each other and that the results for a given sample may not agree with the other experimental observations.

Moreover, it has been remarked that it is necessary to use more than one adsorbate for a correct characterization of the narrow porosity. Thus, in the case of CMSs and other carbon materials (i.e., highly activated carbons) with narrow micropores, N_2 at 77 K is not a suitable adsorbate due to diffusion problems. Other adsorptives and conditions, like CO_2 at 273 or 298 K, avoid such problems. From all of these, it can be concluded that for a suitable characterization of the porosity of carbon materials by physical adsorption, the use of more than one adsorbate and the application of several theories and methods to the adsorption–desorption isotherms are recommended.

Additionally, it has been remarked that a combination of different techniques is convenient to obtain a complete characterization of porous materials. Thus, in addition to the gas adsorption techniques, we have presented some examples of results obtained using SAXS for the characterization of ACF, remarking the interesting research that can be carried out to characterize porous materials by this technique. The examples presented in this work illustrate the suitability of μSAXS technique to characterize single ACFs, instead of a bundle of fibers, due to the existence of microbeams with size down to 0.5 μm. The availability of an area detector allows the two-dimensional scattering patterns to be measured. The isotropic and anisotropic features in carbon fibers from different precursors (pitch- and PAN-based carbon fibers, respectively) and in ACF prepared from these precursors and using different activating agents (CO_2 and steam; KOH and NaOH) have been shown. In addition, the existence of a "scanning setup" has made possible obtaining scattering measurements across the fiber diameter, which allows maps of pores distribution to be obtained. As it has been shown, the type of information obtained by μSAXS (isotropy/anisotropy, maps of pores distribution, etc.) and other techniques, such as gas adsorption, is complementary. Examples showing the use of SAXS and gas adsorption corroborate the limitations of nitrogen adsorption at 77 K for the characterization of narrow micropores. Moreover, the scans across the fiber diameters give the first direct proof confirming the previous results obtained by our research group combining other techniques such as gas adsorption, mechanical properties, and SEM.

ACKNOWLEDGMENTS

The authors thank MEC and FEDER for financial support (project CTQ2006-08958/PPQ) and Generalitat Valenciana (ARVIV/2007/063). F. Suárez-García thanks MCYT for his contract "Juan de la Cierva."

REFERENCES

1. Bansal RC, Donnet JB, and Stoeckli F. *Active Carbon*, New York: Marcel Dekker. 1988.
2. Jankowska H, Swiatkowski A, and Choma J. *Active Carbon*, New York: Ellis Horwood. 1991: 29–38.
3. Marsh H and Rodríguez-Reinoso F. *Activated Carbon*, the Netherlands: Elsevier Science. 2005.
4. Rouquerol F, Rouquerol J, and Sing KSW. *Adsorption by Powders & Porous Solids*, San Diego: Academic Press. 1999.
5. Webb PA and Orr C. *Analytical Methods in Fine Particle Technology*, Norcross, GA: Micromeritics Instrument Co. 1997.
6. Lowell S and Shields JE. *Powder, Surface Area and Porosity*, London: Chapman and Hall, 3rd edn. 1991.
7. Gregg SJ and Sing KSW. *Adsorption, Surface Area and Porosimetry*, London: Academic Press, 2nd edn. 1982.
8. Sing KSW, Everett DH, Haul RAW, Moscou L, Pierotti RA, Rouquerol J, and Siemieniewska T. Reporting physisorption data for gas solid systems with special reference to the determination of surface-area and porosity (recommendations 1984). *Pure Appl. Chem.*, 1985; 57(4): 603–619.
9. Rodríguez-Reinoso F and Linares-Solano A. Microporous structure of activated carbons as revealed by adsorption methods. In: Thrower PA, ed. *Chemistry and Physics of Carbon*, vol. 28, New York: Marcel Dekker. 1988: pp. 1–146.
10. Cazorla-Amorós D, Alcañiz-Monge J, Casa-Lillo MA, and Linares-Solano A. CO_2 as an adsorptive to characterize carbon molecular sieves and activated carbons. *Langmuir*, 1998; 14(16): 4589–4596.
11. Lozano-Castelló D, Cazorla-Amorós D, and Linares-Solano A. Usefulness of CO_2 adsorption at 273 K for the characterization of porous carbons. *Carbon*, 2004; 42(7): 1233–1242.
12. Villar-Rodil S, Martínez-Alonso A, and Tascón JMD. Carbon molecular sieves for air separation from Nomex aramid fibers. *J. Colloid Interface Sci.*, 2002; 254(2): 414–416.
13. Villar-Rodil S, Denoyel R, Rouquerol J, Martínez-Alonso A, and Tascón JMD. Porous texture evolution in nomex-derived activated carbon fibers. *J. Colloid Interface Sci.*, 2002; 252(1): 169–176.
14. Kaneko K, Setoyama N, and Suzuki T. Ultramicropore characterization by He adsorption. In: Rouquerol J, et al., eds. *Characterization of Porous Solids III. Studies in Surface Science and Catalysis*, vol. 87, the Netherlands: Elsevier Science. 1994: pp. 593–602.
15. Setoyama N, Ruike M, Kasu T, Suzuki T, and Kaneko K. Surface characterization of microporous solids with He adsorption and small angle x-ray scattering. *Langmuir*, 1993; 9(10): 2612–2617.
16. Setoyama N, Kaneko K, and Rodríguez-Reinoso F. Ultramicropore characterization of microporous carbons by low temperature helium adsorption. *J. Phys. Chem.*, 1996; 100(24): 10331–10336.

17. Cazorla-Amorós D, Alcañiz-Monge J, and Linares-Solano A. Characterization of activated carbon fibers by CO_2 adsorption. *Langmuir*, 1996; 12(11): 2820–2824.

18. Sweatman MB and Quirke N. Characterization of porous materials by gas adsorption at ambient temperatures and high pressure. *J. Phys. Chem. B*, 2001; 105(7): 1403–1411.

19. Guillot A, Stoeckli F, and Bauguil Y. The microporosity of activated carbon fibre KF1500 assessed by combined CO_2 adsorption and calorimetry techniques and by immersion calorimetry. *Adsorpt. Sci. Technol.*, 2000; 18(1): 1–14.

20. Ravikovitch PI, Vishnyakov A, Russo R, and Neimark AV. Unified approach to pore size characterization of microporous carbonaceous materials from N_2, Ar, and CO_2 adsorption isotherms. *Langmuir*, 2000; 16(5): 2311–2320.

21. Lozano-Castelló D, Cazorla-Amorós D, Linares-Solano A, and Quinn DF. Micropore size distributions of activated carbons and carbon molecular sieves assessed by high-pressure methane and carbon dioxide adsorption isotherms. *J. Phys. Chem. B*, 2002; 106(36): 9372–9379.

22. Puziy AM, Poddubnaya OI, Martínez-Alonso A, Suárez-García F, and Tascón JMD. Synthetic carbons activated with phosphoric acid–II. Porous structure. *Carbon*, 2002; 40(9): 1507–1519.

23. Linares-Solano A, Salinas-Martínez de Lecea C, Alcañiz-Monge J, and Cazorla-Amorós D. Further advances in the characterization of microporous carbons by physical adsorption of gases. *Tanso*. 1998; 185: 316–325.

24. Langmuir I. The constitution and fundamental properties of solids and liquids. Part I. Solids. *J. Am. Chem. Soc.*, 1916; 38(11): 2221–2295.

25. Langmuir I. The adsorption of gases on plane surfaces of glass, mica and platinum. *J. Am. Chem. Soc.*, 1918; 40(9): 1361–1403.

26. Brunauer S, Emmett PH, and Teller E. Adsorption of gases in multimolecular layers. *J. Am. Chem. Soc.*, 1938; 60(2): 309–319.

27. Lippens BC and de Boer JH. Studies on pore systems in catalysts: V. The *t* method. *J. Catal.*, 1965; 4(3): 319–323.

28. Sing KSW. Empirical method for analysis of adsorption isotherms. *Chem. Ind.*, 1968; 44: 1520.

29. Suárez-García F, Paredes JI, Martínez-Alonso A, and Tascón JMD. Preparation and porous texture characteristics of fibrous ultrahigh surface area carbons. *J. Mater. Chem.*, 2002; 12(11): 3213–3219.

30. Kaneko K, Ishii C, Ruike M, and Kuwabara H. Origin of superhigh surface-area and microcrystalline graphitic structures of activated carbons. *Carbon*, 1992; 30(7): 1075–1088.

31. Dubinin MM. Porous structure and adsorption properties of active carbons. In: Walker PL, ed. *Chemistry and Physics of Carbon*, vol. 2, New York: Marcel Dekker. 1966: p. 51.

32. Dubinin MM. Physical adsorption of gases and vapors in micropores. In: Cadenhead DA, Danielli JF, Rosenberg MD, eds. *Progress in Surface and Membranes Science*, vol. 9, New York: Academic Press. 1975: pp. 1–70.

33. Dubinin MM and Stoeckli HF. Homogeneous and heterogeneous micropore structures in carbonaceous adsorbents. *J. Colloid Interface Sci.*, 1980; 75(1): 34–42.

34. Dubinin MM. Generalization of the theory of volume filling of micropores to nonhomogeneous microporous structures. *Carbon*, 1985; 23(4): 373–380.

35. Ehrburger-Dolle E. Analysis of the derived curves of adsorption isotherms. *Langmuir*, 1997; 13: 1189–1198.

36. Linares-Solano A. Textural characterization of porous carbons by physical adsorption of gases. In: Figueredo JL, Moulijn JA, eds. *Carbon and Coal Gasification*, the Netherlands: Martinus Nijhoff Publishers. 1986; pp. 137–180.

37. Dubinin MM and Astakhov VA. Description of adsorption equilibria of vapors on zeolites over wide ranges of temperature and pressure. *Adv. Chem. Ser.*, 1971; 102: 69.

38. Stoeckli HF. Generalization of Dubinin–Radushkevich equation for filling of heterogeneous micropore systems. *J. Colloid Interface Sci.*, 1977; 59(1): 184–185.

39. Stoeckli F and Ballerini L. Evolution of microporosity during activation of carbon. *Fuel*, 1991; 70(4): 557–559.

40. Stoeckli HF, Rebstein P, and Ballerini L. On the assessment of microporosity in active carbons, a comparison of theoretical and experimental-data. *Carbon*, 1990; 28(6): 907–909.

41. Stoeckli HF, Kraehenbuehl F, Ballerini L, and Debernardini S. Recent developments in the Dubinin equation. *Carbon*, 1989; 27(1): 125–128.

42. Stoeckli HF, Ballerini L, and Debernardini S. On the evolution of micropore widths and areas in the course of activation. *Carbon*, 1989; 27(3): 501–502.

43. Stoeckli HF. On the description of micropore distributions by various mathematical-models. *Carbon*, 1989; 27(6): 962–964.

44. Villar-Rodil S, Denoyel R, Rouquerol J, Martínez-Alonso A, and Tascón JMD. Fibrous carbon molecular sieves by chemical vapor deposition of benzene. Gas separation ability. *Chem. Mater.*, 2002; 14(10): 4328–4333.
45. Casa-Lillo M, Lamari-Darkrim F, Cazorla-Amorós D, and Linares-Solano A. Hydrogen storage in activated carbons and activated carbon fibers. *J. Phys. Chem. B*, 2002; 106(42): 10930–10934.
46. Jordá-Beneyto M, Suárez-García F, Lozano-Castelló D, Cazorla-Amorós D, and Linares-Solano A. Hydrogen storage on chemically activated carbons and carbon nanomaterials at high pressures. *Carbon*, 2007; 45: 293–303.
47. Lozano-Castelló D, Cazorla-Amorós D, Linares-Solano A, Shiraishi S, Kurihara H, and Oya A. Influence of pore structure and surface chemistry on electric double layer capacitance in non-aqueous electrolyte. *Carbon*, 2003; 41(9): 1765–1775.
48. Raymundo-Piñero E, Kierzek K, Machnikowski J, and Béguin F. Relationship between the nanoporous texture of activated carbons and their capacitance properties in different electrolytes. *Carbon*, 2006; 44(12): 2498–2507.
49. Barrett EP, Joyner LG, and Halenda PP. The determination of pore volume and area distributions in porous substances. I. Computations from nitrogen isotherms. *J. Am. Chem. Soc.*, 1951; 73(1): 373–380.
50. Cranston RW and Inkley FA. The determination of pore structures from nitrogen adsorption isotherms. *Adv. Catal.*, 1957; 9: 143–154.
51. Dollimore D and Heal GR. Improved method for calculation of pore size distribution from adsorption data. *J. Appl. Chem. USSR*, 1964; 14(3): 109.
52. Roberts BF. A procedure for estimating pore volume and area distributions from sorption isotherms. *J. Colloid Interface Sci.*, 1967; 23(2): 266–273.
53. Horvath G and Kawazoe K. Method for the calculation of effective pore-size distribution in molecular-sieve carbon. *J. Chem. Eng. Jpn.*, 1983; 16(6): 470–475.
54. Gauden PA, Terzyk AP, Rychlicki G, Kowalczyk P, Cwiertnia MS, and Garbacz JK. Estimating the pore size distribution of activated carbons from adsorption data of different adsorbates by various methods. *J. Colloid Interface Sci.*, 2004; 273(1): 39–63.
55. Grebennikov SF and Udal'tsova NN. Analysis of micropore volume size distributions of activated carbon fibers. *Colloid J.*, 2006; 68(5): 541–547.
56. Nishiyama N, Zheng T, Yamane Y, Egashira Y, and Ueyama K. Microporous carbons prepared from cationic surfactant-resorcinol/formaldehyde composites. *Carbon*, 2005; 43(2): 269–274.
57. Pinto ML, Pires J, Carvalho AP, and de Carvalho MB. On the difficulties of predicting the adsorption of volatile organic compounds at low pressures in microporous solid: The example of ethyl benzene. *J. Phys. Chem. B*, 2006; 110(1): 250–257.
58. McEnaney B, Mays TJ, and Causton PD. Heterogeneous adsorption on microporous carbons. *Langmuir*, 1987; 3(5): 695–699.
59. Sosin KA and Quinn DF. Using the high pressure methane isotherm for determination of pore size distribution of carbon adsorbents. *J. Porous Mat.*, 1995; 1(1): 111–119.
60. Do DD and Do HD. Pore characterization of carbonaceous materials by DFT and GCMC simulations: A review. *Adsorpt. Sci. Technol.*, 2003; 21(5): 389–423.
61. Seaton NA, Walton JPRB, and Quirke N. A new analysis method for the determination of the pore size distribution of porous carbons from nitrogen adsorption measurements. *Carbon*, 1989; 27(6): 853–861.
62. Tarazona P. Free-energy density functional for hard-spheres. *Phys. Rev. A*, 1985; 31(4): 2672–2679.
63. Tarazona P. Correction. *Phys. Rev. A*, 1985; 32(5): 3148–3148.
64. Tarazona P, Marconi UMB, and Evans R. Phase-equilibria of fluid interfaces and confined fluids—nonlocal versus local density functionals. *Mol. Phys.*, 1987; 60(3): 573–595.
65. Evans R. Nature of the liquid-vapor interface and other topics in the statistical-mechanics of nonuniform, classical fluids. *Adv. Phys.*, 1979; 28(2): 143–200.
66. Olivier JP. Modeling physical adsorption on porous and nonporous solids using density functional theory. *J. Porous Mat.*, 1995; 2(1): 9–17.
67. Do DD and Do HD. Modeling of adsorption on nongraphitized carbon surface: GCMC simulation studies and comparison with experimental data. *J. Phys. Chem. B*, 2006; 110(35): 17531–17538.
68. Do DD, Nicholson D, and Do HD. Heat of adsorption and density distribution in slit pores with defective walls: GCMC simulation studies and comparison with experimental data. *Appl. Surf. Sci.*, 2007; 253(13 SPEC. ISS.): 5580–5586.
69. Konstantakou M, Samios S, Steriotis T, Kainourgiakis M, Papadopoulos GK, Kikkinides ES, and Stubos AK. Determination of pore size distribution in microporous carbons based on CO_2 and H_2 sorption data. *Stud. Surf. Sci. Catal.*, 2006; 160: 543–550.

70. Kowalczyk P, Tanaka H, Kaneko K, Terzyk AP, and Do DD. Grand canonical Monte Carlo simulation study of methane adsorption at an open graphite surface and in slit like carbon pores at 273 K. *Langmuir*, 2005; 21(12): 5639–5646.

71. Pantatosaki E, Psomadopoulos D, Steriotis T, Stubos AK, Papaioannou A, and Papadopoulos GK. Micropore size distributions from CO_2 using grand canonical Monte Carlo at ambient temperatures: Cylindrical versus slit pore geometries. *Colloids Surf. A: Physicochem. Eng. Aspects*, 2004; 241(1–3): 127–135.

72. Samios S, Papadopoulos GK, Steriotis T, and Stubos AK. Simulation study of sorption of CO_2 and N_2 with application to the characterization of carbon adsorbents. *Mol. Simul.*, 2001; 27(5–6): 441–456.

73. Terzyk AP, Furmaniak S, Harris PJF, Gauden PA, Och J, Kowalczyk P, and Rychlicki G. How realistic is the pore size distribution calculated from adsorption isotherms if activated carbon is composed of fullerene-like fragments? *Phys. Chem. Chem. Phys.*, 2007; 9(44): 5919–5927.

74. Wongkoblap A and Do DD. Characterization of Cabot non-graphitized carbon blacks with a defective surface model: Adsorption of argon and nitrogen. *Carbon*, 2007; 45(7): 1527–1534.

75. Pérez-Mendoza M, Schumacher C, Suárez-García F, Almazán-Almazán MC, Domingo-García M, López-Garzón FJ, and Seaton NA. Analysis of the microporous texture of a glassy carbon by adsorption measurements and Monte Carlo simulation. Evolution with chemical and physical activation. *Carbon*, 2006; 44(4): 638–645.

76. Olivier JP. Improving the models used for calculating the size distribution of micropore volume of activated carbons from adsorption data. *Carbon*, 1998; 36(10): 1469–1472.

77. Lozano-Castelló D, Cazorla-Amorós D, and Linares-Solano A. Microporous solid characterization: Use of classical and "New" techniques. *Chem. Eng. Technol.*, 2003; 26: 852–857.

78. Lozano-Castelló D, Cazorla-Amorós D, Linares-Solano A, Oshida K, Miyazaki T, Kim YJ, Hayashi T, and Endo M. Comparative characterization study of microporous carbons by HRTEM image analysis and gas adsorption. *J. Phys. Chem. B*, 2005; 109: 15032–15036.

79. Hoinkis E. Small-angle scattering of neutrons and x-rays from carbons and graphites. In: Thrower PA, ed. *Chemistry and Physics of Carbon*, vol. 25, New York: Marcel Dekker. 1997: 71–241.

80. Ridgen JS, Dore JC, and North AN. Determination of anisotropic features in porous materials by small-angle x-ray scattering. In: Rouquerol J, et al., eds. *Studies in Surface Science and Catalysis, Characterisation of Porous Solids III*, vol. 87, the Netherlands: Elsevier Science. 1994: pp. 263–271.

81. Calo JM, Hall PJ, Houtmann S, Lozano-Castelló D, Winans RE, and Seifert S. "Real time" determination of porosity development in carbons: A combined SAXS/TGA approach. In: Rodriguez-Reinoso F, et al., eds. *Studies in Surface Science and Catalysis, Characterisation of Porous Solids IV*, vol. 144, the Netherlands: Elsevier Science. 2002: pp. 59–66.

82. Guinier A, Fournet G, and Walker CB. *Small Angle Scattering of X-Rays*. New York: Wiley. 1955: pp. 5–78.

83. Glatter O and Kratky O. *Small Angle X-Ray Scattering*. London: Academic Press Inc. 1982.

84. Ciccariello S and Benedetti A. Oscillatory deviations from Porod's law in the intensities scattered by some glasses. *J. Appl. Crystallogr.*, 1986; 19:195–197.

85. Alcañiz-Monge J, Cazorla-Amorós D, Linares-Solano A, Yoshida S, and Oya A. Effect of the activating gas on tensile strength and pore structure of pitch-based carbon fibres. *Carbon*, 1994; 32(7): 1277–1283.

86. Dubinin MM. The potential theory of adsorption of gases and vapors for adsorbents with energetically nonuniform surfaces. *Chem. Rev.*, 1960; 60: 235–241.

87. Cazorla-Amorós D, Salinas-Martínez de Lecea C, Alcañiz-Monge J, Gardner M, North A, and Dore J. Characterization of activated carbon fibers by small angle x-ray scattering. *Carbon*, 1998; 36(4), 309–312.

88. Lozano-Castelló D, Cazorla-Amorós D, Linares-Solano A, Hall PJ, and Fernandez JJ. Characterization of activated carbon fibers by positron annihilation lifetime spectroscopy (PALS). In: Unger KK, et al., eds. *Studies in Surface Science and Catalysis, Characterisation of Porous Solids V*, vol. 128, the Netherlands: Elsevier Science. 2000: 523–532.

89. Müller M, Czihak C, Vogl G, Fratzl P, Schober H, and Riekel C. Direct observation of microfibril arrangement in a single native cellulose fiber by microbeam small-angle x-ray scattering. *Macromolecules*, 1998; 31: 3953–3957.

90. Müller M, Burghammer M, and Riekel C. Fiber diffraction and small-angle scattering on single cellulose fibers. *ESRF Newsletter*, 1999; 33: 12–13.

91. Müller M, Riekel C, Vuong R, and Chanzy H. Skin/core micro-structure in viscose rayon fibers analysed by x-ray microbeam and electron diffraction mapping. *Polymer*, 2000; 41: 2627–2632.

92. Müller M, Czihak C, Burghammer M, and Riekel C. Combined x-ray microbeam small-angle scattering and fibre diffraction experiments on single native cellulose fibres. *J. Appl. Crystallogr.*, 2000; 33: 817–819.
93. Paris O, Loidl D, Peterlik H, Müller M, Lichtenegger H, and Fratzl P. The internal structure of single carbon fibers determined by simultaneous small- and wide-angle x-ray scattering. *J. Appl. Crystallogr.*, 2000; 33: 695–699.
94. Paris O, Loidl D, Müller M, Lichtenegger H, and Peterlik H. Cross-sectional texture of carbon fibers analysed by scanning microbeam x-ray diffraction. *J. Appl. Crystallogr.*, 2001; 34: 473–479.
95. Lozano-Castelló D, Raymundo-Piñero E, Cazorla-Amorós D, Linares-Solano A, Müller M, and Riekel C. Characterization of pore distribution in activated carbon fibers by microbeam small angle x-ray scattering. *Carbon*, 2002; 40: 2727–2735.
96. Lozano-Castelló D, Raymundo-Piñero E, Cazorla-Amorós D, Linares-Solano A, Müller M, and Riekel C. Microbeam small angle x-ray scattering (µSAXS): A novel technique for the characterization of activated carbon fibers. In: Rodríguez-Reinoso F. et al., eds. *Studies in Surface Science and Catalysis, Characterisation of Porous Solids VI*, vol. 144, the Netherlands: Elsevier Science. 2002: pp. 51–58.
97. Lozano-Castelló D, Maciá-Agulló JA, Cazorla-Amorós D, Linares-Solano A, Müller M, Burghammer M, and Riekel C. Isotropic ad anisotropic microporosity development upon chemical activation of carbon fibers, revealed by microbeam small angle x-ray scattering. *Carbon*, 2006; 44: 1121–1129.
98. Alcañiz-Monge J, Cazorla-Amoros D, and Linares-Solano A. Production of activated carbons: Use of CO_2 versus H_2O as activating agent. A reply to a letter from P.L. Walker, Jr. *Carbon*, 1997; 35(10–11): 1665–1668.
99. Lozano-Castelló D, Lillo-Ródenas MA, Cazorla-Amorós D, and Linares-Solano A. Preparation of activated carbons from Spanish anthracite I. Activation by KOH. *Carbon*, 2001; 39(5): 741–749.
100. Reynolds WN. Structure and physical properties of carbon fibers. In: Walker Jr. PL, Thrower PA,. *Chemistry and Physics of Carbon*, vol. 11, New York: Dekker. 1973: pp. 1–67.
101. Ruland, W. X-ray determination of crystallinity and diffuse disorder scattering. *Acta Crystallogr.*, 1961; 14: 1180–1185.
102. Raymundo-Piñero E, Azaïs P, Cacciaguerra T, Cazorla-Amorós D, Linares-Solano A, and Béguin F. KOH and NaOH activation mechanisms of multiwalled carbon nanotubes with different structural organisation. *Carbon*, 2005; 43(4): 786–795.

5 Surface Chemical and Electrochemical Properties of Carbons

Ljubisa R. Radovic

CONTENTS

5.1 INTRODUCTION

Over the past decades, the unique and remarkable surface properties of carbon materials have been responsible for the market success of many new carbon-based products, including some with very high value added and indeed some for electrochemical applications. In the electrochemical field, the opportunities for both vastly improved and novel products are very attractive, as evidenced by the information provided in the other chapters of this book. And it is not surprising that carbon electrochemistry research has reblossomed in the last decade, as illustrated by the ISI citation frequency of the works by Kinoshita [1] and Randin [2], summarized in Figure 5.1 (see also Figure 5.5). The latter is a more practically oriented review, and it is intriguing that it seems to have been "forgotten." The former was the first attempt to place carbon electrochemistry into the context of carbon's physicochemical properties; it remains necessary though not sufficient reading.

The general issue explored in this chapter is the following: in earlier works, when carbon electrochemistry was critically reviewed [3], a lack of fundamental knowledge was obvious. In large measure that was due to an absence of understanding of some of the essential details of carbon surface chemistry. Even worse, in the interim it had been concluded [4] that "carbon chemists and electrochemists generally appear to have difficulties in finding common ground for discussion."

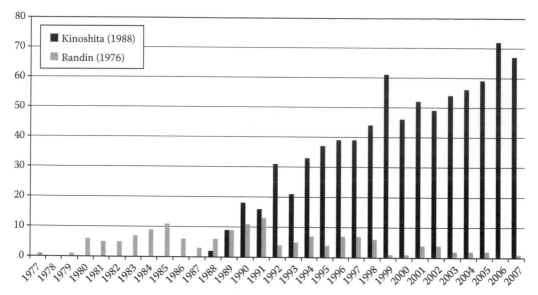

FIGURE 5.1 SCI citation statistics (Web of Science) for reviews by Kinoshita [1] and Randin [2]. Updated December 2007.

Now that carbon surface chemistry is arguably understood much better [4–8], has this led, or can it lead, to improvements in our understanding of carbon surface electrochemistry? More specifically, have we made progress in constructing "realistic speciation diagrams [in which the] participating species would be represented on a potential-pH plot" [4]? And are the surface properties of carbon, which have a "dramatic effect on electrocatalytic activity," still "uncertain and mostly a matter of conjecture" [9]? Can we, finally, say that a "greater understanding of electrocatalytic properties of carbon" is not "hindered [any more] by a lack of basic information concerning the surface state of the edge plane" [9]? The desirable breakthroughs in the performance of carbon-based fuel cells, batteries, capacitors, and electrosensors would certainly be facilitated by such developments. Alas, at least once before such an ambitious agenda formulated at the highest level of expertise [9], with the intent to bridge the gap between the surface chemistry and the electrochemistry of carbon surfaces, had failed to materialize. Fortunately, such state of affairs has not hampered the ubiquity of carbon electrodes in both industrial practice and research laboratories (e.g., even in the form of graphite pencils for a highly didactic instruction of electrochemistry to undergraduate students [10]).

Before proceeding, however, it should be noted that there is no other way to describe the carbon electrochemistry literature than overwhelming and even bewildering (see Section 5.3.1), to the extent that "seeing the forest for the trees" is still, and increasingly so, a monumental challenge. Therefore, if this chapter allows the reader to see in sharper focus some of the key questions, its goal will have been accomplished. A more extensive summary of the proposed answers even to some of the basic questions—including the historical development of the key concepts and their applications—is in preparation. The topics reviewed here are limited to the surface chemistry effects, and therefore, some of the important additional issues are not discussed. (See also the limitations mentioned in Section 5.3.1).

The study by Hsieh and Teng [11] can conveniently set the stage for the discussion that follows. It illustrates well how challenging it is to interpret the electrochemical effects in terms of specific changes in carbon surface chemistry. The authors argued that the following three processes may be responsible for these effects in an acidic solution:

$$>C_x + H^+ = >C_x /\!/ H^+ \tag{5.1}$$

$$>C_xO + H^+ = >C_xO/\!/H^+ \qquad (5.2)$$

$$>C_xO + H^+ + e^- = >C_xOH \qquad (5.3)$$

Reaction 5.1 is meant to represent a nonspecific electrostatic interaction (presumably responsible for double-layer charge accumulation); Reaction 5.2 symbolizes specific adsorption (e.g., ion/dipole interaction); Reaction 5.3 represents electron transfer across the double layer. Together, these three reactions in fact symbolize the entire field of carbon electrochemistry: electric double layer (EDL) formation (see Section 5.3.3), electrosorption (see Section 5.3.4), and oxidation/reduction processes (see Section 5.3.5). The authors did not discuss what exactly $>C_x$ represents, and they did not attempt to clarify how and why, for example, the quinone surface groups (represented by $>C_xO$) sometimes engage in proton transfer only and other times in electron transfer as well. In this chapter, the available literature is scrutinized and the current state of knowledge on carbon surface (electro)chemistry is assessed in search of answers to such questions.

5.2 CARBON SURFACE CHEMISTRY

Here only the most relevant issues will be summarized, because several "mainstream" reviews [4,5,7,8,12], as well as a more "personal" account by Sherwood [13] based on his extensive experience with x-ray photoelectron spectroscopy of carbon fibers are available. Of special interest are the mechanisms of transfer of oxygen, protons, and electrons to and from the carbon surface. The first two mechanisms have been clarified significantly over the past few decades; unfortunately, however, the transfer of electrons—the most relevant one for the applications of interest here—is not well understood; therefore, here we can offer more questions than answers. But based on a careful scrutiny of the literature, we can at least formulate some of the more plausible mechanisms as an attempt to guide future research in this area.

A special feature briefly explored here is a commentary on the literature from the former Soviet Union, which has long suffered from mutual disregard. When it comes to electrochemistry (to which, for example, Frumkin and his collaborators have contributed so much), it is obviously a great disadvantage to ignore it, as many of us still do; of particular interest here is whether the extensive reports on carbon electrochemistry in the Soviet journals have produced, or can yield, insights into carbon surface chemistry that heretofore have not been considered. As an illustration, the early study of Davtyan and Tkach [14] was published very soon after the seminal work of Walker and coworkers [15,16], in which the authors studied "the dependence of the number of active centers of carbon on the temperature of its activation" using both anodic and cathodic polarization of powdered sucrose-derived electrode material for which they assumed the zero charge potential to be 0.05 V. Their results are reproduced in Table 5.1, and they are basically in agreement with well-established facts

TABLE 5.1

Effect of Thermal Treatment of Carbon on Its Chemical Surface Properties as Determined by an Electrochemical Method[a]

Activation Temperature in CO_2 (°C)	Surface Area (m^2/g)	Number of Active Centers per g (anodic)	Number of Active Centers per g (cathodic)
600	974	7.56×10^{20}	6.58×10^{20}
800	1289	3.49×10^{20}	3.34×10^{20}
1100	976	1.83×10^{20}	1.87×10^{20}

Source: Davtyan, O.K. and Tkach, Y.A., *Soviet Electrochem.*, 1, 172, 1965.

[a] In 31% H_2SO_4, at a mean current density of 1.25×10^{-8} A/cm^2.

regarding (i) the effect of heat treatment on carbon active site concentration and (ii) the relationship of the active sites with the total number of surface sites (as determined, for example, by the BET surface area). The authors neither cite any non-Soviet literature, nor has any reference to this study been recorded by the Science Citation Index (SCI). Clearly, however, this is very much undeserved oblivion, because the conclusion that "it is possible to use the electrochemical method for determining the number of active centers" has its merits. In particular, any claims to an understanding of the relationship between surface chemistry and electrochemical properties must be substantiated by an explanation of these authors' findings reported in columns 3 and 4 of Table 5.1: "[t]he number[s] of active centers calculated using data from anodic polarization and from cathodic polarization are in fairly good agreement." Gas-phase oxygen chemisorption on carbon active sites is known to be irreversible and dissociative; the measurements reported here on the "active centers" are electrochemically reversible and "agree fairly well with data obtained by the method of paramagnetic resonance." Are the same active sites involved in the two processes? If not, why not?

And when it comes to "active sites," the issue of adsorption or ion exchange of anions on a positively charged carbon surface has been, and still is to some extent, highly controversial (see Sections 5.2.1 and 5.3.4). Thus, for example, in a paper that cites only three Soviet papers and has never been cited in the non-Soviet literature, Matskevich et al. [17] argued in favor of its treatment "from the electrochemical point of view." These authors did not identify the ion-exchangeable sites on which presumably up to 360 meq/g of I^- adsorbs on the surface of what they called a "positive oxygenous charcoal"—and, based on what we know today about carbon surface chemistry, it is highly unlikely that charcoal activated in CO_2 at 850°C–1000°C would contain any ion-exchangeable oxygen functionalities (see Section 5.2.1), although it would indeed be positively charged over a wide pH range—but clearly this is two orders of magnitude higher than the true (cat)ion exchange capacity of even the best carbons (e.g., those containing ca. 3 meq/g of carboxyl groups). The carbon surface (electro)chemistry literature is littered with such inconsistencies that have remained unaddressed for so many years.

5.2.1 SURFACE FUNCTIONALITIES

At the edges of graphene sheets in carbon materials, especially the ones with high surface area, oxygen-containing surface groups are ubiquitous, whether introduced deliberately (see Section 5.2.2) or accidentally (e.g., by simple exposure to ambient atmosphere). Functional groups containing other heteroatoms, such as sulfur or nitrogen, can be introduced as well, the latter being increasingly popular. The presence of hydrogen is most often the residue from the carbonization process, unless it is introduced in conjunction with the other heteroatoms (e.g., as COOH or NH_2 groups). These surface functionalities are responsible for much of the "chemical activity" of carbons (especially at low temperatures) because the sp^2-hybridized aromatic carbon atoms with a delocalized π-electron system, especially the ones in the basal plane (i.e., within a graphene sheet), are much less (re)active. They are also responsible for the development of surface charge. In this sense, carbons are unique (zwitterionic!?) amphoteric solids [8]: not only can they be acidic (e.g., by virtue of dissociation of carboxyl groups) or basic (see Section 5.2.3), but their surface charge (or electrophoretic mobility) vs. pH plots are not symmetric. The key parameter here is the point of zero charge, pH_{PZC}, with a net positive surface charge developing at $pH < pH_{PZC}$ and a net negative surface charge developing at $pH > pH_{PZC}$.

When it comes to basic carbons—i.e., those carbons exhibiting $pH_{PZC} > 7$ (in the older literature also called H-carbons, following the nomenclature introduced by Steenberg [18,19])—it is convenient to begin the discussion by recalling that a very concrete (and electrochemically relevant!) proposal has been formulated half a century ago by Garten and Weiss [20,21] to "account for the ability of an H-carbon electrode to catalyze the reduction of oxygen [in alkaline solution] at a sufficient rate for it to operate at reasonable current densities." It involves quinone functionalities and the "olefinic bonds" associated with them, which "add on oxygen to form a hydroperoxide or moloxide" and which in turn "accepts an electron from the cathode to form the monovalent perox-ide anion radical which may remain attached to the carbon by the ability of the quinone to form a semiquinone"; subsequently, the oxygen molecule "accepts a second electron and splits off as the peroxide anion." The mechanistic discussion of (the inconsistencies of) such a scheme is offered in Section 5.3.5. Here it is worth emphasizing that one portion of the authors' argument, regarding the quinone–hydroquinone character of activated carbon (AC), carbon black [21], and other carbon materials, has withstood the test of time remarkably well [19].

In addition to quinone groups, and to some extent to pyrones (which undoubtedly confer basicity to carbons), the possibility of existence of an oxygen-free, positively charged basic site on the carbon surface was acknowledged in the elegant early studies by Rivin [22]:

$$Ar + H^+ = ArH^+ \tag{5.4}$$

$$ArH^+ + O_2 = Ar^{2+} + HO_2^- \tag{5.5}$$

$$Ar^{2+} + 2HX = Ar^{2+}\}2X^- + 2H^+ \tag{5.6}$$

A zigzag carbene atom is a very promising candidate for this heretofore unidentified site. This is illustrated in Figure 5.2 (see also Section 5.2.3) and is discussed in more detail elsewhere [6].

5.2.2 SURFACE MODIFICATION AND FUNCTIONALIZATION

Modification of the surface chemistry of carbon materials is still a very active area of research, lately motivated in large measure by the opportunities offered by functionalized nanotubes and fullerenes. A voluminous literature is available on the use of HNO_3 or O_2 (air) to introduce oxygen surface functionalities, especially for ACs, carbon fibers, and carbon blacks. The former treatment introduces primarily carboxyl groups, which upon subsequent inert-gas treatment (most often by temperature-programmed desorption) yield CO_2; the latter introduces primarily the CO-yielding quinone (or phenolic) groups. The extent of their introduction (typically up to 10% by weight) and the distribution among the various groups depend very sensitively on the oxidation conditions used and the nature (i.e., reactivity) of the carbon. For example, in the case of carbon nanotubes (CNTs) (or vapor-grown carbon fibers?!), it was shown—as had been well established for ACs and carbon blacks—that the zeta potential (or electrophoretic mobility) vs. pH plots are shifted from positive to negative values upon acid treatment over increasing periods of time [23]. Other oxidizing agents commonly used are H_2O_2, O_3, as well as the more traditional ones such as $KMnO_4$ or $HClO_4$. Careful perusal of the literature, which is beyond the scope of this chapter, should make it possible

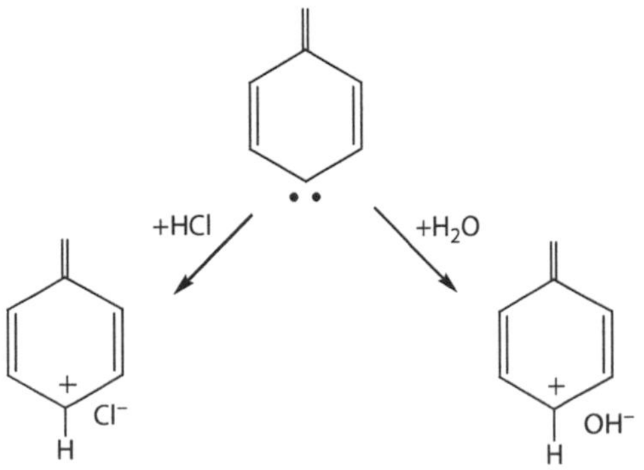

FIGURE 5.2 Chemical identification of an oxygen-free, positively charged basic site on the carbon surface. (See also Figure 5.4.)

to select suitable conditions for tailoring the carbon surface chemistry to specific needs of the user. Under more stringent conditions, these same oxidative treatments, instead of introducing surface oxygen, consume the more reactive carbon and this selective gasification is thus used to purify CNTs [24–33], not only single-walled (SWCNT) but also multi-walled (MWCNT).

Introduction of nitrogen functionalities is not nearly as well documented, but its effects are increasingly attracting the attention of carbon researchers. Thus, for example, a key study in this field is that of Pels et al. [34], with some 170 SCI citations to date, in which the various nitrogen functionalities present in carbons derived from both coal and model compounds have been analyzed in detail; one of its more recent citing papers is that of Holzinger et al. [35], with more than 100 citations of its own, and its "popularity" is undoubtedly due to the analogous use of the XPS technique to characterize nitrogen-containing surface groups in CNTs.

Incorporation of N into carbons is practiced either prior to or after pyrolysis of the carbon precursor. Typical postpyrolysis treatment is a reaction with ammonia at elevated temperature, e.g., as high as 800°C–900°C for an AC [36,37], or as low as 600°C for a carbon black [38,39]. A large variety of N-containing compounds has been used for cocarbonization with the carbon precursor, most often ammonia or urea derivatives [40].

A critical analysis of the currently available evidence for nitrogen functional group assignments, which has been accumulating at a fast pace over the last decade, is lacking. Figure 5.3 is a tentative summary of what are considered to be the most likely ones. Structures 1 (pyridine-like groups), 2 (pyrrole-type groups), and 4 (amine-type groups) are apparently not controversial, even though the number—and thus the relative importance—of five-member rings in any given flat graphene layer is uncertain. The contribution of the so-called "quaternary nitrogen" [41] is much more questionable, as recognized by the authors who originally made the relevant suggestion [34], and the term itself is arguably inappropriate when applied to carbon materials [34,42–45] instead of coals [41]. As emphasized by Kelemen et al. [41], the presence of structure 6 in Figure 5.3 (which indeed contains a quaternary N) becomes much less prevalent as the coal rank increases, as it is presumably converted to a pyridinic structure in chars and other flat-graphene-containing carbons. Furthermore,

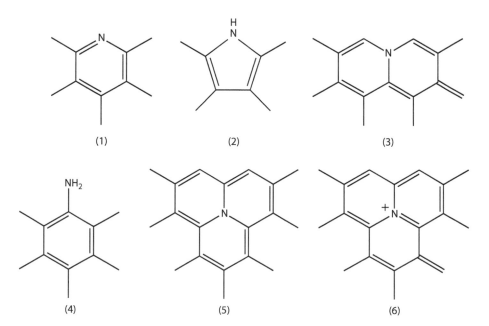

FIGURE 5.3 Examples of nitrogen-containing surface functional groups in carbons: (1) pyridine-type N, (2) pyrrole-type N, (3) basal-plane N (edge), (4) amine-type N, (5) substitutional N, and (6) quaternary N. (The top "layer" shown represents the graphene edge.)

in a highly delocalized graphene sheet, structures 3 and 5 are arguably more plausible (and much more straightforward!) than the cationic quaternary N. This is a substitutional N "functionality," not quaternary and not necessarily a surface group, and its feasibility remains to be demonstrated; this task will perhaps be easier for less ordered carbons (e.g., chars, ACs, and carbon blacks), because for the more ordered graphitic carbons, such a presence (in contrast with substitutional boron) has been explicitly ruled out [46].

A very popular recent research topic has been the functionalization of curved graphenes, in fullerenes and especially in CNTs [47–50]. Not only surface functional groups but also organic compounds and biomaterials have been used [49,51]; in the latter case, a more appropriate term is grafting, and the SCI identified 26 papers explicitly devoted to this procedure for nanotubes, all of them published in the last 3 years. The following two issues—one more fundamental and the other more utilitarian— merit our attention here: (i) Are only the graphene terminations (edges) functionalized, or is sidewall functionalization feasible as well? (ii) Which procedures are of special interest for electrochemical applications?

Banerjee and Wong [28] proposed a "rational sidewall functionalization and purification [procedure for] single-walled carbon nanotubes by solution-phase ozonolysis." They boldly assumed that ozone at subambient temperatures is capable of cleaving C–C bonds within graphene, as well as forming sidewall oxygen functionalities in addition to those at the open ends of SWCNTs [52]. The former is, of course, known not to be possible in flat sp^2-hybridized carbon structures, and it remains to be shown that basal-plane carbon reactivity increase—presumably due to strain introduced by curvature—is sufficient to account for this difference. Similar proposals were offered by other investigators as well (see, for example, Refs. [47,53]), but these authors also failed to address the contradiction mentioned above. Even oxycarbonyl nitrenes have been postulated to interact chemically with the curved graphene sheet, in this case through N bridge-bonding with two carbon atoms [35]. Until and unless such contradictions are resolved, a more reasonable assumption is that "sidewall" functionalization actually occurs at the ubiquitous sidewall defect sites, chemically equivalent to edge sites (see Section 5.2.3), and not at the trigonally bonded (albeit somewhat strained) basal plane sites. This issue has been explicitly addressed in the early study of Smalley and coworkers [54]: "In contrast to the open tube ends, the side-wall of the SWNT's, by virtue of their aromatic nature, possess a chemical stability akin to that of the basal plane of graphite... The chemistry available for modification of the nanotube side-wall (without disruption of the tubular structure) is thus significantly more restrictive." These authors concluded that "reaction temperatures in excess of 150°C are necessary to covalently add significant amounts of fluorine to the tube wall."

Direct proof of sidewall thiolation has been claimed recently [33], based on analyses of elemental maps in high-resolution TEM images. The authors also proposed that such sulfur functionalities replace the carboxyl and phenolic surface groups upon treatment of previously oxidized MWCNT with phosphorus pentasulfide, as follows:

$$MWCNT\text{-}COOH \xrightarrow{+P_4S_{10}} MWCNT\text{-}CSOH \tag{5.7}$$

$$MWCNT\text{-}OH \xrightarrow{+P_4S_{10}} MWCNT\text{-}SH \tag{5.8}$$

It would be important to confirm the feasibility of such interesting surface reactions, because the chemistry of sulfur heteroatoms in carbons has been neglected—relative to that of oxygen and nitrogen—since the pioneering studies of the groups of Boehm [55], Puri [56], and Walker [57], and also because of reports [58] that sulfur addition to carbon electrodes (e.g., in the form of C–S–C groups) may have interesting electrochemical effects.

In a study directly relevant to electrochemical applications of functionalized carbons, Kim et al. [59] reported a "drastic change of electric double layer capacitance by surface functionalization of carbon nanotubes." They subjected MWCNTs to H_2SO_4/HNO_3 in 3:1 ratio and noted that it "is well known that [such] treatment readily introduces the oxygen-containing functional groups, mainly carboxyl groups, on both the open ends and the sidewalls by disrupting graphene structure." The authors' conclusion was that the "introduction of surface carboxyl groups created a 3.2 times larger capacitance due to the increased hydrophilicity," but it was also proposed that the increased capacitance was due "in part to the redox reaction of the surface functional group itself" (see Section 5.3.3.2):

$$-COOH = -COO + H^+ + e^- \qquad (5.9)$$

$$>C-OH = >C=O + H^+ + e^- \qquad (5.10)$$

$$>C=O + e^- = >C-O^- \qquad (5.11)$$

The authors did not discuss these proposed proton and electron transfer reactions [60], but several of their features are noteworthy: (i) quite appropriately, there is a distinction between phenolic or quinone (>C) and carboxyl groups (–C), in the sense that carbon in the former is within the graphene's aromatic system; (ii) the reasons for the absence of charge on the deprotonated carboxyl group (Reaction 5.9) and the presence of charge on the quinone group (Reaction 5.11) require justification and are presumably amenable to experimental verification.

The effect of (presumably covalent) sidewall functionalization on the conductivity of SWCNTs was also reported recently in a theoretical study [61]. The authors found it important (and possible!?) to distinguish between "monovalent" and "divalent" sidewall additions, the latter with or without sidewall opening. They concluded that the "conductance of the monovalently functionalized tube decreases rapidly with the increasing addend concentration, and the nanotube loses its metallicity around 25% of modification," whereas "for the divalent case, the CNTs remain substantially conductive."

In a related study [62], similar effects on conductivity of SWCNTs were reported, but here a comparison was also made between the effects of nitric acid reflux and air plasma treatment, and an attempt was made to relate the changes observed "to the creation of defect sites." The authors did not offer a more concrete proposal regarding the nature of the sites involved in these treatments. After the acid treatment, Raman microscopy results indicated "a dramatic change in SWNT electronic structure," and both treatments enhanced "the electron transfer kinetics for the oxidation of inner-sphere dopamine." By contrast, both treatments had "a negligible effect on the voltammetric response of a simple outer-sphere electron-transfer redox process $Ru(NH_3)_6^{3+/2+}$."

The recent study by Lakshmi et al. [63] is just one example of the very extensive research efforts devoted to the improvement of performance of fuel cells by increasing the dispersion of the electrocatalyst on the carbon support by virtue of carbon surface functionalization [64]. Without acknowledging their familiarity with the most relevant prior studies, they did note that the "point of neutral charge evaluation helps in identifying the platinum complex to be used for electrocatalyst synthesis based on their charge," in the sense that, for example, "if the carbon surface is positively charged an anionic platinum complex is needed" (see Section 5.2.1).

5.2.3 Edge Carbon Atoms

Figure 5.4 summarizes in a schematic way the current state of knowledge of carbon surface chemistry, with an emphasis on the details of graphene edges. As discussed in Section 5.2.1, it shows the dominant surface functionalities, all containing oxygen because of the often inevitable contact of realistic carbon materials with O_2 from air. The main features of interest here are (i) the existence of *free* edge sites and (ii) the notion that the basal plane is not as chemically inert as is often

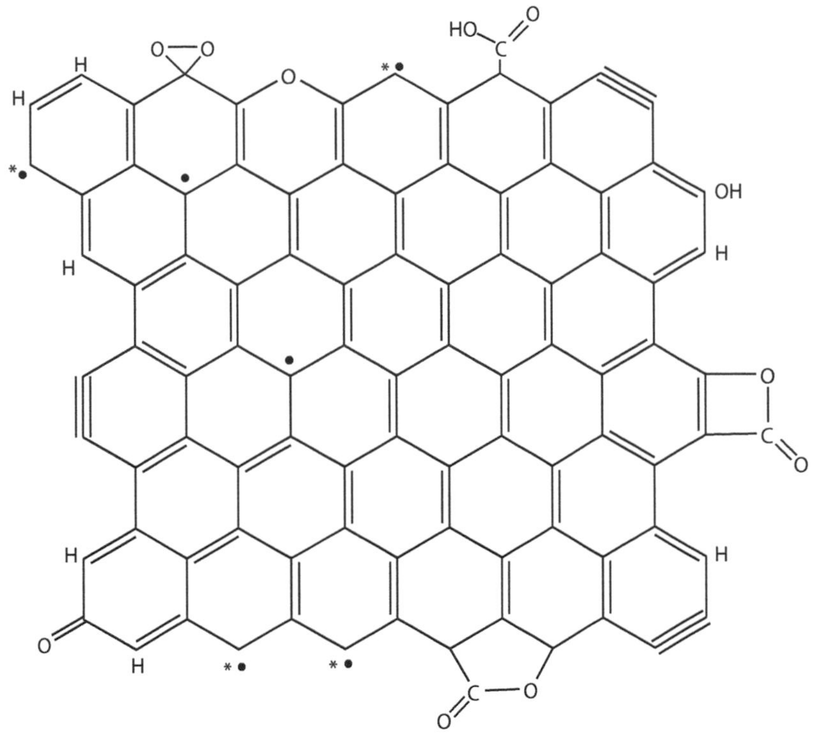

FIGURE 5.4 Schematic representation of the surface chemistry of a graphene layer, including carbene- and carbyne-type free edge sites, carboxyl-, lactone-, quinone-, pyrone-, and phenolic-type oxygen functionalities, delocalized unpaired π electrons, as well as nondissociatively adsorbed O_2 on the carbene-type sites.

envisaged, due to the presence of (delocalized) unpaired electrons, a common consequence of the presence of uneven number of C–H bonds in the condensed polyaromatic structure of graphene.

A detailed justification for the existence of carbene-like zigzag edge sites and carbyne-like armchair edge sites has been provided elsewhere [6]. This proposal is not only theoretically sound, as well as consistent with the key experimental observations regarding the behavior of carbon surfaces, but also provides a straightforward explanation for ferromagnetism in certain impurity-free carbons. It is, therefore, of interest to explore its potential usefulness in advancing our knowledge of electrochemical behavior of carbons.

5.3   IMPACT OF SURFACE CHEMISTRY ON ELECTROCHEMICAL BEHAVIOR

The electrochemistry of carbon surfaces has been a very popular research subject for many years, but from a fundamental viewpoint a very difficult one as well. It is an interdisciplinary topic *par excellence*. At a joint meeting of the electrochemical and carbon societies in 1983, Brodd [3,9] noted that the "interaction between electrochemists and carbon materials scientists has not been strong." In 1994, Leon y Leon and Radovic [4] concluded that the situation had not improved significantly, in spite of the resurgence of industrial interest and the then decade-old scientific stimulus offered by the Cleveland workshop [3]. The goal pursued in this section is to analyze especially the more recent scientific investigations in search of evidence that the gap is finally being bridged, now that the drive toward electric or hybrid vehicles has led to an explosive growth of directly relevant research; some of the important early studies, whose messages had not been given the deserved attention, are also analyzed briefly.

5.3.1 SCOPE

Carbon electrode science [2] does not have as long a history as the carbon electrode technology [65,66]. In a very early paper describing carbon electrodes, Beilby et al. [67] sounded surprised as they concluded that the "problem of the nature of carbon surfaces will not be so simple as the problem for a metal since the properties of carbon and the nature of the surface can vary with the mode of preparation." Almost half a century later, we are not surprised any more, but perhaps surprisingly we remain puzzled by many of the same issues.

The papers collected for analysis in this chapter can be classified, based on their titles, into the topics illustrated in Figure 5.5. Because of the explosive growth of carbon electrochemistry research, particularly for battery and fuel cell applications, and as mentioned earlier, the discussion is necessarily restrictive. It focuses primarily on carbon's EDL and on electrochemical capacitors [68], but the very much related topics of electrosorption and O_2 reduction are also reviewed, albeit to a more limited extent. Notable examples of topics that fall outside the scope of this chapter are the much investigated issue of the relationship between carbon's porous structure and charge storage capacity [69–71], the impact of surface chemistry on the electrochemistry of graphite intercalation compounds [72,73], the effects on Li insertion for battery applications [74–76], and the roles of both porosity and surface chemistry in the performance of carbons as electrocatalyst supports, including their corrosion in fuel cells, both the "old" ones using phosphoric acid [77] and the currently very popular ones using a polymer electrolyte membrane [78].

5.3.2 SUMMARY OF PREVIOUS REVIEWS

Both the mechanistic and the practical aspects of the use of carbons in electrochemical applications have been reviewed many times over the past five decades. As early as 1957, in a seminal

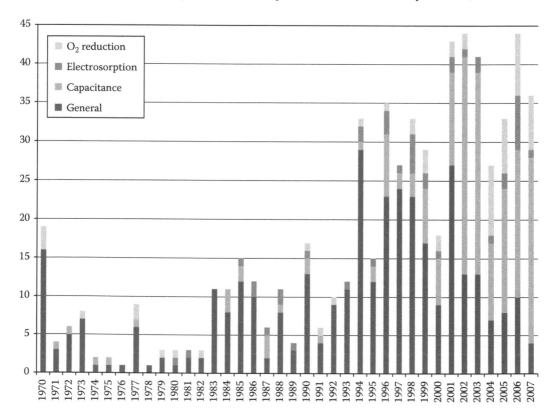

FIGURE 5.5 Distribution by topic of papers analyzed in the present review. (The year "1970" includes papers published during or before 1970.)

contribution whose many convincing messages have been either ignored or misunderstood for many years, Garten and Weiss [79] wrote with (excessive?) optimism that "sufficient data are now available to present a working hypothesis which accounts for and correlates the principal facts concerning the behaviour of carbon as an adsorbent and catalyst." They expressed hope that "manufacturers of activated carbon could, on the basis of what is now known, provide information concerning the nature and concentration of functional groups which would enable their products to be used in a less empirical manner." Well, 50 years later, this has not happened and in this chapter I explore the reason for such unfounded optimism. At the heart of it are concepts such as "hydrolytic adsorption" [18] and Frumkin's "electrochemical theory" of adsorption according to which "the adsorption of electrolytes is attributed to the ability of carbon to function as an oxygen and as a hydrogen electrode." Garten and Weiss had concluded that "the ability of carbon to function as a truly catalytic oxygen electrode is thoroughly established" but its functioning as a hydrogen electrode is much more controversial. Thus, for example, "[h]ydrogen gas was supposed to donate electrons to the carbon, charging it negatively, and this resulted in the consumption of alkali" [79]. Garten and Weiss correctly argued that this theory, while "plausible for platinized carbon," is not acceptable for "non-platinized carbons, which are inert toward gaseous hydrogen at room temperature," but pointed out that the "concept of electrochemically active hydrogen in carbon is not unreasonable in the light of present knowledge. At least some of the acidity of an L-carbon may be due to hydroquinone structures, the hydrogen of which might be regarded as being electrochemically active."

For the better or for worse, Frumkin's electrochemical theory of adsorption has played an important role in carbon electrochemistry research, especially in the prolific literature from the former Soviet Union. It can be summarized as follows [80,81]:

$$C_xO + 2H^+ + 2A^- \rightarrow C_x^{2+} + 2A^- + H_2O \qquad (5.12)$$

$$C_xO_2 + 2H^+ + 2A^- \rightarrow C_x^{2+} + 2A^- + H_2O_2 \qquad (5.13)$$

It is true, for example, that oxygen chemisorbs on the surface of the carbon electrode and is reduced to hydrogen peroxide by "consuming" two electrons (supplied, for example, by zinc in zinc-air batteries). It is in general not true, however, that "in oxygen atmosphere, carbon functions as an oxygen electrode, and its surface is positively charged" and that "[u]nder these conditions, the carbon sends hydroxyl ions into the solution and adsorbs anions" [82]. Garten and Weiss [79] discuss this in some detail. They first emphasize the finding [83] that "an H-carbon, outgassed in a high vacuum, does not adsorb acid from a deaerated solution unless molecular oxygen is admitted" and conclude that this "part of the theory ... is sound, being based on considerable experimental evidence, much of which has been confirmed by others." They do not provide any references for the latter statement, however. As noted above, they also reject Frumkin's claim that "hydrogen, chemisorbed by the unplatinized carbon, is 'electrochemically active'". Instead, Garten and Weiss [20,21,79] proposed the quinone/hydroquinone mechanism of carbon redox reactions, with "H-carbons" being characterized by the presence of quinone groups and "L-carbons" by the presence of phenolic (hydroquinone) groups.

Another pioneering study is the encyclopedic review by Randin [2] (see Figure 5.1) that understandably provided only a few insights. Besenhard and Fritz [84] reviewed the properties and behavior of carbon blacks, with "particular emphasis... on detailed results obtained from chemically modified carbon electrodes and also from surface oxides and fluorides"; a noteworthy conclusion was that the "electrochemistry of black carbons can, to a large extent, be generalized as the 'electrochemistry of polymers with conjugated π-systems,'" but they cautioned that the "incomparable stability of the graphite matrix has not been achieved using 'polyenes,'" and that "irreversible deterioration [has] in some cases been observed after only a few oxidation/reduction cycles."

The monograph by Kinoshita [1] (see Figure 5.1), with 244 references to the electrochemical behavior of carbons, is the most comprehensive review to date. Understandably, it contains few, if any, simple take-home messages and general conclusions. The author emphasizes "more diverse applications requiring high-surface-area carbon" and "the evolving technology of porous carbon electrodes," but an even greater emphasis is placed on research needs, echoing the conclusions and recommendations [9] of the landmark carbon electrochemistry workshop held in 1983 [3]. This transition in emphasis from nonporous, basal-plane-dominated carbon electrochemistry (e.g., in graphite) to porous, edge-plane-dominated electrochemistry (e.g., in ACs) did not receive its deserved scrutiny. A landmark study in this regard is that by Koresh and Soffer [85] whose citation history is reasonably distinguished but arguably quite superficial. Kinoshita's passing reference to it is exemplary in this regard: "[T]he double-layer capacitance of oxidized carbon increases at negative potentials... while at positive potentials... it decreases relative to that for the unoxidized sample" [1]. This statement falls short of summarizing the key arguments offered by Koresh and Soffer [85] from their comparison of cyclic voltammetry (CV) behavior of a "nonporous graphitized carbon" with that of an "ultramicroporous" carbon cloth: (a) that the "high surface ultramicropores take part in double layer charging"; (b) that "the higher capacitance at negative potentials is connected with chemisorbed oxygen"; (c) that "the very low capacity at positive potentials originates from repulsive interaction of the anions with the negative dipoles of the surface oxides which... is not compensated by the attractive positive electronic charge"; and (d) that "discrimination between faradic and double layer processes has to be made regarding the various surface groups existing on carbon." Even Conway [68], who did identify this work as "an important one," missed the opportunity to explain, for example, why "the surface dipoles of chemisorbed oxygen were supposed to lead to repulsive interactions with the electrolyte anions." These issues will be discussed in Section 5.3.3.

Tarasevich and Khrushcheva [86] reviewed the electrocatalytic properties of carbons and noted that "the electrochemistry of carbon materials is now in a period of intensive development"; while they emphasized applications, they concluded (rather optimistically?) that "carbon materials can be synthesized from organic raw materials specifically selected on the basis of their chemical structure in such a way to utilize those groups (elements, fragments) which lead to electrocatalytic properties." McCreery first reviewed the structural effects on electron transfer kinetics at carbon electrodes [87] and then presented a brief overview of electrochemical properties of carbon surfaces [88], whereas McCreery and Cline [89] also offered suggestions "for choosing and applying carbon electrodes for analytical applications"; a brief commentary on some of the fundamental insights and proposals offered here is presented in the subsequent sections. Leon y Leon and Radovic [4] made an initial attempt to unify the relevant concepts, but noted the urgent need for a "development of... an all-inclusive view of the chemistry and electrochemistry of carbon surfaces"; in this sense, the present review is an updated effort in the same direction. Dahn and coworkers [74,90] offered an authoritative view of the applications of carbon in lithium-ion batteries and concluded that "[c]arbon structure influences the electrochemical behavior dramatically" [90]; in particular, they noted that "[s]ubstitutional boron increases specific capacity but raises voltage" while "[n]itrogen additions decrease the average voltage, but induce large amounts of irreversible capacity," without offering convincing reasons for these fascinating effects. An ambitious review of the performance of "graphite, carbonaceous materials and organic solids [including fullerenes] as active electrodes in metal-free batteries" was also offered by Beck [91]; he emphasized the factors that influence the achievement of high energy density, i.e., high theoretical capacity in the normal voltage range from −3 to +2 V (vs. SHE), and concluded, for example, that positively and negatively charged carbon blacks (with a surface area of $1500\,m^2/g$) can achieve (at 1 V) 63.6 and 83.1 Ah/kg, respectively, as compared to 339 for a LiC_6 graphite intercalation compound. (The electrochemistry of fullerenes was also reviewed by Echegoyen et al. [92].) The intriguing review by Kavan [93] focuses on "electrochemical" carbons, defined by the author as " 'synthetic solids consisting mainly of atoms of the element carbon,' which can be prepared electrochemically from suitable precursors"; while its scope transcends the (electrochemical) carbonization of fluoropolymers and thus includes carbon precursors such as C_2H_2, CO_2, and pyridine, the author acknowledges that this is "primarily a subject of academic interest" although he does anticipate that "[e]lectrochemical carbonization of organic precursors is a useful strategy in molecular engineering of all-carbon networks."

An entire chapter in Conway's monograph [68] is devoted to "the double layer and surface functionalities at carbon," and is thus of special interest; a brief commentary is offered here, and a more detailed one is in preparation.

Fialkov [94] has discussed "carbon application in chemical power sources," emphasizing the "influence exerted by a three-dimensional ordering of the structure, dispersion degree, surface properties, and porous structure of carbon particles on the current-producing reactions." His analysis of the effect of surface properties on the behavior of carbons in the oxygen reduction reaction is a typical example of the huge gap that still separates surface electrochemistry from surface chemistry researchers (see Sections 5.3.3.2 and 5.3.5): his proposed mechanistic scheme "suggests that the active centers on activated carbon are carbon atoms directly bound to the base functional groups," but an inquisitive reader is left to wonder whether and how the proposed "base groups" are compatible with the well-known surface functionalities depicted in Figure 5.4, and what evidence, if any, is available for such a proposal.

The brief and authoritative review by Frackowiak and Beguin [60], as well as a more recent one [95], focused on the electrochemical storage of energy in capacitors (see Section 5.3.3). Lukaszewicz [96] has recently reviewed the use of carbon materials ("including graphite, polycrystalline carbon, carbon black, and carbon nanotubes") as chemical sensors in an attempt to "show how the chemical properties of the carbon surface can be controlled, and how the properties influence possible application of carbons to the construction of chemical sensors." It is symptomatic that most of his references for the invoked structure of surface functional groups are more than 30 years old, while "graphene bonded to different oxygen containing functional groups" is misrepresented. It is not surprising that, despite repeated claims regarding the relationship between carbon surface chemistry and (electro)chemical behavior, no clear or consistent correlations are to be found here.

One section of the recent review by Béguin et al. [97] is devoted to surface functionality and chemical and electrochemical reactivity of carbons. Two of the intriguing arguments propagated here are whether the sidewall carbon atoms in a perfect nanotube are really susceptible to attack by an oxidizing agent such as nitric acid (see Section 5.2.2), with the consequent formation of carboxyl functionalities, and whether quinone groups can be considered to have a basic character in the context of the development of positive surface charge; be that as it may, and perhaps of greater relevance to the present discussion, these authors emphasized the fact that irreversible capacity is directly related to carbon surface chemistry in the form of active surface area, which—according to the conventional [98] and arguably outdated [6] view—corresponds to the "[cumulative] surface area of the different types of defects present on the carbon surface (stacking faults, single and multiple vacancies, dislocations)." Intriguingly, however, they noted that ball-milled SWNTs achieve "[s]ignificant values of reversible capacity," of the order of 1000 mAh/g [99]; it is a well-known fact that ball-milling induces "disorder within the bundles" [99], and it also "fractured the nanotubes" [99], but the authors did not comment on the apparent contradiction in this argument. The authors of the cited reference [99] did address the apparent inconsistency, by first echoing [100] the controversial argument that in the case of "ball-milled sugar-carbons… the excess reversible capacity [might arise] from lithium reaction with carbon radicals at the edges of the fractured graphene sheets and with H- and O-containing surface functional groups." But they then dismissed such an interpretation for the nanotubes—because "the oxygen concentration remained at ~2 at.% level after ball-milling, indicating that H- and O-containing functional groups do not play important roles"—in favor of an arguably vague "enhanced… degree of disorder within the bundle" and "lithium diffusion into the inner cores of the fractured nanotubes," acknowledging at the same time that the "exact mechanism for the enhanced Li capacity in ball-milled SWNTs is not clear."

Similar issues were raised by Béguin et al. [97] regarding the capacitance properties of carbons in general and nanotubes in particular, because "the higher the… oxygen content of the nanotubes, the higher the capacitance value," whereas "after a few cycles, the capacitance of nanotubular materials rich in surface groups noticeably decreases"; the authors did not discuss the reasons for such behavior.

Finally, the most recent review by Lee and Pyun [101] discusses the "synthesis and characterization of nanoporous carbon and its electrochemical application to electrode material for supercapacitors." Among the topics covered are the effects of "geometric heterogeneity" and "surface inhomogeneity" on ion penetration into the pores during double layer charging or discharging. The latter issue is discussed primarily in terms of the presence of surface acidic functional groups (SAFGs) that "significantly influence the electrochemical performance of the porous carbon electrodes," e.g., by contributing to "the increase in the total capacitance of EDLC" or by improving electrolyte wettability, but also by increasing the leakage current.

What emerges from the analysis of previous reviews of this eminently interdisciplinary topic is, on the one hand, an increasing awareness that the surface chemistry of carbons and their electrochemical properties and behavior need to be reconciled; but, on the other hand, it is clear that there are many difficulties along the way, and one often senses the temptation to "sweep under the rug" the controversies and the inconsistencies. For example, the goal of McCreery's contribution [87], surely among the most enlightened within the vast literature available, has been nothing less than "the correlation of surface structure with electrochemical behavior." In particular, he discussed the origins of voltammetric background current and its interpretation in terms of surface structure and offered a "working hypothesis" for the relationships between surface structure, on one hand, and electron transfer reactivity, capacitance, and adsorption properties of carbon electrodes on the other; these will be subjected to some scrutiny in Section 5.3.3, and the residual controversies or inconsistencies will be explicitly acknowledged and indeed, emphasized.

5.3.3 CHARGE STORAGE

Recent interest in this topic [102] has been tremendous, spurred not only by the opportunity (and indeed desperate need!) to further enhance the performance of batteries but, especially so, to develop novel supercapacitors [68]. Among the 41 papers published only in 2007 (through October)—a remarkable number, indeed—and identified as directly relevant to this section of the chapter, 24 were devoted primarily to carbon capacitance issues [71,95,103–124], 7 to redox behavior [125–131], 6 to electrosorption [132–137], and the others to more general electrochemical properties and behavior.

One practical and one fundamental question are of interest here: (a) How much charge can be stored in carbons? (b) How does the amount of charge stored depend on the nature of the carbon and thus on its surface chemistry? Their answer(s) should lead to the resolution of an apparent contradiction that is implicit in the following statements from recent authoritative reviews: "[T]he preferred carbon materials [for electrochemical capacitors] should be free from... surface quinonoid structures that can set up self-discharge processes that must be minimized" [68]; "[s]ubstitutional heteroatoms in the carbon network (nitrogen, oxygen) are a promising way to enhance the capacitance" [95].

The pioneering study by Fabish and Hair [138] emphasized the fact that the surface potential "reflects the charge state and electronic structure of the surface region" of carbon materials. Indeed, the connection between electrochemistry and solid-state physics is a laudable example of a mature interdisciplinary subject [139], thanks to the rapid development of the semiconductor industry in the second half of the twentieth century. On the other hand, it has been known at least since the early twentieth century [140–142] that the pH of a carbon suspension reflects carbon's proton transfer characteristics. Therefore, the connection between proton and electron transfer can and indeed should be a good "starting point in the molecular interpretation" [138] of chemical and electrochemical properties and behavior of carbon materials. In this section, the causes and effects of the existence of the surface potential (or work function) of carbons are briefly summarized and illustrated in Figure 5.6. Here both the electrochemical and the solid-state physics approach should be presented; in the former, the interfacial (electrode/electrolyte) double layer is of primary interest (C_{dl}), while in the latter there is also the "space charge" region within the solid (C_{sc}), which typically extends 10–1000 nm away from the surface:

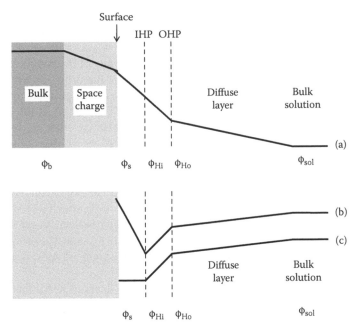

FIGURE 5.6 Schematic representation of the interfacial potential gradients established in the vicinity of the carbon surface. Cases (a)–(c) depend on the relative contributions of proton and electron transfer. (From Duval, J., et al., *Langmuir*, 17, 7573, 2001.)

$$\frac{1}{C} = \frac{1}{C_{sc}} + \frac{1}{C_{dl}} \tag{5.14}$$

Which of these capacities represents the charge-transfer rate-determining step [143], and when is an important issue [144] that deserves much more attention than it is receiving in the recent literature.

A capacitance of ca. 100 F/g develops as a consequence of simple immersion of an AC in an electrolyte and the formation of an electrochemical double layer. In such a case, what exactly does "charging" mean? Converted to millimoles of charge carriers per gram (say, at 2 V in an organic electrolyte), this is of the same order of magnitude as the concentration of oxygen functional groups in an AC (ca. 1 mmol/g). Is that a coincidence? Soffer and Folman [145] make this conversion (using 10 μF/cm² and 500 m²/g), but they do not use it to draw inferences about the origin of the capacitance. Converted to a number of charge carriers, it is ca. 10^{21}/g, which is of the same order of magnitude as the highest spin concentration values measured for ACs. Again, a coincidence? From semiconductor theory, the space charge capacity of graphite [143]

$$C_{sc} = \left(2\varepsilon\varepsilon_0 e^2 d/kT\right)^{0.5} \cosh(\phi_s/2kT) \tag{5.15}$$

can be estimated. Its minimum value (at zero electrode surface potential ϕ_s), using 6×10^{18} carriers/cm³ as the charge density d and $\varepsilon = 3$ as the dielectric constant, is ca. 4.5 μF/cm². The ranges of values that these parameters can take as a function of carbon surface chemistry also deserve more scrutiny than they typically get in the prolific yet arguably superficial recent literature.

A study that illustrates some of these issues most dramatically is that of Koresh and Soffer [85], already mentioned above (see Section 5.3.2). It is instructive to analyze what the literature says about this key aspect of the electrochemical behavior of carbons possessing different surface chemical properties. So far, this paper has been cited close to 50 times in SCI publications, less than its messages deserve. For example, even though their paper deals specifically with the influence of surface oxides on the electrochemical (impedance) behavior of AC fabrics, Nian and Teng [146] cite the paper of Koresh and Soffer only in the Introduction and in support (or illustration) of a statement regarding the well-known formation of oxygen functional groups as a consequence of carbon treatment with oxidizing gases. But, when they discuss the low-scan-rate (3 mV/s) potential sweep CV results for their microporous samples, which do not show any of the effects reported by Koresh and Soffer, they do not take the opportunity to point out this important difference; instead, they simply state that "[o]xygen enhanced the capacitance" even though there was no correlation between the oxygen content and the capacitance. Even less convincing is the statement that the presence of "CO-desorbing complexes would enhance the affinity of the carbon surface toward protons, leading to an excess double layer capacitance"; first, the content of such groups is not the highest for the highest-capacity sample, and also the Faradaic quinone/hydroquinone contribution to the capacitance should be important only in a narrow voltage range. In an attempt to address this inconsistency, the authors postulated both a "negative role" of CO_2-complexes (e.g., carboxyl and lactone groups) and an "aggregation of ionic species or water molecules to hinder the migration of electrolytes in micropores" of highly oxidized electrodes. No (convincing) evidence was provided for these postulates, however. Lozano-Castelló et al. [147] also analyzed the influence of surface chemistry on EDL capacitance of ACs, but they did not cite the Koresh and Soffer paper.

The performance of carbon as a supercapacitor is thus deemed to be the most appropriate topic for an attempt to clarify the connection between carbon surface chemistry and carbon electrochemistry. In the ideal case, it does not involve electron transfer, just the accumulation of charge. In that sense, it is perhaps a necessary stepping stone toward a more fundamental investigation of carbon-based batteries (where electron transfer redox reactions must be understood as well) and fuel cells (where carbon's multiple roles in electrocatalysis—as support, catalyst and conductor—must be clarified as well). The key point to be analyzed is the elusive link between the electrochemical surface potential (e.g., at zero charge) and the electrokinetic point of zero charge (or isoelectric point) of carbon materials. But a major obstacle is the following: much of the early literature on these issues—and a disconcertingly large fraction of the current literature—does not even recognize, and certainly does not accommodate, the now well established fact that the carbon surface develops a charge as a function of pH. Thus, for example, Koresh and Soffer [85] state the following: "Highly oxidized ultramicroporous carbons show much lower double layer capacitance at positive potentials than nonporous carbon surfaces. This is interpreted as repulsive ion-dipole interaction of the anion with the dipole of the chemisorbed oxygen." Such considerations in turn raise the question about the difference in double layer capacitance of carbons before and after charging: what exactly the charging does that prior immersion into an electrolyte has not done? In a recent study, Yang et al. [148] do not help to clarify this issue when they say that "a strong electrical double layer is formed" on the surface of ACs "when an electric field is applied," as if such a layer did not exist in the absence of an external field. They go on to present an EDL model of electrosorption of ions from aqueous solution, where the basic concept is presumably that "charged ions [are forced] to move toward electrodes of opposite charge by imposing an electric field," but they do not take into account the existence of both positive and negative sites on the carbon surface and their dependence on the pH of the solution; instead, they emphasize the importance of the pore size distribution of the carbon adsorbent. They cite the Koresh and Soffer paper [85] for the arguably less important observation that "at positive potential the highly oxidized microporous carbon shows much lower capacity than nonporous carbon" and state that "[t]his effect further prevents ions from entering the micropores inside the porous electrode."

In his recent authoritative monograph on electrochemical capacitors [68], Conway indeed identifies the Koresh and Soffer paper as "an important one," and highlights the following results: (a) the "conductivity of the solution in pores less than 0.7 nm in diameter was found to be several orders of magnitude lower than in a supporting 0.1 M NaCl freely in contact with the outer interface of a porous carbon electrode"; (b) "[h]ighly oxidized, ultramicroporous carbon materials show substantially lower double-layer capacitance at positive polarizations than nonporous carbon surfaces"; (c) "ions can penetrate the smallest pores after depletion of their hydration shells"; (d) "the double-layer charging rate in the microporous carbon is much lower if the NaCl is replaced by 0.1 M LiCl"; and (e) the "surface dipoles of chemisorbed oxygen were supposed to lead to repulsive interactions with the electrolyte anions." He did not elaborate, however, why exactly these findings are important. Indeed, the entire chapter entitled "The Double Layer and Surface Functionalities at Carbon" is arguably woefully outdated and oblivious to some of the key issues in carbon surface (electro)chemistry. Nevertheless, it is instructive to scrutinize in more detail Conway's otherwise insightful writings on these as well as related general issues (see, for example, electrochem.cwru.edu/ed/encycl/art-c03-elchem-cap.htm, accessed 11/02/07); the most relevant ones are reproduced below, so that their applicability to carbon electrodes can be elaborated.

To clarify the analogy between electrochemical and conventional capacitors, the key point is the role of the dielectric in the former kind. Here is what Conway says in the Electrochemistry Encyclopedia (see Figure 5.6):

> Helmholtz envisaged a capacitor-like separation of anionic and cationic charges across the interface of colloidal particles with an electrolyte. For electrode interfaces with an electrolyte solution, this concept was extended to model the separation of 'electronic' charges residing at the electrode surfaces (manifested as an excess of negative charge densities under negative polarization with respect to electrolyte solution or as a deficiency of electron charge density under positive polarization), depending in each case on the corresponding potential difference between the electrode and the solution boundary at the electrode.

> For zero net charge, the corresponding potential is referred to as the 'potential of zero charge'... In response to positive or negative electric polarization of the electrode relative accumulations of cations or anions develop, respectively, at the solution side of the charged electrode. If, for energetic reasons, the ions of the electrolyte are not faradaically dischargeable [that is, no electron transfer can occur across the interface ('ideally polarizable electrode')], then an electrostatic electrical equilibrium is established at the interface, resulting in a 'double layer' of separated charges (electrons or electron deficiency at the metal side and cations or anions at the solution side of the interface boundary), negative and positive, across the interface.

In Conway's view, "it is the solvent of the electrolyte solution that constitutes locally the dielectric of the double layer," and not the electrode/electrolyte interface. He goes on to explain that the "double layer at an electrode/solution interface consists of one real, electronically conducting plate (metal, semiconductor, oxide, or carbon surface) and a second virtual plate that is the inner interfacial limit of a conducting electrolyte solution phase." He explains that "a relatively large specific capacitance of 20–50 μF/cm² can arise" because the "double layer distribution of charges is established across this interphasial region, which is composed of a compact layer having dimensions of about 0.5–0.6 nm, corresponding to the diameters of the solvent molecules and ions that occupy it, and a wider region of thermally distributed ions over 1–100 nm, depending on ionic concentration."

An apparently unresolved issue from the above analysis is the breakdown of the analogy, say, between carbon and metal electrodes: for the latter (e.g., $Zn|Zn^{2+}$), it is easy to define an electrolytic "solution pressure" and an "osmotic pressure," as the pioneers of electrochemistry (Nernst, Ostwald, Arrhenius, and Le Blanc) have done (see books.google.com) more than a century ago; but I am aware of no such discussion (e.g., $C|C^+$) on carbons in which the exact nature of C^+ would have been addressed in those terms.

As discussed above, and in the absence of such detailed scrutiny, it is instructive to estimate the expected density of charges. Here is what Conway says in Chapter 6 of his monograph [68]:

The range of values of accumulated electronic charges on the metal side of interfaces of double layers at electrodes in aqueous solutions extends from about $+30\,mC/cm^2$ at potentials positive to the pzc, to about $-20\,mC/cm^2$ at corresponding negative potentials... For the above figures, it is of interest to calculate the range of excess charge per atom of an assumed planar interface. We can deduce that the number n of atoms having a diameter of 0.3 nm in, say, a square array [i.e., in a (100) lattice configuration of particles in the surface of the electrode], is $n = 1/(3 \times 10^{-8})^2 = 1.1 \times 10^{15}$. Per atom, the excess charge densities in C/cm^2 correspond then to $(+30 \times 10^{-6})$ $(6.022 \times 10^{23})/(1.1 \times 10^{15})/96500 = +0.17$ electrons or $(-0.11$ electrons)... In an electrochemical capacitor, the respective plus and minus charge densities on the two plates are matched by net equal and opposite accumulations of respective negative (anion) and positive (cation) charge densities in the interphasial regions of the solution... It is of interest chemically that the above extents of controllable variation of surface electron excess or deficiency are comparable with the local changes of electronic charge density that can arise chemically on various conjugated aromatic ring structures such as benzene or naphthalene, owing to the presence of substituents such as $-OH$, $-SO_3H-$, $-CH_3$, $-NH_2$, and $-COOH$, where charge density changes arise on account of electronic inductive and resonance effects.

The intrigued reader is left with the essential task—and arguably a rather difficult but very rewarding one—to verify this vague and imprecise reference to carbon surface chemistry.

A related analysis follows from a more general consideration of the semiconductor–electrolyte interphase [149] (see Equation 5.14). Because of the usually much larger double-layer capacitance ($C_{dl} > 20\,\mu F/cm^2$), the overall capacitance is typically of the same order of magnitude as the space charge capacitance ($C_{sc} < 1\,\mu F/cm^2$). In ACs, the latter is expected to be much larger because of the contribution from surface states, and both contributions are expected to be important.

It is intriguing that the most recent review of the potential of zero charge (PZC) [150], like its predecessors (see references therein), contains no reference to carbon, even though carbon electrodes are ubiquitous in so many electrochemical applications. In studies with graphite electrodes immersed in aqueous electrolytes, the double layer capacity at the zero point of charge (zero potential) was found to be quite low, $3–4\,\mu F/cm^2$ [143,151]. In the pioneering work of Randin and Yeager [151–154], the properties and behavior of nonporous carbons were studied. It is instructive to summarize the main findings in these studies and to analyze briefly their citation history, which is illustrated in Figure 5.7.

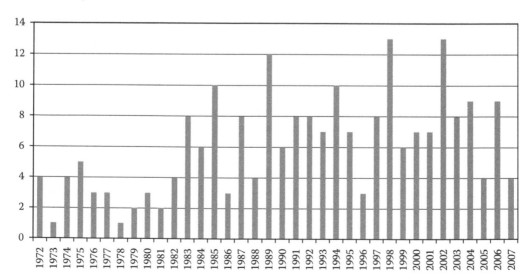

FIGURE 5.7   SCI citation statistics (Web of Science) for the four pioneering papers by Randin and Yeager [151–154]. Updated December 2007.

Using stress-annealed pyrolytic graphite (SAPG), Randin and Yeager [152,154] showed that the edge plane has "at least one order of magnitude" higher capacitance than the basal plane (e.g., 50–70 vs. ca. $3\,\mu F/cm^2$) [151]. They also made an attempt [154] "to establish the extent to which the surface chemistry influences the capacity." Rather than attributing the significant differences (e.g., pH dependence of the electrode potential) to "the presence of surface microfissures," both at the edges of SAPG and in a glassy carbon (GC), they concluded that "the more likely explanation for [the higher capacitance] is slow surface states [associated with surface functional groups] involving changes in the oxidation state of the electrode surface." Understandably, at the time they could not say much more about these "surface states," but they did speculate that the "quinone-hydroquinone surface group is a likely expectation," echoing the proposals of Garten and Weiss [79].

In an attempt to rationalize the measured capacitance values, and especially the low value for the basal plane (ca. $3\,\mu F/cm^2$), these authors first concluded that space charge within the electrode is the dominant contribution (rather than the compact double layer with ca. 15–20 $\mu F/cm^2$, or the diffuse double layer with $>100\,\mu F/cm^2$). They then applied the theory of semiconductor electrodes to confirm this and obtained a good agreement by assuming for SAPG a charge carrier density of $6 \times 10^{18}/cm^3$ and a dielectric constant of 3; for GC, they obtained $13\,\mu F/cm^2$ with the same dielectric constant and 10^{19} carriers per cubic centimeter.

Finally, Randin and Yeager [153] also analyzed the effect of boron addition (at $0.3\,w\%$) on the differential capacitance of SAPG, in order to "test the explanation... concerning the space charge contribution to the capacitance of the graphite/aqueous electrolyte interface," because boron doping causes an increase in the charge carrier concentration and "a change from an intrinsic to p-type semiconductor." For "boronated graphite the potential of the minimum in the capacity-potential curve shifts negative by about $0.5\,V$ relative to that for the non-boronated material," which they concluded to be "in agreement with the semiconductor interpretation." But a similar calculation to the one outlined above indicated that "the Helmholtz capacitance... makes a significant contribution to the measured values" and suggested that "[b]oronation apparently introduces surface chemical groups which contribute to the observed capacity"; here the authors did not venture to speculate on the necessary mechanism or the nature of such groups.

So, 30 years later, what is "posterity's verdict" about these pioneering studies and their reasonably committing conclusions? In particular, is the behavior of the basal plane of SAPG really "abnormal" [151] and have we taken seriously the recommendation that "the space charge characteristics of graphite must be taken into account in examining the electrochemical properties of this material" [151]? And, of course, to what extent do all these apply to nongraphitic and high-surface-area carbon supercapacitors? To facilitate the discussion, three interdependent issues are highlighted in Sections 5.3.3.1 through 5.3.3.3. Before proceeding, however, it should be noted that a large number of studies published over the past decade are devoted primarily to the role of surface area, porosity, and pore size distribution [155–170], or to composite (hybrid) capacitors [170–182]; their analysis is beyond the scope of this chapter.

5.3.3.1 Influence of pH

The pH effect is arguably a very complicated one, even though it has been studied for semiconductor–electrolyte interfaces as early as the 1960s. The origin of the complexity is the fact that accumulation of charge is both the cause and a consequence of proton transfer in the electrical double layer.

According to the Nernst equation, the theory predicts $59\,mV/pH$ units for the flatband potential of an oxide semiconductor. However, Gerischer [183] has analyzed both upward and downward deviations of this slope in the potential vs. pH plots. Thus, for example, he argued that values $<59\,mV/pH$ are to be expected with "decreasing surface charge in adsorbed ions or ionized surface groups," because "the chemical activity will change considerably with the amount of charge on the surface."

It is necessary to analyze first the origin of surface charge as suggested (or revealed?) by pH measurements (see also Figure 5.14). When an "acidic" (or "basic") carbon is immersed in water, its low (or high) isoelectric point (pH$_{IEP}$, or point of zero charge, pH$_{PZC}$), as measured by electrophoresis (or titration) in the absence of specific adsorption, is a consequence of the following phenomena: (a) at low pH (pH < pH$_{IEP}$ = pH$_{PZC}$), the surface is positively charged, OH$^-$ ions are preferentially adsorbed, and there is thus an excess of H$^+$ ions in solution; (b) at high pH (pH > pH$_{IEP}$ = pH$_{PZC}$), the surface is negatively charged, H$^+$ ions are preferentially adsorbed, and there is thus an excess of OH$^-$ ions in solution; (c) adsorption of H$^+$ ions is a consequence of their electrostatic attraction by the dissociated surface OH groups (e.g., carboxylic or phenolic) at the edges of graphene layers; (d) adsorption of OH$^-$ ions is a consequence of the formation of C$_\pi$–H$_3$O$^+$ donor–acceptor sites on the (basal-plane) surface of the graphene layer, as well as of the presence of proton-attracting carbene sites at zigzag graphene edges or of positively charged pyrone sites. The extent to which the recent literature reflects, confirms, or casts doubts on these straightforward facts (or postulates?!) is analyzed below.

Swiatkowski et al. [184] measured the "capacitance current at potentials where no faradic reactions occur" (from −0.2 to 0.9 V) for a series of carefully selected ACs whose suspension pH varied monotonically from 3.0 to 10.2. As is too often the case, they also cite (see Section 5.3.3) the Koresh and Soffer paper [85] only in the Introduction, to support the observation that the "studies of various carbon electrodes have not shown clear evidence to indicate an enhancement or increase in the double layer capacitance resulting from the capacitive contribution of surface oxide groups." But they do arrive at the very important and intriguingly unappreciated conclusion that "the double-layer capacitance of carbon electrodes depends both on the thermal history (surface chemical structure) of the carbon material and, in an aqueous environment, on the pH of the bulk electrolyte solution." Thus, for example, they show that the dependence of the capacitance on the pH is significant at low sweep rates (<0.01 V/s), at which, according to Koresh and Soffer, the contribution of "slowly charging sites" is the greatest. By extrapolating the capacitance results to quasiequilibrium conditions, they then show that the highest capacitances are observed in a strongly acidic environment, which they interpret by emphasizing that "surface oxygen groups are non-ionised" and speculating that "protons can adsorb to a carbon electrode surface by hydrogen bonds." Their argument regarding the low capacitances in strongly basic solutions, "the almost complete ionisation of acidic functional groups and the very low proton concentration," while obviously correct, is not convincing either, because they do not elaborate how exactly these facts are "responsible for the small double-layer capacity observed." Clearly, the very interesting Figure 6 in their paper (not reproduced here) requires a more detailed discussion. For example, at pH = 1.21, the high capacitance may be a consequence of the fact that the double layer is formed by the positive surface sites on the entire carbon surface—i.e., basal plane sites [7,185,186]—and the correspondingly high concentration of counteranions. At the other, low-capacitance extreme, at pH > 10.2, some surface sites (typically not more than ca. 1 meq/g) are negatively charged, as a consequence of the dissociation of carboxyl and phenolic groups, and double layer formation should then be attributed to the accumulation of electrolyte cations at these sites. Indeed, this is a well-known effect in liquid-phase adsorption on carbon surfaces: the adsorption capacity for anions is much greater than the adsorption capacity for cations [7].

A persistent question regarding carbon capacitance is related to the relative contributions of Faradaic ("pseudocapacitance") and non-Faradaic (i.e., double-layer) processes [85,87,95,187]. A practical issue that may help resolve the uncertainties regarding DL- and pseudo-capacitance is the relationship between the PZC (or the point of zero potential) [150] and the point of zero charge (or isoelectric point) of carbons [4]. The former corresponds to the electrode potential at which the surface charge density is zero. The latter is the pH value for which the zeta potential (or electrophoretic mobility) and the net surface charge is zero. At a more fundamental level (see Figure 5.6), the discussion here focuses on the coupling of an externally imposed double layer (an electrically polarized interface) and a double layer formed spontaneously by preferential adsorption/desorption of ions (an electrically relaxed interface). This issue has been discussed extensively (and authoritatively!) by Lyklema and coworkers [188–191] for "amphifunctionally" electrified interfaces [188]. In a particularly illuminating way, Duval et al. [188] tackled head on the difference between what they call the "pristine point of zero charge," which in our terminology is pH_{IEP} or pH_{PZC}, and the "point of zero potential," which, in the terminology adopted in this review, is the PZC; the main distinction for them is that the former arises mainly on a "relaxed" oxide/electrolyte surface, whereas the latter is a characteristic of a "polarized" interface. In a follow-up paper, the authors emphasize [189] that "the way in which the simultaneous functioning of two disparate [double layer] charging mechanisms affects specific adsorption has not yet been addressed" and that "[u]nderstanding the relevant charging mechanisms provides a new route in describing ion adsorption as a function of applied potential and pH." The key practical challenge is to understand the potential vs. pH diagrams. Two such diagrams are reproduced here. In Figure 5.8a, the DL potential (ψ^d) is shown as a function of both externally applied potentials $\Delta\varphi$ and pH: at $pH < pH_{IEP}$ (with a consequent excess of surface positive charge), to achieve the same value of ψ^d, a more negative external potential must be applied than at $pH > pH_{IEP}$. In Figure 5.8b the potential necessary to apply across the interface to reach the isoelectric point is shown for more and less acidic surface: for $pH < pH_{IEP}$, $\Delta\varphi_{IEP} < 0$, and for $pH > pH_{IEP}$, $\Delta\varphi_{IEP} > 0$. Although such theoretical analyses have been performed and limited comparisons have been made between theory and experiment for some metals or oxides, such comparisons have yet to be performed for carbon surfaces, which are "amphifunctional" *par excellence*.

In an earlier work, Jankowska et al. [192] addressed this issue by showing several linear potential vs. pH plots in the pH range from 2.0 to 7.0 with the expected negative slopes for both an AC and a carbon black that "behave alike despite the significant differences in their crystal and pore structure." The measured potentials were in the 0–600 mV range, and yet they suggested that the "most probable" explanation is the "process of electroreduction of [adsorbed] oxygen." They noted that "only in some cases" they obtained the theoretically expected slope of 0.059 V, and they attributed this discrepancy to "a parallel reaction of oxygen electroreduction for which the tangent of the function $E = f(pH)$ is 0.0285 V/pH, and by the dependence of ion adsorption at the carbon surface on pH."

The effect of pH on the CV plots is also of interest. McCreery [87] has discussed this in some detail, and has emphasized both the high sensitivity of this technique in studies of redox processes (e.g., ca. 10^{-12} A of oxidation current produced by 10^{-17} mol of reduced species oxidizing per second) and the opportunities that background current provides for revealing important information about carbon surface structure (e.g., relative importance of basal and edge planes). He identified "at least two and possibly three mechanisms responsible for voltammetric background current at carbon electrodes": (i) double layer capacitance, (ii) "pseudocapacitance" due to potential- and pH-dependent surface Faradaic reactions (e.g., quinone/hydroquinone), and (iii) slow capacitance attributable either to pore-diffusion-controlled double layer charging or surface Faradaic reaction or to "superficial intercalation... in which redox reactions of the graphitic system occur, with accompanying ion transport." Obviously, then, it is essential not only to monitor the pH of the solution or suspension in which carbon's electrochemical behavior is studied, but also to control it to the extent possible considering carbon's surface chemistry as reflected in its isoelectric point (or point of zero charge). Below are just two literature examples, one dealing explicitly with this issue [193] and the other [194] ignoring it even though it is undoubtedly important.

Runnels et al. [193] noted that "during cyclic voltammetry with carbon-fiber electrodes the current varies with changes in concentration of some inorganic cations as a result of their interaction with surface functional groups." Therefore, they "[investigated] which oxide functional groups play a direct role in the electrode's current responses to changes in pH" and "chemically modified" the "surface confined carbonyl and alcohol functionalities," but concluded that "[i]n both cases, the modification did not affect the carbon-fiber electrode's responsiveness to changes in pH." The relevant results are reproduced in Figure 5.9: "The peak potential of the cathodic feature... exhibited a greater dependence on pH than that of the oxidative peak... especially at higher pH values"; the cathodic slope increased from −66 to −134 mV/pH. Based on these trends, the authors surmised

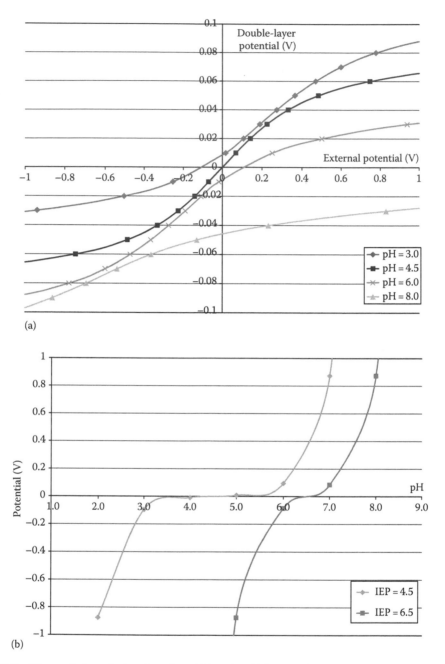

(a)

(b)

FIGURE 5.8 Theoretical analysis of an "amphifunctionally" electrified interface: (a) effect of pH and externally applied potential on the interfacial double-layer potential and (b) potential necessary to apply across the interface to reach the isoelectric point as a function of pH. Electrolyte concentration, 0.01 M; protolytic site density, $3 \times 10^{18}/m^2$; point of zero charge, $pH_{PZC} = 4.5$; inner-layer capacitance, 0.05 F/m²; outer-layer capacitance, 0.30 F/m². (Adapted from Duval, J., et al., *Langmuir*, 17, 7573, 2001.)

that "[i]f a surface-confined species is the origin of the surface wave, it must be one that is less acidic than a solution catechol, such as an alcohol or a carbonyl group," and suggested the following sequential electron- and proton-transfer events (albeit without justifying the existence of the presumed AH_2 moieties on the carbon surface, see Figure 5.4):

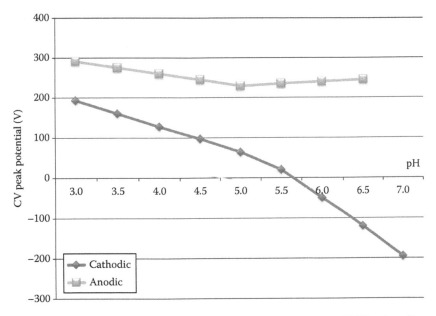

FIGURE 5.9 CV peak potential vs. pH for a microelectrode made from Thornel P55 carbon fibers. (Adapted from Runnels, P.L., et al., *Anal. Chem.*, 7, 2782, 1999.)

TABLE 5.2

Effects of Air Exposure of a Pyrolytic Carbon Electrode and Multiwalled Carbon Nanotubes on Their Electrochemical Performance in Terms of Fe^{2+}/Fe^{3+} Redox Behavior

Time (min)	0	5	10	30	60	120
Basal-Plane PG (mV)	227	269	317	360	407	596
Time (h)	0	1	3	6	12	24
Edge-plane PG (mV)	89	95	99	109	118	137
MWCNT (mV)	66	—	—	—	66	66

Source: Ji, X.B., et al., *Chem. Phys. Chem.*, 7, 1337, 2006.

$$AH_2 \longleftrightarrow AH_2^+ \longleftrightarrow AH^{\cdot} \longleftrightarrow AH^+ \longleftrightarrow A^{\cdot} \qquad (5.16)$$

Table 5.2 reproduces the results of Ji et al. [194], in terms of peak-to-peak separation of the Fe^{2+}/Fe^{3+} redox couple in 0.1 M KCl (except for the MWCNT sample, for which the pH was also controlled at 7, using 0.05 M KCl and 0.05 M NaH_2SO_4). The authors' objective was to "[relate] surface characteristics to electrochemical performance" and "obtain a better understanding of the role of surface oxides/functional groups on [sic] the electrode kinetics." It is puzzling, therefore, that they do not (even attempt to) explain why the presumably more inert "basal-plane-dominated" pyrolytic carbon becomes oxidized in room-temperature air much faster than the "edge-plane-dominated" pyrolytic carbon; this is exactly the opposite of what one would expect based on well-known facts about carbon surface chemistry. Furthermore, as the extent of oxidation increases, the point of zero charge of the carbons is expected to shift from a high to low value, and their electrochemical performance is expected to change accordingly; yet the authors do not even mention this, even though in the case of MWCNTs at pH = 7 there was no potential variation with time. In the light of such fundamental omissions, the authors' interpretations of the intriguing results shown in Table 5.2—in terms of "surface carboxyl groups [being] deprotonated and [repelling] the anion $Fe(CN)_6^{4-/3-}$," or in terms of a "change [in] the nature of the functional groups formed by reaction of oxygen with edge and edge plane-like sites" caused by "structural differences between the graphite and the CNTs," with the effect that "peroxide or other linkages [form] between graphite sheets in pyrolytic graphite which result in the carbon ends becoming capped and unable to participate in electron transfer"—do not sound convincing at all.

Andreas and Conway [187] have "evaluat[ed] the cyclic voltammetry (CV) responses as a function of the pH of an aqueous electrolyte over the range of 0–14," a task that they claimed "few, if any" studies have accomplished. Although they seem to be oblivious to much of the directly relevant literature mentioned above [85,87,184,192,193], their ambitious conclusions deserve close scrutiny. Figure 5.10 summarizes the finding that the "CV shape of the C-cloth depends on whether the titration is initiated at a pH of 0 or at 14." Based on these results and the intriguing argument that "[d]ouble-layer capacitance has the means for storing only 0.18 electrons per surface atom of carbon" (see Section 5.3.3), the authors argued that in acid the "peaks are pseudocapacitive in nature and may be either attributable to electrolyte anion adsorption/desorption... or the oxidation/reduction of some surface functionality"; in base, the lower observed currents were attributed to the absence of "pseudocapacitive peaks, which [were] so prominent in the CV recorded in acid." Unfortunately, the authors did not deem it important to characterize the surface chemical properties of the carbon used (a commercial AC cloth), even though the same electrode material was used repeatedly in several studies by Conway's group [195–198]; the only relevant information is that "it originates by pyrolysis of phenolic polymer fibers followed by heat treatment in O_2-free N_2 between 800°C and 900°C for some hours" [196]. The authors failed to appreciate or explore how such thermal history may be consistent with the observation [195] that "a significant quantity of ions is provided to the solution by the C-felt itself" and the puzzling speculation [195] that the

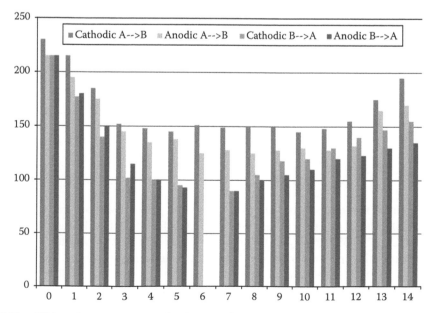

FIGURE 5.10 CV-based average charge density (y-axis, in C/g) vs. pH for an activated carbon cloth in the potential window 0.0–1.0 V: A→B, titration from acid (H_2SO_4) to base (NaOH); B→A, titration from base (NaOH) to acid (H_2SO_4). (Adapted from Andreas, H.A. and Conway, B.E., *Electrochim. Acta*, 51, 6510, 2006.)

"origin of such ions is possibly the surface functional groups of C-felt or of any ions adventitiously introduced in the preparation of the C-felt." In a more recent study devoted not to electrochemical but to adsorption behavior of the same material [199], much more relevant information is available, however: 4.49% O, 0.37% H; pH_{PZC} = 7.4; 0.25 mmol/g acidic groups (0.093 corresponding to carboxyls, 0.14 to phenolics, and the rest to lactones), 0.28 mmol/g basic groups. It remains to be shown how consistent such information is with the authors' interpretation of Figure 5.10: (a) "[B]etween pHs of 0 and 3, there is a 30% decline of capacitance which is attributed to the loss of pseudocapacitance related to the deactivation of the quinone functionalities due to the insufficient concentration of protons in the pores at higher pHs. In the mid-pH range, the charge density is essentially constant as the loss of charge density which results from the decrease in the peak at 0.35 V... is countered by the increase in charge density due to the redox wave centered at 0.75 V (the redox reaction of the pyrone functionalities). At a pH of ca. 11, the charge density begins to increase again... and is attributable to the pseudocapacitance of the newly activated surface functionalities (both the unidentified base-active species and the pyrone groups)." (b) "In changing the pH from ca. 14 to 5, a charge density loss of ca. 40% is seen due to the loss of the pseudocapacitance of the base-active surface functionalities. However, as the pH is decreased even further and sufficient protons become available for the quinone oxidation/reduction, a significant gain in charge density [is seen], resulting in a charge density almost double that of the value at pH 8 and 30% larger than the initial value, due to the newly developed pseudocapacitance." On one hand, the insightful stoichiometric calculations carried out by the authors, starting from a reasonable density of edge surface sites (0.1 nm^2/site), do not necessarily agree (e.g., 0.1% of surface sites having quinone groups, i.e., 0.08 mmol/g) with the surface chemical characteristics mentioned above; and whether these in turn support the prominent role of the pyrone redox reaction (shown below) deserves further scrutiny.

And perhaps even more important is to explicitly analyze the impact of the electrode's pH_{PZC} (presumably ca. 7.4) on the relative contributions of double-layer vs. pseudocapacitance charging.

5.3.3.2 Influence of Heteroatoms

The role of heteroatoms has had a very long research history, and some well-established facts have thus emerged. First, oxygen cannot be considered to be present substitutionally in the graphene lattice. Second, because of its ubiquity in surface functionalities, some of which can be deprotonated, its influence is intimately connected to that of pH, as discussed in the previous section.

Gunasingham and Fleet [200] emphasized the practical implications, associated with the use of carbon electrodes: "[T]here are a large variety of carbons available from various manufacturers that vary in their bulk structure... [which] affects surface properties... [which in turn] has a bearing on the formation of surface carbon–oxygen functionalities, which influence the electrochemical response of the carbon." While inappropriately attributing to Randin and Yeager [151,152] the notion that these functionalities are "formed at unsatisfied valence sites at the exposed edges of graphite planes," the authors fail to either explicitly support or cast doubt on, e.g., the quinone/hydroquinone electrochemical process; instead, they indiscriminately and noncommittingly provide a long list of possible surface chemical effects.

The two issues that are of greatest relevance for our discussion here were emphasized in the very recent review by Frackowiak [95]: "[O]nly the electrochemically available surface area is useful for charging the electrical double layer," and "[s]ubstitutional heteroatoms in the carbon network (nitrogen, oxygen) are a promising way to enhance the capacitance." The author does not take the opportunity to discuss the former issue in any detail. She does emphasize the well-established fact that "capacitance is not directly related to the BET specific surface area," but the possibility is left open that a correlation exists with the total accessible surface area of the electrode, i.e., when the geometry of the pores and of the potential-determining ions is taken into account properly. Such a doubt then brings to the fore the issues of the chemical origin of the surface charge and its relationship to "space charge," as well as the exact role of heteroatoms and the origin of pseudocapacitance. Frackowiak does note that the "presence of a large amount of oxygenated groups can be advantageous, either because they improve the wettability on the carbon surface or they contribute to an additional pseudocapacitance," but does not elaborate on the latter effect. Of course, in this context, oxygen is *not* present substitutionally in the carbon lattice, but as graphene edge functionalities; and whether nitrogen is indeed substitutional in ACs (e.g., as the so-called "quaternary N") can also be debated (see Section 5.2.1), based on the well-known fact that such substitution is not possible either in graphite [46] or, presumably also, in ordered carbons in general. In this context, it is interesting to analyze the proposed mechanisms [95,201,202] by which N-containing carbons [71,116,201–205] enhance the (pseudo)capacitance of carbons (see Figure 5.11a)

$$C* = NH + 2e^- + 2H^+ = C*-H-NH_2 \qquad (5.17)$$

$$C*-NHOH + 2e^- + 2H^+ = C*-NH_2 + H_2O \qquad (5.18)$$

Although some important details of the polyaromatic structure shown in Figure 5.11a are not clear [6] (e.g., the nature of sites marked with an asterisk), the main feature of this proposal is the role of pyridinic N, and not of the presumably substitutional N. It is expected to facilitate the creation of

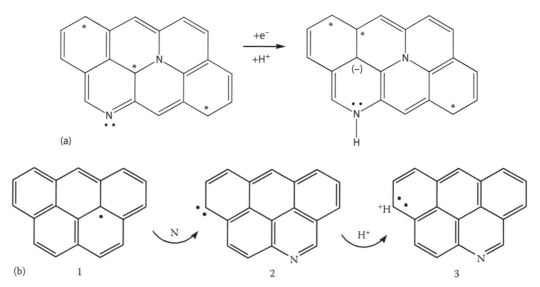

FIGURE 5.11 Proposed mechanisms of participation of N heteroatoms in electrochemical processes on carbon surfaces: (a) "possible pseudo-Faradaic reaction of the pyridinic group in aqueous medium" [95]; (b) N-induced stabilization of carbene-like edge carbon atoms, sites responsible for the development of positive charge in acidic solution.

(a delocalized) negative charge in the graphene layer, presumably under acidic conditions; but the experimental observation that such a situation is meant to support remains unclarified, especially in light of the well-known fact that "the basic character of carbons is also proportional to the nitrogen content" [95], and so a more straightforward explanation is that the double layer capacitance, and not the pseudocapacitance, is enhanced. Similarly, the likelihood of existence of $Ar–H–NH_2$ and $Ar–NHOH$ on the carbon surface remains to be substantiated. In an attempt to further clarify the potentially different roles of nitrogen and oxygen surface functionalities, Ania et al. [71] compared the elemental analyses and cyclic voltammograms of two zeolite Y-templated carbons, one derived from carbonization of acrylonitrile (Y-AN) and the other from deposition of acetylene (Y-Ac). The former data are reproduced in Table 5.3.

The larger charge storage capacity of sample Y-AN (340 vs. 240 F/g) was attributed to "the presence of nitrogen in the carbon matrix," presumably because both materials "possess somewhat similar... oxygen content"; the numbers shown in Table 5.3 obviously do not support the latter statement, especially if, as shown in Figure 5.11a, the quaternary N does not participate in redox processes. Furthermore, the alternative mechanism shown in Figure 5.11b seems equally plausible, and has the additional virtue of being consistent with the emerging detailed knowledge of the edge chemistry of the graphene layers [6]. The availability of nitrogen during carbon preparation (step 1—> 2) is shown here to contribute to the formation of a more stable graphene structure (2): once a pyridine-type group is formed, the stabilization of a divalent carbon atom, with a triplet ground state is favored [6]; and this leads to the formation of a basic site (3) by proton addition.

Another authoritative review, certainly from an electrochemist's point of view, has been offered by Conway [68], as mentioned in Section 5.3.3. In particular, he reinforced the clear distinction between the important concepts of DL- and pseudo-capacitance [206], an issue mentioned previously (see Section 5.3.3.1) and one that Biniak et al. [207] have identified as in need of further exploration "because the results obtained in various works for different carbon materials are divergent and difficult to explain and reconcile." The former is due to potential-dependent surface density of charges stored electrostatically at the electrode interfaces; the accumulated charge is a combination of excess or deficit of conduction-based electrons at or in the near-surface region of the interface, together with counterbalancing charge density of accumulated cations or anions of electrolyte on the solution side of the EDL. The latter is due to a very different charge storage mechanism: passage of charge across the EDL, as in a battery, as a consequence of processes such as oxidation/reduction (redox),

TABLE 5.3

Relative Abundance of Oxygen and Nitrogen Surface Functionalities in Carbon Materials Obtained by Acrylonitrile Carbonization (Y-AN) or Acetylene Deposition (Y-Ac)

	N-Q	N-X	N-6	C–O	C=O
Position of XPS peak (eV)	400.7	403.6	398.5	532.6	530.7
Carbon Y-AN					
Atomic concentration (%)	2.3	0.6	2.3	4.3	1.1
Carbon Y-Ac					
Atomic concentration (%)	—	—	—	3.84	0.33

Source: Ania, C.O., et al., *Adv. Funct. Mater.*, 17, 1828, 2007.

Note: N-Q, presumably quaternary (or substitutional) nitrogen; N-X, six-ring N-surface functionality; N-6, pyridine nitrogen.

intercalation, specific ion adsorption, or partial charge transfer. From the perspective of a carbon scientist, however, Conway's discussion of surface functionalities and their role in such charge storage processes leaves much to be desired. Thus, for example, the reference used to illustrate the presence of "titratable acidic and/or basic functionalities related to some of the surface structures" is, amazingly, a 1938 study of carbon oxidation; and the noncontroversial ketone (or, more precisely, quinone) and lactone structures are misrepresented, while invoking at the same time the mysterious "carbinol" functionalities. It is thus not surprising that the key aspects of carbon surface (electro)chemistry and carbon oxidation are also arguably misrepresented; as an example, the ambitious but largely hypothetical proposals of Sihvonen [208] are elevated to the status of presumably demonstrated (or accepted) fact regarding the anodic oxidation of a quinone functionality at low current density in H_2SO_4:

$$(C_xO)_2 + OH = C_{2x-2}COCOOH \tag{5.19}$$

$$C_xCOCOOH - H = C_x + CO + CO_2 \tag{5.20}$$

How much Conway appears to be out of touch with carbon (electro)chemistry research is confirmed by his "discussion" of a 1972 paper by Thrower, presumably published in *J Electroanal Chem*; anyone familiar with Peter Thrower's expertise—and that means any serious carbon researcher—knows that such a paper cannot (and does not!) exist. Finally, in the light of recent proposals regarding the edge chemistry of carbon surfaces [6], Conway's conclusion that "[m]any porous or powder[ed] carbon materials have 'dangling' surface bonds which are associated with free-radical behavior" should also be viewed with caution (see Figure 5.4). Therefore, perhaps the most (and only?) reliable take-home message from this review is that "much basic research of a substantive kind is required to relate the electrochemical behavior of various preparations more quantitatively to... (5) the surface chemistry of carbon preparations and their shelf-life stability, cycle life, and self-discharge characteristics."

From a practical standpoint, therefore, a reconciliation of the apparently contradicting conclusions regarding the role of carbon surface functionalities in charge storage has been an elusive goal. On the one hand, for example, Conway [68] concluded that the "presence of oxygen-containing functionalities is undesirable because of their involvement in cycle-life stability or self-discharge processes... that probably take place through Faradaic redox reactions in the matrix, although the details of such effects are sparse." (It should be pointed out that the recent study by Andreas and Conway [187]—see Section 5.3.3.1—argues to the contrary.) On the other hand, Frackowiak [95] argued that the "functional groups present on the surface... can considerably enhance the capacitance through additional faradaic reactions" and that the "pseudo-capacitance induced by oxygen functionalities due to close association of oxygen with the carbon surface is the most frequent phenomenon." An earlier authoritative and widely cited review [60] (see also [209]), as well as other studies [11], had already documented the "effect of pseudocapacitance for maximizing the total capacitance" of carbon materials.

Apart from the cycle-life stability and self-discharge processes, a potentially negative effect of the electron-withdrawing oxygen functionalities (e.g., carboxyl groups) is intuitively obvious: in addition to a wettability loss, adsorption of anions (as well as that of aromatics; see Section 5.3.4) may be suppressed and electrical conductivity may be lost by virtue of a decrease in the availability of delocalized π-electrons. The latter issue was recently discussed by Cazorla-Amorós and coworkers [210] and the expected effect was demonstrated by Nian and Teng [146,211], as summarized in Figure 5.12: as the heat-treatment temperature of the previously nitric-acid-treated carbon fabric increased from 150°C to 750°C, the monotonic decrease in the concentration of CO_2-yielding groups and the maximum in the difference between CO- and CO_2-yielding groups resulted in a corresponding capacitance maximum, which the authors

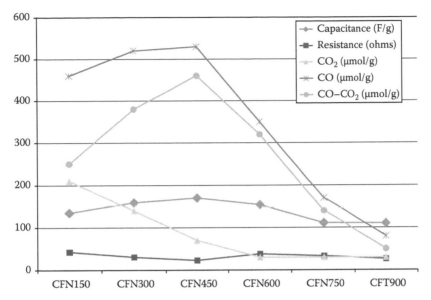

FIGURE 5.12 Effects of heat treatment (*x*-axis, CFNZ, with Z representing treatment temperature in °C) on the CO- and CO_2-yielding oxygen surface functionalities, and on the electrochemical behavior of activated carbon fibers (CFN). (Adapted from Teng, H., et al., *Carbon*, 39, 1981, 2001; Nian, Y.-R. and Teng, H., *J. Electrochem. Soc.*, 149, A1008, 2002.)

interpreted as evidence for "the positive and the negative role that the CO- and the CO_2-desorbing complexes, respectively" [211].

Figure 5.13 reproduces the results of Bleda-Martinez et al. [210], based on which these authors argued that carboxylic anhydride groups (A) exhibit a two-electron redox process and thus have a capacitance-enhancing effect, similar to that of the CO-yielding quinone groups:

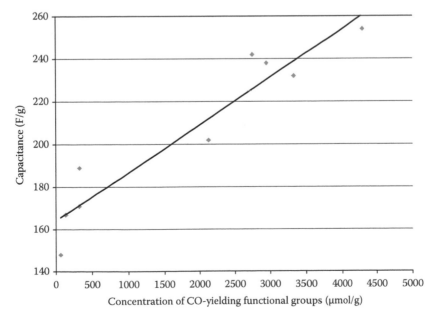

FIGURE 5.13 Correlation between the capacitance of an anthracite-derived activated carbon and its content of CO-yielding surface functionalities. (Adapted from Bleda-Martinez, M.J., et al., *Carbon*, 44, 2642, 2006.)

$$A + e^- <\longrightarrow A^{\cdot-} \qquad (5.21)$$

$$A^{\cdot-} + e^- <\longrightarrow A^{2-} \qquad (5.22)$$

It would be interesting to explore the proton-transfer (i.e., pH) dependence of the above electron transfer process. In the already mentioned study by Hsieh and Teng [11] (see Equations 5.1 through 5.3, Section 5.1), presumably Reaction 5.2, $>C_xO + H^+ = >C_xO//H^+$—described as specific adsorption "basically induced by ion-dipole attraction"—was proposed to be responsible for the observed "excess specific double layer capacitance due to the local changes of electronic charge density"; the invoked changes were not discussed in any detail, however.

An ambitious review of carbon applications in chemical power sources (see also Section 5.3.5) was offered by Fialkov [94]. It is disconcerting, however, that the author discusses the "influence exerted by... surface properties" without citing even one of the well-known—or well-cited or more recent—studies on carbon surface chemistry. And yet, in conjunction with the use of carbon in air (oxygen) electrodes, he speculates that the "oxygen electroreduction kinetics depend on... the degree to which side faces of carbon crystallites are developed" because "base groups" are formed there and presumably interact (e.g., with H_2O_2) in the following manner:

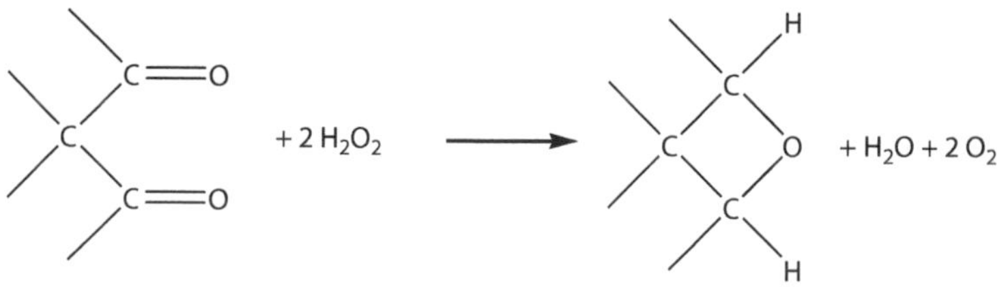

This scheme presumably "suggests that the active centers on AC are carbon atoms directly bound to the base functional groups" or, in mainstream surface chemistry terminology, to the edge active sites. Fialkov does not elaborate how the latter might be "connected with a higher concentration of acid functional groups," which in turn strongly affect the rate of cathodic oxygen reduction. Finally, the author also claims that the quinone groups are "proton-generation centers" according to the following scheme:

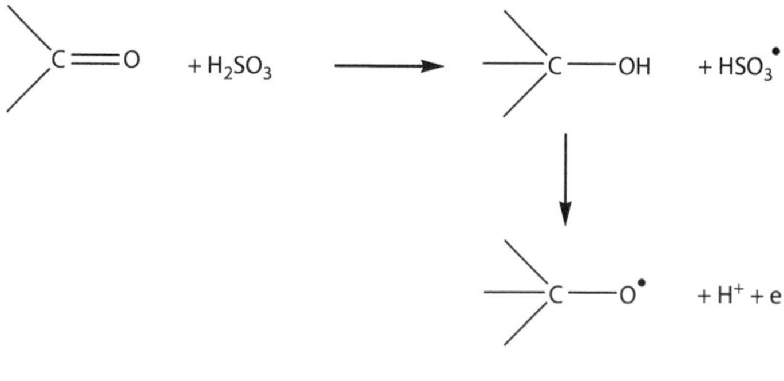

In another recent review, more focused on carbons as supercapacitors, Pandolfo and Hollenkamp [212] emphasized the fact that the "power and energy-storage capabilities of these devices are closely linked to the physical and chemical characteristics of the carbon electrodes" and argued that further improvements are needed in "surface treatments to promote wettability." Interestingly, only the following "chemical and physical properties" are highlighted: high conductivity, high surface area (up to $2000\,m^2/g$ and above) with controlled pore structure, good corrosion resistance, and processability and compatibility in composite materials. And yet, while the "capacitance of a device is largely dependent on the characteristics of the electrode material, particularly the surface area and pore size distribution," these reviewers [212] also note an indirect role of surface functionalities, as they affect the electrical resistivity—presumably by increasing "the barrier for electrons to transfer from one microcrystalline element to the next"—as well as the interfacial resistance between the electrode and the collector. In the section specifically devoted to carbon surface functionalities, the authors list the following factors as being "known to influence the electrochemical interfacial state of the carbon surface and its double-layer properties": wettability, point of zero charge, electrical contact resistance, adsorption of ions (capacitance), and self-discharge characteristics. This last factor is the only one that was elaborated: the "extent of oxygen retention as physically adsorbed molecular oxygen, or as surface complexes, is believed to influence strongly the rate and mechanism of capacitor self-discharge (or leakage)" because the "oxygen functional groups may serve as active sites, which can catalyze the electrochemical oxidation or reduction of the carbon, or the decomposition of the electrolyte components." This interesting argument deserves further scrutiny.

The authors do conclude that "charge storage on carbon electrodes is predominantly capacitive" but do not explain how carbon's surface functionalities (reflected, for example, in its point of zero charge) influence such storage. This is intriguing, because the "cartoon" intended to portray the molecular-level interfacial interactions (see Figure 2 in [212]) suggests the existence of specific sites of positive and negative charges on the carbon surface (see Figure 5.4 and Section 5.3.3.3). Instead, they devote some attention to the effect of oxidative treatments on carbon's pseudocapacitance, and simply summarize the findings of Hsieh and Teng [11] illustrated in Table 5.4: Faradaic current increased significantly with the extent of oxygen treatment, while the change in double layer capacitance was only minor, for reasons that are yet to be discussed (and understood!) more thoroughly. As mentioned previously, in the absence of a more detailed discussion of the chemical origin of surface charge, arguments regarding "an excess specific double layer capacitance due to the local changes of electronic charge density" and "enhanced dipole affinity towards protons in an acid solution" are not very convincing.

Pollak et al. [118,119] recently examined in some detail the source and the possibilities of measurement of pseudocapacitance. They prepared "carbons of different surface chemistry in order to determine the main cause" of unusual potential dependence of carbon electrode's conductivity in acidic media [119]. Oxygen-containing surface groups had little effect, and the anomalous behavior was thus attributed to nitrogen-containing groups, introduced by cellulose carbonization in the presence of ammonium chloride. The mechanism proposed for the observed behavior, disruption of the "conjugated system between the nitrogen surface groups and the carbon backbone," implicated pyridine-type groups and the following proposed redox processes:

$$C^* = NH + 2e^- + 2H^+ \longleftrightarrow C^*HNH_2 \tag{5.23}$$

TABLE 5.4

Effect of Oxidative Treatment on the Capacitance of Activated Carbon Fabrics

Sample	CO$_2$ Evolved (mmol/g)	CO Evolved (mmol/g)	Double-Layer Capacitance (F/g)	Faradaic Current (mA/g)
CF	0.17	0.23	121	0.082
CFO1	0.19	0.41	123	5.7
CFO2	0.20	0.58	123	7.5
CFO3	0.25	0.81	130	22

Source: Hsieh, C.-T. and Teng, H., *Carbon*, 40, 667, 2002.

$$HC^* = N + H_2O \ <\!\!-\!\!> \ C^*HN = O + 2e^- + 2H^+ \hspace{2cm} (5.24)$$

It remains to be seen why similar putative disruptions of π-electron conjugation by the much more ubiquitous oxygen functionalities (e.g., quinones) do not produce analogous effects on the electronic conductivity.

5.3.3.3 Chemical Origin of Surface Charge

Given the state of our fundamental knowledge of carbon surface (electro)chemistry, it is unfortunately still premature to offer sufficiently detailed and thus hopefully convincing arguments about the chemical origin of the electrically induced charges on carbon surfaces. What follows, therefore, is an attempt to indicate the points of departure for such a discussion and set the stage for the much needed "chemical" analysis of the key electrochemical concepts (e.g., PZC, flat-band potential, immersion potential, space charge, and work function).

A convenient point of departure is to recall what Grahame [213] stated in his 1947 review of the electrical double layer and the theory of "electrocapillarity":

> [T]here is a range of potentials for which a current does not flow across the interface [..., which is] electrically similar to a condenser of large specific capacity. The capacity of this condenser gives a fairly direct measure of the electronic charge on the... surface.

As mentioned in Section 5.3.3.2, of special interest in this regard is the exact nature of the relationship between the maximum point on the interfacial tension vs. potential curve (the so-called "electrocapillarity curve") known as the PZC, at which the electric charge density of the surface is zero or the concentrations of ions in solution are adjusted such that adsorption of anions and cations is balanced to produce a maximum interfacial energy [214], and the point of zero charge (pH_{PZC}) or the isoelectric point (pH_{IEP}), at which the concentrations of ions in solution are so adjusted that cations and anions are equally adsorbed or unadsorbed. Indeed, detailed knowledge of the seemingly complex relationship [215] between the electronic structure of carbon (as quantified, for example, by its work function), the electron transfer characteristics of the carbon/liquid interface (quantified by the PZC), and the proton transfer characteristics of the interfacial region (as quantified by the point of zero charge or isoelectric point) is not only of primary interest here, but probably holds the key to many novel and improved applications of carbon materials.

In this context, the "lack of space charge [in carbon pores of size below 1 nm]" [95], as the reason for an efficient EDL formation, deserves closer scrutiny as well; it is intriguing that this argument was presented only in the abstract, suggesting that presumably this is a well-known fact. Is it, really? In earlier reviews by Frackowiak and Beguin [60,97], this issue was not even mentioned, but it was discussed by Conway [68] and more recently by Pandolfo and Hollenkamp [212]. In the former review, only the case of "graphitic carbon materials" is anachronically analyzed, to the extent that their "free electron density is sufficiently large... to behave in an almost metallic way," even though "there are some experimental indications—and here the early studies by Randin and Yeager [152,154] are recalled—for the basal-plane from double-layer capacity measurements that semiconductor properties are in fact exhibited" [68]. In the latter case, the analysis is along the same lines except for the following additional arguments [216]: "[Both capacitance and conductivity] are essentially limited by the number of available charge carriers within the carbon space-charge region," but this can be "alleviated at potentials [on] either side of the pzc because charging of the carbon electrode increases the number of available charge carriers... [, and] as carbons are activated to greater degrees [and] pore walls become thinner[,] space-charge regions begin to overlap and less capacitance is generated per unit electrode area" [212]. This is a key point. Its validity hinges on the assumption that the thickness of pore walls "becomes similar to the dimensions of the space-charge region (defined by the Fermi length, and approximately equivalent to one graphene layer)" [212].

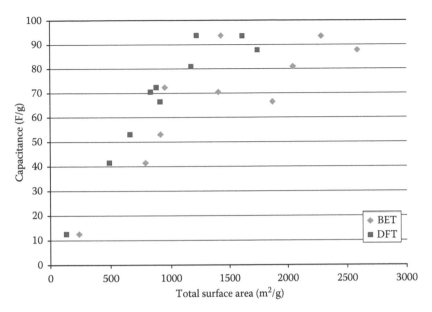

FIGURE 5.14 Correlation between the capacitance of a carbon black and a wood-derived activated carbon and their total surface area. (From Barbieri, O., et al., *Carbon*, 43, 1303, 2005.)

In this context, Barbieri et al. [69] have explored the capacitance limits of high-surface-area ACs. Their results are reproduced in Figure 5.14. In contrast to the above cited assertion regarding the absence of space charge in narrow micropores, here the authors interpreted the apparent saturation effect (when density functional theory is used to determine the specific surface area) to imply that specific capacitance "is limited by the space charge capacitance of the solid." It is puzzling, though, that they claim to have "suggested" this interpretation recently; such an interpretation should be obvious, and even trivial, from basic electrochemical principles of metals and semiconductors, as discussed previously and emphasized below. Even more intriguing is the related argument by Hahn et al. [216] that their "preliminary results... suggest that... the capacitive behavior of... activated carbons is dominated by the electronic properties of the solid, rather than by the properties of the solution side of the double layer," similar to the classical behavior of the basal plane of HOPG; the authors were surprised that the density of electronic states at the Fermi level was calculated to be six times higher for AC ($1.5 \times 10^{21}/cm^3/eV$ at the PZC) than for HOPG. Analysis of the elementary concept of resistances in series (see Figure 5.6 and Equation 5.14), together with the too often evaded consideration of the physicochemical origin of surface charge, must be invoked at this point if the apparent confusion and uncertainties are to be clarified.

It is well known that the "conductivity even of doped semiconductors is usually well below that of an electrolyte solution [217]," and therefore, for most carbons "practically all of the potential drop occurs in the boundary layer of the electrode, and very little on the solution side of the interface [217]." For example, carbon's total charge density in a positively charged space-charge region would be the static positive charge (e.g., protonated carbene sites; see Figures 5.2 and 5.4) plus the charge corresponding to holes created by the localization of π electrons in the process of formation of carbene sites plus the mobile negative charge of the conduction electrons. Its rigorous estimation using the Poisson equation has yet to be carried out; a simplified Mott–Schottky equation [218] has been used by Randin and Yeager [152,154]. In this regard, a potentially fruitful area of research would be to explore the usefulness of Mott–Schottky plots in determining the concentration of carbene sites in different carbon materials (assuming the contribution of the solution to the total capacitance can be neglected).

It is instructive to recall here the argument offered early by Oren et al. [144] as an explanation for potential-independent space charge capacitance:

> In chemical terms, it may be proposed that the π electrons of the first graphite layer at the surface are localized at states which are separate from the bands of the bulk. Significant contribution of these states to the double layer capacitance necessitates an overpotential which is beyond the applied potential range (limited by water electrolysis). In terms of semiconductor theory, it may be assumed that the charge carrier density at the first and probably the second graphite layer is much lower than the bulk charge density value of 6×10^{18} carriers per cm^3.

Observation of distinct, slow and fast electrode charging processes by the same research group [219], presumably not related to micropore diffusion or surface reactions involving OH^- and H^+ ions, was interpreted using the same argument:

> Slow and fast charging entities... may result from layered structure or transversal inhomogeneity... Thus the first and probably also the second graphitic surface layers were supposed to have localized π electrons that are not conjugated with the bulk charge carriers... The population of these [partially localized] states is an activated process, namely it requires some overvoltage, hence the slow charging sections in the [differential voltammogram].

It is tempting to identify these localized π electrons with the recently (re)discovered phenomenon of π-electron localization at carbene zigzag sites [6].

The pioneering studies of capacitance exhibited by basal and edge planes of carbons by Yeager and coworkers showed "abnormal" behavior, which led Randin and Yeager [151] to conclude "that the space charge characteristics of graphite must be taken into account in examining the electrochemical properties of this material."

The concept of electrochemically active surface area holds many clues regarding the chemical origin of surface charge. It is intriguing that an electron spin resonance (ESR) study of the behavior of a carbon black [220] has received very little attention (only two SCI citations up to October 2007); in this study, the behavior of the capacitor—two polymer-blended Ketjen EC electrodes and propylene carbonate containing one $MLiClO_4$ as the electrolyte—was studied using an *in situ* electrolysis/ESR cell. Thus, Liang et al. [179] cite it in the rather prosaic context of an increase in capacitance with increasing surface area, while the key (and unique?) point of this work is that there was "formation of radical sites in the carbon black in the charge [reduction] process" and that the "intensity of the ESR signal increased with the extent of electrochemical reaction." The authors suggested that the "unpaired electron sites (probably anion radicals)" are formed during charging and that "the diffusion of Li^+ ions into the carbon [which presumably stabilizes the anion radical] precedes reduction." Has this been confirmed/substantiated?

No one has addressed the issue of origin of the surface charge in carbon capacitors more explicitly than Bonhomme et al. [221], based on the potential dependence of Raman spectra as well as variations of salt concentration in the electrode vicinity during galvanostatic cycling. These authors noted that the "two double-layer interfaces do not work in a symmetrical way"; and they observed a strong intensity decrease in the Raman spectra of AC cloth electrodes when these were charged positively or negatively, which they interpreted as "the corresponding conductivity increase of this material which behaves as a n- or p-doped semiconductor." This prompted them to offer the following mechanistic explanation for the measured electrolyte (salt) depletion effects during cycling at a given positive or negative current density:

> [T]he salt concentration variations are not the same in the vicinity of the C^+ and C^- interfaces. During the charge, the concentration near C^+ first increases, goes through a maximum and then decreases, whereas near C^- it decreases continuously. A similar difference is found at the end of the discharge... [I]n the discharged state there is a weak but negative potential drop at the two interfaces with a small

excess of negative charge on the two carbon electrodes and compensating cations in the double layer. When a charging current is applied to the system, the C^+ electrode reaches rapidly its pzc as electrons are withdrawn from it while C^- sees its negative charge to increase by a roughly equal amount. Thus, in the vicinity of C^+, ions have been liberated toward the electrolyte, giving a small increase of concentration at the investigated point. On the contrary, the continuous accumulation of cations near C^-, whose negative electronic charge continuously rises, induces an increasing salt depletion at the investigated point. Once the first step is passed through, the C^+ electrode takes an increasingly positive charge with the corresponding amount of anions accumulating in the double layer while the C^- electrode continues to increase its negative charge. The amount of these negative charges always remains higher than the amount of positive charges on C^+ in such a way that the salt concentration depletion near C^- is always more important than near C^+ after the latter has passed the pzc. However, this concentration gradient is permanently compensated by ionic diffusion and, at the end of the charge; the salt concentration within the electrolyte tends to similar values. The same kind of explanation holds for the discharge process as the C^+ electrode again crosses its pzc before the potential of the cell reaches zero whereas the charge of C^- remains always negative.

An analogous explanation was offered for charging under constant potential:

[A]t the beginning of the charge, the C^+ interface evolves first from a negative potential to the pzc and then to a positive potential while the C^+ interface experiences a continuous decrease of negative potential. It follows that the electronic conductivity of the C^+ electrode first decreases, while that of the C^- electrode increases continuously. The response of the latter would be faster and its higher conductivity would allow the electrons to reach quickly the interface. The corresponding amount of electronic depletion in C^+ would be more broadly distributed in the bulk of the electrode, leaving a limited concentration of positive charge at the very interface.

Regarding this "small excess of negative charge" in the discharged state, Bonhomme et al. [221] postulate that

[s]ome kind of [formatting] of the carbon surface occurs after a single charge/discharge process, leading to a situation where the electrical double layer at the ACC/organic electrolyte interface in the absence of applied potential is characterized by a negative excess charge on the carbon surface, compensated by accumulation of Et_4N^+ cations on the electrolyte side... It is not possible to decide whether the negative charge on the carbon comes from some kind of oxygenated surface group, for example, a $C=O^-$ group, from irreversible adsorption of anions in the pores or simply from an excess of electronic charge on the carbon after complete wetting by the electrolyte.

Of course, the straightforward explanation and the one that is consistent with what we know now—and what we have known then—about the surface chemistry of AC cloth, is that its negative surface charge is a consequence of the dissociation of (mostly) carboxyl groups above its pH_{IEP} (or pH_{PZC}).

5.3.4 Electrosorption

Here the focus will be, for the most part, on organic solutes. The huge field of inorganic solutes—related primarily to the removal of cations and anions from solution (e.g., in desalination [136,222,223] or water purification [224]) and the preparation and optimization of the properties (e.g., catalytically active surface area) of carbon-supported electrocatalysts—is beyond the scope of the present discussion. It will be interesting to see whether in the presence of an electric field the controlling factors that have been elucidated for carbon-supported catalysts [225] and adsorbents [7] are different in important ways. In this context, and recalling the discussion in Section 5.3.3.3, it is instructive to keep in mind the conclusions of McDermott and McCreery [226], based on a scanning tunneling microscopy study of quinone adsorption on graphite and GC surfaces:

[W]e should reconsider the definition of the word 'site' at carbon electrodes. The traditional meaning based on functional groups or a particular edge plane geometry is certainly valid for specific chemical interactions such as chemisorption and electrocatalysis. However, for many outer-sphere electron transfer reactions and... for quinone adsorption, the 'site' must include a defect-induced electronic perturbation of the surface. In these cases, it is not the edge itself that promotes charge transfer and adsorption, but rather the electronic and partial charges caused by the step edge.

A similar argument [7] was presented by Müller et al. [227] in their incisive analysis of adsorption of weak electrolytes from aqueous solution on ACs: "The solid surface charges in response to solution pH and ionic strength; the resulting (smeared) surface electrostatic potential influences the adsorption affinity of the ionized solute."

A convenient starting point for this analysis is a recent study by Yang et al. [148], who used an electrical double layer model to rationalize the electrosorption of ions from aqueous solutions by a carbon aerogel (BET surface area, $412 \, m^2/g$; pore volume, $0.6 \, cm^3/g$), even though this model "focuses on the electrosorption of sodium fluoride, which exhibits minimal specific adsorption." The main premise of the authors is that "[f]or electrodes of good electrical conductivity and high surface area, such as AC and carbon aerogel, a strong electrical double layer is formed on their surfaces when an electric field is applied." A similar misconception (or at least misleading statement) is echoed in another recent summary of the "commonly accepted" basic mechanism of electrosorption [228]: "[W]hen an electrical potential is applied to the surface of an active carbon fiber immersed in an aqueous electrolyte solution, charge separation occurs across the interface and an electric double layer (EDL) develops as a result."

The key point of interest in electrosorption (which the above cited arguments do not take into account) is whether—because such a double layer is formed even when an electric field is not applied—the application of an electric field can enhance or reverse the double-layer-controlled interactions between the charged adsorbent and the adsorbate ions in solution (see Figure 5.6). Therefore, the key issue is whether the following arguments are really applicable when it comes to determining the "total capacity of electrosorption" [148]:

A cation-responsive electrode can associate cations with chemical bonds; therefore, its cation capacity is always higher than its anion capacity. In this case, the total capacity is limited solely by the anion capacity. On the other hand, the total capacity of an anion-responsive electrode is limited by its cation capacity.

Is this really true? Does this "responsiveness" of carbon not depend on its isoelectric point (pH_{IEP}) and on the pH of the solution? And is this not subject to (hopefully major) changes in the presence of an applied field? The answer to the latter question requires an explicit dependence of inner-layer capacitance (C_{IHP}), and thus the extent of specific ion adsorption, on surface charge density, which the authors do not provide; the answer to the former requires knowledge of the relationship between the adsorbent's isoelectric point and its PZC, which the authors also do not include in their model. Instead, they use C_{IHP} and PZC as adjustable model parameters, and it is instructive to analyze some of their qualitative and quantitative results. The most obvious (and disconcerting?) finding is that the equilibrium electrosorption capacity of Na^+ (both observed and predicted) is quite low, ca. $0.25 \, mmol/g$ under the most favorable conditions. This could be easily accounted for by adsorption in the absence of an electric field (the so-called "open-circuit" uptake [229]): an air-exposed high-surface-area carbon typically contains several percent of surface oxygen in the form of cation-exchangeable carboxyl groups, and $0.25 \, mmol \, COO^-/g$ corresponds to only 0.8% oxygen, as well as to a surface charge of ca. $0.025 \, C/m^2$; assuming a realistic value of $3 \, \mu m/s$ per V/cm for the electrophoretic mobility of such carbon [7], the corresponding maximum value of surface potential is ca. $70 \, mV$ using the Gouy–Chapman theory. Such a realistic "open circuit" situation [7,227]—for an "acidic," a "basic," and an "amphoteric" carbon surface—is summarized in Figure 5.15, and the interpretation of polarization-driven electrosorption phenomena should be consistent with it.

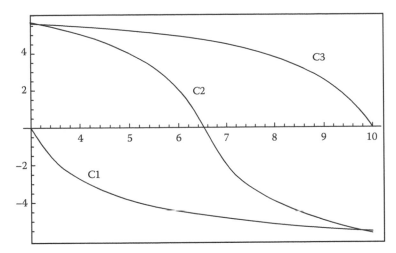

FIGURE 5.15 Dimensionless surface potential (y-axis, $F\phi_s/RT$) vs. pH (x-axis) for three types of carbon electrode materials: C1, "acidic" carbon ($pH_{IEP} = pH_{PZC} = 3.0$); C2, typical as-received ("amphoteric") carbon ($pH_{IEP} = pH_{PZC} = 6.5$); C3, "basic" carbon ($pH_{IEP} = pH_{PZC} = 10.0$). Based on Gouy–Chapman double-layer theory, for a maximum surface charge of $0.03\,C/m^2$ and ionic strength of 10^{-3} M.

It is of interest to analyze some of the follow-up studies that cited the work by Yang et al. [148]. Han et al. [134] investigated the kinetics and equilibrium of m-cresol uptake by AC fibers and reported that "electrochemical polarization could effectively improve the adsorption rate and capacity of ACFs," even though "the surface chemical properties of ACFs hardly changed under… polarization of less than 600 mV." They did not make any comments about the model proposed by Yang et al. [148], but did claim instead that their study "was beneficial to understand the mechanism of electrochemically enhanced adsorption"; intriguingly, however, the only explanation they offered toward such presumed understanding is "the improvement of BET surface areas and average pore sizes besides the affinity between the surface of the ACFs and m-cresol," leaving it up to the reader to wonder why adsorption affinity would be "enhanced by the polarization under electric field." An earlier report by the same group of electrochemically enhanced adsorption of phenol (pK_a = 9.9) on AC fibers in basic aqueous solution [230]—in which the study by Yang et al. [148] was cited, amazingly, for "good efficiency of removal of metal ions by activated carbon"—showed, by contrast, "a nearly 10-fold enhancement of maximum adsorption capacity," the maximum uptake of 3.72 mmol/g (at 700 mV) being close to the monolayer capacity for typical ACs [7]. The arguments offered toward an explanation were again puzzling: the electrosorption enthalpy value of −20 kJ/mol was interpreted as being in agreement "with the results of dipole-dipole interaction" and this presumably "indicated that dipole–dipole interaction, besides electrostatic interaction, might be the main adsorption mechanism [of interaction] between the positive ACF and phenolate ions." No information was provided regarding surface charge, which is likely to be negative in basic solution [7], and no evidence was offered—this arguably being the crucial point—that polarization was able to reverse this situation. A comparison with the behavior of aniline (pK_a = 4.6) is of great interest in this context [231]: here the authors report "a twofold enhancement of adsorption capacity," the maximum uptake being 3.39 mmol/g at 600 mV. Their mechanistic discussion is more informative (though arguably no less confusing), because the effect of pH was studied as well, albeit for an unknown pH_{IEP} of the adsorbent [7]: positive polarization was again assumed to result in a positive surface charge, but the increasing aniline uptake as pH increased from 3.1 to 10.6 was not attributed simply to lesser electrostatic repulsion [7]. Instead, several additional and highly speculative arguments are invoked to explain why "the affinity between the aniline and the surface of ACFs was increased by the polarization," including "hydrogen-bonding between amino group and the surface groups containing oxygen" and an enhancement of "electron density on the benzene ring" (with respect to what?) which was presumably "beneficial to the π–π interaction between aniline and the surface."

It is obvious from the above analysis that the inability to link the concepts of pH_{IEP} (or pH_{PZC}) and PZC—see Section 5.3.3.3—continues to plague [7] the interpretation of electrosorption phenomena on carbon surfaces. The "bottom line" issue should be clear, however, as was illustrated in Figure 5.15 [7].

The influence of electric potential on adsorption of organic compounds has been a topic of interest since at least the 1980s [7,232]. Thus, for example, Eisinger and Alkire [232] analyzed this phenomenon using β-naphthol (pK_a = 9.51) electrosorption on graphite (BET surface area, 1.86 m²/g), which they inappropriately classified as "an adsorbent of moderately high surface area." Their findings are summarized in Figure 5.16. There was no "maximum adsorption near the adsorbent's potential of zero charge"; graphite, presumably, "has been reported to be near −0.7 V (with respect to the mercury/mercurous sulfate reference electrode)." Therefore, the monotonic increase in the uptake as the potential became more positive was "simply viewed as an electrostatic attraction for the negative charge density of the aromatic structure of β-naphthol by the positively charged graphite," despite the fact that the pH was held at 6.6 in order to focus on the effect of dispersion, rather than the coulombic (electrostatic) interactions [7].

Ban et al. [233] reported on an ambitious study of the fundamentals of electrosorption for wastewater treatment of industrial effluents. They emphasize the expectation that "[p]otentials anodic to E_{PZC} decrease the amount of cations adsorbed at the surface and inversely increase the amount of adsorbed anions," whereas "[u]ncharged molecules are most strongly adsorbed at, and close to, the potential of zero charge, E_{PZC}." For a commercial AC (Norit ROW 0.8 Supra), they determined E_{PZC} to be 310 mV (vs. saturated Ag/AgCl), whereas titration experiments yielded 0.3 and 0.2 mmol/g of acidic and basic surface groups, respectively. The potential dependence of the pH (in 0.1 M KCl) is reproduced in Figure 5.17; a similar trend, with the expected slope of ca. 60 mV per pH unit, was observed when pH was the independent parameter. Their results on the potential dependence of adsorbability are reproduced in Figure 5.18, e.g., the trend for the

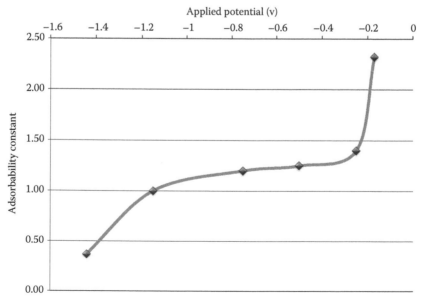

FIGURE 5.16 Correlation between the uptake of β-naphthol on graphite (expressed as a dimensionless "adsorbability constant") and the applied electric potential. (From Eisinger, R.S. and Alkire, R.C., *J. Electroanal. Chem.*, 112, 327, 1980.)

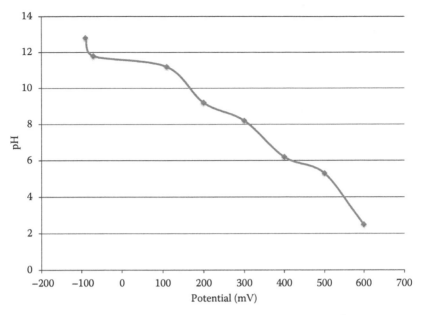

FIGURE 5.17   Potential (vs. Ag/AgCl) vs. pH for an activated carbon electrode. (From Ban, A., et al., *J. Appl. Electrochem.*, 28, 227, 1998.)

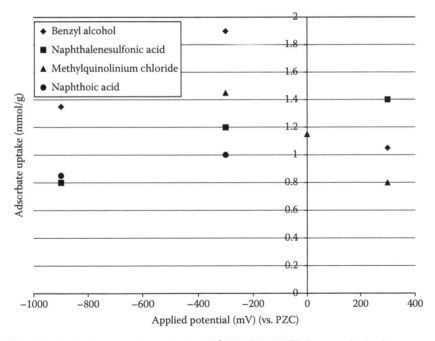

FIGURE 5.18   Correlation between the uptake (at 10^{-3} M in 0.1 M KCl) for several adsorbates on an activated carbon and the applied potential. (From Ban, A., et al., *J. Appl. Electrochem.*, 28, 227, 1998.)

anionic substances (naphthalene sulfonate and naphtoate) was "as expected: going from negative to positive potentials, the adsorption becomes stronger due to coulombic interactions between the charged surface and the ions."

Although in charge storage pseudocapacitance due to a Faradaic process is considered beneficial, it has been argued [234] that in electrosorption "Faradaic processes… complicate the electrosorption process, rendering it unpredictable." These authors also discussed a more fundamental issue: for carbon and graphite electrodes (in contrast to, e.g., a well-behaved mercury electrode), not a sharp but a flat minimum is obtained when the capacitance is plotted as a function of potential and electrolyte concentration; this inconsistency was attributed to "the heterogeneity and the existence of localized π-electron states of the graphite;" intriguingly the reference for this statement was only an earlier paper by Oren and Soffer [219], while the authors acknowledged that "this assumption has not been verified by other experiments." Several decades later, it is of interest whether there is evidence for the existence and explanation of the nature of such "localized π-electron states" (see Figure 5.4).

Chen and Hu [228] found that the maximum electrosorption capacity of an AC felt for thiocyanate anions was obtained at a pH of 3, following the same trend with pH variations as adsorption in the absence of polarization. The explanation offered, in somewhat vague terms, was "the change of the charge state under various pHs." It is intriguing, however, that upon reuse the amount adsorbed increased.

Han et al. [230] reported a 10-fold enhancement of phenol adsorption on AC fibers in a basic solution by application of a 700 mV bias potential; they attributed this to "dipole–dipole interaction with $\Delta H_{ads} = -20$ kJ/mol, besides electrostatic interaction." They did characterize the chemical surface properties of the adsorbent—"acidic and basic groups were 0.82 and 0.54 mmol/g [according to Boehm titration, indicating] that ACFs had a large contribution of surface basic groups, which benefited the adsorption of acidic pollutants, such as phenol"—but they did not discuss how this might explain its electrosorption behavior.

In the recent study by Ania and Béguin [132], such a mechanism was offered for the uptake of bentazone by AC cloth before and after its oxidation with ammonium persulfate, and this is illustrated in Figure 5.19. Although the point of zero charge of the cloth was not determined, the expected suppression of adsorption under acidic conditions and upon surface oxidation [7] was observed. Upon anodic polarization, which induced (albeit in an unspecified way) a positive charge on the carbon surface, there was an enhancement in adsorbate uptake.

In a series of studies by Conway and coworkers [195–197,235–238], both sorption and electrosorption were examined for a variety of adsorbates and carbon adsorbents. The following conclusion is symptomatic of their appreciation of the phenomena governing these processes (and of the persistent absence of "dialogue" between electrochemists and carbon surface chemists): "Conversion of PhOH to PhO⁻ by ionization at alkaline pH causes a major decrease of adsorption of the phenolic structure due to hydration of the resulting anion" [196]; the arguably more obvious explanation, and one that has been shown to be valid in many similar studies [7]—repulsion of PhO⁻ by the negatively charged carbon surface above its pH_{IEP}—is not even considered, let alone dismissed by the authors. In a related study of pyridine (Py) adsorption [236], the authors noted that "the passage of charge gives rise to pH changes in solution, which can appreciably affect the adsorption removal of pollutants... due to base/acid ionization," but did not address the specifics of such interactions, except to note that Py becomes positively charged in acidic solution and to speculate that "hydration of the PyH⁺ ion [makes] the 'Py' moiety more hydrophilic with conversion of the N lone-pair to a quaternary-ion type center (NH⁺) having less affinity for the C-surface." It is left to the reader to reconcile these arguments with the statement that "Py contains an N atom which is more electronegative than an sp^2-hybridized C, so that its preferred adsorption is on a positively charged surface."

Finally, it should be noted that electrosorption of biologically relevant molecules, including bacteria [239], has long been of interest. An in-depth discussion cannot be offered here. Suffice it to mention, for example, that "both the capacitance and dopamine adsorption [on a glassy carbon electrode] decreased as the pH of the electrolyte was increased" [240]; by contrast, the "equilibrium amount of adsorbed material [2,6-diamino-8-purinol on a rough pyrolytic graphite electrode] is the same at high [13.0] and low [1.0] pH" [241] even though "the adsorption rate increases with the

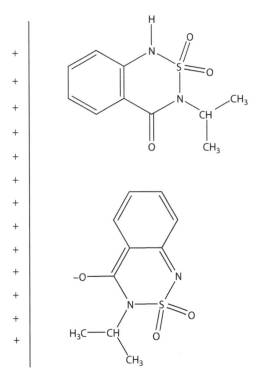

FIGURE 5.19 Postulated interaction between (positively charged) activated carbon cloth surface and (negatively charged) bentazone under anodic polarization. (From Ania, C.O. and Beguin, F., *Water Res.* 41, 3372, 2007.)

decrease in pH" [241]. Clearly, much progress should be expected in the coming years to clarify such important details, not only of the effect of carbon electrochemistry on the adsorption of both simple and complex molecules, but also of the "opposite" effect of the adsorbed species on the electrochemistry of carbon surfaces (e.g., see [242]).

5.3.5 OXYGEN REDUCTION

Carbon is a reducing agent *par excellence.* Its affinity for oxygen is very high. Therefore, carbon in its various material forms is used extensively as a source of electrons or as an electron-transfer medium in O_2 reduction. From a practical point of view, the rate of the O_2 reduction reaction (ORR) is at the heart of the electrocatalysis problem in fuel cells; from a fundamental standpoint, it is a challenging issue because of coupled electron and proton transfer processes [243]. The very much related more general topic of electrolytic reduction of organic and inorganic substances has a long history [244] and is beyond the scope of the present review. The focus here is the more recent literature on the most popular, and arguably the most difficult, redox process: reduction of molecular oxygen in the absence of metal electrocatalysts and as influenced by carbon's surface chemistry.

A tremendous amount of work has been devoted to ORR on all kinds of surfaces; but Bockris and Khan [245] were forced to conclude that "no generalizations as to mechanism have been made, and, correspondingly, no key to the treasure chest of fast catalysis… has been found." Despite decades of research and even though the "effect of electrode material on the O_2 reduction reaction (ORR) has been reviewed extensively," Yang and McCreery [246] concluded even more recently that "detailed mechanisms remain elusive" and that for carbon surfaces "there is no consensus on the mechanism or even the identity of adsorbed intermediates." In spite of the early attempts to generalize the "one-electron mechanism" [243], the same conclusion has been reached by Sljukic et al. [247]: "To date, there is no agreement on the mechanism, rate determining step, or adsorbed intermediates of the electrochemical reduction of oxygen on carbon surfaces."

The surface chemical effects of interest do not go as far as those induced in (extensively) modified carbon electrodes [248], e.g., by pyrolyzed phthalocyanines or macrocycles [249–255], by anthraquinone or its derivatives [126,247,256–259], or by aryl groups [125], or those of "stable and efficient sonoelectrocatalysts" by modifying GC electrodes with 9,10-phenanthraquinone or 1,2-naphthoquinone [260]. Instead, it is explored here whether and how a seemingly simple but crucial issue has been addressed or resolved: what makes O_2 adsorption in ORR nondissociative? The isotopic labeling evidence for this experimental fact has been presented half a century ago [261], and it has not been challenged [262]. The implication, based on the equally noncontroversial literature that O_2 chemisorption on carbons (even at room temperature) is dissociative, is summarized below:

$$O_2 + 2C_{f1} = 2C-O \tag{5.25}$$

$$O_2 + e^- + C_{f2} = C-O_2^{*-} \tag{5.26}$$

Subsequent to Reaction 5.25, in carbon gasification or combustion the adsorbed oxygen invariably wrestles the carbon reactive site (C_{f1}) from the carbon structure to desorb as CO or CO_2; during ORR, however, subsequent to Reaction 5.26 [243] the adsorbed superoxide anion radical picks up a proton from the electrolyte solution to form and desorb a hydroperoxide ion HO_2^-, thus regenerating the same carbon active site (C_{f2}). In what way, if any, are the sites C_{f1} and C_{f2} different? It seems that this line of reasoning had not been explored before and, arguably, this has been the main obstacle in understanding the ORR mechanism and in suitably tailoring carbon surfaces. It is not surprising to encounter the relevant misconceptions in the older electrochemistry literature, even the most authoritative one. Thus, for example, Morcos and Yeager [263] argued that O_2 adsorbs on the carbon surface prior to the charge-transfer step and justified this assumption by invoking a relatively obscure study by Singer and Spry [264]—and thus ignoring Singer's subsequent and better known contributions, as well as the mainstream carbon adsorption literature—according to which "reversible chemisorption of O_2 [occurs] on carbon from the gas phase, presumably without bond cleavage, as revealed by ESR studies." Many years of intense scrutiny [265–267] have not (yet!) clarified the effects of physisorption vs. chemisorption of O_2 on the ESR of carbon surfaces, but there is no question that on the graphene edge sites O_2 adsorbs both irreversibly and dissociatively. In light of the fact that, obviously, "the reaction must involve adsorption" [268], it is surprising that even the recently proposed mechanistic schemes [269] insist on identifying graphene edges as the "active sites" [129,270–273], but fail to realize, let alone address, this basic inconsistency.

A recent study that deals with the "mechanism of reductive oxygen adsorption on active carbons with various surface chemistry" [274] is exactly the topic of our interest here and warrants a detailed commentary. The authors used an electrochemical cell design that allowed them to "investigate quantitatively the processes of oxygen reduction by various types of carbons, measuring a current in an external circuit formed by carbon electrons, captured by O_2 molecules on the previously platinuminated electrode (spongy platinum)." The various types of carbons included those experimentally modified by the introduction of oxygen and nitrogen functionalities, as well as several theoretically analyzed prototypical graphene structures with information about their respective HOMO energy levels. Thus, for example, 37-ring graphenes C_{96}, $C_{92}O_4$, and $C_{93}N_3$ are quoted as having their HOMO levels at −7.20, −5.67, and −5.65 eV (using AM1 level of theory), respectively, presumably reflecting their increasing ease of electron donation, even though the issues of edge termination and the presence of unpaired electrons [6] were not discussed. Although the O_2 reduction kinetic results seemed ambiguous, the authors' main mechanistic conclusion, intriguingly based on elemental analyses and measurements of acidic rather than basic surface functionalities, appears to be that their "results suggest... the key role of furan, chromene and pyrone forms of oxygen in the manifestation of electron-donor (reductive) ability of carbons in relation to strong oxidizers and oxygen." Such a role is presumably consistent with the following key mechanistic step:

As in other electrochemical applications, the influence of N-doping is of great interest, and it has been analyzed in a large number of very recent studies [127,131,269,275–278]. By comparing the behavior of carbon nanofibers (CNF) prepared in the presence (using pyridine and ferrocene) and absence (using xylene and ferrocene) of nitrogen, Maldonado and Stevenson [275] reported "significant catalytic activity" for the former and concluded that both "the exposed edge plane defects and nitrogen doping are important factors for influencing adsorption of reactive intermediates (i.e., superoxide, hydroperoxide) and for enhancing electrocatalysis for the ORR." More specifically, these authors emphasize the "importance of the basic nature of N-doped CNFs" that confers them "stability against oxidation of the carbon which generally forms acidic oxygen functionalities" and propose a "prevalence of adsorbed [superoxide anions] on the surface." Simultaneous doping of boron and nitrogen was also reported [278] to be beneficial; the reasons were not clarified, but the authors speculated that the "roles of the nitrogen surface species and the B–N–C moieties [may be] to change the surface properties such as hydrophilicity and basicity and to form active centers." The same group further discussed the "nature of the active site" for carbons derived from ferrocene-poly(furfuryl alcohol) mixtures [273]. Intriguingly, a recent theoretical study [131]—in which "[n]itrided graphite edges, with N atoms substituting one or two CH groups, [were] modeled to establish some of the effects of N on edges with and without Co added"—suggested that "a bare graphite edge with one N atom... is not active for O_2 reduction because OOH bonds too weakly."

Ambiguity persists regarding the basic role in ORR of oxygen surface functionalities at graphene edges. A typical recent example is the "suggested mechanism for the electrocatalytic reduction of oxygen on MWCNTs surface" where "quinone-type groups on nanotube edges are considered to be the active sites" [279]. As reproduced below, this scheme leaves unacceptably vague all the important mechanistic details of ORR, despite the claim that the "significance of various surface sites for oxygen reduction on carbon electrodes has been thoroughly studied."

Admittedly, identification of the quinone groups as the active sites in ORR is a very popular proposal [130,247], presumably going all the way back [247] to the work of Garten and Weiss [21,79], and more recently extended to their importance in CNTs [280]; but the key supporting arguments have been missing, especially ones that would sound convincing to a carbon surface chemist. Obviously, the enhancement cannot be due to the more favorable adsorption of O_2 onto the quinone sites.

The early proposal of Yeager [245,281] for carbon surfaces, based on the work of Garten and Weiss [21,79], is reproduced in Figure 5.20. On the basis of the experimental fact that the "current densities for the reduction of O_2 to HO_2^- are far lower on the basal plane of well-ordered graphite than on ordinary pyrolytic graphite or glassy carbon," Yeager [281] intriguingly concluded that "[t]his implies that the O_2 reduction on carbon/graphite ordinarily involves a strong interaction of O_2 with functional groups on the surface." Figure 5.20 hardly represents an example of such interaction, but the first step in the sequence shown arguably has the merit from today's perspective [6], assuming that the armchair site of O_2 adsorption is of the carbyne type. Of course, the second step shown makes very little sense, because the expected fate of the O_2 adsorbed on the armchair site is its dissociation and formation of two quinone functional groups, rather than the formation of a peroxy anion.

According to Kinoshita [282], "[e]xperimental evidence suggests that oxygen molecules adsorb strongly on the edge sites of carbonaceous materials, interacting with the surface functional groups that are present and resulting in higher activity than at the basal plane." Echoing the suggestions of Yeager and coworkers [263] and Appleby and Marie [283], the mechanism of oxygen reduction differs on the surfaces of different carbons. Thus, on graphite, it is presumably

$$O_2 \rightarrow O_2\left(ads\right) \tag{5.27}$$

$$O_2\left(ads\right) + e^- \rightarrow O_2\cdot^-\left(ads\right) \tag{5.28}$$

$$2O_2\cdot^-\left(ads\right) + H_2O \rightarrow O_2 + HO_2^- + OH^- \tag{5.29}$$

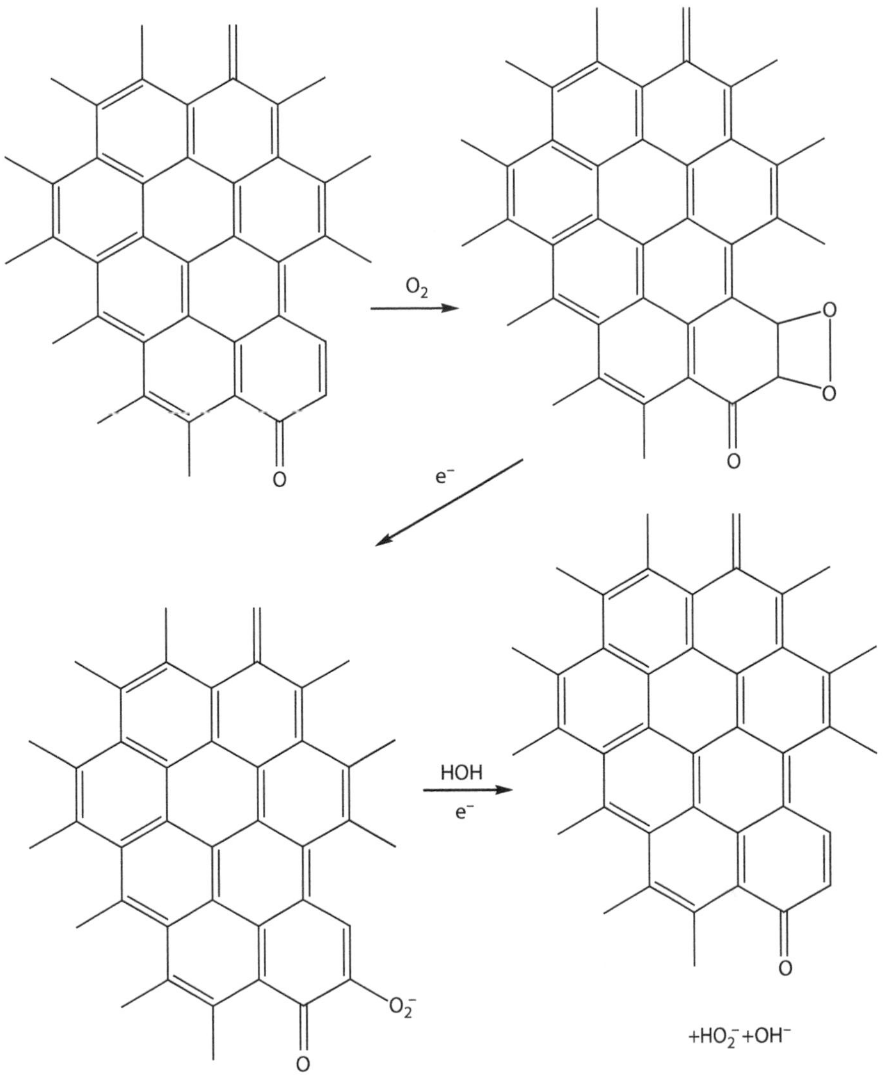

FIGURE 5.20 Mechanism of O_2 reduction on carbon surfaces proposed by Garten and Weiss in the 1950s and endorsed by Yeager [281].

On GC, the mechanism is proposed to be

$$O_2 \rightarrow O_2(ads) \tag{5.30}$$

$$O_2(ads) + e^- \rightarrow [O_2(ads)]^- \tag{5.31}$$

$$[O_2(ads)]^- \rightarrow O_2(ads)^- \tag{5.32}$$

$$O_2(ads)^- + H_2O \rightarrow HO_2(ads) + OH^- \tag{5.33}$$

$$HO_2(ads) + e^- \rightarrow HO_2^-(ads) \tag{5.34}$$

$$HO_2^-(ads) \rightarrow HO_2^- \tag{5.35}$$

And on carbon black, the proposed mechanism is as follows:

$$O_2 + e^- \rightarrow O_2^- \tag{5.36}$$

$$O_2^- + H_2O \rightarrow HO_2^- + OH \tag{5.37}$$

$$OH + e^- \rightarrow OH^- \tag{5.38}$$

Kinoshita did not discuss the possible reasons for these differences, nor has apparently anyone else, at least not convincingly. For example, based on everything known about carbon surface chemistry today, it is not easy to explain why electron transfer would occur on carbon blacks prior to O_2 adsorption, and on graphite or GCs subsequent to O_2 adsorption.

Very recently, Qu [129] recognized this problem and made a valiant effort to tackle the key issue head-on: "The nature of the so-called 'active sites,' which facilitate the electron transfer and possibly also provide adsorption sites for the superoxide ions, has not been reported" and the author thus proposed to investigate "the relationship between the surface structure of activated carbon and its catalytic activity for the oxygen reduction." He reported that the "specific catalytic activity... is determined by the percentage of the edge orientation on the surface" and that, not surprisingly (by analogy with carbon oxidation reaction [284]), the "higher the content of the edge plane, the lower the oxygen reduction potential." The author further argued, rather generally, that "a strong interaction between the reactant and the electrode surface is required" and then anticlimactically echoed Yeager's early ideas [285] that the "functional groups, e.g., quinine [sic] groups, may participate in the course of oxygen reduction," without explaining how exactly such participation might occur. The main result is summarized in Figure 5.21. The slope of the current density vs. potential plot was taken to be a measure of electrocatalytic activity, whereas the parameter R ("measure of the number of [graphene] sheets arranged as a single layer" [286]), obtained from x-ray diffraction, was interpreted to be a measure of the active site concentration. The justification of the latter assumption is very confusing, however, apart from the fact that the correlation obtained is quite weak. What the author calls "the surface content of the edge orientation" is more appropriately obtained from the crystallite width (e.g., the two-dimensional 10 reflection) rather than from the crystallite height, and the existence of a qualitative correlation between this edge-site-concentration parameter and carbon reactivity is well known [287]. Furthermore, the exactly opposite trend from that claimed by the author is much more commonly observed in carbon materials: as the intensity of the 002 peak increases (i.e., as R increases), the edge site concentration decreases. Therefore, the author's conclusion that the "edge orientation on the surface of an activated carbon serves as an 'active site' for the reduction of oxygen," while a reasonable hypothesis (whose verification is indeed the "holy grail" of electrocatalysis research) remains unsubstantiated.

Clearly, despite a tremendous body of available literature, the essential details of the mechanism of oxygen reduction on carbon surfaces remain obscure and, unfortunately, even more experimental work will be needed to elucidate it. However, a judicious and yet unexplored guide for future studies is readily available, based on the principle of microscopic reversibility: the reaction of carbon surfaces with hydrogen peroxide (in the absence of an electric field). Even a cursory analysis of this arguably fruitful topic is beyond the scope of the present review, however. Suffice it to say that the pH dependence of ORR is of interest. It was studied, for example, by Yang and McCreery [246].

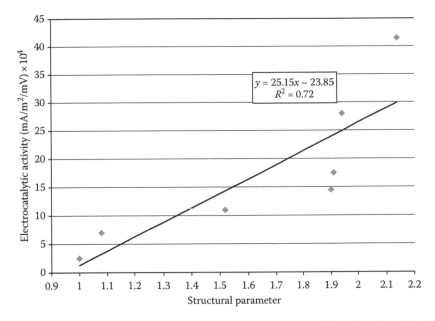

FIGURE 5.21 Presumed correlations between ORR electrocatalytic activity of activated carbon and a "structural parameter" (see text) determined from x-ray diffraction. (Adapted from Qu, D., *Carbon*, 45, 1296, 2007.)

A well-established fact is that, depending on the pH and subsequent to the transfer of the first electron, the reduction product is H_2O_2 or HO_2^-. But a more difficult (and much more controversial!) mechanistic issue is "the nature of the interaction between carbon surfaces and oxygen which leads to electrocatalytic formation of HO_2^- in the absence of transition metals." After examining the behavior of pristine and variously modified or pretreated GC surfaces, they argued that on "clean, unmodified GC, the pH dependence of the O_2 reduction mechanism is consistent with control of the reduction process by adsorbed O_2^{*-} or $^*O_2H$": upon O_2^{*-} adsorption there is an increase in pK_a of O_2^{*-} (ads) compared to O_2^{*-} (aq) and the presumable result is a change in the rate-determining step from O_2^{*-} protonation [O_2^{*-} (ads) $+ H_2O = HO_2^*$(ads) $+ OH^-$] for pH > 12 to O_2 reduction [$O_2 + e = O_2^{*-}$ (ads)] for pH < 10. These authors recognized that the "nature of the adsorption site for O_2^{*-} is not obvious. They speculated that a "likely candidate for the adsorption site is a surface radical or 'dangling bond.'" Such uncertainties further illustrate the problems mentioned earlier in this section: it is not productive, and indeed more often than not it is misleading, to represent the first electron transfer step in the ORR mechanism as shown above, instead of explicitly showing that a surface site is involved, and one that somehow prevents O_2 dissociation. This is especially important if indeed, as stated by Kornienko et al. [271], the "most acceptable explanation of the whole set of experimental data suggests that the slow stage of the reduction is the transfer of the first electron to an adsorbed oxygen molecule."

An early intriguing argument by Balej et al. [288] has been essentially ignored in the published literature, although it does deserve careful scrutiny. These authors explored a "relation between the activity of electrodes from various carbonaceous materials for reduction of oxygen to hydrogen peroxide and [the] concentration of their paramagnetic centers."

Pirjamali and Kiros [289] examined the relationship between the "pH values of the aqueous slurries, [which] provides an indication of the surface functional groups of the carbon blacks and [are] equivalent to the point[s] of zero charge of the materials," and the ORR activities in alkaline electrolyte, and the only conclusion possible, without an explanation, was that "increased temperature treatment has [a] more significant effect on better electrode performances than low temperature treatments." By contrast, Biniak et al. [290] found a "clear correlation… between the basicity of the carbon surface, the catalytic activity of the heterogeneous decomposition of hydrogen peroxide and the activity with respect to dioxygen reduction on powdered activated carbon electrodes."

The early mechanistic proposal by Tarasevich [86,291], and one that has been practically ignored in the "mainstream" ORR literature, arguably comes closest to a "compromise" that should be appealing to both electrochemists and carbon surface chemists. It is reproduced in Figure 5.22 and it is "modernized," as well as made much more specific, in Figure 5.23.

The main idea taken from the Tarasevich proposal is the indirect role of the quinone oxygen that facilitates the formation of either adjacent or remote carbene edge site in the graphene layer [6] and on which nondissociative O_2 adsorption readily occurs. (Such nondissociative adsorption also readily occurs on the basal plane and not only on the carbene edge sites; whether electron transfer necessary for O_2 reduction occurs there as readily as on the carbene sites is worthy of greater scrutiny.) This electron-rich (Lewis base) site also facilitates proton adsorption and the consequent formation and desorption of the hydroperoxide anion. This leaves behind a regenerated albeit electron-deficient site, whose electron-donating ability can be recovered by virtue of the delocalized π system, as illustrated in Figure 5.23. A more detailed quantum-chemical analysis of this process will be presented elsewhere.

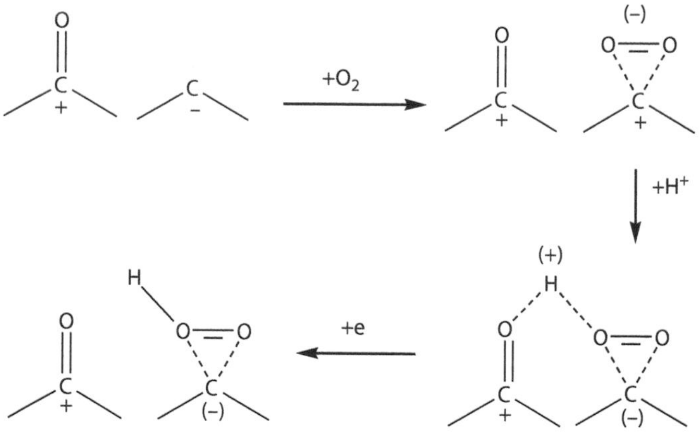

FIGURE 5.22 Mechanism proposed by Tarasevich [291] for O_2 reduction on the carbon surface.

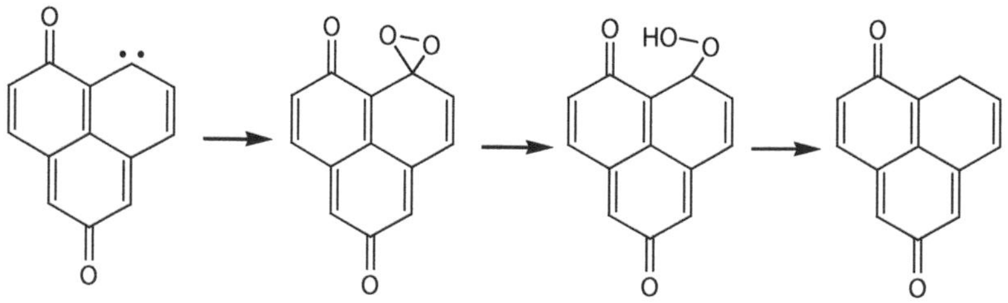

$$C_{13}H_6O_2 + O_2 + H_2O = C_{13}H_6O_2 + HO_2^- + OH^-$$

(100 electrons)   (98 electrons)

FIGURE 5.23 Updated proposal for O_2 reduction that is consistent with current knowledge of carbon surface chemistry.

5.4 SUMMARY AND OUTLOOK

As anticipated, this chapter has posed many questions and has provided few answers or "take-home messages." The justification for such "conclusions" is easy, however: such is the state of knowledge in this eminently interdisciplinary and thus very challenging field. Nevertheless, hopefully the path forward is now (much?) clearer. Surely among the most valuable pieces of information about the behavior of carbon materials in electrochemical applications is a plot of surface potential vs. pH. Such a Pourbaix diagram would be a powerful and convenient summary of carbon's electron and proton transfer characteristics and its differences and similarities with respect to metal and other semiconductor electrodes. Once such equilibrium information becomes available and understood, one can begin to gain a fundamental understanding of the even more important kinetic and catalytic factors that control the relevant electrochemical processes. It is anticipated that, in addition to (or instead of) the proton/electron transfer reactions (Equations 5.1 through 5.3 and 5.9 through 5.11) the following reactions will be the key mechanistic steps in such processes:

$$>C_x + e^- = >C_x^- \qquad (5.39)$$

$$>C_x^- + H^+ = >C_x^- /\!/ H^+ \qquad (5.1a)$$

$$>C_x - e^- = >C_x^+ \qquad (5.40)$$

$$>C_x^+ + OH^- = >C_x^+ /\!/ OH^- \qquad (5.41)$$

$$>C_x O^- + H^+ = >C_x O^- /\!/ H^+ \qquad (5.2a)$$

Quantifying their relative importance as a function of carbon surface properties and of solution electrochemical conditions will undoubtedly keep carbon electrochemists occupied for the next quarter of a century.

During the past quarter of a century, it had already been acknowledged [292] that "relatively little use has been made of [Pourbaix diagrams] in the field of organic chemistry." At the same time, initial insight was provided by Fabish and coworkers [138,293,294] as they interpreted a characteristic minimum observed for carbon black: the work function increase at pH < 6 was due to the increasing contribution by acidic surface groups; the increase at pH > 6 was attributed to "progressive localization of itinerant basal plane electrons" by the chemisorbed oxygen. Since then, a huge wealth of additional information has become available, but not much progress has been made in advancing our understanding of these and other relevant phenomena. Indeed, it could be argued that further experimental work is warranted only to the extent that it is carefully designed to explicitly answer some of the so many fundamental questions that remain unanswered. In particular, future research should focus on the Pourbaix diagram differences among different carbons, and on these differences for a given carbon and a variety of carefully selected redox reactions. For the O_2 reduction reaction, a heretofore unavailable concrete proposal is advanced here that attempts to resolve some of the identified inconsistencies in the carbon electrochemistry and electrocatalysis literature.

ACKNOWLEDGMENTS

In compiling the experimental evidences for the "story" presented in this review, I have strived to take advantage of the entire body of the relevant literature—both the definitive and the half-baked or preliminary, both the illuminating and the speculative or bewildering—and I am grateful to the cited authors for sharing, even if sometimes prematurely, their findings, their insights, their speculations, and occasionally, their confusions. In particular, I dedicate the review to Prof. H. Peter Boehm, as his 80th birthday is being celebrated; his remarkably insightful writings on carbon surface chemistry and catalysis, spanning half a century, have been a constant driving force and a most reliable source of fruitful ideas and inspiration. Financial support provided by FONDECYT-Chile, projects No. 1060950 and 1080334, is gratefully acknowledged.

REFERENCES

1. Kinoshita, K. *Carbon: Electrochemical and Physicochemical Properties*, 1988, New York: Wiley-Interscience.
2. Randin, J.-P. Carbon, in Bard, A. J. (ed.), *Encyclopedia of the Electrochemistry of Elements*, 1976, New York: Marcel Dekker, Inc., pp. 1–251.
3. Sarangapani, S., Akridge, J. R., and Schumm, B. (eds), *Proceedings of the Workshop on the Electrochemistry of Carbon*, 1984, Pennington, NJ: The Electrochemical Society.
4. Leon y Leon, C. A. and Radovic, L. R. Interfacial chemistry and electrochemistry of carbon surfaces, in Thrower, P. A. (ed.), *Chemistry and Physics of Carbon*, Vol 24, 1994, New York: Marcel Dekker, Inc., pp. 213–310.
5. Boehm, H. P. *Carbon*, 1994; **32**: 759–769.
6. Radovic, L. R. and Bockrath, B. *J Am Chem Soc*, 2005; **127**: 5917–5927.
7. Radovic, L. R., Moreno-Castilla, C., and Rivera-Utrilla, J. Carbon materials as adsorbents in aqueous solutions, in Radovic, L. R. (ed.), *Chemistry and Physics of Carbon*, Vol. 27, 2001, New York: Marcel Dekker, pp. 227–405.
8. Radovic, L. R. Surface chemistry of activated carbon materials: State of the art and implications for adsorption, in Schwarz, J. A. and Contescu, C. I. (eds), *Surfaces of Nanoparticles and Porous Materials*, 1999, New York: Marcel Dekker, pp. 529–565.
9. Brodd, R. J. Carbon research needs, in Sarangapani, S., Akridge, J. R., and Schumm, B. (eds), *Proceedings of the Workshop on the Electrochemistry of Carbon*, 1984, Pennington, NJ: The Electrochemical Society, pp. 608–622.
10. Martel, D., Sojic, N., and Kuhn, A. *J Chem Educ*, 2002; **79**(3): 349–352.
11. Hsieh, C.-T. and Teng, H. *Carbon*, 2002; **40**(5): 667–674.
12. Burg, P. and Cagniant, D. Characterization of carbon surface chemistry, in Radovic, L. R. (ed.), *Chemistry and Physics of Carbon*, Vol. 30, 2008, UK: Taylor & Francis (CRC Press), pp. 129–175.
13. Sherwood, P. M. A. Surface properties of carbons and graphites, in Buschow, K. H. J., Cahn, R. W., Flemings, M. C., Ilschner, B., Kramer, E. J., and Mahajan, S. (eds), *Encyclopedia of Materials: Science and Technology*, 2001, Oxford: Pergamon Press, pp. 985–995.
14. Davtyan, O. K. and Tkach, Y. A. *Soviet Electrochem*, 1965; **1**(2): 172–175.
15. Laine, N. R., Vastola, F. J., and Walker, P., Jr. *J Phys Chem*, 1963; **67**: 2030–2034.
16. Laine, N. R., Vastola, F. J., and Walker, P. L., Jr. *Proc 5th Conf Carbon*, 1961; **2**: 211–217.
17. Matskevich, E. S., Ivanova, L. S., and Strazhesko, D. N. *Soviet Electrochem*, 1970; **6**(5): 617–621.
18. Steenberg, B. *Adsorption and Exchange of Ions on Activated Charcoal*, 1944, Uppsala, Sweden: Almqvist & Wiksells.
19. Epstein, B., Dalle-Molle, E., and Mattson, J. S. *Carbon*, 1971; **9**: 609–615.
20. Garten, V. A. and Weiss, D. E. *Third Conference on Carbon*, 1959, Buffalo, NY: Pergamon Press.
21. Garten, V. A. and Weiss, D. E. *Austral J Chem*, 1955; **8**(1): 68–95.
22. Rivin, D. *Fifth Conference on Carbon*, 1963, University Park, PA: Pergamon Press.
23. Esumi, K., Ishigami, M., Nakajima, A., Sawada, K., and Honda, H. *Carbon*, 1996; **34**: 279–281.
24. Shi, Z., Lian, Y., Liao, F., Zhou, X., Gu, Z., Zhang, Y., and Iijima, S. *Solid State Commun*, 1999; **112**: 35–37.
25. Hernadi, K., Siska, A., Thien-Nga, L., Forro, L., and Kiricsi, I. *Solid State Ionics*, 2001; **141–142**: 203–209.
26. Moon, J.-M., An, K. H., Lee, Y. H., Park, Y. S., Bae, D. J., and Park, G.-S. *J Phys Chem B*, 2001; **105**(24): 5677–5681.
27. Sato, Y., Ogawa, T., Motomiya, K., Shinoda, K., Jeyadevan, B., Tohji, K., Kasuya, A., and Nishina, Y. *J Phys Chem B*, 2001; **105**: 3387–3392.
28. Banerjee, S. and Wong, S. S. *J Phys Chem B*, 2002; **106**: 12144–12151.
29. Sen, R., Rickard, S. M., Itkis, M. E., and Haddon, R. C. *Chem Mater*, 2003; **15**(22): 4273–4279.

30. Hu, H., Zhao, B., Itkis, M. E., and Haddon, R. C. *J Phys Chem B*, 2003; **107**(50): 13838–13842.
31. Gajewski, S., Maneck, H.-E., Knoll, U., Neubert, D., Dörfel, I., Mach, R., Strauss, B., and Friedrich, J. F. *Diamond Relat Mater*, 2003; **12**: 816–820.
32. Wiltshire, J. G., Khlobystov, A. N., Li, L. J., Lyapin, S. G., Briggs, G. A. D., and Nicholas, R. J. *Chem Phys Lett*, 2004; **386**(4–6): 239–243.
33. Cech, J., Curran, S. A., Zhang, D. H., Dewald, J. L., Avadhanula, A., Kandadai, M., and Roth, S. *Phys Status Solidi B*, 2006; **243**(13): 3221–3225.
34. Pels, J. R., Kapteijn, F., Moulijn, J. A., Zhu, Q., and Thomas, K. M. *Carbon*, 1995; **33**: 1641–1653.
35. Holzinger, M., Abraha, J., Whelan, P., Graupner, R., Ley, L., Hennrich, F., Kappes, M., and Hirsch, A. *J Am Chem Soc*, 2003; **125**(28): 8566–8580.
36. Grant, K. A., Zhu, Q., and Thomas, K. M. *Carbon*, 1994; **32**(5): 883–895.
37. Biniak, S., Szymanski, G., Siedlewski, J., and Swiatkowski, A. *Carbon*, 1997; **35**(12): 1799–1810.
38. Boehm, H. P., Mair, G., Stoehr, T., de Rincon, A. R., and Tereczki, B. *Fuel*, 1984; **63**: 1061–1063.
39. Solar, J. M., Leon y Leon, C. A., Osseo-Asare, K., and Radovic, L. R. *Carbon*, 1990; **28**(2/3): 369–375.
40. Bimer, J., Salbut, P. D., Berlozecki, S., Boudou, J.-P., Broniek, E., and Siemieniewska, T. *Fuel*, 1998; **77**(6): 519–525.
41. Kelemen, S. R., Gorbaty, M. L., and Kwiatek, P. J. *Energy Fuels*, 1994; **8**(4): 896–906.
42. Kelemen, S. R., Afeworki, M., Gorbaty, M. L., Kwiatek, P. J., Solum, M. S., Hu, J. Z., and Pugmire, R. J. *Energy Fuels*, 2002; **16**(6): 1507–1515.
43. Maldonado, S., Morin, S., and Stevenson, K. J. *Carbon*, 2006; **44**(8): 1429–1437.
44. Jurewicz, K., Babel, K., Pietrzak, R., Delpeux, S., and Wachowska, H. *Carbon*, 2006; **44**(12): 2368–2375.
45. Shalagina, A. E., Ismagilov, Z. R., Podyacheva, O. Y., Kvon, R. I., and Ushakov, V. A. *Carbon*, 2007; **45**(9): 1808–1820.
46. Marinkovic, S. Substitutional solid solubility in carbon and graphite, in Thrower, P. A. (ed.), *Chemistry and Physics of Carbon*, Vol 19, 1986, New York: Marcel Dekker, pp. 1–64.
47. Sun, Y.-P., Fu, K., Lin, Y., and Huang, W. *Acc Chem Res*, 2002; **35**: 1096–1104.
48. Kuzmany, H., Kukovecz, A., Simon, F., Holzweber, A., Kramberger, C., and Pichler, T. *Synthetic Metals*, 2004; **141**(1–2): 113–122.
49. Hirsch, A. and Vostrowsky, O. *Topics Curr Chem*, 2005; **245**: 193–237.
50. Hirsch, A. *Phys Status Solidi B*, 2006; **243**(13): 3209–3212.
51. Gong, K. P., Yan, Y. M., Zhang, M. N., Su, L., Xiong, S. X., and Mao, L. Q. *Anal Sci*, 2005; **21**(12): 1383–1393.
52. Banerjee, S., Kahn, M. G. C., and Wong, S. S. *Chem Eur J*, 2003; **9**(9): 1899–1908.
53. Peng, H. Q., Alemany, L. B., Margrave, J. L., and Khabashesku, V. N. *J Am Chem Soc*, 2003; **125**(49): 15174–15182.
54. Mickelson, E. T., Huffman, C. B., Rinzler, A. G., Smalley, R. E., Hauge, R. H., and Margrave, J. L. *Chem Phys Lett*, 1998; **296**: 188–194.
55. Boehm, H. P., Hofmann, U., and Clauss, A. *Third Conference on Carbon*, 1957, Buffalo, NY: Pergamon Press.
56. Puri, B. R. Surface complexes on carbon, in Walker, P. L., Jr. (ed.), *Chemistry and Physics of Carbon*, Vol 6, 1970, New York: Marcel Dekker, pp. 191–282.
57. Sinha, R. K. and Walker, P. L., Jr. *Carbon*, 1972; **10**: 754–756.
58. Wu, Y. P., Fang, S., Jiang, Y., and Holze, R. *J Power Sources*, 2002; **108**(1–2): 245–249.
59. Kim, Y. T., Ito, Y., Tadai, K., Mitani, T., Kim, U. S., Kim, H. S., and Cho, B. W. *Appl Phys Lett*, 2005; **87**(23): Artn 234106.
60. Frackowiak, E. and Beguin, F. *Carbon*, 2001; **39**(6): 937–950.
61. Park, H., Zhao, J. J., and Lu, J. P. *Nano Lett*, 2006; **6**(5): 916–919.
62. Dumitrescu, L., Wilson, N. R., and Macpherson, J. V. *J Phys Chem C*, 2007; **111**(35): 12944–12953.
63. Lakshmi, N., Rajalakshmi, N., and Dhathathreyan, K. S. *J Phys D*, 2006; **39**(13): 2785–2790.
64. Radovic, L. R. and Rodriguez-Reinoso, F. *Chem Phys Carbon*, 1997; **25**: 243–358.
65. Ollinger, C. G. *J Electrochem Soc*, 1952; **99**(3): 54C–56C.
66. Midgley, D. and Mulcahy, D. E. *Ion Selective Electrode Rev*, 1983; **5**(2): 165–242.
67. Beilby, A. L., Brooks, W., Jr., and Lawrence, G. L. *Anal Chem*, 1964; **36**: 22–26.
68. Conway, B. E. *Electrochemical Supercapacitors: Scientific Fundamentals and Technological Applications*, 1999, New York: Kluwer.
69. Barbieri, O., Hahn, M., Herzog, A., and Kotz, R. *Carbon*, 2005; **43**(6): 1303–1310.

70. Zhao, J. C., Lai, C. Y., Dai, Y., and Xie, J. Y. *Mater Lett*, 2007; **61**(23–24): 4639–4642.
71. Ania, C. O., Khomenko, V., Raymundo-Piñero, E., Parra, J. B., and Beguin, F. *Adv Funct Mater*, 2007; **17**(11): 1828–1836.
72. Noel, M. and Santhanam, R. *J Power Sources*, 1998; **72**(1): 53–65.
73. Noel, M. and Suryanarayanan, V. *J Power Sources*, 2002; **111**(2): 193–209.
74. Zheng, T. and Dahn, J. R. Applications of carbon in lithium-ion batteries, in Burchell, T. D. (ed.), *Carbon Materials for Advanced Technologies*, 1999, Amsterdam, The Netherlands: Elsevier, pp. 341–387.
75. Flandrois, S. Battery carbons, in Buschow, K. H. J., Cahn, R. W., Flemings, M. C., Ilschner, B., Kramer, E. J., and Mahajan, S. (eds.), *Encyclopedia of Materials: Science and Technology*, 2001, Oxford: Pergamon Press, pp. 484–488.
76. Béguin, F., Chevallier, F., Vix, C., Saadallah, S., Rouzaud, J. N., and Frackowiak, E. *J Phys Chem Solids*, 2004; **65**: 211–217.
77. Kinoshita, K. and Bett, J. *Carbon*, 1973; **11**: 237–247.
78. Borup, R., Meyers, J., Pivovar, B., Kim, Y. S., Mukundan, R., Garland, N., Myers, D., Wilson, M., Garzon, F., Wood, D., Zelenay, P., More, K., Stroh, K., Zawodzinski, T., Boncella, J., McGrath, J. E., Inaba, M., Miyatake, K., Hori, M., Ota, K., Ogumi, Z., Miyata, S., Nishikata, A., Siroma, Z., Uchimoto, Y., Yasuda, K., Kimijima, K. I., and Iwashita, N. *Chem Rev*, 2007; **107**(10): 3904–3951.
79. Garten, V. A. and Weiss, D. E. *Rev Pure Appl Chem*, 1957; **7**: 69–122.
80. Burshtein, R. K. and Miller, N. B. *Zhurnal Fizicheskoi Khimii*, 1949; **23**(1): 43–49.
81. Strazhesko, D. N. and Matskevich, E. S. *Soviet Electrochem*, 1965; **1**(3): 250–253.
82. Burshtein, R. K., Vilinskaya, V. S., Zagudaeva, N. M., and Tarasevich, M. R. *Soviet Electrochem*, 1974; **10**(7): 1039–1041.
83. Burshtein, R. K. and Frumkin, A. N. *Z Physik Chem*, 1929; **A141**: 219–230.
84. Besenhard, J. O. and Fritz, H. P. *Angew Chem Internat Edit*, 1983; **22**: 950–975.
85. Koresh, J. and Soffer, A. *J Electrochem Soc*, 1977; **124**: 1379–1385.
86. Tarasevich, M. R. and Khrushcheva, E. I. Electrocatalytic properties of carbon materials, in Conway, B. E., Bockris, J. O. M., and White, R. E. (eds.), *Modern Aspects of Electrochemistry*, Vol 19, 1989, New York: Plenum Press, pp. 295–357.
87. McCreery, R. L. Carbon electrodes: structural effects on electron transfer kinetics, in Bard, A. L. (ed.), *Electroanalytical Chemistry*, Vol 17, 1991, New York: Marcel Dekker, Inc., pp. 221–374.
88. McCreery, R. L. Electrochemical properties of carbon surfaces, in Wieckowski, A. (ed.), *Interfacial Electrochemistry: Theory, Experiment, and Applications*, 1999, New York: Marcel Dekker, Inc., pp. 631–647.
89. McCreery, R. L. and Cline, K. K. Carbon electrodes, in Kissinger, P. T. and Heineman, W. R. (eds.), *Laboratory Techniques in Electroanalytical Chemistry*, 1996, New York: Marcel Dekker, Inc., pp. 293–332.
90. Dahn, J. R., Sleigh, A. K., Shi, H., Way, B. M., Weydanz, W. J., Reimers, J. N., Zhong, Q., and von Sacken, U. Carbons and graphites as substitutes for the lithium anode, in Pistoia, G. (ed.), *Lithium Batteries: New Materials, Developments and Perspectives*, 1994, Amsterdam, The Netherlands: Elsevier, pp. 1–47.
91. Beck, F. Graphite, carbonaceous materials and organic solids as active electrodes in metal-free batteries, in Alkire, R. C., Gerischer, H., Kolb, D. M., and Tobias, C. W. (eds.), *Advances in Electrochemical Science and Engineering*, 1997, Weinheim, Germany: Wiley-VCH, pp. 303–411.
92. Echegoyen, L., Diederich, F., and Echegoyen, L. E. Electrochemistry of fullerenes, in Kadish, K. M. and Ruoff, R. S. (eds.), *Fullerenes: Chemistry, Physics, and Technology*, 2000, New York: John Wiley & Sons, Inc., pp. 1–51.
93. Kavan, L. *Chem Rev*, 1997; **97**: 3061–3082.
94. Fialkov, A. S. *Russ J Electrochem*, 2000; **36**(4): 345–366.
95. Frackowiak, E. *Phys Chem Chem Phys*, 2007; **9**(15): 1774–1785.
96. Lukaszewicz, J. P. *Sensor Lett*, 2006; **4**(2): 53–98.
97. Béguin, F., Flahaut, E., Linares-Solano, A., and Pinson, J. Surface properties, porosity, chemical and electrochemical applications, in Loiseau, A. (ed.), *Understanding Carbon Nanotubes*, 2006, Springer, pp. 495–549.
98. Lahaye, J. and Ehrburger, P. (eds.), *Fundamental Issues in Control of Carbon Gasification Reactivity*. NATO ASI Series E 192. 1991, Dordrecht, The Netherlands: Kluwer Academic Publishers.
99. Gao, B., Bower, C., Lorentzen, J. D., Fleming, L., Kleinhammes, A., Tang, X. P., McNeil, L. E., Wu, Y., and Zhou, O. *Chem Phys Lett*, 2000; **327**(1–2): 69–75.
100. Xing, W. B., Dunlap, R. A., and Dahn, J. R. *J Electrochem Soc*, 1998; **145**(1): 62–70.

101. Lee, G.-J. and Pyun, S.-I. Synthesis and characterization of nanoporous carbon and its electrochemical application to electrode material for supercapacitors, in Vayenas, C., White, R. E., and Gamboa-Aldeco, M. E. (eds.), *Modern Aspects of Electrochemistry*, No 41, 2007, New York: Springer, pp. 139–195.
102. Haas, O. and Cairns, E. J. *Annu Rep Prog Chem Sect C*, 1999; **95**: 163–197.
103. Bakhmatyuk, B. P., Venhryn, B. Y., Grygorchak, I. I., Micov, M. M., and Kulyk, Y. O. *Electrochimica Acta*, 2007; **52**(24): 6604–6610.
104. Bakhmatyuk, B. P., Venhryn, B. Y., Grygorchak, I. I., Micov, M. M., and Mudry, S. I. *Rev Adv Mater Sci*, 2007; **14**(2): 151–156.
105. Balducci, A., Dugas, R., Taberna, P. L., Simon, P., Plee, D., Mastragostino, M., and Passerini, S. *J Power Sources*, 2007; **165**(2): 922–927.
106. Barpanda, P., Fanchini, G., and Amatucci, G. G. *Electrochimica Acta*, 2007; **52**(24): 7136–7147.
107. Barpanda, P., Fanchini, G., and Amatucci, G. G. *J Electrochem Soc*, 2007; **154**(5): A467–A476.
108. Bleda-Martinez, M. J., Morallon, E., and Cazorla-Amoros, D. *Electrochimica Acta*, 2007; **52**(15): 4962–4968.
109. Centeno, T. A., Hahn, M., Fernandez, J. A., Kotz, R., and Stoeckli, F. *Electrochem Commun*, 2007; **9**(6): 1242–1246.
110. Janes, A., Kurig, H., and Lust, E. *Carbon*, 2007; **45**(6): 1226–1233.
111. Kodama, M., Yamashita, J., Soneda, Y., Hatori, H., and Kamegawa, K. *Carbon*, 2007; **45**(5): 1105–1107.
112. Lazzari, M., Mastragostino, M., and Soavi, F. *Electrochem Commun*, 2007; **9**(7): 1567–1572.
113. Li, H.-Q., Liu, R.-L., Zhao, D.-Y., and Xia, Y.-Y. *Carbon*, 2007; **45**(13): 2628–2635.
114. Li, H. Q., Luo, J. Y., Zhou, X. F., Yu, C. Z., and Xia, Y. Y. *J Electrochem Soc*, 2007; **154**(8): A731–A736.
115. Li, W., Chen, D., Li, Z., Shi, Y., Wan, Y., Wang, G., Jiang, Z., and Zhao, D. *Carbon*, 2007; **45**(9): 1757–1763.
116. Li, W., Chen, D., Li, Z., Shi, Y., Wan, Y., Huang, J., Yang, J., Zhao, D., and Jiang, Z. *Electrochem Commun*, 2007; **9**(4): 569–573.
117. Pico, F., Pecharroman, C., Anson, A., Martinez, M. T., and Rojo, J. M. *J Electrochem Soc*, 2007; **154**(6): A579–A586.
118. Pollak, E., Anderson, A., Salitra, G., Soffer, A., and Aurbach, D. *J Electroanal Chem*, 2007; **601**(1–2): 47–52.
119. Pollak, E., Salitra, G., and Aurbach, D. *J Electroanal Chem*, 2007; **602**(2): 195–202.
120. Portet, C., Yushin, G., and Gogotsi, Y. *Carbon*, 2007; **45**(13): 2511–2518.
121. Ruiz, V., Blanco, C., Raymundo-Piñero, E., Khomenko, V., Beguin, F., and Santamaria, R. *Electrochimica Acta*, 2007; **52**(15): 4969–4973.
122. Wang, H. Y., Yoshio, M., Thapa, A. K., and Nakamura, H. *J Power Sources*, 2007; **169**(2): 375–380.
123. Yamada, H., Nakamura, H., Nakahara, F., Moriguchi, I., and Kudo, T. *J Phys Chem C*, 2007; **111**(1): 227–233.
124. Yang, C. M., Kim, Y. J., Endo, M., Kanoh, H., Yudasaka, M., Iijima, S., and Kaneko, K. *J Am Chem Soc*, 2007; **129**(1): 20–21.
125. Kullapere, M., Jurmann, G., Tenno, T. T., Paprotny, J. J., Mirkhalaf, F., and Tammeveski, K. *J Electroanal Chem*, 2007; **599**(2): 183–193.
126. Maia, G., Maschion, F., Tanimoto, S., Vaik, K., Mäeorg, U., and Tammeveski, K. *J Solid State Electrochem*, 2007; **11**(10): 1411–1420.
127. Matter, P. H., Wang, E., Arias, M., Biddinger, E. J., and Ozkan, U. S. *J Mol Catalysis A*, 2007; **264**(1–2): 73–81.
128. Pupkevich, V., Glibin, V., and Karamanev, D. *Electrochem Commun*, 2007; **9**(8): 1924–1930.
129. Qu, D. *Carbon*, 2007; **45**(6): 1296–1301.
130. Saleh, M. M., Awad, M. I., Okajima, T., Suga, K., and Ohsaka, T. *Electrochimica Acta*, 2007; **52**(9): 3095–3104.
131. Vayner, E. and Anderson, A. B. *J Phys Chem C*, 2007; **111**(26): 9330–9336.
132. Ania, C. O. and Beguin, F. *Water Res*, 2007; **41**(15): 3372–3380.
133. Chai, X., He, Y., Ying, D., Jia, J., and Sun, T. *J Chromatogr A*, 2007; **1165**(1–2): 26–31.
134. Han, Y. H., Quan, X., Chen, S., Wang, S. B., and Zhang, Y. B. *Electrochimica Acta*, 2007; **52**(9): 3075–3081.
135. Pakula, M., Biniak, S., Swiatkowski, A., and Derylo-Marczewska, A. *Appl Surf Sci*, 2007; **253**(11): 5143–5148.

136. Park, K.-K., Lee, J.-B., Park, P.-Y., Yoon, S.-W., Moon, J.-S., Eum, H.-M., and Lee, C.-W. *Desalination*, 2007; **206**(1–3): 86–91.
137. Sheveleva, I., Zemskova, L., Zheleznov, S., Voit, A., Barinov, N., Sukhoverstov, S., and Sergienko, V. *Russ J Appl Chem*, 2007; **80**(6): 924–929.
138. Fabish, T. J. and Hair, M. L. *J Coll Interf Sci*, 1977; **62**(1): 16–23.
139. Gerischer, H. *Electrochim Acta*, 1990; **35**: 1677–1699.
140. King, A. *J Chem Soc*, 1935: 889–894.
141. Wiegand, W. B. *J Phys Chem*, 1937; **29**: 953–956.
142. Frampton, V. L. and Gortner, R. A. *J Phys Chem*, 1937; **41**: 567–582.
143. Gerischer, H. *J Phys Chem*, 1985; **89**: 4249–4251.
144. Oren, Y., Tobias, H., and Soffer, A. *J Electroanal Chem*, 1984; **162**: 87–99.
145. Soffer, A. and Folman, M. *J Electroanal Chem Interf Electrochem*, 1972; **38**: 25–43.
146. Nian, Y.-R. and Teng, H. *J Electroanal Chem*, 2003; **540**: 119–127.
147. Lozano-Castelló, D., Cazorla-Amorós, D., Linares-Solano, A., Shiraishi, S., Kurihara, H., and Oya, A. *Carbon*, 2003; **41**(9): 1765–1775.
148. Yang, K.-L., Ying, T.-Y., Yiacoumi, S., Tsouris, C., and Vittoratos, E. S. *Langmuir*, 2001; **17**: 1961–1969.
149. Vijh, A. K. *Electrochemistry of Metals and Semiconductors*, 1973, New York: Marcel Dekker.
150. Trasatti, S. and Lust, E. The potential of zero charge, in White, R. E. (ed.), *Modern Aspects of Electrochemistry*, 1999, New York: Kluwer Academic.
151. Randin, J.-P. and Yeager, E. *J Electrochem Soc*, 1971; **118**(5): 711–714.
152. Randin, J.-P. and Yeager, E. *Electroanal Chem Interf Electrochem*, 1972; **36**: 257–276.
153. Randin, J.-P. and Yeager, E. *J Electroanal Chem*, 1974; **54**(1): 93–100.
154. Randin, J.-P. and Yeager, E. *Electroanal Chem Interfac Electrochem*, 1975; **58**: 313–322.
155. Salitra, G., Soffer, A., Eliad, L., Cohen, Y., and Aurbach, D. *J Electrochem Soc*, 2000; **147**: 2486–2493.
156. Choi, Y. O., Yang, K. S., and Kim, J. H. *Electrochemistry*, 2001; **69**(11): 837–842.
157. Gamby, J., Taberna, P. L., Simon, P., Fauvarque, J. F., and Chesneau, M. *J Power Sources*, 2001; **101**(1): 109–116.
158. Shiraishi, S., Kurihara, H., and Oya, A. *Electrochemistry*, 2001; **69**(6): 440–443.
159. Teng, H., Chang, Y.-J., and Hsieh, C.-T. *Carbon*, 2001; **39**(13): 1981–1987.
160. Weng, T.-C. and Teng, H. *J Electrochem Soc*, 2001; **148**(4): A368–A373.
161. Celzard, A., Collas, F., Mareche, J. F., Furdin, G., and Rey, I. *J Power Sources*, 2002; **108**(1–2): 153–162.
162. Frackowiak, E., Delpeux, S., Jurewicz, K., Szostak, K., Cazorla-Amoros, D., and Beguin, F. *Chem Phys Lett*, 2002; **361**(1–2): 35–41.
163. He, Y. D., Liu, H. B., and Zhang, H. B. *New Carbon Mater*, 2002; **17**(4): 18–22.
164. Jiang, Q., Qu, M. Z., Zhou, G. M., Zhang, B. L., and Yu, Z. L. *Mater Lett*, 2002; **57**(4): 988–991.
165. Pyun, S.-I., Kim, C. H., Kim, S. W., and Kim, J. H. *J New Mater Electrochem Sys*, 2002; **5**(4): 289–295.
166. Tamai, H., Kouzu, M., Morita, M., and Yasuda, H. *Electrochem Solid State Lett*, 2003; **6**(10): A214–A217.
167. Zhuang, X. G., Yang, Y. S., Ji, Y. J., Yang, D. P., and Tang, Z. Y. *Acta Phys Chim Sinica*, 2003; **19**(8): 689–694.
168. Zhou, H., Zhu, S., Hibino, M., and Honma, I. *J Power Sources*, 2003; **122**(2): 219–223.
169. Vix-Guterl, C., Saadallah, S., Jurewicz, K., Frackowiak, E., Reda, M., Parmentier, J., Patarin, J., and Beguin, F. *Mater Sci Eng B*, 2004; **108**(1–2): 148–155.
170. Zhou, C., Liu T., Wang, T., and Kumar, S. *Polymer*, 2006; **47**(16): 5831–5837.
171. Arbizzani, C., Mastragostino, M., and Soavi, F. *J Power Sources*, 2001; **100**(1–2): 164–170.
172. An, K. H., Jeon, K. K., Heo, J. K., Lim, S. C., Bae, D. J., and Lee, Y. H. *J Electrochem Soc*, 2002; **149**(8): A1058–A1062.
173. Park, J. H. and Park, O. O. *J Power Sources*, 2002; **111**(1): 185–190.
174. Shiraishi, S., Kurihara, H., Shi, L., Nakayama, T., and Oya, A. *J Electrochem Soc*, 2002; **149**(7): A855–A861.
175. Wang, X. F., Wang, D. Z., and Liang, J. *J Inorg Mater*, 2002; **17**(6): 1167–1173.
176. Chen, W.-C., Wen, T.-C., and Teng, H. *Electrochimica Acta*, 2003; **48**(6): 641–649.
177. Chen, W.-C. and Wen, T.-C. *J Power Sources*, 2003; **117**(1–2): 273–282.

178. Laforgue, A., Simon, P., Fauvarque, J. F., Mastragostino, M., Soavi, F., Sarrau, J. F., Lailler, P., Conte, M., Rossi, E., and Saguatti, S. *J Electrochem Soc*, 2003; **150**(5): A645–A651.

179. Liang, H. C., Chen, F., Li, R. G., Wang, L. and Deng, Z. H. *Electrochimica Acta*, 2004; **49**(21): 3463–3467.

180. Chen, W. C., Hu, C. C., Wang, C. C., and Min, C. K. *J Power Sources*, 2004; **125**(2): 292–298.

181. Zhang, Q. W., Zhou, X., and Yang, H. S. *J Power Sources*, 2004; **125**(1): 141–147.

182. Hou, C.-H., Liang, C., Yiacoumi, S., Dai, S., and Tsouris, C. *J Coll Interface Sci*, 2006; **302**(1): 54–61.

183. Gerischer, H. *Electrochim Acta*, 1989; **34**(8): 1005–1009.

184. Swiatkowski, A., Grajek, H., Pakula, M., Biniak, S., and Witkiewicz, Z. *Colloids Surf A*, 2002; **208**: 313–320.

185. Leon y Leon, C. A., Solar, J. M., Calemma, V., and Radovic, L. R. *Carbon*, 1992; **30**: 797–811.

186. Carrasco-Marín, F., Solar, J. M., and Radovic, L. R. *International Carbon Conference*, Paris, 1990.

187. Andreas, H. A. and Conway, B. E. *Electrochimica Acta*, 2006; **51**(28): 6510–6520.

188. Duval, J., Lyklema, J., Kleijn, J. M., and van Leeuwen, H. P. *Langmuir*, 2001; **17**(24): 7573–7581.

189. Duval, J., Kleijn, J. M., Lyklema, J., and van Leeuwen, H. P. *J Electroanal Chem*, 2002; **532**(1–2): 337–352.

190. Duval, J. F. L. *J Colloid Interf Sci*, 2004; **269**(1): 211–223.

191. Duval, J. F. L., Sorrenti, E., Waldvogel, Y., Gorner, T., and De Donato, P. *Phys Chem Chem Phys*, 2007; **9**(14): 1713–1729.

192. Jankowska, H., Neffe, S., and Swiatkowski, A. *Electrochim Acta*, 1981; **26**(12): 1861–1866.

193. Runnels, P. L., Joseph, J. D., Logman, M. J., and Wightman, R. M. *Anal Chem*, 1999; **71**(14): 2782–2789.

194. Ji, X. B., Banks, C. E., Crossley, A., and Compton, R. G. *Chem Phys Chem*, 2006; **7**(6): 1337–1344.

195. Ayranci, E. and Conway, B. E. *J Appl Electrochem*, 2001; **31**(3): 257–266.

196. Ayranci, E. and Conway, B. E. *J Electroanal Chem*, 2001; **513**(2): 100–110.

197. Niu, J. and Conway, B. E. *J Electroanal Chem*, 2002; **529**(2): 84–96.

198. Niu, J. J., Pell, W. G., and Conway, B. E. *J Power Sources*, 2006; **156**(2): 725–740.

199. Ayranci, E. and Duman, O. *J Hazard Mater*, 2007; **148**(1–2): 75–82.

200. Gunasingham, H. and Fleet, B. *Analyst*, 1982; **107**(1277): 896–902.

201. Beguin, F., Szostak, K., Lota, G., and Frackowiak, E. *Adv Mater*, 2005; **17**(19): 2380–2384.

202. Lota, G., Lota, K., and Frackowiak, E. *Electrochem Commun*, 2007; **9**(7): 1828–1832.

203. Lota, G., Grzyb, B., Machnikowska, H., Machnikowski, J., and Frackowiak, E. *Chem Phys Lett*, 2005; **404**(1–3): 53–58.

204. Hulicova, D., Kodama, M., and Hatori, H. *Chem Mater*, 2006; **18**(9): 2318–2326.

205. Frackowiak, E., Lota, G., Machnikowski, J., Vix-Guterl, C., and Beguin, F. *Electrochimica Acta*, 2006; **51**(11): 2209–2214.

206. Grahame, D. C. *J Am Chem Soc*, 1941; **63**: 1207–1215.

207. Biniak, S., Swiatkowski, A., and Pakula, M. Electrochemical studies of phenomena at active carbon-electrolyte solution interfaces, in Radovic, L. R. (ed.), *Chemistry and Physics of Carbon*, Vol 27, 2001, New York: Marcel Dekker, pp. 125–225.

208. Sihvonen, V. *Trans Faraday Soc*, 1938; **34**(2): 1062–1072.

209. Beguin F. and Yazami, R. *Actualite Chimique* 2006: 86–90.

210. Bleda-Martinez, M. J., Lozano-Castello, D., Morallon, E., Cazorla-Amoros, D., and Linares-Solano, A. *Carbon*, 2006; **44**(13): 2642–2651.

211. Nian, Y.-R. and Teng, H. *J Electrochem Soc*, 2002; **149**(8): A1008–A1014.

212. Pandolfo, A. G. and Hollenkamp, A. F. *J Power Sources*, 2006; **157**(1): 11–27.

213. Grahame, D. C. *Chem Rev*, 1947; **41**: 441–501.

214. Ruetschi, P. *J Electrochem Soc*, 1959; **106**(9): 819–827.

215. Trasatti, S. *J Electroanal Chem Interf Electrochem*, 1971; **33**: 351–378.

216. Hahn, M., Baertschi, M., Barbieri, O., Sauter, J. C., Kotz, R., and Gallay, R. *Electrochem Solid State Lett*, 2004; **7**(2): A33–A36.

217. Schmickler W. *Interfacial Electrochemistry*, 1996, New York: Oxford University Press.

218. Gelderman, K., Lee, L., and Donne, S. W. *J Chem Educ*, 2007; **84**(4): 685–688.

219. Oren, Y. and Soffer, A. *J Electroanal Chem*, 1985; **186**: 63–77.

220. Watanabe, A., Ishikawa, H., Mori, K., and Ito, O. *Carbon*, 1989; **27**(6): 863–867.

221. Bonhomme, F., Lassègues, J. C., and Servant, L. *J Electrochem Soc*, 2001; **148**(11): E450-E458.

222. Oren, Y. and Soffer, A. *J Appl Electrochem*, 1983; **13**(4): 473–487.

223. Ahn, H. J., Lee, J. H., Jeong, Y., Lee, J. H., Chi, C. S., and Oh, H. J. *Mater Sci Eng A*, 2007; **449**: 841–845.
224. Wang, X. Z., Li, M. G., Chen, Y. W., Cheng, R. M., Huang, S. M., Pan, L. K., and Sun, Z. *Appl Phys Lett*, 2006; **89**(5): Artn 053127.
225. Radovic, L. R. and Rodríguez-Reinoso, F. Carbon materials in catalysis, in Thrower, P. A. (ed.), *Chemistry and Physics of Carbon*, Vol 25, 1997, New York: Marcel Dekker, pp. 243–358.
226. McDermott, M. T. and McCreery, R. L. *Langmuir*, 1994; **10**: 4307–4314.
227. Müller, G., Radke, C. J., and Prausnitz, J. M. *J Phys Chem*, 1980; **84**: 369–376.
228. Rong, C. and Xien, H. *J Colloid Interface Sci*, 2005; **290**(1): 190–195.
229. Woodard, F. E., McMackins, D. E., and Jansson, R. E. W. *J Electroanal Chem*, 1986; **214**: 303–330.
230. Han, Y., Quan, X., Chen, S., Zhao, H., Cui, C., and Zhao, Y. *J Colloid Interface Sci*, 2006; **299**(2): 766–771.
231. Han, Y. H., Quan, X., Chen, S., Zhao, H. M., Cui, C. Y., and Zhao, Y. Z. *Sep Purif Technol*, 2006; **50**(3): 365–372.
232. Eisinger, R. S. and Alkire, R. C. *J Electroanal Chem*, 1980; **112**: 327–337.
233. Ban, A., Schäfer, A., and Wendt, H. *J Appl Electrochem*, 1998; **28**: 227–236.
234. Yang, K.-L., Yiacoumi, S., and Tsouris, C. *J Electroanal Chem*, 2003; **540**: 159–167.
235. Ayranci, E. and Conway, B. E. *Anal Chem*, 2001; **73**(6): 1181–1189.
236. Niu, J. and Conway, B. E. *J Electroanal Chem*, 2002; **521**(1–2): 16–28.
237. Conway, B. E., Ayranci, G., and Ayranci, E. *Zeitschrift Für Physikalische Chemie*, 2003; **217**(4): 315–331.
238. Niu, J. and Conway, B. E. *J Electroanal Chem*, 2003; **546**: 59–72.
239. Oren, Y., Tobias, H., and Soffer, A. *Bioelectrochem Bioenerg*, 1983; **11**(4–6): 347–351.
240. Anjo, D. M., Kahr, M., Khodabaksh, M. M., Nowinski, S., and Wanger, M. *Anal Chem*, 1989; **61**: 2603–2608.
241. Bodalbhai, L. and Brajter-Toth, A. *Anal Chem*, 1988; **60**: 2557–2561.
242. Kavan, L., Novak, P., and Dousek, F. P. *Electrochim Acta*, 1988; **33**: 1605–1612.
243. Sawyer, D. T. and Seo, E. T. *Inorg Chem*, 1977; **16**: 499–501.
244. Swann, S., Jr., Chen, C.-Y., and Kerfman, H. D. *J Electrochem Soc*, 1952; **99**(11): 460–466.
245. Bockris, J. O. M. and Khan, S. U. M. *Surface Electrochemistry: A Molecular Level Approach*, 1993, New York: Plenum Press.
246. Yang, H.-H. and McCreery, R. L. *J Electrochem Soc*, 2000; **147**(9): 3420–3428.
247. Sljukic, B., Banks, C. E., and Compton, R. G. *J Iranian Chem Soc*, 2005; **2**(1): 1–25.
248. Murray, R. W. *Acc Chem Res*, 1980; **13**(5): 135–141.
249. Ohms, D., Gupta, S., Tryk, D. A., Yeager, E., and Wiesener, K. *Z Phys Chem (Leipzig)* 1990; **271**: 451–459.
250. Ladouceur, M., Lalande, G., Guay, D., Dodelet, J. P., Dignard-Bailey, L., Trudeau, M. L., and Schulz, R. *J Electrochem Soc*, 1993; **140**(7): 1974–1981.
251. Biloul, A., Contamin, O., Scarbeck, G., Savy, M., Palys, B., Riga, J., and Verbist, J. *J Electroanal Chem*, 1994; **365**: 239–246.
252. Dignard-Bailey, L., Trudeau, M. L., Joly, A., Schulz, R., Lalande, G., Guay, D., and Dodelet, J. P. *J Mater Res*, 1994; **9**: 3203–3209.
253. Gouerec, P., Biloul, A., Contamin, O., Scarbeck, G., Savy, M., Riga, J., Weng, L. T., and Bertrand, P. *J Electroanal Chem*, 1997; **422**: 61–75.
254. Gouérec, P. and Savy, M. *Electrochim Acta*, 1999; **44**: 2653–2661.
255. Gouérec, P., Savy, M., and Riga, J. *Electrochim Acta*, 1998; **43**(7): 743–753.
256. Sarapuu, A., Vaik, K., Schiffrin, D. J., and Tammeveski, K. *J Electroanal Chem*, 2003; **541**: 23–29.
257. Mirkhalaf, F., Tammeveski, K., and Schiffrin, D. J. *Phys Chem Chem Phys*, 2004; **6**(6): 1321–1327.
258. Vaik, K., Maeorg, U., Maschion, F. C., Maia, G., Schiffrin, D. J., and Tammeveski, K. *Electrochimica Acta*, 2005; **50**(25–26): 5126–5131.
259. Manisankar, P., Gomathi, A., and Velayutham, D. *J Solid State Electrochem*, 2005; **9**(9): 601–608.
260. Sljukic, B., Banks, C. E., Mentus, S., and Compton, R. G. *Phys Chem Chem Phys*, 2004; **6**(5): 992–997.
261. Davies, M. O., Clark, M., Yeager, E., and Hovorka, F. *J Electrochem Soc*, 1959; **106**(1): 56–61.
262. Damjanovic, A. Mechanistic analysis of oxygen reactions, in Bockris, J. O. M. and Conway, B. E. (eds.), *Modern Aspects of Electrochemistry*, No 5, 1969, New York: Plenum Press, pp. 369–483.
263. Morcos, I. and Yeager, E. *Electrochim Acta*, 1970; **15**: 953–975.
264. Singer, L. S. and Spry, W. J. *Bull Am Phys Soc*, 1956; **1**: 214–215.

265. Singer, L. S. *Proc 5th Conf Carbon*, Vol 2, 1963, pp. 37–64.
266. Mrozowski, S. *Carbon*, 1988; **26**(4): 521–529.
267. Wind, R. A., Li, L., Maciel, G. E., and Wooten, J. B. *Appl Magn Reson*, 1993; **5**: 161–176.
268. Appel, M. and Appleby, A. J. *Electrochim Acta*, 1978; **23**: 1243–1246.
269. Matter, P. H. and Ozkan, U. S. *Catalysis Lett*, 2006; **109**(3–4): 115–123.
270. Paliteiro, C., Hamnett, A., and Goodenough, J. B. *J Electroanal Chem*, 1987; **233**(1–2): 147–159.
271. Kornienko, V. L., Kolyagin, G. A., and Saltykov, Y. V. *Russ J Appl Chem*, 1999; **72**(3): 363–371.
272. Goeringer, S., de Tacconi, N. R., Chenthamarakshan, C. R., Rajeshwar, K., and Wampler W. A. *Carbon*, 2001; **39**(4): 515–522.
273. Ozaki, J.-i., Nozawa, K., Yamada, K., Uchiyama, Y., Yoshimoto, Y., Furuichi, A., Yokoyama, T. Oya, A., Brown, L. J., and Cashion, J. D. *J Appl Electrochem*, 2006; **36**(2): 239–247.
274. Strelko, V. V., Kartel, N. T., Dukhno, I. N., Kuts, V. S., Clarkson, R. B., and Odintsov, B. M. *Surf Sci*, 2004; **548**(1–3): 281–290.
275. Maldonado, S. and Stevenson, K. J. *J Phys Chem B*, 2005; **109**(10): 4707–4716.
276. Matter, P. H., Wang, E., Arias, M., Biddinger, E. J., and Ozkan, U. S. *J Phys Chem B*, 2006; **110**(37): 18374–18384.
277. Matter, P. H., Zhang, L., and Ozkan, U. S. *J Catal*, 2006; **239**(1): 83–96.
278. Ozaki, J., Anahara, T., Kimura, N., and Oya, A. *Carbon*, 2006; **44**(15): 3358–3361.
279. Jurmann, G. and Tammeveski, K. *J Electroanal Chem*, 2006; **597**(2): 119–126.
280. Zhang, M. N., Yan, Y. M., Gong, K. P., Mao, L. Q., Guo, Z. X., and Chen, Y. *Langmuir*, 2004; **20**(20): 8781–8785.
281. Yeager, E. *J Mol Catal*, 1986; **38**: 5–25.
282. Kinoshita, K. *Electrochemical Oxygen Technology*, 1992, New York: Wiley-Interscience.
283. Appleby, A. J. and Marie, J. *Electrochim Acta*, 1979; **24**: 195–202.
284. Radovic, L. R., Lizzio, A. A., and Jiang, H. Reactive surface area: an old but new concept in carbon gasification, in Lahaye, J. and Ehrburger, P. (eds.), *Fundamental Issues in Control of Carbon Gasification Reactivity*, 1991, Dordrecht, The Netherlands: Kluwer Academic Publishers, pp. 235–255.
285. Yeager, E. *Electrochim Acta*, 1984; **29**: 1527–1537.
286. Liu, Y., Xue, J. S., Zheng, T., and Dahn, J. R. *Carbon*, 1996; **34**: 193–200.
287. Radovic, L. R., Walker, P. L., Jr., and Jenkins, R. G. *Fuel*, 1983; **62**: 849–856.
288. Balej, J., Balogh, K., Stopka, P., and Spalek, O. *Coll Czech Chem Commun*, 1980; **45**: 3249–3253.
289. Pirjamali, M. and Kiros, Y. *J Power Sources*, 2002; **109**(2): 446–451.
290. Biniak, S., Walczyk, M., and Szymanski, G. S. *Fuel Process Technol*, 2002; **79**(3): 251–257.
291. Tarasevich, M. R. *Russ J Electrochem*, 1981; **17**(8): 988–992.
292. Bailey, S. I., Ritchie, I. M., and Hewgill, F. R. *J Chem Soc Perkin Trans 2*, 1983; **5**: 645–652.
293. Fabish, T. J. and Schleifer, D. E. *Carbon*, 1984; **22**: 19–38.
294. Fabish, T. J. and Schleifer, D. E. Surface chemistry and the carbon black work function, in Sarangapani, S., Akridge, J. R., and Schumm, B. (eds.), *Proceedings of the Workshop on the Electrochemistry of Carbon*, 1984, Pennington, NJ: The Electrochemical Society, pp. 79–109.

6 Electronic Structures of Graphite and Related Materials

Toshiaki Enoki

CONTENTS

6.1   INTRODUCTION

There are two important features in the structure and electronic properties of graphite: the two-dimensional (2D) layered structure and the amphoteric character.[1] The basic unit of graphite, which is called graphene, is an extreme of condensed aromatic hydrocarbons with an infinite in-plane size, where an infinite number of hexagonal rings are fused to form a rigid planar sheet. In a graphene sheet, π-electrons form a 2D extended electronic structure, in which the top of the highest occupied molecular orbital (HOMO) level featured by the bonding π-band is touched at the Fermi energy, E_F, to the bottom of the lowest unoccupied molecular orbital (LUMO) level featured with the antibonding π^*-band, being stabilized in a zero-gap semiconductor state as shown in Figure 6.1a. The *AB* stacking of graphene sheets gives graphite as shown in Figure 6.2, where the weak intersheet interaction modifies the electronic structure into semimetallic one having a quasi-2D nature, as exhibited in Figure 6.1b. Therefore, graphite is featured with a 2D system from both structural and electronic aspects. The amphoteric character is featured by the fact that graphite works not only as oxidizer but also as reducer in chemical reactions. This characteristics stem from the zero-gap-semiconductor-type or semimetallic electronic structure, in which the ionization potential and the electron affinity have the same value of 4.6 eV.[1] Here, the ionization potential is defined as the energy required when one electron is taken from the top of the bonding π-band to the vacuum level, whereas the electron affinity corresponds to the energy produced by taking one electron from the vacuum level to the bottom of the antibonding π^*-band. The amphoteric character gives graphite (or graphene) a unique feature in the charge transfer reactions with a variety of materials. Namely,

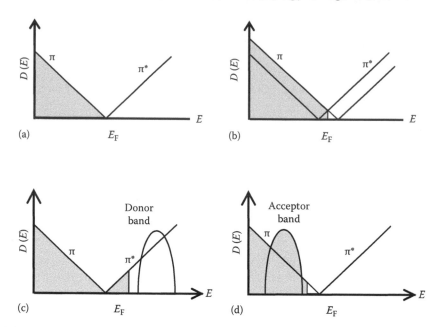

FIGURE 6.1 The electronic structures of graphene (a), graphite (b), donor-GICs (c), and acceptor-GICs (d), where $D(E)$ is the density of states and E_F the Fermi energy.

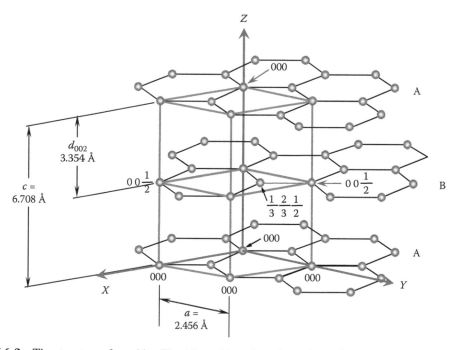

FIGURE 6.2 The structure of graphite. The *AB* stacking of graphene sheets forms the structure of graphite.

not only electron donors but also electron acceptors give charge transfer complexes with graphite as shown in the following reactions:

$$xC + D \rightarrow D^+C_x^-$$

$$xC + A \rightarrow C_x^+ A^-$$

where C, D, and A are graphite, donor, and acceptor, respectively.

In the charge transfer reactions, the layered structure of graphite plays an important role: reacted donor or acceptor species are accommodated in the galleries between graphite layers, since the reaction proceeds by expanding the intersheet distance against the intersheet interaction. The reaction described above is named as an intercalation reaction, where the donor or acceptor is called intercalate. The charge transfer complex with graphite obtained in the intercalation reaction is called graphite intercalation compound (GIC).[2–6]

In recent research trends on graphite-related materials, there are important issues from the viewpoints of chemistry, physics, and applications. One is the issue on graphene that is infinite in in-plane size.[7–16] The discovery of making a single sheet graphite (graphene) using a simple technique[8–10] has recently led us to recognize the importance of understanding graphite as a fundamental issue in nanoscience and cutting-edge device applications in nanotechnology. Indeed, current works[7–16] clarify unconventional electronic features of graphene, which obey the Dirac equation with the kinetic energy being linear wave number dependent. Interestingly, the feature of massless Dirac electrons has been successfully found using a single layer or few graphene layers prepared merely by cleaving bulk graphite flakes.[8,9] In addition, these have provided new basic issues of condensed matter physics, such as unusual half-integer quantum Hall effect,[7,10,13] quantum spin Hall effect,[11] quantum dots,[16] etc.

The second important issue is nanographene and nanographite,[17–22] the latter of which is a stack of nanographene sheets. Nano-sized graphite (nanographite) has attracted attention both from basic science and applications, such as batteries, capacitors, gas adsorbents, and catalysts. Nano-sized graphene (nanographene) and nanographite have electronic features different from those of infinite size graphene and bulk graphite due to the presence of edges and finiteness in size. From chemistry aspect, the issue on nanographene is related to aromaticity or stability of the Kekulé structure. This gives another important factor apart from the 2D structure and amphoteric nature in discussing the electronic properties of nanographene and nanographite. In addition, when nanoporous graphite consists of nanographite domains, a variety of electronic functions is created in relation to the host–guest interaction. Activated carbon fibers (ACFs) are one of the useful materials among these.[20,21] They consist of a 3D disordered network of nanographite domains, each of them comprising a stack of three to four nanographene sheets. The nanopores are created as a space surrounded by nanographite domains.

As explained above, the electronic structure of nanographite or nanographene should also be clarified when we discuss the electronic properties of graphitic materials comprehensively.

In this chapter, we discuss the electronic structures of graphite, graphene, GICs, and nanographene/nanographite. Section 6.2 is devoted to the electronic structure of graphene and graphite. The electronic structure of GICs is discussed in Section 6.3. The issue on nanographene and nanographite is presented in Section 6.4.

6.2 BAND STRUCTURE OF GRAPHITE

In the structure of graphite, graphene hexagon sheets are stacked in an *AB* stacking mode (Figure 6.2), which is characterized with strong in-plane C–C bonds and weak interlayer interactions. This structural feature gives two-dimensionality in its electronic structure. In the in-plane hexagon carbon network, the combination of sp^2 σ and π bonds is the origin of the strong intralayer interaction, whereas the overlap of π-bonds between adjacent graphene sheets contributes to the weak interlayer interaction. In discussing the electronic properties around the Fermi energy, which is particularly important for the electronic structure of graphite, the π-electron orbitals play an essential role, giving uniqueness to graphite. On the contrary, σ-bands, which have larger bonding energy than

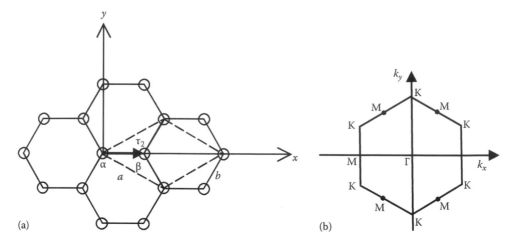

FIGURE 6.3 (a) The unit cell (dashed line) of a graphene sheet consisting of two carbon sites α and β. $a = 2.46\,\text{Å}$ and $b = 1.42\,\text{Å}$ are the in-plane lattice constant and the nearest neighbor C–C distance, respectively. τ_2 denotes the positional vector from α to β atoms. (b) Hexagonal BZ of a graphene.

the π-bands, are located far from the Fermi energy. Thus, they do not contribute seriously to the electronic properties of graphite. Here, we start with the electronic structure of graphite on the basis of the tight-binding model. First, the discussion is devoted to a graphene sheet that is considered to be an infinite 2D hexagonal conjugated π-electron system.[23]

Figure 6.3a presents a unit cell of graphene sheet comprising two kinds of carbon atoms α and β, where $a = 2.46\,\text{Å}$ and τ_2 are the lattice constant and the vector connecting two carbon atoms, respectively. The corresponding Brillouin zone (BZ) in the reciprocal lattice is shown in Figure 6.3b. Assuming that only the nearest neighbors overlap and resonance integrals work in the system, the π-band energies are calculated from the secular equation as expressed by

$$\begin{vmatrix} -E & -(\gamma_0 + SE)g(\mathbf{k}) \\ -(\gamma_0 + SE)g^*(\mathbf{k}) & -E \end{vmatrix} = 0, \tag{6.1}$$

where $\mathbf{k} = (k_x, k_y)$ is a 2D wave vector. The overlap S and resonance integrals γ_0 between the α and β atoms connected by τ_2 are defined in the following equations:

$$S = \int \phi_z(\mathbf{r}) \phi_z(\mathbf{r} - \tau_2) \mathrm{d}^3 r, \tag{6.2}$$

$$\gamma_0 = -\int \phi_z(\mathbf{r})\left[V(\mathbf{r}) - V_0\right]\phi_z(\mathbf{r} - \tau_2)\mathrm{d}^3 r + S\int \phi_z(\mathbf{r})\left[V(\mathbf{r}) - V_0\right]\phi_z(\mathbf{r})\mathrm{d}^3 r, \tag{6.3}$$

where
  $\phi_z(\mathbf{r})$ is the carbon atomic $2p_z$ orbital
  $V(\mathbf{r})$ and V_0 are the crystal and atomic potentials, respectively

A value of γ_0 is estimated at 3.16 eV for graphite, which represents the strength of in-plane interaction.[24] According to the threefold symmetry of the unit cell, $g(\mathbf{k})$ is expressed as follows:

$$g(\mathbf{k}) = \exp(i\mathbf{k}\tau_2) + \exp(ikD_3\tau_2) + \exp(ikD_3^{-1}\tau_2), \tag{6.4}$$

where D_3 is the operator of the $2\pi/3$ rotation. Assuming that the term SE is neglected in Equation 6.1 for simplicity, the energy eigenvalues and wave functions are derived for the π-bands:

$$E_{C,V}(\mathbf{k}) = \pm\gamma_0\left|g(\mathbf{k})\right| = \pm\gamma_0\sqrt{1 + 4\cos\left(\frac{k_y a}{2}\right)\cos\left(\frac{\sqrt{3}k_x a}{2}\right) + 4\cos^2\left(\frac{k_y a}{2}\right)}, \quad (6.5)$$

$$\psi_{C,V}(\mathbf{k}) = \frac{1}{\sqrt{2}}\left[U_\alpha(\mathbf{k}) \pm \frac{g^*(\mathbf{k})}{g(\mathbf{k})}U_\beta(\mathbf{k})\right]. \quad (6.6)$$

$U_{\alpha(\beta)}(\mathbf{k})$ denotes the Bloch function associated with the $2p$ orbital at site $\alpha(\beta)$. E_V with minus sign gives the valence band or the bonding π-band, whereas E_C with + sign corresponds to the conduction band or the antibonding π^*-band. E_F is located at $E = 0$. There is a mirror symmetry between the π- and π^*-bands with respect to the Fermi level. These two π-bands in the hexagonal BZ are depicted in Figure 6.4a.[25] At the Γ point, the π- and π^*-bands take a minimum and maximum, respectively. The difference between them is the total π-band width having $2\gamma_0\sim6\,\text{eV}$. There is a point of contact between the π- and π^*-bands at the K point. There is a trigonal symmetry in the band structure with respect to the K point. In the vicinity of point K near the Fermi energy in the BZ, we can obtain the following approximate expression from Equation 6.5:

$$E_{C,V}(\kappa) = \pm\frac{\sqrt{3}}{2}\gamma_0 a\kappa, \quad (6.7)$$

where κ is the distance from the K point in the BZ. Here, we approximately express the wave function as an isotropic one without angular dependence in terms of the wave vector measured from the K point in the BZ. The density of states derived from the dispersion curves Equation 6.7 is given by

$$D(E) = \frac{2|E|}{\sqrt{3}\pi\gamma_0^2}. \quad (6.8)$$

The energy dispersion curves given from Equation 6.7 are shown in Figure 6.5a. A remarkable feature is that the energies are proportional to the magnitude of the wave vector κ for both valence (V) and conduction bands (C), in contrast to the parabolic energy dispersion in the free electron system. Moreover, the valence and conduction bands touch to each other at the Fermi level. Interestingly, this electronic structure with a linear wave number dependence is effectively reproduced by the Dirac equation in a relativistic quantum mechanical wave equation. Here, Equation 6.7 indicates that the effective mass, which is defined by $m^* = \hbar^2/(\partial^2 E/\partial\kappa^2)$, becomes zero at E_F. The massless state thus given makes the electronic properties of graphene essentially different from those in other ordinary materials.[7–16]

The density of states is a linear function with the energy as shown in Figure 6.5b, which is reflected by the linear energy dispersions. The density of states decreases linearly as the energy approaches the Fermi level for both the conduction and valence bands, resulting in the absence of the density of states just at the Fermi level. This feature in the electronic state makes graphene a zero-gap semiconductor.

In real graphite, the interlayer interaction plays an important role in its electronic structure. The number of resonance integrals, which have to be taken into account, becomes considerably increased since the number of carbon atoms in the unit cell is doubled due to the AB stacking of graphene sheets. Figure 6.6 shows the schematic of resonance integrals in 3D graphite which are named Slonczewski–Weiss–McClure band parameters (Slonczewski–Weiss–McClure band parameters

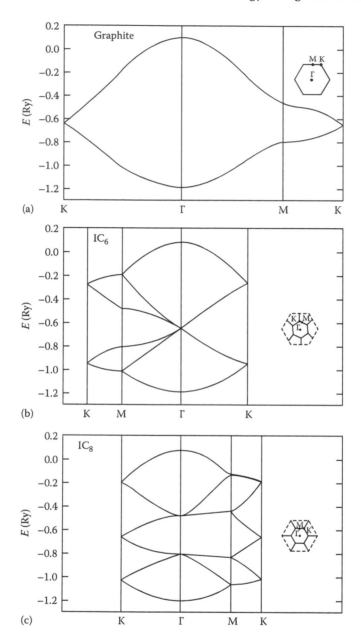

FIGURE 6.4 (a) The π-bands of a 2D graphene layer derived on the basis of the tight-binding method. The lower and upper bands are the bonding π-and antibonding π^*-band, respectively. (b) The folded π-bands in IC_6 having the $\sqrt{3} \times \sqrt{3}$ superlattice. (c) The folded π-bands in IC_8 having the 2×2 superlattice. I denotes intercalate. Insets show the BZ, where those for IC_6 and IC_8 are folded from the BZ of 2D graphene. (From Holzwarth, N.A.W., *Graphite Intercalation Compounds II*, Springer-Verlag, Berlin, 71–152, 1992. With permission.)

model).[26-28] It should be noted that the in-plane interaction γ_0 is more than 10 times larger than the interlayer interactions γ_1–γ_5, suggesting two-dimensionality of the graphite structure. The tight-binding calculation using the Slonczewski–Weiss–McClure parameters gives the electronic structure of π-bands in graphite, as exhibited in Figure 6.7. There are four π-bands E_1, E_2, and E_3 due to the AB stacking of graphite, where the E_3 band is twofold degenerate along the K–H line. The band

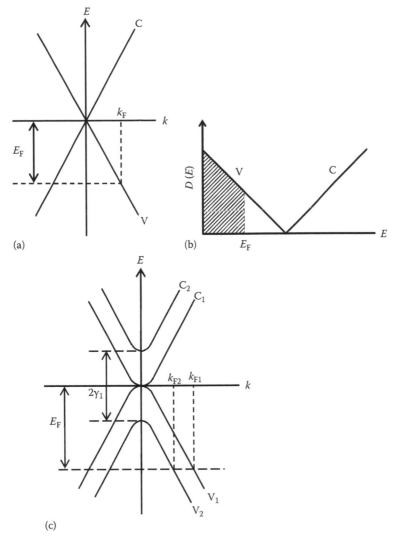

FIGURE 6.5 The energy dispersion (a) and the density of states $D(E)$ (b) for a graphene sheet, and the energy dispersion curves for double graphenc layers coupled to each other in an AB stacking manner (c). They are derived from the tight-binding method, in which (a) and (c) represent the electronic structures of the π- and π^*-bands for stage-1 and stage-2 GICs, respectively. The Fermi level of a single and double graphene sheets are located at the origin. The location of the Fermi level, E_F, is shifted when charge transfer between intercalates and graphene takes place. In the figure, cases for acceptor GICs are shown, where charge transfer from graphene to acceptor makes E_F downshifted.

structure derived from the tight-binding method is compared to that obtained with more elaborated calculation. Figure 6.8 presents the band structure calculated on the basis of a mixed-basis pseudopotential technique, which involves the σ-bands locating far from the Fermi level.[29] The energy dispersions of the π-bands tracks roughly those obtained from the tight-binding method, as we can see the features in the vicinity of K and H points. As a consequence, the use of the tight-binding π-band structure is justified to explain the electronic structure of graphite. Namely, on the basis of the tight-binding model, we can explain experimental results related to the electronic structure of graphite, the electron transport, quantum oscillation, optical, and magneto-optical properties. The participation of interlayer interactions, γ_1–γ_5, modifies the electronic structure from zero-gap

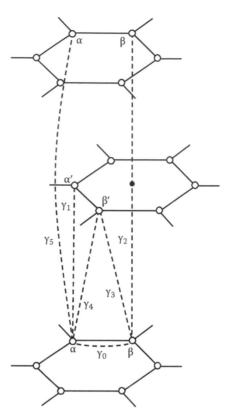

FIGURE 6.6   Schematic view of resonance integrals in graphite. The Slonczewski–Weiss–McClure band parameters (resonance integrals) are given as follows: $\gamma_0 = 3.16$, $\gamma_1 = 0.39$, $\gamma_2 = -0.020$, $\gamma_3 = 0.315$, $\gamma_4 = 0.044$, $\gamma_5 = 0.038$, $\Delta = -0.008\,\mathrm{eV}$, where Δ is the difference in crystalline fields experienced by inequivalent carbon sites in layer planes.

semiconductor to semimetal with the coexistence of electrons and holes, as easily seen in the feature of the energy dispersions and Fermi surfaces. Actually, two hole bands and one electron pocket are present around the H and K points, respectively, at the edge of the BZ, as shown in Figures 6.7a and 6.9a. The number of carriers is estimated at $3 \times 10^{18}\ \mathrm{cm^{-3}}$, being considerably small in comparison with ordinary metals. In addition, the electrons and holes have considerably small effective masses, $m^* = 0.057m$ and $0.039m$, respectively,[1] which are caused by the linear energy dispersion feature around the Fermi energy.

6.3   ELECTRONIC STRUCTURE OF GRAPHITE INTERCALATION COMPOUNDS

6.3.1   TIGHT-BINDING MODEL FOR GRAPHITE INTERCALATION COMPOUNDS

GICs have been first discovered from the reaction of graphite with sulfuric acid more than 150 years ago.[30] In the long history of GICs research, a huge number of compounds have been yielded with a large variety of donors and acceptors, in which alkali metals, alkaline earth metals, transition metal chlorides, acids, and halogens are involved as typical intercalates.

From the electronic structure aspect, the electron transferred from donor to graphite occupies the antibonding π^*-band in donor-GICs, whereas the charge transfer from graphite to acceptor makes the top of the bonding π-band empty in acceptor-GICs. The electronic structures thus produced in donor- and acceptor-GICs are presented in Figure 6.1c and d, respectively. The increase in the number of electrons around E_F after the charge transfer between intercalate and graphite makes GICs

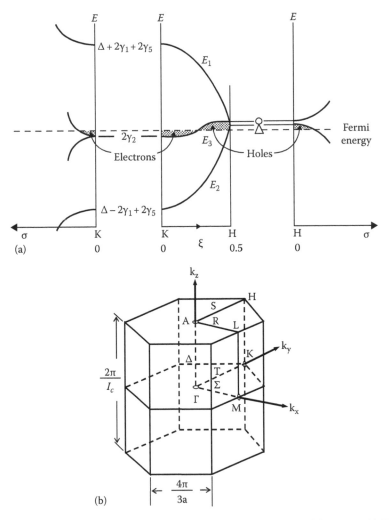

FIGURE 6.7 The electronic structure of 3D graphite. (a) π-band structure of Slonczewski–Weiss–McClure model for pristine graphite. There are four π-bands due to the AB stacking of graphite, where the E_3 band is twofold degenerate along the K–H line. The Fermi level crosses the E_3 band, two hole and one electron pockets being produced. σ denotes the direction to the BZ center; K $\rightarrow$ Γ, H $\rightarrow$ A. (From Dresselhaus, M.S. and Dresselhaus, G., *Adv. Phys.*, 30, 139, 1981.) (b) Hexagonal BZ, in which the electronic structure of graphite is defined.

metallic, in remarkable contrast to the semimetal or zero-gap semiconductor structure of the host graphite or graphene. Therefore, GICs can be called indeed synthetic metals. Actually, GICs have 2D metallic structures. In donor-GICs, the electrons added to the antibonding π^*-band behave as carriers, whereas the vacancies appearing in the bonding π-band produce hole carriers in acceptor GICs. The 2D-featured metallic electronic structures of GICs have been intensively investigated as attractive targets in the physics of low-dimensional electronic systems. The combination of graphitic π-electrons and electrons from the intercalates brings about a variety of novel electronic structures and related electron transport phenomena. In some donor-GICs, superconductivity appears having unconventional features.

The simplest model of the electronic structure of GICs can be obtained from the π-band structure of graphite on the basis of the Slonczewski–Weiss–McClure tight-binding model with the rigid band scheme. Here, the intercalate bands are superimposed on the graphite π-bands, their relative

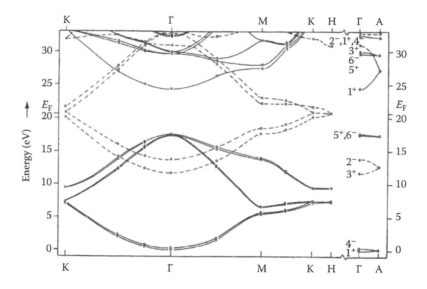

FIGURE 6.8 Band structure of graphite calculated by a mixed-basis pseudopotential method. The solid and dotted lines denote the σ- and π-bands, respectively. The Fermi level is indicated by E_F. (From Holzwarth, N.A.W., et al., *Phys. Rev. B*, 26, 5382, 1982. With permission.)

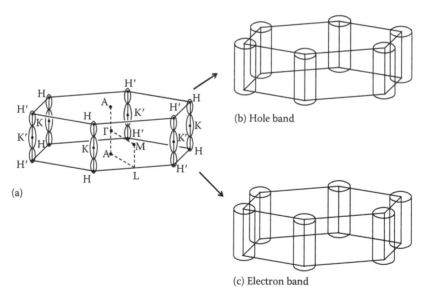

FIGURE 6.9 (a) Fermi surfaces for pristine graphite. One electron and two hole pockets are placed at the K and H points, respectively, around the edge of the BZ. One of the hole pockets at the H point is neglected for simplicity, whose size is considerably smaller than the other. (b) Hole Fermi surfaces of the π-band in acceptor GICs, where the hole pocket at the H point in pristine graphite grows to a cylindrical Fermi surface. (c) Electron Fermi surfaces of the π*-band in donor-GICs, where the electron pocket at the K point in pristine graphite grows to a cylindrical Fermi surface.

positions being adjusted so as to get consistency with the charge transfer rate between graphite and intercalate. In addition, the interaction between graphene sheets and intercalate is neglected. This is justified as a first approximation by more sophisticated theoretical calculations, which will be discussed in the next section. In the Blinowski and Rigaux model,[31,32] the electronic structure of the stage-1 GIC is considered to consist of the π-bands of a single graphene sheet and the intercalate band, the former being given by Equation 6.7. The π-bands of n graphene sheets stacked in the AB stacking manner are introduced in addition to the intercalate band for stage-n GICs. Acceptor GICs are the simplest cases, where the electronic structure can be represented only by the π-bands of graphene sheet(s), since the electronic states of acceptor intercalates are well below the Fermi level. The electrons in the valence π-bands are transferred to the acceptor bands, so that vacancies are produced around the top of the π-band, giving a large number of hole carriers through the charge transfer. This makes the Fermi level shifted down as exhibited in Figure 6.5a and b for stage-1 GICs. Consequently, the Fermi surfaces are different from those of pristine graphite in such a manner that the hole pocket at the H point grows, resulting in the formation of a large cylindrical Fermi surface of hole origin as shown in Figure 6.9b. It should be noted that the Fermi surface cross section is circular with no trigonal warping since we neglect the angular dependence of the wave function in the in-plane directions as discussed in Equation 6.7. In this model, we assume that the formation of an intercalate superlattice is neglected as well, so that we take the same hexagonal BZ as for pristine graphite. Therefore, the volume of the reciprocal space enclosed by the Fermi surface corresponds to the number of hole carriers generated after the charge transfer, which is given by

$$N = \frac{2 \times 2}{(2\pi)^3} \int_0^{k_F} \int_0^{k_F} \int_0^{k_F} dk_x dk_y dk_z = \frac{4E_F^2}{3\pi I_c \gamma_0^2 a^2},\tag{6.9}$$

where I_c is the c-axis repeat distance. The Fermi wave number k_F is, hereafter, defined with respect to the K point as the origin, and it is related to the Fermi level from Equation 6.7 as given by

$$k_F = -\frac{2E_F}{\sqrt{3}\gamma_0 a}.\tag{6.10}$$

Here, E_F takes a negative value for acceptor GICs. Factors 2 in the numerator of Equation 6.9 come from the spin degree of freedom and the number of cylinders in the BZ. In the hexagonal BZ, there are two cylindrical Fermi surfaces, in which holes having up and down spins are accommodated. The charge transfer rate from the graphene sheet to the acceptor intercalate is derived in the following equation:

$$f_C = \frac{1}{\sqrt{3}\pi}\left(\frac{|E_F|}{\gamma_0}\right)^2,\tag{6.11}$$

where f_C is defined to be the charge per carbon atom. The presence of a large concentration of holes modifies the electronic structure to metallic when acceptor intercalates are introduced in the graphitic galleries, where the number of carriers increases more than four orders of magnitude upon charge transfer.

Here, it should be noted that the formation of an in-plane superlattice in the intercalate affects the electronic structure due to the periodic potential of the intercalates. $\sqrt{3} \times \sqrt{3}$ and 2×2 intercalate superlattices are typically observed in GICs. The BZ of GICs having $\sqrt{3} \times \sqrt{3}$ and 2×2 are obtained by folding the BZ of the host graphite as shown in Figure 6.4b and c, respectively. The BZ sizes are reduced to 1/3 and 1/4 of the unfolded BZ of the host graphite for $\sqrt{3} \times \sqrt{3}$ and 2×2 lattices, respectively; the absolute values of the reciprocal lattice vectors become $1/\sqrt{3}$ and 1/2 with respect to the reciprocal lattice vector of host graphite, respectively. Therefore, the Fermi surface is modified by the zone folding effect, if the Fermi wave number estimated with the unfolded BZ is larger than the absolute value of the reciprocal lattice vector of the folded BZ.

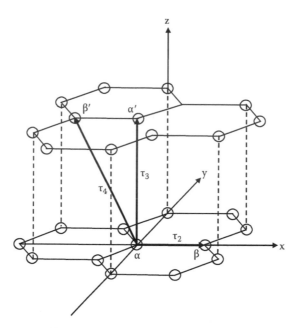

FIGURE 6.10 Lattice structure of graphite. The unit cell has four carbon atoms (α, β, α', β'). The positional vectors measured from atom α are denoted by τ_2, τ_3, and τ_4 for β, α', and β' atoms, respectively.

Next, we discuss the electronic structure of stage-2 GICs on the basis of the Blinowski–Rigaux model, where the stacking structure is represented as *AB-I-BA-I-* for the graphene layers *A*, *B*, and the intercalate layer *I*. The coexistence of the two types of graphene layers facing each other requires interlayer interactions between *A* and *B* sheets. As a reasonable simplification, only a resonance integral, γ_1, is employed for representing the interlayer interaction and other interlayer resonance integrals are neglected (see Figure 6.6). The interaction between graphene sheets and intercalate is neglected as well, as we mentioned before. The unit cell of stage-2 GICs having two graphene layers involves four carbon atoms as shown in Figure 6.10. Similar to the tight-binding treatment for stage-1 GICs or a single graphene sheet, the energies are obtained from the following secular equation:

$$\begin{vmatrix} -E & -\gamma_0 g(\kappa) & \gamma_1 & 0 \\ -\gamma_0 g*(\kappa) & -E & 0 & 0 \\ \gamma_1 & 0 & -E & -\gamma_0 g*(\kappa) \\ 0 & 0 & -\gamma_0 g(\kappa) & -E \end{vmatrix} = 0, \tag{6.12}$$

where the interlayer resonance integral γ_1 is defined as follows:

$$\gamma_1 = \int \phi_z(\mathbf{r})\left[V(\mathbf{r}) - V_0(\mathbf{r})\right]\phi_z(\mathbf{r} - \tau_3)\mathrm{d}^3r - S_1\int \phi_z(\mathbf{r})\left[V(\mathbf{r}) - V_0(\mathbf{r})\right]\phi_z(\mathbf{r})\mathrm{d}^3r. \tag{6.13}$$

Here S_1 is the overlap integral between two adjacent carbon layers (α–α') (see Figure 6.10) and is defined similar to Equation 6.2. The difference between α and β atoms and the terms SE and S_1E in the crystal potentials are neglected in Equation 6.12. From Equation 6.12, the energies of the π-bands are derived

$$-E_{V1} = E_{C1} = \frac{1}{2}\left(\sqrt{\gamma_1^2 + 3\gamma_0^2 a^2 k^2} - \gamma_1\right), \tag{6.14}$$

$$-E_{V_2} = E_{C_2} = \frac{1}{2}\left(\sqrt{\gamma_1{}^2 + 3\gamma_0{}^2 a^2 k^2} + \gamma_1\right), \tag{6.15}$$

where the conduction band is split into E_{C1} and E_{C2}, whereas the valence band is split into E_{V1} and E_{V2}. The band dispersions of stage-2 GICs thus obtained are illustrated in Figure 6.5c. The split between the bands V_2 and C_2 is given by the interlayer resonance integral $2\gamma_1$. A mirror symmetry between the valence and conduction bands exists with respect to $E = 0$, similar to the case for a single graphene sheet or stage-1 case. The location of the Fermi level is determined by the amount of charge transferred from the graphene sheets to the acceptor intercalate. In addition, two Fermi surfaces are produced after the charge transfer, where the Fermi wave numbers for the two Fermi surfaces are given in the following:

$$\frac{E_F + \gamma_1}{E_F - \gamma_1} = \left(\frac{k_{F1}}{k_{F2}}\right)^2, \tag{6.16}$$

where k_{F1} and k_{F2} correspond to the Fermi wave numbers for bands V_1 and V_2, respectively. Equation 6.11 is still valid in the estimation of the charge transfer ratio for stage-2 GICs. For higher stage GICs, the same treatment is taken with n graphene sheets stacked in the AB stacking manner.

More elaborated tight-binding models were presented by Holzwarth[33] and Saito (Saito–Kamimura model)[34] for better quantitative analysis of electronic structures. The Holzwarth model is based on the linear combination of atomic orbitals (LCAO) of the π-bands of the graphitic layers,[35] which includes four in-plane and four interplane interactions and a difference in energy for nonequivalent carbon sites. Here, the introduction of higher order interactions, which are neglected in the Blinowski–Rigaux model, produces trigonal warping of the cylindrical Fermi surfaces and destroys the exact mirror symmetry between the π- and π^*-bands. This makes the Holzwarth model more realistic, especially for the case of large charge transfer rate that gives large Fermi surfaces. However, only stage-1 GICs can be numerically treated in the Holzwarth model, where the LCAO parameters are determined by fitting the π-bands of 2D graphite calculated by Painter and Ellis.[36] In the Saito–Kamimura model, the tight-binding calculation is carried out with one in-plane interaction γ_0, three interplane interactions γ_1, γ_3, γ_4, and energy differences for nonequivalent carbon sites (see Figure 6.6). The bands are least-square-fitted to the self-consistent band structures of Ohno and Kamimura for K GICs.[37] Elaboration of the Saito–Kamimura model is carried out by Yang,[38] where the k-dependence is introduced in γ_0.

In the case of donor-GICs such as alkali metal GICs, the intercalate band (for example, 4s level in potassium) is located around the Fermi level. Consequently, the intercalate band has to be taken into account in addition to the graphitic π^*-bands in the electronic structure of donor-GICs as shown in Figure 6.1c. This makes the electronic structure rather complicated compared to that of acceptor GICs. Therefore, using more sophisticated methods is necessary for detailed information on the electronic structure of donor-GICs. However, the treatment with the Blinowski–Rigaux model still gives important information on the graphitic π-bands even for donor-GICs. It is worth noting that a complete charge transfer takes place from alkali metal to graphite in some of alkali metal GICs, the intercalate bands being well above the Fermi energy. Accordingly, in this case, the tight-binding π-bands are particularly useful in interpreting the electronic properties. For donor-GICs, the electron pocket at the K point grows due to the charge transfer from donor to graphite, resulting in the formation of a large cylindrical electron Fermi surface occupied by a large number of electron carriers, as exhibited in Figure 6.9c. We can estimate the Fermi energy and wave numbers using Equations 6.7, 6.10, and 6.16, where the Fermi energy becomes positive; we omit minus sign in Equation 6.10 and put minus sign ahead of E_F in Equation 6.16. Equation 6.11 is valid in the estimation of the charge transfer rate.

Electronic structures of GICs, thus theoretically characterized, are investigated experimentally by means of various techniques, such as x-ray photoemission spectra, ultraviolet photoelectron spectra, electron energy loss spectra, magneto-oscillation, optical reflectance, Raman spectra, Pauli paramagnetic susceptibility, electronic specific heat coefficient, NMR, positron annihilation, etc. Comparisons between theoretical treatments and experimental characterizations will be discussed in the Sections 6.3.2 and 6.3.3 of this chapter for actual GICs.

6.3.2 ACCEPTOR GRAPHITE INTERCALATION COMPOUNDS

The electronic structures of acceptor GICs are well represented by the rigid band model. The lowest unoccupied molecular orbital (LUMO) of acceptor intercalates, to which electrons are transferred from graphitic π-bands, is considerably far from the Fermi level in general, so that the interaction between graphene and acceptor intercalates becomes weak. In addition, the large interlayer distance ranging typically around $9\,\text{Å}$, which is considerably larger than that of donor-GICs, effectively reduces the three-dimensionality in the electronic structures. This means that the acceptor intercalate works merely as a spacer introduced between graphene sheets, which expands the interlayer distance. Here, as a typical example of applicability of the rigid band model, we discuss the electronic structures of acceptor GICs in relation to optical reflectance spectra that are well explained in terms of the tight-binding model presented in Section 6.3.1.

In general, optical properties are derived from the dielectric function as expressed by the following equation for 2D band model with the light propagating along the c-axis:

$$\varepsilon_\perp = \varepsilon_{\text{inter}} - \frac{\omega_p^{\,2}}{\omega\left(\omega + i\gamma\right)}, \tag{6.17}$$

where

$\varepsilon_{\text{inter}}$ is the dielectric function due to the interband transitions

γ is the damping factor of free carriers that is inversely proportional to the carrier relaxation time τ

ω_p is the plasma frequency, which consists of the plasma frequencies of all the constituent conduction bands:

$$\omega_p^{\,2} = \sum_j \omega_j^{\,2}, \tag{6.18}$$

$$\omega_j^{\,2} = 4\pi \frac{n_j}{I_c + (n-1)d_G} \frac{1}{\hbar^2} \left\langle \frac{\partial^2 E_j(\mathbf{k})}{\partial k_x^2} \right\rangle, \tag{6.19}$$

where I_c, n, and d_G ($=3.35\,\text{Å}$) are the c-axis repeat distance, stage index, and the interlayer distance of pristine graphite, respectively. The actual expressions of the plasma frequency are given as follows:

$$\omega_p^{\,2} = \frac{4e^2 E_F}{I_c \hbar^2}, \tag{6.20}$$

$$\omega_p^{\,2} = \frac{8e^2 E_F}{(I_c + d_G)\hbar^2} \left(\frac{E_F^{\,2} - \gamma_1^{\,2}/2}{E_F^{\,2} - \gamma_1^{\,2}/4} \right), \tag{6.21}$$

for stage-1 and stage-2 GICs, respectively, where e is the charge of electron and $\hbar = h/2\pi$. The dielectric constant due to interband transition is expressed as follows:

$$\varepsilon_{inter} = 1 - \frac{4\pi e^2}{m^2 \omega \left[I_c + (n-1)d_G \right]} \lim_{s \to 0^+} \sum_{j,j'} \frac{2}{(2\pi)^2} \int_{BZ} d^3k f\left(E_j(\mathbf{k})\right)\left[1 - f\left(E_{j'}(\mathbf{k})\right)\right]$$

$$\times \frac{\left|P_{j \to j'}\right|^2}{\omega_{jj'k}} \left[\frac{1}{\hbar\omega_{jj'k} - \hbar\omega - is} - \frac{1}{\hbar\omega_{jj'k} + \hbar\omega + is} \right],$$

(6.22)

where the integration is taken within the BZ, and

$$\hbar\omega_{jj'k} = E_j(\mathbf{k}) - E_{j'}(\mathbf{k}),$$

(6.23)

$$\left|P_{j \to j'}\right|^2 = \frac{1}{2}\left[\left|\langle j'k|P_x|jk\rangle\right|^2 + \left|\langle j'k|P_y|jk\rangle\right|^2\right],$$

(6.24)

where j and j' are the band indices. P_x (P_y) is the $x(y)$-component of the electric dipole moment. All the interband transitions have to be taken into account for the calculation with Equation 6.22. In stage-1 GICs, only the transition from the π-band (V) to the π^*-band (C) contributes to ε_{inter}, where the transition takes place for $\hbar\omega > 2E_F$, as can be seen easily in Figure 6.5a. Stage-2 GICs have five interband transitions according to their electronic structure (see Figure 6.5c): $V_2 \to V_1$, $V_2 \to C_1$, $V_2 \to C_2$, $V_1 \to C_1$, $V_1 \to C_2$.

Optical reflectance studies have been widely employed in investigating the electronic structures of a large number of acceptor GICs on the basis of the Blinowski–Rigaux tight-binding model, its modification,[34] and LCAO calculations.[35] Due to the absence of the intercalate band in the vicinity of the Fermi energy, the tight-binding model only for the graphitic π-electrons can be used to obtain convincing information on the charge transfer and the Fermi energy. Among acceptor GICs, GICs with H_2SO_4,[39,40] $FeCl_3$,[41] $SbCl_5$,[42] $AuCl_3$,[43] AsF_5,[44] OsF_6, MoF_6,[45] BF_3,[46] and F[47] are typical cases for optical studies. In the optical studies, the electronic structures have been well characterized with the aid of the tight-binding models. Figure 6.11 shows the optical reflectance of stage-1 and stage-2 AsF_5-GICs. The theoretical fitting with the Blinowski–Rigaux model leads to the experimental estimation of the Fermi energy, the carrier relaxation time and the charge transfer rate: $E_F = 1.3\,eV$, $\tau = 7 \times 10^{-15}\,s$, $f_C = 0.033$–0.055 for stage-1; $E_F = 1\,eV$, $\tau = 3 \times 10^{-14}\,s$, $f_C = 0.019$–0.032 for stage-2.

Raman spectra are also informative for characterizing the electronic structures of GICs. The Raman active mode E_{2g2}, which is ascribed to the intralayer graphitic vibration and located at $1582\,cm^{-1}$ for pristine graphite, is sensitive to the charge transfer between graphite and intercalate. Actually, the charge transfer from graphite to intercalate in acceptor GICs gives a vacancy at the top of the bonding π-band, whereas electrons from intercalate to graphite in donor-GICs partially occupy the antibonding π^*-band. This makes the intralayer C–C bond strengthened/weakened after charge transfer for acceptor/donor-GICs, resulting in the contraction/elongation of the C–C distance. This means that the detailed x-ray investigation of the intralayer C–C distance provides information on the charge transfer rate or the Fermi energy. Moreover, according to Pietronero and Strassler,[48] the charge-transfer-induced change in the strength of the C–C bond works to upshift/downshift the frequency of the zone-center graphitic phonon modes in acceptor/donor-GICs. From their analysis, the shift in the frequency of the E_{2g2} mode is given by the following relation with stage index n:

$$\frac{\delta\left(\Delta\omega/\omega\right)}{\delta\left(1/n\right)} = \pm 0.007$$

(6.25)

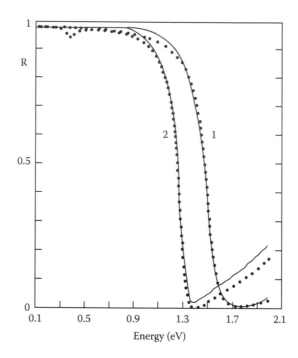

FIGURE 6.11 Optical reflectance spectra of stage-1 C_8AsF_5 and stage-2 $C_{16}AsF_5$. Solid lines represent theoretical fittings with the Fermi energy, the carrier relaxation time, and the charge transfer; $E_F = 1.3$ eV, $\tau = 7 \times 10^{-15}$ s, $f_C = 0.033$–0.055 for stage-1; $E_F = 1$ eV, $\tau = 3 \times 10^{-14}$ s, $f_C = 0.019$–0.032 for stage-2. (From Blinowski, J., et al., *J. Phys. (Paris)*, 41, 47, 1980. With permission.)

for acceptor (+) and donor (−) GICs, since the charge per carbon atom is reduced with increasing stage index. Chan et al. give correlations between the Raman frequency, charge transfer rate and the in-plane lattice constant on the basis of first-principle calculations as given in Figure 6.12, where positive/negative charge transfer corresponds to donor/acceptor systems.[49] It should be noted here that the Raman frequency vs. charge transfer is nonsymmetric between donor and acceptor. This is a consequence of anharmonicity in the potential; graphene hexagons can be stretched more easily than compressed. Here, in Figure 6.13, we show the profiles of the E_{2g2} Raman spectra for $FeCl_3$ GICs as a function of the stage index.[50] A successive upshift in the Raman peak is observed as the stage index decreases from stage-∞ to stage-1, where HOPG or pristine graphite corresponds to stage-∞. The presence of two Raman peaks for GICs having a stage index higher than 3 is associated with the presence of two kinds of graphene layers: the layer faced to intercalate (bounding layer) and the layer not faced to intercalate (interior layer). The bounding layer is easily subjected to charge transfer to intercalate due to the direct interaction, the charge transfer rate being large, whereas the screening effect from the bounding layer works against charge transfer from the interior layer to intercalate. Taking into account this situation, the Raman shift gives the estimate of the Fermi energy or the charge transfer rate.

The electronic structures of acceptor GICs have been characterized well not only by optical spectra but also by various experimental techniques such as Shubnikov-de Haas oscillation, magnetic susceptibility, etc., with the aid of tight-binding electronic structure models. Table 6.1 summarizes the Fermi energies and the charge transfer rates f_C per carbon atom for typical acceptor GICs.[6] The Fermi energies, E_F, are in the range of −1 eV, where the charge transfer rates are estimated at $f_C = 0.01$–0.05. The values are smaller than those of typical donor-GICs such as alkali metal GICs, suggesting weaker charge transfer in acceptor GICs. In the case of H_2SO_4 GICs, $C_p^+ HSO_4^-(H_2SO_4)_x$,

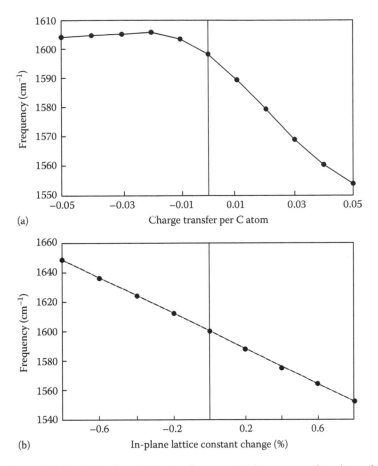

FIGURE 6.12 First-principle calculations of the E_{2g} phonon mode frequency as functions of (a) charge transfer and (b) in-plane lattice-constant change. (From Chan, C.T., et al., *Phys. Rev. B*, 36, 3499, 1987. With permission.)

the composition ratio, p, varies in a wide range: $21 < p < 28$ for stage-1 and $48 < p < 60$ for stage-2. The variation of p contributes to a change in the Fermi energy. Interestingly, strong oxidation with H_2SO_4 in the electrochemical process works beyond the intercalation reaction and induces the formation of C–O bonds on graphene sheets.[51,52] In the case of $SbCl_5$ GICs, the disproportionation reaction into $SbCl_3$ and $SbCl_6^-$ taking place in the intercalates affects the charge transfer or the location of the Fermi energy.

6.3.3 ALKALI, ALKALINE EARTH, AND RARE EARTH METAL GRAPHITE INTERCALATION COMPOUNDS

Alkali metal GICs are the most typical donor-GICs, whose electronic structures have been intensively investigated so far. We can use the tight-binding models such as the Blinowski–Rigaux,[31,32] Holzwarth,[33] Saito–Kaminura[34] models, as a first approximation that are useful to some extent for explaining the experimental results, such as optical reflectance, electronic specific heat, magnetooscillations, Pauli paramagnetic susceptibility, orbital susceptibility, NMR Knight shift, etc. In these models, the electronic structure is analyzed on the basis of the procedures given in Sections 6.3.1 and 6.3.2.

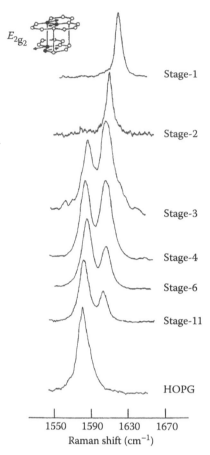

FIGURE 6.13 The Raman spectra of E_{2g_2} graphitic intralayer mode in $FeCl_3$ GICs depending on stage index (From Underhill, C., et al., *Solid State Commun.* 29, 769, 1979. With permission.)

For detailed quantitative analyses, we need more elaborated calculations. Li GICs have been a target of most intensive studies with various calculation methods, such as Green function method (KKR), pseudopotential, LCAO, full-potential self-consistent linearized augumented plane-wave method, etc., due to their simplest electronic structures among donor-GICs.[53–58] All the theoretical calculations suggest a complete charge transfer from Li to graphite, so that the Li 2s level becomes empty. As a consequence, only the graphitic π-bands need to be involved in discussing the electronic structure around the Fermi level, which is most relevant to the electronic properties of Li GICs. In this context, the band structure derived from the tight-binding model is particularly informative in characterizing the electronic structure, since the tight-binding treatment only with the graphitic π-bands gives semiquantitatively the same result as that from more accurate theoretical treatment. The stage-1 Li GIC, LiC_6, forms an in-plane $\sqrt{3} \times \sqrt{3}$ superstructure, which makes the BZ folded on the c-plane and reduces it to 1/3 of the BZ of the pristine graphite, as depicted in Figure 6.4b, where the K point of the pristine graphite is folded to the Γ point in the new BZ. Meanwhile, in the c-axis direction, the change in the c-axis repeat distance I_c upon intercalation from 3.35 to 3.70 Å reduces the reciprocal lattice vector by 10%. Assuming that the interaction between the graphene sheet and the Li intercalate is neglected as we suggested in Section 6.3.1, the structure of π-bands is given as shown in Figure 6.4b. The point of contact between the π- and π*-bands, which appears at the K point in the pristine graphite, is placed at the zone center, the Γ point, after the zone

TABLE 6.1

Fermi Energies E_F and Charge Transfer Rates f_C for Typical Acceptor GICs

GICs (Stage Index)	E_F (eV)	f_C (Charge per Carbon Atom)
BF_4 (1)	−1.21	0.032
BF_4 (2)	−0.96	0.024
AsF_5 (1)	−1.28	0.051–0.060
AsF_5 (2)	−1.02	0.033
SbF_5 (1)	−1.14 through −1.28	
MoF_6 (1)	−1.19	0.029
$SbCl_5$ (1)		0.033
$SbCl_5$ (2)	−0.88 through −0.89	0.013–0.0205
$AuCl_3$ (1)	−0.97	0.017
H_2SO_4 (1)[a]	−1.17 through −1.30	0.036–0.048
H_2SO_4 (2)[b]	−0.86 through −0.92	0.017–0.021

Note: The estimations are carried out using optical reflectance (*o*), magneto-oscillation (*m*), spin susceptibility (*s*), and Raman scattering (*r*).

Source: Enoki, T., et al., *Graphite Intercalation Compounds and Applications*, Oxford University Press,
New York, 2003. With permission.

[a] Changes in E_F and f_C are associated with a change in the carbon-to-intercalate ratio from 28 to 21.

[b] Changes in E_F and f_C are associated with a change in the carbon-to-intercalate ratio from 60 to 48.

folding effect. The charge transfer from Li to graphene sheet elevates the Fermi level from the point of contact and place it in the antibonding π^*-band, where the rate of elevation depends on the degree of charge transfer as we discussed in Section 6.3.1.

The band calculation carried out with a more elaborated model gives roughly similar results concerning the π-band structure to that obtained by the tight-binding model. Figure 6.14 gives the band structure of the stage-1 LiC_6 calculated on the basis of the mixed-basis pseudopotential method.[54] There is no appreciable change in the σ-bands upon Li intercalation except the zone folding effect, which are located far from the Fermi level, judging from the comparison in the electronic structures between pristine graphite and LiC_6. The energy dispersions associated with π-bands roughly track those obtained by the tight-binding method in the Γ-M and Γ-K directions in the 2D plane, as we can see by comparing Figures 6.4b and 6.14. There are deviations from the tight-binding model in the directions A–L and A–H, which are ascribed to the presence of interactions between the graphene sheet and the Li intercalate. This can be clearly seen in the dispersion curves along the *c*-axis direction: the π-band energies, which otherwise are constant with respect to the wave number, depend on the wave vector in the Γ–A direction, indicating a role of interlayer interaction which contributes to adding three-dimensionality into the electronic structure. The shortest *c*-axis repeat distance of 3.70 Å among alkali metal GICs is consistent with the 3D modification of the electronic structure in LiC_6. The Fermi energy is located at 0.85 eV above the contact point between the π- and π^*-bands. Around 1.3 eV above the Fermi energy at the Γ point, there is a contribution from the Li intercalate. The Fermi surfaces of LiC_6 shown in Figure 6.15 are obtained using empirical pseudopotentials and a self-consistent determination of the charge transfer.[56] There are two Fermi surfaces, as we can see from the presence of two energy dispersion curves crossing the Fermi level in Figure 6.14. Due to the presence of interlayer interactions, with a same order of strength as in pristine graphite, the Fermi surfaces deviate from those expected from the 2D tight-binding model. As a consequence, the Fermi surfaces are warping along the *c*-axis direction with minimum cross sections in the Γ-plane.

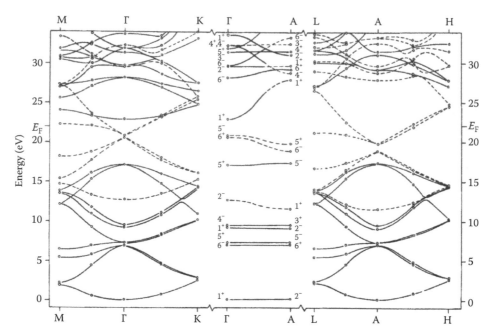

FIGURE 6.14 Band structure of stage-1 LiC_6. The dashed lines denote the π-bands and the solid lines the σ-bands. Zero of energy is taken at the bottom of the σ-band. (From Holzwarth, N.A.W., et al., *Phys. Rev. B*, 28, 1013, 1983. With permission.)

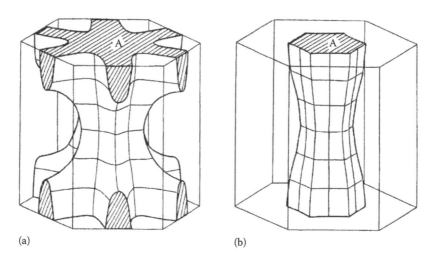

FIGURE 6.15 Fermi surfaces of LiC_6 calculated using empirical pseudo-potentials and a self-consistent determination of the charge transfer; the Fermi surfaces for the lower (a) and upper (b) bands. (From Ohno, T., *J. Phys. Soc. Jpn.* 49(Suppl. A), 899, 1980. With permission.)

In higher stage Li GICs LiC_{6n} ($n \geq 2$), band calculations have been as well carried out using the same techniques.[54,58] Only the stage-2 LiC_{12}, which has an in-plane $\sqrt{3} \times \sqrt{3}$ superlattice, was known from the experiments so far. The number of graphene planes between adjacent intercalate layers, which is the same as the stage index, increases with increasing stage index. Then, there are $2n$ Fermi surfaces for stage-n GICs. It should be noted that the screening effect plays an important role in the modification of the electronic structure for stages higher than 2, as we discussed in the Section 6.3.2

with acceptor GICs. Namely, positively charged Li ions in the graphitic galleries strongly induce polarization of electrons around carbon atoms on a graphene layer (bounding layer) facing Li intercalates, which screens the charges of the intercalates. This weakens the intercalate interaction with an interior graphene layer not facing the intercalates directly in GICs having stages higher than 2. As a consequence, the additional splitting of the π-bands appears due to the difference between the bounding and interior graphitic layers.

Next, we discuss the electronic structures of heavy alkali metal GICs. The stage-1 MC_8 (M = K, Rb, Cs) has a 2 × 2 intercalate superlattice, whereas higher stage GICs have a discommensurated intercalate superlattice with respect to the host graphite lattice. Here, the rigid band model gives a good perspective in characterizing the electronic structures after taking into account the zone folding effect. However, particularly for stage-1 GICs, the close proximity of the interlayer-intercalate state and the graphitic π-bands makes the situation rather complicated. In this connection, there has been a controversy on the electronic structure in relation to the participation of intercalate-related energy state and charge transfer for stage-1 GICs. The experimental estimation of the charge transfer rate claims $f_C \sim 0.6$ for stage-1 systems, while higher stage ones have a completed charge transfer $f_C \sim 1$.[59,60] The partial charge transfer for the former is confirmed by early calculation on the basis of the extended Hückel technique, where the intercalate band partially filled crosses the Fermi level.[61] This is a natural consequence of the rigid band picture. However, this possibility is excluded in later works although the charge transfer rate is still fractional. Therefore, elaborated electronic structure calculation is needed in understanding the complicated situation in the electronic structure in the vicinity of the Fermi energy. The nonself-consistent calculations[62] and the self-consistent calculations[63–65] find that the interlayer-intercalate band is well above the Fermi level and zone center contributions in the Fermi surface appear due to small deviations from the rigid band behavior of the graphite π-bands.

Figure 6.16 shows the band structure diagram for KC_8 and RbC_8, calculated using (a) the self-consistent LCAO method[65] and (b) the self-consistent mixed-basis pseudopotential method.[64] Around the Fermi energy region, we can see the presence of a minimum of the zone-center contribution of the graphitic π^*-electron origin at 1.5 eV below E_F at the Γ point, in addition to the graphitic π^*-band at the zone edge that crosses E_F on the way from the K/H point. Therefore, there are two kinds of graphitic π-character Fermi surfaces that play an important role in the electronic properties of stage-1 heavy alkali metal GICs. Moreover, the interlayer-intercalate band is located around 3.6 eV above the Fermi level, being suggested to be less important in the electronic properties. This seems inconsistent with the partial charge transfer experimentally evidenced. However, consistency is achieved by considering the hybridization of the graphitic π-band and the K 4s band. According to the calculation, the graphitic π^*-bands are mixed with the K 4s band in the vicinity of the Fermi level. From the charge density calculation, the total occupied potassium charge is estimated at 0.4 electrons, the potassium being in a partially ionized state $K^{+0.6}$ in the graphitic galleries. This can well explain the experimental results and suggests the complicated electronic structure of KC_8 in contrast to a simple electronic structure of LiC_6. Other heavy alkali metal GICs, RbC_8 and CsC_8, are considered to have a similar electronic structure to KC_8.[63,64] The interlayer-intercalate band moves to higher energy in the sequence of K < Rb < Cs, and the size of the zone-center contribution to the Fermi surface decreases in the same sequence.

Angle-resolved photoemission,[66–68] X-ray photoelectron spectroscopy (XPS),[69] electron energy loss spectroscopy,[70,71] optical reflectance,[60] magnetooscillation,[72] etc. experimentally confirm the electronic structure of heavy alkali metal GICs thus theoretically predicted. Figure 6.16b presents the angle-resolved photoemission of RbC_8.[68] The energy-vs.-wave number relations experimentally obtained are in good agreement with the theoretical results,[64] especially around the K point. Around the Fermi level, a dispersionless feature appears, in contrast to the presence of a 3D band at the Γ point in the theory. The dispersionless feature is assigned to the zone-center contribution at the Γ point, on the basis of a careful consideration of the experimental finding. The electron energy loss spectra confirm the presence of the interlayer-intercalate band.[71]

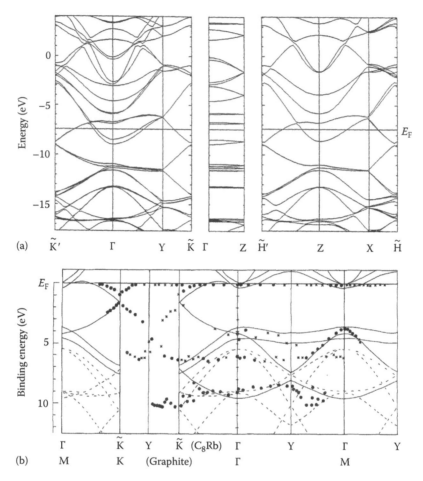

FIGURE 6.16 (a) Band structure diagram for KC_8 calculated using the self-consistent LCAO method. Lowest σ bands are not shown. (From Mizuno, S., et al., *J. Phys. Soc. Jpn.*, 56, 4466, 1987. With permission.) (b) Experimental band structure of RbC_8 compared with the theoretical band calculation of Tatar.[47] In the calculation, solid and broken lines represent π (including π^*) and σ bands, respectively. The calculation was not reported between $\tilde{K}$ and Y points. (From Gunasekara, N., et al., *Z. Phys. B*, 70, 349, 1988. With permission.)

In connection to alkali metal GICs, the presence of unconventional superdense phases should be commented. Application of high pressure in the preparation of alkali metal GICs gives superdense GICs, where a large concentration of alkali metal intercalates is accommodated in the graphitic galleries. Among these superdense alkali metal GICs, LiC_2,[73,74] NaC_{2-3},[75] KC_4,[76] $RbC_{4.5}$,[77] and CsC_4[76] are reported. LiC_2 having the highest intercalate density is the most interesting among the superdense alkali metal GICs from the viewpoints of basic science and applications to batteries, although it is decomposed into less dense phases $Li_{11}C_{24} \rightarrow Li_9C_{24} \rightarrow Li_7C_{24}$ after the pressure is released.[73] The structure of LiC_2 consists of a Li monolayer with a sandwich thickness of 3.7 Å in each graphitic gallery. The in-plane structure is formed by locating Li atoms at the center of every carbon hexagon ring. The Li–Li interatomic distance of 2.49 Å is equal to the length of the in-plane unit cell and is indeed 0.6 Å shorter than that of pristine Li metal. The density of LiC_2 is 10% higher than that of lithium metal. The final product Li_7C_{24} after the pressure is released is stable even in ambient pressure. It has a hexagonal Li_7-cluster based $2\sqrt{3} \times 2\sqrt{3}$ in-plane superlattice.[74] Other lower density phases $Li_{11}C_{24}$ and Li_9C_{24} have also Li_7 cluster-based superlattice with modifications. The Li–Li bondings in the superdense LiC_2 and those in the Li clusters of the lower density phases are reported to be predominated by covalent bonds.[78,79] Indeed, the charge transfer rate f_C from lithium to carbon has almost the same value as that of ordinary Li GIC (LiC_6), whose electronic structure features with Li^+ ion intercalated in the graphitic galleries. Taking into account the considerable increase in the Li concentration in superdense Li GICs, the transferred charge per Li atom is strongly reduced in comparison with that for LiC_6.[80,81] This suggests that the feature of the bonding is considerably modified by the participation of covalent bonds having p-character in the formation of Li_7 clusters and also Li bonding network of LiC_2, as confirmed by NMR experiments.[82,83]

Alkaline earth GICs have a $\sqrt{3} \times \sqrt{3}$ superlattice similar to LiC_6. Their electronic structures have not been well investigated except BaC_6[69,84,85] and CaC_6.[86,87] The electronic structure of BaC_6 considerably deviates from that expected from the tight-binding model. In the rigid band model, the rate of charge transfer is indicated by the amount of electrons transferred to the graphitic π-band, which merely elevates the Fermi energy according to Equation 6.11. However, this is not the case for BaC_6. There are three bands that cross the Fermi level. Two of them are ascribed to the graphitic π-bands, whereas the third one results from the interlayer-intercalate band. The partially occupied interlayer-intercalate bands, which has already been discussed before with alkali metal GICs, play an important role in characterizing the electronic structure in the vicinity of the Fermi energy. It contains contributions of Ba 6s and 5d orbitals where the d-electron nature is considerably enhanced compared to ordinary Ba compounds. The hybridization between Ba 5d and graphite π-bands affects importantly the electronic structure as well. Interestingly, one of two electrons in the Ba 6s band is transferred to the graphitic π-bands, whereas the other stays in the intercalate-interlayer band.[69] This is in contrast to the ordinary Ba compounds that tend to have divalent Ba^{2+} state. Optical reflectance[84] and XPS[69] experiments reveal this anomalous feature in the electronic structure.

The electronic structure of CaC_6 has been intensively investigated under the stimulation triggered by the discovery of high T_c superconducting states of CaC_6 ($T_c = 11.5\,K$) and YbC_6 ($T_c = 6.5\,K$).[88] Density functional theory calculation[87] clarifies the detailed contributions of graphitic π-electrons and Ca intercalate electrons to the band structure. Around the Fermi energy, the electronic band structure consists of a free electron-like Ca band with 3D nature and 2D graphitic π-band as shown in Figure 6.17. The theoretical analysis suggests that the Ca 4s, Ca 3d, Ca 4p, C 2s, C 2pσ, and C 2pπ projections of the density of states at E_F are 0.124, 0.368, 0.086, 0.019, 0.003, and 0.860 states/(cell eV), respectively, as exhibited in Figure 6.17a. The charge transfer rate is estimated to be $f_C = 0.053$. The occurrence of superconductivity is a consequence of the incomplete ionization of the Ca intercalate.

In rare earth GICs that have the $\sqrt{3} \times \sqrt{3}$ superlattice, the electronic structures of La, Eu, and Yb GICs have been investigated from both experimental and theoretical aspects.[86,89–94] Among them, EuC_6 is of particular interest for magnetism in which the interaction between Eu 4f and graphitic π-electrons plays an essential role, giving rise to novel interplay between metallic electron transport and magnetic phase transition. The basic electron configurations for rare earth elements are interestingly featured with 4f and 6s electrons. The 4f electrons form localized states with localized magnetic moments, whereas 6s electrons are delocalized, contributing to the formation of bonds with the graphitic π-electrons upon intercalation. It should be noted that the 6s state tends to be hybridized with 5d and 6p states, resulting in features different from alkali metal GICs. Figure 6.18a and b shows the electronic structures of stage-1 and 2 Eu-GICs (EuC_6, EuC_{12}) calculated on the basis of local-density-approximation with the Hedin–Lundqvist exchange potentials. The calculated electronic structure of stage-2 GIC is compared to the experimental results from angle-resolved photoelectron spectra presented in Figure 6.18c.[94] The band structure is found to be considerably modified from that obtained from the tight-binding model except the fact that the downward shift of graphitic π-bands is the consequence of charge transfer from rare earth metal to graphene layer. The charge transfer makes the $π^*$-band partially filled. Deviations from the tight-binding model appear around the Fermi energy. A splitting is produced between the bonding π- and antibonding $π^*$-bands, which is remarkable in stage-1 GIC. This is seriously affected by the spatial overlap between the graphitic π-band and Eu spd orbitals, in contrast to alkali metal GICs. Actually, Eu-derived sd-hybrid band emerges as a partially filled state between the Fermi level and the $π^*$-band, which contains 30%

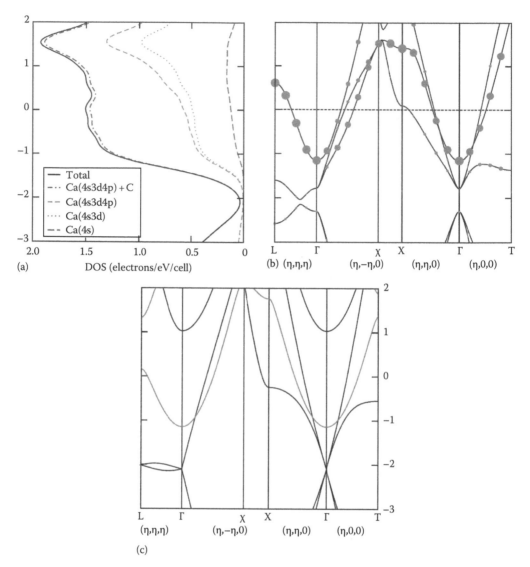

FIGURE 6.17 (a) Total density of states and the partial densities of states projected on selected atomic wave functions in CaC_6. (b) Band structure of CaC_6. The size of the dots represents the percentage of Ca component. As a reference, the dot at ~0.6 eV at the L point represents the 0.95 Ca component. (c) Band structure of CaC_6 with C atoms removed and that with Ca atoms removed. The parabolic band with its bottom located at the Γ point and the bands located around the top of the panel at the L and T points are assigned to the former, while the others are for the latter. The bands have been shifted to compare with the CaC_6 band structure. The directions are given in terms of the rhombohedral reciprocal lattice vectors. χ is the interception between the $(\eta,-\eta,0)$ line with the border of the first BZ ($\eta = 0.3581$). The points L, Γ, X, and T have $\eta = 0.5$. (From Calandra, M. and Mauri, F., *Phys. Rev. Lett.*, 96, 237002-1-4, 2006. With permission.)

d-character at the Γ point for stage-1 compound. Moreover, the π^*-band is modified with the partial contribution of d-character, which grows to 50% at the M' point in the zone boundary. For the stage-2 Eu-GIC, a fully occupied pd-symmetry branch is present in the region of M' point. The amount of charge transferred from Eu to graphite is estimated at 0.5, which is smaller than the one expected for alkali metal GICs. This may be ascribed to covalent admixtures present in Eu GICs as well as large electronegativity of Eu as compared to that of alkali metals. The features of the electronic structures are faithfully tracked by the experiments of angle-resolved photoelectron spectra

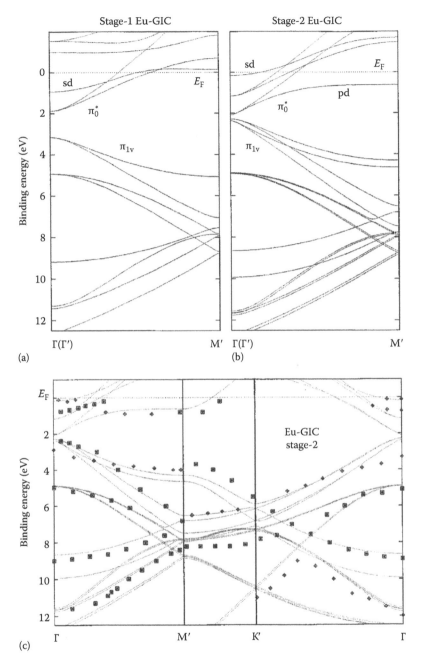

FIGURE 6.18 Band structures of Eu GICs. (a) stage-1 EuC_6, (b) stage-2 EuC_{12}, and (c) comparison with the experiment of angle-resolved photoelectron spectra. Theoretical calculations for (a) and (b) are carried out by means of local-density-approximation. Symmetry points denoted with characters with prime are defined for a $\sqrt{3} \times \sqrt{3}$ superlattice of GICs, whereas Γ is the zone center of the graphite unfolded BZ. (From Molodtsov, S.L., et al., *Phys. Rev. B*, 53, 16621, 1996. With permission.)

shown in Figure 6.18c. The hybridization between Eu spd orbitals and graphitic π-orbitals works to mediate strong magnetic interaction ($J_{\pi-f} \sim 0.15\,\mathrm{eV}$) between Eu f localized magnetic moments and conduction π-carriers on graphitic sheets, giving rise to novel magnetic behavior in EuC_6.[95] Yb GICs have similar electronic features to Eu GICs.

In the electronic structure of YbC$_6$, density functional calculations demonstrate[93] that the Yb f states are fully occupied with nonmixed valence feature. The 4f states are considerably hybridized both with the Yb 5d and C 2p states resulting in a nonnegligible admixture of Yb f at the Fermi level.

6.3.4 Fluorinated Graphite

Fluorine gives unique GICs in which a competition between charge transfer and covalent bond formation governs their structural and electronic features. In the low fluorine concentration range, charge transfer from graphite to the intercalate produces ordinary acceptor GICs having unique stages. In contrast, the increase in the fluorine concentration induces covalent bond formation between fluorine and carbon atoms on the graphitic layer, which brings about a deformation of graphitic plane into corrugated plane due to the conversion of sp^2 bonds to sp^3 bonds.

In relation to the competing two processes, F GICs C$_x$F show a unique electronic structures evidenced by electrical resistivity,[96] optical reflectance,[47] etc. Figure 6.19 shows the room temperature in-plane resistivity for different fluorine concentrations.[96] The increase in the fluorine concentration 1/x increases electrical conductivity in the low fluorine concentration range up to $x{\sim}6$ in stage-2 compound. This is the ordinary trend observed in acceptor GICs related to the increase in the number of hole carriers due to charge transfer. However, the conductivity shows a steep decrease in the higher concentration range after a conductivity maximum around $x{\sim}6$, and, around $x{\sim}3$, it reaches a value two orders of magnitudes lower than the value at $x{\sim}6$, although the stage remains unchanged.[97] Optical reflectance spectra faithfully track the anomalous conductivity behavior upon the increase in the fluorine concentration. Namely, as shown in Figure 6.20, the absolute value of the

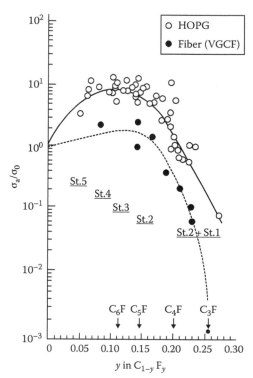

FIGURE 6.19 Room temperature in-plane resistivity of F GIC C$_{1-y}$F$_y$ for different fluorine concentrations. σ_0 is the in-plane conductivity of host graphite; pristine graphite (HOPG) and graphite fiber (VGCF). Note that y is defined as $y = (1 + x)^{-1}$ for C$_x$F. (From Vaknin, D., et al., *Synth. Metals*, 16, 349, 1986. With permission.)

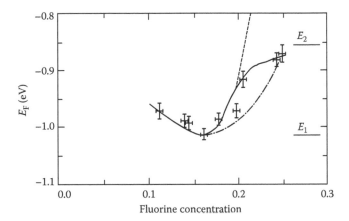

FIGURE 6.20 The Fermi energy, E_F, as a function of fluorine concentration $1/x$ for stage-2 C_xF. The solid line represents the fit using the two-acceptor-stage model, where the two acceptor levels are estimated at $E_1 = -1.034\,eV$ and $E_2 = -0.84\,eV$. The dashed and the dashed-dotted lines represent the model that fluorine atoms added beyond C_6F contribute to the formation of covalent bonds with neighboring carbon atoms and the corresponding region becomes insulating. (From Ohana, I., et al., *Phys. Rev. B*, 38, 12627, 1988. With permission.)

Fermi energy increases ordinarily as the fluorine concentration $(1/x)$ increases in the low concentration range up to $x\sim6$ where the charge transfer rate per fluorine atom becomes 0.125. After that, it shows a decrease at high concentrations, suggesting that the number of holes generated by charge transfer tends to be lowered. The charge transfer rate per fluorine atom becomes 0.063 at $x \sim 4$. The two-step change in the electronic structure thus observed is explained in terms of the presence of two acceptor levels produced from the intercalate fluorine species. In the low fluorine concentration regime, the first acceptor levels, which are well below the Fermi level ($E_1 = -1.034\,eV$), are successively occupied by the electrons transferred from graphene sheets up to the fluorine concentration C_6F, at which the first levels become full. Beyond the concentration, the second acceptor levels ($E_2 = -0.84\,eV$), which are above the first acceptor levels, are created so that with each additional fluorine a total of $A + 1$ fluorines are in the new acceptor levels at the expense of A first acceptor levels. The change in the electronic structure associated with the participation of the second acceptor levels induces the reduction in the conductivity. This model is appropriate in explaining the electronic structure in the fluorine concentration region $8 > x > 4$ for stage-2 F GICs, where fluorine molecules lie along the graphitic planes with a small tilted angle ($\sim12°$) and the sandwich thickness is estimated at $6.14\,Å$. At higher fluorine concentrations $x < 4$, more fluorine atoms cannot be added unless fluorine molecules stand upright perpendicular to the graphitic plane. X-ray studies confirm this structural change, suggesting a remarkable increase in the sandwich thickness by $1.61\,Å$ at $x\sim2$. At this concentration, each fluorine atom faces one neighboring carbon atom, whose concentration is a half of the total fluorine concentration. This structural feature, which is similar to that of fluorinated graphite C_2F, is favorable for the formation of covalent C–F bonds. The drastic decrease in the conductivity in that concentration region, hereby, is explained in terms of the generation of insulating regions networked with fluorine-induced sp^3 bondings.

6.4 NANOGRAPHENE AND NANOGRAPHITE

6.4.1 Electronic Structures of Nanographene

The discoveries of fullerenes and carbon nanotubes have contributed to enriching the variety in the world of carbons, where the new comers have opened a new realm bridging between traditional carbon and nanomaterials. Ball-shaped fullerenes and cylindrical-shaped carbon nanotubes form closed π-electron systems with the participation of pentagon rings or occasionally heptagon rings. Here, the electronic structure depends on their shapes, sizes, and helicities. In such a nanocarbon family, nanographene is included, which is featured with nano-sized 2D flat hexagon ring network with open edges, in contrast to its counterparts having closed π-electron systems, such as fullerenes and carbon nanotubes. According to the theoretical suggestions,[17–19,98,99] the presence of edges adds an extra-electronic state to nanographene depending on the shape of the edges. Namely, edge-inherited nonbonding π-state (edge state) appears around the Fermi level in addition to the bonding π- and antibonding π*-states, when nanographene has zigzag edges. Taking into account that the electronic properties are governed mainly by the electronic states around the Fermi level, the presence of the edge states is considered to make the electronic properties entirely different from those of bulk graphite. In addition, from the magnetism aspects, the magnetic moments created in the edge states bring about unconventional carbon-only magnetism having features different from traditional magnets. The electronic activity of the edge state populated around the edge region is expected to play an important role in chemical reactivity as well. This section is devoted to discussing the electronic structures of nanographene and nanographite, the latter of which is a stacked assembly of nanographene sheets.

Let us start with chemistry on condensed polycyclic hydrocarbon molecules. The electronic structure of benzene is described with six π-orbitals that are split into three occupied bonding π- and three unoccupied antibonding π*-levels located on the opposite sides of the Fermi level E_F with the level splitting being finite between the HOMO and the LUMO. The condensed polycyclic aromatic hydrocarbons formed by the fusion of benzene rings such as naphthalene, anthracene, etc. also have similar electronic structures with the number of π-orbitals and the HOMO–LUMO level splitting increasing and decreasing, respectively, upon the increase of the number of benzene rings associated. These condensed polycyclic aromatic hydrocarbons are called Kekulé molecules, in which the presence of a large HOMO–LUMO level splitting stabilizes the molecules. In the extreme of an infinite size, graphene sheet has a semimetallic electronic structure featured with the bonding π- and antibonding π*-bands that are touched to each other at E_F with no HOMO–LUMO splitting as discussed in Sections 6.2 and 6.3. A group of these condensed polycyclic aromatic hydrocarbons and graphene therefore have common electronic properties based on these π-orbitals. However, among these, there is a subfamily particularly featured with the presence of nonbonding π-electron state. Typical examples are phenalenyl and triangulene free radicals consisting of three and six benzene rings fused, respectively, as shown in Figure 6.21a and b, in which unpaired electron(s) having localized magnetic moment $S = 1/2$ and $S = 1$ exist(s) as a consequence of the presence of a singly occupied nonbonding π-electron state(s) at E_F in addition to the π- and π*-levels. Phenalenyl and triangulene radicals are chemically unstable owing to the presence of the singly occupied nonbonding π-electron state.[100–102] The polycyclic aromatic molecules featured with the presence of nonbonding π-electron state are called non-Kekulé molecules. In general, we can group all the carbon sites of a condensed polycyclic aromatic hydrocarbon into two subgroups, where the neighboring sites directly bonded to a site belonging a subgroup ("starred" subgroup) belong to another subgroup ("unstarred" subgroup). According to Liebs theorem,[103] the unpaired electrons are produced when the numbers of sites belonging to these two subgroups are different. Actually, the difference in the numbers corresponds to the number of unpaired electrons. In the cases of phenalenyl and triangulene free radicals, the differences give one and two unpaired electrons, respectively. Non-Kekulé molecules have a finite value in the difference.

The same theorem is also applicable to nanosized graphene sheet (nanographene). Theory has suggested the presence of nonbonding states in a nanographene dependent particularly on its shape. In general, the circumference of an arbitrary shaped nanographene sheet can be described in terms of a combination of zigzag and armchair edges, which mimic the structures of trans- and cis-polyacetylenes, respectively, as shown in Figure 6.21c and d. According to theoretical predictions,[17–19,98,99] zigzag edges give nonbonding π-electron state (edge state) of edge origin, whose energy level appears at the contact point of the π- and π*-bands. Figure 6.22 presents one

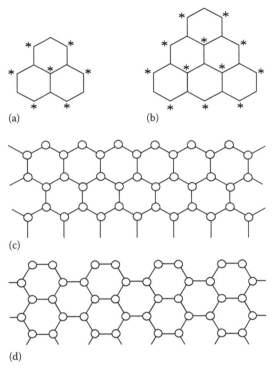

FIGURE 6.21 Phenalenyl (a) and triangulene (b) free radicals where up and down spins are placed on starred and unstarred sites, (c) zigzag edge, and (d) armchair edge. The carbon sites, which are directly bonded to a site belonging to a subgroup (starred), belong to another subgroup (unstarred).

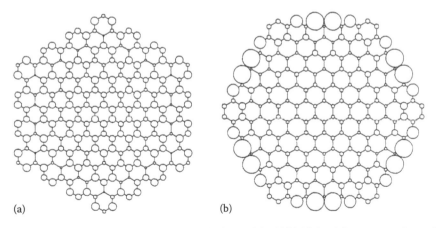

FIGURE 6.22 The spatial distribution of the populations of the HOMO level for nanographene sheets with their edges having (a) armchair and (b) zigzag structures. (From Stein, S.E. and Brown, R.L., *J. Am. Chem. Soc.*, 109, 3721, 1987. With permission.)

of the examples for the electronic structures of armchair- and zigzag-edged nanographene sheets. The HOMO level has the largest populations around the carbon atoms in the zigzag edges in zigzag-edged nanographene sheet, though it distributes uniformly in the entire area of armchair-edged nanographene sheet. The electronic states in the strongly populated zigzag carbon sites are assigned to the nonbonding edge states, which are singly occupied. The presence of the edge state at E_F also makes zigzag-edged nanographene less stable than armchair-edged one.[17]

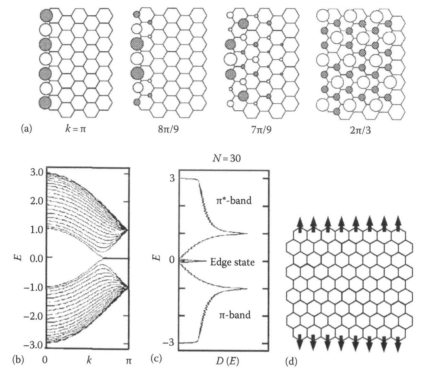

FIGURE 6.23 (a) The wave number dependence of the populations of the edge state, (b) the energy dispersions of a nanographene ribbon having zigzag edges with a width of 30 unit cells, (c) the density of states, and (d) ferromagnetic spin arrangement at the zigzag edges. All the edge carbon atoms are terminated with hydrogen atoms. (From Fujita, M., et al., *J. Phys. Soc. Jpn.*, 65, 1920, 1996. With permission.)

Another example is a nanographene ribbon[19] shown in Figure 6.23. In a nanographene ribbon having zigzag edges, the edge state is localized around the zigzag edges in the wave number region from the BZ edge to $2\pi/3$ of the zone, while it becomes delocalized in the central region $0 \leq k \leq 2\pi/3$. The presence of the edge states, whose density of states has a sharp peak with its occupation being half-filled around E_F, gives rise to localized magnetic moments. Therefore, nanographene is of particular interest from magnetism viewpoint, where even ferromagnetism happens to be produced.

Modifications of edges by a substitution with foreign species are predicted theoretically to give a variation of magnetism.[104–106] When all the edge carbon atoms of zigzag edges on one side of a nanographene ribbon are dihydrogenated with those on the opposite side remaining to be monohydrogenated, a completely localized nonbonding state appears around E_F, where all the carbon atoms are spin polarized even in the interior of nanographene ribbon in a ferromagnetically ordered state.[105] In contrast, fluorination of edges tends to reduce magnetism due to the tendency of forming a closed shell in fluorine. An interesting example is the oxidation of carbon atoms on one zigzag edge side. The oxidized zigzag edge forms an electron conduction path with the hydrogen-terminated carbon atoms on another zigzag edge side being magnetic. This means that the chemical modification can give different roles to these two edges.

The electronic structure of nanographene can be investigated experimentally using scanning tunneling microscopy (STM) and spectroscopy (STS). Graphene edges in the pristine state appear with disordered features, when they are observed in ambient STM experiments. This suggests that the graphene edges are covered completely by oxygen-including functional groups, which are created as a result of oxidation. Therefore, ultra-high vacuum (UHV) STM/STS experiments with graphene edges well defined by edge termination with hydrogen atoms are necessary to characterize the electronic structure of nanographene. According to UHV STM measurements with hydrogen terminated graphene edges,[20–22,107,108] armchair edges are generally long and defect free, whereas zigzag edges tend to be short and defective, as can be seen in comparison between Figures 6.24 and 6.25. Indeed, long well-defined armchair edges are frequently observed, whereas short zigzag edges are embedded between armchair edges. These experimental findings are in good agreement with theoretical suggestions that the armchair edge is energetically more stable than the zigzag edge that has an edge state at the Fermi level.[17] Figure 6.24b shows that the local density of states, which is represented by the dI/dV_s vs. V_s curve (STS curve), consists of the bonding and antibonding bands. These two bands are found to touch each other at E_F, same as the zero-gap-semiconductor-type

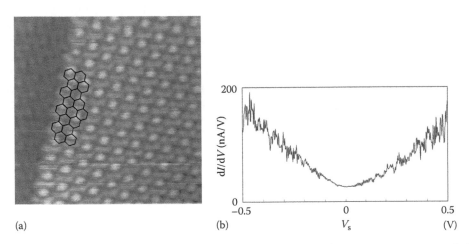

(a)   (b)   (V)

FIGURE 6.24   (a) Atomically resolved UHV STM images ($5.6 \times 5.6\,nm^2$) of a homogeneous armchair edge in constant-height mode with bias voltage $V_s = 0.02\,V$ and current $I = 0.7\,nA$. For clarity of edge structures, a model of the honeycomb lattice is drawn on the image. (b) A dI/dV_s curve from STS measurements taken at the edge in (a). (From Kobayashi, Y., et al., *Phys. Rev. B*, 71, 193406-1-4, 2005. With permission.)

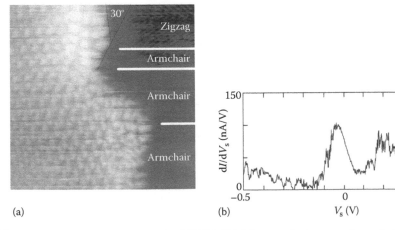

(a)   (b)

FIGURE 6.25   (a) An atomically resolved UHV STM image of zigzag and armchair edges ($9 \times 9\,nm^2$) observed in constant-height mode with bias voltage $V_s = 0.02\,V$ and current $I = 0.7\,nA$. (b) The dI/dV_s curve from STS data at a zigzag edge. (From Kobayashi, Y., et al., *Phys. Rev. B*, 71, 193406-1-4, 2005. With permission.)

electronic structure of a graphene. In contrast, the electronic structure of a zigzag edge has a large density-of-states peak around E_F, which is assigned to the edge state, as shown in Figure 6.25b. The bright spots distributed around the zigzag edge region in Figure 6.25a correspond to the large local density of states in the edge state.

6.4.2 ELECTRONIC STRUCTURES OF NANOGRAPHENE WITH ITS EDGES CHEMICALLY MODIFIED AND CHEMICAL REACTIVITY

The chemical activities of nanographene are different from those of infinite size graphene or bulk graphite due to the presence of edges and the finiteness in its size. In connection to the edge effect, nonbonding edge states appearing in the zigzag edges are one of the important contributions to the chemical activities of nanographene. Indeed, zigzag edges are energetically unstable in comparison with armchair edges, as discussed in Section 6.4.1. In this respect, we can expect a difference in chemical reactivity between armchair and zigzag edges. An early work reports that the oxidation rate of edge carbon atoms is larger by 10%–20% in zigzag edges than armchair edges.[109] This means that zigzag edges are more easily oxidized, in good agreement with their energetically unstable structure. In addition to the difference in the geometry of edge shapes, how to terminate the edge carbon atoms is responsible for creating a variety of chemical activities depending on the electronic features of the functional groups bonded to the edge carbon atoms. Therefore, edge-inherent chemical reactivity should be taken into consideration in discussing nanographene, in addition to charge transfer that is associated with the amphoteric nature of graphene and graphite.

The size of nanographene sheets, which is restricted in nanodimension, plays an important role in modifying the charge transfer reaction as well. Indeed, in case that the in-plane size is smaller than the coherence length of the density correlation function of intercalates in the intercalation process, an intercalation structure with a unique stage is not completed due to the structural fluctuations in the intercalate layers, the charge transfer rate deviating from that of bulk GICs. Actually, alkali-metal intercalation compounds of nanographite has a mixed stage structure with stage-1, stage-2, and nonreacted nanographene sheets.[110,111]

Here, we first discuss the chemical activities of zigzag edges, at which the edge states are created. A theoretical approach indicates differences in the dissociation reactions of several functional groups from the carbon atoms between the zigzag and armchair regions of nanographene ribbons.[112] The bond dissociation energy of a hydrogen atom from a dehydrogenated carbon site of the zigzag edge is in the range of 2.82–2.87 eV, whereas it is estimated as 1.55 eV in the armchair edge. This suggests that the unstable structure of zigzag edge is more stabilized by the dihydrogenation of the edge carbon atoms. The bond dissociation energy is lowered upon the decrease in the ribbon width, narrower ribbons being suggested to be more chemically active. The reactivity depends also on the coverage of the edges. The dissociation energy is lowered as the concentration of dihydrogenated sites increases. In other words, the zigzag edge structure with edge carbon atoms completely dehydrogenated is more unstable in comparison with that with carbon atoms partially dehydrogenated as shown in Figure 6.26, in which the dissociation energy is estimated as 2.86, 2.73, and 1.93 eV for the coverages of 1/6, 1/3, and 1, respectively. The bond dissociation energies are listed in Table 6.2 for several functional groups in comparison with those of substituted ethane molecules.

Next we discuss the charge transfer reaction with intercalates (or guest molecules) in nanographene. This issue is particularly important in discussing the electrochemical process. An important example is charge transfer with lithium, which is used for high performance secondary battery. As we briefed in Section 6.3.3, the maximum guest/host composition ratio of conventional Li-GIC is Li/C = 6 that is achieved for stage-1 GIC. In the super-dense Li-GIC, which is prepared in ultra-high pressure, the ratio reaches Li/C = 2. Interestingly, the ratio can reach Li/C = 2 for nanographene/nanographite in a conventional preparation condition. This is considered to be associated with the hydrogen-termination of the edge carbon atoms in a nanographene sheet.[113–116] Therefore, nanographene has a good advantage to achieve high performance in the secondary batteries.

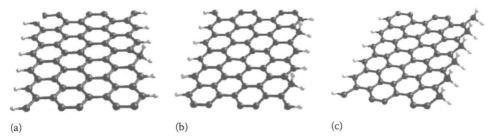

(a)   (b)   (c)

FIGURE 6.26 Chemical reaction of hydrogen atoms with a nanographene ribbon's monohydrogenated zigzag edge at different edge coverage: (a) 1/6, (b) 1/3, and (c) 1 for 5 unit cell width. Coverage is expressed as reacted hydrogen per edge carbon. (From Jiang, D.-E., et al., *J. Chem. Phys.*, 126, 134701-1-6, 2007. With permission.)

TABLE 6.2

Comparison of Bond Dissociation Energy (BDE, in eV) of Zigzag Edge-X Bonds with Experimental BDE of C_2H_5-X

Radical: X	H	OH	CH$_3$	F	Cl	Br	I
BDE (edge X)	2.86	2.76	2.22	3.71	2.18	1.65	1.18
BDE (C_2H_5-X)	4.358	4.055	3.838	4.904	3.651	3.036	2.420

Note: The coverage is at 1/6 X/edge C; zero-point-energy corrections not included.
Source: Jiang, D.-E., et al., *J. Chem. Phys.*, 126, 134701-1-6, 2007. With permission.

The charge transfer reaction of nanographene with Li goes differently from that in bulk graphite, as a consequence of the presence of edges and the finiteness in size.[117–119] Indeed, the way of charge transfer and the arrangement of intercalates crucially depend on whether the edge carbon atoms are terminate with hydrogen atoms[113] or have σ-dangling bonds with no termination. When the edge carbon atoms are bare with dangling bonds, there is no stable site for Li atom guests in the interior of the nanographene sheet after charge transfer reaction. Instead, a Li atom, which is positively charged with +0.29, tends to form covalent bonds with the edge carbon atoms that have σ-dangling bonds as shown in Figure 6.27. Interestingly, the charges transferred from the Li atom to nanographene are distributed in the edge region. In contrast to the bare edged nanographene sheet with σ-dangling bonds, a nanographene sheet with its all edge carbon atoms completely hydrogen terminated has Li guest atoms stabilized through charge transfer in the interior of the sheet. In this case, the $\sqrt{3} \times \sqrt{3}$ superlattice in-plane structure of Li atoms can be formed similar to bulk stage-1 Li-GIC (LiC$_6$). However, the charges in Li atoms depend on their locations. It should be noted that charge fluctuations give a variety of charged species, where not only positively charged Li atoms but also negatively charge species are present. These charge fluctuations are the consequence of the finite size effect, which does not appear in bulk Li-GIC. In addition, the hydrogen-terminated edge sites are available for accommodating extra Li species. Figure 6.28 summarizes the distribution of the Li guest species on a nanographene sheet. Charge fluctuations can be seen in the Li atoms stabilized in the interior of the nanographene sheet. The hydrogenated edges provide sites in which the Li atoms are stabilized as well. Interestingly, the bond formation of a Li atom with the edge carbon atoms is accompanied by the change in the local structure at which the Li atom is reacted.[117] Availability of the hydrogenated edges for accommodating Li atoms contributes to increase the Li/C ratio. This is favorable for enhancing the performance of the Li secondary batteries.[113]

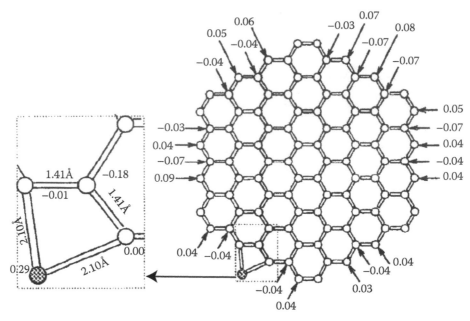

FIGURE 6.27 An optimized Li geometry in nanographene sheet (C_{96} cluster) in which a $Li^{+0.29}$ ion is mostly covalently bonded to the two nearest neighbor carbon atoms. The spotted circle denotes a Li atom. The charges transferred from the Li ion to the nanographene are distributed in the edge region. (From Nakadaira, M., et al., *J. Mater. Res.*, 12, 1367, 1997. With permission.)

Partially hydrogenated nanographene is situated in its properties between the bare nanographene and completely hydrogenated nanographene sheets. The $\sqrt{3} \times \sqrt{3}$ superstructure is deformed due to finite size effect with the aid of the low structural symmetry. Here, what interests us is the presence of dimerized (Li_2) or trimerized Li (Li_3) species created in partially hydrogenated nanographene sheet.

The theoretical works shown above on the electronic structures of Li-doped nanographene sheets give an important clue to explain the electronic features of Li ion batteries.

Interactions of acceptors with nanographene are also different from those in bulk acceptor GICs. Here, iodine is one of the typical examples. In bulk graphite, iodine molecules are not intercalated in the graphitic galleries.[6] In contrast, iodine molecules are intercalated in nanographite with a slight charge transfer ($f_C = 0.0027$–0.0079), where I_2 molecules and I_3^- ions coexist.[120,121] Taking into account the fact that the polycyclic aromatic molecules, such as perylene, pyrene form charge transfer complexes with iodine,[122] this suggests that nanographene has electronic features intermediate between small polycyclic aromatic molecules and bulk graphite. Bromine, which is chemically more active, has different features from iodine in reaction with nanographene, as a consequence of the presence of edges.[123] Bromine is bonded to the edge carbon atoms as covalent species in the substitution reaction, in addition to that bromine molecules form charge transfer species. The charge transfer rate, which is estimated as $f_C \sim 0.0004$, is considerably smaller than that of bulk Br-GICs ($f_C \sim 0.002$).[123]

Interesting is fluorine that is most chemically active in halogen elements.[124,125] Fluorine atoms tend to form covalent bonds with carbon atoms as explained in Section 6.3.4. There are two kinds of carbon sites with which fluorine atoms interact; that is, sites in the interior and edge of a nanographene sheet. The formation of covalent bonds takes place accompanied with structural distortions.[123] Fluorine atoms first attack the edge carbon atoms since the edge region is more chemically active and more easily distorted than the interior in the reaction with fluorine atoms. After all the edge carbon atoms are bonded to fluorine atoms, the carbon atoms in the interior of nanographene

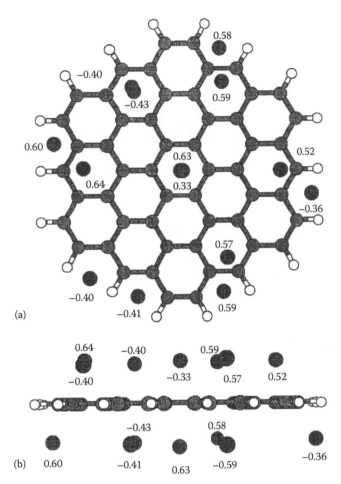

FIGURE 6.28 (a) Top and (b) side views of a completely hydrogenated nanographene with 14 Li atoms ($C_{54}H_{18}Li_{14}$). The dark large circles are Li atoms, and the open circles are H atoms. The charge is labeled for each Li atom. Six Li atoms and eight Li atoms are negatively and positively charged. (From Yagi, M., et al., *J. Mater. Res.*, 14, 3799, 1999. With permission.)

sheet are subjected to fluorination. This two-step change is clearly seen in the fluorine concentration dependence of the localized spin concentration for ACFs shown in Figure 6.29, where the first step is completed in the low fluorine concentration range up to F/C ~0.4, with the second step following the high concentration range up to F/C ~1.2. ACFs consist of a 3D disordered network of nanographite domains, each of which is a stack of 3–4 nanographene sheets with an in-plane size of 2–3 nm.[121,126,127]

In the first step, fluorination of the edge carbon atoms reduces the spin concentration of the edge states as a consequence of a deformation of the edge structure induced by difluorination of the edge carbon atoms, as shown in Figure 6.29b. Indeed, the difluorination of the edge carbon atoms converts the flat sp² structure into the tetrahedral sp³ one, causing the disappearance of the edge-state spins.[105] The structural deformation around the edge region results in the shrinkage of the π-conjugated area in the interior of nanographene sheet, as can been seen in the declining trend of the absolute value of the orbital susceptibility upon fluorine uptake in Figure 6.29a. The second step starting from F/C ~0.4 shows an increase in the spin concentration with a maximum value around F/C ~0.8. The creation of localized spins is associated with the formation of an sp³ covalent bond between fluorine atoms and carbon atoms in the interior of the nanographene sheet. Since the C–F

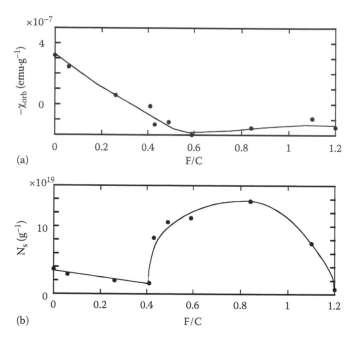

FIGURE 6.29   Orbital susceptibility χ_{orb} (a) and localized spin concentration N_s (b) for fluorinated ACFs with F/C = 0–1.2. Solid lines are guides for the eyes. (From Takai, K., et al., *J. Phys. Soc. Jpn.*, 70, 175, 2001. With permission.)

bond formation requires a larger energy in the interior than in the edge region due to a difficulty in the structural deformation,[124] the fluorination in the interior takes place as the second step after that in the edge region, as evidenced in the experiments. The formation of a covalent bond between a fluorine atom and an interior carbon atom creates a σ-dangling bond at the adjacent carbon atom at the expense of the π-bond between the two carbon atoms. The number of σ-dangling bonds having a localized spin increases as the F/C concentration is elevated up to the value at which a half of the interior carbon atoms are covalently bonded to fluorine atoms. After the F/C concentration at which the spin concentration becomes maximized, the spin concentration decreases due to the formation of C–F covalent bonds at the dangling bond sites. The localized spins completely disappear after all the carbon atoms are completely bonded with fluorine atoms, where the structure of corrugated sheet with an sp³ tetrahedral local structure is stabilized as shown in Figure 6.30. Here, it should be noted that the saturated fluorine concentration exceeds unity. This is contrasted to that of bulk graphite or infinite graphene sheet, which has a saturation concentration at F/C = 1. The deviation from unity is a consequence of the presence of edges. Each edge carbon atom takes two fluorine atoms in the reaction whereas the interior one takes one fluorine atom. The presence of excess fluorine concentration indicates the importance of the edges in the fluorination reaction in nanographene sheets.

6.5   SUMMARY

Graphite, which is one of the allotropes in carbon, has unique features characterized by two-dimensionality and amphoteric nature. This is based on the 2D layered structure with the π-electronic structure extended in the constituent graphene sheets. The electronic structure of graphene, which is described in terms of the extrapolation of condensed polycyclic aromatic molecules to infinite size,

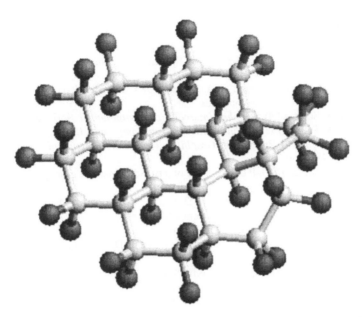

FIGURE 6.30 The structure of a nanographene sheet whose carbon atoms are completely fluorinated. (From Saito, R., et al., *J. Phys. Chem. Solids*, 60, 715, 1999. With permission.)

is given as zero-gap semiconductor with bonding π- and antibonding π^*-electronic states touching to each other at the Fermi level. The stacking of graphene sheets in the *AB* fashion imparts bulk 3D graphite, whose electronic structure is characterized with semimetal having small concentrations of electrons and holes as conduction carriers. The amphoteric nature of grapahene and graphite is a consequence of electronic features of zero-gap semiconductor and semimetal.

The two-dimensionality and amphoteric nature of graphite play an important role in creating a large variety of graphite-based materials, which are called GICs. Guest (intercalates) molecules of electron donors and acceptors are accommodated in the graphitic galleries through charge transfer between host graphite and the guests. A huge number of GICs have been produced so far with a variety of guest species such as alkali metal, alkali earth metal, rare earth metal for donors and metal halides, halogen, and acids for acceptors. The charge transfer from donor to graphite or from graphite to acceptor creates large concentrations of electron or hole carriers, respectively, making GICs metallic in nature.

When the size of graphene or graphite is reduced to nanodimension, the electronic structure is seriously affected by its shape. The circumference of an arbitrary-shaped nanographene sheet is described in terms of a combination of zigzag and armchair edges. The electronic structure of an armchair-edged nanographene sheet is given merely with bonding π- and antibonding π^*-electronic states, similar to an infinite graphene sheet. In contrast, a nonbonding π-electron state appears around the Fermi level in a zigzag-edged nanographene sheet. This state called edge state is well localized around the zigzag edge region. The presence of singly occupied edge state makes zigzag-edged nanographene less stabilized in comparison with armchair-edged nanographene.

The presence of edges and the finiteness in size plays an important role in the chemical activities of nanographene sheets. Chemical reactions in the chemically active edge region and the charge transfer reaction modified by the smallness in size give a rich variety of chemical activities in nanographene and nanographite.

REFERENCES

1. Kelly, B. T. 1981. *Physics of Graphite*. Applied Science, London.
2. Dresselhaus, M. S. and G. Dresselhaus. 1981. Intercalation compounds of graphite. *Adv. Phys.*, 30: 139–326.
3. Brandt, N. B., S. M. Ponomarev, and O. A. Zilbert. 1988. *Semimetals-Graphite and Its Compounds*. North Holland, Amsterdam, the Netherlands.
4. Zabel, H. and S. A. Solin. 1990. *Graphite Intercalation Compounds I*. Springer-Verlag, Berlin.
5. Zabel, H. and S. A. Solin. 1992. *Graphite Intercalation Compounds II*. Springer-Verlag, Berlin.
6. Enoki, T., M. Suzuki, and M. Endo. 2003. *Graphite Intercalation Compounds and Applications*. Oxford University Press, New York.
7. Zheng, Y. and T. Ando. 2002. Hall conductivity of a two-dimensional graphite system. *Phys. Rev.* B65: 245420-1-11.
8. Novoselov, K. S., A. K. Geim, S. V. Morozov, D. Jiang, Y. Zhang, S. V. Dubonos, I. V. Grigorieva, and A. A. Firsov. 2004. Electric field effect in atomically thin carbon films. *Science* 306: 666–669.
9. Novoselov, K. S., A. K. Geim, S. V. Morozov, D. Jiang, M. I. Katsnelson, I. V. Grigorieva, S. V. Dubonos, and A. A. Firsov. 2005. Two-dimensional gas of massless Dirac fermions in graphene. *Nature* 438: 197.
10. Zhang, Y., Y.-W. Tan, H. L. Stormer, and P. Kim. 2005. Experimental observation of the quantum Hall effect and Berry's phase in graphene. *Nature* 438: 201.
11. Kane, C. L. and E. J. Mele. 2005. Quantum spin hall effect in graphene. *Phys. Rev. Lett.* 95: 226801-1-4.
12. Ohta, T., A. Bostwich, T. Seyller, K. Horn, and E. Rotenberg. 2006. Controlling the electronic structure of bilayer graphene. *Science* 313: 951–954.
13. Zhang, Y., Z. Jiang, J. P. Small, M. S. Purewal, Y.-W. Tan, M. Fazlollahi, J. D. Chudow, J. A. Jaszczak, H. L. Stomer, and P. Kim. 2006. Landau-level splitting in graphene in high magnetic fields. *Phys. Rev. Lett.* 96: 136806-1-4.
14. Novoselov, K. S., Z. Jiang, Y. Zhang, S. V. Morozov, H. L. Stormer, U. Zeitler, J. C. Maan, G. S. Boebinger, P. Kim, and A. K. Geim. 2007. Room-temperature quantum hall effect in graphene. *Science* 315: 1379.
15. Meyer, J. C., A. K. Geim, M. I. Katsnelson, K. S. Novoselov, T. J. Booth, and S. Roth. 2007. The structure of suspended graphene sheets. *Nature* 446: 60.
16. Silvestrov, P. G. and K. B. Efetov. 2007. Quantum dots in graphene. *Phys. Rev. Lett.* 98: 016802-1-4.
17. Stein, S. E. and R. L. Brown. 1987. π Electron properties of large condensed polyaromatic hydrocarbons. *J. Am. Chem. Soc.* 109: 3721–3729.
18. Yoshizawa, K., K. Okahara, T. Sato, K. Tanaka, and T. Yamabe. 1994. Molecular orbital study of pyrolytic carbons based on small cluster models. *Carbon* 32: 1517–1522.
19. Fujita, M., K. Wakabayashi, K. Nakada, and K. Kusakabe. 1996. Peculiar localized state at zigzag graphite edge. *J. Phys. Soc. Jpn.* 65: 1920–1923.
20. Enoki, T. and Y. Kobayashi. 2005. Magnetic nanographite: An approach to molecular magnetism. *J. Mater. Chem.* 15(37): 3999–4002.
21. Enoki, T. and K. Takai. 2006. Unconventional magnetic properties of nanographite. In F. Palacio and T. Makorova (eds.), *Carbon-Based Magnetism: An Overview of the Magnetism of Metal Free Carbon-Based Compounds and Materials*, pp. 397–416, Elsevier Science, London.
22. Enoki, T., Y. Kobayashi, and K. Fukui. 2007. Electronic structures of graphene edges and nanographene. *Inter. Rev. Phys. Chem.* 26(4): 609–645.
23. Wallece, P. R. 1947. The band theory of graphite. *Phys. Rev.* 71: 622–634.
24. Toy, W. W., M. S. Dresselhaus, and G. Dresselhaus. 1977. Minority carriers in graphite and the *H*-point magnetoreflection spectra. *Phys. Rev. B* 15: 4077–4090.
25. Holzwarth, N. A. W. 1992. Chapter 2: Electronic band structure of graphite intercalations compounds. In Zabel, H. and S. A. Solin (eds.), *Graphite Intercalation Compounds II*, pp. 7–52, Springer-Verlag, Berlin.
26. Slonczewski, J. C. and P. R. Weiss. 1955. A combined tight binding and perturbation theory of graphite π-states. *Phys. Rev.* 99: 636(A).
27. Slonczewski, J. C. and P. R. Weiss. 1958. Band structure of graphite. *Phys. Rev.* 109: 272–279.
28. McClure, J. W. 1957. Band structure of graphite and de Haas-van Alphen effect. *Phys. Rev.* 108: 612–618.

29. Holzwarth, N. A. W., S. G. Louie, and S. Rabii. 1982. X-ray form factors and the electronic structure of graphite. *Phys. Rev. B* 26: 5382–5390.
30. Schafhaeutl, C. 1840. Ueber die Verbindungen des Kohlenstoffes mit Silicium, Eisen und anderen Metallen, welche die verschiedenen Gallungen von Roheisen, Stahl und Schmiedeeisen bilden. *J. Prakt. Chem.* 21: 129–157.
31. Blinowski, J., N. Hy Hau, C. Rigaux, J. P. Vieren, R. Toullec, G. Furdin, A. Hérold, and J. Melin. 1980. Band structure model and dynamical dielectric function in lowest stages of graphite acceptor compounds. *J. Phys. (Paris)* 41: 47–58.
32. Blinowski, J. and C. Rigaux. 1980. Band structure model and electrostatic effects in third and forth stages of graphite acceptor compounds. *J. Phys. (Paris)* 41: 667–676.
33. Holzwarth, N. A. W. 1980. Graphite intercalation compounds: A simple model of Fermi surface and transport properties. *Phys. Rev. B* 21: 3665–3674.
34. Saito, R. and H. Kaminura. 1985. Orbital susceptibility of higher stage GICs. *Synth. Metals* 12: 295–300.
35. Johnson, L. G. and G. Dresselhaus. 1973. Optical properties of graphite. *Phys. Rev. B* 7: 22752285.
36. Painter, G. S. and J. E. Ellis. 1970. Electronic band structure and optical properties of graphite from a variational approach. *Phys. Rev. B* 1: 4747–4752.
37. Ohno, T. and H. Kamimura. 1983. Band structures and charge distributions along the *c*-axis of higher stage graphite intercalation compounds. *J. Phys. Soc. Jpn.* 52: 223–232.
38. Yang, M. H. and P. C. Eklund. 1988. Optical dielectric function of high-stage potassium graphite intercalation compounds: Experiment and theory. *Phys. Rev. B* 38: 3505–3516.
39. St. Jean, M., M. Menant, N. Hy Hau, C. Rigaux, and A. Metrot. 1983. In situ optical study of H_2SO_4-graphite intercalation compounds. *Synth. Metals* 8: 189–193.
40. Zhang, J. M., D. M. Hoffman, and P. C. Eklund. 1986. Optical study of the *K*-point π-band dispersion in graphite-H_2SO_4. *Phys. Rev. B* 34: 4316–4322.
41. Smith, D. S. and P. C. Eklund. 1983. Optical reflectance studies of stage 1–6 graphite-$FeCl_3$ intercalation compounds. In M. S. Dresselhaus, G. Dresselhaus, J. E. Fischer, and M. J. Moran (eds.), *Materials Research Society Symposium Proceedings, Intercalated Graphite*, Vol. 20, pp. 99–104, Materials Research Society, Pittsburgh, PA.
42. Hoffman, D. M., R. E. Heinz, G. L. Doll, and P. C. Eklund. 1985. Optical reflectance study of the electronic structure of acceptor-type graphite intercalation compounds. *Phys. Rev. B* 32: 1278–1288.
43. Ishii, T., Y. Komatsu, K. Suzuki, T. Enoki, A. Ugawa, K. Yakushi, and S. Bandow. 1994. Electronic structure and transport properties of $AuCl_3$-GIC. *Mol. Cryst. Liq. Crsyt.* 245: 1–6.
44. St. Jean, M., H. N. Hau, C. Rigaux, and G. Furdin. 1983. Optical determination of the charge transfer in AsF_5-graphite intercalation compounds. *Solid State Commun.* 46: 55–58.
45. Ohana, I., D. Vaknin, H. Selig, Y. Yacoby, and D. Davidov. 1987. Charge transfer in stage-1 OsF_6- and MoF_6-intercalated graphite compounds. *Phys. Rev. B* 35: 4522–4525.
46. Brusilovsky, D., H. Selig, D. Vaknin, I. Ohana, and D. Davidov. 1988. Graphite compound with BF_3/F_2: Conductivity and its relation to charge transfer. *Synth. Metals* 23: 377–384.
47. Ohana, I., I. Palchan, Y. Yacoby, and D. Davidov. 1988. Electronic charge transfer in stage-2 fluorine-intercalated graphite compounds. *Phys. Rev. B* 38: 12627–12632.
48. Pietronero, L. and S. Strässler. 1981. Bond length, bond strength and electrical conductivity in carbon based systems. In L. Pietronero, E. Tossati (eds.), *Physics of Intercalation Compounds* pp. 23–32. Springer-Verlag, Berlin.
49. Chan, C. T., K. M. Ho, and W. Kamitakahara. 1987. Zone-center phonon frequencies for graphite and graphite intercalation compounds: Charge-transfer and intercalate-coupling effects. *Phys. Rev. B* 36: 3499–3502.
50. Underhill, C., S. Y. Leung, G. Dresselhaus, and M. S. Dresselhaus. 1979. Infrared and Raman spectroscopy of graphite-ferric chloride. *Solid State Commun.* 29: 769–774.
51. Metrot, A. and J. Fischer. 1981. Charge transfer reactions during anodic oxidation of graphite in H_2SO_4. *Synth. Metals* 3: 201–207.
52. Salaneck, W. R., C. F. Brucker, J. E. Fischer, and A. Metrot. 1981. X-ray photoelectron spectroscopy of graphite intercalated with H_2SO_4. *Phys. Rev. B* 24: 5037–5046.
53. Holzwarth, N. A. W., S. Rabii, and A. Girifalco. 1978. Theoretical study of lithium graphite. I. Band structure, density of states, and Fermi-surface properties. *Phys. Rev. B* 18: 5190–5205.
54. Holzwarth, N. A. W., S. G. Louie, and S. Rabii. 1983. Lithium-intercalated graphite: Self-consistent electronic structure for stages one, two, and three. *Phys. Rev. B* 28: 1013–1025.

55. Kasowski, R. V. 1982. Self-consistent extended-muffin-tin orbital energy-band method: Application to LiC$_6$. *Phys. Rev. B* 25: 4189–4195.
56. Ohno, T. 1980. Electronic band structure of lithium-graphite intercalation compound C$_6$Li. *J. Phys. Soc. Jpn.* 49(Suppl. A): 899–902.
57. Posternak, M., A. Balderschi, A. J. Freeman, and E. Wimmer. 1983. Prediction of electronic interlayer states in graphite and reinterpretation of alkali bands in graphite intercalation compounds. *Phys. Rev. Lett.* 50: 761–764.
58. Samuelson, L. and I. P. Batra. 1980. Electronic properties of various stages of lithium intercalated graphite. *J. Phys. C* 13: 5105–5124.
59. Carver, G. P. 1970. Nuclear magnetic resonance in the cesium–graphite intercalation compounds. *Phys. Rev. B* 2: 2284–2295.
60. Alstrom, P. 1986. Electronic properties of first-stage heavy alkali metal graphite intercalation compounds. *Synth. Metals* 15: 311–322.
61. Inoshita, T., K. Nakao, and H. Kaminura. 1977. Electronic structure of potassium–graphite intercalation compound: C$_8$K. *J. Phys. Soc. Jpn.* 43: 1237–1243.
62. DiVincenzo, D. P. and S. Rabii. 1982. Theoretical investigation of the electronic properties of potassium graphite. *Phys. Rev. B* 25: 4110–4125.
63. Saito, M. and A. Oshiyama. 1986. Self-consistent band structures of first-stage alkali-metal graphite intercalation compounds. *J. Phys. Soc. Jpn.* 55: 4341–4348.
64. Tatar, R. C. 1985. A theoretical study of the electronic structure of binary and ternary first stage alkali intercalation compounds of graphite. PhD Thesis, University of Pennsylvania, PA.
65. Mizuno, S., H. Hiramoto, and K. Nakao. 1987. A new self-consistent band structure of C$_8$K. *J. Phys. Soc. Jpn.* 56: 4466–4476.
66. Takahashi, T., H. Tokailin, T. Sagawa, and H. Suematsu. 1985. Angle-resolved ultraviolet photoemission study of potassium–graphite intercalation compound; C$_8$K. *Synth. Metals* 12: 239–244.
67. Gunasekara, N., T. Takahashi, F. Maeda, T. Sagawa, and H. Suematsu. 1987. Electronic band structure of C$_8$Cs studied by highly-angle-resolved ultraviolet photoelectron spectroscopy. *J. Phys. Soc. Jpn.* 56: 2581–2581.
68. Gunasekara, N., T. Takahashi, F. Maeda, T. Sagawa, and H. Suematsu. 1988. Angle-resolved ultraviolet photoemission study of first stage alkali-metal graphite intercalation compounds. *Z. Phys. B* 70: 349–355.
69. Preil, M. E., J. E. Fischer, S. B. DiCenzo, and G. K. Werthheim. 1984. Barium intra-atomic reconfiguration in BaC$_6$. *Phys. Rev. B* 30: 3536–3538.
70. Guérard, D., C. Takoudjou, and F. Rousseaux. 1983. Insertion d'hydrure de potassium dans le graphite. *Synth. Metals* 7: 43–48.
71. Koma, A., K. Miki, H. Suematsu, T. Ohno, and H. Kaminura. 1986. Density-of-states investigation of C$_8$K and occurrence of the interlayer band. *Phys. Rev. B* 34: 2434–2438.
72. Wang, G., P. K. Ummat, and W. R. Datars. 1988. De Haas-van Alphen effect of potassium intercalated graphite. *Extended Abstracts, Graphite Intercalation Compounds*, 217–219, Materials Research Society, Fall Meeting, Pittsburg, PA.
73. Guérard, D. and V. A. Nalimova. 1994. Crystalline structure of Li and Cs-graphite superdense phases. *Mol. Cryst. Liq. Cryst.* 244: 263–268.
74. Nalimova, V. A., D. Guérard, M. Lelaurain, and O. V. Festeev. 1995. X-ray investigation of highly saturated Li-graphite intercalation compound. *Carbon* 33: 177–181.
75. Avdeev, V. V., V. A. Nalimova, and K. N. Semenenko. 1990. Sodium–graphite system at high pressures. *Synth. Metals* 38: 363–369.
76. Avdeev, V. V., V. A. Nalimova, and K. N. Semenenko. 1990. The alkali metals in graphite matrices—New aspects of metallic state chemistry. *High Press. Res.* 6: 11–25.
77. Nalimova, V. A., S. N. Chepurko, V. V. Avdeev, and K. N. Semenenko. 1991. Intercalation in the graphite-rubidium system under high pressure. *Synth. Metals* 40: 267–273.
78. Krueger, C. and Y.-H. Tsay. 1973. Molekülstructur eines π-Distickstoff-Nickel-Komplexes. *Angew. Chem.* 85: 1051–1052.
79. Nalimova, V. A. 1998. High pressure for synthesis and study of superdense alkali metal-carbon compounds. *Mol. Cryst. Liq. Cryst.* 310: 5–18.
80. Bindra, C. Nalimova, V. A., and J. E. Fischer. 1998. In-plane structure and thermal (in)stability of LiC$_{2.18}$ based on boron-doped graphite. *Mol. Cryst. Liq. Cryst.* 310: 19–25.
81. Morkovich, V. Z. 1996. Synthesis and XPS investigation of superdense lithium–graphite intercalation compound, LiC$_2$. *Synth. Metals* 80: 243–247.

82. Schirmer, A., P. Heitjans, and V. A. Nalimova. 1998. Conduction-electron induced spin-lattice relaxation of ^{8}Li in the high-pressure phase LiC_2. *Mol. Cryst. Liq. Cryst.* 310: 291–296.

83. Conard, J., V. A. Nalimova, and D. Guérard. 1994. NMR study of LiC_X graphite intercalation compounds prepared under high pressure. *Mol. Cryst. Liq. Cryst.* 245: 25–30.

84. Woo, K. C., P. J. Flanders, and J. E. Fischer. 1982. Synthesis and properties of stage 1 barium–graphite. *Bull. Am. Phys. Soc.* 27: 272.

85. Holzwarth, N. A. W., D. P. DiVincenzo, R. C. Tatar, and S. Rabii. 1983. Energy-band structure and charge distribution for BaC_6. *Int. J. Quantum Chem.* 23: 1223–1230.

86. Mazin, I. I. 2005. Intercalant-driven superconductivity in YbC_6 and CaC_6. *Phys. Rev. Lett.* 95: 227001-1-4.

87. Calandra, M. and F. Mauri. 2006. Theoretical explanation of superconductivity in CaC_6. *Phys. Rev. Lett.* 96: 237002-1-4.

88. Weller, T. E., M. Ellerby, S. S. Saxena, R. P. Smith, and N. T. Skipper. 2005. Superconductivity in the intercalated graphite compounds C_6Yb and C_6Ca. *Nature Phys.* 1: 39–41.

89. Prudnikova, G. V., A. Gjatkin, A. V. Ermakov, A. M. Shikin, and V. K. Adamchuk. 1994. Surface intercalation of graphite by lanthanum. *J. Electron Spectros. Relat. Phenomena* 68: 427–430.

90. Shinkin, A. M., S. L. Molodtsov, C. Laubschat, G. Kaindl, G. V. Prudnikova, and V. K. Adamchuk. 1995. Electronic structure of La-intercalated graphite. *Phys. Rev. B* 51: 13586–13591.

91. Shinkin, A. M., G. V. Prudnikova, V. K. Adamchuk, S. L. Molodtsov, C. Laubschat, and G. Kaindl. 1995. Photoemission study of La/fullerite and La/graphite interfaces. *Surf. Sci.* 331–333: 517–521.

92. Molodtsov, S. L., Th. Gantz, C. Laubschat, A. G. Viatkine, J. Avila, C. Cassdo, and M. C. Asensio. 1996. Electron-energy bands in single-crystalline La-intercalated graphite. *Z. Phys. B* 100: 381–385.

93. Mazin, I. I. and S. L. Molodtsov. 2005. Electrons and phonons in YbC_6: Density functional calculations and angle-resolved photoemission measurements. *Phys. Rev. B* 72: 172504-1-4.

94. Molodtsov, S. L., C. Laubschat, M. Richter, Th. Gantz, and A. M. Shikin. 1996. Electronic structure of Eu and Yb graphite intercalation compounds. *Phys. Rev. B* 53: 16621–16630.

95. Sugihara, K., S. T. Chen, and G. Dresselhaus. 1985. Theory of electrical resistivity, magnetoresistance and magnon drag effect in graphite intercalation compound C_6Eu. *Synth. Metals* 12: 383–388.

96. Vaknin, D., I. Palchan, D. Davidov, H. Selig, and D. Moses. 1986. Resistivity and E.S.R. studies of graphite HOPG/fluorine intercalation compounds. *Synth. Metals* 16: 349–365.

97. Ohana, I. 1989. Metal–nonmetal transition induced by reorientation of the fluorine molecules in stage-2 graphite–fluorine compounds. *Phys. Rev. B* 39: 1914–1918.

98. Nakada, K., M. Fujita, G. Dresselhaus, and M. S. Dresselhaus. 1997. Edge state in graphene ribbons: Nanometer size effect and edge shape dependence. *Phys. Rev. B* 54: 17954–17961.

99. Wakabayashi, K., M. Fujita, H. Ajiki, and M. Sigrist. 1999. Electronic and magnetic properties of nanographite ribbons. *Phys. Rev. B* 59: 8271–8282.

100. Goto, K., T. Kubo, K. Yamamoto, K. Nakasuji, K. Sato, D. Shiomi, T. Takui, T. Kobayashi, K. Yakushi, and J. Qutang. 1999. A stable neutral hydrocarbon radical: synthesis, crystal structure, and physical properties of 2,5,8-tri-*tert*-butyl-phenalenyl. *J. Am. Chem. Soc.* 121: 1619–1620.

101. Fukui, K., K. Sato, D. Shiomi, T. Takui, K. Itoh, K. Gotoh, T. Kubo, K. Yamamoto, K. Nakasuji, and A. Naito. 1999. Electronic structure of a stable phenalenyl radical in crystalline state as studied by SQUID measurements, cw-ESR, and ^{13}C CP/MAS NMR spectroscopy. *Synth. Metals* 103: 2257–2258.

102. Inoue, J., K. Fukui, D. Shiomi, Y. Morita, K. Yamamoto, K. Nakasuji, T. Takui, and K. Yamaguchi. 2001. The first detection of a Clar's hydrocarbon, 2,6,10-tri-*tert*-butyltriangulene: A ground-state triplet of non-Kekulé polynuclear benzenoid hydrocarbon. *J. Am. Chem. Soc.* 123: 12702–12703.

103. Lieb, E. 1989. Two theorems on the Hubbard model. *Phys. Rev. Lett.* 62: 1201–1204.

104. Klein, D. J. 1994. Graphitic polymer strips with edge states. *Chem. Phys. Lett.* 217: 261–265.

105. Kusakabe, K. and M. Maruyama, 2003. Magnetic nanographite. *Phys. Rev. B* 67: 092406-1-4.

106. Maruyama, M. and K. Kusakabe, 2004. Theoretical prediction of synthesis methods to create magnetic nanographite. *J. Phys. Soc. Jpn.* 73: 656–663.

107. Kobayashi, Y., K. Kusakabe, K. Fukui, T. Enoki, and Y. Kaburagi. 2005. Observation of zigzag and armchair edges of graphite using scanning tunneling microscopy and spectroscopy. *Phys. Rev. B* 71: 193406-1-4.

108. Kobayashi, Y., K. Fukui, T. Enoki, and K. Kusakabe. 2006. Edge state on hydrogen-terminated graphite edges investigated by scanning tunneling microscopy. *Phys. Rev. B* 73: 125415-1-8.

109. Thomas, J. M. 1965. Microscopic studies of graphite oxidation. In P. L. Walker (ed.), *Chemistry and Physics of Carbon*, Vol. 1, pp. 122–204, Marcel Dekker, New York.

110. Prasad, B. L. V., H. Sato, T. Enoki, Y. Hishiyama, Y. Kaburagi, A. M. Rao, G. U. Sumanasekera, and P. C. Eklund. 2001. Intercalated nanographite: Structures and electronic properties. *Phys. Rev. B* 64: 235407–235416.

111. Takai, K., S. Eto, M. Inaguma, T. Enoki, H. Ogata, M. Tokita, and J. Watanabe. 2007. Magnetic potassium clusters in the nanographite host system. *Phys. Rev. Lett.* 98: 017203-1-4.

112. Jiang, D.-E., B. G. Sumpter, and S. Dai. 2007. Unique chemical reactivity of a graphene nanoribbon's zigzag edge. *J. Chem. Phys.* 126: 134701-1-6.

113. Dahn, J. R., T. Zheng, Y. Liu, and J. S. Xue. 1995. Mechanisms for lithium insertion in carbonaceous materials. *Science* 270: 590–593.

114. Sato, K., M. Noguchi, A. Demachi, N. Oki, and M. Endo. 1994. A mechanism of lithium storage in disordered carbons. *Science* 263: 556–558.

115. Endo, M., K. Takeuchi, S. Igarashi, K. Kobori, M. Shiraishi, and H. W. Kroto. 1994. The production and structure of pyrolytic carbon nanotubes (PCNTs). *J. Phys. Chem. Solids* 54: 1841–1848.

116. Yata, S., Y. Hato, H. Kinoshita, N. Ando, A. Anekawa, T. Hashimoto, M. Yamaguchi, K. Tanaka, and T. Yamabe. 1995. Characteristics of deeply Li-doped polyacenic semiconductor material and fabrication of a Li secondary battery. *Synth. Metals* 73: 273–277.

117. Zhou, P., P. Papanek, C. Bindra, R. Lee, and J. E. Fischer. 1997. High capacity carbon anode materials: Structure, hydrogen effect, and stability. *J. Power Sources* 68: 296–300.

118. Nakadaira, M., R. Saito, T. Kimura, G. Dresselhaus, and M. S. Dresselhaus. 1997. Excess Li ions in a small graphite cluster. *J. Mater. Res.* 12: 1367–1375.

119. Yagi, M., R. Saito, T. Kimura, G. Dresselhaus, and M. S. Dresselhaus. 1999. Electronic states in heavily Li-doped graphite nanoclusters. *J. Mater. Res.* 14: 3799–3804.

120. Shibayama, Y., H. Sato, T. Enoki, M. Seto, Y. Kobayashi, Y. Maeda, and M. Endo. 2000. ^{129}I Mössbauer effect of iodine adsorbed in activated carbon fibers. *Mol. Cryst. Liq. Cryst.* 340: 301–306.

121. Shibayama, Y., H. Sato, T. Enoki, X. X. Bi, M. S. Dresselhaus, and M. Endo. 2000. Novel electronic properties of a nano-graphite disordered network and their iodine doping effects. *J. Phys. Soc. Jpn.* 69: 754–767.

122. Akamatsu, H., H. Inokuchi, and Y. Matsunaga. 1954. Electrical conductivity of the perylene–bromine complex. *Nature* 173: 168.

123. Takai, K., H. Kumagai, H. Sato, and T. Enoki. 2006. Bromine-adsorption-induced change in the electronic and magnetic properties of nanographite network systems. *Phys. Rev. B* 73: 035435-1-13.

124. Saito, R., M. Yagi, T. Kimura, G. Dresselhaus, and M. S. Dresselhaus. 1999. Electronic structure of fluorine doped graphite nanoclusters. *J. Phys. Chem. Solids* 60: 715–721.

125. Takai, K., H. Sato, T. Enoki, N. Yoshida, F. Okino, H. Touhara, and M. Endo. 2001. Effect of fluorination on nano-sized π-electron systems. *J. Phys. Soc. Jpn.* 70: 175–185.

126. Fung, A. W. P., Z. H. Wang, M. S. Dresselhaus, G. Dresselhaus, R. W. Pekala, and M. Endo. 1994. Coulomb-gap magnetotransport in granular and porous carbon structures. *Phys. Rev.* B49: 17325–17335.

127. Shibayama, Y., H. Sato, T. Enoki, and M. Endo. 2000. Disordered magnetism at the metal-insulator threshold in nano-graphite-based carbon material. *Phys. Rev. Lett.* 84: 1744–1747.

7 Carbon Materials in Lithium-Ion Batteries

Petr Novák, Dietrich Goers, and Michael E. Spahr

CONTENTS

7.1 INTRODUCTION AND THE MARKET DIMENSION FOR CARBONS IN LITHIUM-ION BATTERIES

The importance of carbons in secondary lithium batteries considerably increased when Sony introduced the first lithium-ion battery to the market in 1991.[1] Up to that point, metallic lithium was used as the negative electrode in various types of lithium battery systems.[2] The most attractive feature of metallic lithium is the high specific charge capacity of 3.86 Ah g^{-1} combined with a very negative electrode potential of about −3.0 V vs. NHE. However, the coulombic efficiency of the repeated anodic dissolution and cathodic deposition of lithium is normally well below 100% so that at least a threefold lithium excess is required for an acceptable cycle life of rechargeable lithium cells.[2,3] Other drawbacks of the metallic lithium electrode are the grave changes of both volume and morphology during cycling, resulting in a limited cycle life due to a crumbling of the electrode into electrically isolated, chemically unstable metal particles, as well as the dendrite formation on the electrode during the cathodic lithium deposition.[3,4] These lithium dendrites could locally cause internal short circuits between both electrodes of the cell via metallic bridges through the separator, thus significantly increasing the self-discharge rate of the cell. Even worse, the potentially high short-circuit currents through the separator could heat the lithium metal electrode which could melt and exothermally react with the organic electrolyte of the battery. This thermal runaway is a likely cause of cell venting, fire, and even explosions of secondary lithium batteries containing lithium metal electrodes.

The technical breakthrough of the new lithium-ion battery technology was the substitution of the lithium metal negative electrode by a lithium intercalation electrode, in particular by a carbon negative electrode, which significantly improved the safety of the lithium battery system during cycling.[5-12] In the case of a carbon-based negative electrode, during the electrochemical reduction of the negative electrode, lithium ions from the electrolyte are electrochemically reduced and inserted into the carbon host structure. As a consequence, the typical dendritic lithium deposition occurring during cell charging at the carbon negative electrode can be suppressed.[13,14] Normally, the lithium insertion process into the carbon structure is highly reversible as the inserted lithium can be completely de-inserted from the optimized carbon host structure during the electrochemical reoxidation of the carbon negative electrode in the discharge process of the cell.[15,16]

Nowadays, in the secondary battery market, portable lithium-ion batteries almost completely have replaced nickel cadmium batteries and nickel metal hydride batteries in laptop and notebook portable computers, mobile phones, personal digital assistants, and video cameras which have become the typical applications for lithium-ion batteries.[8,17] The increasing global mobility has increased the worldwide distribution of these devices with the consequence of an increasing demand for mobile energy. With annual market growth rates of more than 10% in the past years, the number of globally produced lithium-ion cells has grown to more than 2.5 billion cells per year in 2007.[17] Combined with these growth rates for the worldwide cell production, the carbon annual consumption for lithium-ion batteries has reached a level of more than 10,000 tons of carbon. With the identification and worldwide diffusion of new portable energy demanding applications like video games and digital still cameras, as well as the introduction of small, intermediate, and large-scale lithium-ion power batteries for power tools, telecommunication, power backup, special military applications, and automotive applications like hybrid electric vehicles and electric vehicles, independent market research companies predict another potential doubling of the carbon consumption in the next 5 years until 2012.[17-20]

Carbon is used in lithium-ion cells for different functions: conductive carbon black and/or graphite additives are applied in both the negative and the positive electrode to improve the electronic conductivity of the electrodes. These conductive additives constitute a fraction of up to about 10% of the total carbon consumption. The major fraction is represented by the active carbon materials which are electrochemically reduced and oxidized in the negative electrode during the battery charge and discharge process, respectively.

Since 1991, the cell charge capacity of lithium-ion batteries has been remarkably improved.[21–23] As an example, the first cylindrical 18650 size battery of Sony introduced to the battery market showed a nominal cell capacity of 700 mAh. Today's high-end industrial 18650 size cylindrical batteries show cell capacities of about 3 Ah, and further enhancements still seem possible. These remarkable capacity improvements achieved for all cylindrical and prismatic lithium-ion batteries have been driven by the strong requirement of the original equipment manufacturer (OEM) for higher energy density batteries for newly developed electronic devices with increasing demand of energy and, at the same time, decreasing device size. This increase of energy density in the lithium-ion battery systems was considerably dragged, besides an optimized cell design, by improving the charge density of the carbon negative electrodes. In contrast, the charge density of the positive electrode consisting normally of a lithium cobalt oxide-based active material could not considerably be improved so far.

The development of carbon negative electrode materials with higher crystallinity and compressed densities opened the way for negative electrodes with high density and high specific charge. The application of more crystalline materials has become feasible by identifying suitable, especially tailored electrolyte systems for these negative electrodes and by adapting other components like the binder system as well as of the electrode manufacturing process to the modified carbon active material.[24–28]

In parallel to the improvement of the cell performance, the drastic price decay of lithium-ion cells from above \$3/Wh to below \$0.5/Wh has put strong pressure on battery material and manufacturing costs. The development of new higher performing but lower cost carbon active materials produced with more efficient manufacturing processes and the identification of suitable technology that allowed to apply the improved carbon materials in the battery have been the prerequisites to achieve cell improvements in combination with cost reductions. However, battery producers so far have prioritized the improvement of the cell performance and cell safety to cost aspects when substituting a negative active material in the lithium-ion battery.

In the following sections, the material science and manufacturing technology of the most important industrially used carbon negative electrode materials, their manufacturing processes, as well as the application technology of these materials in the lithium-ion battery will be discussed in more detail. A separate section focuses on carbon materials used as conductive additives in the positive electrode. Even though only contributing to about 10% of the relevant total carbon consumption, carbon conductive additives are essential for the performance of the positive electrodes providing sufficient electronic conductivity and, thus, influence the cell performance at high current rates. Conductive carbon materials are also applied as conductivity enhancer in the negative electrode.[29] A particular case is conductive graphite materials with high degree of crystallinity that are blended as minor component to the actual active material in the negative electrode mass. These conductive graphite components act both as conductivity enhancer and as lithium insertion host. The section about carbons in the negative electrode will discuss the bifunctionality of these conductive graphite components together with the typical active negative electrode materials.

In view of the bright commercial future that is predicted to carbons in the lithium-ion batteries, it must be considered that the race for higher energy density of lithium-ion batteries for portable applications like mobile phones, portable computers, and video cameras accelerates the search for electrode materials with higher charge densities. New higher charge capacity materials could substitute carbon as an active material in the negative electrode in future. Battery companies already have claimed lithium-ion battery prototypes containing noncarbon or carbon composite electrodes. The first lithium-ion battery containing a carbon-amorphous cobalt doped tin metal composite negative electrode was industrialized by Sony under the name "Nexelion," specifically for video cameras in 2006.[30] Giving higher energy densities to lithium-ion cells compared to carbon-based ones, there is, however, still a long way to go for those high-capacity negative electrodes until they fulfill all requirements of a negative electrode material. The substitution process also will be slowed down by the fact that the electronic devices have to be adapted to the modified voltage profile which is accompanied by the replacement of the electrode materials in the lithium-ion battery system. In this context, a short outlook at the end of this chapter throws light on some new research directions in the field of negative electrode materials as well as in the field of new conductive carbon additives which could be found in future lithium battery system.

7.2 ELECTROCHEMICAL PRINCIPLES OF THE LITHIUM-ION CELL

The industrial lithium-ion cell consists of a positive electrode mass containing a lithium transition metal oxide (e.g., $LiCoO_2$, $LiNi_{1-x}Co_xO_2$, $LiMn_2O_4$, $LiFePO_4$, $LiNi_{1-x-y}Mn_xCo_yO_2$) which is coated as a layer of up to about 100 μm on both sides of an aluminum current collector foil. Carbon conductive additives are added to improve the electronic conductivity of the electrode. The adhesion of the electrode particles on the current collector and the cohesion of the electrode film are achieved by a polymer binder, e.g., polyvinylidene difluoride (PVDF). The negative carbon electrode mass which typically forms a layer of 30–100 μm on both sides of a copper current collector foil contains normally PVDF or styrene butadiene rubber (SBR)/carboxy methyl cellulose (CMC) binder and, if required, a conductive carbon additive. Both electrodes are separated by a porous separator (usually a microporous polyethylene/polypropylene film of 10–20 μm thickness) which is soaked with an electrolyte solution. The latter consists normally of the $LiPF_6$ salt (ca. 1 M) dissolved in a mixture of liquid organic carbonates. The sandwich assembly of the electrodes and separators is either rolled for a cylindrical cell design or wrapped or stacked in the case of a prismatic cell design. The cells are in a metallic can or are laminated using an aluminum pouch packaging.

The electrolyte must be a pure ionic conductor, preferably with a high transport number for lithium ions, as an electronic conductivity of the electrolyte would create short-circuit ("leakage") currents between the electrodes. Both electrodes must have a high electronic conductivity and a sufficient ionic conductivity for lithium. The metal current collectors foils (current collectors) are pure electron conductors that allow only electrons to migrate to the external electric leads to the consumer or charger unit.

During the charging process of a lithium-ion cell (Figure 7.1), lithium ions originating from the lithium metal oxide $LiMeO_2$ (Me = Co, Ni, Mn) migrate through the electrolyte and are reversibly inserted into the carbon structure of the negative electrode. During cell discharge, the direction of this lithium migration is reversed. The electrochemical reactions can be described as follows:

$$\text{Positive electrode}: \quad LiMeO_2 \underset{\text{discharge}}{\overset{\text{charge}}{\rightleftarrows}} Li_{1-x}MeO_2 + xLi^+ + xe^-$$

$$\text{Negative electrode}: \quad C_6 + xLi^+ + xe^- \underset{\text{discharge}}{\overset{\text{charge}}{\rightleftarrows}} Li_xC_6$$

$$\text{Total cell reaction}: \quad LiMeO_2 + C_6 \underset{\text{discharge}}{\overset{\text{charge}}{\rightleftarrows}} Li_{1-x}MeO_2 + Li_xC_6$$

The average cell voltage on discharge is around 3.5 V. Thus, in comparison to cells utilizing aqueous electrolytes, very high average voltage is the reason for the high energy densities and power densities achievable with lithium-ion batteries. Normally, the energy density (energy per cell volume, W_E/V) and power density (power per cell volume, P/V) are calculated using the following equations:

$$\text{Energy density}: \quad \frac{W_E}{V} = \frac{\overline{U}\int I\,dt}{V} \quad (7.1)$$

$$\text{Power density}: \quad \frac{P(t)}{V} = \frac{\overline{U}I(t)}{V} \quad (7.2)$$

where
$\overline{U}$ is the average cell voltage
I is the electric current

Alternatively, the energy and power are given with respect to the cell or battery mass and are called specific energy and specific power, respectively.

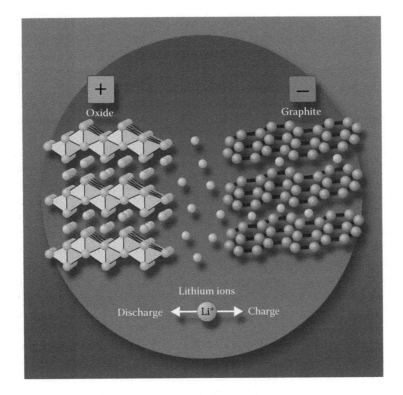

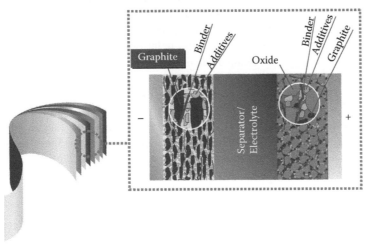

FIGURE 7.1 Schematics of the lithium-ion cell (top) and the electrode assembly (bottom).

7.3 CARBONS AS CONDUCTIVE ADDITIVES IN POSITIVE ELECTRODES

7.3.1 ELECTROCHEMICAL LIMITATIONS OF POROUS ELECTRODES AND THE CONSEQUENCES FOR THE CONDUCTIVE ADDITIVE

To ensure a sufficient electronic and thermal conductivity of the positive electrode during the charge and discharge process, conductive additives are required in the positive, transition metal oxide-based electrode.[31–33] Compared to metal powders as potential conductivity enhancer, carbon materials combine high electronic and thermal conductivity with low weight, low costs, relatively high chemical inertness, and nontoxicity. Conductive carbons optimize the electrical resistivity of the positive electrode mass but are not involved in the electrochemical redox process which delivers the energy of the electrochemical cell. Consequently, to maximize the energy density of the lithium-ion cell, the amount of conductive carbon needs to be minimized. Nowadays the quantity of the conductive carbon typically used in positive electrodes of commercial batteries is significantly below 10% of the total electrode mass. This carbon concentration is kept usually above the critical value at which the electrode conductivity sharply decreases.[34,35] At and above this critical concentration, neighboring carbon particles are close enough to assure electrical contact to each other either by electron tunneling or direct contact to form, in the whole electrode volume, a conductive matrix in which the positive active material is embedded.

The percolation theory which mathematically describes the conductivity of a conductor–insulator blend at different volume ratios in bulk gives only an approximate description of the performance of mixtures of carbon with the positive electrode material, as the well-known limitations of porous electrodes apply.[36–41] Note that the performance of (industrial) porous electrodes at high currents depends on multiple parameters, including the electrode thickness, the electrode porosity and tortuosity, the pore size distribution, the quality of electrolyte percolation (which is related to the surface wetting properties of the used material/electrolyte combination), the ion transfer rate in the electrolyte, the particle orientation in the electrode, and the amount and nature of binder material(s).[42,43] The ion transfer across the porous electrode depends, apart from the electrode's properties, on the conductivity of the electrolyte in the pores and the diffusion and migration rate of lithium ions in the pores.[44,45] The diffusion is driven by concentration gradients generated by the insertion of lithium ions into or de-insertion from the electroactive material, creating also concentration gradients of the anions due to the fact that the transport number of lithium ions is much less than unity. The thicker the electrode the more important the electrode pore structure becomes to achieve a sufficiently high ionic flow. Optimized conductive additives help by enhancing the electrolyte penetration and retention in the electrode by influencing the electrode density, the size and shape of the pores (electrode tortuosity), and the electrolyte absorption properties.[46,47]

The purity of the used conductive carbon plays an important role with regard to cell performance and safety. Trace element impurities in the battery deteriorate the energy density, cycling stability, and calendar life of the cell by provoking undesired electrochemical side reactions that lead to irreversible charge losses in the cell. In addition, transition metal impurities, which once dissolved in the electrolyte due to the high oxidative potential of the positive electrode, are reductively deposited on the negative electrode in form of dendrites. These dendrites may generate local short circuits between the electrodes that cause self-discharge of the cell and create a risk of cell fire of even explosion. Note that, if present in the electrode mass, particles larger than the desired electrode thickness ("oversized particles") can puncture the battery separator with the same consequence. A minimum amount of trace element impurities and low contamination with metal particles and oversized particles are therefore prerequisites for good battery properties, long battery cycle and calendar life, and safety.

Besides the electrochemical parameters, the carbon additive influences the process parameters during the electrode manufacturing. A homogeneous dispersion of the conductive carbon in the electrode mass, which is necessary to take full advantage out of the conductive carbon for the electrode performance,[48] is controlled by a homogeneous dispersion of the powders in the solvent-based slurry used for the coating process of the electrode mass on the current collector. The carbon additive influences the slurry rheology as well as the mechanical stability and flexibility of the produced film electrodes. The polymer binder in the electrode mass works as an adhesive but flexible link between the electrode particles as well as between the particles and the current collector foil leading to sufficiently high film cohesion and adhesion on the current collector foil. As insulating and electrochemically inactive material, the amount of polymer binder must be minimized in the electrode. A correlation between the oil adsorption number of carbon materials with their polymer (i.e., binder) absorption and electrolyte absorption capability has been shown making, thus, the oil adsorption number, an important characterization parameter of carbons for battery applications.

In modern commercial lithium-ion batteries, a variety of graphite powder and fibers, as well as carbon black, can be found as conductive additive in the positive electrode. Due to the variety of different battery formulations and chemistries which are applied, so far no standardization of materials has occurred. Every individual active electrode material and electrode formulation imposes special requirements on the conductive additive for an optimum battery performance. In addition, varying battery manufacturing processes implement differences in the electrode formulations. In this context, it is noteworthy that electrodes of lithium-ion batteries with a gelled or polymer electrolyte require the use of carbon black to attach the electrolyte to the active electrode materials.[49–54] In the following, the characteristic material and battery-related properties of graphite, carbon black, and other specific carbon conductive additives are described.

7.3.2 GRAPHITE CONDUCTIVE ADDITIVES

The types of graphitic carbon powders which primarily are applied as conductive additive belong to the family of highly crystalline graphite materials. These graphite materials show real densities of 2.24–2.27 g cm^{-3} (values based on the xylene density according to DIN 12 797 and DIN 51 901-X) and average interlayer distances of $c/2 = 0.3354$–0.3360 nm.[55]

The crystal structure of graphite was first described by Hull.[56] Carbon atoms with sp^2 hybridization are covalently bound to condensed six-membered rings forming planar layers. The covalent carbon bonds have a bond length of 0.1421 nm. Besides the three localized σ-bonds to its nearest neighbors, every carbon atom participates with one valence electron to a delocalized π molecular orbital along the graphene layer being responsible for the metallic behavior parallel to graphene layers. The graphene layer stacks are bound by relatively weak van der Waals forces being the reason for a relatively large theoretical interlayer distance of 0.3354 nm. The specific electrical conductivity parallel to the layer is found to be $2.6 \times 10^4\,\Omega^{-1}$cm^{-1} at room temperature and decreases with increasing temperature.[57] Perpendicular to the layer, the electrical conductivity is smaller by a factor of 10^{-4} and increases with increasing temperature, which is typical for a semiconductor.

As shown in Figure 7.2, the graphene layers can be stacked in a hexagonal symmetry. In this case, every third layer has an identical position to the first layer resulting in a stacking sequence A, B, A, B, A, B, etc. This structure type represents the thermodynamically stable hexagonal crystal structure. Besides, a rhombohedral structure exists in which only every fourth graphite layer has an identical position to the first layer resulting in a sequence of A, B, C, A, B, C, etc.[58,59] This rhombohedral structure appears as statistical stacking defects and can be formed by mechanical deformation of hexagonal crystals by shear forces.[60,61] Rhombohedral defects can be highly dispersed in the hexagonal graphite crystal or segregated to isolated rhombohedral phase.[62] Heat treatment completely transforms the rhombohedral into the hexagonal structure.

To achieve a sufficient electronic conductivity in the positive electrode the required amount of a high-crystallinity graphite is generally lower the lower the apparent (volume) density of the graphite powder is. The high apparent density, besides the requirements given by the dimensions of the electrode film, is the reason why mainly fine graphite powders are used as conductive additive. An average particle size below 10 μm is typical for graphite conductive additives in the positive electrode. Besides the particle size distribution, also the graphite texture and morphology influence the conductivity of the electrode mass. Graphite powders consist of polycrystalline particles having the shape of platelets. These platelets are agglomerates of intergrown single crystals. The graphite texture describes the orientation of the single crystal in the particle (mosaicity). As an example, the transmission electron microscopy (TEM) pictures in Figure 7.3 show the mosaicity of TIMREX MX15 graphite in comparison to TIMREX KS15. The average particle size of both graphite powders is about 8 μm. The graphite MX15 shows large single crystals aligned along the xy-plane of the platelet giving rise to anisotropy of the graphite properties. In contrast, the graphite KS15 shows a more isotropic texture.

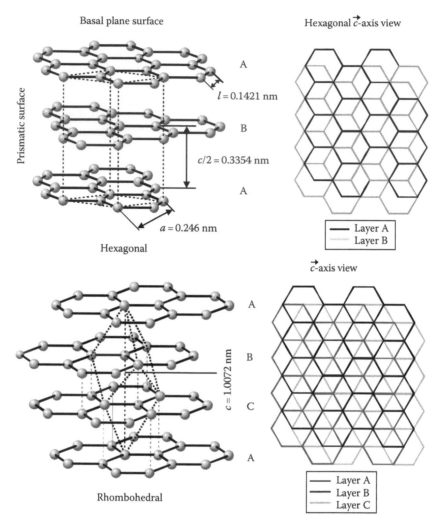

FIGURE 7.2 The hexagonal and rhombohedral graphite crystal structures.

As shown in Figure 7.4, the higher amount of anisometric particles cause a higher conductivity of $LiCoO_2$ mixtures at a given graphite concentration, at concentrations below 10 wt.%. Usually, the percolation threshold of graphite powders with anisometric particle shape is at a lower concentration than the percolation threshold of graphite materials with isometric particle shape.

Besides the electrochemical performance, the selection of graphite additives depends on electrode processing aspects. Flaky graphite particles have a higher dibutyl phthalate absorption (DBPA) and, thus, require more polymer binder than isometric graphite particles to achieve a sufficiently high mechanical electrode stability, as shown in Table 7.1.

Graphite conductive additives are manufactured from natural sources or synthesized in an industrial process. Primary synthetic graphite powders are derived from mixtures of selected high-purity carbon precursors like petroleum cokes and coals from natural sources. These amorphous carbons are graphitized above 2500°C under the exclusion of oxygen. During this heat treatment, the amorphous coke material is transformed into crystalline carbon. Franklin, and later Méring and Maire from the same group established the fundamental basis of the graphitization process.[55,63] During the graphitization process, a progressive decrease of d_{002} spacing occurs from ca. 344 pm down to the ultimate graphite value, 335.4 pm. In addition to the ordering of perfect graphene-like layers to a

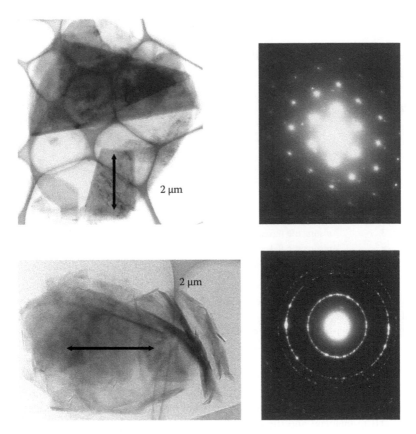

FIGURE 7.3 TEM pictures of graphites TIMREX MX15 (top) and TIMREX KS15 (bottom).

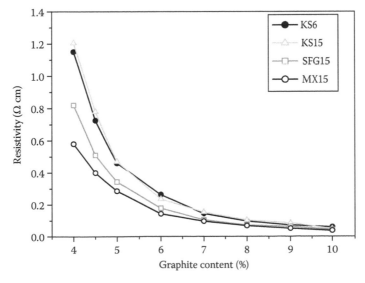

FIGURE 7.4 Resistivity of $LiCoO_2$–graphite mixtures as a function of the concentration of the graphites TIMREX MX15, TIMREX SFG15, TIMREX KS15, and TIMREX KS6.

TABLE 7.1

Typical Material Parameters of Fine TIMREX Graphite Powders Used as Conductive Additives

Material Parameter	Average Particle Size (μm)	Particle Shape	BET SSA (m² g⁻¹)	DBPA (g DBP/100 g Graphite) (ASTM 281)	OAN (ml DBP/100 g graphite) (ASTM D2414)	Scott Density (g cm⁻¹) (ASTM B329)
KS4	2	Isometric irregular spheroids	26	200	116	0.05
KS6	3	Isometric irregular spheroids	20	170	114	0.07
KS15	8	Isometric irregular spheroids	12	140	104	0.10
SFG6	3	Anisometric flakes	16	180	117	0.06
SFG15	8	Anisometric flakes	7	150	110	0.09
MX15	8	Strongly anisometric flakes	9	190	120	0.06

BET SSA, Brunauer Emmett Teller specific surface area; SSA, specific surface area; DBPA, dibutyl phthalate absorption; OAN, oil absorption number.

The Scott Density is a Standardized Method to Measure the Apparent Density of a Powder (ASTM B329).

3D structure, healing of defects within the graphene layers occurs.[63] Especially, the Acheson furnace technology is known to provide synthetic graphite materials with high crystallinity.[64–66]

Whereas in the case of the primary synthetic graphite process, the final product purity is controlled by the starting material and the graphitization, the natural graphite process primarily is the purification of the graphite ore taken from a mine.[67] Geological graphite deposits in nature are the result of the transformation of organic matter under high metamorphic pressure combined with high temperatures. The graphite contents of graphite ores available from exploited graphite mines can vary between 3% and more than 90% and mainly depend on the location and quality of the mine. The graphite is retrieved from the ore in various subsequent crushing, grinding, and flotation steps. Modern flotation processes achieve purities of more than 98%. To achieve purities above 99.9% being required for lithium-ion batteries, the graphite material is further purified either by thermal or chemical purification. Thermal purification processes are performed at high temperatures, mostly above 1500°C. Depending on the nature of the impurities, reactive gases like chlorine may be added. Chemical treatment is performed by either acid or alkaline leaching. The efforts to purify natural graphite increase with the purity level required. Nowadays, the costs for natural graphite powders with purities above 99.9% are in a similar range as for synthetic graphite with the same purity. Mechanical treatments like specific grinding, sieving, or classification processes are applied to achieve the final particle size distribution of the graphite material.

7.3.3 FIBROUS GRAPHITE MATERIALS

Vapor-grown carbon fibers like VGCF from Showa Denko K.K. or similar carbon filaments are materials with low bulk densities below 0.05 g cm⁻³ that show both high intrinsic electronic (typically ca. 50 Ω⁻¹ cm⁻¹ at 0.8 g cm⁻³ compaction density) and thermal conductivity. These properties explain the excellent performance of the vapor-grown carbon fibers as conductive additive in battery electrodes.[35,68–70] Besides a strong conductivity enhancement, vapor grown carbon fiber (VGCF) containing electrodes show an improved dissipation of the heat locally evolved at high current rates. Typical practical fiber concentrations range below 1 wt% of the total electrode mass. The texture of the VGCF carbon fibers consisting of concentrically wound graphite layers is shown in the tunneling electron microscope picture in Figure 7.5. This texture explains the high electronic and thermal conductivity along the carbon fiber axis.

50 nm

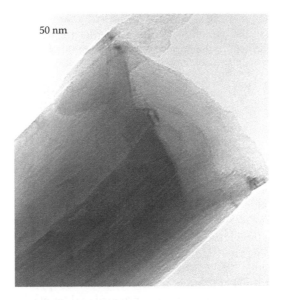

FIGURE 7.5 Tunneling microscope picture of a VGCF sample showing the concentrically ordered graphene sheets resulting in a fibrous morphology. (Courtesy of ICSI-CNRS, Mulhouse, France.)

In addition to the electronic and thermal properties, the fibrous morphology results in a high flexural modulus and a low coefficient of thermal expansion which explains the increased flexibility and mechanical stability of VGCF containing electrode films.[71] The vapor-grown fibers are produced by exposing a metallic catalyst to a mixture of hydrocarbon gases and hydrogen at temperatures above 1000°C. Decomposition of the hydrocarbon gas provides carbon from which filaments grow.[71–73] The complex manufacturing process is the reason for the high material costs which have been a considerable drawback for vapor-grown fibers as conductive additive in lithium batteries.

7.3.4 CONDUCTIVE CARBON BLACKS

Among the 9 million tons of carbon black which are produced globally per year, only a small fraction of very specific, high-purity conductive carbon blacks can be used as conductive additive in lithium-ion batteries. A traditional conductive carbon is acetylene black, a special form of a thermal black produced by the thermal decomposition of hydrocarbon feedstock.[74,75] The particularity of acetylene black to other thermal carbon black production is that the starting hydrocarbon, acetylene, exothermally decomposes above 800°C.[75–77] Once the reaction is started, the acetylene decomposition autogenously provides the energy required for the cracking of acetylene to carbon followed by the synthesis of the carbon black:

$$n\mathrm{C_2H_2} \rightarrow 2n\mathrm{C} + n\mathrm{H_2} + \text{Energy}$$

The reaction generates temperatures exceeding 2500°C at the carbon black surface. The carbon black formation takes place in the temperature region below 2000°C; above 2000°C, a partial graphitization occurs.[78] The Shawinigan process is a typical example of an acetylene-based process. Due to the gaseous acetylene precursor and the process conditions, acetylene blacks are known for high purities and large crystalline domains in the primary particles.[76,77,79] A decreasing availability of acetylene from the petrochemical industry has limited the global production capacity of acetylene blacks.

Carbon blacks with similar purity can be produced by partial combustion of hydrocarbon feedstock with high purity. A partial combustion of the hydrocarbon feedstock by co-injecting an oxidant like air in the reaction compartment provides the energy required for the thermal decomposition of the hydrocarbon.[74,75] The generated energy cracks the hydrocarbon at temperatures above 1000°C followed by the synthesis of the carbon black. Usually, after the carbon black synthesis a cooling step is applied to stop the growth of the particles. The chemical reactions are

$$\text{Combustion:} \quad C_xH_{2x+z} + \left(\frac{3X}{2} + \frac{z}{4}\right)O_2 \rightarrow xCO_2 + \left(x + \frac{z}{2}\right)H_2O + \text{Energy}$$

$$\text{Cracking:} \quad C_xH_{2x+z} + \text{Energy} \rightarrow C_xH_y + \left(x + \frac{z}{2} - \frac{y}{4}\right)H_2 \quad (z > y)$$

$$\text{Synthesis:} \quad C_xH_y \rightarrow xC + \frac{y}{2}H_2$$

Battery grade carbon blacks produced by a specially designed partial combustion process include the ENSACO/SUPER P carbon black and the by-products of the Shell gasification process, known as Ketjen Black.[74–78]

The carbon black formation mechanism, as derived from the carbon black structure, includes (1) the vaporization of the feedstock (in case of an oil feedstock) and pyrolysis down to C_1 or C_2 units, (2) formation of nuclei or growth centers for the primary carbon particles, (3) growth and fusion of the nuclei to concentric primary particles, (4) aggregation of the primary particles to primary aggregates, (5) in some cases a step of secondary growth which relates to the formation of a pyrolytic deposit on the surface of the aggregate, (6) agglomeration of the aggregates by van der Waals forces, and (7) possibly aggregation of the aggregates followed by subsequent coating of carbon (as sometimes observed in the plasma process).[77,79–84]

Conductive carbon blacks are highly structured aggregates of primary carbon particles with several tens of nanometers in size. Typical primary particles have a concentric structure composed of repeating carbon layers.[79,85–87] Figure 7.6 shows a TEM picture of a highly structured aggregate of SUPER P Li conductive carbon black as well as an enlarged picture of the almost spherical primary particles. The higher the carbon black structure the lower is the required concentration of carbon black in mixtures with insulating or semiconducting materials.[87,88]

Packing of the primary carbon particle aggregates creates voids. The resulting void volume depends on the size and shape of the aggregates, the aggregate agglomeration, and the porosity of the primary particles. Therefore, the carbon black structure can be considered as the sum of a number of accessible voids by unit weight, namely

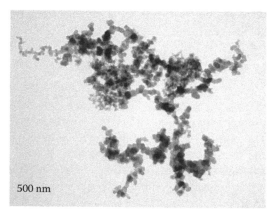
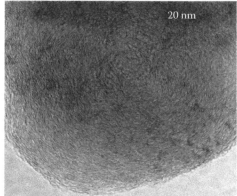

FIGURE 7.6 TEM picture of a SUPER P Li carbon black. Left: highly structured aggregates of fused primary particles. Right: primary particle.

TABLE 7.2

Typical OAN Numbers from DBPA, BET Specific Surface Area, Relative Conductivity, Dispersibility, and Purity of Various Conductive Carbon Blacks from Different Manufacturing Processes

Carbon Black Type	BET SSA (m² g⁻¹)		OAN (mL DBP/100 g Carbon)		Conductivity	Dispersibility	Purity
Acetylene black	80	Low	250	High	+++	+++	++++
ENSACO 250G	65	Low	190	High	+++	+++	++++
SUPER P Li	60	Low	290	High	++++	++++	++++
ENSACO 260G	70	Low	190	High	++++	+++	++++
ENSACO 350G	800	Very high	320	Very high	+++++	++	+++
Ketjenblack EC300	800	Very high	320	Very high	+++++	+	++
Ketjenblack EC-600 JD	1270	Super high	495	Very high	+++++	+	++

1. Interaggregate space
2. Interstices within the aggregates
3. Interparticle porosity and particle pores

The higher the structure level of the aggregates is, the higher is the volume of the voids. DBPA, today replaced by oil absorption number (OAN; ASTM D2414), is employed to measure the void volume and, thus, the average structure level. The higher the OAN the more complex is the structure of the aggregates. Table 7.2 gives an overview about the OAN of the typical conductive carbon black types obtained from different manufacturing processes. Note that conductive carbon blacks typically show the OAN of more than 170 mL/100 g carbon.

The Brunaurer Emmett Teller specific surface area (BET SSA) of carbon black is influenced by the size, the porosity, and the surface microstructure of the primary particles. Surface heterogeneities given by graphitic planes at the surface, amorphous carbon, crystallite edges, and slit-shaped cavities representing adsorption sites of different energies describe the surface microstructure.

High shear forces that are generated by intense mixer or extruder units during the dispersion of the carbon black in a liquid or in elastomers do not change the primary particles but may break down the primary agglomerates of carbon black. This breakdown is accompanied by a decrease of the OAN resulting in an increase in the electrode resistivity. Recent investigations have experimentally proven that the mechanical properties of carbon black are mainly governed by its electrical state.[89] The collapse of the carbon black structure provoked by shear energy could be interpreted as de-agglomeration of electrically charged aggregates. According to this approach, the comparison of the densification behavior and the resistivity vs. press density of SUPER P Li with acetylene black could explain the higher stability of SUPER P Li toward shear energy generally observed in the electrode manufacturing process.

7.3.5 CONDUCTIVE GRAPHITE VS. CARBON BLACK

When applied as conductive additive in the positive electrode, graphite and carbon black show complementary properties which are summarized in Table 7.3. The decision which carbon type should be selected depends on the cell requirements and the type of active electrode materials used in the electrodes. The TEM pictures in Figure 7.7 compare the morphology of a typical conductive carbon black and a graphite powder and illustrate the dimensional differences of the primary particles of a factor of about 10.

TABLE 7.3

Effect of Carbon Black and Graphite Conductive Additives on the Electrode Parameters

Electrode Parameter	Carbon Black	Graphite
Electronic electrode conductivity	Particle–particle contact and contact to current collector	Conductive paths through electrode; higher compressibility causes higher electrode density and particle contacts
Ionic electrode conductivity	Electrolyte absorption	Porosity control
Cycling stability	Flexible network Optimal SEI formation	Low BET → Low side reactions' rate
Energy density of the cell	Low amounts required	Compressibility Low swelling
Electrode manufacturing process	Stable dispersions High binder compatibility	Low binder amounts Low slurry viscosity

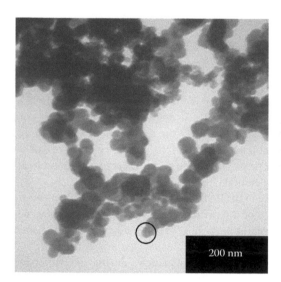 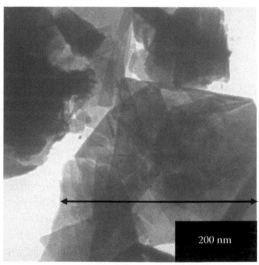

FIGURE 7.7   TEM pictures of SUPER P Li carbon black (left) and TIMREX KS4 graphite (right).

Compared to graphite, usually a lower amount of carbon black conductive additive is required in the positive electrode. The reason could be the lower volumetric density and the highly structured aggregates of primary particles of the conductive carbon black which generates a conductive network in the positive electrode. However, the use of highly structured conductive carbon blacks is limited because of their relatively high BET SSA. The higher the carbon surface area, the higher is the overall reaction rate for the electrochemical and chemical side reactions that can be expected at the positive electrode, especially at highly positive potentials. High electrolyte decomposition rates at highly oxidative potentials are reported for $LiMn_2O_4$/C positive electrodes.[49,50,90] The oxidation of high surface area carbon black deteriorating the electrical contacts between the electrode particles is reported as possible reason for the capacity fading of $LiMn_2O_4$/C electrodes during cycling.[34] The drawback of graphite additives is the risk of insertion of solvated anions from the battery electrolyte at highly oxidative potentials, which has been identified by *in situ* x-ray diffraction (XRD).[91] This anion insertion expands the graphite particles and deteriorates the electrical contacts. On the other hand, the higher thermal conductivity of graphite might facilitate the dissipation of heat locally generated at high current rates. The control of the electrolyte percolation in the electrode is different for the two materials: carbon black absorbs the liquid electrolyte; thus, the OAN of carbon black can be used to estimate the capability to absorb electrolyte. In the case of graphite, variation of the graphite particle size influences the size of the electrode pores filled with the electrolyte.

Besides the electrical and thermal aspects, carbon conductive additives influence the mechanical properties of the electrodes. In particular, due to its compressibility, graphite improves the electrode density and mechanical stability. The generally lower DBPA of graphite is the reason for the lower amount of binder material necessary to achieve a suitable mechanical stability of the electrode. Further, a more facile spreadability of graphitic filaments in the electrode mass is reported for primary lithium cells.[92]

Figure 7.8 (left) shows the scanning electron microscope (SEM) picture of a positive electrode containing $LiCoO_2$. A mixture of graphite and carbon black is used for the conductive matrix. The carbon black is mainly attached at the surface of the active electrode material, whereas the fine graphite particles fill the voids between the coarser active material particles. It can be concluded that carbon black and graphite fulfill complementary electrical functions in the electrode. The carbon black improves the contact between the particles of the active electrode material, whereas the graphite creates the electronic conductive path through the electrode, as it already has been reported.[93] The beneficial synergy of both carbon types for an optimal conductive matrix which lead to an improved cell impedance, resulting in an improved cycling stability at high discharge currents, is shown in Figure 7.9 for the

LiCoO₂ positive electrode

Graphite negative electrode

FIGURE 7.8 SEM pictures of a $LiCoO_2$ positive electrode (left) and of a surface-treated graphite negative electrode (right) both containing TIMREX KS6 graphite and SUPER P Li carbon black conductive additives.

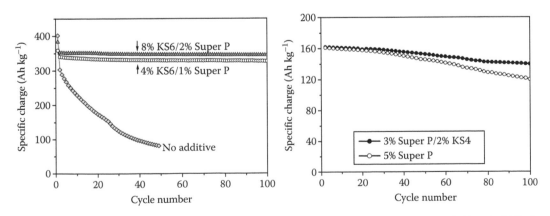

FIGURE 7.9 Cycling behavior of half cells containing a surface-treated graphite negative (left) and a $LiCoO_2$ positive (right) electrode, respectively, with different fractions of TIMREX KS6 or KS4 graphite, and SUPER P Li carbon black as conductive additives. (Electrode porosity ca. 35%; electrolyte 1 M $LiPF_6$ in ethylene carbonate (EC)/ethyl methyl carbonate (EMC) 1:3 (v:v); metallic lithium as counter-electrode.)

cases of a positive and a negative electrode, respectively. In addition, Cheon et al.[47] recently reported optimized cell performance with regard to the energy and power density by using a binary mixtures of TIMREX KS6 graphite and SUPER P Li (1:1) as a conductive matrix in a positive $LiCoO_2$ electrode. High rate performance was obtained in the case of a $LiMn_2O_4$ spinel electrode containing TIMREX SFG6 graphite and ENSACO 250 P.[94] The positive effect of a mixed graphite/carbon black conductive additive has been claimed in several patents.

7.4 STRUCTURE AND ELECTROCHEMICAL PERFORMANCE OF ELECTROACTIVE CARBONS IN THE NEGATIVE ELECTRODE

Until now, in the field of lithium-ion batteries, carbon was investigated primarily as electroactive material for the negative electrode. It is therefore not surprising that several comprehensive reviews partly or entirely focusing on this particular field were published.[12,22,95–98] In this chapter, some basic information on electroactive carbons for the negative electrode will be provided and recent developments in the field will be highlighted. For more detail on earlier works, the previous reviews and the primary literature listed here are recommended.[12,22,95–98]

7.4.1 EARLY WORK

Since the pioneering work of Hérold and coworkers[99,100] about graphite intercalation compounds (GICs) in the 1950s, the intercalation of lithium into graphite up to a chemical composition of LiC_6 was known. With the proposal of the "rocking chair battery" concept by Armand in 1980,[5] first trials to use graphite as host structure in the negative electrode failed due to the strong reactivity of the negative electrode with the electrolyte.[101,102] The electrochemical reduction of graphite in aprotic (usually organic) electrolytes is clearly visible in cyclic voltammograms of graphite electrodes and occurs at more positive potentials than the potential at which the plating of metallic lithium occurs. The reversible reduction process was accompanied by huge swelling and macroscopic disintegration of the electrode, an effect interpreted by Besenhard and coworkers[103,104] as solvent decomposition. The first successful electrochemical intercalation of lithium into graphite was achieved at high temperatures using a solid polymer electrolyte.[105]

Kanno et al. studied less crystalline carbons like pyrolyzed polymers and commercial carbon fibers. They reported an irreversible charge capacity in the first cycle related to the carbon surface. However, the reversible charge capacity which they could observe over several cycles was independent of the surface reaction.[106] Mohri et al.[107] demonstrated only 20% capacity fading over 500 cycles with a lithium-ion cell containing a lithium metal oxide and low crystallinity pyrolytic carbon electrode.

The first commercial lithium-ion battery marketed by Sony Energytec Inc. in 1991 contained a nongraphitizable carbon negative electrode produced by thermal decomposition of thermosetting polyfurfuryl alcohol (PFA) resin.[1,108,109] Due to the hardness of the carbon particles, these amorphous nongraphitizable materials are also called hard carbons. The steeply sloping galvanostatic curve (showing the potential of the electrode under constant current conditions, plotted either vs. time or vs. total electric charge) of PFA-based carbon at low lithium content changes with increasing lithium insertion level into a long plateau slightly positive to the potential of metallic lithium, as shown for a typical hard carbon electrode in Figure 7.10. The maximum reversible charge capacity of 350 mAh g^{-1} obtained for this electrode could be improved by doping with phosphorous atoms up to 550 mAh g^{-1}.[110,111] Other hard carbons prepared from solid phase by decomposition of polyacrylonitrile (PAN), polyvinyl acetate, or phenylic resins showed similar electrochemical performance.[112] Although the reversible charge capacity of such carbons is not confined by the theoretical reversible capacity of graphite of 372 mAh g^{-1} (being equivalent to a chemical composition of LiC_6; based on the carbon mass), usually only a part of the maximum charge capacity of hard carbons can be utilized in electrochemical cells. It was shown that for nongraphitizable carbon electrodes,

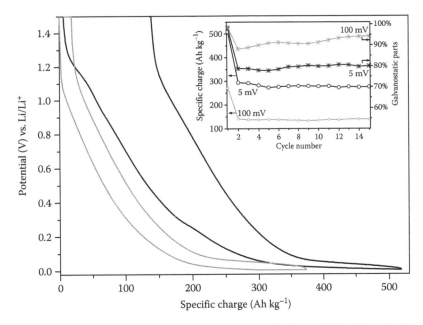

FIGURE 7.10 First (black line) and second (grey line) galvanostatic charge (lithium insertion) and discharge (lithium de-insertion) cycle of a nongraphitizing carbon (pitch-derived hard carbon) at 10 mA g^{-1} in the 1 M LiPF$_6$, EC:EMC 1:3 (*v:v*) electrolyte. The counter-electrode was metallic lithium. The inset shows the cycling behavior of this electrode when charged to 5 mV vs. Li/Li$^+$ and 100 mV vs. Li/Li$^+$.

with increasing utilization of the possible reversible capacity, the cycling stability of the electrode decreases and the first cycle charge "loss" (irreversible capacity) increases.[112]

The success of Sony's hard carbon electrode initiated numerous research activities on carbons from alternative raw materials of natural and synthetic sources.[109,113–119] The large variety of low crystallinity carbon materials showed different electrochemical behaviors depending on their differences in macrostructure and microstructure. The suggested classification of carbons with different degree of crystallinity, according to their electrochemical potential as a function of the lithium content and the amount of lithium per mass unit of carbons which can be reversibly inserted into these carbons, is discussed in Section 7.4.2. The achievable reversible charge capacity as a function of the electrochemical potential is used to select the optimal carbon negative electrode for a lithium-ion cell with maximum energy per mass unit of carbon.

With the development of more suitable electrolyte formulations based on ethylene carbonate (EC), the use of more graphitic carbon negative electrodes became possible.[25,120,121] Yamamoto et al.[122] reported about the increase of reversible capacity and surface reactivity of heat-treated mesophase pitch-based carbon fibers with increasing treatment temperature. Several groups compared the electrochemical performance of natural graphite to other carbon materials.[22,113,123,124] Nishio showed that the discharge capacities of graphite vs. acetylene black, petroleum coke (1400°C), pitch coke (1200°C), and artificial graphitic carbons heat-treated at different temperatures were found to be inversely proportional to their d_{002} values.[125] The monotonic decrease of the specific charge with decreasing crystallinity from graphite up to crystallite thickness $L_{c002} < 10$ nm for graphitizable carbons heat-treated above 1500°C was confirmed by several groups.[21,96,126–128] Even though the theoretical maximum reversible specific charge of 372 mAh g^{-1} achievable for graphite is smaller compared to other carbons with lower crystallinity, it significantly contributes to a high energy density at the cell level due to the fact that the redox potential for the lithium insertion is close to that of metallic lithium over almost the entire charge/discharge curve, as shown in Figures 7.11 and 7.12.

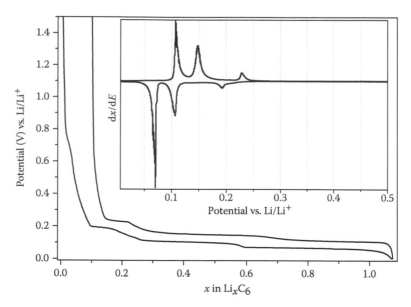

FIGURE 7.11 First galvanostatic charge (lithium insertion) and discharge (lithium de-insertion) curve of a highly crystalline graphite (TIMREX SFG44) at 10 mA g^{-1} in the 1 M LiPF$_6$, EC:EMC 1:3 (v:v) electrolyte. The counter-electrode was metallic lithium. The derivative of the composition (x in Li$_x$C$_6$) with respect to the potential (dx/dE) is shown in the inset. The phase transitions between stage-1 and stage-2 and between dilute stage-1 and stage-4 are observed at ca. 70 mV and 190 mV vs. Li/Li$^+$ during the lithium intercalation. The broader peak at ca. 105 mV vs. Li/Li$^+$ during the lithium intercalation indicates the formation of stage-2, most probably by the intermediate diluted stage-2 and stage-3 phases. The experimental conditions do not allow all possible phases to be observed.

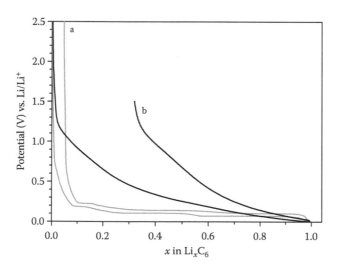

FIGURE 7.12 First galvanostatic charge (lithium insertion) and discharge (lithium de-insertion) curve of (a) graphitized mesocarbon (MCMB, Osaka Gas Chemical Co.) and (b) a graphitizable petroleum coke heat-treated at 1500°C at 10 mA g^{-1} in the 1 M LiPF$_6$, EC:EMC 1:3 (v:v) electrolyte.

7.4.2 LITHIUM INTERCALATION INTO GRAPHITE

Generally speaking, various species can be chemically and/or electrochemically intercalated into graphite and form graphite intercalation compounds (GICs). A characteristic structural feature of GICs is the staging phenomenon which corresponds to a periodical arrangement of intercalated layers within the graphite layer matrix. GICs are classified by the stage index n which characterizes the number of graphene layers which separate two nearest intercalated layers. The intercalation of species into layered host structures in its strict meaning occurs without a change of the chemical structure of the host material, whereas the insertion of species describes the more frequent case where the host structure is changed.[5] Note that, nowadays the term "insertion" is often used to describe a general case of the insertion of species into the crystal structure of a solid (in this respect, even the alloying of elements like silicon with lithium is sometimes termed "insertion"); the term "intercalation" is normally used for insertion into 2D crystal structures like graphite. Without doubt the terminology in the scientific and technical literature is not consequent here.

The staging phenomena and the resulting phase diagram for lithium intercalation into graphite have been investigated by XRD and neutron diffraction.[129] The chemical intercalation of lithium into graphite from the vapor phase proceeds via the successive formation of the following GICs: stage-4 (composition not well defined), stage-3 (LiC_{24}), a dilute lattice gas disordered stage-2 (LiC_{18}), stage-2 (LiC_{12}), and stage-1 (LiC_6). The intercalation reaction proceeds via the prismatic graphite surfaces of the host carbon; the intercalation through the basal plane can only occur at defect sites.[130] During the intercalation reaction, the stacking order of the ideal (hexagonal) graphite changes from ABAB to AAAA. The final order of the intercalated rhombohedral graphite is also AAAA. In the stage-1 compound LiC_6, the interlayer stacking order of lithium atoms between adjacent graphite layers is $\alpha\alpha$.[131,132] The lithium insertion causes an increase of the interlayer distance of the graphene layers by ca. 10%.[133,134] Between the graphene layers the lithium atoms are distributed in a way that the nearest neighbor sites are not occupied resulting in a Li–Li distance of 0.430 nm, as shown in Figure 7.13. The occupation of the nearest neighbor sites can be found in the lithium-rich intercalation phases with the composition of LiC_4 and LiC_2, so far stabilized only at high lithium partial pressure of 60 kbar and elevated temperatures.[135–140] In LiC_2, the Li–Li distance of adjacent Li atoms within one layer decreases to 0.250 nm. This distance is lower than the Li–Li distance observed for

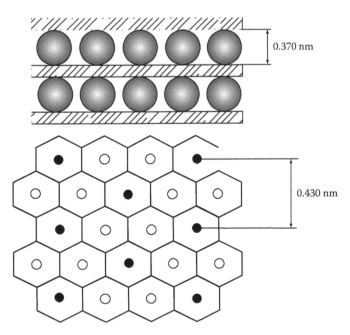

FIGURE 7.13 The structure of LiC_6. Top: Schematic representation showing the AAAA stacking of graphene layers and the $\alpha\alpha$ ordering of the adjacent lithium layers between the graphene sheets. Bottom: Simplified schematic representation of the in-plane ordering of LiC_6 (black circles: lithium atoms). Open circles indicate the in-plane ordering of lithium atoms in LiC_2. (Redrawn from Winter, M., Besenhard, J.O., Spahr, M.E., and Novák, P., *Adv. Mater.*, 10, 725, 1998.)

metallic lithium (0.267 nm) which could explain a negative open circuit voltage of <-10 mV vs. Li/Li$^+$ measured for LiC$_2$. Under ambient conditions, LiC$_2$ decomposes slowly via different metastable intermediate phases to LiC$_6$ and lithium.[136,139]

The electrochemical insertion of unsolvated lithium ions into the graphite structure proceeds reversibly in the potential range between 0 and 0.25 V vs. Li/Li$^+$. For graphite materials with high structural order, provided that the insertion current density is low enough, the theoretical charge capacity (specific charge) of 372 mAh g^{-1} of graphite, which corresponds to the stoichiometry of the stage-1 lithium-graphite intercalation compound (Li–GIC) LiC$_6$, can be obtained. The redox potential vs. composition curve exhibits several reversible plateaus, as it is clear from Figure 7.11 where a charge/discharge curve at a low rate is shown for a graphite/lithium cell. The potential plateaus describe two-phase regions of two co-existing single phase, in which the overall lithium activity is constant according to the Gibbs' phase rule. The existence of single phases is expected in regions in which with increasing lithium content the lithium activity in the graphite electrode is increasing and, thus, the potential is decreasing. The derivative of the stoichiometric index x in Li$_x$C$_6$ with respect to the potential, plotted against the potential, allows to determine the two-phase regions by peaks, as illustrated in Figure 7.11.

Several groups have determined the electrochemical Li–C phase diagram using *ex situ* or *in situ* XRD[120,121,141,142] and Raman spectroscopy.[143] At room temperature, the transition between stage-1 (LiC$_6$) and stage-2 (LiC$_{12}$) occurs at ca. 0.09 V vs. Li/Li$^+$, between stage-2 and stage-2L (LiC$_{18}$) at ca. 0.12 V vs. Li/Li$^+$, between stage-2L and stage-3 as well as stage-3 and stage-4 at ca. 0.14 vs. Li/Li$^+$, and stage-4 and the dilute stage-1 at ca. 0.20 V vs. Li/Li$^+$. Some disagreement can be found in the literature especially for the identified Li–GIC in the low lithium concentration range which could be explained by the experimental conditions, nature of the graphite samples, and the weak X-ray scattering of the light lithium atoms. The composition of the stage-3 and stage-4 compounds is not well defined and indicate the chemical composition ranges between LiC$_{25}$ and LiC$_{30}$ for stage-3 and between LiC$_{44}$ and LiC$_{50}$ for stage-4 compound. The transition between stage-3 and stage-4 seems to be continuous as XRD indicates a smooth shift of the (001) diffraction peak from the stage-3 to the stage-4 positions, and the electrode potential steadily varies during the transition. Safran proposed mixed sandwiches of stage-3 and stage-4 in the compound; however, the continuous transition between the two pure phases has not yet been understood.[144] At low lithium content, the intercalated lithium is randomly distributed throughout the graphite host in a dilute stage-1 phase. The potential plateau at about 0.2 V vs. Li/Li$^+$ generally corresponds to the transition between the diluted stage-1 phase and the stage-4 phase. In contrast, Ohzuku et al.[120] showed evidence for the existence of a stage-8 phase.

7.4.3 Graphitizable and Nongraphitizable Carbons

As shown above, there are differences in the electrochemical intercalation of lithium into hard carbons and highly graphitized carbons (cf. Figures 7.10 and 7.11), which depend on the degree of graphitization. Graphitization is defined as a solid state transformation of thermodynamically unstable carbon into graphite by thermal activation.[145] The existence of graphitizable carbons, non-graphitizable carbons, and intermediates was first recognized by Franklin.[55] With increasing heat treatment temperatures, graphitizable carbons go through a progressive structural ordering process. During the graphitization process, a progressive decrease of the medium d_{002} interlayer spacing ($c/2$ value) occurs from 0.3440 nm down to the ultimate value for graphite, 0.3354 nm. The value of 0.3440 nm is considered as the lowest value possible in a turbostratic structure. The turbostratic order of graphitic carbons was introduced by Warren and Biscoe as a roughly parallel and equidistant stacking of graphene layers, but with each layer having a completely random orientation about the layer normal.[146] According to Franklin, the degree of graphitization g of a graphitizing carbon is defined as

$$g = \frac{(0.3440 - d_{002})}{(0.3440 - 0.3354)} \tag{7.3}$$

During the graphitization process, the crystal size (both the L_c and L_a values) increases from about 5 nm at 1200°C to the order of 10 nm at 2000°C and to more than 100 nm at higher graphitization temperatures.

Méring and Maire expanded the Franklin model for the graphitization process suggesting not only an ordering of disordered stacks of graphitic atomic layers (graphene layers) to perfect graphene layers with a 3D structure but also the healing of defects within the elementary graphite layers.[63,147] According to this work, the fundamental process of graphitization lies in the transformation of elemental layer states which are disordered and contain carbon defects to the ideal graphene layers. The transformation of disordered stacks of graphitic atomic layers into an ordered arrangement, as the graphitization was described by Franklin, is only a consequence of the internal transformation.

The pyrolytic conversion of organic compounds to carbon proceeds at temperatures below 1000°C and involves an immense number of parallel and sequential complex reactions, depending on the starting materials.[145] After the primary carbonization step up to about 600°C which involves a resolidification from a fluid phase, the resulting carbon residues consist of basic structural units of planar arrangements of 10–20 aromatic rings which are piled up more or less in parallel by groups of two to four.[148] In the case of graphitizable carbons, with increasing temperature, the structural units form distorted columns which coalesce during the graphitization process. In the case of non-graphitizing carbons, the formation of columns and their coalescence is hindered by cross-linking groups between the structural units, being present especially in the case of oxygen-containing precursors. The mobility of the carbon structural units, which determines the degree of microstructural ordering of the carbonaceous materials, is dependent on the aggregation state of the intermediate phase during pyrolysis. Commonly, nongraphitizing carbons are prepared by solid state pyrolysis, whereas graphitizable carbons are prepared in a liquid or gas phase process.[149] Due to their degree of particle hardness, graphitizing carbons are called "soft carbons" and nongraphitizing carbons "hard carbons."[55]

Common features of carbons carbonized between 500°C and 1000°C are the low crystallinity with interlayer spacings larger than 0.344 nm, a large amount of microporosity and nanoporosity (observable by small-angle X-ray scattering), low real densities between 1.4 and 2.0 g cm^{-3},[150] and a large amount of heteroatoms left from the organic precursor (which can be essentially hydrogen, nitrogen, oxygen, and sulfur). Most part of the hydrogen is cleaved up to 1000°C,[151] oxygen and nitrogen at about 1500°C,[152] and sulfur around 2000°C.[153] During the heat treatment, the electrical conductivity increases by several orders of magnitude for which a disorder-induced nonmetal–metal transition was suggested as an explanation.[154]

7.4.4 Lithium Intercalation into Graphitizable Carbons

The electrochemical intercalation of lithium into graphitizable carbons like mesophase carbon or petroleum coke, heat-treated at temperatures between 1000°C and 3000°C, exhibit common features being extensively described in numerous publications.[12,25,96,126,155–166] The electrochemical double layer capacitance of the carbons which is related to the electrode's surface being in contact with the electrolyte is decreasing with increasing heat treatment temperature and degree of graphitization. Correlating with the double layer capacitance, the faradaic "losses" during the first electrochemical lithium intercalation half-cycle (due to reactions other than lithium intercalation) decrease independently from the type of the electrolyte system, up to a certain degree of graphitization. This is presumably due to the decrease of the carbon's active surface area (ASA) resulting from structural reorganization and elimination of microporosity.[167] Above a heat treatment temperature of 2800°C and a graphitization degree of about 0.8, the electrochemical performance depends on the electrolyte system. For example, in propylene carbonate (PC)- and butylene carbonate-based electrolytes, strong gas formation due to the electrolyte decomposition at the carbon surface combined with the exfoliation of the graphitic carbon structure results in high charge "losses" and a failure of the lithium-ion cell, whereas a good carbon surface passivation can be achieved in EC-based electrolytes.[25,159]

The reversible charge capacity of graphitized soft carbons decreases with increasing graphitization degree up to a minimum value at a heat treatment temperature of about 2000°C and then, with increasing graphitization degree, increases until the theoretical charge capacity for graphite is reached.[12,96,128,168] With increasing treatment temperatures above 2000°C, under galvanostatic conditions, the potential curve (vs. Li/Li+) for the electrochemical lithium intercalation into graphite shows a progressive appearance of plateaus indicating defined two-phase regions. On the other hand, carbons treated below 2000°C show in galvanostatic experiments sloping cell voltages in lithium half cells with two domains of different voltage behaviors, depending on the treatment temperature.

The progressive appearance of distinct voltage plateaus and the increasing reversible charge capacity for heat treatment temperatures above 2000°C can be explained with the increasing crystallinity of the carbon. For the behavior below 2000°C, several, sometimes controversial, models have been proposed. Endo et al.[126] showed that the minimum charge capacity occurred if the crystallite thickness L_c (calculated from the (002) XRD) was in the range of 10 nm. For $L_c > 10$ nm, a classical intercalation should occur according to the mechanism as proposed for the intercalation of sulfuric acid into graphite.[21,169] For $L_c < 10$ nm, a different process of doping and undoping was proposed to occur considering the formation of covalent Li_2 molecules. With decreasing size of the crystalline domains, this process would be enhanced.[126] Tatsumi introduced an index value, P_1 for the fractional ratio of graphite stacking in a carbon material and an index $(1 - P_1)$ for a fractional ratio of the turbostratic structure. He showed that, for heat treatment temperatures of mesocarbon microbeads (MCMB) above 2000°C (region I), the reversible charge capacity is proportional to the P_1 index. The maximum possible lithium composition of the fully turbostratic structure was estimated to be around $Li_{0.2}C_6$. To explain the different behavior in the low-temperature region II, Tatsumi proposed two different mechanisms for the storage of lithium. Compared to ideal graphite, they are based on the larger interplanar spacing of the turbostratic structure and the existence of crystal defects.

A similar model proposed by Dahn et al.[12,128,156] is based on the graphitization model of Franklin which considers graphitic carbons as mixtures of organized and unorganized regions.[55] Organized regions are constituted of parallel layers of graphene sheets with ABAB stacking or turbostratically structured carbon. The unorganized regions consist of groups of tetrahedrally bonded sp3-carbon atoms or highly buckled graphene sheets. XRD were used to calculate the amount of organized and unorganized domains. Dahn et al. introduced a P-factor describing the probability of random stacking between adjacent graphene layers and a g-factor describing the probability of parallel stacking. Based on some earlier works, they assumed that buckled layers being characteristic of unorganized carbon ($P = 1$, $g = 0$) can intercalate an amount of $x_{uc} = 0.9$ per six carbon atoms in Li_xC_6, that randomly stacked parallel layers of turbostratically structured layers ($g = 1$, $P = 1$) can intercalate $x_r = 0.25$ per six carbon atoms for treatment temperatures below 2200°C and $x_r = 0$ for treatment temperatures above 2200°C, and that parallel layers with registered stacking as in graphite ($g = 1$, $P = 0$) can intercalate $x_g = 1$ per six carbon atoms. The variation of the maximum lithium content x_{max} in $Li_{x max}C_6$ with the treatment temperature (and the related structural parameters) was fitted with the following expression:

$$x_{max} = g[(1-P)+Px_r]+(1-g)x_{uc} \tag{7.4}$$

For soft carbons heat-treated above 2000°C, the experimentally determined variation of the maximum reversible charge capacity with the P-factor could be well fitted for $P < 0.4$ with the linear relationship deducted from Equation 7.4 considering that no lithium can be intercalated between randomly stacked parallel layers ($x_r = 0$). To be able to apply the relationship over the whole possible range of P between 0 and 1, they concluded from the fact that the interlayer spacing d_{002} depends on the heating temperature that x_r varies with d_{002}, being $x_r \cong 0$ for $d_{002} < 0.338$ nm and $x_r \cong 0.25$ for $d_{002} > 0.338$ nm. For heat treatment temperatures above 2200°C ($d_{002} < 0.338$ nm), the interlayer spaces or galleries between randomly stacked layers would not host lithium ions and these "blocked" galleries would frustrate the formation of the regular sequence of full and empty galleries (staged phases) and therefore the appearance of voltage plateaus.[163,170] Materials with $P > 0.3$ only showed stage-1 and stage-2 phases, whereas those with $P < 0.25$ showed evidence for stages up to stage 4.[171]

The probably most comprehensive model was proposed by Flandrois et al.[96,166] who showed that the dependence of charge capacities and voltage behavior on the heating temperature can be understood in the framework of the Méring's model of graphitization.[147] This model describes the fundamental process of graphitization by the transformation of elementary layer states. Méring concluded from XRD data that each face of an elemental graphene layer could be in either α or β state. The surface of a perfect graphene layer is in the β state, the presence of interstitial carbon atoms grafted onto the graphene layer face is a sign of the α state. During graphitization, every single graphene layer is converted from the α state to the β state which can be described as cleaning of every graphene layer's face. As this cleaning occurs independently from a particular graphene face, the ordered arrangement for a single graphene interlayer space is only possible between two adjacent $\beta\beta$ faces. In a partially graphitized carbon, three types of interlayer spaces ($\alpha\alpha$, $\alpha\beta$, and $\beta\beta$) can be considered. Considering the graphitization degree, g, as a fraction of the layer faces in the β state, the relative proportion of each type can be expressed as $(1 - g)^2$ for $\alpha\alpha$, $2g(1 - g)$ for $\alpha\beta$, and g^2 for $\beta\beta$. It was assumed that the exclusive presence of interlayer spaces in the $\alpha\alpha$ state results in the mean d_{002} value of 0.3440 nm for turbostratic carbon and that the graphene interlayer spaces in the $\beta\beta$ state correspond to the d_{002} value of 0.3354 nm. Therefore, the mean interlayer distance of partially graphitized carbon could be expressed as

$$d_{002} = 0.3354g^2 + [(0.3354 + 0.3440)/2]2g(1-g) + 0.3440(1-g)^2 \qquad (7.5)$$

being equivalent to Equation 7.3 derived by Franklin.

Flandrois et al. considered that the stoichiometric coefficient x in Li_xC_6 (describing the amount of electrochemically intercalated lithium atoms per six carbon atoms) should be a sum of the contributions of $\alpha\alpha$, $\alpha\beta$, and $\beta\beta$ spaces, with $x_{\beta\beta} = 1$ for the ideal graphite:

$$x = (1-g)2x_{\alpha\alpha} + 2g(1-g)x_{\alpha\beta} + g^2 x_{\beta\beta} \qquad (7.6)$$

They showed that the reversible charge capacity of heat-treated cokes as a function of their degree of graphitization is in remarkable agreement with the expression (Equation 7.6) if $x_{\alpha\alpha} = 0.75$, $x_{\alpha\beta} = 0.20$, and $x_{\beta\beta} = 1$. The lower value of $x_{\alpha\beta}$ was qualitatively confirmed by STM observations on pyrocarbons heat-treated at 2000°C.[172] In the $\alpha\beta$ state, interstitial clusters of carbon atoms which commensurate with the graphite lattice hindered the intercalation of lithium and, thus, decreased the stoichiometric coefficient x with respect to a random distribution of interstitials.

Note that the appearance of potential plateaus on the galvanostatic curves of carbons treated at temperatures higher than 2000°C can be explained with the above model. For $g = 0.5$ ($g^2 = 0.25$), on average, one out of four interlayer spaces should be of the $\beta\beta$ type and could be completely filled with lithium resulting in the formation of a stage-4 Li-GIC. A plateau at a potential of 0.18 V vs. Li/Li$^+$ was observed for a coke heat-treated at 2200°C which could correspond to the transition between stage-4 and dilute stage-1.[142] A coke sample (heat-treated at 2400°C) with $g^2 \cong 0.5$ indicating one out of two interlayer spacings in the $\beta\beta$ state showed a potential plateau in which the stage-2 graphite intercalation compound (GIC) compound was formed.

7.4.5 LITHIUM INSERTION INTO LOW-TEMPERATURE AND NONGRAPHITIZABLE CARBONS

The electrochemical performance of a large number of carbonaceous materials carbonized at temperatures below 1000°C like pitch cokes,[173–175] petroleum cokes,[115] mesophase carbons,[157,173,174] coals,[114] cellulose,[176] phenolic resins,[116,117,177–179] sucrose and other sugars,[179–183] mixtures of pyrene and dimethyl-*p*-xylene glycol[184] and other polyaromatic compounds,[185] as well as polymers like poly-*p*-phenylene,[127,168,186–190] PAN,[191,192] polyvinyl chloride (PVC), PVDF, and polyphenylene sulfide[115] has been investigated. Due to the low crystallinity, the existence of internal porosity, and the presence of heteroatoms and functional groups (being typical for low-temperature carbons), the electrochemical lithium insertion mechanism of these materials is different compared to soft carbons heat-treated above 1300°C. In contrast to the latter, for low-temperature and nongraphitizing carbons with basic structural units of 1–1.5 nm, it does not seem justifiable to talk about lithium intercalation being defined as introduction of guest species into the interspace of host layers. The electrochemical performance of these carbons shows many similarities which include (i) high reversible charge capacities, (ii) high irreversible capacities in the first cycle, and (iii) large hysteresis between the galvanostatic curves of the lithium insertion and de-insertion. Usually, a poor cycling stability and a reduced energy efficiency of the lithium-ion cell are observed. These features are more pronounced for lower pyrolysis temperatures.

Several mechanisms have been proposed to explain this behavior. To explain the high reversible charge capacity, Sato et al.[127] concluded from [7]Li nuclear magnetic resonance (NMR) measurements the existence of covalent Li_2 molecules in addition to lithium regularly inserted between the graphene layers. The ^{7}Li NMR investigations revealed two bands with chemical shifts of 0 and 10 ppm, an indication of two different lithium sites. More recent NMR investigations interpreted the band close to 0 ppm as due to lithium located at the edges of the carbon layers.[193] Xiang et al.[178] emphasized the importance of the carbon layer edges, as they observed a linear relationship between the charge consumed in the plateau near 1 V vs. Li/Li[+] during the charge of the carbon electrode and the total edge length per mass unit deduced from L_a and L_c measurements. The electrochemically estimated charge capacity of various polymers pyrolyzed below 1000°C linearly correlated with the H/C atomic ratio.[115,177,186,194] Therefore, the effect of the edges on the high electrochemical charge capacity of carbon materials heat-treated below 1000°C could be a result of the presence of hydrogen atoms bonded at the periphery of the aromatic molecules, which according to Oberlin constitute the basic structural units.[148] The influence of covalently bonded hydrogen in the carbon material could explain the voltage hysteresis observed between lithium insertion and removal. Zheng et al.[194–196] proposed that the lithium atoms may bind on hydrogen-terminated edges of hexagonal carbon fragments, causing a bond change of the terminated carbon atom from sp[2] to sp[3].

However, the latter model could not be used to explain the high reversible charge capacities of more than 400 mAh g[−1] being obtained for some hard carbons whose hydrogen contents were significantly reduced by heat treatment above 1000°C.[179,197] In addition, such heated nongraphitizing carbons show in electrochemical experiments only little hysteresis and exhibit a pronounced part of the galvanostatic curve with significant capacity below 0.1 V vs. Li/Li[+].[116] Heat treatment at temperatures considerably above 1000°C closed the micropores and further decreased the hysteresis of the electrochemical galvanostatic curve; however, it also resulted in drastically reduced reversible charge capacities.[198–200] The existence of the low-voltage potential plateau is not well understood yet. Based on XRD and small-angle X-ray scattering experiments,[179,201] it was proposed that lithium was adsorbed at internal surfaces of nanopores formed by single-layer, bi-layer, and tri-layer groups of graphene sheets which are arranged like a *house of cards*.[128,197,202] This model was extended to the *falling card model* considering also the storage of lithium in micropores.[203] The adsorption of lithium at the surface of the nanopore walls and lithium storage in the micropores proposed by Dahn and coworkers is similar to the pore-filling mechanism proposed by Sonobe and coworkers[204] and the cavity filling mechanism proposed by Tokumitsu et al.[184] Yazami and Deschamps[173,174] proposed a multilayer model in which each (a,b) graphene face of a structural building unit would be covered by two or three lithium layers. Electrochemical *ex situ* and *in situ* ^{7}Li NMR measurements of half cells containing hard carbon fibers indicated that lithium is first introduced in the small openings between nanometer-size graphitic-type layers and then penetrates the nanopores where growing quasi-metallic lithium clusters are formed.[176,205–207]

The electrochemical formation of a passivation layer at the carbon's surface as it is observed in the case of graphite could explain only part of the irreversible capacity during the first electrochemical lithium insertion into low-temperature carbons and hard carbons. These charge "losses" increase if the passivation layer is also formed within the micropores of the carbons.[96] Additional charge "losses" occur by the reaction of lithium with surface defects, surface groups, and other species, especially water molecules adsorbed at the carbon surface and absorbed inside the micropores.[208] Matsumura et al.[209] showed by Fourier transform infrared spectroscopy-attenuated total reflection method (FTIR-ATR), secondary ion mass spectrometry (SIMS), and X-ray photo electron spectroscopy (XPS) that a large portion of the irreversible capacity could be caused by the reaction of lithium with active sites like hydroxyl groups or carbon radicals being present in the bulk of the carbon electrode. In the case of disordered carbons, such active sites could be present in high concentrations giving rise to surface complexes with oxygen and other elements.[210] Kikuchi et al.[211] suppressed the irreversible process observed in the cyclic voltammogram of a pitch-based carbon fiber electrode by heating the carbonaceous material at 980°C under vacuum. Xing and Dahn[181] could considerably reduce the irreversible capacity of pyrolyzed sugars by exposure to argon and nitrogen for several days, this in contrary of opposite effects of air exposure[180] or oxidation.[212] Treatments in CO_2, CO, water, and steam increase the irreversible capacity.[181] Water steam exposure resulted in an additional potential plateau close to 2 V vs. Li/Li$^+$ which could be explained by the reaction of lithium with water absorbed in the micropores of the sugar-based carbon samples. In addition to the reactions at the carbon surface and inside the pores, charge "losses" are generated by trapping a part of the inserted lithium in the carbon structure. This fraction of trapped lithium was reported to be about 20%.[213]

7.4.6 DOPING OF CARBONS

Depending on the carbon precursor(s) and heat treatment temperature, carbonaceous materials commonly contain heteroatoms. Usually, these heteroatoms are located at dislocations and/or at the edges of the graphene layers. Intentional doping of carbon materials has been used to change the distribution of electrons in the energy levels of the carbon material, to affect the graphitization process and to modify the chemical state of the carbon particles' surface.[214] Doping of carbon has been performed (1) by co-deposition of carbon and foreign atoms, e.g., by chemical vapor deposition (CVD), (2) by pyrolysis of organic molecules containing foreign atoms, and (3) by chemical treatment of carbon.

Doping with boron, nitrogen, phosphorus, and silicon atoms: Boron can substitute up to 2.3% of the carbon atoms near the thermodynamical equilibrium in the carbon lattice. Higher amounts of boron in the carbon lattice segregate as B_4C during heat treatment above 2350°C.[215] At lower temperatures, higher boron amounts could be included up to a composition close to BC_3[216] although the maximum amount of boron which can substitute carbon atoms in the carbon structure is lower.[217,218] Way and Dahn[219] prepared boron substituted carbons $C_{1-z}B_z$ by CVD using benzene and BCl_3 at 900°C. Lithium half cells with $C_{1-z}B_z$ showed increase of the reversible charge capacity with the increase in the boron content. Cells with boron-doped carbon showed considerably higher lithium intercalation potentials of about 0.5 V vs. Li/Li$^+$ in comparison to cells with boron-free carbon, due to the acceptor character of boron which lowers the Fermi level and allows more lithium to be intercalated.[220,221] Kwon et al.[222] confirmed these findings for boron-doped pyrolytic carbon. In addition, the presence of boron strengthens the chemical bond between the intercalated lithium and the boron–carbon host when compared to the boron-free carbon. Besides the acceptor effect of boron dopants on the carbon band structure, boron is well known as graphitization catalyst which is added to soft carbons like coal tar-based mesocarbons, petroleum coke, and pitch in the graphitization process to improve crystallinity and therefore the reversible specific charge of the final graphitized carbon product. For graphitized mesocarbon pitch-based fibers containing 1% of boron, the doping increased the reversible charge capacity up to 360 mAh g^{-1} (compared to 300 mAh g^{-1} for the nondoped fibers), resulted in a slightly more positive lithium intercalation potential, and the cycling efficiency was improved.[188,223–227] Similar effects have been found for boron-doped MCMB, graphite, and graphitized coke.[228–236]

Contrary to boron, nitrogen acts as a donor weakening the bond of lithium to the carbon host structure which causes less positive lithium intercalation potentials in comparison to pure carbon. C_xN ($7.3 < x < 62$) and BC_xN ($x = 2, 3, 7, 10$) compounds have been prepared by CVD of acetylene and ammonia or by pyrolysis of nitrogen-containing organic molecules like acetonitrile, acrylonitrile, or pyridine with BCl_3 between 400°C and 1000°C.[237-239] The reversible charge capacities found for these materials were low, especially for the boron containing ternary compounds. The irreversible capacity increased with increasing nitrogen content due to the reaction of lithium with nitrogen chemically bound at aromatic molecules being present besides the nitrogen substituted in the carbon lattice. More promising results were reported for $C_{7.3}N$ and $C_{28}S$ synthesized by CVD at 800°C from pyridine and thiophene, respectively, showing charge capacities higher than 500 mAh g^{-1}.[240] Most of the charge . . . extracted at relatively high potentials, and the irreversible capacity is high. An increased amount of substitutional incorporated nitrogen and a reduced amount of chemically bound nitrogen could be achieved for compositions $C_{14}N$ and $C_{62}N$ prepared by thermal decomposition of acetonitrile at around 1100°C in the presence of a nickel catalyst. Reversible charge capacities of about 400 mAh g^{-1} and an improved cyclability in electrochemical cells were reported.[241] The nickel catalyst improved the crystallinity of the synthesized filaments or particles, also for boron containing carbons BC_x.[242-244]

Phosphorus-doping of petroleum green cokes[245] and hard carbons from epoxy resin[111] or PFA[110] by phosphorous acid resulted in significantly increased reversible capacities and decreased irreversible charge. However, the charge capacity increase occurred at potentials above 0.9 V vs. Li/Li$^+$. The voltage profile was similar to those reported for H-containing carbons, which indicates a similar lithium insertion mechanism.

Silicon–oxygen-containing carbonaceous materials with different Si–O–C composition have been prepared by CVD of benzene and silicon containing precursors,[246,247] pyrolysis around 1000°C of epoxy silane composites,[248] of pitch/polysilane blends,[249,250] and of polysiloxanes.[251-253] The resulting disordered carbon materials contained nano-dispersed amorphous silicon oxycarbide clusters whose amount and Si–O–C compositions depended on the precursor. Materials with the highest identified reversible charge capacity for lithium (about 900 mAh g^{-1}) contained about 43 at% carbon, 32 at% oxygen, and 25 at% silicon. It was suggested that the amorphous glass could reversibly react with lithium, provided carbon is present to provide a path for electrons and Li ions. However, the hysteresis between charge and discharge potentials and the irreversible capacity increase almost linearly with the oxygen content of the materials.

7.4.7 Modifications of the Carbon Surface Morphology

The surface morphology of the carbons is an important material parameter relevant to both the engineering of the electrodes and the irreversible charge consumption due to side reactions, in particular, due to the electrochemical passivation of the surface (i.e., the formation of the so-called solid electrolyte interphase layer, solid electrolyte interphase (SEI)). The modification of the carbon surface morphology can be done by mechanical treatments like grinding, by oxidation in air, oxygen, or other oxidizing agents, or by postheat treatments in an inert gas atmosphere. The type and extent of these treatments affect the particle size and shape, the surface defect concentration, the surface crystallinity, the shape and dimensions of the pores, as well as the surface group chemistry.

Peled et al.[254-256] showed that mild oxidation of graphite at around 600°C increases the reversible charge capacity by 10%–30%. In the case of small burn-offs up to 6%, the irreversible capacity decreased by 10%–20%. The authors assigned the reasons for the performance improvement (1) to the formation of carboxylic groups from hydrogen groups at the prismatic edges at low burn-off providing hence an improved chemical bonding of the SEI for a more efficient passivation and, thus, suppression of the solvent co-intercalation and (2) to the accommodation of extra lithium at edge sites and inside the nanochannels formed by oxidation (the latter were shown by STM to have an opening of a few nanometers). Ein-Ely and Koch[257] found similar performance improvements for a synthetic graphite oxidized by strong chemical reagents like ammonium peroxy disulfate or hot concentrated nitric acid. Buqa et al.[258–260] confirmed the influence of surface oxidation on the charge losses by reacting the graphite surface with various oxidation agents like CO, CO_2, and oxygen.

In contrast, low-temperature carbons from epoxy resins with significant amount of micropores showed a dramatic deterioration of the electrochemical performance after oxidation in air at temperatures from 300°C to 600°C.[212] As shown by small angle x-ray scattering (SAX), the oxidation did not considerably increase the volume of the micropores but the size of the pore openings increased allowing, thus, the electrolyte solution to penetrate into the pores. The consequence was the enhancement of the irreversible electrolyte decomposition reactions during the first electrochemical reduction of the electrode. Surface oxides resulting from oxygen chemisorption contributed to the irreversible capacity and resulted in voltage hysteresis of the galvanostatic charge/discharge curves as well.

Mechanical grinding of carbons generates active sites with a considerable reactivity toward oxygen and/or other elements affecting, thus, the reactivity of the carbon toward the electrolyte solution.[261] Both air jet milling and turbo milling showed their potential to positively influence the electrochemical performance of graphite negative electrode materials by influencing both particle shape and surface morphology.[262] Important is that mild grinding does not alter the interlayer spacing in graphite while severe grinding converts pristine graphite into disordered carbons.[263] Severe grinding increased the reversible charge capacity up to 700 mAh g^{-1} but this positive effect was combined with high irreversible capacity and large hysteresis of the galvanostatic insertion/de-insertion curve.[264] The increase in reversible charge capacity was interpreted as an adsorption of additional lithium at single layer surfaces. Xing et al.[179,180,182] showed by studying the electrochemical behavior of sugar-based carbons before and after milling that the lithium insertion mechanism of severely ground graphite differed to disordered microporous carbons. It was shown that the milling created broken carbon–carbon bonds which are highly reactive and that the air-milled carbons contained substantial amounts of oxygen. The authors proposed a mechanism for the quasi-reversible lithium insertion into milled carbons involving (1) the reaction of lithium at the edge of small graphene sheets, (2) the intercalation in remaining graphene stacks, and (3) the reaction with surface functional groups.

The postheat treatment of graphite reduces the amount of surface defects and surface oxides. It has been shown that heat-treated graphites have a lower ASA and a modified surface reactivity toward the electrolyte. The heat treatment affects, thus, the passivation mechanism, as discussed in more detail in Section 7.4.8.[265,266]

7.4.8 CARBON SURFACE PASSIVATION (SEI FORMATION)

A sufficient stability of the electrolyte is crucial for the shelf life of any kind of electrochemical energy storage cells. The voltage of rechargeable lithium-ion cells is typically around 4 V. Hence, aqueous as well as many nonaqueous electrolytes cannot be used because their thermodynamic or kinetic potential windows of electrochemical stability are not wide enough.[267,268] Actually, no thermodynamically stable electrolyte solutions working in such a broad potential window were identified so far. Only a few classes of aprotic electrolytes possessing sufficient kinetic stability can be used. The consequence is that lithium-ion cells function far beyond the thermodynamic stability limits of their electrolyte. Thus, the surface of both the negative and the positive electrode is normally covered with reduction and oxidation products, respectively, of the electrolyte.[269–271] In the following sections, the interaction with carbons in the negative electrode of typical liquid electrolyte solutions used in lithium-ion batteries based on EC and PC will be discussed. Note that the so-called "polymer electrolyte cells" normally utilize similar liquid electrolytes immobilized in an appropriate matrix, thus, the discussion below is valid also for this electrolyte class. True polymer electrolytes and inorganic solid electrolytes are out of the scope of this chapter.

From thermodynamic reasons, the reduction of the electrolyte takes place on the electrolyte-wetted surface of all components of a charged negative electrode (inclusive current collector) because the potential of lithiated carbon is close to that of metallic lithium (about −3 V vs. normal hydrogen electrode [NHE]). Fortunately, in suitable electrolytes, films of electrolyte decomposition products are formed on the electrolyte-wetted surface. These surface films can cause passivation of the carbon surface. In an ideal case, the surface passivation films kinetically protect the electrolyte from further reduction. The passivation film is normally called the SEI, a term introduced by Peled who developed the concept about three decades ago.[4,256,272–280] Good SEI films possess unique characteristics which are essential for the kinetic stability of the electrolyte solutions and allow the utilization of highly crystalline carbons in negative electrodes. Good SEI films must be electronically insulating to inhibit the electrolyte reduction at the interface SEI/electrolyte. At the same time, they should act as a "sieve" permeable only for the charge-carrier Li^+ but impermeable to other electrolyte components to avoid reduction of the electrolyte components at the carbon/SEI interface and/or in the bulk of the electroactive carbon. Note that a comprehensive book on the properties of the SEI was published a few years ago.[281] In this section, only the basics of SEI will be discussed; for deeper insight into this topic, the aforementioned book and the numerous references therein are recommended.[281]

The composition and morphology of the SEI layers depend on the kind of the electrolyte used.[24,258,282–329] Electrochemical parameters like current density during the first reduction (called "formation") of the carbonaceous negative electrode and the temperature during the formation process influence the quality of the formed SEI. The SEI films are typically nonhomogeneous and composed of two interpenetrating components[95,280,330] (Figure 7.14). At the surface of the electrode, there is a rather thin and more compact film of mostly inorganic decomposition products. Further, to the electrolyte side there is a thicker, possibly porous and electrolyte permeable film of organic (polymeric and oligomeric) decomposition products.

FIGURE 7.14 (Top) The structure of the SEI film. The spheres on the electrode surface symbolize inorganic precipitations (e.g., LiF, Li_2O), the polymeric components of the SEI were drawn as fibrils. The gradually fading grey background symbolizes the decreased material density toward the electrolyte. (Bottom) SEM image of a graphite electrode surface showing the morphology of the SEI film after galvanostatic reduction with 50 mA g^{-1} graphite and stabilization at 300 mV vs. Li/Li^+ for 2 days. The electrode was taken out from a half cell (metallic lithium as counter-electrode, EC:DMC = 1:1, 5% vinylene carbonate (VC) additive, 1 M LiPF$_6$) and rinsed with dimethyl carbonate (DMC) solvent before the SEM examination. (Adapted from Peled, E., Golodnitsky, D., and Ardel, G., *J. Electrochem. Soc.*, 144, L208, 1997.)

The SEI film formation reactions on carbon electrodes in organic electrolytes are generally irreversible, and the film formation follows the nucleation and growth mechanism.[331–333] The SEI films are typically rough; their average thickness increases from about 0.02 μm for freshly formed SEI films up to about 0.1 μm for films on strongly aged electrodes.[332–334] In some electrolytes, the SEI films are partially soluble. Further, the protective SEI films can be damaged under certain conditions. This typically happens when there is an excursion of the potential of the negative electrode to a region >1.5 V vs. Li/Li⁺, and/or the temperature increases too much. The critical temperature, which is normally slightly above 100°C, strongly depends on the electrolyte composition.[292,316,335–342]

Needless to say that a deep understanding of both the formation mechanism of the SEI layer and the underlaying question of carbon's surface chemistry in a particular electrolyte solution is of utmost importance for battery developers. Clearly, the surface chemistry of graphite electrodes plays a key role in their performance.[259,312,325,343–352] A lot of work was devoted to decipher this very complicated surface chemistry. It is therefore not surprising that the advancement in the understanding of surface chemistry of carbon electrodes in nonaqueous electrolytes correlates well with the worldwide production rate of lithium-ion batteries.

Usually, in a given electrolyte solution, there is a similarity in the mechanism of SEI formation on carbon and metallic lithium.[285,353,354] The mechanisms of SEI formation on lithium in numerous electrolytes are investigated since about three decades. In about the last 15 years, the focus continuously shifted from metallic lithium to carbon. There are a huge number of publications covering manifold aspects of the carbon's reactivity with the electrolytes and/or the SEI formation. The reader of this chapter is referred to the books published in this field recently and especially to the primary literature listed therein. Examples include *Nonaqueous Electrochemistry* from 1999 edited by Aurbach,[355] *Advances in Lithium-Ion Batteries* from 2002 edited by van Schalkwijk and Scrosati,[356] and *Lithium-Ion Batteries: Solid-Electrolyte Interphase* from 2004 edited by Balbuena and Wang.[281]

Carbonaceous materials are covered by surface oxide groups such as hydroxyl, phenyl, carbonyl, ether, carboxyl, carboxylic ester, and lactone due to the oxidation of the reactive carbon atoms at their surfaces.[211] The amount and the nature of these groups (and the temperature as well) influence the reactivity of the carbon surface toward the electrolyte. Of course, the distribution of these groups along the carbon surface depends on the carbon surface type (basal, prismatic) and the local concentration of defects. At the carbon electrode surface, there is a competition among many parallel and follow-up reduction reactions of surface groups, electrolyte salts, electrolyte solvents, electrolyte additives (if present), and impurities. The rates of all these reactions are functions of intrinsic physicochemical parameters of involved species and surfaces, local temperature (which, at the interface, is influenced due to the local heat of reaction and the heat dissipation due to the shift of the electrochemical potential from the thermodynamic conditions), and concentrations of both starting materials and reaction products. It is clear that these numerous possible reaction paths are complicated, interdependent, and the lateral distribution of their rates among the carbon surface wetted with the electrolyte can be very inhomogeneous. An additional parameter is the influence of the reaction products precipitating on the surface of the electrode which introduces further normally unknown parameters to be considered including local ohmic resistance. Diffusion and migration rates of numerous species in the electrolyte, the SEI layer, and the electrode must be considered as well.[357] Finally, it must be considered that the reactions of some electrolyte additives (vinylene carbonate [VC] is a prominent example) are initiated electrochemically, but then, the reaction can proceed chemically further without significant charge consumption, the reaction products forming layers on the electrode surface.[358]

The products of reduction of salt anions are typically inorganic compounds like LiF, LiCl, Li₂O, which precipitate on the electrode surface. Reduction of solvents results, apart from the formation of volatile reaction products like ethylene, propylene, hydrogen, carbon dioxide, etc., in the formation of both insoluble (or partially soluble) components like Li_2CO_3, semicarbonates, oligomers, and polymers.[281,283,359] A combination of a variety of advanced surface (and bulk) analytical tools (both *ex situ* and *in situ*) is used[286,321,332,344,352,353,360–377] to gain a comprehensive characterization of processes acurring at the carbon/electrolyte interface, but a more detailed discussion of the proposed reaction mechanisms would be out of the scope of this chapter.

Theoretically, the Li$^+$ intercalation into carbons is fully reversible. In practical cases, however, the charge consumed in the first reduction half-cycles of carbon electrodes apparently exceeds the theoretical charge capacity of the respective carbon type (e.g., 372 mAh g^{-1} for highly crystalline graphites). This additional reduction charge is consumed due to the formation of the passivation layer which can be considered as a side reaction.[25,209,258,285,303,345,378–421] Occasionally other side reactions must be considered, the most important ones are the reduction of traces of water, oxygen, and/or solvents (like N-methyl-pyrrolidone) used during the manufacture of the electrode.[422] In the second and subsequent cycles, however, this additional reduction charge normally decreases rapidly but 20–40 full charge/discharge cycles are needed to complete the formation of the SEI layer.[297] Because of the irreversible consumption of material (lithium and electrolyte), the "irreversible charge" corresponding to the formation of the SEI is in the technical literature frequently called charge "loss" or "irreversible charge capacity" in contrast to the charge associated with the reversible lithium insertion which is called "reversible charge capacity" or simply "charge capacity." Note that the correct term for the latter is "specific charge."[423] Since the positive electrode is the only lithium source in a lithium-ion cell, the "losses" of charge and lithium are detrimental to the specific energy of the whole cell and have to be minimized.

During the last two decades, a vast amount of highly contrary experimental results have been published on the irreversible charge capacity of carbonaceous negative electrodes in lithium-ion cells.[96,181,350,374,379,386–390,403,407,415,419,421,424–454] The differences in the published numbers are immense, even for identical materials in the same electrolyte solution. It is our experience that both the preparation technique and the drying procedure of carbon working electrodes are crucial for their electrochemical performance. A low value of irreversible charge as well as a good cyclability can only be obtained when water contamination of the cell is avoided. For example, the irreversible charge consumption doubles from about 20% in "dry" electrolytes (<10 ppm H$_2$O) to about 40% when the water concentration is ≈1500 ppm H$_2$O (data for a high surface area graphite TIMREX SFG6 in a LiClO$_4$-based electrolyte, with respect to the reversible charge capacity which does not change).[455] Moreover, the scatter of the results rapidly increases with increasing water content.

In a simple model,[333] we distinguish between three types of irreversible charge consumption (Figure 7.15): (1) the irreversible charge related primarily to the reduction of surface groups of the carbon between ca. 3 V and ca. 0.8 V vs. Li/Li$^+$, (2) the irreversible charge primarily due to the SEI formation via electrolyte decomposition between ca. 0.8 V and ca. 0.2 V vs. Li/Li$^+$, and (3) the irreversible charge due to side reactions parallel to the reversible intercalation and de-intercalation of lithium ions (corrosion-like reactions of Li$_x$C$_6$ which contribute to the SEI film growth). It can be clearly seen from Figure 7.15 that, under typical experimental conditions, the main contribution to the irreversible charge is in the region (2) where the SEI is essentially formed. Note that the exfoliation of highly crystalline carbons which is discussed further below is not considered in this simple model.

It should be noted that the SEI formation is strictly a surface-related process (which includes the surface of the pores wetted with the electrolyte, of course). Subject to the conditions that (1) the carbon electrodes are fully wetted with the electrolyte, (2) the influence of surface groups can be considered as minor (which is the case of graphites with comparatively low BET SSA), and (3) there are no significant effects due to the exfoliation of the graphite, there is a linear dependence of the irreversible charge consumption on the BET SSA of the carbon.[333,394] A similar dependence exists between the irreversible charge consumption and the double-layer capacity of graphite electrodes.[394] Thus, from the BET SSA of a particular graphite type, the irreversible charge consumption in a lithium-ion cell can be directly estimated using, e.g., for electrodes utilizing a PVDF binder the correlations shown in Figure 7.16. Actually, there is a small irreversible charge of about 2.5% when extrapolating the correlations to zero BET SSA. This is an indication that minor volume-based reactions must be also considered. The latter are normally associated with the reduction of the

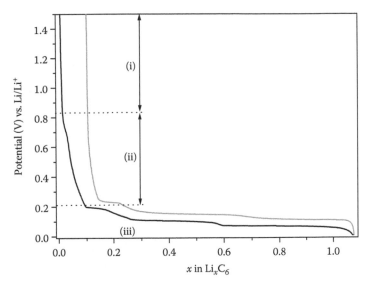

FIGURE 7.15   First galvanostatic charge/discharge cycle at ≈C/40 of a graphite TIMREX SLX50 electrode in the 1 M LiPF$_6$ + EC:DMC electrolyte. (i) to (iii): see text.

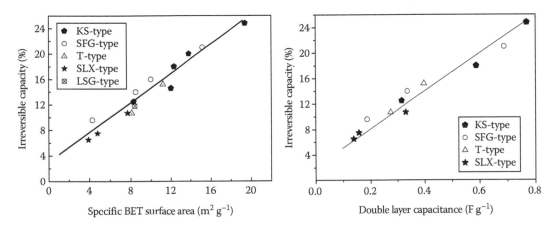

FIGURE 7.16   Irreversible charge consumption in the first cycle at ≈C/40 of TIMREX graphites in the 1 M LiPF$_6$ + EC:DMC electrolyte as a function of the graphite BET SSA.

binder,[456] reduction of impurities in the electrolyte solution, and/or reduction of some oxygen-containing groups located on defect sites within the graphite structure.

The electrochemical performance of disordered (hard) carbons is rather independent of the electrolyte composition. In contrast, the electrochemical performance of highly crystalline carbons (graphites) strongly depends on the composition of the used electrolyte solution. A prominent example is the incompatibility of PC-based electrolytes with some graphite materials.[280,374,457–471] During the electrochemical insertion of lithium from PC-based electrolytes, the graphite structure might exfoliate leading to severe battery failure, whereas graphite exfoliation can be suppressed if PC is (partially) replaced by the lower cyclic carbonate homologue, EC.[347,348,461,462,464,472–479] One reason for this different performance is the quality of the SEI layer formed on the carbon's surface. The PC-based SEI does not suppress the exfoliation of the graphite structure, whereas the EC-based SEI does. Note that the exfoliation of graphite results in a drastic enlargement of the surface area in contact with the electrolyte (Figure 7.17) leading, thus, to a significant increase in the irreversible

FIGURE 7.17   SEM picture of electrochemically exfoliated graphite.

charge capacity and, in the last consequence, to the failure of the lithium-ion cell. The exfoliation of graphite can be suppressed or reduced by the use of suitable electrolyte additives, the most popular among them being VC.[413,467,473,474,479–484] When VC is used as an additive, the SEI formation is completed prior to the beginning of the exfoliation. Thus, to obtain an efficient and high-quality passivation SEI film on the surface of the graphite negative electrode material, the electrolyte composition needs to be adjusted to the graphite material. Interestingly, electrochemical exfoliation of graphite can be observed also in EC- and EC/PC-based electrolyte systems with low PC concentration.[477] Normally, only a few particles from many exfoliate.

Two controversial models have been proposed to explain the exfoliation of graphite. The first model[368,482,485–492] postulates solvent ("solv") co-intercalation and the formation of ternary GICs $Li_x(solv)_yC_6$. The way how these GICs further decompose determines whether the film formation on the graphite is optimized to avoid the exfoliation of graphite and further electrolyte decomposition. The second model interprets graphite exfoliation as the destruction of the electrode by gaseous reaction products which are continuously formed on the graphite surface due to the poor passivation properties of the SEI layer which is formed initially.[474,493] Recently, experiments based on *in situ* XRD[368] and *in situ* Raman microscopy[486,492,494] confirmed the existence of the ternary GICs $Li_x(solv)_yC_6$ making, thus, the first model more likely.

There is an influence of the particle size, the mechanical pressure, and the porosity of the electrode on the tendency of the graphite particles to exfoliate.[483] However, graphite bulk properties like crystallite size, the rhombohedral stacking defects of the graphite layer structure, the graphite texture, and graphite porosity only play an indirect role on both the course of the SEI formation and the graphite tendency to exfoliate since some of these bulk parameters are linked to the surface properties of the graphite material.[265,266,347,348,394,422,476–478,484,495–497] It has been shown that the surface crystallinity and morphology as well as the nature of the surface group chemistry of polycrystalline graphite influence the SEI forming process on the graphite surface in the first electrochemical reduction. Clearly, the exfoliation of graphite is a complex process which is closely related to the mechanism and kinetics of the SEI formation reactions. For example, heat treatment of a highly crystalline graphite TIMREX SLX50 at temperatures above 1200°C in an inert gas atmosphere results in a significant increase of the irreversible charge during the first electrochemical lithium insertion, due to the exfoliation of the graphite.[347,478] Several parameters were identified as being responsible. Among them, the surface groups' chemistry and the amount of surface defects are prominent, both parameters being related. The amount of acidic surface groups seems to be an important factor that influences graphite exfoliation during electrochemical lithium insertion in EC containing electrolytes. Furthermore, apparently, graphite particles containing single crystals with low amount of defects tend to exfoliate easier than graphite powders containing a higher amount of defects. By using a procedure based on oxygen chemisorption developed for carbon materials, the quantification of these defects is possible.[498–500] In fact, the cumulated surface area of the different types of defects present . . . at the carbon surface (stacking faults, single and multiple atoms, vacancies, and dislocations) corresponds to what is called the "active surface area (ASA)" which is an intrinsic characteristic of the carbon which does not depend on the nature of the organic surface groups. In short, the elimination of surface defects by, e.g., heat treatment (decrease of the ASA) hinders the formation of the SEI and consequently favors the exfoliation of graphite. On the contrary, the increase in the ASA results in faster electrolyte decomposition and subsequent graphite surface passivation at potentials which are more positive than the potential at which the electrochemical exfoliation of graphite can be observed. It clearly appears that the passivation is kinetically enhanced at graphite surfaces with high ASA. The concept of active sites is therefore a suitable tool to predict the passivation behavior of the graphite.[266]

The interaction between the electrolyte and graphite particle surface is not always sufficient to explain the exfoliation effects of graphite electrodes. The distribution of the local electric potential inside the porous electrode assembly and, as a consequence, the distribution of the local current densities at every single particle of the electrode during the first galvanostatic reduction process could be an explanation of the exfoliation of parts of the electrode. Variations from the average electrode potential can be created by high interparticle resistivity causing high local IR drops. During the galvanostatic electrode passivation process, the local current density for the passivation at badly contacted particles is not high enough to accomplish the formation of an effective SEI layer. Thus, during the further reduction of the electrode, badly contacted particles in the electrode assembly tend to exfoliate. The high exfoliation sensitivity of graphite particles with low compressibility and presumably high contact resistance could be an indication for the importance of good interparticle contacts in the electrode. This interparticle contact can be improved by applying pressure on the graphite electrode during the electrochemical lithium insertion. In technical electrodes, the interparticle contact can be improved by the use of an optimal binder system and conductive additives.

Another important issue is the development of gas on the negative carbonaceous electrode, due to the reductive decomposition of the electrolyte during the formation of the protective SEI layer. Gas development is an unwanted process in a sealed lithium-ion cell system affecting the dimensional stability and safety of the cell.[270,290,317,331,372,484,501–504] Fortunately, the SEI, once formed, normally prevents further reductive electrolyte decomposition and gas formation. The prominent gaseous reaction products are ethylene and propylene from EC and PC electrolyte components, respectively, as well as hydrogen. The amount of gas depends on the electrolyte composition and the type of carbon used.[266] Interestingly, the formation of ethylene and propylene can be suppressed when electrolyte additives like VC are used due to the mostly polymeric and hence involatile nature of the VC decomposition products.[484] However, the use of VC additive does not suppress the reductive formation of hydrogen. In fact, the detection of the type and quantity of gas formed during the formation of the passivation layer on the carbon negative electrode is used to explain the mechanism of the graphite passivation as well as the composition of the passivation layer.[266,290,306,317,331,333,372,484,503,505–507]

The quality of the SEI passivation layer on the carbon electrode has a significant influence on the safety of the entire lithium-ion battery.[23,326,340,342,508–526] Needless to say that safety is one of the most important requirements of lithium-ion batteries. As a general rule, the more energy is stored the more hazardous the energy storage system potentially would be. One of the potentially most dangerous events is self-heating of a lithium-ion cell followed by a thermal runaway. The potentially most dangerous component in a lithium-ion battery is the Li–metal oxide in the positive electrode mass[300,386,509–514,517,519,521–523,526–552] (typically $LiCoO_2$) because of the reactivity of the delithiated (fully charged or overcharged) oxides with the electrolyte solution. This reaction, which may cause a thermal runaway of the cell, normally starts at temperatures well above 200°C where oxygen evolution from the lithium metal oxide occurs. Cycled carbon electrodes have an exothermal effect at temperatures of about 140°C. This effect is much smaller than that observed with the positive electrode but the heat evolved might still bring the cell to the temperature above 200°C. Therefore, the carbon negative electrode could be a potential fuse for the thermal runaway of a lithium-ion battery, and some attention must be paid to its thermal behavior. The heat evolution of graphite negative electrodes at temperatures between 80°C and 220°C is associated with the decomposition of the SEI layer on the graphite negative electrode material.[342] The relative mass of the SEI in a given electrode mass is closely related to the interface area wetted with the electrolyte (which is higher for carbons with higher BET SSA). Therefore, there is less heat evolution from carbons with lower BET SSA.

The quality (above all the flexibility) of the passivation SEI layer on the carbon has also a significant influence on the lifetime (during both cycling and storage) of the entire lithium-ion cell. There is a rather reversible expansion/contraction of the carbon electrode during the electrochemical cycling leading to pronounced mechanical fatigue of the carbon electrode upon prolonged cycling.[457,505,553–555] The carbon particles can therefore break up (and become possibly noncontacted, the consequence of which is cell capacity fading) and new surface is thus created. Moreover, during cycling of a carbon electrode, the SEI periodically swells and shrinks (as proven by *in situ* atomic force microscopy (AFM) experiments).[556] Cracks in the SEI layer can thus be created by both processes. The consequence is that a new SEI layer is formed on continuously created "defect" areas, which is a process consuming both lithium and charge. Therefore, during the lifetime of the cell, the average thickness of the SEI layer increases (the consequence of which is cell capacity fading) and the interfacial resistance increases (the consequence of which is power fading). In addition to the increasing interfacial resistance, an SEI layer increasing in thickness during the cycling of the cell could diminish the electrolyte penetration into the electrode pores or even clog the pores. A decreasing electrolyte penetration into the porous electrode decreases the apparent migration and diffusion rates of lithium ions in the electrode, i.e., increases the apparent ionic resistivity of the electrode. The resistivity increase during cycling of the cell decreases the usable charge capacity of the electrode and therefore the lifetime of the cell.[557–561]

7.5 ELECTRODE ENGINEERING

7.5.1 THEORETICAL ASPECTS

The intrinsic maximum electrochemical performance of an electroactive carbon is not always achieved in industrial electrodes as limitations of the electrodes can occur.[41] For example, model microelectrodes consisting of a single MCMB carbon particle demonstrated high charge capacity retention during cycling and extreme high discharge rates up to 600 C.[562–567] In fact, both electrodes of commercial lithium-ion cells belong to the category of porous electrodes being an assembly of a multitude of particles of the electrode material. Porous electrodes find wide application in applied electrochemistry and have been therefore subject of intense research in the past. The reason for the utilization of porous electrodes is to increase the surface area in contact with the electrolyte. Thus, although the apparent current density pertaining to the geometric surface area of the porous electrode is high, the local current density (pertaining to the interface area of the porous electrode in contact with the electrolyte) is comparatively low. This enables electrochemical reactions to proceed at sufficiently low polarization, i.e., the sum of all overpotentials at the electrode is low.

The structure and composition of porous electrodes must ensure a sufficient electrical conductivity, both in the solid and liquid phases, and at the same time minimum overpotentials for the electrochemical reaction(s). Electrochemical reactions like the intercalation and de-intercalation of lithium into/from carbon consume or generate, respectively, electroactive species (lithium ions) in the electrolyte solution and, thus, generate concentration gradients that are compensated by mass transport in the solution. Therefore, the conditions in a porous electrode can be described with the theory of transport of charge and mass in porous systems. Two mechanisms are relevant, namely diffusion and migration, and the mass transport in porous electrodes is the main object of theoretical calculations and numerous experiments.[444,568-576] Needless to say that there are also economic considerations such as long life and acceptable costs, so the development of porous electrodes is by no means simple.[42,577-579]

Good electronic conductivity of the solid phase is achieved by the use of conductive additives (cf. Chapter 3). Good electrical conductivity of the liquid phase (which means its good ionic conductivity) can be achieved by a proper choice of the electrolyte solution. As far as the choice of the solvent is considered, there are almost no effective degrees of freedom because, apart from the cost issue, the choice is normally based on the inherent necessity to consider the formation of an effective SEI passivation layer, to pay attention to the temperate window in which the battery will be operated, and to the safety of the battery. Generally, only the concentration of the electrolyte salt (in industrial lithium-ion batteries normally $LiPF_6$) can be varied to come close to the maximum of the ionic conductivity which is typically about 1 mol of the salt per liter depending on the expected average operating temperature of the battery.

The most serious challenges arise from the condition of minimum overpotential for the electrode processes. Needless to say that, depending on the application of the electrode (high power or high energy systems), the conductivity of both the solid phase and the electrolyte, the porosity of the electrode, and the pore tortuosity must be optimized to reach maximum performance. Only the most important rules for electrode optimization will be mentioned here: (1) for a given power requirement, the porosity of the electrode should be as low as possible in order to maximize the energy density of the battery; of course, the risk of lithium depletion in the pores (with the consequence of gravely increased concentration overpotential) must be carefully considered; (2) the tortuosity of the pores should be minimized to decrease the ohmic potential drop in the electrolyte and to enhance the mass transport in the pores; and (3) the electronic conductivity of the solid phase and the ionic conductivity of the liquid phase should be well balanced to achieve nearly uniform utilization of the entire electrode mass. Provided that the apparent ionic conductivity of the porous electrode is much higher than its electronic conductivity, the part of the electrode close to the current collector will work faster. If the electronic conductivity of the porous electrode is much higher than its apparent ionic conductivity, the part of the electrode close to the electrolyte-soaked separator will work faster. Both cases are suboptimal from the point of view of electrode performance.

7.5.2 Practical Aspects

Basic performance characteristics of industrial carbon electrodes are the maximum (low current) charge capacity per unit of volume and mass, respectively, the coulombic efficiency of both the first charge/discharge cycle and during the subsequent cycling, the practically achievable charge capacity at high charge and discharge rates, the stability of the charge capacity during cycling and with time (called capacity fading), and the resistance (correctly speaking, overpotential) increase during cycling and with time. The charge "losses" and the resistance increase during the lifetime of the electrode (which implicitly means the number of electrochemical cycles, their depth, the influence of possible temperature excursions, the calendar age of the electrode, etc.) will influence the energy density, the power density, and the cycle life at both the cell level and the battery module level. A satisfying electrode performance can be obtained by optimizing the electrode parameters like electronic and ionic resistivity, electrode thickness and density (i.e., its porosity), material loading, as well as the orientation of anisometric graphitic particles in the electrode.[245,580,581] In many cases, limitations from the (not optimal) engineering of carbon electrodes have pretended limitations of the carbon electroactive material.

Good contacts between the carbon particles are the prerequisite for a low electronic resistivity of the electrode, high reversible specific charge, and an optimal formation of the SEI passivation layer. This can be achieved by designing an electrode with a comparatively high density and the use of conductive additive(s).[580] The (apparent) ionic resistivity of the electrode is influenced by the conductivity of the electrolyte solution in the pores of the electrode as well as by the overall transport rate of lithium ions (due to diffusion and migration) in the pores. The driving forces for the latter are the concentration gradients built up during the insertion of lithium ions into the carbon electrode and the local electrical field, both depending on the size and shape of the electrode pores.[44,45] Note that the lithium transport in the solid state (i.e., in the graphite) is neglected in this simple consideration.

From the parameters influencing the electronic and ionic conductivity of the electrode, the shape of the electrode particles plays an important role: hardened spherical graphite particles result in higher current rate performance than soft flaky graphite particles.[582–584] Pore forming additives may help to tailor the electrode porosity.[577] The wetting with the electrolyte solution of the pores in the graphite electrode can be improved by special electrolyte additives or by the use of optimized carbonate electrolyte formulations with low viscosity and surface tensions.[580,585] The right choice of the electrolyte formulation for the individual graphite surface optimizes the SEI formation and, thus, the coulombic efficiency, cycling stability, and performance at high currents.[477,586,587] The thickness of the electrode determines the maximum particle size of the electrode material. The electrode thickness significantly influences its high current performance because the electrode particles in thin electrodes have an ideal wetting by the electrolyte. In addition, the smaller diffusion lengths in smaller particles may improve the electrode polarization behavior at high current rates, both for charge and discharge.[44,497,580,588,589] Thin electrodes are especially suitable for high power cells, whereas the electrode thickness is commonly maximized in energy-oriented cell designs—thick electrodes increase the ratio electrode vs. current collector volume and, thus, the amount of the active material in the given cell volume.

The choice of the binder influences the electrode performance too. Sufficient mechanical electrode stability has to be reached with a minimized binder amount since the insulating polymer binder increases the electrode resistivity and at the same time reduces the specific charge of the entire electrode.[586] An insulating polymer binder system may form films on the carbon electrode and reduce the charge losses by partially covering the electrochemically active carbon surface.[590] Film forming binder polymers may also affect the reversibility of the negative electrode by protecting the filmed particles from solvent co-intercalation.[591] Polymeric binders can be involved in an electrochemical reaction at negative potentials; perhalogenated binder materials known to be reduced by lithium amalgams[592] are also reactive with respect to metallic lithium and lithiated carbons. This is the reason why, in the case of polytetrafluoroethylene, the irreversible charge capacity increases with the increasing binder content in the negative electrode.[593,594] Of course, the reaction rate in practical electrodes of lithium-ion batteries is much slower compared to the well-investigated case of lithium amalgam. In contrast, the PVDF binder is much less reactive at the very negative potentials and, thus, shows no significant variation of the irreversible charge capacity with the increasing binder concentration.[595] The swelling of the binder material by the electrolyte solvents improves the electrolyte retention of the electrode but at the same time increases the electrode thickness and, thus, decreases the energy density of the cell.[596]

Very often a compromise has to be found between the achievable energy density and the power density of the cell. The energy density of the cell can be maximized by a high-electrode loading and density. This might be achieved by highly compressed electrode materials. However, the increase in the electrode density means a decrease of the porosity and, thus, it lowers the electrolyte retention. The decreased overall ionic conductivity deteriorates the high rate performance of the electrode and therefore the power density at the cell level.

7.6 COMMERCIAL NEGATIVE ELECTRODE MATERIALS

7.6.1 Hard Carbons

Being used as negative electrode material in Sony's first generations of commercial cylindrical lithium-ion cells at the beginning of the 1990s,[597] hard carbons in the meantime completely have disappeared from energy-oriented lithium-ion cells and have been substituted by graphitic carbons. The reason has been the growing demand for batteries with higher energy density. Recently, however, hard carbons have attracted attention for their use in large-scale, power-oriented lithium-ion batteries, especially those lithium-ion batteries which could be applied in mild hybrid vehicles (HEV). Mild HEV require batteries with comparatively low energy densities (but still sufficient for the typically shallow depths of discharge of only 5%–10%).[18,598–605] The advantages of hard carbons to graphitic materials as active material in this battery application are (1) the sloping discharge curve allowing an easy monitoring of the cell state-of-charge during charge and discharge; (2) the high stability of hard carbons in PC-based electrolyte systems (which, below −20°C, exhibit better low-temperature performance than EC-based systems, as required for automotive applications); and (3) a higher solid state lithium diffusion coefficient due to the amorphous structure and isotropic texture of the hard carbons allowing, thus, faster lithium acceptance.[606] In contrast, power-oriented lithium-ion battery systems requiring higher energy densities like power tool batteries and batteries for strong hybrid electric vehicles, plug-in hybrid electric vehicles, and vehicles with pure electric drive drain do not allow for the use of hard carbons but require graphitic carbon based negative electrodes.

Although some hard carbons exhibit faradaic capacities significantly above the theoretical charge capacity of graphite, the specific charge of hard carbons is limited in practical applications. The main disadvantage of hard carbons is the large charge "loss" during the first reduction which is generally attributed to surface reactions, to the passivation layer (SEI) formation at the carbon surface and in the micropores, as well as to the trapping of some lithium in the carbon network structure.[213,346] In addition, a limited charge capacity retention on cycling was reported when the whole stoichiometric range was used for the electrochemical lithium insertion.[346,607] To avoid a large hysteresis between the charge and discharge cell voltage, hard carbons heat-treated above 1100°C with low H/C-ratios are used.[198,608] It was claimed that posttreatments and/or carbon coatings of the hard carbon surface could further improve the electrochemical performance.[198,207,208,213,609–612] However, treatments above 1000°C significantly have decreased the available reversible charge capacity well below the values obtained for the graphite. In addition, a significant charge capacity is available negative to 0.1 V vs. Li/Li$^+$ in the flat part of the galvanostatic curve (cf. Figure 7.10). Unfortunately, this charge capacity cannot be utilized in an industrial battery due to the following reasons: (1) there is a serious safety risk due to easy lithium plating during high current charging at low temperatures (cold cranking); (2) the reversibility of the lithium insertion in this potential range is limited due to the partial trapping of lithium in the carbon structure which leads to "charge losses" during cycling and, thus, to a reduced cycling stability; (3) the flat shape of the galvanostatic curve makes it difficult to control the cut-off voltage for an overcharge protection of the cell; and (4) the kinetically hindered acceptance of lithium in this low-voltage region results in a reduced high current charging performance and a limited cycling stability at high currents.[213,346]

Commercial hard carbons like Carbotron P from Kureha Chemical Co. being used in Sony's lithium-ion batteries[204,597] are produced by pyrolysis of petroleum pitch which includes several steps: (1) cross-linking of the basic structural units of the pitch by oxygen functional group introduced by an oxidation treatment, (2) calcination of the treated pitch, (3) grinding of the calcined pitch to the required particle size, and (4) final heat treatment. Other preparation methods of hard carbons include pyrolysis of cross-linked phenolic resins or polymers like PVC, PAN, or PFA.[108,113,115,116,197,613,614] Doping with phosphorus atoms allowed to increase the reversible charge capacity.[110,111] Recently, a graphite–hard carbon hybrid material has been proposed by Sanyo as promising negative electrode material.[615] From Figure 7.10 (Chapter 4.1) showing the first galvanostatic lithium insertion/de-insertion cycles of a typical pitch-based hard carbon, the kinetic limitation of the lithium insertion process in the potential range negative to 0.1 V vs. Li/Li$^+$ becomes obvious. The latter could be a possible reason for the slightly reduced cycling stability of the cells at high currents if the whole achievable voltage range is utilized.

Table 7.4 compares typical properties of a commercial pitch-derived hard carbon product with some other types of commercial negative electrode materials, namely graphitized mesocarbon, coated natural graphite, and synthetic graphite. The typical microporosity of hard carbon is indicated by the relatively low real density (xylene density, which also indicates the amorphous state of the material) and much more by the significant volume of adsorbed nitrogen at low partial pressures as it is clear from the gas adsorption isotherms measured at 77 K, as illustrated in Figure 7.18. The morphology of the materials presented in Table 7.4 is illustrated by SEM pictures shown in Figure 7.19.

TABLE 7.4

Comparison of Material Properties and Typical Electrochemical Charge Capacities of Some Commercial Carbonaceous Electroactive Materials (Examples)

Material Property	Hard Carbon (Pitch-Derived)	Graphitized Mesocarbon (MCMB 25-28)	Coated Natural Graphite	Synthetic Graphite (TIMREX SLG5)
Xylene density (g cm^{-3})	1.60	2.10	2.21	2.26
L_c(002) (nm)	1	70	150	150
$c/2$(002) (nm)	0.377	0.338	0.336	0.336
BET SSA (m^2 g^{-1})	4.3	1	1.5	1.5
Average particle size (μm)	9	25	18	22
Bulk density (Scott density) (g cm^{-3})	0.35	0.90	0.83	0.60
Typical reversible charge capacity (mAh g^{-1})	400	335	360	360

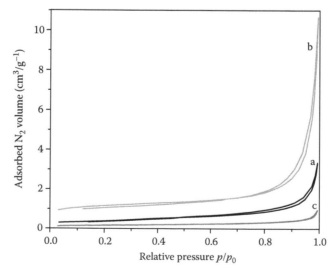

FIGURE 7.18 Nitrogen adsorption isotherms at 77 K of synthetic graphite (a), pitch-derived hard carbon (b), and graphitized MCMB 25-28 (c).

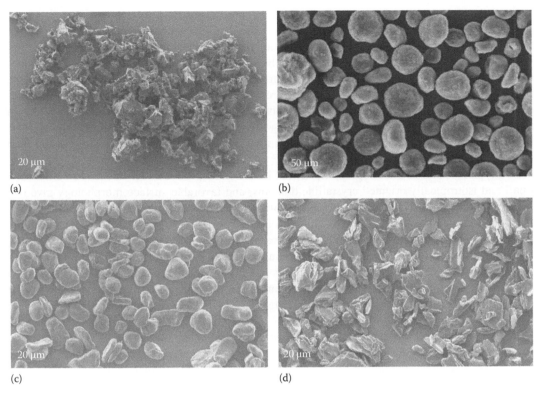

(a)

(b)

(c)

(d)

FIGURE 7.19 Scanning electron microscope pictures of a pitch-based hard carbon (a), graphitized MCMB 25-28 (courtesy of Osaka Gas Chemical Co.) (b), coated natural graphite (c), and synthetic graphite (d).

7.6.2 Graphitized Mesocarbons

The spherically shaped graphitized mesocarbon microbeads (MCMB) 25-28 by Osaka Gas Chemical Co.[157,164,425,616,617] and the chopped graphitized pitch-based mesophase carbon fibers (MCF)[21,162,231,618–621] by Petoca Co. were the first graphitic carbons used in commercial lithium-ion cells as a substitution for hard carbon and coke negative electrode materials. Besides these graphitic materials, other spherical or block-shaped graphitized mesocarbon-based electrode materials have been homologated in lithium-ion batteries. Examples of commercial graphitized mesocarbon products are KMCF by JFE, CMS by Shanghai Shanshan, graphized mesocarbons by Nippon Carbon Co. (e.g., P15B-HG) and Sumitomo Metals Co. At the end of the last decade, graphitized mesocarbons were the mostly used carbonaceous negative electrode materials in lithium-ion batteries representing, thus, a kind of a standard material in lithium-ion battery technology.

Mesophase carbon discovered by Brooks and Taylor in the 1960s is an intermediate isotropic phase between isotropic pitch and anisotropic semicoke generated by thermal reaction of organic pitch components between 400°C and 500°C.[622,623] The mesophase appears as initial stage during the carbonization of aromatic molecules as optically anisotropic spheres in the isotropic organic matrix. The spheres grow in diameter, coalesce with each other, and finally convert together with the carbonizing matrix to a solidified anisotropic semicoke.[624] The carbon mesophase is composed of lamellar macromolecules of different molecular size bonded by the van der Waals forces in a parallel stacking, with a long crystallographic order similar to liquid crystals.[625] By further heat treatment to ca. 1000°C, the stacking of aromatic planes of the mesophase grows inside the carbon planes forming the typical graphitizable carbons.[624]

The spherical MCMB is formed in the mesophase of coal tar during carbonization. The heat treatment is stopped at temperatures at which the mesophase spheres formed in the coal tar pitch have reached the desired particle size. After cooling, the solid carbon spheres are separated from the isotropic matrix by extraction with quinoline or pyridine. Stabilization of the spheres by oxidation in air at elevated temperatures, carbonization, and subsequent graphitization are the process steps to the final product.[624] Spinning of molten, basically thermoplastic mesophase pitch, provides the pitch fibers after resolidification, with structures depending on their flow properties. The shape and molecular orientations in fibers, intimately related to their mechanical properties, are principally determined by the spinning process.[624] Stabilization of the fibers is achieved in an oxidation process in which the fibers become thermosetting in an aromatization and cross-linking process.[626] The mesophase carbon fibers are carbonized, milled, and then graphitized resulting in the final powdered electrode materials.

The typical characteristics of graphitized MCMB and MCF products such as their low BET SSA, small and isotropically oriented crystalline domains, and favorable surface morphology give rise to their low irreversible capacity, good safety characteristics and capacity retention at high current, and a relatively high stability in PC-based electrolytes.[627–635] A stable passivation (SEI) layer at the MCMB and MCF surface, respectively, is the reason for their good compatibility with $LiMn_2O_4$ spinel positive electrodes. In contrast, the use of a graphite negative electrode in combination with positive manganese oxide spinel electrode results in a limited cycling stability of the cells due to a mechanism which includes the dissolution of manganese ions into the electrolyte and subsequent manganese plating on the carbon surface. The latter disturbs the effective passivation of the graphite surface.[526,636]

The high bulk and tap density, as well as the isotropic texture and shape of the mesocarbon particles are the reason for their good charge capacity retention at high currents, the low polymer binder and solvent absorption, and the advantageous pore size distribution of compacted electrodes allowing, thus, easy processing of graphitized mesocarbon based electrodes[637] and especially the application of MCMB type negative electrodes in lithium polymer batteries.[638] The good performance at high current drains is the reason why graphitized fine MCMB 6-28 has been suggested as negative electrode material in lithium-ion batteries for power tools.[616] Electrode resistivity problems due to bad particle contacts can be solved by adding fine conductive graphite additives (cf. Chapter 6.6). Typical reversible charge capacities obtained with graphitized mesophase carbons reach about 320 mAh g⁻¹. Improved graphitization at temperatures above 3000°C and the use of graphitization promoters like boron during the graphitization (incorporated by co-pyrolysis of coal tar pitch and boron compounds) as well as surface oxidation allow to increase the maximum obtainable specific charge capacity to about 340 mAh g⁻¹.[224–227,229,231,235,630,639,640] A further surface modification may establish a favorable surface morphology for a tailored electrolyte compatibility as well as an improved energy density and safety of the lithium-ion cell (cf. Figure 7.20). The high production costs of mesophase carbon based electrode materials are a drawback of mesocarbon negative electrode materials. The reasons are the low yield of the mesophase spheres production as well as environmental and health issues of some polyaromatic mesophase carbon components requiring a strict process control. High productions costs as well as the limitations in reversible charge capacity are the reasons for the technical substitution of graphitized mesophase carbons by graphite materials, which has been occurring recently in the lithium-ion battery technology. Nowadays, the market share of mesophase carbon-based electrode materials in lithium-ion batteries has dropped below the total market share of natural and synthetic graphite-based negative electrode materials.

7.6.3 Coated Natural Graphites

High-crystallinity natural graphite materials are attractive active materials in the negative electrode of lithium-ion batteries due to their high theoretical reversible charge capacity of 372 mAh g⁻¹ as well as the low and flat potential profile below 0.2 V vs. Li/Li⁺, which are important features which are needed to improve the energy density of portable lithium-ion batteries. The drawbacks for the

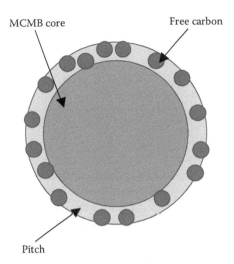

FIGURE 7.20   Schematic of the cross-section of a MCMB particle. (Courtesy of Osaka Gas Chemical Co.)

use of natural graphite in negative electrodes are the relatively high BET SSA leading to high faradaic "losses" during the first electrochemical reduction, high gassing rates during the passivation (SEI formation), limitations in the cell safety and storage, and the high sensitiveness of natural graphite to exfoliation in certain electrolytes, especially based on PC.[25,101,102,641] In addition to the high reactivity of the high crystalline surface, the orientation of the flaky-shaped particles in the electrode causes unfavorable pore size distribution and, thus, limitations of the electrolyte retention which in turn decreases the charge capacity achievable at high current drains and worsens the cycling stability of electrodes utilizing natural graphite.[642,643] Due to the softness of the particles, high compaction pressure during the roll-pressing of the electrodes results in electrodes with low porosity which (in addition to the orientation effect) negatively influences the electrode tortuosity and, thus, both cycling and high current performance.

To decrease the SSA, to decrease the surface reactivity with the electrolyte system, and to overcome the particle anisometry and softness, particles of most commercially available natural graphite electrode materials are spherically shaped by mechanical treatments and are surface-coated with a layer of hard carbon.[395,582,644–647] The core-shell concept of hard carbon coated graphite was introduced by Kuribayashi et al.[648] in 1994. The coating can be achieved by a treatment with pitch, phenol thermosetting resins, or thermosetting polymers like PVC and subsequent heat treatment in an inert gas atmosphere[414,584,643,648–653] or by chemical vapor deposition or thermal vapor deposition of gaseous carbon hydrides at the graphite surface.[582–584,643,645,646,654,655] With increasing amount of hard carbon material on the graphite surface, the real density of the material decreases.[584] In addition, the hard carbon coating increases the contact resistance between the particles and decreases the compressibility of the graphite material.[448] An important aspect is the stability of the hard carbon coating. The hard carbon layer has to be flexible enough to withstand the repeated dimensional changes of the particles during lithium insertion and de-insertion. In addition, the high compaction pressure applied during the electrode manufacturing process to reach high electrode densities can crack the coating and, thus, can increase the irreversible capacity of the electrode. A SEM picture of hard carbon coated spherical graphite particles is shown in Figure 7.19c.

Some examples of commercial products are MPG and ICG by Mitsubishi Chemical Co., GDA by Mitsui Mining Co., OMAC by Osaka Gas Chemical Co., 818 by BTR Energy Materials Co., NG-7 by Kansai Thermochemical Co., and DAG by Sodiff.[656] Hard carbon coated cokes and soft carbon coated hard carbons are other carbons with core-shell structure reported in the literature.[610,657]

7.6.4 SYNTHETIC GRAPHITES

Synthetic graphite active electrode materials are typically graphitized isotropic soft carbons with high degree of graphitization. Typical examples of commercial synthetic graphite products are MAG by Hitachi Chemical Co., TIMREX SLG5 by TIMCAL Ltd., or the product of Showa Denko Co.[658] The graphitization process applied to the manufacturing of these carbons is especially designed to reach large crystallite domains and low porosity in the particle bulk and small surface defect concentrations at the particle surface, which results in a high crystallinity and high real density combined with a low BET SSA. The optimized particle morphology (an example is shown in the SEM picture in Figure 7.19d) facilitates the electrode engineering. The high surface crystallinity of such graphite materials is the reason for a relatively high sensitiveness to exfoliation that requires the use of tailored electrolyte compositions with SEI film forming additives. However, the advantages of highly graphitized materials include their high reversible charge capacity, good compaction behavior and particle stability, a good compatibility with PVDF and SBR/CMC binder systems, and comparatively low manufacturing costs. Variations of the material properties achieved by modifications of the raw materials and/or the manufacturing process allow a versatile use of synthetic graphite materials in both energy and power-oriented lithium-ion battery systems. The use of conductive additives decreases the electrode resistance and improves the high rate cycling performance as well as the uniformity of the SEI formation process on synthetic graphite active materials. As an example, the first electrochemical lithium intercalation/de-intercalation process of TIMREX E-SLG5 is shown in Figure 7.21, demonstrating charge capacity of ca. 360 mAh g^{-1} of graphite and a coulombic efficiency in the first cycle of 95%.

7.6.5 CARBON-BASED HYBRID MATERIALS

Nowadays, industrial lithium-ion batteries almost have reached the energy density limits of the LiCoO$_2$/graphite electrochemical system. To further increase the energy density of lithium-ion batteries, other electrode materials with higher specific charge upon the electrochemical lithium insertion are needed. Dragged by the increasing demand of the OEMs for high energy density battery systems, the research activities for high charge capacity electrode materials have increased exponentially in the last years. A total replacement has been proposed of the graphite negative electrode

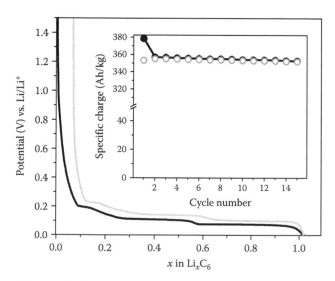

FIGURE 7.21 First galvanostatic intercalation/de-intercalation of lithium into/from TIMREX E-SLG5 graphite in a 1 M LiPF$_6$, EC:EMC (1:3) electrolyte containing VC as a SEI forming additive. The inset shows the charge capacity of the graphite (●: reduction, ○: oxidation). The counter-electrode was metallic lithium.

material by either silicon or metallic tin or alloys. With respect to their lithiation, all these materials theoretically can achieve considerably higher reversible charge capacities. However, these materials generally suffer from large dimensional changes during the lithium insertion process with the consequence of crumbling of the electrode. This crumbling is the reason for the limited cycling stabilities and high charge "losses," both unsolved challenges for the use of such materials in commercial lithium-ion batteries.

A promising intermediate step could be a partial replacement of the carbon by finely dispersed nanosized metals or alloys of tin, silicon, or other elements. Numerous publications on this topic can be found in the contemporary scientific and technical literature. The following includes a few examples. To increase both the reversible charge capacity and the electronic conductivity of the negative electrode material for power batteries, a silver coating of graphite (or other carbons) by aqueous co-precipitation was suggested by Hitachi in the frame of the *LIBES* program.[659–662] A dispersion of nanosized silicon in the carbon has been achieved by mechanical co-milling or mechanical alloying of graphite and other carbons with a silicon metal or an alloy component[663–672] or by pyrolysis of carbon precursors in the presence of silicon particles.[673] Other groups suggest simple mixing or impregnation processes to incorporate into carbon nanosized silicon particles which are produced by laser or chemical vapor decomposition.[505,674,675] Silicon-coated carbon materials have been prepared by chemical vapor deposition, laser-induced vapor deposition, thermal vapor decomposition, a gas suspension spray method, and spray pyrolysis.[246,676–684] To decrease the irreversible capacity of these materials, they could be coated by a layer of hard carbon at the particle surface.[685] Typically, the increase of the reversible charge capacity depends on the amount of silicon (or other elements) in the carbon based composite. However, with the increasing amount of silicon, the specific charge "loss" increases and the cycling stability of the electrode decreases.[420,686] The deterioration of the contacts of the silicon particles and the trapping of lithium in the silicon structure (due to irreversibilities of the lithium–silicon alloying process at some composition ranges of the Li_xSi phase diagram) have been suggested as an explanation in addition to a problematic SEI film formation at the surface of the alloy.[670,679] The SEI formation process and, thus, the cycling stability of the electrode could be significantly improved by the use of VC as a film forming additive.[674] The performance of the silicon/carbon composite electrodes can be influenced by the electrode engineering and the binder selection.[590,687] To maintain the dimensional integrity of the hybrid materials, highly flexible, conductive binders were proposed.[688]

Recently, Morita and Takami[689] proposed a nanosilicon cluster-SiO_x–C composite negative electrode material including nanosize silicon particles which were prepared by disproportionation of silicon monoxide in a polymerized furfuryl alcohol based carbon. TEM analysis showed that Si clusters in the range of 2–10 nm were distributed homogeneously within a silicon oxide matrix. The nanosilicon composite anode had a large capacity of ca. 700 mAh g⁻¹ and a long cycle life of more than 200 cycles. The improvement of cyclability was explained by the nanosize silicon particles, and their uniform dispersion within the silicon oxide phase retained by the carbon matrix, which could effectively suppress the pulverizing of Si particles by the volume change during lithium insertion and extraction.

The first commercial lithium-ion cell containing a carbon based hybrid material is the Sony Nexelion cell which was introduced in 2006.[30] Nexelion cells contain a graphite/cobalt-doped amorphous tin hybrid electrode. Batteries of this cell type are used in video cameras which require high energy density but can accept the lower cycling stability of the Nexelion batteries compared to conventional lithium-ion battery systems.

7.6.6 Conductive Carbon Components in the Negative Electrode Mass

At the first glance, the use of conductive additives does not seem to be necessary for negative carbon electrodes due to their significant intrinsic electronic conductivity (especially of the crystalline carbon active materials). However, in most of the commercial carbon negative electrodes, conductivity enhancers are used to control the electronic and thermal conductivity. The application at the surface-sputtered metal films significantly improved the electrochemical performance of graphitized carbon fibers and other graphitic materials.[690–699] In the literature, the use of carbon black, graphite, and carbon fibers can be found frequently in combination with commercial electrode materials. Conductive carbon blacks like acetylene black, SUPER P Li, and VGCF are used to improve the interparticle contacts in all types of active electrode materials.[29,68,700] The homogeneous dispersion of carbon black in the electrode is a prerequisite for a good electrode performance.[48] The quantity of the additives used in the electrode is limited to small amounts below 1%–2% due to their relatively high BET SSA which has an impact on the specific charge "losses" of the negative electrode. High crystallinity graphite materials like TIMREX SFG6 or SFG15 are used as minor electrode components in electrodes mainly based on graphitized mesocarbons like MCMB or MCF.[162,618,621] Usually, the amount of graphite in these electrodes can reach more than 10 wt%.

The use of conductive additives in combination with typical commercial carbonaceous electrode materials exhibiting low BET SSA (like graphitized mesocarbons, hard carbon-coated graphite materials, and synthetic graphite materials) is required. This is because these electroactive materials show, due to a low defect concentration at the surface, relatively high contact resistances between the particles. The function of the carbon black conductive additives in the negative electrode resembles those in the positive electrode. In contrast to carbon black, graphite additives in the negative electrode have a double function—they act as conductivity enhancers and at the same time as electrochemically active electrode materials. Due to their high crystallinity (xylene densities of $2.260\,g\,cm^{-3}$ and higher), they provide a high reversible charge capacity for lithium intercalation: Typical graphite additives like the TIMREX SFG15 reach their theoretical charge capacity at low current drains enhancing, thus, the overall reversible charge capacity of the graphitized mesocarbon-based electrode. The positive effect of a graphite conductive additive on the electrode impedance, cycling stability, and rate capability of a mesocarbon-based negative electrode has been reported.[566,701] As shown in Figure 7.22, the capacity fading of a half cell containing an MCMB negative electrode could be significantly improved by adding 10% of SFG6. The graphite SFG6 improves the impedance of the electrode, improving, thus, the capacity retention of the cell especially at high currents.

In addition to the beneficial effect on the electrochemical performance, the graphite conductive additives influence the electrode density positively. Typically, the graphitized mesocarbon or

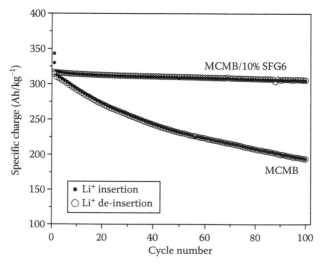

FIGURE 7.22 Capacity fading of a lithium-ion cell with a MCMB electrode with and without the addition of 10% graphite SFG6. Electrolyt 1 M $LiPF_6$, EC:EMC (1:3).

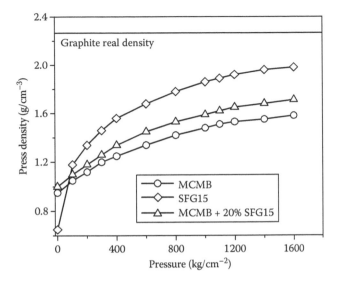

FIGURE 7.23 Electrode density as a function of the compaction pressure of a pure MCMB electrode, a pure TIMREX SFG15 electrode, and an electrode containing MCMB blended with 20% graphite SFG15.

surface-treated graphite electrodes show a relatively large spring back after pressure release being the reason for their relatively low compressibility. In addition, high compaction pressure during the electrode manufacturing process may lead to particle breakage.[702] In contrast, a conductive graphite powder with high crystallinity shows high compressibility. Adding conductive graphite to the meso-carbon electrode mass can improve the electrode density as shown in Figure 7.23.

The positive effect of TIMREX SFG15 and SFG6 on the electrochemical performance of the negative carbon electrodes was explained by the high bulk crystallinity and a high ASA indicating a high surface defect concentration which could favorably influence the contact resistance between the carbon particles.[266] These properties may be the reason for the high capacity retention at high discharge rates of up to 30C and high lithium acceptance rates which were reported for pure SFG electrodes.[588]

The use of fine high crystallinity graphites as additives to the negative electrode mass based on other electroactive materials is limited due to the increased BET SSA which may lead, if the electrolyte is not optimized in the appropriate way, to relatively high charge "losses" in the first cycle combined with gassing due to the electrolyte decomposition. There are also possible safety issues related to the heating of the cell. The comparatively high polymer binder absorption of these graphite additives may cause delamination of the negative electrode from the current collector foil when the binder content is low.

7.7 REMARKS TO THE TECHNOLOGICAL ASPECTS OF COMPLETE CELLS

The aim of this chapter is to call the attention of the reader to a few technology-related issues which are implicit to any kind of complete electrochemical cells. In the following, the case of the carbon electrode in a complete lithium-ion cell will be emphasized. Note that under a "cell" a single cell is understood; a "battery" is strictly speaking an assembly of two or more cells.

First of all, it is important to pay attention to the fact that most of the available literature data describes the behavior of a carbon working electrode against a laboratory counter- electrode which is normally metallic lithium. This counter-electrode is normally oversized and acts therefore as a nearly infinite source (and sink) of lithium ions. If the coulombic efficiency during cycling of the carbon electrode is less than 100% (e.g., due to side reactions, trapping of lithium in the carbon, growth of the SEI, etc.), the so-caused consumption of lithium is masked in half cell experiments. Therefore, the cycling behavior of a carbon electrode in a half cell will appear better that it would be in a complete cell where the only source of lithium is the electroactive oxide in the counter-electrode (cf. Chapter 2.0).

The second point is to raise the awareness of the trivial relationship between the potentials of the carbon and oxide electrode, respectively, and the voltage of the complete cell. Note that the cell voltage, U_{cell}, is the measurable difference of the potential of the positive electrode, E_{oxide}, and the potential of the negative electrode, E_{carbon}:

$$U_{cell} = E_{oxide} - E_{carbon} \qquad (7.7)$$

If a complete cell is charged to, e.g., 4.1 V, then the potential E_{carbon} of the fully lithiated negative electrode will be about 0.1 V vs. Li/Li$^+$. Therefore, the potential E_{oxide} of the fully charged positive electrode in this example will be 4.2 V vs. Li/Li$^+$. Needless to say that this trivial relationship must be remembered when data for half cells (vs. metallic lithium) are compared to the data for complete cells. An important consequence of this trivial relationship is the potential excursion of the counter-electrode in the case of an anomalous behavior of the carbon electrode (and vice versa). Imagine that in the previous example the potential of the carbon would shift to 0.3 V vs. Li/Li$^+$ due to a malfunction of the carbon electrode. If the end-of-charge voltage of the complete cell would be the same, namely 4.1 V, then the potential of the positive electrode would be 4.4 V vs. Li/Li$^+$. In such a case, the safety of the entire cell could be compromised.

The next point is the balancing of complete cells.[703] Obviously, the highest energy density of a complete cell will be achieved when the charge capacity of both electrodes will match. (Note that in energy-optimized industrial batteries the carbon electrode is moderately oversized for safety reasons, to avoid plating of metallic lithium.) But, the crucial point is that the charge capacity of both electrodes must match *after* the SEI is formed on the carbon electrode.[704] The consumption of lithium during the formation of the SEI layer on the carbon (the so called "irreversible capacity") must therefore be carefully taken into account when calculating the optimal amounts of carbon and oxide in the negative and positive electrodes, respectively. Because the charge consumption during the SEI formation depends on many factors including the carbon type, the carbon's BET SSA, and the electrolyte composition (along with additives, if any), it is clear that the optimal balancing of the electrodes will be different for any particular case.

The final point is the dynamics of the lithium intercalation process. Provided that the engineering of a carbon composite electrode is optimal, there is a natural limit of the electrode's dynamics which is related to the magnitude of the lithium diffusion coefficient, D_{Li}, in the carbon. For example, with an apparent diffusion coefficient of 10^{-9} cm^2 s^{-1}, it can be calculated from $dx = (2D_{Li}t)^{1/2}$ that the lithium ions diffuse about 27 μm within 1 h.[705] For a 1-h discharge, rate the grain size therefore should be of the same order of magnitude. Clearly, the dynamics of both negative and positive electrode should be comparable to achieve the maximum possible performance of the complete cell.

7.8 OUTLOOK

The demand for increased energy and power density of batteries for all kinds of mobile applications will certainly not cease in the future. It is obvious that the industrial development lines already split into two, high energy and high power batteries. The former development line maximizes the energy density of the cell even at the cost of the cell lifetime which is expected to be a few years only, which seems to be acceptable for consumer electronics like mobile phones, cameras, personal computers, etc. The latter development line focuses on high power (both during charge and discharge), very high reliability, wide temperature window, and long service time as required for automotive applications where HEV are the most prominent example. Of course, the highest achievable safety of the battery must be of utmost importance for the battery developers, especially in the light of the economic demand to reduce the material and productions cost. Naturally, the increasing public awareness of environmental issues imposes on the battery manufacturers the restriction to use environmentally benign materials and processes. In this regard, carbon materials have their advantage. New improved carbon electrode materials in energy-oriented lithium-ion batteries will show high charge capacities, may be above the theoretical charge capacity of graphite, and will allow high compaction for high carbon electrode densities. Major challenges for the development of carbon electrode materials in high power battery applications will be to increase the crystallinity and, thus, the lithium insertion capacity without loosing the compatibility with electrolyte systems covering broad operation temperature windows (like PC-based electrolytes) as well as to further increase the lithium acceptance rate for a fast charging process.

The lithium-ion battery based on $LiCoO_2$ and graphite has become an industrial standard which certainly cannot be easily substituted by higher performing systems in electronic applications. The worldwide trend to buy at the lowest cost will affect the battery manufacturer's selection of the carbons but the battery safety must not be sacrificed. The selection of materials and battery components with further decreasing costs will allow a further decrease of the costs per energy unit for the lithium-ion battery. This cost decay could finally lead in some applications to an "all purpose lithium-ion battery" complementing or even substituting current lead acid battery systems and primary batteries. This substitution effect of primary batteries by secondary lithium-ion batteries becomes already visible in the consumer electronics, e.g., in digital still cameras.

Apart from the low cost demand which goes to another direction, the never ending demand for higher energy density could be the driving force for the substitution of the electroactive carbon by metals or alloys in the energy density oriented battery market segments in the future. Moth probably, this substitution will be a gradual one, examples include attempts to use carbon-based composite electrodes with alloys, nano silicon, etc. In a longer term perspective nanostructured carbons (prepared, e.g., by template synthesis), carbon nanotubes, and carbon nano fibers could be utilized in industrial products especially when the high power demand will be the decisive factor.

ABBREVIATIONS

AFM	atomic force microscopy
ASA	active surface area
BET SSA	Brunauer-Emmett-Teller specific surface area
CMC	carboxy methyl cellulose
CVD	chemical vapor deposition
DBPA	dibutyl phthalate absorption
DMC	dimethyl carbonate
EC	ethylene carbonate
EMC	ethyl methyl carbonate
FTIR-ATR	Fourier transform infrared spectroscopy-attenuated total reflection method
GIC	graphite intercalation compound
HEV	hybrid electrical vehicles
MCF	mesophase carbon fibers
MCMB	mesocarbon microbeads
NHE	normal hydrogen electrode
NMR	nuclear magnetic resonance
OAN	oil absorption number
OEM	original equipment manufacturer
PAN	polyacrylonitrile
PC	propylene carbonate

PFA   polyfurfuryl alcohol
PVC   polyvinyl chloride
PVDF   polyvinylidene difluoride
SAX   small angle x-ray scattering
SBR   styrene butadiene rubber
SEI   solid electrolyte interphase
SEM   scanning electron microscopy
SIMS   secondary ion mass spectrometry
TEM   transmission electron microscopy
VC   vinylene carbonate
VGCF   vapor-grown carbon fibers
XPS   x-ray photoelectron spectroscopy
XRD   x-ray diffraction

REFERENCES

1. T. Nagaura and K. Tozawa, *Progress in Batteries and Solar Cells* **1990**, *9*, 209.
2. K. M. Abraham, *Electrochimical Acta* **1993**, *38*, 1233.
3. Y. Matsuda, *Nihon Kagaku Kaishi* **1989**, 110, 1.
4. E. Peled, *Journal of the Electrochemical Society* **1979**, *126*, 2047.
5. M. Armand, in J. Broadhead and B. C. H. Steele (eds), *Materials for Advanced Batteries*, Plenum Press, New York, **1980**, pp. 145–150.
6. Y. Nishi, H. Azuma, and A. Omaru, EP, Sony Corp., Japan, EP89-115940, **1990**, p. 9.
7. Y. Nishi, *Chemical Record* **2001**, *1*, 406.
8. Y. Nishi, *Journal of Power Sources* **2001**, *100*, 101.
9. Y. Nishi, *Macromolecular Symposia* **2000**, *156*, 187.
10. Y. Nishi, in M. Wakihara and O. Yamamoto (eds), *Lithium Ion Batteries*, Wiley-VCH, New York, **1998**, p. 181.
11. S. Megahead and B. Scrosati, *Journal of Power Sources* **1994**, *51*, 79.
12. J. R. Dahn, A. K. Sleigh, H. Shi, B. M. Way, W. J. Weydanz, J. N. Reimers, Q. Zhong, and U. v. Sacken, in G. Pistoia (ed.), *Lithium Batteries–New Materials, Developments, and Perspectives*, Vol. 5, Industrial Chemistry Library, Elsevier, Amsterdam, London, New York, Tokyo, **1994**, pp. 1–47.
13. M. Broussely, P. Biensan, and B. Simon, *Electrochimica Acta* **1999**, *45*, 3.
14. U. von Sacken, E. Nodwell, A. Sundher, and J. R. Dahn, *Solid State Ionics* **1994**, *69*, 284.
15. R. Yazami and P. Touzain, *Journal of Power Sources* **1983**, *9*, 365.
16. R. Yazami and D. Guerard, *Journal of Power Sources* **1993**, *43*, 39.
17. H. Takeshita, *The 24th International Battery Seminar and Exhibit*, Fort Lauderdale, FL, **2007**.
18. Y. Nishi, K. Katayama, J. Shigetomi, and H. Horie, *13th Annual Battery Conference on Applications and Advances*, Jan. 13–16, 1998, Long Beach, CA, **1998**, p. 31.
19. M. Broussely and G. Archdale, *Journal of Power Sources* **2004**, *136*, 386.
20. M. Broussely, *Lithium Batteries: Science and Technology* **2004**, 645.
21. M. Endo, C. Kim, K. Nishimura, T. Fujino, and K. Miyashita, *Carbon* **2000**, *38*, 183.
22. N. Imanishi, Y. Takeda, and O. Yamamoto, *Lithium Ion Batteries* **1998**, 98.
23. R. J. Brodd, *Proceedings—Electrochemical Society* **2001**, *2000–36*, 14.
24. K. Abe, H. Yoshitake, T. Kitakura, T. Hattori, H. Wang, and M. Yoshio, *Electrochimica Acta* **2004**, *49*, 4613.
25. R. Fong, U. Von Sacken, and J. R. Dahn, *Journal of the Electrochemical Society* **1990**, *137*, 2009.
26. H. Ota, Y. Sakata, A. Inoue, and S. Yamaguchi, *Journal of the Electrochemical Society* **2004**, *151*, A1659.
27. K. Abe, T. Takaya, H. Yoshitake, Y. Ushigoe, M. Yoshio, and H. Wang, *Electrochemical and Solid-State Letters* **2004**, *7*, A462.
28. K. Abe, H. Yoshitake, T. Tsubura, H. Nakamura, and M. Yoshio, *Electrochemistry (Tokyo, Japan)* **2004**, *72*, 487.
29. T. Takamura, M. Saito, A. Shimokawa, C. Nakahara, K. Sekine, S. Maeno, and N. Kibayashi, *Journal of Power Sources* **2000**, *90*, 45.

30. H. Inoue, *International Meeting on Lithium Batteries 2006*, Abstract No. 228, Claude Delmas, Biarritz, France, **2006**.
31. M. Nishizawa, H. Koshika, T. Itoh, M. Mohamedi, T. Abe, and I. Uchida, *Electrochemistry Communications* **1999**, *1*, 375.
32. K. Tachibana, T. Nishina, T. Endo, and K. Matsuki, *Progress in Batteries and Battery Materials* **1998**, *17*, 242.
33. K. Tachibana, T. Nishina, K. Matsuki, and A. Kozawa, *ITE Battery Letters* **1999**, *1*, 33.
34. S. Mandal, J. M. Amarilla, J. Ibanez, and J. M. Rojo, *Journal of the Electrochemical Society* **2001**, *148*, A24.
35. F. Mizuno, A. Hayashi, K. Tadanaga, and M. Tatsumisago, *Journal of the Electrochemical Society* **2005**, *152*, A1499.
36. K.-J. Euler, *Elektrotechnische Zeitschrift. Ausg. B* **1976**, *28*, 45.
37. S. R. Broadbent and J. M. Hammersley, *Proceedings of the Cambridge Philosophical Society, Mat. Phys. Sci.* **1957**, *53*, 629.
38. K.-J. Euler, R. Kirchhof, and H. Metzendorf, *Journal of Power Sources* **1980**, *5*, 255.
39. D. A. G. Bruggemann, *Annals of Physics (Leipzig)* **1935**, *24*, 636.
40. D. Adler, L. P. Flora, and S. D. Senturla, *Solid State Communication* **1973**, *12*, 9.
41. M. Broussely, J. P. Planchat, G. Rigobert, D. Virey, and G. Sarre, *Journal of Power Sources* **1997**, *68*, 8.
42. I. Roušar, K. Micka, and A. Kimla, *Electrochemical Engineering*, Vol. 2, Academia, Prague, **1986**.
43. L. G. Austin and H. Lerner, *Electrochimica. Acta* **1964**, *9*, 1469.
44. P. Arora, M. Doyle, and R. E. White, *Journal of the Electrochemical Society* **1999**, *146*, 3543.
45. K. Sawai and T. Ohzuku, *Journal of the Electrochemical Society* **2003**, *150*, A674.
46. L.-Z. Wang, H.-F. Wang, S.-H. Gu, H.-Z. Ren, J.-C. Li, R.-F. Li, Z.-P. Yuan, and Y. Yuan, *Dianchi Gongye* **2005**, *10*, 262.
47. S. E. Cheon, C. W. Kwon, D. B. Kim, S. J. Hong, H. T. Kim, and S. W. Kim, *Electrochimica Acta* **2000**, *46*, 599.
48. K. Endo, J. Suzuki, K. Sekine, and T. Takamura, *Electrochemistry (Tokyo, Japan)* **2006**, *74*, 374.
49. D. Guyomard and J. M. Tarascon, *Proceedings—Electrochemical Society* **1993**, *93-24*, 75.
50. D. Guyomard and J. M. Tarascon, *Journal of the Electrochemical Society* **1993**, *140*, 3071.
51. J. M. Tarascon and D. Guyomard, *Solid State Ionics* **1994**, *69*, 293.
52. H. S. Kim, B. W. Cho, J. T. Kim, K. S. Yun, and H. S. Chun, *Journal of Power Sources* **1996**, *62*, 21.
53. K. Kezuka, T. Hatazawa, and K. Nakajima, *Journal of Power Sources* **2001**, *97–98*, 755.
54. Y. Nishi, in W. A. van Schalkwijik and B. Scrosati (eds), *Advances in Lithium-Ion Batteries*, Kluwer Academic/Plenum Publishers, New York, **2002**, p. 233.
55. R. E. Franklin, *Acta Crystallographica* **1951**, *4*, 253.
56. A. W. Hull, *Physical Review* **1917**, *10*, 661.
57. W. Primak and L. H. Fuchs, *Physical Review* **1954**, *95*, 1.
58. H. Lipson and A. R. Stokes, *Proceedings of the Royal Society (London)* **1942**, *A181*, 101.
59. A. Kochanovska, *Czechoslovak Journal of Physics* **1953**, *3*, 193.
60. F. Laves and Y. Baskin, *Zeitschrift fuer Kristallographie, Kristallgeometrie, Kristallphysik, Kristallchemie* **1956**, *107*, 337.
61. H. Gasparoux, *Carbon* **1967**, *5*, 441.
62. H. A. Wilhelm, B. Croset, and G. Medjahdi, *Carbon* **2007**, *45*, 2356.
63. J. Méring and J. Maire, *Journal of Chemical Physics* **1960**, *57*, 803.
64. E. G. Acheson, US Patent 702758, **1902**.
65. E. G. Acheson, US Patent 711031, **1902**.
66. E. G. Acheson, US Patent 645285, **1900**.
67. H. A. Wilhelm and J. L'Heureux, *Actualité de Chimie* **2006**, *19*, 295–296.
68. B.-S. Jin, C.-H. Doh, S.-I. Moon, M.-S. Yun, J.-K. Jeong, H.-D. Nam, and H.-G. Park, *Journal of the Korean Electrochemical Society* **2004**, *7*, 143.
69. K. Tatsumi, K. Zaghib, Y. Sawada, H. Abe, and T. Ohsaka, *Proceedings—Electrochemical Society* **1995**, *94(28)*, 97.
70. C. A. Frysz, X. Shui, and D. D. L. Chung, *Journal of Power Sources* **1996**, *58*, 41.
71. D. D. Eddie and R. J. Diefendorf, *Carbon–Carbon Materials and Composites*, Vol. 1254, NASA, **1992**, p. 19.
72. N. Murphie, in H. Marsch, E. A. Heintz, and F. Rodriguez-Reinoso (eds), *Introduction to Carbon Technologies*, Universidad Alicante, Alicante, **1997**, p. 603.
73. J. B. Donnet and R. C. Bansal, *Carbon Fibers*, Marcel Dekker Inc., New York, **1984**.

74. Anonymous, *Ullmann's Encyclopedia of Industrial Chemistry*, Vol. 5, Weinheim, **1986**.
75. G. Kühner and M. Voll, in J.-B. Donnet, R. C. Bansal, and M.-J. Wang (eds), *Carbon Black—Science and Technology*, 2nd ed., Marcel Dekker Inc., New York, **2003**, pp. 1–66.
76. J. B. Donnet, R. P. Bansal, and M. J. Wang, *Carbon Black Science and Technology*, 2nd ed., Marcel Dekker Inc., New York, **1993**.
77. R. Taylor, in H. Marsh, E. A. Heintz, and F. Rodriguez-Reinoso (eds), *Introduction to Carbon Technologies*, Universidad Alicante, Alicante, **1997**, Chapter 4.
78. V. Schwob, in P. L. Walker and P. A. Thrower (eds), *Chemistry and Physics of Carbon*, Vol. 15, Marcel Dekker, New York, **1982**, p. 109.
79. X. Bourrat, in H. Marsh and F. Rodriguez-Reinoso (eds), *Science of Carbon Materials*, Publicaciones de la Universidad de Alicante, Alicante, **2000**, pp. 1–97.
80. J. Lahaye, *Carbon* **1992**, *30*(3), 309.
81. J. Lahaye and G. Prado, in P. L. Walker and P. A. Thrower (eds), *Chemistry and Physics of Carbon*, Vol. 14, Marcel Dekker Inc., New York, **1978**, pp. 168–200.
82. E. H. Homann, *Combustion and Flame* **1967**, *11*, 265.
83. H. F. Calcote, *Combustion and Flame* **1981**, *42*, 215.
84. R. C. Bansal and J.-B. Donnet, in J.-B. Donnet, R. C. Bansal, and M.-J. Wang (eds), *Carbon Black-Science and Technology*, 2nd ed., Marcel Dekker, Inc., New York, **2003**, pp. 67–88.
85. X. Bourrat, *21st Biennal Conference on Carbon*, American Chemical Society, Buffalo, NY, **1993**, pp. 229–230.
86. W. M. Hess and C. R. Herd, in J.-B. Donnet, R. C. Bansal, and M.-J. Wang (eds), *Carbon Black—Science and Technology*, 2nd ed., Marcel Dekker, Inc., New York, **2003**, pp. 89–173.
87. N. Probst, in J.-B. Donnet, R. C. Bansal, and M.-J. Wang (eds), *Carbon Black—Science and Technology*, 2nd ed., Marcel Dekker Inc., New York, **2003**, pp. 271–288.
88. F. Carmona, *Annals of Chimica, France* **1988**, *13*, 395.
89. E. Grivei and N. Probst, *Belg. KGK, Kautschuk, Gummi, Kunststoffe* **2003**, *56*, 460.
90. D. Guyomard and J. M. Tarascon, *Solid State Ionics* **1994**, *69*, 222.
91. T. Ishihara, M. Koga, H. Matsumoto, and M. Yoshio, *Electrochemical and Solid-State Letters* **2007**, *10*, A74.
92. C. A. Frysz, X. Shui, and D. D. L. Chung, *Journal of Power Sources* **1996**, *58*, 41.
93. H. Wang, T. Umeno, K. Mizuma, and M. Yoshio, *Journal of Power Sources* **2008**, *175*, 886–890.
94. M. Lanz, C. Kormann, H. Steininger, G. Heil, O. Haas, and P. Novák, *Journal of the Electrochemical Society* **2000**, *147*, 3997.
95. M. Winter, J. O. Besenhard, M. E. Spahr, and P. Novák, *Advanced Materials (Weinheim, Germany)* **1998**, *10*, 725.
96. S. Flandrois and B. Simon, *Carbon* **1999**, *37*, 165.
97. M. Winter and J. O. Besenhard, in J. O. Besenhard (ed.), *Handbook of Battery Materials*, Wiley-VCH, Weinheim, **1998**, pp. 383–418.
98. M. Winter and J. O. Besenhard, in M. Wakihara and O. Yamamoto (eds), *Lithium Ion Batteries*, Kodansha Scientific Ltd., Tokyo, Wiley-VCH, Weinheim **1998**, p. 127.
99. A. Hérold, *Bulletin de la societe Chimique* **1955**, *187*, 999.
100. D. Guérard and A. Hérold, *Carbon* **1975**, *13*, 337.
101. A. N. Dey and B. P. Sullivan, *Journal of the Electrochemical Society* **1970**, *117*, 222.
102. M. Arakawa and J. Yamaki, *Journal of Electroanalytical Chemistry* **1987**, *219*, 273.
103. J. O. Besenhard and H. P. Fritz, *Journal of Electroanalytical Chemistry* **1974**, *53*, 329.
104. J. O. Besenhard and G. Eichinger, *Journal of Electroanalytical Chemistry* **1976**, *68*, 1.
105. R. Yazami and P. Touzain, *Journal of Power Sources* **1985**, *56*, 81.
106. R. Kanno, Y. Takeda, T. Ichikawa, K. Nakanishi, and O. Yamamoto, *Journal of Power Sources* **1989**, *26*, 534.
107. M. Mohri, N. Yanagisawa, Y. Tajima, H. Tanaka, T. Mitate, S. Nakajima, M. Yoshida, Y. Yoshimoto, T. Suzuki, and H. Wada, *Journal of Power Sources* **1989**, *26*, 545.
108. A. Omaru, H. Azuma, and Y. Nishi (Sony Corp., Japan). Patent Application: WO 92-JP238 9216026, **1992**.
109. K. Sekai, H. Azuma, A. Omaru, S. Fujita, H. Imoto, T. Endo, K. Yamaura, Y. Nishi, S. Mashiko, and M. Yokogawa, *Journal of Power Sources* **1993**, *43*, 241.
110. A. Omaru, H. Azuma, M. Aoki, A. Kita, and Y. Nishi, *Proceedings—Electrochemical Society* **1993**, *93(24)*, 21.
111. H. H. Schonfelder, K. Kitoh, and H. Nemoto, *Journal of Power Sources* **1997**, *68*, 258.

112. M. Nagamine, *Electrochemical Society Japan, Spring Meeting.* Extended Abstract, p. 267, Tokyo, Japan, **1993**.
113. H. Imoto, A. Omaru, H. Azuma, and Y. Nishi, *Proceedings—Electrochemical Society* **1993**, *93-24*, 9.
114. T. Zheng, W. Xing, and J. R. Dahn, *Carbon* **1996**, *34*, 1501.
115. T. Zheng, Y. Liu, E. W. Fuller, S. Tseng, U. von Sacken, and J. R. Dahn, *Journal of the Electrochemical Society* **1995**, *142*, 2581.
116. T. Zheng, Q. Zhong, and J. R. Dahn, *Journal of the Electrochemical Society* **1995**, *142*, L211.
117. T. Zheng and J. R. Dahn, *Condensed Matter News* **1996**, *5*, 11.
118. S. I. Yamada, H. Kajiura, K. Seka, H. Ooki, H. Imoto, and M. Nagamine, *Proceedings of the Sony Research Forum* **1999**, *8*, 328.
119. Y. Nishi, *Molecular Crystals and Liquid Crystals Science and Technology, Section A: Molecular Crystals and Liquid Crystals* **2000**, *340*, 419.
120. T. Ohzuku, Y. Iwakoshi, and K. Sawai, *Journal of the Electrochemical Society* **1993**, *140*, 2490.
121. D. Billaud, F. X. Henry, and P. Willmann, *Materials Research Bulletin* **1993**, *28*, 477.
122. O. Yamamoto, N. Imanishi, Y. Takeda, and H. Kashiwagi, *Journal of Power Sources* **1995**, *54*, 72.
123. H. Imoto, M. Nagamine, and Y. Nishi, *Proceedings—Electrochemical Society* **1995**, *94-28*, 43.
124. M. Fujimoto, K. Ueno, T. Nohma, M. Takahashi, K. Nishio, and T. Saito, *Proceedings—Electrochemical Society* **1993**, *93-23*, 280.
125. K. Nishio, M. Fujimoto, Y. Shoji, R. Ohshita, T. Noohma, K. Moriwaki, S. Narukawa, and T. Saito, *8th International Meeting on Lithium Batteries,* Extented Abstract, p. 93, Nagoya, Japan, **1996**.
126. M. Endo, Y. Nishimura, T. Takahashi, K. Takeuchi, and M. S. Dresselhaus, *Journal of Physical Chemistry Solids* **1996**, *57*, 725.
127. K. Sato, M. Noguchi, A. Demachi, N. Oki, and M. Endo, *Science* **1994**, *264*, 556.
128. J. R. Dahn, T. Zheng, Y. Liu, and J. S. Xue, *Science (Washington, DC)* **1995**, *270*, 590.
129. J. E. Fischer, in A. P. Legrand and S. Flandrois (eds), *Chemical Physics of Intercalation*, Plenum Press, New York, **1987**, pp. 59–70.
130. T. Tran and K. Kinoshita, *Journal Electroanalytical Chemistry,* **1995**, *386*, 221.
131. R. Moret, in M. S. Dresselhaus (ed.), *Intercalation in Layered Materials, Vol. NATO ASI Series B148*, Plenum Press, New York, **1986**, pp. 185–189.
132. J. Rossad-Mignod, D. Fruchart, M. J. Moran, J. W. Milliken, and J. E. Fischer, *Synthetic Metals* **1980**, *2*, 143.
133. D. Billaud, E. McRae, and A. Hérold, *Materials Research Bulletin* **1979**, *14*, 857.
134. X. Y. Song, K. Kinoshita, and T. D. Tran, *Journal of the Electrochemical Society* **1996**, *143*, L120.
135. R. Yazami, A. Cherigui, V. A. Nalimova, and D. Guerard, *Proceedings—Electrochemical Society* **1993**, *93-24*, 1.
136. R. Yazami, in G. Pistoia (ed.), *Lithium Batteries—New Materials, Developments and Perspectives*, Elesevier, London, **1994**, pp. 49–91.
137. V. V. Avdev, V. A. Nalimova, and K. N. Semenenko, *High Pressure Research* **1990**, *6*, 11.
138. D. Guerard and V. A. Nalimova, in G.-A. Nazri, J.-M. Tarascon, and M. Schreiber (eds), *Solid State Ionics IV, Materials Research Society Proceedings*, Vol. 369, **1995**, p. 155.
139. V. A. Nalimova, G. Guerard, M. Lelaurain, and O. V. Fateev, *Carbon* **1995**, *33*, 177.
140. V. A. Nalimova, C. Bindra, and J. E. Fischer, *Solid State Communications* **1996**, *97*, 583.
141. J. R. Dahn, *Physical Review B: Condensed Matter and Materials Physics* **1991**, *44*, 9170.
142. D. Billaud, F. X. Henry, M. Lelaurain, and P. Willmann, *Journal of Physical Chemistry Solids* **1996**, *57*, 775.
143. M. Inaba, H. Yoshida, Z. Ogumi, T. Abe, Y. Mizutani, and M. Asano, *Journal of the Electrochemical Society* **1995**, *142*, 20.
144. S. A. Safran, in A. P. Legrand and S. Flandrois (eds), *Chemical Physics of Intercalation*, Plenum Press, New York, **1987**, p. 47.
145. N. Christu, E. Fitzer, and W. Fritz, *Berichte der Deuschen Keramischen Gesellschaft* **1964**, *41(2)*, 143.
146. J. Biscoe and B. E. Warren, *Journal of Applied Physics* **1942**, *13*, 364.
147. J. Maire and J. Méring, in P. C. Walker (ed.), *Chemistry and Physics of Carbon*, Vol. 6, Marcel Dekker Inc., New York, **1970**, p. 125.
148. A. Oberlin, in P. A. Thrower (ed.), *Chemistry and Physics of Carbon*, Vol. 22, Marcel Dekker Inc., New York, **1989**, pp. 1–143.
149. O. Vohler, F. v. Sturm, E. Wege, H. v. Kienle, M. Voll, and P. Kleinschmidt, in W. Gerhartz (ed.), *Ullmann's Encyclopedia of Industrial Chemistry*, Vol. 5A, 5th ed., Wiley-VCH, Weinheim, **1986**, p. 95.

150. E. Hoinkis, in P. A. Thrower (ed.), *Chemistry and Physics of Carbon*, Vol. 25, Marcel Dekker Inc., New York, **1997**, pp. 71–241.

151. A. Guvenilir, T. M. Breunig, J. H. Kinney, and S. R. Stock, *Acta Materialia* **1997**, *45*, 1977.

152. A. Marchand and J. V. Zanchetta, *Carbon* **1966**, *3*, 483.

153. M. Cerutti, J. Uebersfeld, J. Millet, and J. Parisot, *Journal of Chemical Physics* **1960**, *57*, 907.

154. P. Delhaès and F. Carmona, in P. L. Walker and P. A. Thrower (eds), *Chemistry and Physics of Carbon*, Vol. 17, Marcel Dekker Inc., New York, **1981**, pp. 89–174.

155. J. R. Dahn, R. Fong, and M. J. Spoon, *Physical Review B: Condensed Matter and Materials Physics* **1990**, *42*, 6424.

156. J. R. Dahn, A. K. Sleigh, H. Shi, J. N. Reimers, Q. Zhong, and B. M. Way, *Electrochimica Acta* **1993**, *38*, 1179.

157. A. Mabuchi, K. Tokumitsu, H. Fujimoto, and T. Kasuh, *Journal of the Electrochemical Society* **1995**, *142*, 1041.

158. K. Tatsumi, N. Iwashita, H. Sakaebe, H. Shioyama, S. Higuchi, A. Mabuchi, and H. Fujimoto, *Journal of the Electrochemical Society* **1995**, *142*, 716.

159. K. Tatsumi, K. Zaghib, Y. Sawada, H. Abe, and T. Ohsaki, *Journal of the Electrochemical Society* **1995**, *142*, 1090.

160. A. Mabuchi, H. Fujimoto, K. Tokumitsu, and T. Kasuh, *Journal of the Electrochemical Society* **1995**, *142*, 3049.

161. A. Satoh, N. Takami, and T. Ohsaki, *Solid State Ionics* **1995**, *80*, 291.

162. N. Takami, A. Satoh, M. Hara, and T. Ohsaki, *Journal of the Electrochemical Society* **1995**, *142*, 2564.

163. T. Zheng and J. R. Dahn, *Physical Review B: Condensed Matter* **1996**, *53*, 3061.

164. K. Tatsumi, T. Akai, T. Imamura, K. Zaghib, N. Iwashita, S. Higuchi, and Y. Sawada, *Journal of the Electrochemical Society* **1996**, *143*, 1923.

165. R. Kostecki, T. Tran, X. Song, K. Kinoshita, and F. McLarnon, *Journal of the Electrochemical Society* **1997**, *144*, 3111.

166. S. Flandrois, A. Fevrier-Bouvier, K. Guerin, B. Simon, and P. Biensan, *Molecular Crystals and Liquid Crystals Science and Technology, Section A: Molecular Crystals and Liquid Crystals* **1998**, *310*, 389.

167. F. Béguin, F. Chevalier, C. Vix-Guterl, S. Saadalah, V. Bertagna, M. Letalier, J. N. Rouzaud, and E. Frackowiak, *Journal of Physics and Chemistry of Solids* **2004**, *65*, 211.

168. M. Endo, Y. Nishimura, T. Takahashi, K. Takeuchi, and M. S. Dresselhaus, *Journal of Physics and Chemistry of Solids* **1996**, *57*, 725.

169. M. Inagaki, N. Iwashita, and Y. Hishiyama, *Molecular Crystals and Liquid Crystals* **1994**, *244*, 89.

170. T. Zheng, J. N. Reimers, and J. R. Dahn, *Physical Review B: Condensed Matter* **1995**, *51*, 734.

171. T. Zheng and J. R. Dahn, *Synthetic Metals* **1995**, *73*, 1.

172. H. Saadaoui, J. C. Roux, S. Flandrois, and B. Nysten, *Carbon* **1993**, *31*, 481.

173. R. Yazami and M. Deschamps, *Journal of Power Sources* **1995**, *54*, 411.

174. R. Yazami and M. Deschamps, *Materials Research Society Symposium Proceedings* **1995**, *369*, 165.

175. Y. Mori, T. Iriyama, T. Hashimoto, S. Yamazaki, F. Kawakami, H. Shiroki, and T. Yamabe, *Journal of Power Sources* **1995**, *56*, 205.

176. S. Gautier, F. Leroux, E. Frackowiak, A. M. Faugere, J. N. Rouzaud, and F. Béguin, *Journal of Physical Chemistry A* **2001**, *105*, 5794.

177. S. Yata, H. Kinoshita, M. Komori, N. Ando, T. Kashiwamura, T. Harada, K. Tanaka, and T. Yamabe, *Synthetic Metals* **1994**, *62*, 153.

178. H. Q. Xiang, S. B. Fang, and Y. Y. Jiang, *Journal of the Electrochemical Society* **1997**, *144*, L187.

179. W. Xing, J. S. Xue, T. Zheng, A. Gibaud, and J. R. Dahn, *Journal of the Electrochemical Society* **1996**, *143*, 3482.

180. W. Xing, J. S. Xue, and J. R. Dahn, *Journal of the Electrochemical Society* **1996**, *143*, 3046.

181. W. Xing and J. R. Dahn, *Journal of the Electrochemical Society* **1997**, *144*, 1195.

182. W. Xing, R. A. Dunlap, and J. R. Dahn, *Journal of the Electrochemical Society* **1998**, *145*, 62.

183. T. Zheng and J. R. Dahn, *Carbon Materials for Advanced Technologies* **1999**, 341.

184. K. Tokumitsu, A. Mabuchi, H. Fujimoto, and T. Kasuh, *Journal of the Electrochemical Society* **1996**, *143*, 2235.

185. K. Tokumitsu, A. Mabuchi, H. Fujimoto, and T. Kasuh, *Journal of Power Sources* **1995**, *54*, 444.

186. S. Wang, Y. Zhang, L. Yang, and Q. Liu, *Solid State Ionics* **1996**, *86–88*, 919.

187. M. J. Matthews, M. Endo, T. Takahashi, Y. Nishimura, T. Takamuku, and M. S. Dresselhaus, *Materials Research Society Symposium Proceedings* **1996**, *413*, 701.

188. C. Kim, T. Fujino, T. Hayashi, M. Endo, and M. S. Dresselhaus, *Journal of the Electrochemical Society* **2000**, *147*, 1265.
189. M. Endo, C. Kim, T. Karaki, T. Fujino, M. J. Matthews, S. D. M. Brown, and M. S. Dresslhaus, *Carbon* **1998**, *36*, 1403.
190. M. Endo, C. Kim, T. Karaki, T. Fujino, M. J. Matthews, S. D. M. Brown, and M. S. Dresslhaus, *Synthetic Metals* **1998**, *98*, 17.
191. M. W. Verbrugge and B. J. Koch, *Journal of the Electrochemical Society* **1996**, *143*, 24.
192. Y. Jung, M. C. Suh, H. Lee, M. Kim, S. I. Lee, S. C. Shim, and J. Kwak, *Journal of the Electrochemical Society* **1997**, *144*, 4279.
193. S. Yamazaki, T. Hashimoto, T. Iriyama, Y. Mori, H. Shiroki, and N. Tamura, *Journal of Molecular Structure* **1998**, *441*, 165.
194. T. Zheng, J. S. Xue, and J. R. Dahn, *Chemistry of Materials* **1996**, *8*, 389.
195. T. Zheng, W. R. McKinnon, and J. R. Dahn, *Journal of the Electrochemical Society* **1996**, *143*, 2137.
196. T. Zheng and J. R. Dahn, *Journal of Power Sources* **1997**, *68*, 201.
197. Y. Liu, J. S. Xue, T. Zheng, and J. R. Dahn, *Carbon* **1996**, *34*, 193.
198. E. Buiel, A. E. George, and J. R. Dahn, *Journal of the Electrochemical Society* **1998**, *145*, 2252.
199. E. R. Buiel, A. E. George, and J. R. Dahn, *Carbon* **1999**, *37*, 1399.
200. K. Tokumitsu, A. Mabuchi, H. Fujimoto, and T. Kasuh, in S. Megahead, B. Barnett, and L. Xie (eds), *Rechargeable Lithium and Lithium-Ion Batteries*, The Electrochemical Society, Pennington, NJ, **1995**, p. 136.
201. A. Gibaud, J. S. Xue, and J. R. Dahn, *Carbon* **1996**, *34*, 499.
202. D. A. Stevens and J. R. Dahn, *Journal of the Electrochemical Society* **2001**, *148*, A803.
203. J. R. Dahn, *Carbon* **1997**, *35*, 825.
204. A. Nagai, M. Ishikawa, J. Masuko, N. Sonobe, H. Chuman, and T. Iwasaki, *Materials Research Society Symposium Proceedings* **1995**, *393*, 339.
205. E. Frackowiak, S. Gautier, F. Leroux, J.-N. Rouzaud, and F. Béguin, *Molecular Crystals and Liquid Crystals Science and Technology, Section A: Molecular Crystals and Liquid Crystals* **2000**, *340*, 431.
206. F. Chevallier, M. Letellier, M. Morcrette, J. M. Tarascon, E. Frackowiak, J. N. Rouzaud, and F. Béguin, *Electrochemical and Solid-State Letters* **2003**, *6*, A225.
207. M. Letellier, F. Chevallier, C. Clinard, E. Frackowiak, J.-N. Rouzaud, F. Béguin, M. Morcrette, and J.-M. Tarascon, *Journal of Chemical Physics* **2003**, *118*, 6038.
208. E. Buiel and J. R. Dahn, *Journal of the Electrochemical Society* **1998**, *145*, 1977.
209. Y. Matsumura, S. Wang, and J. Mondori, *Journal of the Electrochemical Society* **1995**, *142*, 2914.
210. B. R. Puri, in *Chemistry and Physics of Carbon*, Vol. 6, Marcel Dekker Inc., New York, **1970**, pp. 191.
211. M. Kikuchi, Y. Ikezawa, and T. Takamura, *Journal of Electroanalytical Chemistry* **1995**, *396*, 451.
212. J. S. Xue and J. R. Dahn, *Journal of the Electrochemical Society* **1995**, *142*, 3668.
213. F. Chevallier, S. Gautier, J. P. Salvetat, C. Clinard, E. Frackowiak, J. N. Rouzaud, and F. Béguin, *Journal of Power Sources* **2001**, *97–98*, 143.
214. A. Marchand, in P. L. Walker (ed.), *Chemistry and Physics of Carbon*, Vol. 7, Marcel Dekker Inc., New York, **1971**, pp. 155–189.
215. C. E. Lowell, *Journal of the American Ceramic Society* **1967**, *50*, 142.
216. J. Kouvetakis, R. B. Kaner, M. L. Sattler, and N. Bartlett, *Journal of Chemical Society Chemical Communications* **1986**, 1758.
217. B. Ottaviani, A. Derré, E. Grivei, O. A. M. Mahmoud, M. F. Guimon, S. Flandrois, and P. Delhaès, *Journal of Materials Chemistry* **1998**, *8*, 197.
218. T. Shirasaki, A. Derre, K. Guerin, and S. Flandrois, *Carbon* **1999**, *37*, 1961.
219. B. M. Way and J. R. Dahn, *Journal of the Electrochemical Society* **1994**, *141*, 907.
220. J. R. Dahn, J. N. Reimers, A. K. Sleigh, and T. Tiedje, *Physical Review B: Condensed Matter and Materials Physics* **1992**, *45*, 3773.
221. J. R. Dahn, J. N. Reimers, T. Tiedje, Y. Gao, A. K. Sleigh, W. R. McKinnon, and S. Cramm, *Physical Review Letters* **1992**, *68*, 835.
222. I.-H. Kwon, M. Y. Song, E. Y. Bang, Y.-S. Han, K.-T. Kim, and J.-Y. Lee, *Journal of the Korean Electrochemical Society* **2002**, *5*, 30.
223. Y. Nishimura, T. Takahashi, T. Tamaki, M. Endo, and M. S. Dresselhaus, *Tanso* **1996**, *172*, 89.
224. M. Endo, C. Kim, T. Karaki, T. Tamaki, Y. Nishimura, M. J. Matthews, S. D. M. Brown, and M. S. Dresselhaus, *Physical Review B: Condensed Matter and Materials Physics* **1998**, *58*, 8991.
225. T. Tamaki, T. Kawamura, and Y. Yamazaki, *Materials Research Society Symposium Proceedings* **1998**, *496*, 569.

226. M. Endo, C. Kim, T. Karaki, Y. Nishimura, M. J. Matthews, S. D. M. Brown, and M. S. Dresselhaus, *Carbon* **1999**, *37*, 561.
227. T. Morita and N. Takami, *Electrochimica Acta* **2004**, *49*, 2591.
228. H. Fujimoto, A. Mabuchi, C. Natarajan, and T. Kasuh, *Carbon* **2002**, *40*, 567.
229. M.-H. Chen, G.-T. Wu, G.-M. Zhu, J.-K. You, and Z.-G. Lin, *Journal of Solid State Electrochemistry* **2002**, *6*, 420.
230. Y. Lee, D.-Y. Han, D. Lee, A. J. Woo, S. H. Lee, D. Lee, and Y. K. Kim, *Carbon* **2002**, *40*, 403.
231. N. Takami, *Materials Chemistry in Lithium Batteries* **2002**, 1.
232. Y.-H. Lee, K.-C. Pan, S.-J. Jeng, F.-S. Li, and S.-T. Chang, *Guoli Taiwan Daxue Gongcheng Xuekan* **2003**, *89*, 121.
233. G. Yin, Y. Gao, P. Shi, X. Cheng, and A. Aramata, *Materials Chemistry and Physics* **2003**, *80*, 94.
234. J. Machnikowski, L. Ziolkowski, and E. Frackowiak, *Karbo* **2000**, *45*, 265.
235. E. Frackowiak, J. Machnikowski, H. Kaczmarska, and F. Béguin, *Journal of Power Sources* **2001**, *97–98*, 140.
236. J. Machnikowski, E. Frackowiak, K. Kierzek, D. Waszak, R. Benoit, and F. Béguin, *Journal of Physics and Chemistry of Solids* **2004**, *65*, 153.
237. M. Morita, T. Hanada, H. Tsutsumi, and Y. Matsuda, *Journal of the Electrochemical Society* **1992**, *139*, 1227.
238. W. J. Weydanz, B. M. Way, T. van Burren, and J. R. Dahn, *Journal of the Electrochemical Society* **1994**, *141*, 900.
239. M. Kawaguchi, *Advanced Materials* **1997**, *9*, 615.
240. S. Ito, T. Murata, M. Hasegawa, Y. Bito, and Y. Toyoguchi, *Journal of Power Sources* **1997**, *68*, 245.
241. T. Nakajima, M. Koh, and M. Takashima, *Electrochimical Acta* **1998**, *43*, 883.
242. T. Nakajima, *Recent Research Developments in Electrochemistry* **1998**, *1*, 57.
243. T. Nakajima, M. Koh, and K. Dan, *Materials Research Society Symposium Proceedings* **1998**, *496*, 581.
244. M. Koh and T. Nakajima, *Electrochimica Acta* **1999**, *44*, 1713.
245. T. D. Tran, J. H. Feikert, X. Song, and K. Kinoshita, *Journal of the Electrochemical Society* **1995**, *142*, 3297.
246. A. M. Wilson, B. M. Way, J. R. Dahn, and T. van Buuren, *Journal of Applied Physics* **1995**, *77*, 2363.
247. A. M. Wilson and J. R. Dahn, *Journal of the Electrochemical Society* **1995**, *142*, 326.
248. J. S. Xue, K. Myrtle, and J. R. Dahn, *Journal of the Electrochemical Society* **1995**, *142*, 2927.
249. A. M. Wilson, X. Weibing, G. Zank, B. Yates, and J. R. Dahn, *Solid State Ionics* **1997**, *100*, 259.
250. W. Xing, A. M. Wilson, G. Zank, and J. R. Dahn, *Solid State Ionics* **1997**, *93*, 239.
251. A. M. Wilson, G. Zank, K. Eguchi, W. Xing, B. Yates, and J. R. Dahn, *Chemistry of Materials* **1997**, *9*, 2139.
252. A. M. Wilson, G. Zank, K. Eguchi, W. Xing, B. Yates, and J. R. Dahn, *Chemistry of Materials* **1997**, *9*, 1601.
253. W. Xing, A. M. Wilson, K. Eguchi, G. Zank, and J. R. Dahn, *Journal of the Electrochemical Society* **1997**, *144*, 2410.
254. E. Peled, C. Menachem, D. Bar-Tow, and A. Melman, *Journal of the Electrochemical Society* **1996**, *143*, L4.
255. C. Menachem, E. Peled, L. Burstein, and Y. Rosenberg, *Journal of Power Sources* **1997**, *68*, 277.
256. E. Peled, D. Golodnitsky, and J. Penciner, in J. O. Besenhard (ed.), *Handbook of Battery Material*, Wiley-VCH, Weinheim, New York, Chichester, Brisbane, Singapore, Toronto **1998**, pp. 419–456.
257. Y. Ein-Ely and V. R. Koch, *Journal of the Electrochemical Society* **1997**, *144*, 2968.
258. H. Buqa, M. Wachtler, G. H. Wrodnigg, J. O. Besenhard, M. Winter, R. Blygh, M. Ramsey, and F. Netzer, *Advances in Science and Technology (Faenza, Italy)* **1999**, *24*, 125.
259. H. Buqa, R. I. R. Blyth, P. Golob, B. Evers, I. Schneider, M. V. S. Alvarez, F. Hofer, F. P. Netzer, M. G. Ramsey, M. Winter, and J. O. Besenhard, *Ionics* **2000**, *6*, 172.
260. H. Buqa, P. Golob, M. Winter, and J. O. Besenhard, *Journal of Power Sources* **2001**, *97–98*, 122.
261. H. Harker, J. B. Horsley, and D. Robson, *Carbon* **1971**, *9*, 1.
262. H. Wang, T. Ikeda, K. Fukuda, and M. Yoshio, *Journal of Power Sources* **1999**, *83*, 141.
263. M. Tidjani, J. Lachter, T. S. Kabre, and R. H. Bragg, *Carbon* **1986**, *24*, 447.
264. F. Disma, L. Aymard, L. Dupont, and J. M. Tarascon, *Journal of the Electrochemical Society* **1996**, *143*, 3959.
265. M. E. Spahr, H. Wilhelm, F. Joho, J.-C. Panitz, J. Wambach, P. Novák, and N. Dupont-Pavlovsky, *Journal of the Electrochemical Society* **2002**, *149*, A960.

266. M. E. Spahr, H. Buqa, A. Wuersig, D. Goers, L. Hardwick, P. Novák, F. Krumeich, J. Dentzer, and C. Vix-Guterl, *Journal of Power Sources* **2006**, *153*, 300.
267. D. Aurbach and I. Weissman, in D. Aurbach (ed.), *Nonaqueous Electrochemistry*, MO, Dekker, New York, **1999**, pp. 1–52.
268. G. Pistoia, *Lithium Batteries—New Materials, Developments and Perspectives*, Elsevier, Amsterdam, the Netherlands, **1994**.
269. K. Edstrom, T. Gustafsson, and J. O. Thomas, *Electrochimica Acta* **2004**, *50*, 397.
270. A. Wuersig, J. Ufheil, and P. Novák, *205th Meeting of the Electrochemical Society*, San Antonio, TX, **2004**, p. 78.
271. A. Wuersig, H. Buqa, M. Holzapfel, F. Krumeich, and P. Novák, *Electrochemical and Solid-State Letters* **2005**, *8*, A34.
272. E. Peled and H. Yamin, *Israel Journal of Chemistry* **1979**, *18*, 131.
273. A. Meitav and E. Peled, *Journal of Electroanalytical Chemistry* **1982**, *134*, 49.
274. H. Yamin and E. Peled, *Journal of the Electrochemical Society* **1983**, *130*, C307.
275. E. Peled, D. Golodnitsky, G. Ardel, and V. Eshkenazy, *Electrochimica Acta* **1995**, *40*, 2197.
276. D. Bar-Tow, E. Peled, and L. Burstein, *Proceedings—Electrochemical Society* **1997**, *97-18*, 324.
277. E. Peled, C. Menachem, D. BarTow, and A. Melman, *Journal of the Electrochemical Society* **1996**, *143*, L4.
278. E. Peled and D. Golodnitsky, in P. B. Balbuena und Y. Wang (eds), *Lithium-Ion Batteries*, Imperial College Press, London, **2004**, Chapter 1.
279. V. Eshkenazi, E. Peled, L. Burstein, and D. Golodnitsky, *Solid State Ionics* **2004**, *170*, 83.
280. H. Buqa, A. Würsig, J. Vetter, M. E. Spahr, F. Krumeich, and P. Novák, *Journal of Power Sources* **2006**, *153*, 385.
281. P. B. Balbuena and Y. Wang, *Lithium-Ion Batteries: Solid–Electrolyte Interphase*, Imperial College Press, London, **2004**.
282. E. Peled and D. Golodnitsky, in C. F. L. A. R. Holmes (ed.), *1st Joint Meeting of the Electrochemical-Society/International-Society-of-Electrochemistry*, Vol. 97, Electrochemical Society Inc., Paris, France, **1997**, pp. 281–286.
283. E. Peled, D. Golodnitsky, C. Menachem, and D. BarTow, *Journal of the Electrochemical Society* **1998**, *145*, 3482.
284. M. Inaba, Y. Kawatate, A. Funabiki, S. K. Jeong, T. Abe, and Z. Ogumi, *Electrochimica Acta* **1999**, *45*, 99.
285. D. Rahner, *Journal of Power Sources* **1999**, *82*, 358.
286. S. I. Pyun, *Metals and Materials-Korea* **1999**, *5*, 101.
287. Z.-y. Xu and H.-h. Zheng, *Dianyuan Jishu* **2000**, *24*, 295.
288. M. Winter, W. K. Appel, B. Evers, T. Hodal, K. C. Moller, I. Schneider, M. Wachtler, M. R. Wagner, G. H. Wrodnigg, and J. O. Besenhard, *Monatshefte Fur Chemie* **2001**, *132*, 473.
289. P. Novák, F. Joho, M. Lanz, B. Rykart, J. C. Panitz, D. Alliata, R. Kötz, and O. Haas, *Journal of Power Sources* **2001**, *97–98*, 39.
290. M. Lanz and P. Novák, *Journal of Power Sources* **2001**, *102*, 277.
291. H. Maleki, G. Deng, A. Anani, and I. Kerzhner-Haller, US Patent Application (Motorola Inc., USA), **2002**.
292. A. M. Andersson, M. Herstedt, A. G. Bishop, and K. Edstrom, *Electrochimica Acta* **2002**, *47*, 1885.
293. C. S. Wang, X. W. Zhang, A. J. Appleby, X. L. Chen, and F. E. Little, *Journal of Power Sources* **2002**, *112*, 98.
294. Y. X. Wang and P. B. Balbuena, *Journal of Physical Chemistry B* **2002**, *106*, 4486.
295. M. Balasubramanian, H. S. Lee, X. Sun, X. Q. Yang, A. R. Moodenbaugh, J. McBreen, D. A. Fischer, and Z. Fu, *Electrochemical and Solid State Letters* **2002**, *5*, A22.
296. K. Araki and N. Sato, *Journal of Power Sources* **2003**, *124*, 124.
297. J. Vetter and P. Novák, *Journal of Power Sources* **2003**, *119–121*, 338.
298. Y. S. Han and J. Y. Lee, *Electrochimica Acta* **2003**, *48*, 1073.
299. K. Xu, S. S. Zhang, and T. R. Jow, *Electrochemical and Solid State Letters* **2003**, *6*, A117.
300. M. Ihara, B. T. Hang, K. Sato, M. Egashira, S. Okada, and J. Yamaki, *Journal of the Electrochemical Society* **2003**, *150*, A1476.
301. X. D. Wu, Z. X. Wang, L. Q. Chen, and X. J. Huang, *Electrochemistry Communications* **2003**, *5*, 935.
302. W. S. Kim, K. I. Chung, J. H. Cho, D. W. Park, C. Y. Kim, and Y. K. Choi, *Journal of Industrial and Engineering Chemistry* **2003**, *9*, 699.
303. S. E. Sloop, J. B. Kerr, and K. Kinoshita, *Journal of Power Sources* **2003**, *119*, 330.

304. S. Komaba, B. Kaplan, T. Ohtsuka, Y. Kataoka, N. Kumagai, and H. Groult, *Journal of Power Sources* **2003**, *119*, 378.
305. A. M. Andersson, A. Henningson, H. Siegbahn, U. Jansson, and K. Edstrom, *Journal of Power Sources* **2003**, *119*, 522.
306. M. R. Wagner, P. R. Raimann, A. Trifonova, K. C. Moeller, J. O. Besenhard, and M. Winter, *Electrochemical and Solid-State Letters* **2004**, *7*, A201.
307. H. H. Zheng, K. L. Zhuo, J. J. Wang, and Z. Y. Xu, *Chemical Journal of Chinese Universities-Chinese* **2004**, *25*, 729.
308. H. H. Lee, C. C. Wan, and Y. Y. Wang, *Journal of the Electrochemical Society* **2004**, *151*, A542.
309. J. T. Lee, Y. W. Lin, and Y. S. Jan, *Journal of Power Sources* **2004**, *132*, 244.
310. H. Tanaka, T. Osawa, Y. Moriyoshi, M. Kurihara, S. Maruyama, and T. Ishigaki, *Thin Solid Films* **2004**, *457*, 209.
311. X. D. Wu, Z. X. Wang, L. Q. Chen, and X. J. Huang, *Surface and Coatings Technology* **2004**, *186*, 412.
312. M. Herstedt, A. M. Andersson, H. Rensmo, H. Siegbahn, and K. Edstrom, *Electrochimica Acta* **2004**, *49*, 4939.
313. K. Xu, U. Lee, S. S. Zhang, and T. R. Jow, *Journal of the Electrochemical Society* **2004**, *151*, A2106.
314. A. Augustsson, M. Herstedt, J. H. Guo, K. Edstrom, G. V. Zhuang, P. N. Ross, J. E. Rubensson, and J. Nordgren, *Physical Chemistry Chemical Physics* **2004**, *6*, 4185.
315. H. J. Santner, C. Korepp, M. Winter, J. O. Besenhard, and K. C. Moller, *Analytical and Bioanalytical Chemistry* **2004**, *379*, 266.
316. M. Herstedt, H. Rensmo, H. Siegbahn, and K. Edstrom, *Electrochimica Acta* **2004**, *49*, 2351.
317. J. Ufheil, M. C. Baertsch, A. Würsig, and P. Novák, *Electrochimica Acta* **2005**, *50*, 1733.
318. K. Xu, S. S. Zhang, and R. Jow, *Journal of Power Sources* **2005**, *143*, 197.
319. H. H. Lee, Y. Y. Wang, C. C. Wan, M. H. Yang, H. C. Wu, and D. T. Shieh, *Journal of Applied Electrochemistry* **2005**, *35*, 615.
320. Q. S. Wang, J. H. Sun, X. L. Yao, and C. H. Chen, *Thermochimica Acta* **2005**, *437*, 12.
321. H. L. Zhang, F. Li, C. Liu, J. Tan, and H. M. Cheng, *Journal of Physical Chemistry B* **2005**, *109*, 22205.
322. Q. S. Wang, J. H. Sun, X. L. Yao, and C. H. Chen, *Journal of the Electrochemical Society* **2006**, *153*, A329.
323. M. Itagaki, S. Yotsuda, N. Kobari, K. Watanabe, S. Kinoshita, and M. Ue, *Electrochimica Acta* **2006**, *51*, 1629.
324. L. J. Fu, H. Liu, C. Li, Y. P. Wu, E. Rahm, R. Holze, and H. Q. Wu, *Solid State Sciences* **2006**, *8*, 113.
325. K. Edstrom, M. Herstedt, and D. P. Abraham, *Journal of Power Sources* **2006**, *153*, 380.
326. I. Watanabe and J. Yamaki, *Journal of Power Sources* **2006**, *153*, 402.
327. M. Q. Xu, X. X. Zuo, W. S. Li, H. J. Zhou, J. S. Liu, and Z. Z. Yuan, *Acta Physico-Chimica Sinica* **2006**, *22*, 335.
328. S. S. Zhang, K. Xu, and T. R. Jow, *Journal of Power Sources* **2006**, *156*, 629.
329. H. Nakahara and S. Nutt, *Journal of Power Sources* **2006**, *160*, 1355.
330. E. Peled, D. Golodnitsky, and G. Ardel, *Journal of the Electrochemical Society* **1997**, *144*, L208.
331. P. Novák, J. C. Panitz, F. Joho, M. Lanz, R. Imhof, and M. Coluccia, *Journal of Power Sources* **2000**, *90*, 52.
332. D. Alliata, R. Kotz, P. Novák, and H. Siegenthaler, *Electrochemistry Communications* **2000**, *2*, 436.
333. P. Novák, F. Joho, M. Lanz, B. Rykart, J. C. Panitz, D. Alliata, R. Kotz, and O. Haas, *Journal of Power Sources* **2001**, *97–98*, 39.
334. R. Kostecki, B. Schnyder, D. Alliata, X. Song, K. Kinoshita, and R. Kotz, *Thin Solid Films* **2001**, *396*, 36.
335. H. Maleki, G. Deng, A. Anani, and J. Howard, *Journal of the Electrochemical Society* **1999**, *146*, 3224.
336. A. M. Andersson, K. Edstrom, and J. O. Thomas, *Journal of Power Sources* **1999**, *82*, 8.
337. E. P. Roth, *Proceedings of the Intersociety Energy Conversion Engineering Conference* **2000**, *35*, 962.
338. E. P. Roth, *Proceedings—Electrochemical Society* **2000**, *99-25*, 763.
339. K. Edstrom, A. M. Andersson, A. Bishop, L. Fransson, J. Lindgren, and A. Hussenius, *Journal of Power Sources* **2001**, *97–98*, 87.
340. J. Yamaki, H. Takatsuji, T. Kawamura, and M. Egashira, *Solid State Ionics* **2002**, *148*, 241.
341. M.-S. Wu, P.-C. J. Chiang, J.-C. Lin, and Y.-S. Jan, *Electrochimica Acta* **2004**, *49*, 1803.
342. F. Joho, P. Novák, and M. E. Spahr, *Journal of the Electrochemical Society* **2002**, *149*, A1020.
343. D. Aurbach, Y. Ein-Eli, O. Chusid, Y. Carmeli, M. Babai, and H. Yamin, *Journal of the Electrochemical Society* **1994**, *141*, 603.

344. D. Aurbach, M. D. Levi, E. Levi, and A. Schechter, *Journal of Physical Chemistry, B* **1997**, *101*, 2195.
345. J. C. Panitz, U. Wietelmann, M. Wachtler, S. Strobele, and M. Wohlfahrt-Mehrens, *Journal of Power Sources* **2006**, *153*, 396.
346. F. Béguin, F. Chevallier, C. Vix, S. Saadallah, J. N. Rouzaud, and E. Frackowiak, *Journal of Physics and Chemistry of Solids* **2004**, *65*, 211.
347. M. E. Spahr, H. Wilhelm, T. Palladino, N. Dupont-Pavlovsky, D. Goers, F. Joho, and P. Novák, *Journal of Power Sources* **2003**, *119–121*, 543.
348. H. Buqa, D. Goers, M. E. Spahr, and P. Novák, *Journal of Solid State Electrochemistry* **2003**, *8*, 79.
349. Z. Ogumi and S. K. Jeong, *Electrochemistry* **2003**, *71*, 1011.
350. M. Inaba, H. Tomiyasu, A. Tasaka, S. K. Jeong, Y. Iriyama, T. Abe, and Z. Ogumi, *Electrochemistry* **2003**, *71*, 1132.
351. C. Liebenow, M. W. Wagner, K. Luehder, P. Lobitz, and J. O. Besenhard, *Journal of Power Sources* **1995**, *54*, 369.
352. S. Yamaguchi, H. Asahina, K. A. Hirasawa, T. Sato, and S. Mori, *Molecular Crystals and Liquid Crystals Science and Technology, Section A: Molecular Crystals and Liquid Crystals* **1998**, *322*, 239.
353. O. Chusid, Y. E. Ely, D. Aurbach, M. Babai, and Y. Carmeli, *Journal of Power Sources* **1993**, *43*, 47.
354. D. Aurbach, B. Markovsky, A. Shechter, Y. Ein-Eli, and H. Cohen, *Journal of Electrochemical Society* **1996**, *143*, 3809.
355. D. Aurbach, *Nonaqueous Electrochemistry*, Marcel Dekker AG, New York, **1999**.
356. W. van Schalkwijk and B. Scrosati, *Advances in Lithium-Ion Batteries*, Kluwer Academic Publishers Group, New York, **2002**.
357. D. Aurbach, M. D. Levi, E. Levi, H. Teller, B. Markovsky, G. Salitra, U. Heider, and L. Heider, *Journal of the Electrochemical Society* **1998**, *145*, 3024.
358. D. Aurbach, K. Gamolsky, B. Markovsky, Y. Gofer, M. Schmidt, and U. Heider, *Electrochimica Acta* **2002**, *47*, 1423.
359. Y. Ein-Eli, S. R. Thomas, V. Koch, D. Aurbach, B. Markovsky, and A. Schechter, *Journal of the Electrochemical Society* **1996**, *143*, L273.
360. Y. Ein-Eli, B. Markovsky, D. Aurbach, Y. Carmeli, H. Yamin, and S. Luski, *Electrochimica Acta* **1994**, *39*, 2559.
361. D. Aurbach and Y. Ein-Eli, *Journal of the Electrochemical Society* **1995**, *142*, 1746.
362. J. S. Gnanaraj, M. D. Levi, E. Levi, G. Salitra, D. Aurbach, J. E. Fischer, and A. Claye, *Journal of the Electrochemical Society* **2001**, *148*, A525.
363. M. Koltypin, Y. S. Cohen, B. Markovsky, Y. Cohen, and D. Aurbach, *Electrochemistry Communications* **2002**, *4*, 17.
364. J. S. Gnanaraj, M. D. Levi, Y. Gofer, D. Aurbach, and M. Schmidt, *Journal of the Electrochemical Society* **2003**, *150*, A445.
365. J. Ufheil, A. Würsig, O. D. Schneider, and P. Novák, *Electrochemistry Communications* **2005**, *7*, 1380.
366. R. Dominko, M. Gaberscek, M. Bele, J. Drofenik, E. M. Skou, A. Würsig, P. Novák, and J. Jamnik, *Journal of the Electrochemical Society* **2004**, *151*, A1058.
367. M. Hirayama, N. Sonoyama, T. Abe, M. Minoura, M. Ito, D. Mori, A. Yamada, R. Kanno, T. Terashima, M. Takano, K. Tamura, and J. Mizuki, *Journal of Power Sources* **2007**, *168*, 493.
368. M. R. Wagner, J. H. Albering, K. C. Moeller, J. O. Besenhard, and M. Winter, *Electrochemistry Communications* **2005**, *7*, 947.
369. H. J. Santner, C. Korepp, M. Winter, J. O. Besenhard, and K. C. Moeller, *Analytical and Bioanalytical Chemistry* **2004**, *379*, 266.
370. J. F. Ni, H. H. Zhou, J. T. Chen, and G. Y. Su, *Progress in Chemistry* **2004**, *16*, 335.
371. M. Inaba, H. Tomiyasu, A. Tasaka, S. K. Jeong, and Z. Ogumi, *Langmuir* **2004**, *20*, 1348.
372. M. R. Wagner, P. R. Raimann, A. Trifonova, K. C. Moller, J. O. Besenhard, and M. Winter, *Analytical and Bioanalytical Chemistry* **2004**, *379*, 272.
373. K. C. Moller, H. J. Santner, W. Kern, S. Yamaguchi, J. O. Besenhard, and M. Winter, *Journal of Power Sources* **2003**, *119–121*, 561.
374. C. S. Wang, A. J. Appleby, and F. E. Little, *Journal of Electroanalytical Chemistry* **2002**, *519*, 9.
375. K. Edstrom and M. Herranen, *Journal of the Electrochemical Society* **2000**, *147*, 3628.
376. C. S. Wang, I. Kakwan, A. J. Appleby, and F. E. Little, *Journal of Electroanalytical Chemistry* **2000**, *489*, 55.
377. S. Yamaguchi, *Tanso* **1999**, *186*, 39.
378. J. M. Chen, C. Y. Yao, C. H. Cheng, W. M. Hurng, and T. H. Kao, *Journal of Power Sources* **1995**, *54*, 494.
379. T. D. Tran, J. H. Feikert, X. Song, and K. Kinoshita, *Journal of the Electrochemical Society* **1995**, *142*, 3297.

380. F. Coowar, D. Billaud, J. Ghanbaja, and P. Baudry, *Journal of Power Sources* **1996**, *62*, 179.
381. Y. Nakagawa, S. Wang, Y. Matsumura, and C. Yamaguchi, *Synthetic Metals* **1997**, *85*, 1343.
382. S. Ahn, *Electrochemical and Solid-State Letters* **1998**, *1*, 111.
383. M. Jean, A. Chausse, and R. Messina, *Electrochimica Acta* **1998**, *43*, 1795.
384. D. J. Derwin, K. Kinoshita, T. D. Tran, and P. Zaleski, *Symposium on Materials for Electrochemical Energy Storage and Conversion II-Batteries, Capacitors and Fuel Cells, at the 1997 MRS Fall Meeting*, Vol. 496, Materials Research Society, Boston, MA, **1998**, pp. 575–580.
385. D. Larcher, C. Mudalige, M. Gharghouri, and J. R. Dahn, *Electrochimica Acta* **1999**, *44*, 4069.
386. G. A. Nazri, B. Yebka, M. Nazri, M. D. Curtis, K. Kinoshita, *Materials Research Society Symposium Proceedings*, Vol. 548 (Solid State Ionics V), Materials Research Society, Warrendale, PA, **1999**, pp. 27–36.
387. K. Guérin, A. Fevrier Bouvier, S. Flandrois, M. Couzi, B. Simon, and P. Biensan, *Journal of the Electrochemical Society* **1999**, *146*, 3660.
388. M. C. Smart, B. V. Ratnakumar, S. Surampudi, Y. Wang, X. Zhang, S. G. Greenbaum, A. Hightower, C. C. Ahn, and B. Fultz, *Journal of the Electrochemical Society* **1999**, *146*, 3963.
389. G. A. Nazri, B. Yebka, M. Nazri, D. Curtis, K. Kinoshita, and D. Derwin, *Symposium on Solid State Ionics at the 1998 MRS Fall Meeting*, Vol. 548, Materials Research Society, Boston, MA, **1999**, pp. 27–36.
390. K. Guérin, A. Fevrier-Bouvier, S. Flandrois, M. Couzi, B. Simon, and P. Biensan, *Journal of the Electrochemical Society* **1999**, *146*, 3660.
391. F. Cao, I. V. Barsukov, H. J. Bang, P. Zaleski, and J. Prakash, *Journal of the Electrochemical Society* **2000**, *147*, 3579.
392. M. Gaberscek, M. Bele, J. Drofenik, R. Dominko, and S. Pejovnik, *Electrochemical and Solid State Letters* **2000**, *3*, 171.
393. K. Zaghib, G. Nadeau, and K. Kinoshita, *Journal of the Electrochemical Society* **2000**, *147*, 2110.
394. F. Joho, B. Rykart, A. Blome, P. Novák, H. Wilhelm, and M. E. Spahr, *Journal of Power Sources* **2001**, *97–98*, 78.
395. S. Yoon, H. Kim, and S. M. Oh, *Journal of Power Sources* **2001**, *94*, 68.
396. S. Pejovnik, M. Bele, R. Dominko, J. Drofenik, and M. Gaberscek, *Acta Chimica Slovenica* **2001**, *48*, 115.
397. K. Zaghib, G. Nadeau, and K. Kinoshita, *Journal of Power Sources* **2001**, *97–98*, 97.
398. K. Yamaguchi, J. Suzuki, M. Saito, K. Sekine, and T. Takamura, *Journal of Power Sources* **2001**, *97–98*, 159.
399. K. Zaghib, F. Brochu, A. Guerfi, and K. Kinoshita, *Journal of Power Sources* **2001**, *103*, 140.
400. Z. W. Xie, X. G. Qu, D. X. Zhu, J. Gong, L. Y. Qu, and D. M. Xie, *Chemical Journal of Chinese Universities-Chinese* **2001**, *22*, 2048.
401. J. Drofenik, M. Gaberscek, R. Dominko, M. Bele, and S. Pejovnik, *Journal of Power Sources* **2001**, *94*, 97.
402. Q. M. Pan, K. K. Guo, L. Z. Wang, and S. B. Fang, *Journal of the Electrochemical Society* **2002**, *149*, A1218.
403. K. Zaghib and K. Kinoshita, *Conference of the NATO Advanced-Study-Institute on New Trends in Intercalation Compounds for Energy Storage*, Vol. 61, Kluwer Academic Publishers, Sozopol, Bulgaria, **2002**, pp. 27–38.
404. S. Komaba, T. Ohtsuka, B. Kaplan, T. Itabashi, N. Kumagai, and H. Groult, *Chemistry Letters* **2002**, 1236.
405. J.-B. Gong, H.-S. Gao, Y. Gao, H. Liang, C.-G. Liu, and D.-G. Wang, *Dianyuan Jishu* **2003**, *27*, 205.
406. J. Drofenik, M. Gaberscek, R. Dominko, F. W. Poulsen, M. Mogensen, S. Pejovnik, and J. Jamnik, *Electrochimica Acta* **2003**, *48*, 883.
407. B. Blizanac, S. Mentus, N. Cvijeticanin, and N. Pavlovic, *Journal of the Serbian Chemical Society* **2003**, *68*, 119.
408. J. T. Chen, H. H. Zhou, W. B. Chang, and Y. X. Ci, *Acta Physico-Chimica Sinica* **2003**, *19*, 278.
409. S. Hossain, Y. K. Kim, Y. Saleh, and R. Loutfy, *Journal of Power Sources* **2003**, *114*, 264.
410. M. Yoo, C. W. Frank, and S. Mori, *Chemistry of Materials* **2003**, *15*, 850.
411. S. Hossain, R. Loutfy, Y. Saleh, and Y.-K. Kim, *Proceedings of the Power Sources Conference* **2004**, *41*, 112.
412. J. Y. Xu, H. Tanaka, M. Kurihara, S. Maruyama, Y. Moriyoshi, and T. Ishigaki, *Journal of Power Sources* **2004**, *133*, 260.
413. M. Holzapfel, C. Jost, A. Prodi-Schwab, F. Krumeich, A. Wuersig, H. Buqa, and P. Novák, *Carbon* **2005**, *43*, 1488.

414. J. K. Baek, H. Y. Lee, S. W. Jang, and S. M. Lee, *Journal of Materials Science* **2005**, *40*, 347.
415. X. M. Wang, H. Naito, Y. Sone, G. Segami, and S. Kuwajima, *Journal of the Electrochemical Society* **2005**, *152*, A1996.
416. J. Christensen and J. Newman, *Journal of the Electrochemical Society* **2005**, *152*, A818.
417. M. K. Datta and P. N. Kumta, *Journal of Power Sources* **2006**, *158*, 557.
418. E. Kim, Y. Kim, M. G. Kim, and J. Cho, *Electrochemical and Solid State Letters* **2006**, *9*, A156.
419. M. Wachtler, M. Wohlfahrt-Mehrens, S. Strobele, J. C. Panitz, and U. Wietelmann, *Journal of Applied Electrochemistry* **2006**, *36*, 1199.
420. V. G. Khomenko and V. Z. Barsukov, *Electrochimica Acta* **2007**, *52*, 2829.
421. J. Shim and K. A. Striebel, *Journal of Power Sources* **2007**, *164*, 862.
422. F. Joho, B. Rykart, R. Imhof, P. Novák, M. E. Spahr, and A. Monnier, *Journal of Power Sources* **1999**, *81–82*, 243.
423. G. Gritzner and G. Kreysa, *Pure and Applied Chemistry* **1993**, *65*, 1009.
424. Y. Q. Chang, H. Li, L. Wu, and T. H. Lu, *Journal of Power Sources* **1997**, *68*, 187.
425. L. Feng, Y. Chang, L. Wu, and T. Lu, *Journal of Power Sources* **1996**, *63*, 149.
426. R. Yazami, *Electrochimica Acta* **1999**, *45*, 87.
427. B. Simon, S. Flandrois, K. Guérin, A. Fevrier-Bouvier, I. Teulat, and P. Biensan, *Journal of Power Sources* **1999**, *81–82*, 312.
428. E. Frackowiak, S. Gautier, H. Gaucher, S. Bonnamy, and F. Béguin, *Carbon* **1999**, *37*, 61.
429. R. S. Rubino and E. S. Takeuchi, *Journal of Power Sources* **1999**, *82*, 373.
430. K. Suzuki, T. Hamada, and T. Sugiura, *Journal of the Electrochemical Society* **1999**, *146*, 890.
431. S. Flandrois and K. Guérin, *Molecular Crystals and Liquid Crystals Science and Technology, Section A: Molecular Crystals and Liquid Crystals* **2000**, *340*, 493.
432. I. Mochida, C. H. Ku, and Y. Korai, *Carbon* **2001**, *39*, 399.
433. Y. K. Choi, K. I. Chung, W. S. Kim, and Y. E. Sung, *Microchemical Journal* **2001**, *68*, 61.
434. K. Chung, M. W. Chung, W. S. Kim, S. K. Kim, Y. E. Sung, and Y. K. Choi, *Bulletin of the Korean Chemical Society* **2001**, *22*, 189.
435. T. L. Kulova, L. S. Kanevskii, A. M. Skundin, D. E. Sklovskii, N. A. Asryan, and G. N. Bondarenko, *Russian Journal of Electrochemistry* **2001**, *37*, 1017.
436. K. Zaghib and K. Kinoshita, *NATO Science Series, II: Mathematics, Physics and Chemistry* **2002**, *61*, 27.
437. G. T. K. Fey, K. L. Chen, and Y. C. Chang, *Materials Chemistry and Physics* **2002**, *76*, 1.
438. Z. H. Yang, H. Q. Wu, and B. Simard, *Electrochemistry Communications* **2002**, *4*, 574.
439. E. Frackowiak and F. Béguin, *Carbon* **2002**, *40*, 1775.
440. K. Kinoshita and K. Zaghib, *Journal of Power Sources* **2002**, *110*, 416.
441. Y. P. Wu, C. Jiang, C. Wan, and R. Holze, *Journal of Power Sources* **2002**, *111*, 329.
442. T. L. Kulova and A. M. Skundin, *Russian Journal of Electrochemistry* **2002**, *38*, 1319.
443. Z. H. Yu and F. Wu, *New Carbon Materials* **2002**, *17*, 29.
444. D. Aurbach, H. Teller, M. Koltypin, and E. Levi, *Journal of Power Sources* **2003**, *119–121*, 2.
445. T. L. Kulova and A. M. Skundin, *Journal of Solid State Electrochemistry* **2003**, *8*, 59.
446. J. P. Shim and K. A. Striebel, *Journal of Power Sources* **2004**, *130*, 247.
447. S. S. Zhang, K. Xu, and T. R. Jow, *Journal of Power Sources* **2004**, *130*, 281.
448. K. A. Striebel, A. Sierra, J. Shim, C. W. Wang, and A. M. Sastry, *Journal of Power Sources* **2004**, *134*, 241.
449. T. L. Kulova, *Russian Journal of Electrochemistry* **2004**, *40*, 1052.
450. M. Yoshio, H. Y. Wang, K. Fukuda, T. Umeno, T. Abe, and Z. Ogumi, *Journal of Materials Chemistry* **2004**, *14*, 1754.
451. L. J. Ning, Y. P. Wu, L. Z. Wang, S. B. Fang, and R. Holze, *Journal of Solid State Electrochemistry* **2005**, *9*, 520.
452. H. J. Guo, X. H. Li, Z. X. Wang, W. J. Peng, and Y. X. Guo, *Journal of Central South University of Technology* **2005**, *12*, 50.
453. K. J. Takeuchi, A. C. Marschilok, G. C. Lau, R. A. Leising, and E. S. Takeuchi, *Journal of Power Sources* **2006**, *157*, 543.
454. B. Zhang, H.-j. Guo, X.-h. Li, Z.-x. Wang, and W.-j. Peng, *Zhongnan Daxue Xuebao, Ziran Kexueban* **2007**, *38*, 454.
455. F. Joho, B. Rykart, R. Imhof, P. Nova, M. E. Spahr, and A. Monnier, *Journal of Power Sources* **1999**, *81–82*, 243.
456. G. B. Li, R. J. Xue, and L. Q. Chen, *Solid State Ionics* **1996**, *90*, 221.

457. M. Winter, G. H. Wrodnigg, J. O. Besenhard, W. Biberacher, and P. Novák, *Journal of the Electrochemical Society* **2000**, *147*, 2427.
458. J. Besenhard and H. P. Fritz, *Journal of the Electrochemical Society* **1972**, *119*, 1697.
459. P. Touzain and A. Jobert, *Electrochimica Acta* **1981**, *26*, 1133.
460. S. S. Zhang, K. Xu, J. L. Allen, and T. R. Jow, *Journal of Power Sources* **2002**, *110*, 216.
461. S. K. Jeong, M. Inaba, Y. Iriyama, T. Abe, and Z. Ogumi, *Electrochemical and Solid State Letters* **2003**, *6*, A13.
462. T. Abe, N. Kawabata, Y. Mizutani, M. Inaba, and Z. Ogumi, *Journal of the Electrochemical Society* **2003**, *150*, A257.
463. Y. S. Hu, W. H. Kong, Z. X. Wang, H. Li, X. J. Huang, and L. Q. Chen, *Electrochemical and Solid State Letters* **2004**, *7*, A442.
464. W. S. Kim, D. W. Park, H. J. Jung, and Y. K. Choi, *Bulletin of the Korean Chemical Society* **2006**, *27*, 82.
465. T. L. Kulova and A. M. Skundin, *Russian Journal of Electrochemistry* **2006**, *42*, 251.
466. R. Chandrasekaran, Y. Ohzawa, T. Nakajima, M. Koh, and H. Aoyama, *Journal of New Materials for Electrochemical Systems* **2006**, *9*, 181.
467. J. S. Gnanaraj, R. W. Thompson, J. F. DiCarlo, and K. M. Abraham, *Journal of the Electrochemical Society* **2007**, *154*, A185.
468. T. Achiha, T. Nakajima, and Y. Ohzawa, *Journal of the Electrochemical Society* **2007**, *154*, A827.
469. J. Gao, H. P. Zhang, T. Zhang, Y. P. Wu, and R. Holze, *Solid State Ionics* **2007**, *178*, 1225.
470. T. Achiha, S. Shibata, T. Nakajima, Y. Ohzawa, A. Tressaud, and E. Durand, *Journal of Power Sources* **2007**, *171*, 932.
471. L. Fu, J. Gao, T. Zhang, Q. Cao, L. C. Yang, Y. P. Wu, and R. Holze, *Journal of Power Sources* **2007**, *171*, 904.
472. M. Inaba, Z. Siroma, A. Funabiki, Z. Ogumi, T. Abe, Y. Mizutani, and M. Asano, *Langmuir* **1996**, *12*, 1535.
473. B. Simon, S. Flandrois, A. Fevrier-Bouvier, and P. Biensan, *Molecular Crystals and Liquid Crystals Science and Technology, Section A: Molecular Crystals and Liquid Crystals* **1998**, *310*, 333.
474. D. Aurbach, B. Markovsky, I. Weissman, E. Levi, and Y. Ein-Eli, *Electrochimica Acta* **1999**, *45*, 67.
475. M. E. Spahr, H. Wilhelm, F. Joho, J. C. Panitz, J. Wambach, P. Novák, and N. Dupont-Pavlovsky, *Journal of the Electrochemical Society* **2002**, *149*, A960.
476. A. Wuersig, H. Buqa, M. E. Spahr, and P. Novák, *GDCh-Monographie* **2004**, *29*, 110.
477. M. E. Spahr, T. Palladino, H. Wilhelm, A. Wuersig, D. Goers, H. Buqa, M. Holzapfel, and P. Novák, *Journal of the Electrochemical Society* **2004**, *151*, A1383.
478. H. Buqa, A. Wuersig, D. Goers, L. J. Hardwick, M. Holzapfel, P. Novák, F. Krumeich, and M. E. Spahr, *Journal of Power Sources* **2005**, *146*, 134.
479. M. E. Spahr, H. Buqa, A. Würsig, D. Goers, L. Hardwick, P. Novák, F. Krumeich, J. Dentzer, and C. Vix-Guterl, *Journal of Power Sources* **2006**, *153*, 300.
480. M. Dubois and D. Billaud, *Electrochimica Acta* **1998**, *44*, 805.
481. G. H. Wrodnigg, L. H. Lie, M. Winter, and J. O. Besenhard, *Advances in Science and Technology (Faenza, Italy)* **1999**, *24*, 131.
482. G. C. Chung, H. J. Kim, S. I. Yu, S. H. Jun, J. W. Choi, and M. H. Kim, *Journal of the Electrochemical Society* **2000**, *147*, 4391.
483. H. Buqa, A. Würsig, A. Goers, L. J. Hardwick, M. Holzapfel, P. Novák, F. Krumeich, and M. E. Spahr, *Journal of Power Sources* **2005**, *146*, 134.
484. H. Buqa, A. Wuersig, J. Vetter, M. E. Spahr, F. Krumeich, and P. Novák, *Journal of Power Sources* **2006**, *153*, 385.
485. Q. M. Pan, K. K. Guo, L. Z. Wang, and S. B. Fang, *Journal of Materials Chemistry* **2002**, *12*, 1833.
486. S. K. Jeong, M. Inaba, Y. Iriyama, T. Abe, and Z. Ogumi, *Electrochimica Acta* **2002**, *47*, 1975.
487. T. Abe, Y. Mizutani, N. Kawabata, M. Inaba, and Z. Ogumi, *Synthetic Metals* **2001**, *125*, 249.
488. G. C. Chung, H. J. Kim, S. H. Jun, and M. H. Kim, *Electrochemistry Communications* **1999**, *1*, 493.
489. R. Yazami and S. Genies, *Denki Kagaku* **1998**, *66*, 1293.
490. J. O. Besenhard, M. Winter, J. Yang, and W. Biberacher, *Journal of Power Sources* **1995**, *54*, 228.
491. J. O. Besenhard, J. Yang, and M. Winter, *Journal of Power Sources* **1997**, *68*, 87.
492. T. Abe, Y. Mizutani, T. Tabuchi, K. Ikeda, M. Asano, T. Harada, M. Inaba, and Z. Ogumi, *Journal of Power Sources* **1997**, *68*, 216.
493. D. Aurbach, M. Koltypin, and H. Teller, *Langmuir* **2002**, *18*, 9000.
494. L. J. Hardwick, PhD thesis no. 16992, ETH Zürich (Zürich), **2007**.

495. M. E. Spahr, P. Novák, O. Haas, and R. Nesper, *Journal of Power Sources* **1995**, *54*, 346.
496. M. E. Spahr, H. Wilhelm, F. Joho, and P. Novák, *ITE Letters on Batteries, New Technologies and Medicine* **2001**, *2*, B53.
497. H. Buqa, D. Goers, M. E. Spahr, and P. Novák, *ITE Letters on Batteries, New Technologies and Medicine* **2003**, *4*, 38.
498. W. P. Hoffman, F. J. Vastola, and P. L. Walker, *Carbon* **1984**, *22*, 585.
499. N. R. Laine, F. J. Vastola, and P. L. Walker Jr., *Journal of Physical Chemistry* **1963**, *67*, 2030.
500. C. Vix-Guterl and Q. Ehrburger, in Q. Delhaes (ed.), *World of Carbon*, Vol. 2, Taylor and Francis, London, **2003**, p. 188.
501. R. Imhof and P. Novák, *Proceedings—Electrochemical Society* **1997**, *97-18*, 313.
502. R. Imhof and P. Novák, *Journal of the Electrochemical Society* **1998**, *145*, 1081.
503. M. Winter, R. Imhof, F. Joho, and P. Nova, *Journal of Power Sources* **1999**, *81–82*, 818.
504. M. Hahn, A. Wuersig, R. Gallay, P. Novák, and R. Kotz, *Electrochemistry Communications* **2005**, *7*, 925.
505. M. Holzapfel, H. Buqa, L. J. Hardwick, M. Hahn, A. Wuersig, W. Scheifele, P. Novák, R. Koetz, C. Veit, and F.-M. Petrat, *Electrochimica Acta* **2006**, *52*, 973.
506. M. Lanz and P. Novák, *Journal of Power Sources* **2001**, *102*, 277.
507. M. E. Spahr, T. Palladino, H. Wilhelm, A. Würsig, D. Goers, H. Buqa, M. Holzapfel, and P. Novák, *Journal of the Electrochemical Society* **2004**, *151*, A1383.
508. G. Au and M. Sulkes, Electronics Power Sources Directorate, Army Research Laboratory, Fort Monmouth, NJ, **1993**, p. 37.
509. S. Hossain and S. Megahed, *EVS-13*, 13th International Electric Vehicle Symposium, Japan Electric Vehicle Association, Tokyo, Japan, **1996**, pp. 45–49.
510. B. A. Johnson, Phd thesis no. DA9841733, Univ. of South Carolina, Columbia, SC **1998**.
511. F. Bis and D. Warburton, *Proceedings of the Power Sources Conference* **1998**, *38*, 33.
512. A. K. Saraswat, C. G. Castledine, and T. G. Messing, *Proceedings of the Power Sources Conference* **1998**, *38*, 37.
513. Z. Zhang, D. Fouchard, and J. R. Rea, *Journal of Power Sources* **1998**, *70*, 16.
514. C. Lampe-Onnerud, J. Shi, S. K. Singh, and B. Barnett, *Annual Battery Conference on Applications and Advances, 14th*, Jan. 12–15, 1999, Long Beach, CA, **1999**, p. 215.
515. P. Biensan, B. Simon, J. P. Pere, A. de Guibert, M. Broussely, J. M. Bodet, and F. Perton, *Journal of Power Sources* **1999**, *81–82*, 906.
516. C. Crafts, T. Borek, and C. Mowry, *Proceedings of the Power Sources Conference* **2000**, *39th*, 52.
517. J. A. Jeevarajan, F. J. Davies, B. J. Bragg, and S. M. Lazaroff, *Proceedings of the Power Sources Conference* **2000**, *39*, 56.
518. H. Maleki, G. P. Deng, I. Kerzhner-Haller, A. Anani, and J. N. Howard, *Journal of the Electrochemical Society* **2000**, *147*, 4470.
519. T. D. Hatchard, D. D. MacNeil, A. Basu, and J. R. Dahn, *Journal of the Electrochemical Society* **2001**, *148*, A755.
520. M. D. Farrington, *Journal of Power Sources* **2001**, *96*, 260.
521. E. P. Roth, *Proceedings of the Power Sources Conference* **2002**, *40*, 44.
522. A. Hammami, N. Raymond, and M. Armand, *Nature (London, U.K.)* **2003**, *424*, 635.
523. J. Yamaki, Y. Baba, N. Katayama, H. Takatsuji, M. Egashira, and S. Okada, *Journal of Power Sources* **2003**, *119–121*, 789.
524. P. G. Balakrishnan, R. Ramesh, and T. P. Kumar, *Journal of Power Sources* **2006**, *155*, 401.
525. T. Yoshida, K. Kitoh, T. Mori, H. Katsukawa, and J. Yamaki, *Electrochemical and Solid State Letters* **2006**, *9*, A458.
526. T. Yoshida, K. Kitoh, S. Ohtsubo, W. Shionoya, H. Katsukawa, and J.-i. Yamaki, *Electrochemical and Solid-State Letters* **2007**, *10*, A60.
527. E. Ferg, R. J. Gummow, A. Dekock, and M. M. Thackeray, *Journal of the Electrochemical Society* **1994**, *141*, L147.
528. M. W. Juzkow and S. T. Mayer, in H. A. Frank and E. T. Seo (eds), *Annual Battery Conference Application Advance, 12th*, Institute of Electrical and Electronics Engineers, New York, **1997**, pp. 189–193.
529. Y. A. Gao, M. V. Yakovleva, and W. B. Ebner, *Electrochemical and Solid State Letters* **1998**, *1*, 117.
530. C. O. Kelly, H. D. Friend, and S. Wilson, *IEEE Aerospace and Electronic Systems Magazine* **1999**, *14*, 37.
531. H. Maleki, G. P. Deng, I. Kerzhner Haller, A. Anani, and J. N. Howard, *Journal of the Electrochemical Society* **2000**, *147*, 4470.

532. R. A. Leising, M. J. Palazzo, E. S. Takeuchi, and K. J. Takeuchi, *Journal of Power Sources* **2001**, *97–98*, 681.

533. B. Ammundsen and J. Paulsen, *Advanced Materials* **2001**, *13*, 943.

534. T. Ohzuku and Y. Makimura, *Chemistry Letters* **2001**, *30*(8), 744.

535. R. A. Leising, M. J. Palazzo, E. S. Takeuchi, and K. J. Takeuchi, *Journal of the Electrochemical Society* **2001**, *148*, A838.

536. A. D'Epifanio, F. Croce, F. Ronci, V. R. Albertini, E. Traversa, and B. Scrosati, *Physical Chemistry Chemical Physics* **2001**, *3*, 4399.

537. C. Delmas and L. Croguennec, *MRS Bulletin* **2002**, *27*, 608.

538. S. Al-Hallaj and J. R. Selman, *Journal of Power Sources* **2002**, *110*, 341.

539. D. H. Doughty, P. C. Butler, R. G. Jungst, and E. P. Roth, *Journal of Power Sources* **2002**, *110*, 357.

540. H. Arai, M. Tsuda, K. Saito, M. Hayashi, and Y. Sakurai, *Journal of the Electrochemical Society* **2002**, *149*, A401.

541. M. Mohamedi, H. Ishikawa, and I. Uchida, *Journal of Applied Electrochemistry* **2004**, *34*, 1103.

542. M. S. Wu, P. C. J. Chiang, J. C. Lin, and Y. S. Jan, *Electrochimica Acta* **2004**, *49*, 1803.

543. H. Omanda, T. Brousse, C. Marhic, and D. M. Schleich, *Journal of the Electrochemical Society* **2004**, *151*, A922.

544. E. P. Roth, D. H. Doughty, and J. Franklin, *Journal of Power Sources* **2004**, *134*, 222.

545. J. W. Jiang and J. R. Dahn, *Electrochimica Acta* **2004**, *49*, 4599.

546. E. S. Hong, S. Okada, T. Sonoda, S. Gopukumar, and J. Yamaki, *Journal of the Electrochemical Society* **2004**, *151*, A1836.

547. J. Jiang and J. R. Dahn, *Electrochemistry Communications* **2004**, *6*, 39.

548. J. Jiang and J. R. Dahn, *Electrochimica Acta* **2005**, *50*, 4778.

549. M. Takahashi, H. Ohtsuka, K. Akuto, and Y. Sakurai, *Journal of the Electrochemical Society* **2005**, *152*, A899.

550. I. Belharouak, W. Q. Lu, D. Vissers, and K. Amine, *Electrochemistry Communications* **2006**, *8*, 329.

551. Y. H. Chen, Z. Y. Tang, X. H. Lu, and C. Y. Tan, *Progress in Chemistry* **2006**, *18*, 823.

552. H. Maleki and J. N. Howard, *Journal of Power Sources* **2006**, *160*, 1395.

553. M. R. Wagner, P. R. Raimann, A. Trifonova, K. C. Moeller, J. O. Besenhard, and M. Winter, *Analytical and Bioanalytical Chemistry* **2004**, *379*, 272.

554. S. B. Yang, H. H. Song, and X. H. Chen, *Carbon* **2006**, *44*, 730.

555. J. Christensen and J. Newman, *Journal of Solid State Electrochemistry* **2006**, *10*, 293.

556. F. P. Campana, R. Koetz, J. Vetter, P. Novák, and H. Siegenthaler, *Electrochemistry Communications* **2005**, *7*, 107.

557. G. Sarre, P. Blanchard, and M. Broussely, *Journal of Power Sources* **2004**, *127*, 65.

558. M. Broussely, P. Biensan, F. Bonhomme, P. Blanchard, S. Herreyre, K. Nechev, and R. J. Staniewicz, *Journal of Power Sources* **2005**, *146*, 90.

559. M. Broussely, S. Herreyre, F. Bonhomme, P. Biensan, P. Blanchard, K. Nechev, and G. Chagnon, *Proceedings—Electrochemical Society* **2003**, *2001–21*, 75.

560. M. Broussely, *Advances in Lithium-Ion Batteries* **2002**, 393.

561. M. Broussely, S. Herreyre, P. Biensan, P. Kasztejna, K. Nechev, and R. J. Staniewicz, *Journal of Power Sources* **2001**, *97–98*, 13.

562. K. Dokko, N. Nakata, and K. Kanamura, *The 48th Battery Symposium in Japan Vol. Extended Abstracts 2B01*, Jun-ichi Yamaki, Fukuoka, **2007**, p. 194.

563. M. Umeda, K. Dokko, Y. Fujita, M. Mohamedi, I. Uchida, and J. R. Selman, *Electrochimica Acta* **2001**, *47*, 885.

564. M. Umeda, K. Dokko, Y. Fujita, M. Mohamedi, I. Uchida, and J. R. Selman, *Electrochimica Acta* **2001**, *47*, 885.

565. K. Dokko, Y. Fujita, M. Mohamedi, M. Umeda, I. Uchida, and J. R. Selman, *Electrochimica Acta* **2001**, *47*, 933.

566. M. Nishizawa, H. Koshika, T. Itoh, M. Mohamedi, T. Abe, and I. Uchida, *Electrochemistry Communication* **1999**, *1*, 375.

567. M. Nishizawa, H. Koshika, R. Hashitani, T. Itoh, T. Abe, and I. Uchida, *Journal of Physical Chemistry B* **1999**, *103*, 4933.

568. R. de Levie, *Electrochimica Acta* **1963**, *8*, 751.

569. R. de Levie, *Electrochimica Acta* **1964**, *9*, 1231.

570. R. de Levie, *Electrochimica Acta* **1965**, *10*, 113.

571. M. D. Levi, Z. Lu, and D. Aurbach, *Solid State Ionics* **2001**, *143*, 309.

572. M. D. Levi, C. Wang, and D. Aurbach, *Journal of Electroanalytical Chemistry* **2004**, *561*, 1.

573. M. D. Levi, E. Markevich, and D. Aurbach, *Journal of Physical Chemistry B* **2005**, *109*, 7420.

574. E. Markevich, M. D. Levi, and D. Aurbach, *Journal of the Electrochemical Society* **2005**, *152*, A778.

575. M. D. Levi, E. Markevich, and D. Aurbach, *Electrochimica Acta* **2005**, *51*, 98.

576. M. D. Levi and D. Aurbach, *Journal of Physical Chemistry B* **2005**, *109*, 2763.

577. P. Novák, W. Scheifele, M. Winter, and O. Haas, *Journal of Power Sources* **1997**, *68*, 267.

578. M. W. Riley, P. S. Fedkiw, and S. A. Khan, *Proceedings—Electrochemical Society* **2001**, *2000-21*, 371.

579. V. Srinivasan and J. Newman, *Journal of the Electrochemical Society* **2004**, *151*, A1530.

580. Z. X. Shu, R. S. McMillan, and J. J. Murray, *Journal of the Electrochemical Society* **1993**, *140*, 922.

581. Q. Liu, T. Zhang, C. Bindra, J. E. Fischer, and J. Y. Josefowicz, *Journal of Power Sources* **1997**, *68*, 287.

582. M. Yoshio, H. Wang, K. Fukuda, T. Umeno, T. Abe, and Z. Ogumi, *Journal of Materials Chemistry* **2004**, *14*, 1754.

583. M. Yoshio, H. Wang, K. Fukuda, Y. Hara, and Y. Adachi, *Journal of the Electrochemical Society* **2000**, *147*, 1245.

584. H. Wang, M. Yoshio, T. Abe, and Z. Ogumi, *Journal of the Electrochemical Society* **2002**, *149*, A499.

585. M.-S. Wu, T.-L. Liao, Y.-Y. Wang, and C.-C. Wan, *Journal of Applied Electrochemistry* **2004**, *34*, 797.

586. C. Siret, F. Castaing, and P. Biensan, *LiBD2003*, Arcachon, France, **2003**.

587. M. Herstedt, L. Fransson, and K. Edström, *Journal of Power Sources* **2003**, *124*, 191.

588. H. Buqa, D. Goers, M. Holzapfel, M. E. Spahr, and P. Novák, *Journal of the Electrochemical Society* **2005**, *152*, A474.

589. K. K. Patel, J. M. Paulsen, and J. Desilvestro, *Journal of Power Sources* **2003**, *122*, 144.

590. H. Buqa, M. Holzapfel, F. Krumeich, C. Veit, and P. Novák, *Journal of Power Sources* **2006**, *161*, 617.

591. R. Santhanam and M. Noel, *Journal of Power Sources* **1996**, *63*, 1.

592. L. Kavan, F. Dousek, and K. Micka, *Solid State Ionics* **1990**, *38*, 109.

593. G. Li, R. Xue, and L. Chen, *Solid State Ionics* **1996**, *90*, 221.

594. Y. Nakagawa, S. Wang, Y. Matsumura, and C. Yamaguchi, *Synthetic Metals* **1997**, *85*, 1343.

595. A. Ohta, H. Koshina, H. Okuno, and H. Murai, *Journal of Power Sources* **1995**, *54*, 6.

596. Z. Chen, L. Christensen, and J. R. Dahn, *Journal of Applied Polymer Science* **2004**, *91*, 2958.

597. Y. Nishi, in M. Wakihara and O. Yamamoto (eds), *Lithium Ion Batteries—Fundamentals and Performance*, Wiley-VCH-Kodansha, Tokyo, **1998**, pp. 181–198.

598. Y. Koga, K. Katayama, and Y. Nishi, *Kagaku Kogyo* **1998**, *49*, 192.

599. D. Bechtold, T. Brohm, M. Maul, and E. Meissner, *Proceedings of the Power Sources Conference* **1998**, *38*, 508.

600. Y. Tanjo, T. Abe, H. Horie, T. Nakagawa, T. Miyamoto, and K. Katayama, *Society of Automotive Engineers, [Special Publication] SP* **1999**, *SP-1417*, 51.

601. T. Horiba, K. Hironaka, T. Matsumura, T. Kai, M. Koseki, and Y. Muranaka, *Journal of Power Sources* **2003**, *119–121*, 893.

602. J. Arai, T. Yamaki, S. Yamauchi, T. Yuasa, T. Maeshima, T. Sakai, M. Koseki, and T. Horiba, *Journal of Power Sources* **2005**, *146*, 788.

603. T. Horiba, T. Maeshima, T. Matsumura, M. Koseki, J. Arai, and Y. Muranaka, *Journal of Power Sources* **2005**, *146*, 107.

604. K. Gotoh, M. Maeda, A. Nagai, A. Goto, M. Tansho, K. Hashi, T. Shimizu, and H. Ishida, *Journal of Power Sources* **2006**, *162*, 1322.

605. R. Elger and G. Lindbergh, *Proceedings—Electrochemical Society* **2004**, *2003-24*, 150.

606. E. Buiel and J. R. Dahn, *Electrochimica Acta* **1999**, *45*, 121.

607. E. Frackowiak, S. Gautier, L. Duclaux, K. Metenier, and F. Béguin, *ITE Battery Letters* **1999**, *1*, 12.

608. K. Tatsumi, T. Kawamura, S. Higuchi, T. Hosotubo, H. Nakajima, and Y. Sawada, *Journal of Power Sources* **1997**, *68*, 263.

609. T. Hashimoto, M. Yamashita, K. Kanekiyo, and H. Shiroki, *Proceedings—Electrochemical Society* **2000**, *99-24*, 315.

610. M. Yoshio, H. Wang, K. Fukuda, T. Abe, and Z. Ogumi, *Chemistry Letters* **2003**, *32*, 1130.

611. Y. Ohzawa, Y. Yamanaka, K. Naga, and T. Nakajima, *Journal of Power Sources* **2005**, *146*, 125.

612. J.-H. Lee, H.-Y. Lee, S.-M. Oh, S.-J. Lee, K.-Y. Lee, and S.-M. Lee, *Journal of Power Sources* **2007**, *166*, 250.

613. S. R. Mukai, T. Tanigawa, T. Harada, H. Tamon, and T. Masuda, *Adsorption Science and Technology, Proceedings of the Pacific Basin Conference, 3rd*, May 25–29, 2003, Kyongju, Republic of Korea, **2003**, p. 313.

614. Y. J. Kim, H. J. Lee, S. W. Lee, B. W. Cho, and C. R. Park, *Carbon* **2004**, *43*, 163.
615. K. Yanagida, A. Yanai, Y. Kida, A. Funahashi, T. Nohma, and I. Yonezu, *Journal of the Electrochemical Society* **2002**, *149*, A804.
616. K. Amine, J. Liu, and I. Belharouak, *Electrochemistry Communications* **2005**, *7*, 669.
617. A. Mabuchi, H. Fujimoto, K. Tokumitsu, and T. Kasuh, *Journal of the Electrochemical Society* **1995**, *142*, 3049.
618. N. Takami, A. Satoh, M. Hara, and T. Ohsaki, *Journal of the Electrochemical Society* **1995**, *142*, 371.
619. T. Ohsaki, N. Takami, M. Kanda, and M. Yamamoto, *Studies in Surface Science and Catalysis* **2001**, *132*, 925.
620. N. Takami, T. Ohsaki, H. Hasebe, and M. Yamamoto, *Journal of the Electrochemical Society* **2002**, *149*, A9.
621. T. Ohsaki, M. Kanda, Y. Aoki, H. Shiroki, and S. Suzuki, *Journal of Power Sources* **1997**, *68*, 102.
622. G. H. Taylor, *Fuel* **1961**, *40*, 465.
623. J. D. Brooks and G. H. Taylor, in P. L. Walker Jr. (ed.), *Chemistry and Physics of Carbon*, Vol. 4, Marcel Dekker Inc., New York, **1968**, pp. 243–286.
624. I. Mochida, S.-H. Yoon, Y. Korai, K. Kanno, Y. Sakai, and M. Komatsu, in H. Marsh and F. Rodríguez-Reinoso (eds), *Sciences of Carbon Materials*, Universidad di Alicante, Alicante, **2000**, pp. 259–285.
625. R. Menéndez, M. Granda, and J. Bermejo, in H. Marsh, E. A. Heintz, and F. Rodríguez-Reinoso (eds), *Introduction to Carbon Technologies*, Universidad de Alicante, Alicante, **1997**, pp. 461–490.
626. S.-H. Yoon, Y. Korai, and I. Mochida, in H. Marsh and F. Rodríguez-Reinoso (eds), *Sciences of Carbon Materials*, Universidad de Alicante, Alicante, **2000**, pp. 287–325.
627. M. N. Richard and J. R. Dahn, *Journal of Power Sources* **1999**, *79*, 135.
628. M. N. Richard and J. R. Dahn, *Journal of the Electrochemical Society* **1999**, *146*, 2068.
629. M. N. Richard and J. R. Dahn, *Journal of Power Sources* **1999**, *83*, 71.
630. G. Nadeau, X. Y. Song, M. Masse, A. Guerfi, K. Kinoshita, and K. Zaghib, *Proceedings—Electrochemical Society* **2000**, *99-24*, 326.
631. H. Maleki, G. Deng, I. Kerzhner-Haller, A. Anani, and J. N. Howard, *Journal of the Electrochemical Society* **2000**, *147*, 4470.
632. H. Maleki and J. N. Howard, *Carbon'01, An International Conference on Carbon*, July 14–19, 2001, Lexington, KY, **2001**, p. 1244.
633. J. Jiang and J. R. Dahn, *Electrochemical and Solid-State Letters* **2003**, *6*, A180.
634. W. Lu and J. Prakash, *Journal of the Electrochemical Society* **2003**, *150*, A262.
635. E. P. Roth and D. H. Doughty, *Journal of Power Sources* **2004**, *128*, 308.
636. Z. Chen and K. Amine, *Journal of the Electrochemical Society* **2006**, *153*, A316.
637. M. Yoo, C. W. Frank, S. Mori, and S. Yamaguchi, *Polymer* **2003**, *44*, 4197.
638. T. Sato, K. Banno, T. Maruo, and R. Nozu, *Journal of Power Sources* **2005**, *152*, 264.
639. A. Mabuchi, *Tanso* **1994**, *165*, 298.
640. H. Fujimoto, A. Mabuchi, K. Fujiwara, K. Tokumitsu, K. Kitaba, and T. Kasuh, *Progress in Batteries & Battery Materials* **1998**, *17*, 169.
641. H. Nakamura, H. Komatsu, and M. Yoshio, *Journal of Power Sources* **1996**, *62*, 219.
642. W. Qiu, G. Zhang, S. Lu, and Q. Liu, *Solid State Ionics* **1999**, *121*, 73.
643. H.-C. Wu, Z.-Z. Guo, H.-P. Wen, and M.-H. Yang, *Journal of Power Sources* **2005**, *146*, 736.
644. H. Y. Wang, M. Yoshio, K. Fukuda, and Y. Adachi, in S. M. R. A. O. Z. P. J. Surampudi (ed.), *International Symposium on Lithium Batteries of the 196th Electrochemical-Society Fall Meeting*, Vol. 99, Electrochemical Society Inc, Honolulu, Hawaii, **2000**, pp. 55–72.
645. H. Wang, M. Yoshio, K. Fukuda, and Y. Adachi, *Proceedings—Electrochemical Society* **2000**, *99-25*, 55.
646. H. Wang and M. Yoshio, *Journal of Power Sources* **2001**, *93*, 123.
647. N. Ohta, H. Nozaki, K. Nagaoka, K. Hoshi, T. Tojo, T. Sogabe, and M. Inagaki, *New Carbon Materials* **2002**, *17*, 61.
648. I. Kuribayashi, M. Yokoyama, and M. Yamashita, *Journal of Power Sources* **1995**, *54*, 1.
649. H. Y. Lee, J. K. Baek, S. W. Jang, S. M. Lee, S. T. Hong, K. Y. Lee, and M. H. Kim, *Journal of Power Sources* **2001**, *101*, 206.
650. N. Ohta, H. Nozaki, K. Nagaoka, K. Hoshi, T. Tojo, T. Sogabe, and M. Inagaki, *Xinxing Tan Cailiao* **2002**, *17*, 61.
651. H. Wang, K. Fukuda, M. Yoshio, T. Abe, and Z. Ogumi, *Chemistry Letters* **2002**, *31*(2), 238.
652. H. Wang, K. Fukuda, and M. Yoshio, *ITE Letters on Batteries, New Technologies and Medicine* **2003**, *4*, 410.

653. H.-Y. Lee, J.-K. Baek, S.-M. Lee, H.-K. Park, K.-Y. Lee, and M.-H. Kim, *Journal of Power Sources* **2004**, *128*, 61.
654. M. Yoshio, H. Wang, and K. Fukuda, *Angewandte Chemie, International Edition* **2003**, *42*, 4203.
655. Y.-S. Ding, W.-N. Li, S. Iaconetti, X.-F. Shen, J. DiCarlo, F. S. Galasso, and S. L. Suib, *Surface and Coatings Technology* **2006**, *200*, 3041.
656. S. Yamaguchi, *Batteries 2007*, Nice, France, **2007**.
657. Y. Sato, Y. Kikuchi, T. Nakano, G. Okuno, K. Kobayakawa, T. Kawai, and A. Yokoyama, *Journal of Power Sources* **1999**, *81–82*, 182.
658. E. P. Roth and D. H. Doughty, *Journal of Power Sources* **2004**, *128*, 308.
659. Y. Muranaka, K. Nishimura, H. Honbo, S. Takeuchi, H. Andou, Y. Kozono, H. Miyadera, M. Oda, M. Koseki, and T. Horiba, *EVS-13, 13th International Electric Vehicle Symposium*, Japan Electric Vehicle Association, Tokyo, Japan, **1996**, pp. 682–687.
660. J. Aragane, K. Matsui, H. Andoh, S. Suzuki, H. Fukuda, H. Ikeya, K. Kitaba, and R. Ishikawa, *Journal of Power Sources* **1997**, *68*, 13.
661. K. Nishimura, H. Honbo, S. Takeuchi, T. Horiba, M. Oda, M. Koseki, Y. Muranaka, Y. Kozono, and H. Miyadera, *Journal of Power Sources* **1997**, *68*, 436.
662. K. Tamura and T. Horiba, *Journal of Power Sources* **1999**, *81–82*, 156.
663. N. Dimov, S. Kugino, and M. Yoshio, *Journal of Power Sources* **2004**, *136*, 108.
664. H. Dong, R. X. Feng, X. P. Ai, Y. L. Cao, and H. X. Yang, *Electrochimica Acta* **2004**, *49*, 5217.
665. H.-Y. Lee and S.-M. Lee, *Electrochemistry Communications* **2004**, *6*, 465.
666. J.-Y. Lee, Y.-M. Kang, K.-T. Kim, Y.-J. Lee, Y.-M. Kim, S.-J. Lee, and K.-Y. Lee, *Advanced Materials for Energy Conversion II, Proceedings of a Symposium held during the TMS Annual Meeting, 2nd*, Mar. 14–18, 2004, Charlotte, NC, **2004**, p. 169.
667. Y.-S. Lee, J.-H. Lee, Y.-W. Kim, Y.-K. Sun, and S.-M. Lee, *Electrochimica Acta* **2006**, *52*, 1523.
668. M.-S. Park, Y.-M. Kang, S. Rajendran, H.-S. Kwon, and J.-Y. Lee, *Materials Chemistry and Physics* **2006**, *100*, 496.
669. P. Wang, Y. NuLi, J. Yang, and Y. Zheng, *International Journal of Electrochemical Science* **2006**, *1*, 122.
670. M. Yoshio, T. Tsumura, and N. Dimov, *Journal of Power Sources* **2006**, *163*, 215.
671. M. K. Datta and P. N. Kumta, *Journal of Power Sources* **2007**, *165*, 368.
672. P. Zuo, G. Yin, J. Zhao, Y. Ma, X. Cheng, P. Shi, and T. Takamura, *Electrochimica Acta* **2006**, *52*, 1527.
673. Y. Liu, T. Matsumura, A. Hirano, T. Ichikawa, N. Imanishi, and Y. Takeda, *Research Reports of the Faculty of Engineering, Mie University* **2004**, *29*, 9.
674. M. Holzapfel, H. Buqa, W. Scheifele, P. Novák, and F.-M. Petrat, *Chemical Communications (Cambridge, U.K.)* **2005**, *12*, 1566.
675. W.-J. Jin, T. R. Kim, S. H. Moon, Y.-S. Lim, and M.-S. Kim, *Materials Science Forum* **2006**, *510–511*, 1078.
676. A. M. Wilson, J. R. Dahn, J. S. Xue, Y. Gao, and X. H. Feng, *Materials Research Society Symposium Proceedings* **1995**, *393*, 305.
677. X.-Q. Yang, J. McBreen, W.-S. Yoon, M. Yoshio, H. Wang, K. Fukuda, and T. Umeno, *Electrochemistry Communications* **2002**, *4*, 893.
678. M. Yoshio, H. Wang, K. Fukuda, T. Umeno, N. Dimov, and Z. Ogumi, *Journal of the Electrochemical Society* **2002**, *149*, A1598.
679. N. Dimov, S. Kugino, and M. Yoshio, *Electrochimica Acta* **2003**, *48*, 1579.
680. U. S. Kasavajjula and C. Wang, *AIChE Annual Meeting, Conference Proceedings*, Nov. 7–12, 2004, Austin, TX, **2004**, p. 332B/1.
681. B. J. Jeon, S. W. Kang, and J. K. Lee, *Korean Journal of Chemical Engineering* **2006**, *23*, 854.
682. S.-H. Ng, J. Wang, D. Wexler, K. Konstantinov, Z.-P. Guo, and H.-K. Liu, *Angewandte Chemie, International Edition* **2006**, *45*, 6896.
683. Z. Zhao, Z. Wu, W. Yang, X. Liu, and W. Jiang, *Beijing Daxue Xuebao, Ziran Kexueban* **2006**, *42*, 39.
684. T. Zhang, J. Gao, L. J. Fu, L. C. Yang, Y. P. Wu, and H. Q. Wu, *Journal of Materials Chemistry* **2007**, *17*, 1321.
685. H.-Y. Lee, Y.-L. Kim, M.-K. Hong, and S.-M. Lee, *Journal of Power Sources* **2005**, *141*, 159.
686. M. Holzapfel, H. Buqa, F. Krumeich, P. Novák, F. M. Petrat, and C. Veit, *Electrochemical and Solid-State Letters* **2005**, *8*, A516.
687. N. Dimov, H. Noguchi, and M. Yoshio, *Journal of Power Sources* **2006**, *156*, 567.

688. Y. Liu, T. Matsumura, N. Imanishi, A. Hirano, T. Ichikawa, and Y. Takeda, *Electrochemical and Solid-State Letters* **2005**, *8*, A599.

689. T. Morita and N. Takami, *Journal of the Electrochemical Society* **2006**, *153*, A425.

690. K. Sekine, T. Shimoyamada, R. Takagi, K. Sumiya, and T. Takamura, *Proceedings—Electrochemical Society* **1997**, *97-18*, 92.

691. K. Sumiya, K. Sekine, and T. Takamura, *Proceedings—Electrochemical Society* **1997**, *97-18*, 523.

692. T. Takamura, K. Sumiya, Y. Nishijima, J. Suzuki, and K. Sekine, *Materials Research Society Symposium Proceedings* **1998**, *496*, 557.

693. T. Takamura, J. Suzuki, C. Yamada, K. Sumiya, and K. Sekine, *Surface Modification Technologies XII, Proceedings of the 12th International Conference on Surface Modification Technologies,* Oct. 12–14, 1998, Rosemont, IL, **1998**, p. 313.

694. T. Takamura, K. Sumiya, J. Suzuki, C. Yamada, and K. Sekine, *Journal of Power Sources* **1999**, *81–82*, 368.

695. T. Takamura, J. Suzuki, C. Yamada, K. Sumiya, and K. Sekine, *Surface Engineering* **1999**, *15*, 225.

696. K. Sumiya, J. Suzuki, R. Takasu, K. Sekine, and T. Takamura, *Journal of Electroanalytical Chemistry* **1999**, *462*, 150.

697. J. Suzuki, M. Yoshida, C. Nakahara, K. Sekine, M. Kikuchi, and T. Takamura, *Electrochemical and Solid-State Letters* **2001**, *4*, A1.

698. J. Suzuki, O. Omae, K. Sekine, and T. Takamura, *Solid State Ionics* **2002**, *152–153*, 111.

699. J. Suzuki, M. Yoshida, Y. Nishijima, K. Sekine, and T. Takamura, *Electrochimica Acta* **2002**, *47*, 3881.

700. T. Takamura, H. Awano, T. Ura, and K. Sumiya, *Journal of Power Sources* **1997**, *68*, 114.

701. T. Takumaura, M. Saito, A. Shimokawa, C. Nakahara, K. Sekine, S. Maeno, and N. Kibayashi, *Journal of Power Sources* **2000**, *90*, 45.

702. C.-W. Wang, Y.-B. Yi, A. M. Sastry, J. Shim, and K. A. Striebel, *Journal of the Electrochemical Society* **2004**, *151*, A1489.

703. W. F. Bentley, in H. A. Frank and E. T. Seo (eds), *Annu Battery Conf Appl Adv, 12th*, Institute of Electrical and Electronics Engineers, New York, **1997**, pp. 223–226.

704. C. H. Doh, H. S. Kim, and S. I. Moon, *Journal of Power Sources* **2001**, *101*, 96.

705. O. Haas, E. Deiss, P. Novák, W. Scheifele, and A. Tsukada, *Proceedings—Electrochemical Society* **1997**, *97-18*, 451.

8 Electrical Double-Layer Capacitors and Pseudocapacitors

François Béguin, Encarnación Raymundo-Piñero, and Elzbieta Frackowiak

CONTENTS

8.1 INTRODUCTION

Supercapacitors (or ultracapacitors or electrochemical capacitors) are a relatively new energy storage system which was the object of an important scientific and industrial development during the last years. Supercapacitors provide a higher energy density than dielectric capacitors and a higher power density than batteries. They are particularly adapted for applications which require energy pulses during short periods of time, e.g., seconds or tens of seconds. They are recommended for automobiles, tramways, buses (being able to boost the system during acceleration and to recover energy during braking, the so-called regenerative braking), cranes, forklifts, wind turbines, electricity load leveling in stationary and transportation systems, etc. In the case of automobiles, they are combined to another power source (lead-acid battery, nickel-metal hydride battery, lithium-ion battery, internal combustion engine, fuel cell) which enables the autonomy of the system (Figure 8.1).

In order to satisfy the industrial demand, the performance of supercapacitors must be improved and new solutions should be proposed. The development of new materials and new concepts has enabled important breakthroughs during the last years. In this forecast, carbon plays a central role. Due to its low cost, versatility of nanotextural and structural properties, high electrical conductivity, it is the main electrode component. Nanoporous carbons are the active electrode material, whereas carbon blacks or nanotubes can be used for improving the conductivity of electrodes or as support of other active materials, e.g., oxides or electrically conducting polymers.

This chapter intends to discuss the fundamental role played by carbons, taking particularly into account their nanotexture and surface functionality. The general properties of supercapacitors are reviewed, and the correlation between the double-layer capacitance and the nanoporous texture of carbons is shown. The contribution of pseudocapacitance through pseudofaradaic charge transfer reactions is introduced and developed for carbons with heteroatoms involved in functionalities able to participate to redox couples, e.g., the quinone/hydroquinone pair. Especially, we present carbons obtained by direct carbonization (without any further activation) of appropriate polymeric precursors containing a high amount of heteroatoms.

The Section 8.4.3 is devoted to hydrogen storage by electrochemical reduction of water with a nanoporous carbon cathode. The mechanism of hydrogen storage and nature of sites is discussed in light of various experiments. It is shown that hydrogen stored electrochemically is weakly chemically bonded with the carbon surface. Consequently, carbons loaded with hydrogen are perfect as electrode of electrochemical cells, as they are able to keep charge in ambient pressure and temperature conditions.

Based on the information developed in the previous sections of the chapter, the last part (8.5) presents hybrid capacitors based on two different electrodes in organic and aqueous media. Especially, novel high-performance asymmetric capacitors are able to operate at voltage values as high as 2 V in water. The power and energy density of these environment-friendly devices are comparable with the values obtained with symmetric capacitors in organic electrolyte. Different constructions with carbon-based electrodes are presented, playing on nanotexture, surface functionality, and electrical conductivity of carbons. The most efficient device associates a negative electrode storing hydrogen and a positive one able to operate at high values of potential.

8.2 PRINCIPLE AND PROPERTIES OF ELECTROCHEMICAL CAPACITORS

8.2.1 TERMINOLOGY AND KINDS OF SUPERCAPACITORS

Depending on the charge storage mechanism, one must distinguish between the electrical double-layer capacitors (EDLC) and the pseudocapacitors. The principles and properties of both types of supercapacitors will be further described.

The most known supercapacitor is the symmetric one, i.e., with two identical electrodes immersed in an aqueous or an organic electrolyte (Figure 8.2). The two electrodes are separated by a porous membrane (paper, glass fiber, polymer) named separator. In the industrial capacitors, the electrode

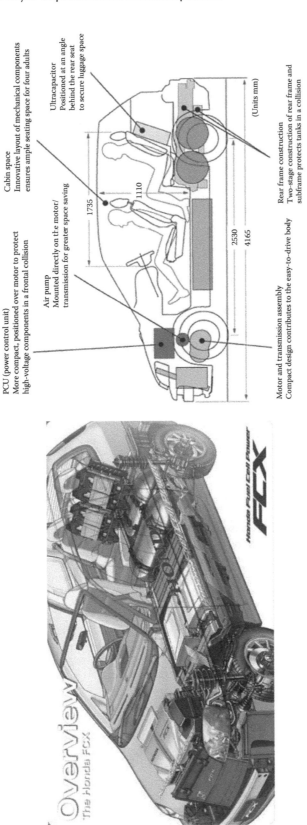

PCU (power control unit)
More compact, positioned over motor to protect high-voltage components in a frontal collision

Cabin space
Innovative layout of mechanical components ensures ample seating space for four adults

Ultracapacitor
Positioned at an angle behind the rear seat to secure luggage space

Air pump
Mounted directly on the motor/transmission for greater space saving

Rear frame construction
Two-stage construction of rear frame and subframe protects tanks in a collision

Motor and transmission assembly
Compact design contributes to the easy-to-drive body

(Units mm)

1735
1110
2530
4165

FIGURE 8.1 An example of utilization of an ultracapacitor in the Honda FCX. The ultracapacitor (at the back of the rear seat) associated to a fuel cell boosts the system during acceleration, recovering energy during braking.

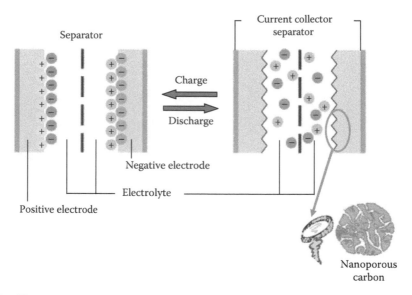

FIGURE 8.2 Charged and discharged states of a supercapacitor.

material is a nanoporous activated carbon (AC), which coats a current collector (aluminum in organic electrolyte, stainless steel in aqueous KOH). A binder (polyvinylidene fluoride, PVdF; carboxymethylcellulose, CMC; polytetrafluoroethylene, PTFE) agglomerates and links the grains of active materials with the current collector. A percolator (carbon black, carbon nanotubes [CNTs]) is added for improving the electrodes conductivity.

Figure 8.2 shows that in its charged state, a supercapacitor is equivalent to two electrodes of capacities C_1 and C_2 in series. The capacity of the total system is given by Equation 8.1:

$$\frac{1}{C} = \frac{1}{C_1} + \frac{1}{C_2} \tag{8.1}$$

As the capacity of the two electrodes is different, even in a symmetric capacitor, Equation 8.1 indicates that the value of C is determined by the electrode with the smallest capacity value. Moreover, the later electrode operates in a larger potential window than the other one, which consequently reduces the voltage range of the device. This drawback can be circumvented by balancing the respective masses of the electrodes or by using, for each electrode, different materials working in their optimal potential range.

8.2.2 ENERGY AND POWER OF SUPERCAPACITORS

The energy density of a supercapacitor is given by Equation 8.2:

$$E = \frac{1}{2}CU^2 \tag{8.2}$$

where U is the operating voltage. Hence, the energy depends essentially on voltage, i.e., on the electrolyte stability window. For symmetric capacitors, the later is limited to 0.7–0.8 V in aqueous electrolyte and 2.7 V in organic medium. Tetraethylammonium tetrafluoroborate, $(C_2H_5)_4N^+BF_4^-$ in acetonitrile is the most frequent industrial electrolyte in Europe and in the United States. The smaller value of voltage window in water medium as compared to the thermodynamic one, i.e., 1.23 V, is related to the uncontrolled potential of each electrode. When the voltage is higher than 0.7–0.8 V, the potential of one of the electrodes may be beyond the thermodynamic limit, leading to the electrolyte decomposition on the corresponding electrode. From this point of view, the organic electrolyte is more interesting, although the electrodes capacity is smaller than that in aqueous medium. However, the voltage window in organic electrolyte is strongly limited by the presence of impurities, for example traces of water. The stability window of high purity acetonitrile determined with a glassy carbon electrode is 5.9 V, and it decreases to 3.8 and 2.7 V in the presence of 14 and 40 ppm of water, respectively (Figure 8.3) [1].

With nanoporous carbon electrodes, even if high purity organic solvents are used, it is necessary to reduce the voltage window in order to limit the electrolyte decomposition on the active surface of carbon. When the supercapacitor operates, the nanopores of carbon are partly filled by decomposition products; the capacity decreases and the resistance of the system increases [2]. Unpaired electrons are assumed to be responsible of electrolyte decomposition. An example of the most frequent functionalities for graphene sheets constituting a nanoporous carbon material is shown in Chapter 5 (Figure 5.4) [3]. In addition to functional groups, free-edge sites and unpaired electrons on the basal planes are observed. The cycle life of supercapacitors is generally improved through special treatments reducing the surface functionality of carbons [4].

Capacity also influences the energy value. Actually, most of the research effort is dedicated to improving the capacity of electrode materials. Since ACs are taken as reference, the cost of the materials is an important criterion at the industrial level. Most of the materials developed for EDLC are of little interest for applications, because of their cost and only slightly improved performance. By contrast, it will be shown that advanced carbon materials can be developed for pseudocapacitors.

The second important parameter for characterizing supercapacitors is their power given by Equation 8.3:

$$P = \frac{U^2}{4R_S} \tag{8.3}$$

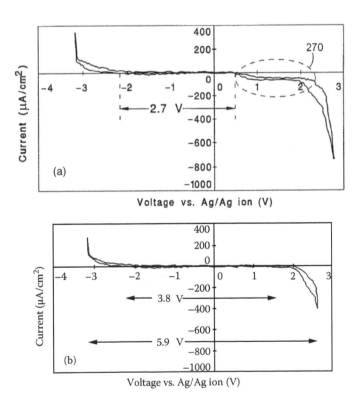

FIGURE 8.3 Stability domains of acetonitrile (a) with 40 ppm water (2.7 V); (b) with 14 ppm water (3.8 V) and pure acetonitrile (5.9 V). (From Farahmendi, J.C., et al., Patents WO9815962, EP0946954 and JP 2001502117T for Maxwell Technologies, 1998. With permission.)

TABLE 8.1

Compared Properties of Supercapacitors in Aqueous and Organic Electrolytes

Property	Organic Electrolyte	Aqueous Electrolyte
Operating voltage	2.3–2.7 V	0.6–0.7 V
Conductivity	~0.02 S cm^{-1}	~1 S cm^{-1}
Maximum capacity	150–200 F g^{-1}	250–300 F g^{-1}
Technical, economic characteristics, and security	Construction in inert atmosphere, expensive, unfriendly for environment	Easy construction, low cost, environment-friendly

where R_S is the internal resistance or equivalent series resistance (ESR) of the supercapacitor. The time constant, i.e., $R_S C$, determines the charge/discharge propagation, in other words the power. R_S includes the resistance of all materials between the current collectors (Figure 8.2), i.e., the substrate, the active material (the nanoporous carbon), the binder, the separator, and electrolyte. A thermal treatment allows the conductivity of ACs to be improved; however, the porous volume of the electrodes may simultaneously decrease. In general, a small amount of percolator (carbon black, nanotubes) improves the conductivity of the electrodes. Even if nanotubes are expensive, they help to improving performance of the active material [5,6], increasing the electrodes resiliency, and consequently the calendar life of systems during cycling. According to Equation 8.3, showing that power depends on voltage, the organic electrolytes seem to be preferable. However, this profit is strongly moderated by their small electrical conductivity as compared to the aqueous medium. Optimizing the electrode composition and the electrical contact between the substrate and the active material is crucial. Table 8.1 shows the respective advantages and drawbacks of aqueous and organic electrolytes.

The performance of electrochemical systems is compared in a Ragone plot (Figure 8.4). Supercapacitors have a higher power density than any battery; by contrast, their energy density is much lower. The main research effort is now oriented to improving the energy.

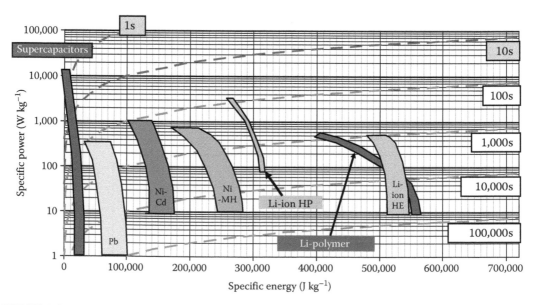

FIGURE 8.4 Ragone plot of electrochemical storage systems.

8.3 ELECTRICAL DOUBLE-LAYER CAPACITORS

8.3.1 INTRODUCTION TO THE ELECTRICAL DOUBLE LAYER

As shown in Figure 8.2, when the system is connected to a power supply, the surface of the electrodes is charged and attracts the ions of opposite charge. The ions are stored in an electrical double layer (EDL) at the surface of electrodes (Figure 8.5). Each electrode is equivalent to a capacitor with capacity given by the classical formula (Equation 8.4) [7]:

$$C_e = \frac{\varepsilon S}{d} \tag{8.4}$$

where

     S is the surface area of the electrode/electrolyte interface
     ε is the permittivity or dielectric constant
     d is the EDL thickness

As d is of the order of $1\,nm$, the specific capacity is very high, e.g., $0.1\,F\,m^{-2}$. Nanoporous carbons are ideal materials for supercapacitor electrodes [8], because of their low cost, good electrical conductivity, and very high specific surface area (between 1000 and $2500\,m^2\,g^{-1}$). The values of capacity are generally ranging from 100 to $200\,F\,g^{-1}$.

Different types of well-defined carbon nanofibers (CNFs), including platelet (PCNFs), herringbone (HCNFs), and tubular (TCNFs) types, have been used as model materials to correlate capacitance with the surface structure [9]. The surface of PCNFs and HCNFs shows graphitic edges, whereas basal planes parallel to the fiber axis are predominantly present at the surface of TCNFs. Capacitance values of 12.5, 23.4, and $4.5\,F\,g^{-1}$ were measured in $0.5\,mol\,L^{-1}$ sulfuric acid solution for PCNFs, HCNFs, and TCNFs, respectively. Hence, the edge surface of PCNFs and HCNFs is 3–5 times more effective in capacitive charging than the basal planes of TCNFs. On the other hand, graphitization of PCNFs transformed their outer surface in dome-like basal planes,

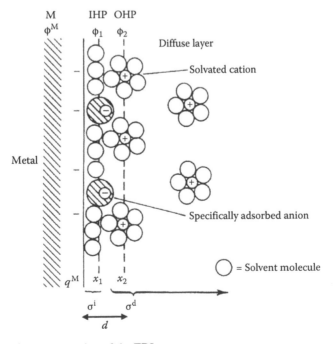

FIGURE 8.5 Schematic representation of the EDL.

and the corresponding capacitance decreased from $12.5\,F\,g^{-1}$ in pristine PCNFs to $3.2\,F\,g^{-1}$ in the graphitized ones. Definitely these data demonstrate that higher electrical charge is built up at graphitic edge surfaces under an electrochemical polarization. The different behavior of the various CNFs is attributed to the higher electrical conduction in the direction parallel to the basal planes than in the perpendicular one [9]. Under polarization of PCNFs and HCNFs, the π electrons move easily along the graphene layers to reach the edge surface of the fibers.

In literature, it is very often assumed that the BET specific surface area (S_{BET}) is equivalent to the surface area of the electrode/electrolyte interface and consequently that some proportionality should exist between capacitance and S_{BET}. Therefore, many authors try to investigate highly activated carbons. A plot of capacitance vs. S_{BET} is shown in Figure 8.6 for a variety of commercial ACs, carbon blacks, and ACs from pyrolyzed wood [10]. In the initial part of the curve, capacitance increases almost linearly with S_{BET}, but it rapidly levels-off for high S_{BET} values. This phenomenon is partly due to the fact that the BET model overestimates the surface area values [11,12]. Therefore, the density functional theory (DFT) model is proposed as a more accurate way to determine the specific surface area. The plot of capacitance vs. S_{DFT} shows a more extended region of proportionality than with the BET model, but saturation is still observed for the high surface area values, i.e., for carbons with increasing activation degree. In order to explain the saturation, Barbieri et al. suggested that the increase of pore volume is associated to a decrease of average pore wall thickness [10]. Consequently, if the walls become too thin, the electric field (and the corresponding charge density) no longer decays to zero within the pore walls, as it would do for thicker walls or the bulk material. For the various carbon materials, Table 8.2 shows the normalized capacitance (gravimetric capacitance divided by the DFT specific surface area) together with the screening length of the electric field (δ_{SC}) and the pore wall thickness (δ_W). The ratio between the pore wall thickness and the field screening length decreases with increasing capacitance. However, for surfaces areas around $1200\,m^2\,g^{-1}$, the average pore wall thickness becomes close to the screening length of the field. Hence, the observed capacitance limitation in highly activated carbons can be also ascribed to a space constriction for charge accommodation inside some pore walls.

In nanoporous carbons, however, the charge storage mechanisms cannot be restricted only to surface effects. For example, a noticeable expansion of the carbon electrode has been observed at >2 V for a typical electrochemical double-layer capacitor based on AC and an aprotic organic solvent [13]. It suggests that ion intercalation (or insertion) processes might play some role for charge storage in EDLCs. The respective sizes of ions and pores and the way the pores are connected with the outer surface of the carbon particles may also strongly influence the values of capacitance, especially under high operating

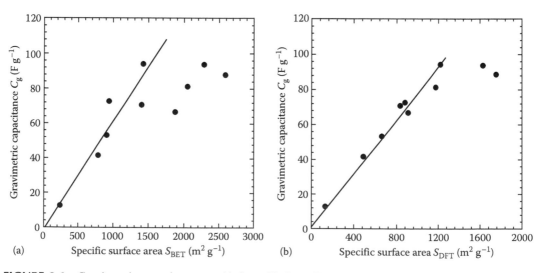

FIGURE 8.6 Gravimetric capacitance vs. (a) S_{BET}; (b) S_{DFT}. (From Barbieri, O., et al., *Carbon*, 43, 1303, 2005. With permission.)

TABLE 8.2

Gravimetric Capacitance (C), BET- and DFT-Specific Surface Area, Normalized Capacitance (C), Field Screening Length (δ_{SC}) and Pore Wall Thickness (δ_W) for Different ACs

	C (F g^{-1})	S_{BET} (m^2 g^{-1})	S_{DFT} (m^2 g^{-1})	C (μF cm^{-2})	δ_{SC} (nm)	δ_W (nm)	δ_W/δ_{SC}
Vulcan XC72	13	240	132	9.6	0.16	6.91	43.42
W30	41	790	491	8.5	0.20	1.85	9.25
W37	53	913	682	7.8	0.23	1.33	5.82
MM192	66	1875	910	7.3	0.25	1.00	3.93
Black pearl 2000	70	1405	832	8.5	0.20	1.09	5.50
W54.5	72	950	850	8.5	0.20	1.07	5.43
Picactif	81	2048	1172	6.9	0.28	0.78	2.80
K1500	94	1429	1219	7.7	0.23	0.75	3.19
K2200	94	2286	1613	5.8	0.36	0.56	1.58
K2600	88	2592	1744	5.0	0.43	0.52	1.20

Source: Barbieri, O., et al., *Carbon*, 43, 1303, 2005. With permission.

current densities of supercapacitors. Moreover, the mobility of ions under polarization and their size completely differ from those of the probes (N_2, CO_2) which are used for determining porosity data by gas adsorption. Consequently, not only the specific surface area but also the porous nanotexture of carbons must be seriously taken into account for understanding their capacitance properties.

8.3.2 INFLUENCE OF THE NANOPOROUS TEXTURE ON THE PERFORMANCE OF IONS ELECTROSORPTION

Lately, it has been unambiguously demonstrated that the EDL capacity of nanoporous carbons is essentially determined by the respective size of pores and ions. Eliad et al. [14–16] showed that a molecular sieving effect appears when the pore size approaches the effective size of ions. It was found that common cations could not enter pores smaller than their hydrated size. Figure 8.7 shows a typical box-like shape of three-electrode cell voltammetry curve with $MgSO_4$ as electrolyte when the average pore size (0.58 nm) is significantly larger than the size of ions. By contrast, for a smaller pore size carbon (0.51 nm), the current is negligible in all the potential range, implying that the Mg^{2+} and SO_4^{2-} ions are larger than 0.51 nm. Figure 8.7 also displays the voltammograms of a small average pore size carbon (C 0.51 nm) when using Li_2SO_4 and $MgCl_2$ electrolytes which combine monovalent and bivalent ions. The triangular shape of the two voltammetry curves confirms the adsorption of the small monovalent ions (Li^+ or Cl^-) into the pores and the absence of adsorption of the larger bivalent ions (SO_4^{2-} or Mg^{2+}, respectively) on the other side of the point of zero charge.

The sieving effect of the carbon host was also demonstrated by measuring the capacitance values of an AC in a series of solvent-free ionic liquids (ILs) of increasing cation size [17]. Since ions are not solvated in pure ILs, it was easy to interpret the electrochemical properties by comparing the nanoporous characteristics of carbon and the size of cations calculated by molecular modeling. It was found that the overall porosity of the carbon is noticeably underused, due to pores smaller than the effective size of the cations. The results with ILs confirm that the optimal pore size depends on the kind of electrolyte, i.e., the dimensions of pores and ions must match each other.

Comparable results were obtained when an IL (ethyl-methylimmidazolium-bis(trifluoro-methane-sulfonyl)imide; EMI-TFSI) was tested using a series of nanoporous carbons with average pore width in the range of 0.65–1.1 nm [18]. The ion sizes, calculated as 0.79 and 0.76 nm in the longest dimension for TFSI and EMI ions, respectively, are within the range of carbons pore size. Figure 8.8 points out that, when the average pore size decreases from 1.1 to 0.7 nm, the normalized capacitance increases; below ~0.7 nm, the normalized capacitance decreases. Since the maximum at 0.7 nm is

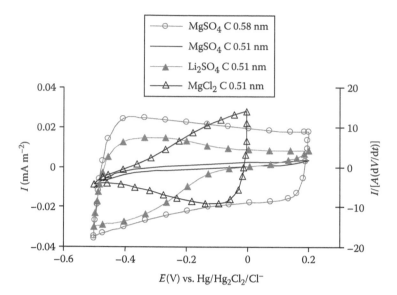

FIGURE 8.7 Voltammetry curves obtained in 0.1 mol L^{-1} MgSO$_4$ solution using AC electrodes with average pore size of 0.58 and 0.51 nm. For comparison, curves are also given for 0.1 mol L^{-1} Li$_2$SO$_4$ and MgCl$_2$ solutions with pores of 0.51 nm. The current is normalized per unit of BET specific surface area. (Adapted from Eliad, L., et al., *J. Phys. Chem. B*, 105, 6880, 2001.)

very close to the size of both types of ions, one must conclude that the maximum capacitance is observed when the ion size approaches the pore size. As average pore sizes are determined from gas adsorption data, it means that at 0.7 nm a maximum number of pores fit with the ion size. It must also be noticed that the results reported in references [17,18] are well interpreted with the assumption that the IL cations stand vertically, i.e., with their largest dimension perpendicular to the pore walls. In this orientation, more ions can occupy the available active surface of the pores, i.e., the electrosorbed charges are more closely packed. Moreover, both papers rule out the fact that mainly pores within the range of 0.8–2 nm contribute to the EDL formation [8,19,20]. Definitely, only one layer of ions is electrosorbed in each pore.

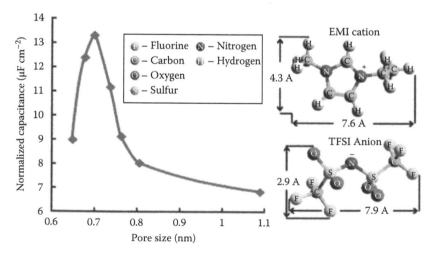

FIGURE 8.8 Normalized capacitance (gravimetric capacitance divided by the BET specific surface area) vs. average pore size of a series of CDCs; models of the structure of EMI and TFSI ions. (From Largeot, C., et al., *J. Am. Chem. Soc.*, 130, 2730, 2008. With permission.)

Raymundo-Piñero et al. have studied the effect of pore size in aqueous ($6 \, mol \, L^{-1}$ KOH, $1 \, mol \, L^{-1}$ H_2SO_4) and organic ($1 \, mol \, L^{-1}$ TEABF$_4$ in acetonitrile) electrolytes using a series of porous carbons prepared by KOH activation of chars obtained by pyrolysis of bituminous coal at $520°C–1000°C$ [21]. Since the materials were all prepared from the same precursor, by simply changing the carbonization temperature, one can reasonably assume that they are texturally identical (same pore shape and pore interconnection) and that they only differ by their average pore width. As the pyrolysis temperature of the coal increases, the BET specific surface area (S_{BET}) decreases from ca. 3000 to $1000 \, m^2 \, g^{-1}$ and the average micropore width (L_{0N_2}) from 1.4 to 0.8 nm. It is interesting to note that the volume of ultramicropores (pores in the range of 0.4–0.8 nm) measured by CO_2 adsorption, V_{DRCO_2}, remains almost unchanged for all the samples at a value around $0.3 \, cm^3 \, g^{-1}$, whereas the volume of the wider micropores measured by N_2 adsorption (V_{DRN_2}) decreases with the heat-treatment temperature. The volumetric capacitance (gravimetric capacitance, in F g^{-1}, divided by the Dubinin–Radushkevich [DR] micropore volume, in $cm^3 \, g^{-1}$) or the normalized capacitance (gravimetric capacitance, in F g^{-1} divided by the BET specific surface area, in $m^2 \, g^{-1}$) of these carbons increase as the average micropore size, L_0, decreases (Figure 8.9) [21]. These data fit well with Equation 8.4, $C = \varepsilon S/d$, showing that capacitance increases when the distance d between pore walls and ions decreases. A similar trend was observed for a series of carbide-derived carbons (CDCs) obtained by chlorination of carbides at various temperatures [22].

The theoretical values of volumetric capacitance plotted in Figure 8.9a were obtained considering that one slit-shape pore of volume L_0^3 is occupied by one ion, allowing the volumetric charge, Q_v in C cm^{-3}, to be calculated. From the equation $Q_v = C_v \times U$, the volumetric capacitance C_v (F cm^{-3}) is obtained by assuming that the voltage window, U, applied in the two-electrode cell is 1 V. From this simple approach, very interesting information is extracted without any assumption on the ions diameter. For average pore diameters larger than ~0.7 and ~0.8 nm in aqueous and organic electrolyte, respectively, the experimental values are higher than the theoretical ones, indicating that each model pore is occupied by more than one ion. By contrast, for average pore diameters smaller

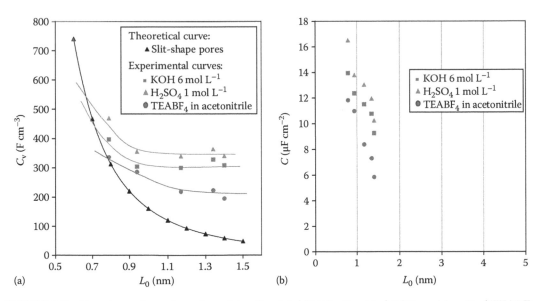

FIGURE 8.9 Capacitance in various electrolytes ($1 \, mol \, L^{-1} \, H_2SO_4$, $6 \, mol \, L^{-1}$ KOH, and $1 \, mol \, L^{-1}$ TEABF$_4$ in acetonitrile) vs. average pore size L_0 for a series of bituminous coal-derived carbons activated by KOH at 800°C. (a) Experimental volumetric capacitance (gravimetric capacitance divided by DR micropore volume) and theoretical volumetric capacitance (calculated for one ion occupying one pore of volume L_0^3); (b) Normalized capacitance (gravimetric capacitance divided by the BET specific surface area). (Adapted from Raymundo-Piñero, E., et al., *Carbon*, 44, 2498, 2006.)

than 0.7 or 0.8 nm, the experimental values are lower than the theoretical ones; in this case only the volume fraction corresponding to pores larger than the ion size is effective for their trapping. Hence, in KOH and H_2SO_4 electrolytes, the pores efficiency for forming the EDL increases when their size decreases from 0.9 to 0.7 nm. In 1 mol L^{-1} TEABF$_4$, this efficiency increases for a pore size decreasing from 1.1 to 0.8 nm, that is, logically considering the higher ion size for the organic electrolyte [21]. Another interesting feature is that for pore diameters higher than 0.9 and 1.1 nm in aqueous and organic electrolyte, respectively, the volumetric capacitance, i.e., the number of ions forming the double layer per volume unit, is nearly constant. Remarking that pores larger than 0.9 and 1.1 nm correspond to values of BET specific surface area larger than ca. 1500 and 2000 m^2 g^{-1}, one can easily understand the saturation phenomenon shown in Figures 8.10 [21] and 8.6 [10].

Since the optimal average pore size of carbon in $(C_2H_5)_4N^+$ BF_4^- solution is closer to the crystallographic diameter of the $(C_2H_5)_4N^+$ cation (0.672 nm) [23] than to the diameter of the cation solvated by acetonitrile (1.30 nm [24]), it seems reasonable to conclude that the ions penetrate in the pores without their solvation shell [21,22,25]. However, whatever the kind of nanoporous carbon, the pore size distribution is sufficiently broad for suggesting a noticeable contribution of pores larger than the average value. Therefore, data obtained from nitrogen or argon adsorption do not provide sufficiently unambiguous proof of the desolvation. In order to circumvent this problem, the porosity of templated carbons was investigated by CO_2 adsorption at 293 K [26]. As shown in Chapter 4, CO_2 is a very accurate probe for ultramicropores, e.g., pores <0.7–0.8 nm [27].

Porous carbons were prepared using MCM-48 and SBA-15 mesoporous templates and various carbon precursors (see Chapter 3 for preparation description). Figure 8.11 displays the nitrogen adsorption isotherms at 77 K of SBA-15 and of the corresponding templated carbon obtained by carbonization of sucrose in the template. Both isotherms show a bimodal porosity: in the templated carbon, mesopores are generated by the removal of the silica walls, and micropores are present in

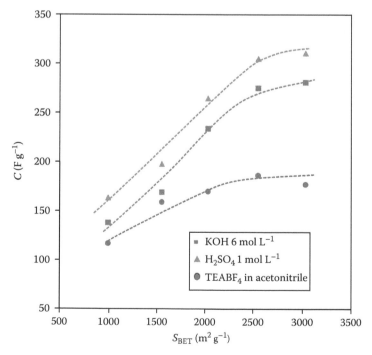

FIGURE 8.10 Gravimetric capacitance in various electrolytes (1 mol L^{-1} H_2SO_4, 6 mol L^{-1} KOH, and 1 mol L^{-1} TEABF$_4$ in acetonitrile) vs. BET specific surface area for a series of bituminous coal-derived carbons activated by KOH at 800°C. (Adapted from Raymundo-Piñero, E., et al., *Carbon*, 44, 2498, 2006.)

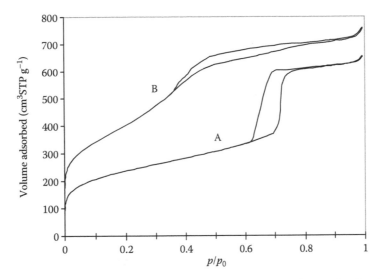

FIGURE 8.11 Nitrogen adsorption isotherm at 77 K of (A) SBA-15 and (B) the templated carbon obtained by carbonization of sucrose in SBA-15.

the walls. The capacitance values, either in $1 \, \text{mol} \, L^{-1} \, H_2SO_4$ or $1 \, \text{mol} \, L^{-1}$ TEABF$_4$ in acetonitrile, increase with the BET specific surface area, however, without showing any proportionality. A linear relationship was found between the capacitance values in aqueous or organic medium and the micropore volume determined by CO_2 adsorption (Figure 8.12) [26], confirming that the ions electrosorbed in ultramicropores mainly contribute to the value of capacitance. Considering the case of the organic electrolyte, it is obvious that ultramicropores are smaller than the diameter of the solvated ions, e.g., 1.3 and 1.16 nm for the solvated $(C_2H_5)_4N^+$ and BF_4^- ions [24], respectively. From molecular modeling calculations, the diameters of the nonsolvated $(C_2H_5)_4N^+$ and BF_4^- ions are estimated to be 0.672 and 0.454 nm, respectively [23], i.e., they fit with the size of ultramicropores. The micropore and mesopore of these templated carbons are perfectly interconnected. Therefore,

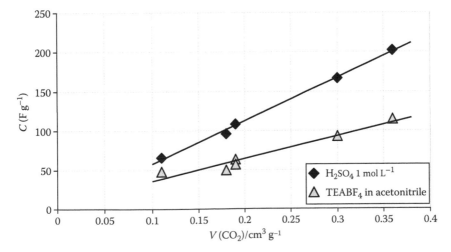

FIGURE 8.12 Capacitance in $1 \, \text{mol} \, L^{-1} \, H_2SO_4$ and $1 \, \text{mol} \, L^{-1}$ TEABF$_4$ in acetonitrile vs. ultramicropore volume for various templated carbons. (Adapted from Vix-Guterl, C., et al., *Carbon*, 43, 1293, 2005.)

under the application of an electrical polarization, the solvated ions easily diffuse in the straight mesopores (playing the role of corridors) until reaching the entrance of ultramicropores where they are finally desolvated [26]. By contrast, in typical ACs, the pathway of solvated ions to reach the active micropores is very tortuous and long, with several bottlenecks (nanogates) before reaching the ultramicropores. The determinant pore size in this case is probably slightly higher than the size of the pores where ions are finally accumulated.

The results presented in this section confirm that an adequate pore size is more important than a high surface area for an optimization of the capacitance values. For the production of compact systems, an important objective is to limit as much as possible the useless porosity in order to enhance the volumetric capacity. Moderately activated carbons, with pores at the boarder of the ultramicropore region, e.g., 0.7–0.9 nm, are the most profitable for ions electrosorption.

Due to the confinement of ions in a situation very close to intercalation/insertion, the term "double layer" seems to be no longer relevant for explaining the charge storage mechanism in electrochemical capacitors based on nanoporous carbons.

8.4 CARBON MATERIALS IN PSEUDOCAPACITORS

8.4.1 INTRODUCTION

In general, two modes of energy storage are combined in electrochemical capacitors: (1) the electrostatic attraction between the surface charges and the ions of opposite charge (EDL) and (2) a pseudocapacitive contribution which is related with quick faradic charge transfer reactions between the electrolyte and the electrode [7,8]. Whereas the redox process occurs at almost constant potential in an accumulator, the electrode potential varies proportionally to the charge utilized dq in a pseudocapacitor, which can be summarized by Equation 8.5:

$$dq = C * dU \qquad (8.5)$$

The electrical response of such a system is comparable to that of a capacitor. Being of faradic origin and nonelectrostatic, this capacity is distinguished from the double layer one and is called pseudocapacity. In summary, the EDL formation is a universal property of a polarized material surface. Pseudocapacity is an additional property which depends both on the type of material and electrolyte. Compared to the double-layer normalized capacity ($\sim$10 μF cm^{-2}), it has generally a high value (100–400 μF cm^{-2}), because it involves the bulk of the electrode and not only the surface. From a practical point of view, pseudocapacity contributes to enhancing the capacity of materials and their energy density.

With nanoporous carbon electrodes in organic electrolyte, the double-layer capacity is the major contribution. An important pseudocapacitive contribution is observed in aqueous medium with three main families of materials: conducting polymers [28–31], metallic oxides [5,32–34], and nanotextured carbons enriched with heteroatoms (nitrogen and oxygen) [35,36]. Nitrogen is either substituted to carbon (lattice nitrogen) or included in the form of functional groups (chemical nitrogen) at the periphery of polyaromatic structural units [37,38].

Reversible hydrogen electrosorption is another kind of pseudofaradic effect which might be also observed with nanoporous carbons in aqueous medium. In this case, nascent hydrogen formed by water reduction under negative polarization of carbon is adsorbed in the material; reversal of polarization provokes its desorption by oxidation [39].

8.4.2 PSEUDOCAPACITANCE OF HETEROATOM-ENRICHED CARBONS

As shown in Chapter 5, functional groups are always present on the surface of nanoporous carbons. Some of them, as the quinone-type groups, may be involved in reversible redox reactions giving a

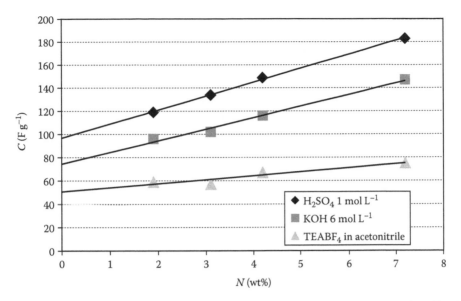

FIGURE 8.13 Capacitance vs. nitrogen content of N-enriched carbon materials. (Adapted from Frackowiak, E., et al., *Electrochim. Acta*, 51, 2209, 2005.)

pseudocapacitive contribution in addition to the double-layer capacity in aqueous electrolyte. As the proportion of these groups is rather small, techniques have been developed in order to enrich nanoporous carbons with heteroatoms.

Nitrogen has been incorporated by ammoxidation of nanoporous carbons [35] or by steam activation of nitrogen-enriched carbons [36,40]. As oxidation reactions are involved in both processes, oxygen is also incorporated in the materials. Consequently, it is difficult to assess completely the values of capacitance to the unique contribution of the nitrogenated functionality. Nevertheless, a series of nitrogen-doped carbons of comparable nanotexture ($S_{BET} \approx 800 \, m^2 \, g^{-1}$) demonstrates a proportional dependence of capacity with the amount of nitrogen (Figure 8.13) [41]. Extrapolating the curves to N% = 0 gives a capacity of 75–100 F g^{-1} in aqueous medium, which is of the order of magnitude expected for a BET surface area of 800 m^2 g^{-1}, assuming a double-layer normalized capacity of 9.4–12.5 μF cm^{-2} [7].

The increase of capacity in H$_2$SO$_4$ medium with the amount of nitrogen is due to pseudofaradic charge transfer reactions as those illustrated by Equations 8.6 through 8.8 [41,42]:

$$> C = NH + 2e^- + 2H^+ \leftrightarrow > CH - NH_2 \tag{8.6}$$

$$> C - NHOH + 2e^- + 2H^+ \leftrightarrow > C - NH_2 + H_2O \tag{8.7}$$

$$\tag{8.8}$$

More effective nanoporous carbons have been obtained by the template technique. A nitrogenated precursor is introduced in a nanoporous scaffold and subsequently pyrolyzed; then the nitrogen-enriched replica is obtained by etching the host with hydrofluoric acid. The first materials of this type for supercapacitors were synthesized by melamine intercalation in a lamellar aluminosilicate, e.g., mica [43,44]. Much better performance was demonstrated by carbons obtained through acrylonitrile pyrolysis in NaY zeolite. Their exceptionally high gravimetric capacity, e.g., 340 F g^{-1} comes from the synergy of different contributions: an ordered superstructure inherited from the scaffold, micropores with a size perfectly adapted to the electrolyte, an important pseudocapacitive effect related with rich functionality [45]. Moreover, a symmetric capacitor from these carbons could operate at 1.2 V, which is much higher than the usual operating voltage for symmetric systems based on activated carbon electrodes in aqueous media [46]. Nitrogen enrichment provides a profitable shift of the operating potential of the electrodes.

Other nitrogen-enriched carbons were formed by one-step carbonization of polyacrylonitrile/CNT blends [47] and melamine formaldehyde/CNTs blends [42]. Although presenting a moderate specific surface area (S_{BET} ~ 400 m^2 g^{-1}), the carbons obtained from the later blend give a gravimetric capacity of 170 F g^{-1} with a discharge current density of 50 mA g^{-1} and 130 F g^{-1} at a high current of 5 A g^{-1} [42]. Such exceptional performance is due to the synergy between the nanotubes and the pseudocapacitive properties related with the presence of heteroatoms. The open mesoporosity of nanotubes and their good electrical conductivity permit a good charge propagation.

A very interesting carbon was obtained by one-step pyrolysis of a seaweed biopolymer, e.g., sodium alginate, without any further activation [48]. Although sodium alginate has a structure very close to the structure of cellulose, its thermal behavior is completely different. Whereas carbonization of cellulose is almost completed at 400°C, a noticeable weight loss associated to CO evolution is observed for sodium alginate between 700°C and 900°C (Figure 8.14). Taking into account this information, a carbon material named ALG600 has been prepared by pyrolysis of sodium alginate at 600°C under argon atmosphere. The resulting material is slightly microporous (S_{BET} = 273 m^2 g^{-1}) and contains a high amount of oxygen (15 at%) retained in the carbon framework. From the deconvolution of the X-ray photoelectron spectroscopy (XPS) C_{1s} peak, the oxygenated functionalities are phenol and ether groups (C–OR; 7.1 at%), keto and quinone groups (C=O; 3.5 at%), and carboxylic groups (COOR; 3.4 at%). Despite the low BET specific surface area of this carbon, the capacitance in 1 mol L^{-1} H$_2$SO$_4$ medium reaches 200 F g^{-1}, i.e., a value comparable to the best ACs available on the market. Voltammetry curves of this material in a three-electrode cell show cathodic and

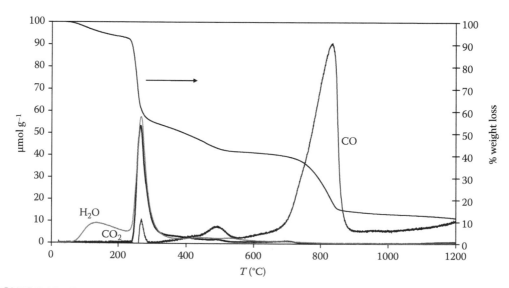

FIGURE 8.14 Thermogravimetric analysis of sodium alginate under helium atmosphere coupled with mass spectrometry analysis of the evolved gases. Heating rate 10°C min^{-1}. (Adapted from Raymundo-Piñero, E., et al., *Adv. Mater.*, 18, 1877, 2006.)

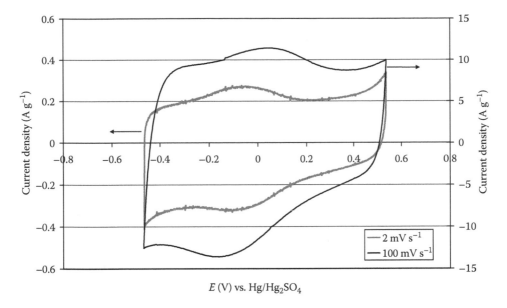

FIGURE 8.15 Cyclic voltammograms in a three-electrode cell where the carbon from alginate—ALG600—is used as working and counter electrodes; reference electrode Hg/Hg$_2$SO$_4$. (Adapted from Raymundo-Piñero, E., et al., *Adv. Mater.*, 18, 1877, 2006.)

anodic humps at around −0.1 and 0.0 V vs. Hg/Hg$_2$SO$_4$, respectively (Figure 8.15) [48]. For ACs, peaks at such positions are traditionally assessed to electrochemical reactions of oxygen surface functionalities such as the quinone/hydroquinone pair [49]. Pyrone-like structures (combinations of nonneighboring carbonyl and ether oxygen atoms at the edge of the graphene layers) can also accept two protons and two electrons in the same electrochemical potential range as the quinone/hydroquinone pair [50]. Consequently, in the carbon obtained from sodium alginate, the high capacitance value is related with charge transfer reactions of the quinone, phenol, and ether groups [48].

Although biopolymers, e.g., alginate, are reasonably cheap precursors, it is worth to investigate the direct carbonization of seaweeds in order to find out a better performance/cost ratio. Seaweeds rich in alginate, due to their intrinsic chemical composition, were found to be fantastic precursors for producing nanotextured carbons with adjustable porosity and surface functionality [51]. The carbon LN600 obtained by heat treating the Lessonia Nigrescens (a seaweed containing 30–40 wt% of alginate in the dried state) at 600°C demonstrates a well-balanced micro/mesoporosity; it has a BET specific surface area of 746 m^2 g^{-1}. The surface oxygen content determined by XPS amounts to 10.1 at% and is distributed in phenol and ether groups (C–OR; 4.6 at%), keto and quinone groups (C=O: 2.4 at%), and carboxylic groups (COOR: 1.7 at%). The relatively higher porosity of LN600 as compared to ALG600 is due to the higher amount of metallic elements in Lessonia Nigrescens. During the thermal treatment under neutral atmosphere, the formed salts are distributed at the nanometric scale in the carbonaceous matrix allowing an effective autoactivation. Due to its unique functionality and nanoporosity, the carbon LN600 exhibits a capacitance of 264 F g^{-1} in 1 mol L^{-1} H$_2$SO$_4$ [51].

Both ALG600 and LN600 carbons were compared to commercial ACs frequently used for supercapacitors, e.g., Maxsorb (AC1, 3487 m^2 g^{-1}) and Norit Super 50 (AC2, 1402 m^2 g^{-1}), giving capacitance of 203 and 119 F g^{-1} in 1 mol L^{-1} H$_2$SO$_4$ electrolyte, respectively. The capacitor from LN600 could operate in a range of 0–1.4 V during 10,000 cycles, with a capacitance loss of only 11%, whereas even in a limited voltage range of 0–0.7 V, the capacitance loss with AC1 was 21% after only 1000 cycles and 13% after 10,000 cycles with AC2. Another interesting property of the biopolymer carbons, related with their moderately developed porosity, is their high electrical conductivity

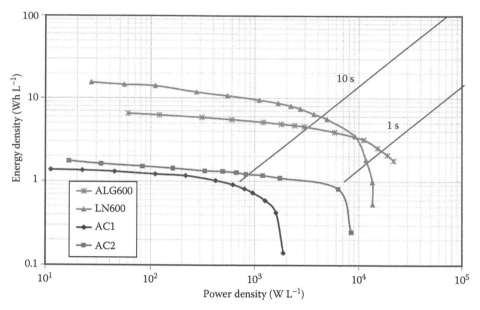

FIGURE 8.16   Compared Ragone plots of capacitors in 1 mol L^{-1} H$_2$SO$_4$. Electrodes from alginate (ALG600) and Lessonia Nigrescens (LN600) carbons and from two commercial ACs AC1 and AC2. The ALG600 and LN600 electrodes are without percolator, whereas AC1 and AC2 require 5 wt% percolator (Pure Black, Graphitized carbon black 205–110, Superior Graphite, Chicago, IL). (Adapted from Raymundo-Piñero, E., et al., *Adv. Funtc. Mat.*, 19, 1, 2009.)

and the high density of pressed electrodes: 0.79 and 0.91 g cm^{-3} for LN600 and ALG600, respectively, against 0.47 and 0.65 g cm^{-3} for AC1 and AC2, respectively. As shown on the volumetric Ragone plot presented in Figure 8.16, LN600 and ALG600 exhibit a much higher volumetric energy than the activated carbons AC1 and AC2, and the capacitors can be charged/discharged at high regime without requiring any conductivity additive in the electrodes [48,51]. The better power performance of ALG600 might be attributed to the remarkably high density of the electrodes from this material.

8.4.3   PSEUDOCAPACITANCE RELATED WITH REVERSIBLE HYDROGEN ELECTROSORPTION IN NANOPOROUS CARBONS

Promising hydrogen capacities, up to ca. 2 wt%, were found by electroreduction of aqueous media on a nanoporous carbon electrode at ambient pressure and temperature conditions [39,52–54]. Interestingly, the electrosorbed hydrogen can be desorbed by anodic oxidation giving rise to a pseudofaradic contribution in addition to the EDL capacity of the material. Although this chapter deals with electrochemical applications for supercapacitors, physical adsorption will be briefly reviewed in order to show the textural analogies between the two processes and the differences concerning the kind of interaction between hydrogen and the carbon network.

8.4.3.1   Optimal Texture for Hydrogen Physisorption and Electrosorption in Nanoporous Carbons

A remarkable adsorption capacity on high surface area ACs under a hydrogen pressure has been reported for the first time at the beginning of the 1980s [55,56]. Whereas hydrogen is absorbed in the interstitial sites of metallic alloys, the main storage mechanism in carbon materials is the adsorption in micropores [57,58]. Depending on the authors, theoretical studies found that the optimum pore size for hydrogen adsorption corresponds to one layer of hydrogen adsorbed (0.35 nm) [59] or two layers (0.56 nm) [60]. Nevertheless, most of the experimental works claimed about some proportionality between the amount of hydrogen adsorbed and the BET specific surface area [61–63].

The hydrogen adsorption capacity of a large variety of microporous activated carbons measured at room temperature and at pressures up to 70 MPa increases with the DR micropore volume measured by nitrogen or CO_2 adsorption [64]. The hydrogen density inside the micropores varies with the average pore size and reaches a maximum at a value of 0.66 nm, which matches fairly well with the theoretical value calculated for a pore which holds two layers of hydrogen (0.56 nm [60]). At room temperature, the H_2 saturation pressure has been estimated empirically to a value of $P_S = 106$ MPa [65]. At 50 MPa, the relative pressure ($P_r = 0.47$) is such that the wider micropores contribute essentially to the total adsorption; the hydrogen adsorption capacity correlates with the total micropore volume, whereas ultramicropores are more involved for low pressure. At 77 K, the estimated saturation pressure of hydrogen is $P_S = 7$ MPa. Consequently, at 4 MPa ($P_r = 0.57$), not only the narrow micropores, but also the wider micropores contribute to the hydrogen adsorption.

To date, Ref. [65] provides probably the most reliable data on a series of chemically activated carbons and also on other types of carbon materials, such as activated carbon fibers, CNTs, and CNFs. The best values measured for adsorption capacity at 298 K were 1.2 and 2.7 wt% at 20 and 50 MPa, respectively. At 77 K, the hydrogen adsorption capacity reached 5.6 wt% at 4 MPa. Such values demonstrate that nanoporous carbons are not worst than other kinds of materials studied at the moment for hydrogen storage.

Although a relatively acceptable proportionality is observed when the amount of hydrogen physisorbed is plotted vs. the DR micropore volume of activated carbons [65,66], this is no longer the case for electrosorption. A trend to increasing the sorbed amount with the DR micropore volume, measured either by nitrogen or by CO_2 adsorption, still exists in the case of electrochemistry but with scarce values [67]. Interestingly, the data show that the correlation between the sorption capacity and the pore volume do not cross the origin, suggesting that a part of electrosorbed hydrogen is irreversibly trapped. By contrast, a series of carbons prepared by using mesoporous silica templates showed a good correlation between the hydrogen capacity and the DR_{CO_2} volume [26,54]. As already mentioned in Section 8.3.2, it is obvious that the mesoporous channels of templated carbons play a fundamental role in the transportation of hydrated ions to ultramicropores. In the case of activated carbons, the diffusion of ions is perturbed by the tortuosity and the presence of bottlenecks. As it will be shown in Section 8.4.3.3, it is also important to keep in mind that hydrogen is weakly chemisorbed, being in a state different of the CO_2 probe used for measuring the ultramicropore volume.

8.4.3.2 Mechanism of Hydrogen Electrosorption in Nanoporous Carbons

Whereas high pressures (at least 10 MPa) and/or low temperatures (77 K) are necessary for enhancing the hydrogen adsorption capacity of carbons, reversible hydrogen storage by cathodic electro-decomposition of water occurs at ambient conditions [39,52–54]. Due to the driving force of the negative polarization, *in situ* produced hydrogen easily penetrates into the nanopores of carbon, where it is adsorbed. By this process, higher pressures (calculated approximately from the Nernst law) than in the gas phase are locally reached [39].

In alkaline solution, for example, water is reduced according to Equation 8.9:

$$H_2O + e^- \rightarrow H + OH^- \tag{8.9}$$

and the formed nascent hydrogen is adsorbed on the surface of nanopores:

$$<C> + xH \rightarrow <CH_x> \tag{8.10}$$

where

<C> represents the carbon host

<CH$_x$> hydrogen adsorbed in the later

The overall charge/discharge phenomenon is summarized by Equation 8.11:

$$<C> + xH_2O + xe^- \leftrightarrow <CH_x> + xOH^- \tag{8.11}$$

A capacity of 272 mAh g^{-1} corresponds to 1 wt% of hydrogen.

Cyclic voltammetry is a well-adapted electrochemical technique to elucidate the mechanism and kinetics of reversible hydrogen storage. An example of voltammetry characteristics, using a microporous activated carbon cloth (ACC) from viscose (ACC; S_{BET} = 1390 m^2 g^{-1}) in 3 mol L^{-1} KOH, is shown in Figure 8.17. The minimum potential is shifted of −100 mV for each cycle, i.e., toward hydrogen evolution.

When the electrode potential is higher than the thermodynamic value which corresponds to water decomposition (the theoretical value of the equilibrium potential (EP) in 3 mol L^{-1} KOH is −0.856 V vs. Normal Hydrogen Electrode (NHE), i.e., −0.908 V vs. Hg/HgO), the voltammetry curves have the rectangular shape typical of charging the EDL. From the fourth loop, as the potential becomes lower than this value, both the double layer is charged and hydrogen is adsorbed in the pores of carbon. During the anodic scan, the reactions (Equations 8.9 and 8.10) run in opposite direction and a peak corresponding to the electrooxidation of sorbed hydrogen is observed. This pseudocapacity contributes to the total charge in addition to the EDLC.

When the negative potential limit decreases, the anodic current increases, and the corresponding hump shifts toward more positive values of potential. When the negative potential reaches −2 V vs. Hg/HgO, the hump is located close to 0 V vs. Hg/HgO, i.e., +0.052 V vs. NHE. This important polarization between the cathodic and the anodic processes indicates that hydrogen stored

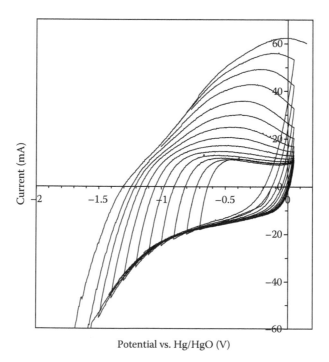

Potential vs. Hg/HgO (V)

FIGURE 8.17 Voltammetry scans on an ACC in 3 mol L^{-1} KOH at various values of negative potential cutoff (−100 mV stepwise shift). Auxiliary electrode: nickel. Reference electrode: Hg/HgO. (Adapted from Jurewicz, K., et al., *Appl. Phys. A*, 78, 981, 2004.)

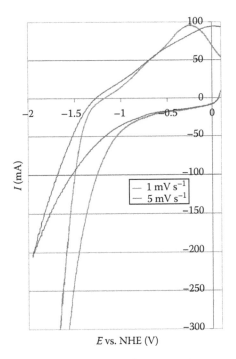

FIGURE 8.18 Influence of the scan rate (1 and 5 mV s^{-1}) on the voltammetry scans of a microporous ACC in 6 mol L^{-1} KOH. (Adapted from Béguin, F., et al., *Carbon*, 44, 2392, 2006.)

electrochemically is trapped more energetically than for a classical physisorption and/or that there are important diffusion limitations.

The contribution of diffusion has been estimated on the microporous ACC by cyclic voltammetry at 1 and 5 mV s^{-1} scan rates (Figure 8.18) [68]. In order to take into account the differences of scan rate and for an easy comparison of the two curves, the current values obtained at 1 mV s^{-1} have been multiplied by 5. As seen from Figure 8.18, the peak related with hydrogen oxidation is noticeably shifted to a lower potential value when the scan rate is diminished to 1 mV s^{-1}. The current at the end of oxidation, close to 0 V vs. NHE, is much lower in the case of the experiment at 1 mV s^{-1}, i.e., when the diffusion limitations are less important. Taking into account that the electrochemical hydrogen trapping takes place essentially in ultramicropores (i.e., pores smaller than 0.7–0.8 nm) and that tortuous voids may slow down the desorption process from these pores, one easily understands the improved tendency to reversibility at low scan rate. However, the anodic peak position observed at 1 mV s^{-1} is still higher than the hydrogen EP in 6 mol L^{-1} KOH, e.g., −0.874 V vs. NHE. Therefore, the observed polarization suggests a weak chemical hydrogen bonding with the carbon surface.

8.4.3.3 State of Hydrogen Electrosorbed in Nanoporous Carbons

The fact that hydrogen is strongly trapped was confirmed by galvanostatic intermittent titration technique (GITT) using a meso/microporous carbon prepared by the template technique [69]. Selecting a carbon with an interconnected network of mesoporous canals and micropores should allow to get rid of an important part of the diffusion limitations [26]. For the GITT experiment presented in Figure 8.19, the galvanostatic cycle includes periods of 1 h at a constant current of ±25 mA g^{-1}, interrupted by 2 h relaxation periods at open circuit. The potentials at the end of each relaxation period are equivalent to equilibrium values. Hence, the important polarization (overpotential) between

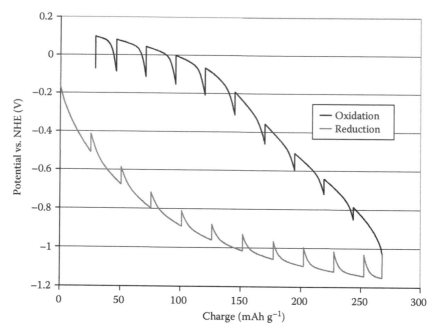

FIGURE 8.19 Galvanostatic intermittent reduction/oxidation in 6 mol L^{-1} KOH of a template carbon obtained by pyrolysis of SBA-15 silica impregnated with a sucrose solution. Charging/discharging current: ±25 mA g^{-1} during 1 h; the vertical segments represent interruption periods of 2 h at opened circuit. (Adapted from Vix-Guterl, C., et al., *Carbon*, 43, 1293, 2005.)

reduction and oxidation EPs indicates that the interaction of sorbed hydrogen with the carbon substrate is stronger than physisorption. However, since hydrogen can be still desorbed in ambient conditions, it is weakly chemically bonded with the carbon network.

This type of hydrogen bonding has been confirmed by thermoprogrammed desorption (TPD) analyses on ACC samples submitted to galvanostatic charge at −500 mA g^{-1} during 15 min or 12 h (Figure 8.20) [68]. In order to avoid any unwanted hydrogen desorption, the samples were not rinsed with water after the electrochemical charging. As a consequence they both present a peak above 400°C which is due to the reaction of excess KOH with carbon, according to Equation 8.12 [70]:

$$6KOH + 2C \rightarrow 2K + 3H_2 + 2K_2CO_3 \tag{8.12}$$

It is the only peak observed for the sample charged 15 min at −500 mA g^{-1}. Within this limited charging time, the Nernst potential for water reduction is not reached, and only the EDL is charged. By contrast, when charging is prolonged during 12 h, an additional peak appears in the TPD curve at around 200°C [68]. From this peak, the activation energy was estimated to 110 kJ mol^{-1} for hydrogen desorption [71]; such a value is characteristic of a weakly chemisorbed state of hydrogen with the carbon surface [72]. The desorption temperature is smaller than the value which is generally observed for covalently bonded hydrogen, e.g., 700°C [73].

The important role of active sites (dangling bonds) in hydrogen chemisorption has been proved by studying nanoporous carbons of comparable texture, but with varying amount of surface-oxygenated functionality, as estimated by TPD measurement of the CO + CO$_2$ amount desorbed at 950°C [74]. This series of carbons was obtained by chemical oxidation of an activated carbon with nitric acid, followed by thermal treatment under nitrogen at temperatures from 300°C to 900°C. Figure 8.21 shows that the hydrogen storage capacity decreases when the amount of oxygenated groups increases, i.e., when the number of available dangling bonds decreases. The formation of

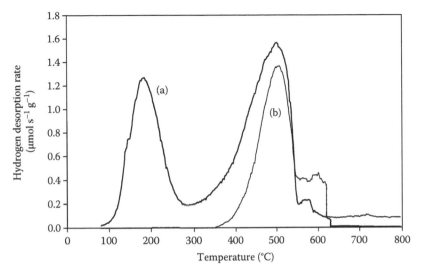

FIGURE 8.20 TPD analysis of hydrogen evolved from a microporous ACC charged at −500 mA g⁻¹ in 6 mol L⁻¹ KOH during (a) 12 h; and (b) 15 min. (Adapted from Béguin, F., et al., *Carbon*, 44, 2392, 2006.)

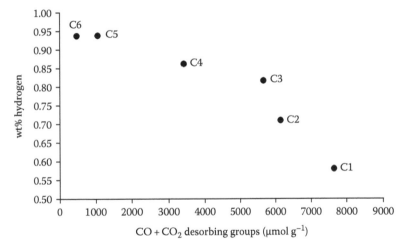

FIGURE 8.21 Hydrogen electrosorption capacity (wt%) vs. the amount of CO + CO₂ desorbing groups. (From Bleda-Martínez, M.J., et al., *Carbon*, 46, 1053, 2008. With permission.)

$C(sp^2)$–H bonds during cathodic reduction and their dissociation during the oxidation step has been confirmed by *in situ* Raman spectroscopy [74].

As a consequence of the chemical type of hydrogen bonding with the carbon surface, it was shown that the electrochemical properties depend on temperature. The voltammetry curves on the ACC recorded at different temperatures from 20°C to 60°C show that the amplitude of the reduction and oxidation peak increases with temperature (i.e., the amount of reversibly sorbed hydrogen increases), while the polarization between the two peaks decreases (Figure 8.22). Similar information is given by the galvanostatic charge/discharge curves (Figure 8.23). As the cell temperature increases, the value of potential reached during the charging step is less negative, because of a reduced polarization by enhancement of the ionic conductivity of the solution. The curves show also a better defined discharge plateau, positioned at lower potential, indicating that the temperature

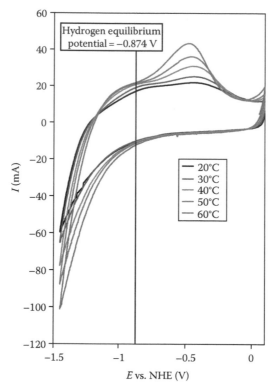

FIGURE 8.22 Cyclic voltammetry on a nanoporous carbon cloth (ACC) at different temperatures in 6 mol L^{-1} KOH. (From Béguin, F., et al., *Carbon*, 44, 2392, 2006.)

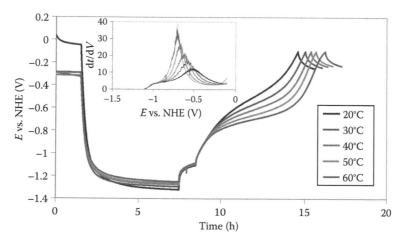

FIGURE 8.23 Effect of temperature on the galvanostatic charge/discharge characteristics of a microporous ACC in 6 mol L^{-1} KOH. Charging current –150 mA g^{-1}; discharging current + 50 mA g^{-1}. Inset: Derivative of the galvanostatic oxidation curves showing a shift of maximum from –0.5 to –0.7 V vs. NHE as temperature increases from 20°C to 60°C. (From Béguin, F., et al., *Carbon*, 44, 2392, 2006. With permission.)

increase reduces the kinetic barrier and facilitates the hydrogen extraction (oxidation) from the micropores. However, the most interesting observation from the curves is an increase of the amount of hydrogen stored reversibly, confirming that hydrogen electrosorption in nanoporous carbons is activated by temperature [68].

The particularity of electrochemical storage compared to the typical adsorption under gas pressure is due to the production of nascent hydrogen through water electroreduction. This very reactive form of hydrogen interacts with the active sites of carbon, being energetically trapped. As hydrogen is stabilized in the carbon substrate by weak chemical bonds, the self-discharge is not important [68]. When temperature is raised to 60°C, one can remark a noticeable enhancement of the reversible capacity and a shift of oxidation potential from about −0.5 to −0.7 V vs. NHE (see the shift of the derivative curve maximum; inset of Figure 8.23). Hence, the voltage (and consequently the energy density) of a cell using activated carbon as negative electrode in aqueous solution will be enhanced by operating at higher values of temperature. Taking into account that the capacity is quite high (up to 500 mAh g^{-1}) and that the process occurs in the negative range of potentials, it will be further shown that nanoporous carbons are very interesting as negative electrode of asymmetric supercapacitors.

8.4.4 NANOTUBES AS BACKBONE OF PSEUDOCAPACITIVE COMPOSITE ELECTRODES

Electrode materials with pseudocapacitive properties, such as oxides [5,32–34,75,76] or electronically conducting polymers (ECPs) [28–31,77] are very promising for enhancing the performance of supercapacitors, e.g., the specific capacitance. However, they are mostly investigated in three-electrode cells where a very thin layer of the active material coats a metallic current collector (very often platinum), which is far from the requirements for industrial applications. Moreover, swelling and shrinkage may occur during doping/dedoping of the active film, leading to mechanical degradation of the electrodes and fading of the electrochemical performance during cycling. Such problems have been partly overcome in the case of ECPs by adding an insulating polymer with good mechanical properties such as poly-N-vinylalcohol [78], but the electrical conductivity of the composite material is lower than in the pristine ECP. Adding carbon [79], and especially CNTs, is the most interesting solution which has been proposed to improving the mechanical and electrochemical properties of the electrodes [5,29]. CNTs, due to their unique morphology and extended graphitic layers, are characterized by exceptional conducting and mechanical properties which allow them to be used directly as three-dimensional support of active materials [80]. With nanotubes, the percolation of the active particles is more efficient than with the traditional carbon blacks which are generally used for the manufacture of electrodes [5]. Additionally, the charge propagation is also improved by the fact that the migration of ions in the electrodes is facilitated by the open mesoporous network formed by nanotubes entanglement. Finally, because of the high resiliency of nanotubes, the composite electrodes can easily adapt to the volumetric changes during charge and discharge, and are consequently able to bear a high number of cycles without a noticeable mechanical degradation. For all these reasons, composites where nanotubes are homogeneously dispersed in an active phase with pseudocapacitive properties represent an interesting breakthrough for developing a new generation of supercapacitors.

8.4.4.1 Conducting Polymer/Carbon Nanotube Composites

ECPs as polypyrrole (PPy), polyaniline (PANI), or polythiophene can store and release charge through redox processes [7,8,28,31,77]. When oxidation occurs (also referred to as "p-doping"), ions from the electrolyte are transferred to the polymer backbone, and upon reduction (dedoping) they are released back into the solution as shown in Figure 8.24. The doping/dedoping process takes place throughout the bulk of the electrodes, offering the opportunity of achieving high values of specific capacitance. However, the insertion/deinsertion of counterions causes volumetric changes of ECPs during cycling and progressive electrode degradation. Therefore, it has been proposed to use a moderate amount of carbon materials (carbon black, CNTs,...) as surface area enhancing component for improving the mechanical properties of the electrodes [29,81–91]. Moreover, the presence of carbon in the bulk of ECPs allows a good electronic conduction to be ensured in the electrode when the polymer is in its insulating state.

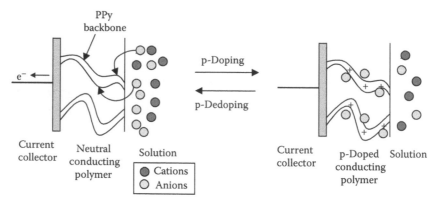

FIGURE 8.24 Schematic representation of the p-doping/dedoping of a conducting polymer (example of PPy). It can be summarized by $[PPy^+ A^-] + e^- \leftrightarrow [PPy] + A^-$.

Chemical and electrochemical polymerization of the monomer has been considered in order to get ECP/CNT nanocomposites. Figure 8.25 shows the scanning electron micrograph (SEM) picture of a PANI/CNT composite with 80 wt% of PANI [92]. Whereas the films from pure PANI are dense and compact, it can be seen that the composite material is porous, keeping the advantage of the entangled network of the nanotubes, that will allow a good access of the electrolyte to the active polymer. There is no doubt that such a porous nanotexture is optimal for a fast ionic diffusion and migration in the polymer so that the electrode performance should be improved.

Coating by a thin layer of PPy has been realized on multiwalled nanotubes (MWNT) [29,91,93], well-aligned MWNT [85] and single-wall nanotubes (SWNT) [88]. When MWNT are oxidized, their surface is covered with oxygenated functionalities, which can be used as anionic dopant of a PPy film electrodeposited on the MWNT [94]. These films are notably less brittle and more adhesive to the electrode than those formed using an aqueous electrolyte as source of counterion.

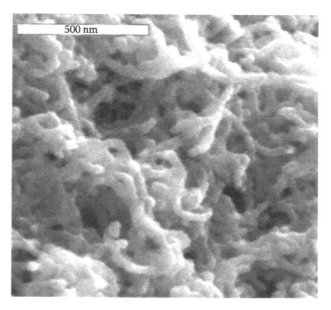

FIGURE 8.25 SEM showing the morphology of a PANI/CNT composite material containing 80 wt% of PANI. (From Khomenko, V., et al., *Electrochim. Acta*, 50, 2499, 2005. With permission.)

The electrochemical behavior of ECP/CNT composites has been studied either in two- or three-electrode cell. Comparing the redox performance of PPy films on aligned MWNT and on flat Ti and Pt surfaces shows a noticeable improvement in the case of the composites with MWNT due to the high accessible surface area of the CNTs in the aligned arrays [85]. By contrast, the results found with SWNT/PPy nanocomposites [88] are probably of limited application because (1) the nickel foam used as current collector supplies an additional capacity in the alkaline solution used for the study; and (2) PPy degrades quickly in alkaline solution. With the MWNT/PPy composite obtained electrochemically, the capacitance value reaches ca. $170 \, F \, g^{-1}$ with a good cyclic performance over 2000 cycles [29]. The high values of capacitance found with the MWNT/ECP composites are due to the unique property of the entangled nanotubes, which supply a perfect three-dimensional volumetric charge distribution and a well-accessible electrode/electrolyte interface. Comparing the result of the two coating techniques, the nonhomogeneous PPy layer deposited chemically is more porous, less compact than electrochemically deposited. Consequently, the diffusion of ions proceeds more easily, giving a better efficiency for charge storage [92].

In fact, it has been shown that the capacitance values for the composites with PANI and PPy strongly depend on the cell construction [92]. With chemically deposited ECP, extremely high values of specific capacitance can be found—from 250 to $1100 \, F \, g^{-1}$—using a three-electrode cell, whereas smaller values of $190 \, F \, g^{-1}$ for MWNT/PPy and $360 \, F \, g^{-1}$ for MWNT/PANI have been measured in a two-electrode cell. It highlights the fact that only two-electrode cells allow the materials' performance to be well estimated in electrochemical capacitors.

The applied voltage was found to be the key factor influencing the specific capacitance of pseudo-capacitors based on ECP/CNT nanocomposites [95]. Figure 8.26 shows the variation of specific capacitance by cycling a PPy/CNT-based symmetric capacitor. The cyclability of this capacitor is excellent when the maximum cell voltage is fixed at $0.4 \, V$. Upon cycling up to $0.6 \, V$, the capacitance loss is 20% of the initial value after 500 cycles, and it reaches almost 50% at $0.8 \, V$. Hence, using the same ECP/CNT material for both electrodes provides a poor cycling stability of the supercapacitor if the voltage range exceeds some limit. Beyond this limit, one electrode reaches a potential at which the ECP is electrochemically unstable, being the reason for the capacity fade [95].

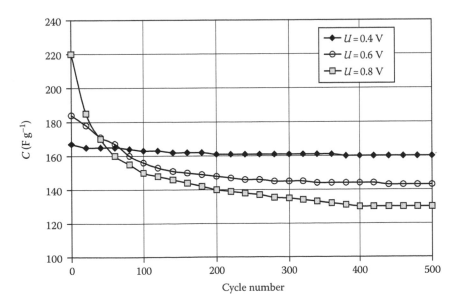

FIGURE 8.26 Influence of the maximum voltage on the specific capacitance (per mass of one electrode) vs. the number of galvanostatic cycles of a symmetric two-electrode system based on a PPy/CNT composite (20 wt% of CNT) in $1 \, mol \, L^{-1} \, H_2SO_4$. (From Khomenko, V., et al., *Appl. Phys. A*, 82, 567, 2006. With permission.)

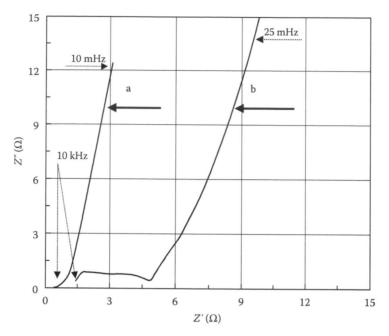

FIGURE 8.27 Impedance spectra of a symmetric capacitor based on PPy/CNT composite electrodes (20 wt% of CNT) in 1 mol L^{-1} H$_2$SO$_4$, before (a) and after (b) performing 500 galvanostatic charge–discharge cycles at a maximum voltage of 0.8 V. (From Khomenko, V., et al., *Appl. Phys. A*, 82, 567, 2006. With permission.)

In Figure 8.27, the impedance spectra of a supercapacitor based on PPy/CNTs composite electrodes, before galvanostatic cycling and after 500 cycles at a maximum voltage of 0.8 V, show that the resistance dramatically increases after cycling. Moreover, the semicircle at high frequency after cycling (curve b) suggests the parallel existence of a charge transfer resistance. Hence, irreversible redox transitions are at the origin of the high values of capacitance observed in Figure 8.26 under high voltage during the first galvanostatic cycles [95]. Generally, the operating voltage of a symmetric capacitor based on ECP electrodes cannot exceed 0.6–0.8 V due to oxygen evolution in the positive range of potentials and switching to an insulating state in the negative values [92].

8.4.4.2 Amorphous Manganese Oxide/Carbon Nanotube Composites

CNTs were also demonstrated to be a perfect support for cheap transition metal oxides of poor electrical conductivity, such as amorphous manganese oxide (a-MnO$_2$·nH$_2$O) [5,96]. The pseudocapacitance properties of hydrous oxides are attributed to the redox exchange of protons and/or cations with the electrolyte as in Equation 8.13 for a-MnO$_2$·nH$_2$O [97]:

$$MnO_a(OH)_b + nH^+ + ne^- \leftrightarrow MnO_{a-n}(OH)_{b+n} \qquad (8.13)$$

where MnO$_a$(OH)$_b$ and MnO$_{a-n}$(OH)$_{b+n}$ indicate interfacial a-MnO$_2$·nH$_2$O in higher and lower oxidation states, respectively. However, due to the low electrical conductivity of a-MnO$_2$·nH$_2$O, a conducting additive is required for the realization of supercapacitor electrodes. Therefore, a-MnO$_2$ has been precipitated on CNTs by adding Mn(OAc)$_2$ 4H$_2$O to a KMnO$_4$ solution which contained predetermined amounts of CNTs [5]. The SEM picture presented in Figure 8.28 shows the morphological aspect of a composite loaded with 15 wt% of nanotubes. The striking fact is a remarkable template effect of the entangled nanotubes framework, with an extremely good adhesion of the coating layer

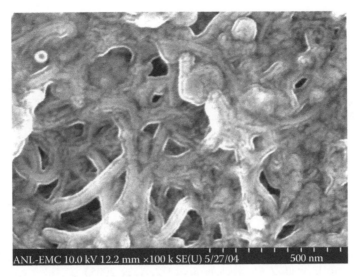

FIGURE 8.28 SEM image of the a-MnO$_2$/CNTs composite with 15 wt% of nanotubes. (Courtesy of M. Thackeray, Argonne National Laboratory, Argonne, IL.)

on the CNTs which were grafted with surface groups by KMnO$_4$ oxidation. These textural characteristics are favorable simultaneously for an easy access of ions to the bulk of the active material, a low electrical resistance and a good resiliency of the composite electrodes.

The values of specific capacitance and cell resistance obtained with symmetric two-electrode capacitors built with a-MnO$_2$/CNTs composite electrodes are presented in Table 8.3. The addition of nanotubes to a-MnO$_2$ causes a drastic decrease of cell resistance and an increase of specific capacitance referred to the mass of a-MnO$_2 \cdot n$H$_2$O. However, when the specific capacitance is referred to the total mass of the composite electrode material, to be realistic, a CNTs loading higher than 10–15 wt% does not improve the electrodes performance. Therefore, 10–15 wt% of CNTs conducting additive seems to be an optimal amount both on the point of view of electrodes capacitance

TABLE 8.3

Cell Resistance and Specific Capacitance in 1 mol L^{-1} Na$_2$SO$_4$ Medium for a-MnO$_2$ Loaded with Different Weight Percentages of CNTs

CNTs Loading (wt%)	Specific Capacitance (F g$^{-1}_{Composite}$)	Specific Capacitance (F g$^{-1}_{Oxide}$)	Cell Resistance (Ω cm^2)
0	0.1	0.1	2000
5	19	20	625
10	137	154	5.2
15	137	161	4.0
20	139	174	3.8
25	141	188	3.5
30	138	197	3.0
40	140	234	2.7
50	138	277	2.5
100	20	—	2.0

Source: Raymundo-Piñero, E., et al., *J. Electrochem. Soc.*, 152, A229, 2005. With permission.

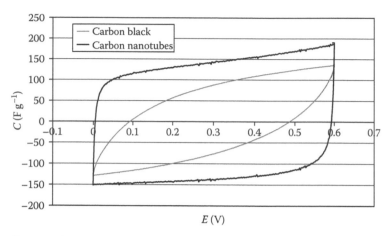

FIGURE 8.29 Comparative voltammograms of two-electrode cells built with a-MnO$_2$/carbon black and a-MnO$_2$/CNT electrodes, respectively, in 1 mol L^{-1} Na$_2$SO$_4$. Carbon black or CNTs loading: 15 wt%. (Adapted from Raymundo-Piñero, E., et al., *J. Electrochem. Soc.*, 152, A229, 2005.)

and cell resistance. Usually, in the battery industry, carbon black is used for the percolation of electrodes. Therefore, in order to appreciate the advantages of nanotubes, two-electrode capacitors have been built with a-MnO$_2 \cdot n$H$_2$O composite electrodes loaded with 15 wt% of CNTs and 15 wt% of carbon black, respectively. The shape of the voltammetry curves presented in Figure 8.29 demonstrates that the capacitive behavior of a-MnO$_2 \cdot n$H$_2$O is remarkably improved when CNTs are used instead of carbon black for the realization of the composite electrodes. Moreover, the values of ESR are 4 and 50 Ω cm^2 for nanotubes and carbon black, respectively, showing that nanotubes are very interesting for the development of high-power pseudocapacitors with a-MnO$_2$ as active material. Similar profitable effects of CNTs as backbones for nanocomposite electrodes have been shown with other pseudocapacitive oxides as nickel oxide [98].

8.5 ASYMMETRIC SYSTEMS

In recent years, a new concept of high energy density electrochemical capacitor, with different materials for the positive and negative electrodes, has been developed. In the literature, this kind of device is named either hybrid or asymmetric. The strategies followed to find the best electrode materials for obtaining the highest cell capacitance and voltage depend mainly on the nature of the electrolyte used. Understanding the charge storage mechanisms, e.g., the faradic reactions involving some materials in aqueous electrolyte or the Li-intercalation capability of other materials in Li-based organic electrolytes, allowed such promising devices to be developed.

8.5.1 ASYMMETRIC SYSTEMS INCLUDING A BATTERY ELECTRODE

The first asymmetric systems presented in the literature used a rechargeable battery-type electrode combined with an activated carbon capacitive electrode, either in aqueous [99,100] or in organic [101–109] electrolytes. The reason which led to follow this direction is the possibility to enhance at the same time the cell voltage and the cell capacitance [110–112].

By definition, a capacitive material is characterized by a potential swing during the charge and discharge of the capacitor. The initial output potential difference of a symmetric cell with two capacitive electrodes is 0 V, and the potentials diverge linearly during charging of the cell. Symmetric capacitors are generally charged up to 0.7–0.8 V in aqueous electrolyte and 2.5–2.7 V in organic one.

By contrast, a battery-like electrode is characterized by an almost constant potential during charging and discharging. Thus, in order to get the largest cell voltage in an asymmetric capacitor, where the capacitive electrode is replaced by a battery-like one, the selected battery-electrode potential must be close to the low or high limit of the potential window.

As it was already described in Section 8.2.1, the total capacitance of a supercapacitor is given by Equation 8.1. In a symmetric capacitor with equal values of capacitance for the positive (C_1) and negative (C_2) electrodes, the total capacitance of the system is half the capacitance of one electrode. In the asymmetric device, as the capacitance of the battery electrode is much higher than the capacitive one, the capacitance of the system approaches that of the electrode with the smallest value. In other words, the capacitance of the asymmetric configuration combining a battery-like electrode with a capacitive one will be close to the value of the capacitive electrode, i.e., twice larger than that for a symmetric configuration with two capacitive electrodes.

Taking into account that the charging characteristics of the battery electrode and the capacitive one are very different, special operating conditions of the asymmetric device have to be adopted in order to optimize its overall performance with respect to [112] (1) Long cycle life; (2) balanced charge characteristics, taking into account the effective equivalent weights of both electrodes. Indeed, the charge of one electrode depends on its mass, capacitance, and working potential. As the charge passing through the positive and the negative electrodes is the same, the masses of the electrodes must be modulated in order to maximize the energy density; and (3) Ensuring an optimal power density. Different battery-like electrode materials have been proposed depending on the electrolyte, as metal oxides for aqueous electrolytes and Li-intercalation compounds for nonaqueous ones.

8.5.1.1 Asymmetric Systems Using a Metal Oxide Battery-Like Electrode in Aqueous Electrolyte

On the basis of the high potential of metal oxides, Razumov et al. [99,100] introduced the first concept of hybrid capacitor in an aqueous electrolyte, using the metal oxide as faradic positive electrode and a capacitive activated carbon as negative electrode The faradic electrodes were either $PbO_2/PbSO_4$ in aqueous H_2SO_4 or $NiOOH/Ni(OH)_2$ in KOH electrolyte. In the capacitive electrode, the charge is stored electrostatically at the electrode/electrolyte interface. The process is very fast and is highly reversible, assuring a high power capability and a long cycle life. By contrast, at the battery-like component, the charge is stored through a redox reaction and an electrochemical phase transformation of the material. Therefore, even if the faradic electrode offers a better capacity than the capacitive one, it presents several disadvantages as [112] (1) smaller power density and (2) shorter cycle life. In particular, the cycle life depends on the operating rate and to a larger extent on the depth-of-discharge.

In such a case, the electrodes mass is balanced in order to ensure appropriate conditions for charge and discharge and to avoid the overcharge or overdischarge of any of the electrodes. In the case of the PbO_2/activated carbon asymmetric system, it has been claimed that the capacitive electrode should store preferably, at the most, one third of the charge of the faradic electrode in order to prevent a high depth-of-discharge or overdischarge of the later. In that case, the effective equivalent weight of the positive electrode must be increased in relation to that of the negative one. Taking into account the previous considerations, these asymmetric configurations can operate with a maximum cell voltage of 1.4–1.6 V. This fact, together with the improved capacitance, makes high energy density devices in comparison with symmetric capacitors built with two capacitive electrodes in aqueous electrolytes. However, even after optimization, the system presents some disadvantages still related with the use of a battery-like electrode: (1) the cycle life is more similar to a battery than to a capacitor because it is limited by the faradic electrode and (2) the power density is also limited by the electrode with the lowest power density.

8.5.1.2 Asymmetric Systems Using a Li-Intercalation Electrode in Organic Electrolyte

Following the idea presented by Razumov et al., Li-intercalation compounds have been proposed for electrodes in asymmetric capacitors working in organic electrolytes [101–110]. Figure 8.30 shows the redox potentials of some intercalation compounds in comparison with the relative potentials of the positive and negative electrodes in an activated carbon EDLC [110]. Some of these compounds are potential candidates for being used as negative electrode combined with a positive activated carbon electrode, in an asymmetric configuration. In theory, the nominal cell voltage (and consequently the energy density) can be increased in comparison to the symmetric EDLC.

In practice, Table 8.4 shows that even if intercalation compounds present very interesting initial potentials and high specific capacity in comparison with a capacitive activated carbon, they demonstrate quite severe disadvantages as low rate capability and limited cycle life. The later is related with the volume changes produced by Li insertion/deinsertion, which results in the formation of microcracks causing the failure of the electrode. Additionally, a high polarization of the negative electrode down to potentials close to 0 V vs. Li should be avoided in order to prevent lithium plating.

Nevertheless, Amatucci et al. [101] have shown the possibility of using $Li_4Ti_5O_{12}$ (LTO) after taking into account some considerations. LTO is one of the few lithium intercalation compounds which exhibit little expansion or contraction during the lithium insertion–deinsertion process. However, in order to obtain an acceptable lithiation rate capability and not to limit the power of the asymmetric cell, it was necessary to develop a nanostructured material. An asymmetric cell with activated carbon as positive electrode, LTO as negative one, and $LiBF_4$ in acetonitrile as electrolyte can reach a maximum voltage around 3.2 V after a good weight balance of the electrodes. For the long-term robustness test, a 500 F plastic prismatic prototype has been made applying the technology developed for Li-ion batteries. A maximum cell voltage of 2.8 V had to be used for assuring correct power characteristics and a good cyclability, e.g., 450,000 cycles [102,110].

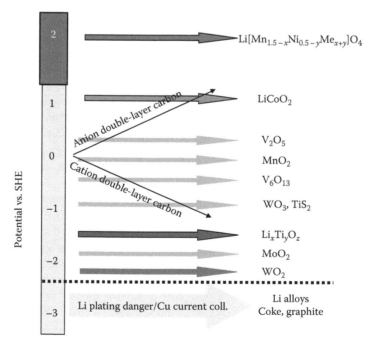

FIGURE 8.30 Schematic three-electrode plot vs. Standard Hydrogen Electrode (SHE). The plot shows the relative potentials of the positive and negative electrodes of an activated carbon EDLC as a function of charge vs. the redox potential of various intercalation compounds. (From Plitz, I., et al., *Appl. Phys. A.*, 82, 615, 2006. With permission.)

TABLE 8.4

Disparity of the Electrochemical Properties of Intercalation Electrode Materials Compared with Nonfaradic Electrodes

Property	Intercalation Material	Nonfaradic Activated Carbon
Specific capacity	100–200 mAh g^{-1}	10–30 mAh g^{-1} (per 1 V)
Rate	1–3C	1000–10,000C
Cycle life	500–1000	500,000–2,000,000
Expansion	2–10%	0%
Initial voltage (range)	−3 to 2 V SHE	0 V SHE

Source: Plitz, I., et al., *Appl. Phys. A*, 82, 615, 2006. With permission.

In order to obtain a higher output voltage, another nanostructured material as WO_2 can be used as negative electrode, considering that Li intercalates at lower potential values than in the lithium titanates (see Figure 8.30) [101,110]. However, even if the cell voltage can be increased up to 3.6 V, the maximum energy is similar to that obtained with the previous system using LTO as negative electrode. The performance of both asymmetric systems is limited by the positive electrode which is underused in the charging/discharging conditions needed to ensure a good cycle life of the faradic-type material.

Some other asymmetric devices have been proposed implying a LTO negative electrode and alternative positive electrode materials as (1) conducting polymers and (2) battery-like material/activated carbon composites.

1. *Conducting polymers*: The positive doping (p-doping) process of conducting polymers is interesting from the point of view of capacity, efficiency, and electrochemical stability. By contrast, negative doping was not found to be stable enough for being used in energy storage devices [103]. Different reports can be found where p-doping polymers, e.g., poly(4-fluorophenyl-3-thiophene) [103,104] or poly(3-methylthiophene) [104,105], are used as positive electrode in an asymmetric configuration with an activated carbon negative electrode using organic electrolytes ($TEABF_4$ in acetonitrile or propylene carbonate). Maximum cell voltages around 3.0 V can be reached, but the cyclability is not good. Consequently, an asymmetric configuration combining a p-doping conducting polymer as positive electrode with a battery-like LTO negative electrode has been tested, using $LiBF_4$ in acetonitrile as electrolyte [106,107]. The devices exhibit a high-capacitance fade due to the degradation of the polymer backbone related with its overoxidation at high voltage. Moreover, the high cost production of conducting polymers compared to activated carbons precludes their practical use.

2. *Battery-like material/activated carbon composites*: Other asymmetric configurations with LTO and another battery-like material as negative and positive electrodes, respectively, have been reported [107,108,110]. For Li-intercalation compounds, Figure 8.30 shows the highest potential differences between LTO and either $LiCoO_2$ or $Li[Mn_{1.5-x}Ni_{0.5-y}Me_{x+y}]O_4$ [107,110]. In practice, the asymmetric device with $LiCoO_2$ (with 10–20 wt% of activated carbon for improving the conductivity and the cycle life of the electrode) and LTO as positive and negative electrodes, respectively, can reach a maximum voltage of 3.0 V (Figure 8.31). However, the cyclability does not go beyond 1000–2000 cycles and the performance is too close to a battery and too far from a supercapacitor. Such devices should be defined as Li-ion batteries with an improved power capability but not as supercapacitors.

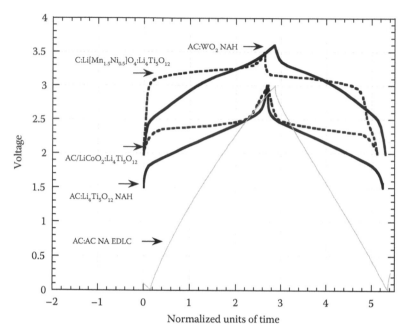

FIGURE 8.31 Voltage profiles of various hybrid systems in nonaqueous electrolytes (NAH) including Li-intercalation electrodes compared with a nonaqueous (NA) EDLC based on activated carbon electrodes cycled to 3 V. All profiles are normalized with respect to each other. (From Plitz, I., et al., *Appl. Phys. A*, 82, 615, 2006. With permission.)

The same occurs when a $Li[Mn_{1.5-x}Ni_{0.5-y}Me_{x+y}]O_4$/carbon composite is used as positive electrode in combination with LTO as negative one [108,110]. Figure 8.31 shows that the maximum cell voltage reaches 3.5 V, but the cyclability of the device is very poor. Even if activated carbon is used as negative electrode in combination with a Li-based compound similar to the previous one ($LiMn_{1.2}Ni_{0.5}O_4$) for the positive electrode [109], the maximum cell voltage is only 2.8 V and the cyclability and power capability are not improved.

In conclusion, among all the presented hybrid capacitors, the most realistic is the LTO/Activated carbon system [101,110]. However, the positive activated carbon electrode must be oversized, making the energy density of the asymmetric system only few times higher than for a symmetric EDLC. Consequently, to achieve a higher energy density, it is desirable to improve the capacity of the positive electrode. In addition, since intercalation/deintercalation in $Li_4Ti_5O_{12}$-based materials takes place at around 1.5 V vs. Li, the hybrid capacitor could be only charged/discharged between 1.2 and 3.2 V, leading to underusing the positive electrode. Therefore, other approaches have been explored for extending the operating voltage.

8.5.2 New Asymmetric Systems Working in Organic Electrolytes

From the Li-ion battery technology, it is known that carbons can be intercalated at more negative potentials than any other Li-intercalation material (see Figure 8.30). In commercial Li-ion batteries, two kinds of carbon materials are mainly used as negative electrode: (1) nongraphitizable or hard carbons (HCs) and (2) graphite.

The common approach which has been recently introduced for realizing asymmetric capacitors using any of these two materials, as negative electrode combined with an activated carbon, consists in extending the working potential window of the positive electrode [113–116]. In this sense, Figure

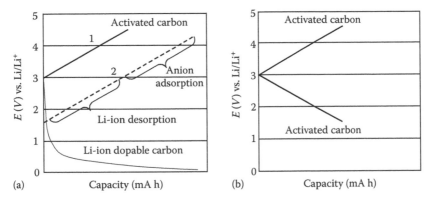

FIGURE 8.32 Typical potential profiles for (a) positive electrode in a conventional asymmetric capacitor built with a non "pre-doped" Li inter calation carbon for the negative electrode (curve 1), an asymmetric capacitor with a "pre-doped" Li-ion intercalation carbon material (curve 2), and (b) positive and negative electrodes of an EDLC during the charging process. (From Aida, T., et al., *Electrochem. Solid-State Lett.*, 9, A534, 2006. With permission.)

8.32 shows a scheme of the potential profiles during charging two different asymmetric configurations (Figure 8.32a) and a symmetric double-layer capacitor (Figure 8.32b) [114]. In the asymmetric systems presented in the previous section, a potential range from about 3.0 to 4.5 V vs. Li has been used for the activated carbon (Figure 8.32a line 1). However, the potential window of the activated carbon could be extended to lower values, as in the EDLC negative electrode (Figure 8.32b). If a wider potential range would be used at the positive electrode, i.e., from 1.5 to 4.5 V vs. Li as presented in Figure 8.32a (dotted line), the charge storage capacity of this electrode could be increased by a factor of 2.

Following this idea, Aida et al. [114–116] proposed an asymmetric capacitor with activated carbon and nongraphitizable carbon as positive and negative electrodes, respectively, using $LiPF_6$ in 1:1 ethylene carbonate/diethyl carbonate as electrolyte. Underusing the positive electrode is avoided by balancing the volumes of both electrodes (volume of positive electrode = 10 × volume of the negative electrode), allowing the cell to be charged/discharged from 1.0 to 4.3 V. However, the system has a very poor cycle life, only few 10 cycles, due to the capacity loss of the nongraphitizable carbon negative electrode. The authors interpret this capacity loss by the fact that the high polarization of the active carbon electrode (over 4.0 V vs. Li) provokes the electrolyte decomposition and hydrofluoric acid (HF) formation. HF further reacts with the solid electrolyte interface at the negative electrode, forming LiF which blocks the electrode. In an attempt to improve the performance, a Li foil has been placed between the two electrodes in order to trap HF in metallic lithium. However, only 1000 cycles are shown, which is not sufficient to validate the system as a capacitor.

More recently, Khomenko et al. [113] have shown that graphite is a better choice for the negative electrode in such kind of asymmetric configuration. Actually, there are some disadvantages of nongraphitizable carbon or HC as compared with graphite: (1) HC has a very poor rate performance related to a slow diffusion of lithium in the internal carbon structure; (2) the HC charge/discharge curve is not flat, and the potential gradually varies during the intercalation/deintercalation process; and (3) HC has a large initial irreversible capacity [117]. Graphite-based electrodes do not have the above-mentioned disadvantages and they are considered as the best choice for the negative electrode of lithium-ion batteries [118–120].

Therefore, a hybrid cell has been designed in $1 \, mol \, L^{-1}$ $LiPF_6$ in 1:1 ethylene carbonate/diethyl carbonate electrolyte by combining graphite and activated carbon as negative and positive electrodes, respectively [113]. The activated carbon electrode is stable in the potential window between 1.0 and 5.0 V vs. Li, whereas the graphite electrode can be polarized down to low potential values. The mass of the electrodes should be balanced to fully take profit of the performance of both materials in their optimal working potential range. Considering that the capacity of the graphite electrode is higher than that of the activated carbon, the mass of the AC electrode should be higher than the graphite one in order to maximize the energy density. However, it has been demonstrated that the time constant decreases when increasing the mass of graphite (due to the decrease of Li-intercalation depth). Then, considering that a capacitor must be essentially a power device, the best compromise between energy and power has been obtained for an electrode mass ratio of 1:1.

Another important consideration is that for obtaining a hybrid cell where the electrochemical behavior will be controlled by the AC electrode, the working potential range of the graphite electrode should be located below 0.2 V vs. Li/Li$^+$, while avoiding lithium plating which would decrease the cycle life of the system. For this purpose, it has been found that the charging of the cell should start from 1.5 V instead of 0.0 V. However, starting from 1.5 V is not enough to keep the graphite electrode in such low potential and a special procedure comprising some "formation cycles" has been designed in order to shift the potential of the graphite electrode to the expected value. The principle of the procedure came from the fact that the overall self-discharge of the hybrid capacitor is essentially determined by the AC-based electrode which has a higher self-discharge rate than the redox graphite one [106]. These differences in the rate of self-discharge allow the desired potential of 0.1 V vs. Li/Li$^+$ to be achieved for the graphite electrode after a series of charge cycles at increasing cell voltage, e.g., 4.0–4.5 V, followed by 3 h relaxation periods (Figure 8.33). After such formation cycles, the graphite electrode potential becomes stable, and the capacitor can be charged/discharged with an almost ideal linear profile as it can be seen for the last cycles in Figure 8.33. Figure 8.34 shows more in detail the charge/discharge profile of the asymmetric cell in the voltage range of 1.5–4.5 V. By comparing Figure 8.34 with the charge/discharge profiles presented in Figure 8.31 for previously proposed asymmetric systems, a very remarkable difference in linearity can be easily observed [113]. The advantage of the present system is that, after the formation cycles, the potential changes for the graphite electrode during charging/discharging the cell are very small and

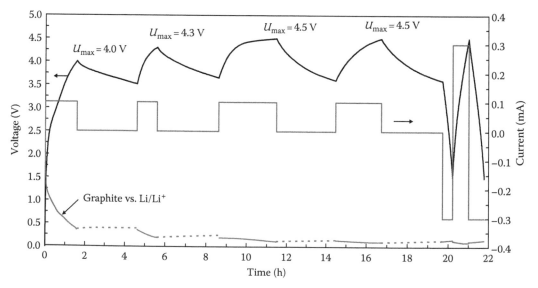

FIGURE 8.33 Charge/discharge voltage (upper curve) and current profile (middle curve) of the hybrid graphite/AC capacitor during formation cycles at a current rate C/20. The continuous down curve represents the potential variation of the graphite electrode during charging, whereas the dotted line corresponds to the relaxation periods between charge cycles. The cell was assembled with a mass ratio: m(activated carbon)/m(graphite) = 1/1. (From Khomenko, V., et al., *J. Power Sources*, 177, 643, 2008. With permission.)

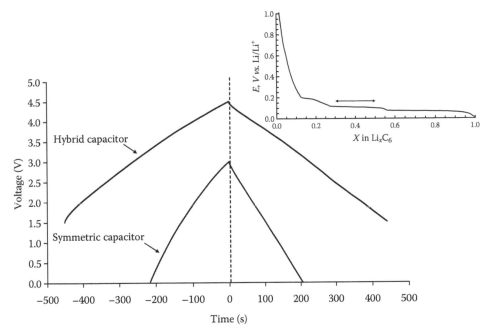

FIGURE 8.34 Cell output voltage comparison for one charge–discharge cycle of symmetric (both electrodes based on activated carbon) and hybrid capacitors using the same electrolyte (1 mol L^{-1} LiPF$_6$ in EC/DMC) and electrodes mass loading. The inset shows the potential and capacity range (arrow) of the graphite electrode during the charge–discharge of the hybrid capacitor between 1.5 and 4.5 V. (From Khomenko, V., et al., *J. Power Sources*, 177, 643, 2008. With permission.)

in the very desirable window of 0.16–0.10 V vs. Li/Li$^+$ (see inset in Figure 8.34). Simultaneously, the potential of the activated carbon electrode changes in the range of 1.66–4.6 V vs. Li/Li$^+$. Therefore, the charge–discharge profile of the hybrid cell is largely regulated by the activated carbon electrode and not by the battery-like electrode as in the other systems.

This graphite/AC asymmetric configuration presents very high values of energy density, while keeping a very good cycle life in a voltage range from 1.5 to 4.5 V [113]. The discharge capacitance after 10,000 cycles was over 85% of the initial value, which is much higher than that of any symmetric capacitor operating at a voltage higher than 2.7 V. Additionally, this asymmetric or hybrid system fulfils all the requisites for being used in high power applications.

In conclusion, it is possible to realize asymmetric systems in organic electrolyte having a better energy density than symmetric EDLCs while keeping an acceptable power density and a good cycle. To date, the best configuration was obtained by associating graphite at the negative electrode and activated carbon at the positive one.

8.5.3 NEW ASYMMETRIC SYSTEMS WORKING IN AQUEOUS ELECTROLYTES

Nowadays, industrial capacitors are mainly based on activated carbon electrodes and an organic electrolyte. The reason for this choice is dictated by Equations 8.2 and 8.3 which demonstrate the dominant influence of voltage on energy and power. The voltage of industrial supercapacitors reaches 2.7 V in organic medium and only 0.6–0.7 V in aqueous solution. Although organic electrolytes are the most currently adopted by companies, Table 8.1 shows that they have a number of disadvantages: (1) their electrical resistance is higher than for an aqueous electrolyte, and consequently the delivered power is smaller (Equation 8.3); (2) the capacitance of nanoporous carbons is smaller in organic medium; (3) the organic electrolytes are more expensive and the supercapacitors have to be built in a moisture-free atmosphere; and (4) during the operation of supercapacitors, the organic electrolytes are decomposed with the emission of environment unfriendly gases and risks of explosion. All these disadvantages could be eliminated by the use of an aqueous electrolyte, provided that the voltage window is enhanced.

Therefore, new research directions were proposed in order to develop cheap and safe supercapacitors of high performance in a more environment-friendly electrolyte. By analogy with the asymmetric capacitors designed in organic electrolytes with Li-intercalation compounds, Wang et al. proposed to combine $LiCoO_2$ [121] or $LiCo_{1/3}Ni_{1/3}Mn_{1/3}O_2$ [122] as positive electrode with activated carbon as negative one in aqueous Li_2SO_4 electrolyte. Although these devices exhibited voltage values between 0.5 and 2.0 V, their poor cyclability demonstrates that this is not the right solution. All these results clearly indicate that the use of a battery-like electrode in aqueous electrolyte is not effective.

The most effective approach is through an asymmetric system with two pseudocapacitive electrodes of high overpotential in water medium, allowing the voltage to be enhanced while keeping a good cyclability. The promising electrode materials are some metal oxides (MnO_2, Fe_3O_4,...) [97,123–126], conducting polymers [92,95,127] or oxygen-rich carbons [128]. In most systems, the negative electrode is from nanoporous carbon which stores charges in the EDL and hydrogen by electroreduction of water [39].

8.5.3.1 Asymmetric Systems Using Pseudocapacitive Oxides or Conducting Polymers

As it was shown in Section 8.4.4, in pseudocapacitive materials as a-MnO_2 or ECPs, the pseudofaradic reactions depend on the pH of the electrolytic medium and are potential dependent. Therefore, the voltage range determines the reactions taking place at each of the electrodes. As a consequence, pseudocapacitive materials as a-MnO_2 or ECPs are precluded in symmetric two-electrode capacitors because the working potential of the positive and negative electrodes cannot be controlled. In particular, the voltage window of a two-electrode capacitor based on a-MnO_2 is limited to 0.6 V by the irreversible reaction $Mn(IV) \rightarrow Mn(II)$ taking place at the negative electrode [5]. In the case of ECPs, the main limitation is also found at the negative electrode, where the ECP reaches its insulating state when the voltage window is higher than 0.6–0.8 V [92].

By contrast, in an asymmetric or hybrid configuration, with two electrodes of different nature, it is possible to completely take profit of the pseudofaradic reactions. The main target, designing these systems, is to optimize the potential range of each electrode, taking into account the need to obtaining a high cell voltage and a good cycle life of the supercapacitor as a whole. Considering that in symmetric systems the problems for a-MnO_2 and ECPs are appearing at the negative electrode, the most realistic alternative is to use them as positive electrode in combination with activated carbon for the negative electrode. As it was deeply discussed in Section 8.4.3, when nanoporous carbons are polarized at enough negative potential in aqueous medium, both the EDL is charged and hydrogen is weakly chemisorbed [39].

The best-performing asymmetric configuration is obtained when using a-MnO_2 for the positive electrode and activated carbon for the negative one [97,123,124]. Figure 8.35 presents the cyclic voltammetry curves of activated carbon and a-MnO_2 in 2 mol L^{-1} KNO_3 using a three-electrode cell. In the case of activated carbon, the overpotential for dihydrogen H_2 evolution reaches 0.5 V (the thermodynamic value in this medium is around −0.37 V vs. NHE). In the same electrolytic medium, a-MnO_2 presents a high reversibility in the positive range of potential, up to 1.2 V. Figure 8.35 also shows that by carefully balancing the respective masses of the two materials in the asymmetric construction, they can operate in the potential ranges represented by the arrows, e.g., $\Delta E_{\text{Activated carbon}}$ and ΔE_{MnO_2}, allowing theoretically a working voltage window of 2.2 V to be obtained. In practice, it is necessary to reduce slightly the voltage value in order to avoid irreversible reactions, such as hydrogen evolution at the negative electrode or the partial dissolution of MnO_2. The optimized

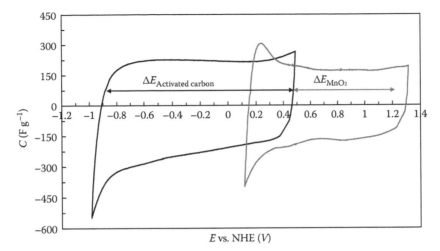

FIGURE 8.35 Comparative cyclic voltammograms of a-MnO$_2$/CNTs composite and of activated carbon (Maxsorb, Kensai, S_{BET} = 3487 m^2 g^{-1}) obtained with a three-electrode cell in 2 mol L^{-1} KNO$_3$ (pH = 6.4). (Adapted from Khomenko, V., et al., *J. Power Sources*, 153, 183, 2006.)

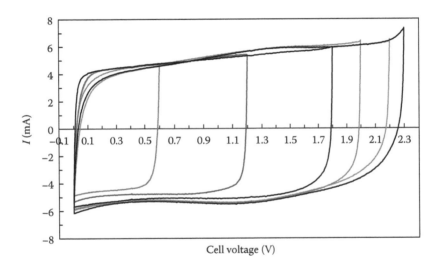

FIGURE 8.36 Cyclic voltammograms on an optimized asymmetric manganese oxide/activated carbon capacitor in 2 mol L^{-1} KNO$_3$ electrolyte. (Adapted from Khomenko, V., et al., *J. Power Sources*, 153, 183, 2006.)

asymmetric capacitor demonstrates a perfect rectangular shape of voltammogram (Figure 8.36) and is able to operate at 2 V with an excellent stability during cycling [124].

Another possibility which has been explored is the combination of a-MnO$_2$ and Fe$_3$O$_4$ as positive and negative electrodes, respectively [126]. However, even if the cell voltage can be extended up to 1.8 V, modest energy densities are obtained because of the low capacity of Fe$_3$O$_4$.

By analogy with the a-MnO$_2$ example, the cycling stability and voltage range of supercapacitors based on ECPs are improved when using an asymmetric configuration with activated carbon and ECP as negative and positive electrode, respectively [92,95,127]. The high stability suggests that undoping does not take place in the operating potential range of ECPs.

Table 8.5 summarizes the electrochemical performance of different types of symmetric and asymmetric supercapacitors in aqueous medium, including the maximum cell voltage (V_{max}), the

TABLE 8.5

Electrochemical Characteristics of Symmetric and Asymmetric Capacitors Based on Different Active Materials. 20 wt% of CNTs Are Used for the Percolation of ECPs and 15 wt% for a-MnO$_2$ Electrodes

Electrode Materials		Supercapacitor Characteristics		
Positive	Negative	V_{max} (V)	E (Wh kg^{-1})	P_{max} (k Wkg^{-1})
Activated carbon	Activated carbon	0.7	3.74	22.4
a-MnO$_2$	a-MnO$_2$	0.6	1.88	3.8
a-MnO$_2$	Activated carbon	2.0	10.50	123.3
PANI	PANI	0.5	3.13	10.9
PPy	PPy	0.6	2.38	19.7
PEDOT	PEDOT	0.6	1.13	23.8
PANI	Activated carbon	1.0	11.46	45.6
PPy	Activated carbon	1.0	7.64	48.3
PEDOT	Activated carbon	1.0	3.82	54.1

Source: Khomenko, V., et al., *Appl. Phys. A*, 82, 567, 2006. With permission.

maximum energy density (E), and the maximum specific power (P_{max}). After analyzing these data, it is possible to conclude that the performance of the asymmetric capacitors is much better than for any of the symmetric systems. It is obvious that combining different active materials for both the negative and positive electrodes allows the practical voltage in aqueous electrolyte to be significantly extended. In addition, the specific energy and maximum power are always higher for the asymmetric configurations than for the corresponding symmetric system using any of the electrode materials, independently of being an ECP, a-MnO$_2$ or activated carbon. The highest energy and power density (10.5 Wh kg^{-1} and 123 kW kg^{-1}, respectively) are for the asymmetric cell with a-MnO$_2$ and activated carbon as positive and negative electrode, respectively, due to the large voltage window.

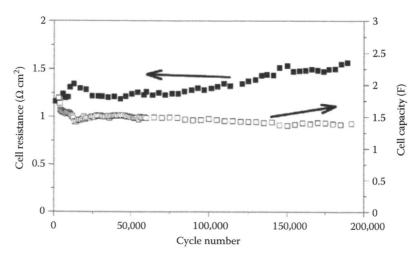

FIGURE 8.37   Capacitance (open squares) and resistance (black squares) of a hybrid AC–MnO$_2$ cell. Charge/ discharge current 40 mA cm^{-2} between 0 and 2 V. (From Brousse, T., et al., *J. Power Sources*, 173, 633, 2007. With permission.)

A recent work [125] has shown that the cycle life of the asymmetric manganese oxide/carbon system is similar or even better than for the double-layer capacitors built with carbon electrodes in organic electrolytes. Figure 8.37 shows the evolution of the capacitance of a laboratory test cell after N_2 bubbling prior to the cell sealing in order to minimize the corrosion of the stainless steel current collectors by the dissolved oxygen present in the electrolyte. The capacitance fading after 195,000 cycles is only of 12.5% and the limited ESR change upon cycling demonstrates the attractive performance of this asymmetric configuration. After scaling up the device by assembling several electrodes in parallel, the capacity and the cell voltage could be maintained over 600 cycles at 380 F and 2 V, respectively.

In conclusion, while being environment-friendly, the asymmetric construction in aqueous electrolyte offers electrochemical characteristics comparable with EDLCs in organic electrolytes. Combining materials with pseudocapacitance properties in an asymmetric cell is a very promising issue for developing a new generation of high-performance supercapacitors.

8.5.3.2 Asymmetric Systems Using Two Different Activated Carbons

A new type of supercapacitor based on only activated carbons for both electrodes in an environment-friendly electrolyte (sulfuric acid) has been recently discovered [128]. The novelty consists in using different activated carbons adapted for the positive and negative electrode, respectively, in a nonsymmetric configuration. According to this simple construction, either the capacitance value or the voltage window of the capacitor is enhanced, resulting in a high energy density device. For adapting the performance of an activated carbon, either as positive or as negative electrode, very simple processes have been followed in order to have materials in which EDL and pseudofaradic properties are combined.

Figure 8.38 compares the cyclic voltammograms in a three-electrode cell (1 mol L^{-1} H_2SO_4 electrolyte) of a commercial activated carbon from Norit (AC2 – $S_{BET} = 1400\,m^2\,g^{-1}$) and of the same carbon after oxidation in 30% nitric acid ($AC2_{ox}$). After oxidation, the oxygen content increases from 4.6 at% in AC2 to 15.0 at% in $AC2_{ox}$ and the carbon capacitance from 110 to 240 F g^{-1}, due to the redox processes involving the oxygenated functionalities. Interestingly, after chemical oxidation,

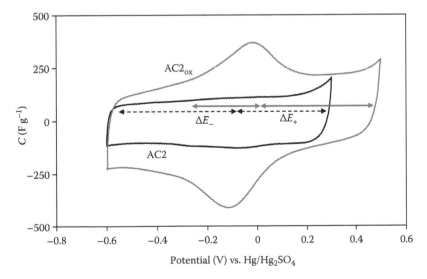

FIGURE 8.38 Cyclic voltammograms of the activated carbons AC2 and $AC2_{ox}$ in a three-electrode cell using 1 mol L^{-1} H_2SO_4. Graphite is the auxiliary electrode. The arrows represent the real potential range ΔE for the positive and the negative electrode when a two-electrode cell is charged up to 0.8 V. The dashed and full arrows are for the AC2 and $AC2_{ox}$ based capacitors, respectively. (Adapted from Khomenko, V., et al., *J. Power Sources*, 2008.)

the oxidized carbon $AC2_{ox}$ may operate in a higher potential range than as-received AC2. However, this carbon is not interesting in a symmetric capacitor, because the redox reactions involving the oxygenated functionalities control simultaneously the EP at 0 V and the potential range ΔE for the positive and negative electrodes (arrows included in Figure 8.38). For the oxidized sample, ΔE_+ of the positive electrode is displaced to very positive potentials. Consequently, the positive electrode of the symmetric capacitor based on $AC2_{ox}$ works only partly in the range where redox reactions take place. Thus, the capacitance of the positive electrode is smaller than for the negative one and, due to the series dependence for the real capacitor (Equation 8.1), the total cell capacitance is determined by the positive electrode. To summarize, the strong chemical oxidation of AC2 provides a carbon perfectly adapted for operating in the positive range of potential.

As in the previous examples, the pseudofaradic behavior of activated carbons related with the redox sorption of hydrogen in the pores under negative polarization has been used in order to obtain a perfectly optimized material as negative electrode. In particular, a nanoporous carbon, $AC1_{ox}$, combining the most interesting features for being used as negative electrode has been obtained by mild oxidation (20% H_2O_2 at room temperature) of an activated carbon (AC1, Maxsorb, Kensai, $S_{BET} = 3487 \, m^2 \, g^{-1}$). The oxygen content is 5.0 and 6.6 at% for AC1 and $AC1_{ox}$, respectively. Such mild oxidized carbon ($AC1_{ox}$) presents: (1) a very high capacitance, e.g., 480 F g^{-1} resulting from the combination of the EDL charging and a large pseudocapacitance related to the reversible hydrogen electrosorption; (2) a very high overpotential for H_2 evolution (see Figure 8.39).

From the foregoing, different pseudocapacitive properties of activated carbon can be used in order to fulfill the requirements of the positive and negative electrodes. Figure 8.39 shows together the cyclic voltammograms of the activated carbons $AC1_{ox}$ and $AC2_{ox}$ obtained in a three-electrode cell. From this figure, it can be expected that combining the two electrodes should allow a total cell voltage of 1.8 V to be reached without electrolyte decomposition or destructive oxidation of the activated carbon. Also the capacitance of the electrochemical capacitor should increase because the redox reactions at the positive and negative electrodes occur in different potential range.

In practice, up to 1.6 V, the asymmetric cell demonstrates a perfect cycle life with a capacitance value of 320 F g^{-1} [128]. The capacity drop of 15% during the first 2000 cycles under a maximum

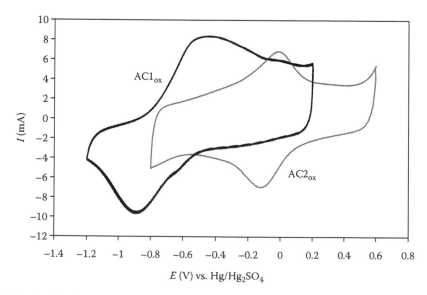

FIGURE 8.39 Cyclic voltammograms of the activated carbons $AC1_{ox}$ and $AC2_{ox}$ in a three-electrode cell using 1 mol L^{-1} H_2SO_4. Graphite is the auxiliary electrode. (Adapted from Khomenko, V., et al., *J. Power Sources*, 2008.)

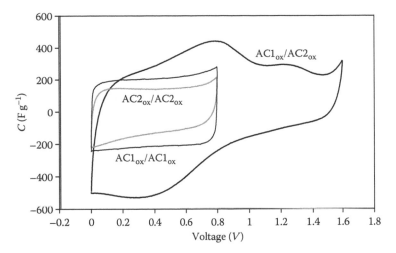

FIGURE 8.40 Cyclic voltammograms of an optimized asymmetric $AC1_{ox}/AC2_{ox}$ capacitor in $1\,\text{mol}\,\text{L}^{-1}$ H_2SO_4 electrolyte in comparison with the symmetric capacitors $AC1_{ox}/AC1_{ox}$ and $AC2_{ox}/AC2_{ox}$. (Adapted from Khomenko, V., et al., *J. Power Sources*, 2008.)

voltage of $1.6\,\text{V}$ and a current density of $1\,\text{A}\,\text{g}^{-1}$ is attributed to conditioning of the system. Thereafter, the specific capacitance remains almost constant during the subsequent 8000 cycles, indicating a high cycling stability of the supercapacitor.

Figure 8.40 shows the cyclic voltammograms obtained in the asymmetric configuration and in symmetric cells designed with each of the materials used for the positive or the negative electrodes. Due to the clear differences in capacitance and cell voltage, the specific energy of the asymmetric capacitor reaches a value of about $30\,\text{Wh}\,\text{kg}^{-1}$ which is very close to the values observed with symmetric carbon-based electrochemical capacitors using an aqueous electrolyte. These results demonstrate that most of the disadvantages of the organic electrolyte can be eliminated by developing asymmetric carbon/carbon supercapacitors in aqueous electrolyte. Due to the use of the water solution, this capacitor is safe and environment-friendly, and it can be produced at a low cost.

8.6 CONCLUSION

During the last years, a tremendous number of papers related with supercapacitors have been published. The largest number is devoted to investigating the electrochemical properties of various types of activated carbons. Since many authors consider that some equivalence exists between the surface area of the electrode/electrolyte interface and the specific surface area of carbons, a great attention has been generally focused to producing carbons with a highly developed porosity. Most experiments on these carbons were realized in aqueous medium, with the believing that only the EDL capacity is measured in this electrolyte, and neglecting the existence of a pseudocapacitive contribution.

From the researches achieved recently, three main breakthroughs might be observed. The first one concerns the traditional concept of EDL in a nanoporous carbon. It has been demonstrated from independent experiments performed by few groups that the highest capacitance related with electrosorption is observed when the nanopores of carbons perfectly fit the dimensions of ions. Some experiments have even shown some swelling of the carbon host which could be attributed to an intercalation/insertion-type penetration of ions. Moreover, the ions could penetrate desolvated in the pores. From the foreword, it is obvious that some computational work is now necessary for a better interpretation of these experimental observations. Especially, the energy calculations should allow the kind of ion–host interaction to be described better. In particular, if some kind of intercalation would be confirmed, it would be worth to study the influence of carbon doping on the EDL properties.

The second breakthrough is related with the development of pseudocapacitive materials using environmental-friendly electrolytes. Definitely, it seems that conducting polymers are not adapted for practical systems, because of their relatively high cost and poor cycle life. On the other hand, attention should be paid to some composites involving low-cost oxides and a carbon framework as electrical conductivity percolator, e.g., MnO_2/CNTs, and particularly to designing materials where the active oxide better operates. Undoubtedly, the novelty of the last years in this domain concerns the development of very interesting pseudocapacitive carbons with in-frame incorporated heteroatoms, e.g., nitrogen and/or oxygen. Optimized nitrogen-doped carbons have been prepared by the template technique or using nanotubes as a support. Although expensive, such materials are very useful for understanding the pseudofaradic properties and the ways to improve them. From the point of view of applications, a new generation of doped carbons has been obtained from seaweeds and from biopolymers. Because of their poorly developed porosity, these high-density carbons exhibit high volumetric capacity, high volumetric energy, and power densities. The reversible chemisorption of hydrogen by water electrodecomposition on a negatively polarized carbon electrode is also an interesting way for providing a pseudofaradic contribution.

Definitely, the symmetric capacitors in aqueous solutions do not fulfill industrial requirements. Based on pseudocapacitive materials, the last important breakthrough concerns the development of asymmetric (or hybrid) capacitors able to operate in aqueous electrolyte at higher voltage than the symmetric systems. In general, the negative electrode is from nanoporous carbon which undergoes both EDL charging and weak hydrogen chemisorption through water reduction. The positive electrode might be oxidized carbon or some oxide, e.g., a-MnO_2. The record to-date was obtained with an asymmetric activated carbon/a-MnO_2 capacitor which presents a good cycle life up to 2 V. Asymmetric capacitors with two different activated carbon electrodes seem to be a very attractive solution able to fulfill the main industrial requirements, such as low cost, safety, and environment compatibility. Taking into account the high versatility of carbons, there are many possibilities for optimizing these systems. The adaptable surface functionality and nanotexture of carbons offers incredible opportunities for developing supercapacitors able to fulfill the industry expectations.

REFERENCES

1. J.C. Farahmendi, J.M. Dispennette, E. Blank, and A.C. Kolb (1998) Patents WO9815962, EP0946954 and JP 2001502117T for Maxwell Technologies.
2. P. Azaïs, L. Duclaux, P. Florian, D. Massiot, M.A. Lillo-Ródenas, A. Linares-Solano, J.P. Peres, C. Jehoulet, and F. Béguin, *J. Power Sources* 171 (2007) 1046.
3. L.R. Radovic and B. Bockrath, *J. Am. Chem. Soc.* 127 (2005) 5917.
4. P. Azaïs, Recherche des causes du vieillissement de supercondensateurs à électrolyte organique à base de carbons activés, PhD thesis, Orléans, France (2004).
5. E. Raymundo-Piñero, V. Khomenko, E. Frackowiak, and F. Béguin, *J. Electrochem. Soc.* 152 (2005) A229.
6. P. Portet, P.L. Taberna, P. Simon, and E. Flahaut, *J. Power Sources* 139 (2005) 371.
7. B.E. Conway, *Electrochemical Supercapacitors–Scientific Fundamentals and Technological Applications*, Kluwer Academic/Plenum, New York (1999).
8. E. Frackowiak and F. Béguin, *Carbon* 39 (2001) 937.
9. T. Kim, S. Lim, K. Kwon, S.H. Hong, W. Qiao, C.K. Rhee, S.H. Yoon, and I. Mochida, *Langmuir* 22 (2006) 9086.
10. O. Barbieri, M. Hahn, A. Herzog, and R. Kötz, *Carbon* 43 (2005) 1303.
11. S.J. Gregg and K.S.W. Sing, *Adsorption, Surface Area and Porosity*, Academic Press, London (1982), pp. 103–104.
12. K. Kaneko and C. Ishii, *Colloids Surf.* 67 (1992) 203.
13. M. Hahn, O. Barbieri, F.P. Campana, R. Kötz, and R. Gallay, *Appl. Phys. A* 82 (2006) 633.
14. G. Salitra, A. Soffer, L. Eliad, Y. Cohen, and D. Aurbach, *J. Electrochem. Soc.* 147 (2000) 2486.

15. L. Eliad, G. Salitra, A. Soffer, and D. Aurbach, *J. Phys. Chem. B* 105 (2001) 6880.
16. L. Eliad, E. Pollak, N. Levy, G. Salitra, A. Soffer, and D. Aurbach, *Appl. Phys. A* 82 (2006) 607.
17. C.O. Ania, J. Pernak, F. Stefaniak, E. Raymundo-Piñero, and F. Béguin, *Carbon* 44 (2006) 3126.
18. C. Largeot, C. Portet, J. Chmiola, P.L. Taberna, Y. Gogotsi, and P. Simon, *J. Am. Chem. Soc.* 130 (2008) 2730.
19. Y.J. Kim, Y. Horie, S. Ozaki, Y. Matsuzawa, H. Suezaki, C. Kim, N. Miyashita, and M. Endo, *Carbon* 42 (2004) 1491.
20. C. Lin, J.A. Ritter, and B.N. Popov, *J. Electrochem. Soc.* 146 (1999) 3639.
21. E. Raymundo-Piñero, K. Kierzek, J. Machnikowski, and F. Béguin, *Carbon* 44 (2006) 2498.
22. J. Chmiola, G. Yushin, Y. Gogotsi, C. Portet, P. Simon, and P.L. Taberna, *Science* 313 (2006) 1760.
23. M. Ue, *J. Electrochem. Soc.* 141 (1994) 3336.
24. C.M. Yang, Y.J. Kim, M. Endo, H. Kanoh, M. Yudasaka, S. Iijima, and K. Kaneko, *J. Am. Chem. Soc.* 129 (2007) 20.
25. J. Chmiola, C. Largeot, P.L. Taberna, P. Simon, and Y. Gogotsi, *Angew. Chem. Int. Ed. Enge.* 120 (2008) 3440.
26. C. Vix-Guterl, E. Frackowiak, K. Jurewicz, M. Friebe, J. Parmentier, and F. Béguin, *Carbon* 43 (2005) 1293.
27. D. Cazorla-Amorós, J. Alcañiz-Monge, M.A. de la Casa-Lillo, and A. Linares-Solano, *Langmuir* 14 (1998) 4589.
28. A. Laforgue, P. Simon, C. Sarrazin, and J.F. Fauvarque, *J. Power Sources* 80 (1999) 142.
29. K. Jurewicz, S. Delpeux, V. Bertagna, F. Béguin, and E. Frackowiak, *Chem. Phys. Lett.* 347 (2001) 36.
30. M. Mastragostino, C. Arbizzani, and F. Soavi, *J. Power Sources* 87–98 (2001) 812.
31. M. Mastragostino, C. Arbizzani, and F. Soavi, *Solid State Ionics* 148 (2002) 493.
32. M. Toupin, T. Brousse, and D. Bélanger, *Chem. Mater.* 16 (2004) 3184.
33. N.L. Wu, *Mater. Chem. Phys.* 75 (2002) 6.
34. J.P. Zheng, P.J. Cygan, and T.R. Jow, *J. Electrochem. Soc.* 142 (1995) 2699.
35. K. Jurewicz, K. Babel, A. Ziolkowski, and H. Wachowska, *Electrochim. Acta* 48 (2003) 1491.
36. G. Lota, B. Grzyb, H. Machnikowska, J. Machnikowski, and E. Frackowiak, *Chem. Phys. Lett.* 404 (2005) 53.
37. J.R. Pels, F. Kapteijn, J.A. Moulijn, Q. Zhu, and K.M. Thomas, *Carbon* 33 (1995) 1641.
38. F. Kapteijn, J.A. Moulijn, S. Matzner, and H.P. Boehm, *Carbon* 37 (1999) 1143.
39. K. Jurewicz, E. Frackowiak, and F. Béguin, *Appl. Phys. A* 78 (2004) 981.
40. B. Grzyb, J. Machnikowski, and J.V. Weber, *J. Anal. Appl. Pyrolysis* 72 (2004) 121.
41. E. Frackowiak, G. Lota, J. Machnikowski, K. Kierzek, C. Vix, and F. Béguin, *Electrochim. Acta* 51 (2005) 2209.
42. G. Lota, K. Lota, and E. Frackowiak, *Electrochem. Comm.* 9 (2007) 1828.
43. D. Hulicova, J. Yamashita, Y. Soneda, H. Hatori, and M. Kodama, *Chem. Mater.* 17 (2005) 1241.
44. D. Hulicova, M. Kodama, and H. Hatori, *Chem. Mater.* 18 (2006) 2318.
45. C.O. Ania, V. Khomenko, E. Raymundo-Piñero, J.B. Parra, and F. Béguin, *Adv. Func. Mater.* 17 (2007) 1828.
46. A. Burke, *J. Power Sources* 91 (2000) 37.
47. F. Béguin, K. Szostak, G. Lota, and E. Frackowiak, *Adv. Mater.* 17 (2005) 2380.
48. E. Raymundo-Piñero, F. Leroux, and F. Béguin, *Adv. Mater.* 18 (2006) 1877.
49. S. Biniak, A. Swiatkowski, and M. Makula, In *Chemistry and Physics of Carbon*, Vol. 27, L.R. Radovic (Ed.), Marcel Dekker, New York (2001) Chapter 3.
50. M.A. Montes-Moran, D. Suarez, J.A. Menendez, and E. Fuente, *Carbon* 42 (2004) 1219.
51. E. Raymundo-Piñero, M. Cadek, and F. Béguin, *Adv. Funtc. Mat.* 19 (2009) 1.
52. K. Jurewicz, E. Frackowiak, and F. Béguin, *Electrochem. Solid State Lett.* 4 (2001) A27.
53. K. Jurewicz, E. Frackowiak, and F. Béguin, *Fuel Process. Tech.* 77–78 (2002) 415.
54. B. Fang, H. Zhou, and I. Honma, *J. Phys. Chem. B* 110 (2006) 4875.
55. C. Carpetis and W. Peschka, *Int. J. Hyd. Energy* 5 (1980) 539.
56. R.K. Agarwal, J.S. Noh, J.A. Schwarz, and P. Davini, *Carbon* 25 (1987) 219.
57. L. Schlapbach and A. Züttel, *Nature* 414 (2001) 353.
58. F. Cuevas, J.M. Joubert, M. Latroche, and A. Percheron-Guégan, *Appl. Phys. A* 72 (2001) 225.
59. M. Rzepka, P. Lamp, and M.A. de la Casa-Lillo, *J. Phys. Chem. B* 102 (1998) 10894.
60. Q. Wang and J.K. Johnson, *J. Chem. Phys.* 110 (1999) 577.
61. A. Züttel, P. Sudan, P. Mauron, T. Kioyobayashi, C. Emmenegger, and L. Schlapbach, *Int. J. Hyd. Energy* 27 (2002) 203.
62. R. Ströbel, L. Jörissen, T. Schliermann, V. Trapp, W. Schütz, K. Bohmhammel, G. Wolf, and J. Garche, *J. Power Sources* 84 (1999) 221.

63. M.G. Nijkamp, J.E.M.G. Raaymakers, A.J. van Dillen, and K.P. de Jong, *Appl. Phys. A* 72 (2001) 619.
64. M.A. de la Casa-Lillo, F. Lamari-Darkrim, D. Cazorla-Amorós, and A. Linares-Solano, *J. Phys. Chem.* 106 (2002) 10930.
65. M. Jorda-Beneyto, F. Suárez-García, D. Lozano-Castelló, D. Cazorla-Amorós, and A. Linares-Solano, *Carbon* 45 (2007) 293.
66. N. Texier-Mandoki, J. Dentzer, T. Piquero, S. Saadallah, P. David, and C. Vix-Guterl, *Carbon* 42 (2004) 2735.
67. F. Béguin, K. Kierzek, M. Friebe, A. Jankowska, J. Machnikowski, K. Jurewicz, and E. Frackowiak, *Electrochim. Acta* 51 (2006) 2161.
68. F. Béguin, M. Friebe, K. Jurewicz, C. Vix-Guterl, J. Dentzer, and E. Frackowiak, *Carbon* 44 (2006) 2392.
69. R. Ryoo, S.H. Joo, and S. Jun, *J. Phys. Chem. B* 103 (1999) 7743.
70. M.A. Lillo-Ródenas, D. Cazorla-Amorós, and A. Linares-Solano, *Carbon* 41 (2003) 267.
71. J.M. Gatica, J.M. Rodriguez-Izquierdo, D. Sanchez, T. Chafik, S. Harti, H. Zaitan, and H. Vidal, *Comptes Rendus Chimie* 9 (2006) 1215.
72. S. Haydar, C. Moreno-Castilla, M.A. Ferro-Garcia, F. Carrasco-Marin, J. Rivera-Utrilla, A. Perrard, and J.P. Joly, *Carbon* 38 (2000) 1297.
73. S. Orimo, T. Matsushima, H. Fujii, T. Fukunaga, and G. Majer, *J. Appl. Phys.* 90 (2001) 1545.
74. M.J. Bleda-Martínez, J.M. Perez, A. Linares-Solano, E. Morallón, and D. Cazorla-Amorós, *Carbon* 46 (2008) 1053.
75. J.M. Miller, B. Dunn, T.D. Tran, and R.W. Pekala, *Langmuir* 15 (1999) 799.
76. M. Toupin, T. Brousse, and D. Bélanger, *Chem. Mater.* 14 (2002) 3946.
77. C. Arbizzani, M. Mastragostino, and F. Soavi, *J. Power Sources* 100 (2001) 164.
78. R.S. Campomanes, E. Bittencourt, and J.S.C. Campos, *Synth. Met.* 102 (1999) 1230.
79. W.A. Wampler, C. Wei, and K. Rajeshwar, *J. Electrochem. Soc.* 141 (1994) L13.
80. J. Sandler, M.S.P. Shaffer, T. Prasse, W. Bauhofer, K. Schulte, and A.H. Windle, *Polymer* 40 (1999) 5967.
81. H. Liang, F. Chen, R. Li, L. Wang, and Z. Deng, *Electrochim. Acta* 49 (2004) 3463.
82. A. Rudge, J. Davey, I. Raistrick, and S. Gottesfeld, *J. Power Sources* 47 (1994) 89.
83. A. Rudge, I. Raistrick, S. Gottesfeld, and J.P. Ferraris, *Electrochim. Acta* 39 (1994) 273.
84. J.H. Park, J.M. Ko, O.O. Park, and D. Kim, *J. Power Sources* 105 (2002) 20.
85. J.H. Chen, Z.D. Huang, D.Z. Wang, S.X. Yang, W.Z. Li, J.G. Wan, and Z.F. Ren, *Synth. Met.* 125 (2002) 289.
86. M. Hughes, G.Z. Chen, M.S.P. Schaffer, D.J. Fray, and A.H. Windle, *Chem. Mater.* 14 (2002) 1610.
87. M. Hughes, M.S.P. Schaffer, A.C. Renouf, C. Singh, G.Z. Chen, D.J. Fray, and A.H. Windle, *Adv. Mater.* 14 (2002) 382.
88. K.H. An, K.K. Jeon, J.K. Heo, S.C. Lim, D.J. Bae, and Y.H. Lee, *J. Electrochem. Soc.* 149 (2002) A1058.
89. Y. Zhou, B. He, W. Zhou, J. Huang, X. Li, B. Wu, and H. Li, *Electrochim. Acta* 49 (2004) 257.
90. J. Huang, X. Li, J. Xu, and H. Li, *Carbon* 41 (2003) 2731.
91. E. Frackowiak, K. Jurewicz, S. Delpeux, and F. Béguin, *J. Power Sources* 97–98 (2001) 822.
92. V. Khomenko, E. Frackowiak, and F. Béguin, *Electrochim. Acta* 50 (2005) 2499.
93. Q. Xiao and X. Zhou, *Electrochim. Acta* 48 (2003) 575.
94. G.Z. Chen, M.S.P. Schaffer, D. Coleby, G. Dixon, W. Zhou, D.J. Fray, and A.H. Windle, *Adv. Mater.* 12 (2000) 522.
95. V. Khomenko, E. Raymundo-Piñero, E. Frackowiak, and F. Béguin, *Appl. Phys. A* 82 (2006) 567.
96. G.X. Wang, B.L. Zhan, Z.L. Yu, and M.Z. Qu, *Solid State Ionics* 176 (2005) 1169.
97. M.S. Hong, S.H. Lee, and S.W. Kim, *Electrochem. Solid State Lett.* 5 (2002) A227.
98. J.Y. Lee, K. Liang, K.H. An, and Y. H. Lee, *Synthetic Metals* 150 (2005) 153.
99. S.N. Razumov, A.D. Klementov, S.V. Litvinenko, and A.I. Beliakov, US Patent 6,222,723 (2001).
100. S.N. Razumov, S.V. Litvenenko, A.D. Klementov, and A.I. Belyakov, European Patent 1,156,500 (2001).
101. G.G. Amatucci, F. Badway, A. Du Pasquier, and T. Zheng, *J. Electrochem. Soc.* 148 (2001) A930.
102. A. Du Pasquier, I. Plitz, J. Gural, S. Menocal, and G.G. Amatucci, *J. Power Sources* 113 (2003) 62.
103. A. Laforgue, P. Simon, J.F. Fauvarque, J.F. Sarrau, and P. Lailler, *J. Electrochem. Soc.* 148 (2001) A1130.
104. A. Laforgue, P. Simon, J.F. Fauvarque, M. Mastragostino, F. Soavi, J.F. Sarrau, P. Lailler, M. Conte, E. Rossi, and S. Saguatti, *J. Electrochem. Soc.* 150 (2003) A645.

105. A. Di Fabio, A. Giorgi, M. Mastragostino, and F. Soavi, *J. Electrochem. Soc.* 148 (2001) A850.
106. A. Du Pasquier, A. Laforgue, P. Simon, G.G. Amatucci, and J.F. Fauvarque, *J. Electrochem. Soc.* 149 (2002) A302.
107. A. Du Pasquier, A. Laforgue, and P. Simon, *J. Power Sources* 125 (2004) 95.
108. K. Ariyoshi, S. Yamamoto, and T. Ohzuku. *J. Power Sources* 119–121 (2003) 959.
109. H. Li, L. Cheng, and Y. Xia, *Electrochem. Solid-State Lett.* 8 (2005) A433.
110. I. Plitz, A. Du Pasquier, F. Badway, J. Gural, N. Pereira, A. Gmitter, and G.G. Amatucci, *Appl. Phys. A* 82 (2006) 615.
111. J.P. Zheng, *J. Electrochem. Soc.* 150 (2003) A484.
112. W.G. Pell and B.E. Conway, *J. Power Sources* 136 (2004) 334.
113. V. Khomenko, E. Raymundo-Piñero, and F. Béguin, *J. Power Sources* 177 (2008) 643.
114. T. Aida, K. Yamada, and M. Morita, *Electrochem. Solid-State Lett.* 9 (2006) A534.
115. T. Aida, I. Murayama, K. Yamada, and M. Morita, *J. Electrochem. Soc.* 154 (2007) A798.
116. T. Aida, I. Murayama, K. Yamada, and M. Morita, *Electrochem. Solid-State Lett.* 10 (2007) A93.
117. J.R. Dahn, T. Zheng, Y. Liu, and J.S. Xue, *Science* 270 (1995) 590.
118. D. Linden, *Handbook of Batteries*, 2nd ed., McGraw-Hill, New York (1995).
119. W.A. Schalkwijk and B. Scrosati, *Advances in Lithium-Ion Batteries*, Kluwer Academic/Plenum Publishers, New York (2002).
120. T. Zheng, J.R. Dahn, Applications of carbon in lithium-ion batteries, In T.D. Burchell (ed.), *Carbon Materials for Advanced Technologies*, Pergamon, Amsterdam, the Netherlands (1999) 341–388.
121. Y.G. Wang and Y.Y. Xia, *J. Electrochem. Soc.* 153 (2006) A450.
122. Y.G. Wang, J.Y. Luo, C.X. Wang, and Y.Y. Xia, *J. Electrochem. Soc.* 153 (2006) A1425.
123. T. Brousse, M. Toupin, and D. Bélanger, *J. Electrochem. Soc.* 151 (2004) A614.
124. V. Khomenko, E. Raymundo-Piñero, and F. Béguin, *J. Power Sources* 153 (2006) 183.
125. T. Brousse, P.L. Taberna, O. Crosnier, R. Dugas, P. Guillemet, Y. Scudeller, Y. Zhou, F. Favier, D. Bélanger, and P. Simon, *J. Power Sources* 173 (2007) 633.
126. T. Brousse and D. Bélanger, *Electrochem. Solid State Lett.* 6 (2003) A244.
127. J.H. Park and O.O. Park, *J. Power Sources* 111 (2002) 185.
128. V. Khomenko, E. Raymundo-Piñero, and F. Béguin, *J. Power Sources* (2008) submitted.

9 Fuel Cell Systems: Which Technological Breakthrough for Industrial Development?

Claude Lamy

CONTENTS

9.1 INTRODUCTION

Discovered in England in 1839 by Sir William Grove [1,2], the fuel cell is an electrochemical device which transforms directly the heat of combustion of a fuel (hydrogen, natural gas, methanol, ethanol, hydrocarbons, etc.) into electricity. The fuel is oxidized electrochemically at the anode, without producing any pollutants (only water and/or carbon dioxide are rejected in the atmosphere), whereas the oxidant (oxygen from the air) is reduced at the cathode. This process does not follow the Carnot's theorem, so that higher energy efficiencies are expected: 40%–50% in electrical energy, 80%–85% in total energy (electricity + heat production).

There is now a great interest in developing different kinds of fuel cells with several applications (besides the first and until now main application in space programs) depending on their nominal power: stationary electric power plants (100 kW to 10 MW), electricity and heat production for buildings and houses (5–200 kW), power train sources (20–200 kW) for the electrical vehicle (bus, truck, and individual car), auxiliary power units (APU) (1–100 kW) for different uses (automobiles, aircrafts, space launchers, space stations, etc.), and also for portable electronic devices (1–100 W), such as cell phones, computers, cam recorders, etc. For the latter applications, low-temperature fuel cells are more convenient, particularly the alkaline fuel cell (AFC) working at 80°C with pure hydrogen and oxygen, the proton-exchange membrane fuel cell (PEMFC), which can operate from ambient temperature to 70°C–80°C with hydrogen as the fuel, either ultrapure hydrogen (produced, e.g., by water electrolysis) or hydrogen-containing reformate gas, and the direct alcohol (methanol or ethanol) fuel cell (DAFC) which needs higher temperatures (up to 120°C–150°C) to electro-oxidize directly methanol (ethanol). In both cases, because of the relatively low working temperature, the kinetics of the electrochemical reactions involved is rather slow. This requires the development of new electrocatalysts, in order to increase the rate of fuel oxidation and oxygen reduction [3–5].

In this chapter, after recalling the working principles and the different kinds of fuel cells, the discussion will be focused on low-temperature fuel cells (AFC, PEMFC, and DAFC), in which several kinds of carbon materials are used (catalyst support, gas-diffusion layer [GDL], bipolar plates [BP], etc.). Then some possible applications in different areas will be presented. Finally the materials used in fuel cells, particularly carbon materials, will be discussed according to the aimed applications. To read more details on the use of carbon in fuel cell technology, see the review paper on "The role of carbon in fuel cell technology" recently published by Dicks [6].

9.2 PRINCIPLE OF FUEL CELLS

The working principle of fuel cells, the different kinds of fuel cells, the main fuels used and their possible applications will be first discussed [7–10].

9.2.1 How Do They Work?

A fuel cell system consists of a fuel cell stack and auxiliary equipments, which produce electric energy (and heat) directly from the electrochemical oxidation of a fuel in an elementary electrochemical cell (Figure 9.1).

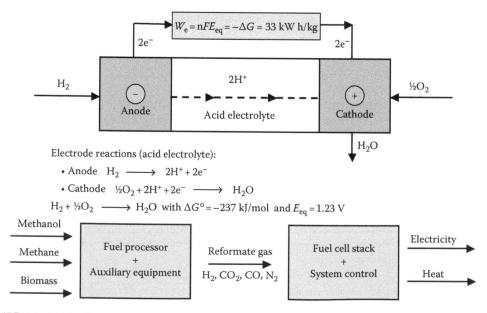

FIGURE 9.1 Fuel cell system.

An elementary electrochemical cell converts directly the chemical energy of combustion in oxygen (i.e., the Gibbs energy change, $-\Delta G$) of a given fuel (hydrogen, natural gas, hydrocarbons, kerosene, alcohols, etc.) into electricity [11–13]. Electrons liberated at the anode (negative pole of the cell) by the electro-oxidation of the fuel pass through the external circuit (producing an electrical energy, nFE_{cell}, equal to $-\Delta G$, where E_{cell} is the cell voltage and $F = 96{,}485$ C, the Faraday constant) and arrive at the cathode (positive pole), where they reduce oxygen (from air). Inside the fuel cell, the electrical current is transported by migration and diffusion of the electrolyte ions (H^+, OH^-, $O^=$, $CO_3^=$).

The electrochemical reactions involved are the electro-oxidation of hydrogen at the anode:

$$H_2 \rightarrow 2H^+ + 2e^- \quad E_1^\circ = 0.000 \text{ V/SHE}$$

and the electroreduction of oxygen at the cathode:

$$O_2 + 4H^+ + 4e^- \rightarrow 2H_2O \quad E_2^\circ = 1.229 \text{ V/SHE}$$

where E_i° are the electrode potentials vs. the standard hydrogen (reference) electrode (SHE). These electrochemical reactions correspond to the combustion reaction of hydrogen in oxygen

$$H_2 + 1/2O_2 \rightarrow H_2O$$

with the thermodynamic data, under standard conditions (liquid water):

$$\Delta G^\circ = -237 \text{ kJ mol}^{-1}; \quad \Delta H^\circ = -286 \text{ kJ mol}^{-1} \text{ of } H_2$$

The cell voltage, E_{eq}°, at equilibrium (no current flowing) under standard conditions, is thus

$$E_{eq}^\circ = -\frac{\Delta G_o}{nF} = \frac{237 \times 10^3}{2 \times 96485} = E_2^\circ - E_1^\circ = 1.229 \text{ V} \approx 1.23 \text{ V}$$

This process produces an electric energy, $W_e = nFE_{eq}^\circ = -\Delta G^\circ$, corresponding to an energy mass density of the fuel, $W_{el} = -\Delta G^\circ/(3600\,M) = 32.9$ kW h kg^{-1}, where $M = 0.002$ kg is the molecular weight of hydrogen, $F = 96{,}485$ C, the Faraday constant (i.e., the absolute value of the electric charge of one mole of electrons), and $n = 2$ the number of electrons involved in the oxidation of one hydrogen molecule.

With gaseous water the cell voltage at equilibrium of a hydrogen–oxygen fuel cell at 25°C under standard conditions is $E_{eq}^\circ = 1.18$ V, since $\Delta G_{gas}^\circ = -227$ kJ mol^{-1} at 25°C.

The working of the cell under reversible thermodynamic conditions does not follow Carnot's theorem so that the theoretical energy efficiency, defined as the ratio of the produced electric energy ($-\Delta G$) to the chemical combustion energy ($-\Delta H$), i.e., $\Delta G/\Delta H$, is very high: 83% for the hydrogen–oxygen fuel cell and 97% for the methanol–oxygen fuel cell (see Section 9.3.3) at 25°C under standard conditions.

This theoretical efficiency is much greater (by a factor of about 2) than that of a thermal combustion engine, producing the reversible work, W_r, according to the Carnot's theorem:

$$\varepsilon_{therm}^{rev} = \frac{W_r}{(-\Delta H)} = 1 - \frac{Q_2}{Q_1} = 1 - \frac{T_2}{T_1} = 0.43 \quad \text{for, e.g., } T_1 = 350°C \text{ and } T_2 = 80°C$$

where Q_1 and Q_2 are the heat exchanged with the hot source and cold source, at temperatures T_1 and T_2, respectively.

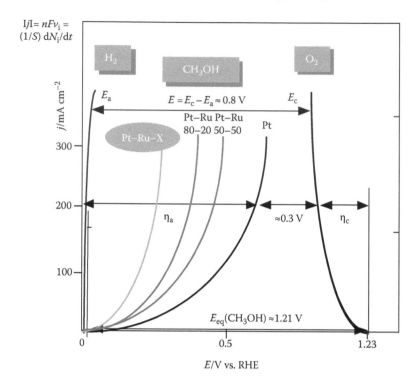

FIGURE 9.2 Current density vs. electrode potential curves for the H_2–O_2 and the CH_3OH–O_2 fuel cells showing the reaction overvoltages η_a and η_c at different catalytic electrodes (Pt, Pt–Ru,…).

However under working conditions with a current density j (different from zero), the cell voltage $E(j)$ becomes smaller than the equilibrium voltage E_{eq} as the result of reaction overvoltages η_a and η_c at both electrodes (η is defined as the difference between the working electrode potential $E_{(i)}$ and the equilibrium potential $E_{eq}^{(i)}$) and ohmic losses (Figure 9.2).

This deviation from the thermodynamic reversible behavior leads to a decrease of the cell voltage by ca. 0.4–0.6 V, which reduces the energy efficiency of the fuel cell by a factor $\varepsilon_E = E(j)/E_{eq}$, called voltage efficiency ($\varepsilon_E = 0.80/1.23 = 0.65$ for a cell voltage of 0.80 V), so that the overall energy efficiency at 25°C for the H_2/O_2 fuel cell becomes $0.83 \times 0.65 = 0.54$.

For the elementary fuel cell, the cell voltage $E(j)$ is the difference between the cathode potential E_c and the anode potential E_a:

$$E(j) = E_c(j) - E_a(j) = E_{eq} - \left(|\eta_a| + |\eta_c| + R_e j \right)$$

where
  E_{eq} is the equilibrium cell voltage under the temperature and pressure working conditions
  R_e is the specific resistance of the cell (electrolyte + interfacial resistances).

In the $E(j)$ characteristics, one may distinguish three zones associated with these energy losses (Figure 9.3)

- Zone I: The E vs. j linear curve corresponds to ohmic losses $R_e j$ in the electrolyte and interface resistances; a decrease of the electric resistance R_e from ca 0.3 to 0.15 Ω cm^2 leads to an increase of the current density j (at 0.7 V) from 0.25 to 0.4 A cm^{-2}, i.e., an increase in the energy efficiency and in the power density by 1.6 times.

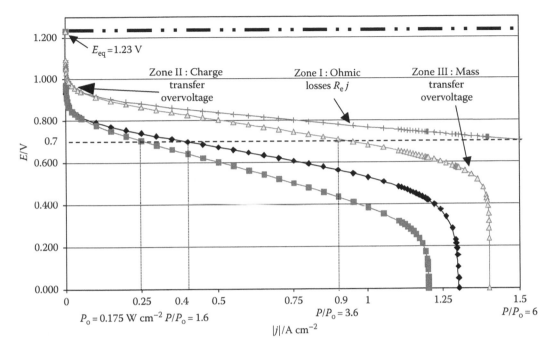

FIGURE 9.3 Theoretical $E(j)$ electric characteristics of a fuel cell. ▬▬ $j_o = 10^{-8}$; $R_e = 0.30$; $j_l = 1.2$ ◆ $j_o = 10^{-8}$; $R_e = 0.15$; $j_l = 1.3$ ▲ $j_o = 10^{-6}$; $R_e = 0.15$; $j_l = 1.4$ + $j_o = 10^{-6}$; $R_e = 0.10$; $j_l = 2.2$ (j_o, j_l, in A cm^{-2}, R_e in Ω cm^2).

- Zone II: The E vs. Ln j/j_o logarithmic curve corresponds to the charge transfer polarization, i.e., to the activation overvoltages due to a relatively low electron exchange rate at the electrode–electrolyte interface, particularly for the oxygen reduction reaction whose exchange current density j_o is much smaller than that of the hydrogen oxidation: an increase of j_o from 10^{-8} to 10^{-6} A cm^{-2} leads to an increase of the current density j (at 0.7 V) from 0.4 to 0.9 A cm^{-2}, i.e., an increase in the energy efficiency and in the power density by 3.6 times compared to the initial curve.

- Zone III: The E vs. Ln $(1-j/j_l)$ logarithmic curve corresponds to concentration polarization which results from the limiting value j_l of the mass transfer limiting current density for the reactive species and reaction products to and/or from the electrode active sites; an increase of j_l from 1.4 to 2.2 A cm^{-2} leads to a further increase of the current density j (at 0.7 V) from 0.9 to 1.5 A cm^{-2}, i.e., an overall increase in the energy efficiency and in the power density by six times compared to the initial curve.

Thus these three key points will determine the energy efficiency and the specific power of the elementary fuel cell: an improvement in each component of the cell will increase the power density from 0.175 to 1.05 W cm^{-2}, i.e., an increase by a factor of 6. As a consequence for the fuel cell systems, the weight and volume will be decreased by a similar factor, for a given power of the system, and presumably the overall cost will be similarly diminished. The improvement in the components of the elementary fuel cell has a direct effect on the system technology and therefore on the overall cost.

For a hydrogen–air fuel cell, the working cell voltage is typically 0.8 V at 500 mA cm^{-2}, which leads to a voltage efficiency ε_E:

$$\varepsilon_E = \frac{E(j)}{E^o_{eq}} = 1 - \frac{\left| \eta_a(j) \right| + \left| \eta_c(j) \right| + R_e |j|}{E^o_{eq}} = \frac{0.8}{1.23} = 0.65$$

The overall energy efficiency (ε_{cell}) thus becomes

$$\varepsilon_{cell} = \frac{W_e}{(-\Delta H^o)} = \frac{n_{exp} FE(|j|)}{(-\Delta H^o)} = \frac{n_{exp}}{n_{th}} \frac{E(|j|)}{E^o_{eq}} \frac{n_{th} FE^o_{eq}}{(-\Delta H^o)} = \varepsilon_F \varepsilon_E \varepsilon^{rev}_{eq}$$

where the Faradaic efficiency $\varepsilon_F = n_{exp}/n_{th}$ is the ratio between the number of electrons n_{exp} effectively exchanged in the overall reaction and the theoretical numbers of electrons n_{th} for complete oxidation of the fuel (to H_2O and CO_2). $\varepsilon_F = 1$ for hydrogen oxidation, but $\varepsilon_F = 4/6 = 0.67$ for methanol oxidation stopping at the formic acid stage (four electrons exchanged instead of six electrons for complete oxidation to CO_2) (see Section 9.3.3).

The overall energy efficiency of an H_2/O_2 fuel cell, working at 0.8 V under 500 mA cm^{-2}, is thus

$$\varepsilon^{H_2/O_2}_{cell} = 1 \times 0.65 \times 0.83 = 0.54$$

9.2.2 Different Kinds of Fuel Cells

Different types of fuel cells have been developed and are classified mainly according to (1) type of fuel, (2) operating temperature range and/or electrolyte, or (3) direct or indirect utilization of fuel (see Table 9.1):

- Solid oxide fuel cell (SOFC) working between 700°C and 1000°C with a solid oxide electrolyte, such as yttria stabilized zirconia (ZrO_2—8% Y_2O_3), conducting by the O$^=$ anion.
- Molten carbonate fuel cell (MCFC) working at about 650°C with a mixture of molten carbonates (Li_2CO_3/K_2CO_3) as electrolyte, conducting by the $CO_3^=$ anion, both of them being high-temperature fuel cells.
- Phosphoric acid fuel cell (PAFC) working at 180°C–200°C with a porous matrix of PTFE-bonded silicon carbide impregnated with phosphoric acid as electrolyte, conducting the H$^+$ cation. This medium temperature fuel cell is now commercialized by ONSI, a division of the United Technologies Inc. (the United States) mainly for stationary applications.
- AFC working at 80°C with concentrated potassium hydroxide as electrolyte, conducting by the OH$^-$ anion. This kind of fuel cell, developed by International Fuel Cells (IFC, the United States), is now used in the space shuttle.
- PEMFC working at around 70°C with a polymer membrane electrolyte, such as Nafion, which is a solid protonic conductor (conducting by the H$^+$ cation). This was first developed in 1990 by Ballard (Canada) [14].
- Direct methanol fuel cell (DMFC) working between 30°C and 110°C with a proton-exchange membrane (such as Nafion) as electrolyte, which realizes the direct oxidation of methanol at the anode.

For fuel cells operating at low (<100°C) and intermediate temperatures (up to 200°C), H_2 and H_2/CO_2 (with minimal amounts of CO) are the ideal fuels; the H_2/CO_2 gas mixture is produced by steam-reforming/water–gas shift conversion, or partial oxidation/shift conversion of the primary or secondary organic fuels. Hydrogen is a secondary fuel and, like electricity, is an energy carrier. On a large scale, hydrogen is produced from the primary sources—natural gas, coal, or oil. For

TABLE 9.1
Status of Fuel Cell Technologies

Types of Fuel Cell Operating fuel and operating temperature	Power Rating kW	Fuel Efficiency %	Power Density mW cm^{-2}	Lifetime h	Capital Cost $/kW	Applications
SOFC; CH_4, Coal ? 800°C–1000°C	25–5,000	50–60	200–400	8,000–40,000	1,500	Base-load and intermediate load, power generation—cogeneration
MCFC; CH_4, Coal ? 650°C	100–5,000	50–55	150–300	10,000–40,000	1250	Base-load and intermediate load, power generation—cogeneration
PAFC; CH_4, CH_3OH, oil 200°C	200–10,000	40–45	200–300	30,000–40,000	200–3,000	On-site integrated energy systems, transportation (fleet vehicles), load-leveling
AFC; H_2 80°C	20–100	65	250–400	3,000–10,000	1,000	Space flights, space stations, transportation, APU
PEMFC; H_2, CH_3OH 25°C–100°C	0.01–250	40–50	500–1,000	10,000–100,000	50–2,000	Transportation, stand-by power, portable power, space stations
DMFC; CH_3OH 25°C–150°C	0.001–10	30–45	50–200	1,000–10,000	1,000	Portable power, stand-by power, transportation (?), APU

SOFC, solid oxide fuel cell; MCFC, molten carbonate fuel cell; PAFC, phosphoric acid fuel cell; AFC, alkaline fuel cell; PEMFC, proton-exchange membrane fuel cell; DMFC, direct methanol fuel cell.

high-temperature fuel cell systems (>650°C), a mixture of H_2, CO, and CO_2, produced by steam reforming, can be used in fuel cells quite efficiently (power plant efficiency of over 50%).

These fuels (pure H_2, H_2/CO_2, and $H_2/CO/CO_2$) can also be produced from renewable energy sources—biomass, solar, windmills, and hydroelectric power. On the other hand, pure H_2 can be generated by water electrolysis using nuclear power plants. Hydrogen is the most electro-reactive fuel for fuel cells operating at low and intermediate temperatures.

Methanol is the most electro-reactive organic fuel, and, when it is electro-oxidized directly at the fuel anode (instead of to be transformed by steam reforming in a hydrogen-rich gas), the fuel cell is called a DMFC. More generally if the direct oxidation of a given fuel (alcohols, borohydrides, etc.)

does occur in the fuel cell, it is called a direct fuel cell, such as the direct ethanol fuel cell (DEFC), the DAFC, the direct borohydride fuel cell, etc.

The state of the art and a survey of the different applications of fuel cells are given in Table 9.1.

9.2.3 WHICH IS THE MOST CONVENIENT FUEL FOR FUEL CELLS?

For most kinds of hydrogen fuel cells natural gas (methane) is the primary fuel, leading to hydrogen by steam reforming, followed by water–gas shift reactions and CO cleaning (Scheme 9.1).

In view of the abundance of natural gas resources found since the 1980s, and natural gas being a considerably cleaner fuel than petroleum or coal for the fuel processors, the main goals of the major worldwide fuel cell programs are to develop fuel cell power plants and portable power sources using natural gas or natural gas-derived fuel cells. A chart of all types of fuel cells, using natural gas as a fuel, is presented in Scheme 9.2; the applications being considered for the different types of fuel

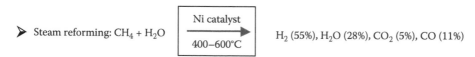

➤ Steam reforming: $CH_4 + H_2O$ — Ni catalyst, $400–600°C$ → H_2 (55%), H_2O (28%), CO_2 (5%), CO (11%)

Followed by water–gas shift reactions at high (HTWGS) and low temperature (LTWGS)

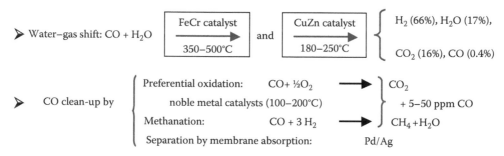

➤ Water–gas shift: $CO + H_2O$ — FeCr catalyst, $350–500°C$ — and — CuZn catalyst, $180–250°C$ → H_2 (66%), H_2O (17%), CO_2 (16%), CO (0.4%)

➤ CO clean-up by
  Preferential oxidation: $CO + ½O_2$ → CO_2
  noble metal catalysts (100–200°C)   + 5–50 ppm CO
  Methanation: $CO + 3 H_2$ → $CH_4 + H_2O$
  Separation by membrane absorption: Pd/Ag

SCHEME 9.1   Schematic representation of hydrogen production from natural gas.

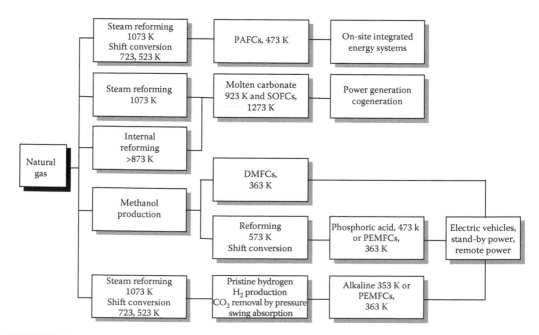

SCHEME 9.2   Schematic representation of the methods of utilization of natural gas in fuel cells [13].

cells are also indicated. The most advanced type of fuel cell is the PAFC system, which operates at about 473 K. However, in order to make this fuel cell system reasonably efficient, it is necessary to steam reform the fuel and also to use a water–gas shift converter to reduce the carbon monoxide levels to about 1%–2%. With natural gas as the fuel, the high-temperature systems (i.e., molten carbonate and solid oxide fuel cells, MCFC, and SOFC, respectively) are more attractive because of the following advantages: (1) carbon monoxide is a reactant and not a poison, (2) noble metal electrocatalysts are no more required, and (3) the waste heat from these fuel cells, which is of sufficiently high quality, can be used for cogeneration or transferred to a bottoming cycle gas turbine to produce more electricity.

For the transportation application, methanol is an ideal fuel. Even though its energy content is half of that of gasoline, a methanol fuel cell vehicle could have the same range (driving distance) as a gasoline internal combustion engine (ICE) or diesel engine-powered vehicle, for the same volume of fuel carried on board of the vehicle; this is because the fuel cell could operate at twice the efficiency as that of the ICE or diesel engine. The increasing concern about global warming has stimulated the utilization of pristine hydrogen as a fuel to be carried on board of the electric vehicles. It is also the unique one for space fuel cells. Pristine hydrogen can be produced either from fossil fuels (natural gas, oil, etc.) or from renewable fuels (biomass resources, energy crops, etc.) by steam reforming, water–gas shift conversion, and pressure-swing absorption for the complete removal of CO_2. It can also be produced from the decomposition of water by high-temperature electrolysis, e.g., using electricity produced by fourth-generation nuclear reactors, or through thermodynamic cycles. However, for transportation applications, hydrogen storage with an acceptable mass ratio (at least 5%) is still a great challenge as illustrated in Table 9.2.

TABLE 9.2
Different Ways of Hydrogen Storage

Storage Procedure	Hydrogen Content		Energy Density	
	Stored Weight (%)	Stored Volume (g cm^{-3})	kW h kg^{-1}	kW h L^{-1}
Gaseous state				
Steel cylinder 60 kg, 50 1200 bar	1.5	0.018	0.49	0.23
Al composite 75 kg, 125 1200 bar	2.6	0.018	0.85	0.40
Glass microsphere	6.0	0.006	1.97	0.20
Zeolite	0.8	0.006	0.26	0.19
Liquid state				
Cryogenic liquid 300 m^3 tank truck	12.5	0.071	4.12	2.29
Solid State (Hydride)				
MgH_2	7.0	0.101	2.29	3.30
$FeTi_2$	1.6	0.096	0.53	3.12
$LaNi_5H_6$	1.4	0.089	0.46	2.95
Mg_2NiH_4	3.2	0.081	1.04	2.65
Chemical state				
Methane	25.0	0.139	13.91	7.72
n-Octane	15.8	0.110	11.00	7.73
Methanol	12.5	0.150	6.09	4.82
Ammoniac	17.6	0.136	5.33	3.98
High surface area carbon	7.2	0.0014	2.36	0.44

If the problems of hydrogen storage, in respect to energy density, specific energy, and safety, can be solved, the AFC or the PEMFC will be ideal candidates for terrestrial applications. An advantage of the alkaline system is that there is no noble metal requirement, but a great disadvantage is that, even with air as a cathodic reactant, a small amount of CO_2 carbonates the electrolyte, leading to the lowering of the conductivity of the electrolyte, altering its pH, and giving a deposition of insoluble carbonates in the porous gas-diffusion electrode.

With a methanol-fed fuel cell, there are two schemes that look most promising for fuel cell-powered vehicles. One is to feed either a PAFC or a PEMFC with hydrogen produced by on-board methanol steam reforming (see Figure 9.6), and the second one is to use it directly as the anodic reactant in a DMFC (see Figure 9.7). The PAFC has the advantage that, because of good thermal matching between the temperatures for the endothermic steam-reforming reaction and the exothermic fuel cell reaction, the heat required for the former process can effectively be provided by the waste heat generated during the fuel cell reaction. The DMFC has a great advantage for transportation, since in such a fuel cell, the heavy and bulky fuel-processing system is not required. The PEMFC has a considerably better prospect of attaining high power densities and a faster start-up time than the PAFC and thus is one of the more favored power source for the electric vehicle.

9.2.4 APPLICATIONS OF FUEL CELLS

Depending on their rating power fuel cells are electrochemical devices able to produce electric energy, together with heat (cogeneration systems), in a wide range of power with a similar electric energy efficiency (between 40% and 60%). This concerns

- Electricity generation in centralized power plant (1–100 MW)
- Power trains for the electric vehicle (10–200 kW)
- APU (1–100 kW) for terrestrial and space applications
- Household power sources (1–10 kW)
- Power sources for portable electronic devices (1–100 W)

For terrestrial applications, the PAFC power plant is the most advanced one and a 200 kW system manufactured by ONSI, a Division of United Technologies Inc., has reached the state of commercialization. Its main applications are focused on on-site integrated energy systems that could provide electricity and heat for apartment buildings, hospitals, shopping centers, etc. Overall efficiencies higher than 85% for fuel consumption (natural gas → electricity plus heat) have been demonstrated in these power plants. In the United States Department of Energy sponsored fuel cell program, PAFC systems were considered as power plants for fleet vehicles (buses, trucks, delivery vehicles, etc.) and 100 kW units have been installed and tested in buses; the performances of these buses are quite promising. Apart from the ONSI program, the most active PAFC development and commercialization programs are in Japan by Fuji, Toshiba, and Mitsubishi.

MCFCs have the advantages of (1) not requiring noble metal electrocatalysts, (2) efficient heat transfer from the exothermic fuel cell reaction to the endothermic fuel-processing reaction (internal reforming), (3) CO being a reactant and not a poison, and (4) the rejected heat from the MCFC power plant being of high enough quality to produce electricity in a gas turbine or steam in a steam turbine (for cogeneration). The active MCFC developers are (1) Energy Research Corporation and M-C Power Corporation in the United States; (2) ECN, Ansaldo and CFC Solutions GmbH (Tognum Group) in Europe; and (3) IHI, Toshiba, Mitsubishi, Hitachi and Sanyo in Japan. Power plants, with power ratings from 100 kW to 2 MW have been constructed and tested.

The SOFCs have practically the same advantages as the MCFCs for applications in electric utility companies and chemical industries. An additional advantage is that, because the SOFC power plant is a two-phase system (gas and solid) whereas all other types of fuel cells are three-phase systems (gas, liquid, and solid), the complex problems associated with liquid electrolytes are eliminated and mass transport problems are thus greatly minimized. The world leader in this technology is Westinghouse Electric Corporation in the United States, which is now part of the Siemens Group (Germany). The materials, used in the electrochemical cell stack, are relatively inexpensive but the costs of fabrication of thin film structures are prohibitive. The Westinghouse power plants (100 kW), with the tubular structure, have reached the performance goals and have shown promise of reaching the lifetime goals. Flat-plate designs are more desirable than the tubular one for the SOFCs from the points of view of (1) scale-up to MW sizes and (2) markedly lowering the capital cost. Siemens in Germany is at the most advanced state in developing the flat-plate design. Other companies, which are working in this area, are IHI in Japan and ZTEK, Allied Signal, SOFCO and MTI in the United States.

A fuel cell system, which has drawn the most attraction since the mid-1980s, is the one with a proton-exchange membrane as the electrolyte (PEMFC). This was the first type of fuel cell to find an application, i.e., as an auxiliary power source for NASA's Gemini Space Flights in the 1960s. This technology was developed by General Electric Company in the United States. Since the AFCs were more energy efficient and could attain higher power densities, as required for the subsequent manned space flights, the PEMFC technology was dormant, after the Gemini flights, until the mid-1980s. Since then, Ballard Power Systems Inc. in Canada made major strides in this technology, particularly for the electric vehicle application. The Ballard/Daimler-Benz/Ford venture for the development and commercialization of PEMFC-powered electric vehicles has provided great enthusiasm for this technology leading to demonstration Fuel Cell Electric Vehicles (FCEVs) (Ford Focus, Daimler A class). Similarly Toyota, Honda, and Nissan developed in Japan several FCEVs. As may be seen from Table 9.1, the significant advantage of this technology is that it has the greatest potential for attaining the highest power densities and the longest lifetime. An attractive feature of PEMFC is that it uses an innocuous electrolyte (i.e., Nafion, which is a perfluorosulfonic polymer membrane). The potential for reducing the platinum loading by more than a 100 times (as compared with that used in the Gemini Fuel Cells) has been demonstrated in high-power density PEMFCs. There are good prospects of further reducing the platinum loading to an even lower value (about $0.2\,g\,kW^{-1}$, as targeted by the DOE program on fuel cells), so that the Pt cost could be as low as €5 per kW. Ballard/Daimler-Benz has demonstrated 60–75 kW PEMFC-powered automobiles, and Ballard had also been sponsored by the California and Illinois State Agencies to develop and to test 120–200 kW PEMFC-powered buses.

The cost target in the Partnership for a New Generation of Vehicles program in the United States—for hybrid electric vehicles with the goals of attaining three times the efficiency for fuel consumption and meeting the same performance characteristics as conventional vehicles—is most challenging; it is US$30–50 per kW, and is thus a factor of 100 lower than that of the ONSI's 200 kW commercialized PAFCs. The R&D projects (to advance the PEMFC technology in industries, universities, and government laboratories) are at a peak level. Practically all the automobile companies (Daimler-Benz, G.M., Ford, Chrysler, Toyota, Honda, Nissan) are collaborating with (1) chemical industries (DuPont, 3 M, W.L. Gore and Assoc., Asahi Chemical, Asahi Glass, Solvay, Hoechst, Johnson Matthey, etc.) to develop advanced proton conducting membranes for electrolytes and (2) high technology engineering companies (Ballard in Canada, IFC, Energy Partners, Allied Signal, MTI, Plug Power in the United States, Siemens in Germany, De Nora in Italy, and Fuji, Mitsubishi, Sanyo in Japan) to develop and manufacture high-performance PEMFCs.

Until recently (i.e., till early 1990s), most of the efforts to develop DMFCs has been with sulfuric acid as the electrolyte. The recent success with a proton conducting membrane (perfluorosulfonic acid membrane) in PEMFCs has steered DMFC research toward the use of this electrolyte. The positive feature of a liquid feed to a DMFC is that it eliminates the humidification subsystem, as required for a PEMFC with gaseous reactants. Another positive point is that the DMFC does not require the heavy and bulky fuel processor. Two problems continue to be nerve-wracking in the projects to develop DMFCs: (1) the exchange current density for methanol oxidation, even on the best electrocatalyst to date (Pt–Ru) is 10^5–10^6 times lower than that for the electro-oxidation of hydrogen and (2) the crossover of methanol through the membrane, from the anode to the cathode side, reduces the Coulombic efficiency for methanol utilization by about 30%. However, a compensating feature of DMFCs, as compared to PEMFCs, is that it eliminates the fuel processor and a lower performance of the electrochemical cell stack may still be acceptable for some applications, e.g., portable power sources.

According to some projections, the energy densities of DMFCs could be considerably higher than those of even lithium-ion batteries, so that DMFCs could find applications for low power applications (laptop computers, backpack power sources for soldiers). IFC in the United States, Siemens and Daimler in Germany have designed, constructed, and tested kW-size DMFC power plants. Research studies in several government laboratories, universities, and industries have shown the prospects of attaining a current density of 400 mA cm^{-2} at a cell potential of 0.5 V.

9.3 WHICH MATERIALS FOR WHICH APPLICATION?

9.3.1 ALKALINE FUEL CELL

The AFC has the highest electrical efficiency of all fuel cells but it only works properly with very pure gases, which is considered a major drawback in most applications. The KOH electrolyte, which is used in AFCs (usually in concentrations of 30–45 wt%), has an advantage over acidic fuel cells, i.e., the oxygen reduction kinetics is much faster than in acid medium.

The first AFC (1950s) was a Bacon fuel cell, provided 5 kW power, and used a Ni anode, a lithiated NiO cathode, and 30 wt% aqueous KOH. Its operating temperature and pressure were 200°C and 50 bar, respectively. For the Apollo program, an AFC which employed a 85% KOH solution at 200°C–230°C was used. In the U.S. Space Shuttles, the fuel cells are used both for producing energy, cooling the Shuttle compartments, and producing drinkable water. Three plant modules are used, each with a maximum power output of 12 kW. AFCs are now normally run at operating temperatures below 100°C, as the high temperature is not needed to improve oxygen reduction kinetics (although higher temperatures are advantageous for the hydrogen oxidation kinetics).

In an alkaline medium, oxygen reduction and hydrogen oxidation involve OH$^-$ ions. The OH$^-$ species formed by the cathodic reduction of oxygen move through the electrolyte to the anode, where recombination with hydrogen, oxidized at the anode, produces water.

Cathode reaction	$\frac{1}{2}O_2 + H_2O + 2e^- \rightarrow 2OH^-$
Anode reaction	$H_2 + 2OH^- \rightarrow 2H_2O + 2e^-$
Overall reaction	$H_2 + \frac{1}{2}O_2 \rightarrow H_2O$

The advantage of AFCs over the other systems lies in the fact that the reduction of oxygen to OH$^-$ is much faster than the acidic equivalent of oxygen to H$_2$O due to a better kinetics, which makes the AFC a more efficient system [15]. The hydrogen oxidation reaction in alkaline medium, however, is slower.

AFC electrodes were Ni-based catalysts sometimes activated with platinum. It should be noted, however, that metal loadings are usually lower (0.3 mg cm^{-2} can be attained) than for a PAFC or a PEM fuel cell. The Pt/C gas-diffusion electrodes are now generally used for both the anode and cathode, although various groups are investigating other possibilities.

Pt/Co alloys have been suggested and have proved to have a superior activity to Pt for oxygen reduction due to a higher exchange current density [16]. The Tafel slope for oxygen reduction for Pt/Co catalysts is the same as for Pt catalysts (60 mV per decade). Further work on Ag and Co catalysts is being undertaken to replace some or the entire Pt in the electrodes.

A Pt/Pd anode was tested for stability characteristics in comparison to Raney nickel. It is known that Raney Ni electrodes have a high activity for hydrogen oxidation but, due to the wettability of the inner pores and changes in chemical structure under operation conditions, decay in performance occurs. The Pt/Pd activity was also seen to have a very rapid decay initially but after a short time the decay stopped and the performance remained constant afterward.

The AFCs used in space had KOH in a stabilized matrix, usually gel-like materials, but this is disadvantageous for most applications. It has since been found that a much longer operating lifetime can be obtained with circulating KOH. A circulating electrolyte provides a good barrier against gas leakage and it can be used as a cooling liquid in the cell or stack. Other advantages include the use of the electrolyte circulation to clean the cell and remove the water produced at the anode.

One of the most controversial issues in AFCs is the formation of carbonates. It is generally accepted that the CO_2, both originally in the air and formed by corrosion reaction of the carbon catalyst support, interacts with the electrolyte according to the following equation:

$$CO_2 + 2OH^- \rightarrow CO_3^{2-} + H_2O$$

The formation of carbonates, via the oxidation of the carbon support material in the electrodes during open circuit conditions, clogs up the pores of the "gas-diffusion" electrodes, and the cell performance rapidly decreases. A circulating electrolyte avoids the build up of carbonates.

9.3.2 PROTON-EXCHANGE MEMBRANE FUEL CELL

PEMFC is now the most advanced fuel cell technology, because it can be used in several applications (space, electric vehicle, stationary power plants, portable electronics), the progress made in one application being beneficial to the others.

An elementary PEMFC comprises several elements and components: the membrane-electrode assembly (MEA), the flow-field plate (bipolar plate, which also ensures electric contact with the next cell), gaskets to ensure tightness to reactants and end plates (Figure 9.4).

The MEA consists of a thin (10–200 µm) solid polymer electrolyte (a protonic membrane, such as Nafion) on both sides of which are pasted the electrode structures (fuel anode and oxygen cathode) (Figure 9.5). The electrode structure comprises several layers: the first layer made of carbon paper (or cloth) to strength the structure, on which are coated the GDL, and then the catalyst layer (CL), directly in contact with the protonic membrane (usually Nafion).

A single cell delivers a cell voltage between 0.5 and 0.9 V (instead of the theoretical electromotive force [emf] of 1.23 V under standard equilibrium conditions) depending on the working current

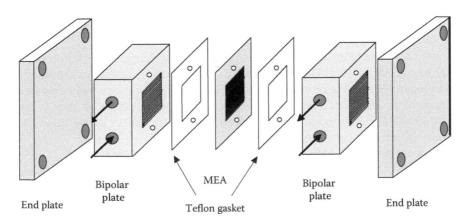

FIGURE 9.4 Schematic representation of a PEMFC elementary cell.

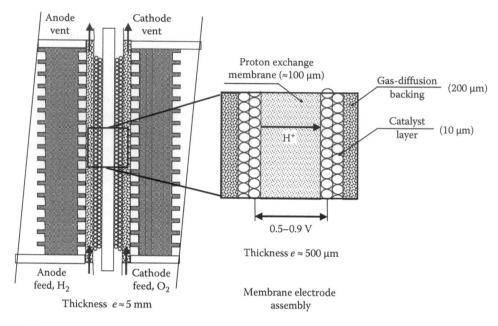

FIGURE 9.5 Schematic representation of a MEA.

density, so that many elementary cells, electrically connected by BP, are assembled together (in series and/or in parallel) to reach the nominal voltage (such as 48 V for electric vehicles) and power of the fuel cell stack.

In PEMFC working at low temperature (20°C–90°C) several problems need to be solved before technological development of fuel cell stacks for different applications. This concerns the components of the elementary cell, i.e., the proton-exchange membrane, the electrode (anode and cathode) catalysts, the membrane—electrode assemblies, and the BP [17].

The proton-exchange membrane, which is a solid polymer electrolyte, plays a key role in the PEMFC. It allows the electrical current to pass through it thanks to its H^+ ionic conductivity and prevents any electronic current through it in order that electrons are obliged to circulate in the external electric circuit to produce the electric energy corresponding to the combustion reaction of the fuel; it must avoid any gas leakage between the anodic and the cathodic compartments, so that no chemical combination between hydrogen and oxygen is directly allowed; it must be stable mechanically, thermally (up to 150°C in order to increase the working temperature of the cell), and chemically; and finally its lifetime must be sufficient for practical applications (e.g., >2000 h for transportation power trains or >40,000 h for stationary power plant).

The actually developed PEMFC have a Nafion membrane, which partially fulfills these requirements, since its thermal stability is limited to 100°C, and its protonic conductivity strongly decreases at higher temperatures because of its dehydration. On the other hand, it is not completely tight to liquid fuels (such as alcohols). This is more important as the membrane is thin (a few 10 μm). Furthermore its actual cost is too high (more than €500 per m^2), so that its use in a PEMFC for a low-cost passenger electric car is actually difficult.

One of the main problems of low-temperature (20°C–80°C) PEMFCs is the relatively low kinetics rate of the electrochemical reactions involved, e.g., the oxygen reduction at the cathode and fuel (hydrogen from a reformate gas, or alcohols) oxidation at the anode. Another problem, particularly crucial for the electric vehicle, is the relatively low-working temperature (70°C–80°C), which prevents efficient exhaust of the excess heat generated by the power fuel cell.

Therefore, new membranes are investigated with improved stability and conductivity at higher temperatures (up to 150°C). For power fuel cell, the increase of temperature will increase the rate of the electrochemical reactions occurring at both electrodes, i.e., the current density at a given cell voltage and the specific power. Furthermore, thermal management and heat utilization will be improved, particularly for residential applications with heat cogeneration and for large-scale applications to exhaust excess heat.

The reaction rates can only be increased by the simultaneous action of the electrode potential and electrode material (this is the field of electrocatalysis) [3–5]. Moreover increasing the working temperature from 80°C to 150°C would strongly increase (by a factor of at least 100–1000) the rates of the electrochemical reactions (thermal activation). All these combined effects would increase the cell voltage by ca. 0.1–0.2 V, since at room temperature the anode and cathode overvoltages are close to 0.2–0.4 V, which decreases the cell voltage by 0.4–0.7 V leading to values close to 0.5–0.7 V, instead of the 1.23 V theoretical cell voltage.

The investigation of new electrocatalysts more active for the oxygen reduction and fuel oxidation (hydrogen from reformate gas, or alcohols) is thus an important point for the development of PEMFCs [15,16].

The realization of the MEA is a crucial point to build a good fuel cell stack. The method currently used consists in hot pressing (at 130°C and 35 kg cm^{-2}) the electrode structures on the polymer membrane (Nafion). This gives nonreproducible results (in terms of interface resistance) and this is difficult to industrialize. New concepts must be elaborated, such as the continuous assembly of the three elements in rolling tape process (like in the magnetic tape industry) or successive deposition of the component layers (microelectronic process), etc.

The BP, which separate both electrodes of neighboring cells (one anode of a cell and one cathode of the other), have a triple role:

- To ensure the electronic conductivity between two neighboring cells
- To allow the distribution of reactants (gases and liquids in the case of alcohols) to the electrode catalytic sites, and to evacuate the reaction products (H_2O, and CO_2, in the case of alcohols)
- To provide thermal management inside the elementary cell by evacuating the excess heat

The BP are usually fabricated with nonporous machined graphite or corrosion-resistant metal plates. Distribution channels are engraved in these plates. Metallic foams can also be used for distributing the reactants.

A more detailed picture of a PEMFC system, with hydrogen produced by methanol reforming, including the auxiliary and control equipments, is given in Figure 9.6.

Fuel supply is usually from liquid hydrogen or pressurized gaseous hydrogen. For other fuels, such as methanol, a fuel processor is needed, which includes a reformer, water–gas shift reactors, and purification reactors (in order to decrease to an acceptable level—a few tens of ppm—the amount of CO, which would otherwise poison the platinum-based catalysts). This equipment is still heavy and bulky and limits the dynamic response of the fuel cell stack, particularly for the electric vehicle in some urban driving cycle.

On the other hand, the other auxiliary equipments depend greatly on the stack characteristics:

- Air compressor, the characteristics of which is related to the differential pressure supported by the proton-exchange membrane
- Humidifiers for the reacting gases with controlled humidification conditions
- Preheating of gases to avoid any condensation phenomena
- Hydrogen recirculation, purging system of the anode compartment
- Cooling system for the MEAs
- Control valves of pressure and/or gas flows
- DC/DC or DC/AC electric converters

The system control must ensure correct working of the system, not only under steady state conditions, but also during power transients. All the elementary cells must be controlled (cell

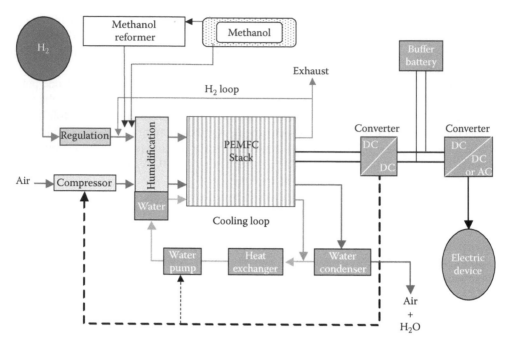

FIGURE 9.6 Detailed scheme of a PEMFC system with its auxiliary and control equipments.

voltage of each elementary cell, if possible) and the purging system must be activated in case of technical hitch.

According to different applications (stationary power plants, power sources for electric vehicle, APU), different specifications and research objectives are required.

9.3.3 DIRECT METHANOL FUEL CELL

DMFC transforms directly the Gibbs energy of combustion of methanol into electricity, without a fuel processor (Figure 9.7). This greatly simplifies the system reducing its volume and cost. Recent development of DMFC is due to the use of proton-exchange membrane as electrolyte, instead of liquid acid electrolyte, as previously done [13].

At the anode (negative pole of the cell), the complete electro-oxidation of methanol takes place as follows:

$$CH_3OH + H_2O \rightarrow CO_2 + 6H^+ + 6e^-   E_1^o = 0.016 \text{ V/SHE}$$

whereas the cathode (positive pole) undergoes the electroreduction of oxygen, i.e.,

$$O_2 + 4H^+ + 4e^- \rightarrow 2H_2O   E_2^o = 1.229 \text{ V/SHE}$$

where E_i^o are the electrode potentials vs. the SHE.

This corresponds to the overall combustion reaction of methanol in oxygen:

$$CH_3OH + 3/2O_2 \rightarrow CO_2 + 2H_2O$$

with the thermodynamic data (under standard conditions):

$$\Delta G^o = -702 \text{ kJ mol}^{-1};   \Delta H^o = -726 \text{ kJ mol}^{-1}$$

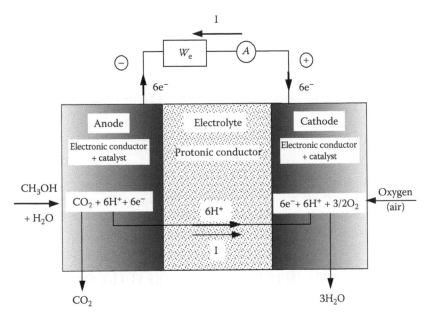

FIGURE 9.7 Principle of a DMFC.

This gives a standard emf at equilibrium:

$$E_{eq}^{o} = -\frac{\Delta G^{o}}{nF} = \frac{702 \times 10^{3}}{6 \times 96,485} = E_{2}^{o} - E_{1}^{o} = 1.21 \text{ V}$$

with $F = 96,485$ C, the Faraday constant, and $n = 6$ the number of electrons involved in the complete oxidation of one methanol molecule.

The protons produced at the anode crossover the membrane, ensuring the electrical conduction inside the electrolyte, whereas the electrons liberated at the anode reach the cathode (where they reduce oxygen) through the external circuit producing an electrical energy, $W_{e} = nFE_{eq}^{o} = -\Delta G^{o}$, with a mass energy density $W_{el} = -\Delta G^{o}/(3600M) = 6.09$ kW h kg^{-1}, where $M = 0.032$ kg is the molecular weight of methanol.

The energy efficiency, under reversible standard conditions, defined as the ratio between the electrical energy $(-\Delta G^{o})$ and the heat of combustion $(-\Delta H^{o})$ at constant pressure, is

$$\varepsilon_{eq}^{rev} = \frac{W_{e}}{(-\Delta H^{o})} = \frac{\Delta G^{o}}{\Delta H^{o}} = 1 - \frac{T\Delta S^{o}}{\Delta H^{o}} = \frac{702}{726} = 0.97$$

Under working conditions, with a current density j, the cell voltage $E(j)$ decreases greatly as the result of three limiting factors: the overvoltages η_{a} and η_{c} at both electrodes due to a rather low reaction rate of the electrochemical processes (activation polarization), the ohmic drop $R_{e}j$ in the electrolyte and interface resistance R_{e}, and mass transfer limitations for reactants and products (concentration polarization).

For a DMFC working at 0.5 V under 400 mA cm^{-2} (with complete oxidation to CO_2) the energy efficiency would be

$$\varepsilon_{cell}^{CH_{3}OH/O_{2}} = 1 \times 0.41 \times 0.97 = 0.40 \quad \text{since} \quad \varepsilon_{E} = \frac{0.5}{1.21} = 0.41$$

which is similar to that of the best thermal engine (diesel engine, see Section 9.2.1).

However, if the combustion reaction is not complete, one has to introduce the Faradaic efficiency $\varepsilon_F = n_{exp}/n_{th}$, defined as the ratio between the number of electrons n_{exp} effectively exchanged in the overall reaction and the theoretical numbers of electrons n_{th} for a complete oxidation of the fuel (to H_2O and CO_2). $\varepsilon_F = 1$ for hydrogen oxidation, whereas $\varepsilon_F = 4/6 = 0.67$ for methanol oxidation stopping at the formic acid stage (four electrons exchanged instead of six electrons for complete oxidation to CO_2). This reduces further the overall energy efficiency, which becomes

$$\varepsilon_{cell}^{CH_3OH/O_2} = 0.67 \times 0.41 \times 0.97 = 0.27$$

An additional problem arises from methanol crossover through the proton-exchange membrane. It results that the platinum cathode experiences a mixed potential, since both oxygen reduction and methanol oxidation take place at the same electrode. The cathode potential is thus lowered, leading to a smaller cell voltage and thus to a decrease of the voltage efficiency.

9.4   CARBON MATERIALS IN LOW-TEMPERATURE FUEL CELLS

Carbon materials have been widely used in fuel cells due to their attractive features, such as high electronic conductivity, good stability under fuel cell operation (acidic/alkaline medium, oxidizing/reducing environment, relatively high temperatures, etc.), and low cost.

Excepting the use as a fuel in the direct carbon fuel cell, or as a part of the feedstock in other cells, the carbon materials play a role in the fuel cell technology. The first huge interest was the use of the carbon as host for the hydrogen storage. Despite enormous scientific efforts, unfortunately until now there is no evidence for an efficient storage of the hydrogen in any carbon or carbon-related materials. This aspect will be not discussed in this chapter, but many investigations are still in progress in this field, showing that hydrogen sorption in carbon materials does not exceed a few wt% at room temperature under reasonable pressure (below 100 bar) [18,19].

For low-temperature fuel cells (AFC, PEMFC, and DAFC) carbon is mainly used

- As material for the BP
- As component of the GDL
- As electrocatalyst support in the CL

For these purposes there are three industrial kinds of carbon materials ready to be used, e.g., machined graphite for BPs, carbon fibers or papers for GDL, and carbon blacks for CL.

9.4.1   STRUCTURE AND PROPERTIES OF CARBON MATERIALS

Many forms of carbon materials do exist with a continuous variation in the structure/nanotexture and properties between the graphite (sp^2 hybrid orbitals) and diamond (sp^3 hybrid orbitals) limiting structures.

9.4.1.1   Graphite

The graphite structure has free π-electrons leading to high electronic conductivity, which can reach a few hundred S cm^{-1}, and which is the essential property in fuel cell applications. The actual annual production is about 1 million of metric tons of natural graphite and about 200 ktons of synthetic graphite [20]. Although natural graphite can have a very high crystallinity, it must be treated at high temperature in order to remove impurities. Synthetic graphite is obtained by the thermal treatment of graphitizable cokes (petroleum, pitch, metallurgical, etc.) at temperatures above 3000°C [21].

The particle size ranges from a few µm to a few millimeters. The crystallinity, the particle size and shape, and the grinding procedure represent the main parameters influencing the final properties. Recently, exfoliated graphite has been found to present a high technological interest. It has a very high aspect ratio, in the range of 1000 or more, presenting high advantage for the percolation in electrochemical systems.

9.4.1.2 Carbon Black and Activated Carbons

The carbon black has a quasigraphitic structure. A few layers form crystallites, which combine to form "spherical" primary particles, the diameter of which is in the order of 50 nm (Figure 9.8).

The primary particles continue to grow into aggregates, with the exception of some special carbon blacks. An aggregate represents the carbon black unit and is indestructible (Figure 9.9). By Van der Waals interactions or by external mechanical energy, the aggregates are more or less agglomerated and the resistance to this agglomeration process is dependent on the charge carrier concentration [23,24]. The elementary particle size is about a few tens of nanometers, while the aggregate size is in the range of 150–200 nm.

The individual graphitic layers are the basic building blocks of carbon black particles. The morphology and particle size distribution of carbon black is dependent on the source material and the process of its thermal decomposition. Particle size and distribution determine directly the specific surface area (SSA) which is one of the most important properties of carbon black for fuel cell applications. High surface area (ranging from a few hundreds to 2000–3000 m² g⁻¹) carbon blacks suitable for fuel cell applications can be obtained from Cabot Corporation (Vulcan XC-72R, Black Pearls BP 2000), Ketjen Black International, Chevron (Shawinigan), Erachem and Denka.

The carbon black is produced by two ways: the partial combustion of different hydrocarbons and the exothermic reaction of acetylene decomposition. About 10 million of tons are produced annually in the world, from which 99% for the tire industry.

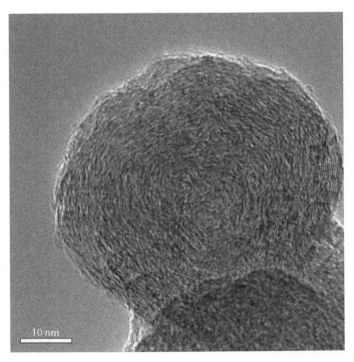

FIGURE 9.8 Primary particle of carbon black. (Grivei, E., TIMCAL Belgium, e.grivei@be.timcal.com, personal communication.)

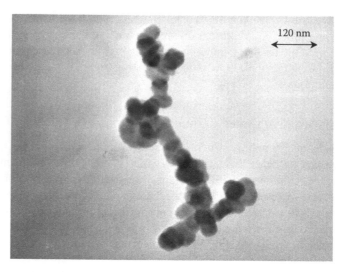

FIGURE 9.9 Few aggregates of carbon black Ensaco 250G. (Grivei, E., TIMCAL Belgium, e.grivei@be.
timcal.com, personal communication.)

Activated carbons are obtained by physical (steam or CO_2 at 800°C–1000°C) or chemical (KOH
at 700°C–800°C) treatment of carbons. The most usual carbon precursors are lignocellulosic derivatives, polymers, pitch, etc. Activated carbons are generally highly microporous (<2 nm) with some
amount of mesopores (2–50 nm); depending on the degree of activation, their BET SSA ranges from
1000 to 3500 $m^2\ g^{-1}$.

Activated carbon has been used as a support for industrial precious metal catalysts for many years
and has been a natural choice for supporting the electrocatalysts in PAFC, AFC, and PEMFC.

9.4.1.3 Carbon Fibers

The carbon fibers present a filamentary shape with a diameter of few tens of nanometers (nanofibers)
to few tens of micrometers. The structure is that of short graphitic segments. The preferred orientation of the graphitic planes is parallel to the fiber axis. The morphology can be radial, cross-sectional,
or circumferential [25]. The carbon fibers present excellent mechanical and thermal properties. The
carbon fibers are produced by two ways [26]:

- The pyrolysis and the graphitization of polymeric fibers
- By chemical vapor deposition from hydrocarbon precursors in the presence of a catalyst
  followed by high-temperature thermal treatment.

9.4.2 CATALYST LAYER

In PEMFC and DAFC working at relatively low temperatures in an acidic medium, electrocatalysts
are essential to increase the rate of electrode reactions (oxidation of hydrogen, methanol, etc., at the
anode and oxygen reduction at the cathode). The most efficient electrocatalyst is platinum or a platinum-based catalyst, which needs to be highly dispersed on an electron conducting support, in order
to utilize at maximum a costly precious metal. Carbon blacks, activated carbons, and graphites are
widely used in industry as catalyst support due to high conductivity of carbon and high stability
in acid or alkaline medium. CLs of a few micrometers (<10 µm) are prepared with high precious
metal loading (>40 wt%) in order to minimize the layer thickness and thus its electric resistance. In
order to extend the three-phase (electrode–electrolyte–reactants) boundary into the whole CL, the
carbon material of the CL is mixed with a soluble form of Nafion, which is the commonly protonic
electrolyte used in PEMFC.

Different methods have been developed to prepare Pt-based nanoparticles supported on carbon powders (particularly Vulcan XC-72R).

9.4.2.1   Catalyst Synthesis by Electrochemical Methods

The direct electrochemical deposition methods for the preparation of electrocatalysts allow to localize the catalyst particles on the top surface of the carbon support, as close as possible to the solid polymer electrolyte and does not need heat (oxidative and/or reducing) treatment, as most of the chemical methods do, in order to clean the catalytic particles from surfactant contamination [27,28]. This will prevent catalyst sintering due to the agglomeration of nanoparticles under thermal treatment.

The deposition of metals on a conductive support by applying a constant potential was extensively studied [29–36]. Platinum, ruthenium, and molybdenum could be deposited in a reproducible way using this method. The electrodeposition of platinum particles by potentiostatic pulses was also described [37,38]. However, the potentiostatic set up requires a three-electrode configuration, and this technique is not convenient for preparing electrodes with large geometric surface area (greater than a few ten square centimeters), as required for industrial applications. Choi et al. [39] have developed a galvanostatic pulse method for the electrodeposition of nanoscaled platinum particles on a conductive carbon support. This technique consists in applying a square wave signal of current (typically $-20\,\mathrm{mA\,cm^{-2}}$ during a time $t_{on} = 0.1$ s) between two carbon electrodes: the metal salts are reduced at the working electrode, so that metals can be directly deposited on the GDL. By dissolving different metal salts with different concentrations in sulfuric acid solution, plurimetallic catalysts could be deposited at the surface of the diffusion layer with a high surface loading suitable for a DAFC (ca. $2\,\mathrm{mg\,cm^{-2}}$).

PtRu catalysts with controlled atomic ratios were prepared by adjusting the nominal concentrations of platinum and ruthenium salts in the solution, whereas different mean particle sizes could be obtained by adjusting some electric parameters of the deposition process, e.g., t_{on} (during which the current pulse is applied) and t_{off} (when no current is applied to the electrode), as determined by different physicochemical methods (XRD, EDX, and TEM) [40]. Characterization by XRD led to determine the crystallite size, the atomic composition and the alloy character of the PtRu catalysts. The atomic composition was confirmed using EDX, and TEM pictures led to evaluate the particle size and to show that PtRu particles formed small aggregates of several tens of nanometers (Figure 9.10).

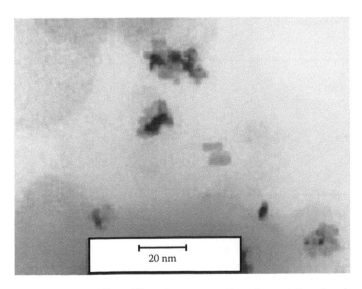

FIGURE 9.10   TEM image of a $Pt_{0.65}Ru_{0.35}$/C catalyst prepared by galvanostatic pulse electrodeposition with $t_{off} = 2.5$ s.

TABLE 9.3

Conditions of Catalyst Preparation by Galvanostatic Pulse Electrodeposition and Characterizations of the PtRu Catalysts Deposited on Vulcan XC-72R
$(j = -20\,\text{mA cm}^{-2}, t_{on} = 0.1\,\text{s})$

	Pt/Ru Atomic Ratio in Solution	Catalyst Atomic Ratio (XRD)	EDX Analyses	Particle Size (nm)
$t_{off} = 0.3\,\text{s}$	50/50	$Pt_{0.57}Ru_{0.43}$	$Pt_{0.53}Ru_{0.47}$	5
	65/35	$Pt_{0.66}Ru_{0.34}$	—	5
	80/20	$Pt_{0.77}Ru_{0.23}$	$Pt_{0.78}Ru_{0.22}$	5
$t_{off} = 2.5\,\text{s}$	50/50	$Pt_{0.55}Ru_{0.45}$	$Pt_{0.60}Ru_{0.40}$	8
	65/35	$Pt_{0.69}Ru_{0.31}$	$Pt_{0.69}Ru_{0.31}$	8
	80/20	$Pt_{0.75}Ru_{0.25}$	$Pt_{0.82}Ru_{0.18}$	8

Table 9.3 gives the physicochemical characterizations of the PtRu/C catalysts obtained under different experimental conditions (nominal metal atomic ratio in solution and value of t_{off}).

These electrodes were tested in a 5 cm² DMFC working at 110°C. First, the effect of the bulk composition (atomic ratio) was evaluated. It appeared that the $Pt_{0.8}Ru_{0.2}$/C catalyst led to the best electrical performance in terms of maximum achieved power density (Figure 9.11). The lower performance was obtained with the ruthenium-rich $Pt_{0.5}Ru_{0.5}$/C catalyst. The effect of t_{off} was also pointed out. Higher maximum power density was obtained with lower t_{off} (Figure 9.12). This latter effect can be related either to a particle size effect (the mean particle size increasing from ca. 5 to 8 nm with increasing t_{off}) or to a deeper penetration of the metal salts inside the diffusion layer. But in the case of platinum, for some authors the surface activity reaches higher values for particle sizes between 5 and 8 nm [41], whereas for others the specific activity decreases for particle sizes lower than 5 nm [42], or the specific activity always increases with the increase of the particle size, the higher specific activity being reached with bulk platinum [43]. However, a mean particle size of platinum catalyst close to 5 nm seems to be a good compromise in terms of surface and specific activity and for an economical point of view. On the other hand, the penetration of platinum in the depth of the diffusion layer can lead to decrease its utilization efficiency (the triple-phase boundary being not achieved for the particles in the depth of the electrode [44–46]).

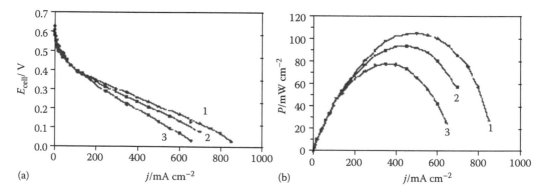

FIGURE 9.11 (a) Cell voltage E_{cell} and (b) power density, P, vs. the current density, j, curves recorded in a single 5 cm² surface area DMFC with PtRu/C anodes of different atomic compositions prepared by pulse electrodeposition with $t_{off} = 0.3$ s (Nafion 117 membrane, 2 M CH_3OH, $P_{O_2} = 2.5$ bar, $P_{MeOH} = 2.0$ bar, $T = 110°C$): (1) $Pt_{0.8}Ru_{0.2}$; (2) $Pt_{0.65}Ru_{0.35}$; (3) $Pt_{0.5}Ru_{0.5}$.

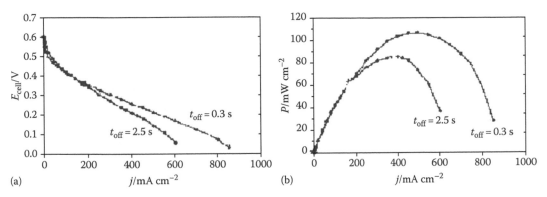

FIGURE 9.12   (a) Cell voltage, E_{cell}, and (b) power density, P, vs. current density, j, curves recorded in a single 5 cm² surface area DMFC with $Pt_{0.8}Ru_{0.2}/C$ anodes prepared by pulse electrodeposition with different t_{off} (Nafion 117, 2 M CH_3OH, P_{O_2} = 2.5 bar, P_{MeOH} = 2.0 bar, T = 110°C).

Thus, the galvanostatic pulse electrodeposition method displays some advantages for the preparation of catalysts. It is a clean process, which does not need or involve any organic solvents or reactants. The process is easy to perform and very reproducible, which is of interest for industrial development. It makes possible to control the atomic composition of the catalysts and allows preparing mono-, bi-, and even trimetallic catalysts with convenient metal loading or surface concentration. It makes also possible to control the particle size by controlling the electric parameters (t_{on} and t_{off}) of the synthesis process. Therefore, this method seems convenient to prepare real fuel cell electrodes for a DAFC. However, this method also displays an important limitation. As soon as some platinum is deposited on the carbon electrode, hydrogen evolution does occur when current pulses are applied. The faradic yield is then drastically decreased, and finally, only 10% of the electric charge is used for metal deposition. Therefore, to obtain electrodes with high metal loading or surface concentration, experiments have to be performed for a long time.

9.4.2.2   Catalyst Synthesis by Chemical Methods

9.4.2.2.1   Impregnation–Reduction Method

The impregnation–reduction method is widely used in the field of heterogeneous catalysis to prepare nanostructured catalysts. Among the different possibilities, the cationic exchange method [47] can be used. It consists in activating a carbon support (typically Vulcan XC-72) with sodium hypochlorite to form surface carboxylic acid groups, which are transformed into ammonium salts after treatment with ammonia. The ammonium groups are exchanged by contact with a metal salt (e.g., $PtCl_4$), and the catalyst precursor is then reduced under pure hydrogen atmosphere to obtain metallic particles.

A Pt/C catalyst prepared by this method displays a good dispersion on the support and a mean particle size close to 2 nm (Figure 9.13). But, the total loading of metal is very limited and does not exceed 10 wt%, which is a too low value for application to fuel cells.

Higher metal loadings could be obtained by slightly modifying the synthesis protocol [48–50]. The modification of the carbon surface using *aqua regia* as reactant leads to form hydroxyl groups, which can be acidified by treatment with an acid. The OH_2^+ groups can fix an anionic metal salt (e.g., $PtCl_6^{2-}$), which after reduction for 4 h at 300°C under pure hydrogen atmosphere leads to the formation of metallic particles. By this way, Pt/C and Pt_xSn_{1-x}/C catalysts with different atomic ratios could be prepared, with a loading up to 30 wt%, which is suitable for a DEFC.

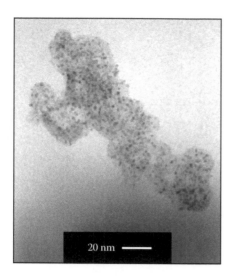 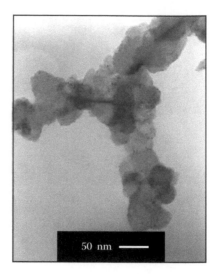

FIGURE 9.13   TEM photographs of a Pt/C catalyst prepared by the cationic exchange method (10 wt% metal loading).

Pt_xSn_{1-x}/C catalysts with different atomic ratios were then characterized electrochemically with respect to the ethanol oxidation reaction (Figure 9.14). All tin-containing catalysts led to a shift of the onset of the oxidation wave toward lower potentials compared with a pure platinum catalyst. Higher current densities were obtained over the whole potential range studied with a $Pt_{0.83}Sn_{0.17}$/C catalyst (Figure 9.14a). This latter catalyst was used in a DEFC working at 90°C. Again, an important enhancement of cell performance was observed with the platinum–tin anodic catalyst in comparison with pure platinum or platinum–rhenium catalysts, in terms of higher open circuit voltage (OCV), which increased from ca. 0.5 to 0.8 V (Figure 9.14b), and of higher power density, from ca. 8 to 33 mW cm^{-2}, with a Pt/C and a $Pt_{0.83}Sn_{0.17}$/C anode, respectively (Figure 9.14c).

However, TEM measurements performed on a $Pt_{0.83}Sn_{0.17}$/C (Figure 9.15) indicated that the increase of the metal loading on the carbon support led to the formation of a multimodal distribution of the particle size. Then, to overcome this problem, colloidal methods were also developed in our laboratory.

9.4.2.2.2   Colloidal Methods

In these methods, the use of a surfactant to form microreactors where metal salts are reduced is expected to limit the size dispersion of the metallic particles.

The procedure described by Bönnemann et al. [51,52] was slightly modified and adapted to the preparation of mono- and multimetallic CLs. The synthesis is carried out under controlled atmosphere (argon) free of oxygen and water, with nonhydrated metal salts (e.g., 99.9% $PtCl_2$, $RuCl_3$, and $SnCl_2$). The first step consists in the preparation of a reducing agent (a tetra-alkyl ammonium triethylborohydride), which, after reducing the metal salt in a tetrahydrofuran medium, will act also as a surfactant, preventing any agglomeration of the metallic nanoparticles. The obtained colloidal solution is mixed with a carbon powder (e.g., Vulcan XC-72) and heated at 300°C for 1 h under air atmosphere to remove the organic surfactant. Then a catalytic powder with a controlled metal loading is obtained [53–58].

One of the main advantage of this method is that it allows different possibilities of synthesis of metallic particles deposited on a carbon support: (1) synthesis of the catalysts with a controlled atomic ratio by coreduction, which consists in mixing different metal salts before their reduction leading to colloid formation and deposition on carbon; (2) synthesis of the catalysts with a controlled atomic ratio by codeposition, which consists in mixing colloids of different

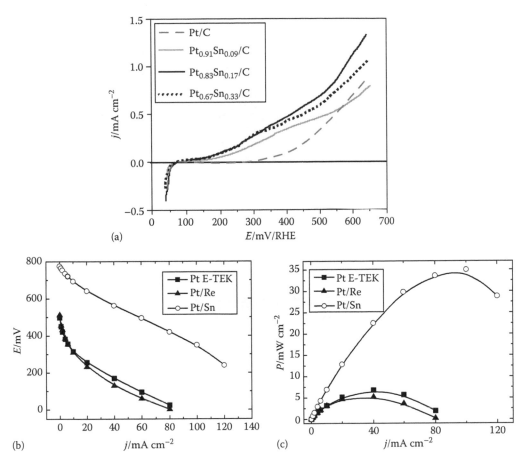

FIGURE 9.14 (a) $j(E)$ polarization curves for ethanol oxidation on Pt/C and $Pt_{(1-x)}Sn_x$/C catalysts (30 wt% metal loading) with different atomic ratio, prepared by the impregnation–reduction method (anionic exchange): 0.1 M $HClO_4$, 1.0 M EtOH, $v = 5$ mV s^{-1}, $T = 293$ K. (b) Cell voltage E. (c) Power density, P, vs. current density, j, curves recorded at 90°C in a single DEFC of 5 cm² surface area using Pt/C, PtSn(83:17)/C, PtRe(83:17)/C anodes (30 wt% metal loading) (cathode E-TEK, Nafion 117 membrane, $C_{CH3CH2OH} = 1.0$ M, $P_{CH3CH2OH} = 1$ bar, $P_{O_2} = 1.0$ bar).

metals before the deposition on carbon; (3) synthesis by simply mixing different M/C catalytic powders. Each method leads to a catalyst having a different structure. For example, it was shown in a previous work [58] that the coreduced $Pt_{0.8}Ru_{0.2}$/C catalyst displayed well-dispersed particles with a narrow size distribution, a mean particle size close to 2 nm and an alloy character as determined by XRD, whereas the codeposited $Pt_{0.8} + Ru_{0.2}$/C catalyst displayed also a mean particle size close to 2 nm, but analysis of the XRD pattern indicated that the compound was not an alloy, but looked like particles of platinum in strong interaction with small ruthenium particles, may be decoration of platinum particles by very small ruthenium particles. Table 9.4 summarizes the obtained structural characteristics for the different catalysts prepared with the Bönnemann method.

The effect of the structure of the catalysts on their electroactivity was first evaluated by CO stripping. The oxidation of a saturated layer of adsorbed CO occurs at lower potentials on the codeposited $Pt_{0.8}+Ru_{0.2}$/C catalyst than on the $Pt_{0.8}Ru_{0.2}$/C coreduced one, and the mixture of a Pt/C catalyst with a Ru/C catalyst gives the worst activity (Figure 9.16a). The activity for methanol electro-oxidation is also higher at the nonalloyed codeposited catalyst than at the alloyed

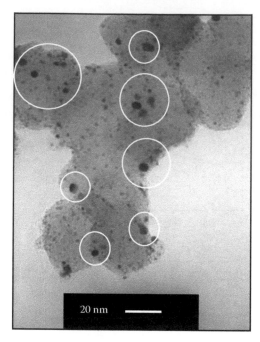

FIGURE 9.15   TEM photograph of a Pt/C catalyst prepared by the impregnation–reduction method of an anionic metal salt (30 wt% metal loading).

TABLE 9.4

Mean Particle Sizes as Determined by TEM and Crystallographic Structure as Determined by XRD Measurements of the Different Supported Catalysts Prepared by the Bönnemann method

Catalyst	Mean Particle Size (nm)	Crystallographic Structure
Pt/C	2.2	fcc
Ru/C	1.5	hcp
$Pt_{0.8}Ru_{0.2}$/C coreduction	1.9	fcc alloy
$Pt_{0.8}$ + $Ru_{0.2}$/C codeposition	2.1	fcc (Pt) + hcp (Ru) in interaction
Pt/C + Ru/C mixture (80 at% Pt + 20 at% Ru)	2.1 (Pt) + 1.5 (Ru)	fcc (Pt) + hcp (Ru)

---

coreduced catalyst, as the former one leads to higher current density over the whole potential range studied (Figure 9.16b).

The Bönnemann method allows also preparing PtSn/C and PtSnRu/C catalysts. Unfortunately, we did not succeed to prepare a tin colloidal solution using this method, so that only coreduced catalysts could be synthesized. Then coreduced $Pt_{0.9}Sn_{0.1}$/C and $Pt_{0.84}Sn_{0.1}Ru_{0.06}$/C catalysts were obtained. The atomic composition was confirmed by EDX measurements in different zones of the catalysts. A more detailed physicochemical characterization was given in previous works [53–56]. In particular, it was shown by TEM measurements that the mean particle size of a $Pt_{0.9}Sn_{0.1}$/C catalyst is close to 2.6 nm for a metal loading on carbon of 30 wt% (Figure 9.17) and 2.8 nm for a metal loading of 60 wt%.

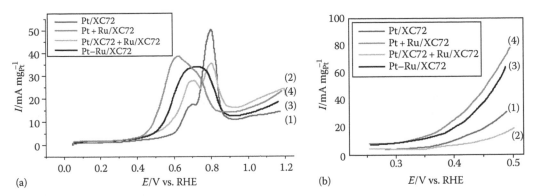

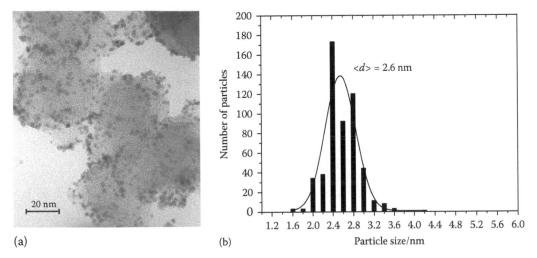

FIGURE 9.16 (a) $I(E)$ polarization curves obtained for the electro-oxidation of a preadsorbed saturated layer of CO (CO stripping) on Pt/C and $Pt_{0.8}$–$Ru_{0.2}$/C. (b) $I(E)$ polarization curves obtained for the electro-oxidation of methanol on Pt/C and $Pt_{0.8}$–$Ru_{0.2}$/C. The catalysts were prepared by codeposition or coreduction using the Bönnemann method (0.5 M H_2SO_4, 1.0 M MeOH, v = 10 mV s^{-1}, T = 298 K): (1) Pt/C; (2) Pt/C + Ru/C; (3) coreduced PtRu/C; and (4) codeposited Pt + Ru/C.

FIGURE 9.17 Physical characterization of a 30% $Pt_{0.9}Sn_{0.1}$/C catalyst with 30 wt% metal loading prepared with the Bönnemann method: (a) TEM image and (b) particle size distribution.

9.4.2.2.3 Water-in-Oil Method

Although the Bönnemann method is very interesting by allowing to vary and to control easily the composition and the nanostructure of the catalyst and is adapted to the preparation of real fuel cell electrodes, it displays also some limitations. For example, bismuth-containing colloids could not be prepared with the Bönnemann method, and even in presence of platinum salts. Moreover, the presence of bismuth hinders the reduction of platinum salts [59]. However, platinum–bismuth is a good catalyst for ethylene glycol electro-oxidation in alkaline medium [59–62]. Moreover, colloid of tin alone could not be obtained, and the reaction was only possible by coreduction in the presence of a platinum salt. Then, other colloidal methods should be developed keeping in mind the necessity of a similar flexibility as that of the Bönnemann method.

Among the different possibilities, the "water-in-oil" microemulsion method [63–73] can be a good approach. This method was derived from that developed by Boutonnet et al. [64]. Two different microemulsions are prepared, one containing the metal salts (e.g., 99.9% H_2PtCl_6, $SnCl_2$, and/or $BiCl_3$) dissolved in ultrapure water, the other one containing a solution of sodium borohydride ($NaBH_4$) as reducing agent. The surfactant is polyethyleneglycol-dodecylether (BRIJ 30), the organic phase is n-heptane. Accurate amount of carbon Vulcan XC72 is added to obtain catalysts with the desired metal loadings.

The microemulsion method offers the possibility of coreduction and codeposition as the Bönnemann method does. It also offers the possibility to prepare catalytic powders with high metal loadings (up to 60 wt%) without significant modifications of the structure [59]. All these characteristics make this method convenient for the preparation of real fuel cell electrodes. Moreover, the synthesis can be carried out without controlling the atmosphere, conversely to the Bönnemann method, which needs a dry argon atmosphere free of dioxygen. However, because small perturbations can easily destabilize the "water-in-oil" microemulsion, the scale-up is more difficult to realize compared to the Bönnemann method, which can be a limitation for large-scale preparation.

As an example, platinum–bismuth catalysts with different atomic ratios have been prepared by the "water-in-oil" method and tested for the ethylene glycol electro-oxidation in alkaline medium [59].

9.4.3 Gas-Diffusion Layer

The GDL is located on the back of the CL in order to improve gas distribution and water management in the cell. This layer has to be porous to the reacting gases, must have good electronic conductivity, and has to be hydrophobic so that the liquid produced water does not saturate the electrode structure and reduce the permeability of gases. The GDL needs to be resilient and the material of choice for the PEMFC is usually carbon fiber, paper or cloth, with a typical thickness of 0.2–0.5 mm [74,75]. This macroporous support layer is coated with a thin layer of carbon black mixed with a dispersed hydrophobic polymer, such as PTFE, in order to make it hydrophobic. This latter compound can, however, reduce the electronic conductivity of the GDL, and limit the three-phase boundary access.

A GDL is also an important component of the AFC. For these fuel cells, the GDLs are usually made by a rolling procedure similar to that used in the paper industry. The activated carbon/PTFE mixture is rolled onto a current collector made of planar nickel mesh. By using GDL and a two-layer electrode structure, the Pt loading in the AFC can be reduced to 0.3 mg cm^{-2} [76].

The structure of the porous GDL is governed by the type of carbon and by the hydrophobic polymer which is used in its preparation. These, in turn, greatly influence the performance of the PEMFC. The most widely used carbons have been an oil furnace carbon (Vulcan XC-72R) and acetylene carbon black.

The amount of hydrophobic polymer used to coat the carbon in the GDL is also important. High polymer content, higher sintering temperatures during sample preparation, or a combination of the two leads to better hydrophobicity but at the expense of electrical conductivity [77]. For example it was shown that a 10 wt% loading of tetrafluoroethylene–hexafluoropropylene (FEP) copolymer in a carbon cloth GDL enables a higher power to be achieved from a PEMFC than a greater polymer loading (40 wt%) [78]. This is attributed to the polymer dissolution, which, at high concentrations, blocks the micropores in the carbon surface and thereby contributes to significantly higher concentration polarization. The marked effect of the polymer loading on the power density is illustrated in Figure 9.18.

With fluorosulfonate ionomers other than Nafion, the specifications for the GDL has to be changed. More recently, Nishikawa et al. [79] developed GDLs for novel organic–inorganic hybrid electrolytes. They mixed Triton and PTFE dispersions with carbon black and coated the mixture on to carbon paper that had been wet-proofed with FEP copolymer. A nanohybrid electrolyte was prepared by a sol–gel method and formed over the GDL. Addition of uncatalyzed carbon black to the cathode GDL was found to enhance the performance at high current densities.

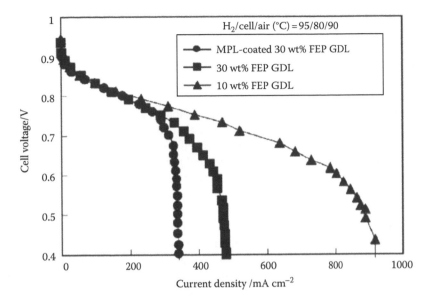

FIGURE 9.18 Cell voltage–current density curves recorded at 80°C in 2 bar H_2/air with MEAs using three different types of cathode GDL under an air stoichiometry of 2.1: (—●—) microporous layer-coated and 30 wt% FEP-impregnated cathode GDL; (—■—) 30 wt% FEP-impregnated cathode GDL; (—▲—) 10 wt% FEP-impregnated cathode GDL. Anode and cathode humidification temperature is 95°C and 90°C, respectively. Pt loading is 0.22 mg cm^{-2} on both electrodes. (From Lim, C. and Wang, C.Y., *Electrochim. Acta*, 49, 4149, 2004.)

9.4.4 INTERCONNECTING MATERIALS (BIPOLAR PLATES)

The bipolar flow-field plate is the main hardware component of the PEMFC and of the DAFC. This has to fulfill several requirements, namely

- Good electrical conductivity
- Nonporous and leak-free channels for the reactants
- Provision for distribution of the fuel and oxidant gases and removal of water and reaction products
- Good thermal conductivity to achieve stack cooling and satisfactory temperature distribution between cells
- Construction to high tolerance in large volumes
- Acceptable corrosion resistance
- Good mechanical stability at low thickness
- Low weight, especially for transport applications
- Low cost

Carbon meets many of these requirements and has been used by fuel cell makers over many years. Nevertheless, the development of low-cost, nonporous, carbon materials continues to be a challenge and the bipolar plate remains one of the most costly components in a PEMFC system.

The design of BP for PEMFCs is dependent on the cell architecture, on the fuel to be used, and on the method of stack cooling (e.g., water or air-cooling). To date, most of the fuel cells have employed traditional filter-press architecture, so that the cells are planar and reactant flow distribution to the cells is provided by the bipolar plate. The bipolar plate therefore incorporates reactant channels machined or etched into the surface. These supply the fuel and oxidant and also provide a means for removing water from the cells. Some alternative designs of plate have been suggested for nonplanar cell architecture and may provide higher volumetric power densities, while maintaining low contact resistance and good structural integrity [80].

Carbon BP for PEMFC applications have been traditionally made by machining graphite sheets that have been impregnated with a polymer resin filler to make them gas-tight, but this is a very expensive process. In order to reduce costs, fuel cell makers are turning to use either metal or carbon composite materials. Metals have the advantage in that they can be made very thin and can be formed by stamping, which is a low-cost process. Plates can also be fabricated by molding carbon composite materials, another low-cost option compared with machined graphite. Carbon composites have been used industrially for several decades and generally comprise carbon fibers to provide strength reinforcement for organic polymers, such as polyethylene and polyvinyl chloride, or epoxy resins. For PEMFC bipolar plates, several composite materials have been investigated, including carbon–carbon composites, graphite/polymer mixtures, carbon fiber/epoxy resin, and carbon cloths impregnated with polymer and various conducting fillers [81–89].

The polymer or resin used to bind the carbon in composite plate material is usually thermoplastic or thermosetting, which implies that the plates can be fabricated either by injection molding [83,85] or by hot pressing. To meet the required electronic conductivity, manufacturers have been using quite high loadings of carbon in the composite material, typically 60%–90% by volume [87]. Unfortunately, this makes the plates brittle, resulting in high scrap rates and an inability to make thin plates, required for high stack power densities.

The structure/nanotexture of carbon used in the composite material also significantly influences the properties of the final material. In the case of carbon blacks in polyvinyl chloride, for example, the surface area and mesopore volume are found to affect significantly the rheology and conductivity of the composite [90]. To enable low loadings of carbon to be used, researchers have been developing specific forms of carbon for use in composite materials [82,89,91–93]. In producing the composite material, the proportion of carbon filler to polymer is found to affect not only the electronic conductivity but also the mechanical properties such as the flexural strength, density, water absorption, and gas permeability of the final plate. Good mechanical properties are afforded by keeping the carbon content low and the polymer content relatively high. Heinzel et al. [87], for example, have shown that 50 wt% carbon additive loadings can achieve bulk conductivities of 50 S cm^{-1} for injection-molded materials. They have also reported conductivities of up to 150 S cm^{-1} for some samples although the nature of the carbon additives was not disclosed.

9.5 CONCLUSIONS

Many fuel cell systems have been developed since the first discovery of Sir William Grove. Fuel cell systems can produce electricity from several fuels (hydrogen, natural gas, alcohols, etc.) for many applications: stationary power plants, power train sources, APU, and electronic portable devices, with nearly the same energy efficiency (around 40% in electric energy), irrespective of their size (from tens of MW for power plants to a few W for portable electronics).

In low-temperature fuel cells (AFC, PEMFC, DAFC, etc.), carbon materials are important since they are involved in the fabrication of BP, GDL, and CL. It appears that no other materials can replace carbon with the same properties (good electronic conductivity, good thermal and chemical stabilities, and low cost). But much work is needed to optimize carbon materials for fuel cell applications and to ensure that they meet the performance targets for conductivity, physical properties, and lifetime within operating stacks.

Furthermore, significant breakthroughs in some components are still missing for a large-scale development of fuel cell systems. This is particularly true for PEMFCs, which need better protonic membranes, allowing working at higher temperatures (120°C–150°C) in order to increase the rate of the electrochemical reactions involved (oxidation of the fuel, reduction of oxygen) and to better utilize (or evacuate) the heat produced. The electrocatalysts need to have better performances to improve the kinetics of the ORR, which is responsible for one third of the energy losses and thus of the power density and fuel utilization decrease. Carbon BP is also a limiting factor, particularly in terms of cost and industrial manufacturing. Therefore one needs breakthroughs in all these components, before an industrial development of fuel cell systems may occur, particularly for the electric vehicle, for which a cost target of €50 per kW is not yet attained.

ACKNOWLEDGMENTS

The author would like to acknowledge both Eusebiu Grivei, TIMCAL Belgium, for personal communication and François Béguin, CRMD, CNRS-Université d'Orléans, for helpful discussions concerning some parts of the chapter on carbon materials in low-temperature fuel cells.

REFERENCES

1. W.R. Grove, *Phil. Mag.* 14 (1839) 127.
2. W.R. Grove, *Phil. Mag.* 15 (1839) 287.
3. J.O'M. Bockris, A.K.N. Reddy, *Modern Electrochemistry*, Vol. 2, Plenum Press, New York, 1972, p. 1141.
4. G.P. Sakellaropoulos, in *Advances in Catalysis*, D.D. Eley, H. Pines, P.B. Weisz, Eds., Academic Press, New York, 1981, p. 218.
5. A.J. Appleby, in *Comprehensive Treatise of Electrochemistry*, Vol. 7, B.E. Conway, J.O'M. Bockris, E. Yeager, S.U.M. Khan, R.E. White, Eds., Plenum Press, New York, 1983, pp. 173–239.
6. A.L. Dicks, *J. Power Sources* 156 (2006) 128–141.
7. A.J. Appleby and F.R. Foulkes, *Fuel Cell Handbook*, Van Nostrand Reinhold, New York, 1989, pp. 203–500.
8. K. Kordesch and G. Simader, *Fuel Cells and Their Applications*, VCH, Weinheim, 1996.
9. P. Stevens, F. Novel-Catin, A. Hammou, C. Lamy, M. Cassir, *Piles à combustible*, Techniques de l'Ingénieur, traité Génie Electrique, D3 340, 2000, pp. 1–28.
10. W. Vielstich, A. Lamm, H. Geisteiger, *Fuel Cells: Fundamental, Technology and Applications*, Wiley, Chichester, 2003.
11. B.V. Tilak, R.S. Yeo, S. Srinivasan, in *Comprehensive Treatise of Electrochemistry*, Vol. 3, J.O'M. Bockris, B.E. Conway, E. Yeager, R.E. White, Eds., Plenum Press, New York, 1981, pp. 39–122.
12. C. Lamy, J.-M. Léger, *J. Phys.* IV (1994) C1–253.
13. C. Lamy, J.-M. Léger, S. Srinivasan, in *Modern Aspects of Electrochemistry*, Vol. 34, J.O'M. Bockris, B.E. Conway, R.E. White, Eds., Kluwer Academic/Plenum Publishers, New York, 2001, pp. 53–118.
14. K. Prater, *J. Power Sources* 29 (1990) 239.
15. M.R. Tarasevich, A. Sadkowski, E. Yeager in *Comprehensive Treatise of Electrochemistry*, Vol. 7, B.E. Conway, J.O'M. Bockris, E. Yeager, S.U.M. Khan, R.E. White, Eds., Plenum Press, New York, 1983, pp. 301–398.
16. T.R. Ralph, M.P. Hogarth, *Platinum Metals Rev.* 46 (2002) 3.
17. G. Hoogers, Ed., *Fuel Cell Technology Handbook*, CRC Press LLC, Boca Raton, FL, 2003.
18. G.G. Tibbetts, G.P. Meisner, C.H. Olk, *Carbon* 39 (2001) 2291–2301.
19. M. Jorda-Beneyto, F. Suarez-Garcia, D. Lozano-Castello, D. Cazorla-Amoros, A. Linares-Solano, *Carbon* 45 (2007) 293–303.
20. U.S. Geological Survey. 2007. USGS minerals information: Graphite. http://minerals.usgs.gov/minerals/pubs/commodity/graphite.
21. E.G. Acheson. 1893. US Patent 492, p. 767.
22. E. Grivei, TIMCAL Belgium, e.grivei@be.timcal.com, personal communication.
23. N. Probst and E. Grivei. Structure and electrical properties of carbon black. *Carbon* 40 (2002) 201–205.
24. E. Grivei and N. Probst. Electrical conductivity and carbon network in polymer composites. *Kautschuk Gummi Kunststoffe* 56, 11 (2003) 460–464.
25. M.S. Dresselhaus, G. Dresselhaus, K. Sugihara, I.L. Spain, H.A. Goldberg. *Graphite Fibers and Filaments*, Vol. 5, *Springer Series in Materials Science*, Springer-Verlag, Berlin, 1988.
26. J. Pilling. 2006. Carbon fibers, carbon-polymer composites and carbon-carbon composites. http://www.mse.mtu.edu/~drjohn/my4150/class14/class14.html

27. T.J. Schmidt, M. Noeske, H.A. Gasteiger, R.J. Behm, P. Britz, H. Bönnemann, *J. Electrochem. Soc.* 145 (1998) 925.
28. T.J. Schmidt, M. Noeske, H.A. Gasteiger, R.J. Behm, P. Britz, W. Brijoux, H. Bönnemann, *Langmuir* 13 (1997) 2591.
29. A.D. Jannakoudakis, E. Theodoridou, D. Jannakoudakis, *Synth. Met.* 10 (1984/85) 131.
30. E.J. Taylor, E.B. Anderson, N.R. Vilambi, *J. Electrochem. Soc.* 139 (1992) 145.
31. D.-L. Lu, K.-I. Tanaka, *Surf. Sci.* 373 (1997) L339.
32. T. Napporn, M.-J. Croissant, J.-M. Léger, C. Lamy, Electrochim. *Acta* 43 (1998) 2447.
33. C. Catteneo, M.I. Sanchez de Pinto, H. Mishima, B.A. Lopez de Mishima, D. Lescano, *J. Electroanal. Chem.* 461 (1999) 32.
34. C. Coutanceau, M.-J. Croissant, T. Napporn, C. Lamy, *Electrochim. Acta* 46 (2000) 579.
35. A.A. Mikhaylova, O.A. Khazova, V.S. Bagotzky, *J. Electroanal. Chem.* 480 (2000) 225.
36. A. Lima, C. Coutanceau, J.-M. Léger, C. Lamy, *J. Appl. Electrochem.* 31 (2001) 379.
37. F. Gloaguen, J.-M. Léger, C. Lamy, A. Marmann, U. Stimming, R. Vogel, *Electrochim. Acta* 44 (1999) 1805.
38. S.D. Thomson, L.R. Jordan, M. Forsyth, *Electrochim. Acta* 46 (2001) 1657.
39. K.H. Choi, H.S. Kim, T.H. Lee, *J. Power Sources* 75 (1998) 230.
40. C. Coutanceau, A.F. Rakotondrainibé, A. Lima, E. Garnier, S. Pronier, J.-M. Léger, C. Lamy, *J. Appl. Electrochem.* 34 (2004) 61.
41. F. Frelink, W. Visscher, J.A. Van Veen, *J. Electroanal. Chem.* 382 (1995) 65.
42. S. Mukerjee, J. McBreen, *J. Electroanal. Chem.* 448 (1998) 163.
43. Y. Takasu, T. Iwazaki, W. Sugimoto, Y. Murakami, *Electrochem. Commun.* 2 (2000) 671.
44. N. Inagaki, S. Tasaka, Y. Horikawa, *J. Polym. Sci. Polym. Chem.* 27 (1989) 3495.
45. S. Litster, G. McLean, *J. Power Sources* 130 (2004) 61.
46. P. Brault, S. Roualdès, A. Caillard, A.-L. Thomann, J. Mathias, J. Durand, C. Coutanceau, J.-M. Léger, C. Charles, R. Boswell, *Eur. Phys. J.* AP 34 (2006) 151.
47. D. Richard, P. Gallezot, in *Preparation of Highly Dispersed Carbon Supported Platinum Catalysts*, B. Delmon, P. Grange, P.A. Jacobs, G. Poncelet (Eds.), Preparation of catalysts IV, 1987, Elsevier Science Publishers B.V., Amsterdam, the Netherlands.
48. F. Vigier, C. Coutanceau, A. Perrard, E.M. Belgsir, C. Lamy, *J. Appl. Electrochem.* 34 (2004) 439.
49. C.L. Pieck, P. Marecot, J. Barbier, *Appl. Catal. A General* 145 (1996) 323.
50. C. Roman-Martinez, D. Cazorla-Amoros, H. Yamashita, S. de Miguel, O.A. Scelza, *Langmuir* 16 (2000) 1123.
51. H. Bönnemann, W. Brijoux, R. Brinkmann, E. Dinjus, T. Joussen, B. Korall, *Angew. Chem. Int. Ed. Engl.* 30 (1991) 1312.
52. H. Bönnemann, W. Brijoux, R. Brinkmann, R. Fretzen, T. Joussen, R. Köppler, B. Korall, P. Neiteler, J. Richter, *J. Mol. Catal.* 86 (1994) 129.
53. F. Vigier, S. Rousseau, C. Coutanceau, J.-M. Léger, C. Lamy, *Topics Catal.* 40 (2006) 111.
54. S. Rousseau, C. Coutanceau, C. Lamy, J.-M. Léger, *J. Power Sources* 158 (2006) 18.
55. J.-M. Léger, S. Rousseau, C. Coutanceau, F. Hahn, C. Lamy, *Electrochim. Acta* 50 (2005) 5118.
56. C. Lamy, S. Rousseau, E.M. Belgsir, C. Coutanceau, J.-M. Léger, *Electrochim. Acta* 49 (2004) 3901.
57. L. Dubau, F. Hahn, C. Coutanceau, J.-M. Léger, C. Lamy, *J. Electroanal. Chem.* 554–555 (2003) 407.
58. L. Dubau, C. Coutanceau, J.-M. Léger, C. Lamy, *J. Appl. Electrochem.* 33 (2003) 419.
59. L. Demarconnay, S. Brimaud, C. Coutanceau, J.-M. Léger. *J. Electroanal. Chem.* 601 (2007) 169.
60. F. Kadirgan, B. Beden, C. Lamy, *J. Electroanal. Chem.* 136 (1982) 119.
61. F. Kadirgan, B. Beden, C. Lamy, *J. Electroanal. Chem.* 143 (1983) 135.
62. H. Cnobloch, D. Groöppel, H. Kohlmüller, D. Kühl, G. Siemsen, in J. Thomson (Ed.), *Power Sources 7, Proceedings of the 11th Symposium, Brighton 1978*, Vol. 24, Academic Press, London, 1979, p. 389.
63. S. Brimaud, C. Coutanceau, E. Garnier, J.-M. Léger, F. Gérard, S. Pronier, M. Leoni, *J. Electroanal. Chem.* 602 (2007) 226.
64. M. Boutonnet, J. Kizling, P. Stenius, G. Maire, *Colloids Surf.* 5 (1982) 209.
65. J.F. Rivadulla, M.C. Vergara, M.C. Blanco, M.A. Lopez-Quintela, J. Rivas, *J. Phys. Chem. B* 101 (1997) 8997.
66. J. Solla-Gullon, A. Rodes, V. Montiel, A. Aldaz, J. Clavillier, *J. Electroanal. Chem.* 554–555 (2003) 273.
67. I. Capek, *Adv. Colloid Interface Sci.* 110 (2004) 49.
68. M.L. Wu, L.B. Lai, *Colloids Surf.* A 224 (2004) 149.
69. S. Eriksson, U. Nylén, S. Rojas, M. Boutonnet, *Appl. Catal. A, General* 265 (2004) 207.

70. Z. Liu, J. Yang Lee, M. Han, W. Chen, L.M. Gan, *J. Mater. Chem.* 12 (2002) 2453.
71. Z. Liu, L. Gan, L. Hong, W. Chen, J. Yang Lee, *J. Power Sources* 139 (2005) 73.
72. L. Xiong, A. Manthiram, *Electrochim. Acta* 50 (2005) 2323.
73. S. Rojas, F.J. García-García, S. Järas, M.V. Martínez-Huerta, J.L. García Fierro, M. Boutonnet, *Appl. Catal. A: General* 285 (2005) 24.
74. H. Chang, P. Koschany, C. Lim, J. Kim, *J. New Mater. Electrochem. Syst.* 3 (2000) 55–59.
75. Gore-Tex Inc., US Patent No. 6,127,059.
76. E. Han, I. Eroglu, L. Turker, *Int. J. Hydrogen Energy* 25 (2000) 157–165.
77. D. Bevers, D. Rogers, M. von Bradke, *J. Power Sources* 63 (1996) 193–201.
78. C. Lim, C.Y. Wang, *Electrochim. Acta* 49 (2004) 4149–4156.
79. O. Nishikawa, K. Doyama, K. Miyatake, H. Uchida, M. Watanabe, *Electrochim. Acta* 50 (13) (2005) 2719–2723.
80. W.R. Merida, G. McLean, N. Djilali, *J. Power Sources* 102 (2001) 178–185.
81. G. Marchetti, World Patent No. WO 00/57500.
82. F. Jousse, J. Granier, *Clefs CEA* 44 (2001) 52.
83. M.H. Oh, Y.S. Yoon, S.G. Park, *Electrochim. Acta* 50 (2004) 773–776.
84. E.A. Cho, U.-S. Jeon, H.Y. Ha, S.-A. Hong, L.-H. Oh, *J. Power Sources* 125 (2004) 178–182.
85. M. Wu, L.L. Shaw, *J. Power Sources* 136 (2004) 37–44.
86. M. Wu, L.L. Shaw, *Int. J. Hydrogen Energy* 30(4) (2005) 373–380.
87. A. Heinzel, F. Mahlendorf, O. Niemzig, C. Kreuz, *J. Power Sources* 131 (2004) 35–40.
88. E. Middelman, W. Kout, B. Vogelaar, J. Lenssen, E. de Waal, *J. Power Sources* 118 (2003) 44–46.
89. R.H.J. Blunk, D.J. Lisi, Y.-E. Yoo, C.L. Tucker III, *AIChE J.* 49 (2003) 18–29.
90. J.R. Nelson, W.K. Wissing, *Carbon* 22(2), 1984, 230.
91. Ticona, Press release, November 24, 2004, http://www.ticona.com
92. Timcal, Timrex graphite powders and dispersions for fuel cells, Data sheet, 2004, available from http://www.timcal.com
93. Z. Iqbal, Honeywell, International Patent No. WO 01/89013, 2001.

10 Carbon in Batteries and Energy Conversion Devices

Ralph J. Brodd

CONTENTS

10.1 INTRODUCTION

Batteries convert chemical energy directly into electrical energy. The energy reactions at the negative (anode) and positive (cathode) poles generate a voltage and occur only on demand when the external device is connected and the circuit is closed. The design of the battery separates the reactants to prevent the reactions from proceeding until the external circuit is closed. The ability to deliver energy and power is dependent on the quality and characteristics of the materials used to construct the battery/cell. There are two general classes of batteries, primary (single use and discard) and secondary or rechargeable (capable of being electrically restored to its original state for reuse). Within each of these general classes are many different systems distinguished by the chemistry that produces the conversion of chemical energy into electrical energy. Although many electrochemical systems can store chemical energy, only a few reactions meet the criteria for use in commercial systems for cost and storage as well as efficient conversion of chemical energy into electrical energy.

This chapter is concerned with primary and secondary batteries. Please refer to the respective chapters for a detailed discussion of fuel cells (Chapter 9), ultracapacitors (Chapter 8), and advanced carbon materials for rechargeable lithium-ion batteries (Chapter 7).

Carbon and carbon materials are essential components in most electrochemical energy conversion devices, and range from carbon black and graphite to amorphous and nanostructured materials and carbon nanotubes [1–4]. The carbon used in an energy source can be tailored and optimized for that particular application. The high purity, low cost, morphology, and physical form are very important factors for its effective use in all these applications. In the final analysis, the high electrical conductivity, inertness and nontoxic characteristics of carbon, as well as its availability in large quantities in a variety of forms, have made carbon a component of essentially all electrochemical energy conversion systems. It is estimated that the battery industry requires an excess of 50,000 t of various carbonaceous materials [5].

Most battery systems employ carbon materials in one form or another, as noted in Table 10.1. The use of carbon materials in batteries stretches across a wide spectrum of battery technologies. The variety of carbon runs the gamut from bituminous materials, used to seal carbon–zinc and carbon black powders in lead acid batteries, to high performance synthetic graphites, used as active materials in lithium-ion cells. The largest use is as a conductive diluent to enhance the performance of cathode materials. In many instances, it is used as a conductive diluent for poorly conducting cathode materials where carbon blacks, such as acetylene black, are preferred. It is essential that

TABLE 10.1
Typical Uses of Carbon and Carbon Materials in Batteries

Battery System, Common Name	Usage
Primary Battery Systems	
Carbon–zinc (Leclanché)	Acetylene/carbon black provides a conductive diluent in the MnO_2 cathode and acts to absorb the electrolyte
	Carbon rod cathode current collector
	Bituminous seal
Alkaline	Graphite for good conductivity in the MnO_2 cathode matrix
	Expanded graphite in high performance cells
	Cathode may also contain acetylene/carbon black to absorb electrolyte
	Can coating for low resistance contact
Zinc–air	Polymer-bonded active carbon cathode
Li–CFx	Active cathode material
Li–FeS$_2$	Conductive graphite/carbon diluent in cathode
Li–MnO$_2$	Conductive graphite diluent in cathode
Li–SOCl$_2$, Li–SO$_2$	Polymer-bonded acetylene black cathode
Fuel cells	Polymer-bonded active carbon anode and cathode
	Intercell connectors and gas flow chambers
Rechargeable Battery Systems	
Lead acid	Carbon black in the anode to act as a conductive diluent and nucleation site for Pb deposition on charge
	Current collector for advanced lead acid batteries
Ni–Cd and Ni–MH	Graphite used as conductive diluent in the cathode
Li-ion	Conductive diluent in the cathode
	Anode-active material component
	Conductive polymer material used for the positive temperature coefficient resistor (PTC) overcurrent protection device
Ultracapacitors	Activated carbon cathode and anode
	Current collector for redox couples in asymmetric configurations

the carbon materials are free from heavy metal impurities and have good ability to absorb and hold cell electrolytes. Graphites and acetylene blacks find application as conductive additives from primary carbon–zinc to advanced lithium-ion batteries. They have high purity with low ash, moisture, sulfur, and volatile contents. They also have a structure that absorbs and holds electrolyte. Acetylene black consists of colloidal-sized particles of carbon black linked together in long chains. Each particle of acetylene black is composed of surface-graphitized crystallites bonded together in a long-chain structure. This chain-like structure provides the highly developed ability to absorb and hold liquid. The linkage of the carbon particles, together with the degree of surface graphitization, provides excellent electrical and thermal conductivity.

10.1.1 CARBON AS CONDUCTIVE DILUENT

The main use of carbon in primary batteries is as a conductive diluent to provide a low resistance path between the active materials and the current collector. The conductive diluent plays an essential role in the design of a battery electrode. Current distribution in electrode structures has a strong influence on the ability to efficiently deliver energy to the external circuit [6]. The design of the electrode differs for a high energy vs. a high power cell. One can have one or the other, but seldom do they have both in one system design.

When carbon is a conductive diluent in an electrode structure, to a first approximation, carbon should coat the outer shell of the active material, whether it is manganese dioxide for carbon–zinc primary cells or lithium cobalt oxide in a lithium-ion rechargeable cell. This amount is usually determined experimentally. Figure 10.1 shows a typical resistivity plot of the active material in the electrode matrix measured in addition of carbon [7]. This dramatic change in resistance indicates the minimum amount of conductive diluent needed for efficient discharge. Once the amount of carbon diluent for a single cell has been determined, the next step is to determine the optimum mixing time for the active mass in the production process. Whereas, an abrupt change in resistivity defines the proper ratio of carbon to active material, the plotted curve for change in conductivity of the production mix vs. time of mixing shown in Figure 10.2, has a bathtub shape. Samples are taken at regular intervals in the production mixing process; conductivity of the samples is measured, and then plotted. The time of mixing is determined by the minimum in resistivity for the production mixer and is specific to the mixer.

A similar effect can be expected for the carbon conductive diluent used in cathode formulations for all commercial cells. In all cases, it is necessary to take into account the particular cell design and the electrical resistivity of the electrode-active mass, perpendicular to the current collector, to optimize cell performance [8]. One cannot standardize on any one type of carbon for all battery environments. Fortunately, since carbon is a versatile material, one can find a unique form for each application.

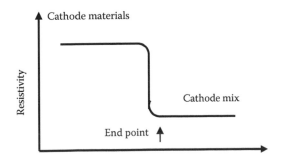

FIGURE 10.1 Resistivity of the cathode-active material with added conductive diluent.

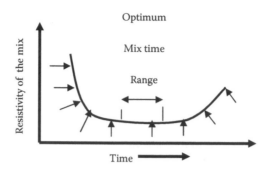

FIGURE 10.2 Resistivity of cathode mix as a function of mixing time. The mixing should be stopped in the optimum range to avoid changes in the distribution in the active mix as time continues. Each mixer will have its own unique shape and time of mixing. The arrows indicate the point in time that a sample was taken to determine resistivity of the mix.

10.2 PRIMARY BATTERIES

Primary batteries are designed to be used once and then discarded. They are good and stable chemical systems. The storage life of each system varies in length according to the chemistry. These systems include aqueous electrolyte with zinc anodes coupled with manganese dioxide, silver oxide, and air (oxygen) cathodes, as well as cells with lithium anodes with iron sulfide, manganese dioxide or carbon monofluoride cathodes, and nonaqueous electrolytes.

10.2.1 Carbon–Zinc Cells

Carbon–zinc is the generic term for the Leclanché and zinc chloride system. Carbon–zinc cells provide an economical source of electrical energy for low drain applications. The service life depends strongly on the discharge rate, the operating schedule and the cutoff voltage, as well as temperature and storage conditions.

Carbon–zinc cells are distinguished by the composition of the electrolyte. The Leclanché cell has an aqueous ammonium chloride–zinc chloride electrolyte. The higher performance zinc chloride cell mainly has zinc chloride electrolyte and may contain a small amount of ammonium chloride. Cells are available in cylindrical and flat plate constructions, as well as combinations of cells for higher voltage applications. Approximately 30 billion carbon–zinc cells are manufactured annually.

10.2.1.1 Leclanché Cells

Leclanché cells are the least expensive primary batteries. The first zinc–manganese dioxide cell was developed by Georges Leclanché in 1866. He developed the primary battery with an ammonium chloride and zinc chloride electrolyte, and with a natural MnO_2 and carbon (usually acetylene black) cathode inserted into a zinc can. His name is still associated with this chemistry today. The battery reactions are given in Equation 10.1.

$$Zn + 2MnO_2 + 2NH_4Cl = Zn(NH_3)_2Cl_2 + H_2O + Mn_2O_3 \tag{10.1}$$

A typical cell construction is shown in Figure 10.3. Natural manganese dioxide ore is blended with acetylene black and Leclanché electrolyte, and molded into a round bobbin shape with a carbon rod current collector in the center. This molded cathode is then inserted into a zinc can. A paste-coated paper separates the zinc can and the MnO_2 cathode. This provides a barrier to prevent solid particles

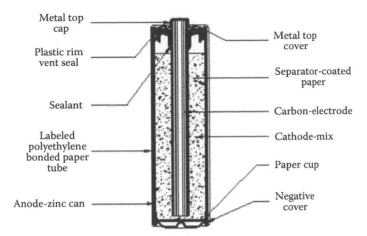

FIGURE 10.3 Construction of the carbon–zinc cells. The zinc chloride and Leclanché have the same construction except for the seal. The zinc chloride has a plastic-compression seal while the Leclanché has a poured hot bituminous seal instead of the plastic-compression seal shown. (Courtesy of Eveready Battery Co., St Louis, MO. With permission.)

from migrating from the cathode to the zinc anode and short the cell. The zinc can function as the anode as well as the cell container. Once the molded cathode is inserted into the can, the cell is sealed with the application of a quantity of hot liquid bituminous material poured into the top of the can. The Leclanché cell is used in applications such as flashlights and portable radios. These cells find wide application, wherever a low-cost, portable power supply is required. Recently, thin flat cells have been developed to power smart radio frequency identification tags (RFID) tags.

Acetylene black is the carbon material of choice for the cathode of Leclanché cells. Acetylene black serves a dual purpose: (1) it provides a conducting path between the carbon rod current collector and the particle of active MnO_2 material and (2) it absorbs and holds, or retains, the electrolyte for ionic conductivity throughout the cathode structure. Carbons, used as the conductive diluent, do not participate in the redox reactions that generate current and voltage.

10.2.1.2 Zinc Chloride Cells

The zinc chloride cell is a high performance version of the carbon–zinc cell. As its name implies, the zinc chloride cell uses a $ZnCl_2$ electrolyte, along with synthetic or electrolytic MnO_2 (EMD). The cell reaction is given in Equation 10.2.

$$4Zn + 8MnO_2 + ZnCl_2 + 9H_2O = 8\, MnOOH + ZnCl_2 \cdot 4ZnO \cdot 5H_2O \qquad (10.2)$$

Because the zinc chloride electrolyte is more acidic, the cell requires a stronger plastic-compression seal. The ingress of oxygen from the air would otherwise react directly with the zinc anode and shorten cell life.

10.2.2 ALKALINE MANGANESE DIOXIDE (MnO₂) CELLS

Alkaline $Zn–MnO_2$ cells set the standard for performance in primary cells. The alkaline cell uses a mixture of high purity EMD and a natural graphite cathode molded into a cylinder and inserted into a nickel-plated steel can. In the past, the cathode had a small percentage of acetylene black to hold electrolyte near the reaction site and to enhance conductivity. However, the oxygen-rich surface groups on acetylene black may contribute to reduced shelf life of the cells and acetylene black's use in this system has diminished since the early 1990s. Purity of all critical materials is held in the parts per million range. The cell reaction is given in Equation 10.3.

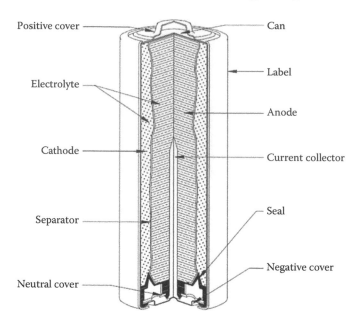

Positive cover — Can
Electrolyte — Label
Cathode — Anode
Current collector
Separator — Seal
Neutral cover — Negative cover

FIGURE 10.4 Construction of the alkaline Zn–MnO_2 cell. (Courtesy of Eveready Battery Co., St Louis, MO. With permission.)

$$Zn + 2MnO_2 + H_2O = ZnO + 2MnOOH \qquad (10.3)$$

The cell construction is shown in Figure 10.4. The cathode is a mixture of EMD, high purity flake, synthetic or expanded graphite, acetylene black, functional additives, and a polymer or portland cement binder. This cathode mix is granulated and then molded into cylinders at high pressure (approx. $3\,t/cm^2$), and inserted into a nickel-plated steel can, or as a pellet for button and coin cell configurations. The good electrical conductivity and lubricity of graphite contribute to the low internal resistance. Unlike some other carbon materials, the graphite mix has minimal "spring-back" after molding. As a result, the close dimensional tolerances critical to reliability and automated cell assembly can be maintained. Advanced performance cells use a graphite coating on the interior surface of the can. This graphite coating serves as a protection against can corrosion and reduces contact resistance between the molded cathode and the can. The anode consists of powdered zinc alloy held by a KOH gel containing a corrosion inhibitor. The cell is closed with a polymer gasket-compression seal.

To meet the competition from rechargeable batteries for portable device applications, it has been necessary to improve the high-rate performance alkaline cells. This has forced a complete redesign of the internal contents of the cell. Developments in the past 5–6 years include new zinc alloy powder and corrosion inhibitors, new gelling agents for the powder, new separator materials and high-purity, nanostructured manganese dioxide, as well as new conductive, high-purity graphite materials, such as expanded graphites, for the conductive diluent. The expanded graphite is produced from natural graphite by treatment in sulfuric acid with added oxidizers, followed by rapid heating to a high temperature and causes exfoliation of the graphite [9]. Photomicrographs of the expanded graphite after exfoliation, in the form suitable for battery use, and natural graphite are shown in Figure 10.5a though c, respectively. These expanded graphite particles, when mixed with MnO_2 at low concentrations (6 wt% or less), form a matrix with higher conductivity than that produced from the same formulation using natural graphite. The new cathode formulation has higher amount of MnO_2 and a lower internal resistance, giving significantly improved performance as noted in Figure 10.6.

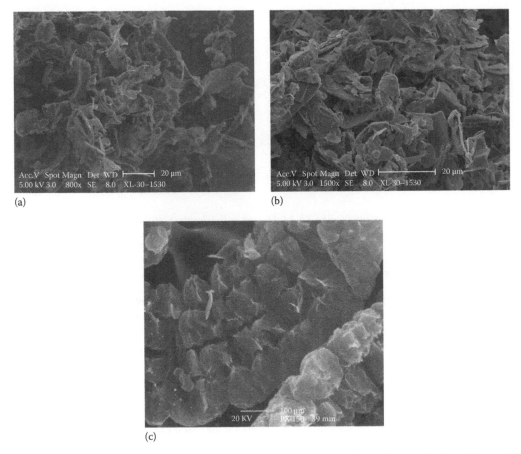

FIGURE 10.5 Photomicrographs of (a) purified exfoliated graphite worms, (b) expanded graphite, and (c) purified natural flake graphite (a precursor to making expanded graphite). (Courtesy of Superior Graphite Co., Chicago, IL. With permission.)

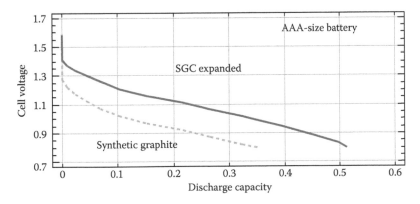

FIGURE 10.6 Continuous discharge of AAA-size alkaline cells with synthetic and expanded graphite in the cathodes at 500 mA. (Courtesy of Superior Graphite Co., Chicago, IL. With permission.)

Production lines typically operate at upwards of 500 cells/min or more. The alkaline cell has higher energy storage, longer shelf life, and lower internal resistance than either of the two carbon–zinc cells. The alkaline cells are also produced in coin/button cell configuration as shown in Figure 10.7.

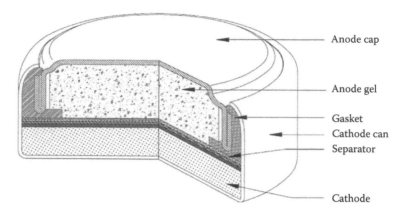

FIGURE 10.7   Construction of an alkaline Zn–MnO$_2$ coin cell. (Courtesy of Eveready Battery Co., St Louis, MO. With permission.)

10.2.3 COMPARISON OF SYSTEMS

Figure 10.8 compares the performance of the three Zn–MnO$_2$ chemistries. The difference in rated capacity is directly related to the type of MnO$_2$ used, as well as the cell construction. Discharges are efficient at low rates, determined by the amount of active MnO$_2$ in the cell. At high drains, the alkaline cell reactions give greater service than those in the carbon–zinc chemistry, where inefficiencies limit performance at rates of discharge.

10.2.4 ZINC–AIR CELLS

The cathode of the zinc–air cell uses a zinc metal anode and an air electrode similar to that for an alkaline fuel cell. The cell reaction is given in Equation 10.4. The cell is constructed with a pull-off

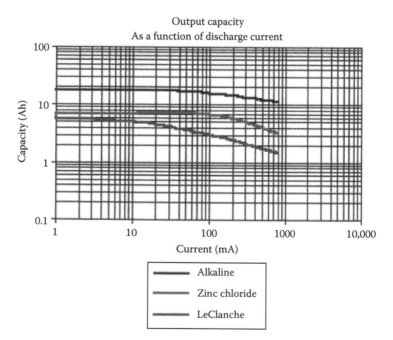

FIGURE 10.8   Comparison of the performance of the three AA-size Leclanché, zinc chloride, and alkaline cells. (Courtesy of Eveready Battery Co., St Louis, MO. With permission.)

$$2Zn + O_2 = ZnO \tag{10.4}$$

tab covering the air inlet holes to the cell. When ready for use, the tab is removed and air flows into the bonded carbon cathode. The air electrode is constructed with two zones, similar in construction to the alkaline fuel cell in Figure 10.7. The outer layer consists of a Teflon sheet or low carbon black, high Teflon content. This provides a waterproof layer, allowing air in, but preventing the leakage of electrolyte. The inner layer is a polymer-bonded high surface area carbon with a catalyst, usually of manganese dioxide. An expanded metal or screen current collector provides structural stability and acts as the current collector for the positive oxygen electrode. The miniature Zn–air cells are used almost exclusively for hearing aids, while large cells find use in military applications. The Zn–air cell has the same construction as the alkaline coin cell, except that a thin air electrode structure replaces the manganese cathode pellet. This essentially doubles the zinc gel anode capacity and gives the zinc–air battery almost double the service life of a zinc–silver oxide cell. The 2007 zinc–air market is estimated at about $250 million.

10.3 LITHIUM BATTERIES

The 1.5 V lithium iron sulfide system competes directly with the alkaline manganese system for high-performance electronic applications. It gives better high-rate performance than the alkaline manganese system. The other main commercial systems are the 3 V lithium carbon monofluoride (Li–CFx) system, and the lithium manganese dioxide (Li–MnO$_2$) system.

The lithium sulfur dioxide and the lithium thionyl chloride systems are specialty batteries. Both have liquid cathode reactants where the electrolyte solvent is the cathode-active material. Both use polymer-bonded carbon cathode constructions. The Li–SO$_2$ is a military battery, and the Li–SOCl$_2$ system is used to power automatic meter readers and for down-hole oil well logging. The lithium primary battery market is estimated to be about $1.5 billion in 2007.

10.3.1 LITHIUM IRON SULFIDE

The lithium cells have high energy density, but usually have poor high rate capability. The internal design, with a wound high surface area construction, thin electrodes and the choice of electrolyte, give the iron sulfide cell good performance. Carbon black is used as a conductive diluent in the cathode to improve the performance. The cell reaction is given in Equation 10.5 and the cell construction is shown in Figure 10.9.

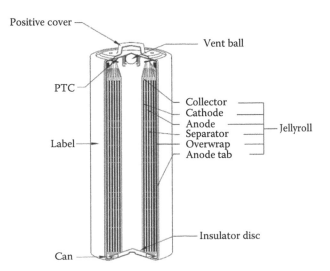

FIGURE 10.9 Construction of the lithium iron sulfide cell. (Courtesy of Eveready Battery Co., St Louis, MO. With permission.)

$$Li + FeS_2 = LiFeS_2 \qquad (10.5)$$

The Li–FeS$_2$ cell has superior low temperature and high-rate performance compared to the alkaline and rechargeable Ni–MH cells, as noted in Figures 10.10 and 10.11.

10.3.2 LITHIUM CARBON MONOFLUORIDE CELLS

Two types of 3 V lithium metal primary cells are in common use. The first commercial lithium primary cell was the cylindrical pin-shaped Li–CFx initially used to power lighted floats for night fishing in Japan [10–13]. The CFx is very stable to over 400°C and gives the cell excellent high temperature and long storage life characteristics. Although the CFx is conductive, it is mixed with acetylene black for use in the cathode. The typical cell electrolyte is γ-butrylactone with LiAsF$_6$. As the cell discharges, LiF accumulates at the surface of the CFx particles in the cathode structure. The cell reaction is given in Equation 10.6. However, the

$$x\, Li + CFx = x\, LiF + C \qquad (10.6)$$

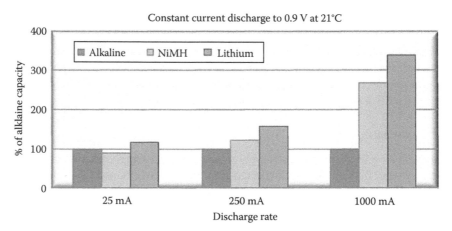

FIGURE 10.10 Comparison of the performance of rechargeable Ni–MH, alkaline, and Li–FeS$_2$ cells. (Courtesy of Eveready Battery Co., St Louis, MO. With permission.)

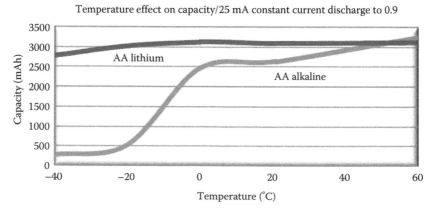

FIGURE 10.11 Comparison of the Li–FeS$_2$ and alkaline cell performance as a function of temperature.

carbon formed is more conductive than the original CFx, and as a result, the discharge is very efficient. Coin cells find application in watches, cameras, and memory protection in electronic equipment. The Li–CFx is one of the highest energy battery systems at over 1000 Wh/L and 400 Wh/kg.

As noted in Figure 10.12, the CFx cathode material is formed by reacting finely sized carbon materials (calcined coke, carbon fibers, carbon black, or even flake graphite) directly with fluorine gas at 400°C. The preferred starting material for Li–CFx cells has been either calcined petroleum coke or coal-derived coke. The reaction time and temperature depend on the particular carbon material used as the precursor. Reaction time can range from 3 to 8 h. Generally, the reaction vessel operates at one atmosphere total pressure, with a ratio of nitrogen to fluorine of 1:1. The carbon fluoride is preferably made from carbons with a lattice constant of 3.40–3.50 Å in the d_{002} plane. As noted in Figure 10.13, these materials result in a lattice constant in the range of 3.50 Å, and a very small grain size, providing maximum performance. Reaction conditions that produce a smaller grain size for the final product are important for good performance [14].

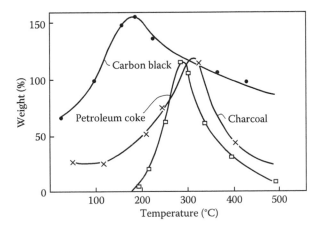

FIGURE 10.12 Effect of temperature on the preparation of CFx materials from various precursors.

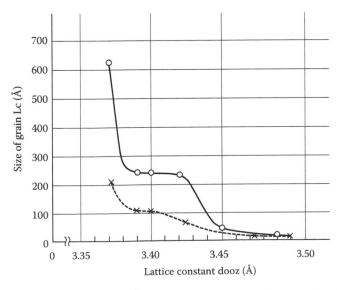

FIGURE 10.13 Effect of the lattice constant of the carbon precursor on the grain size of CFx produced.

In addition to the direct fluorination process described above, two other processes may be noted. A low temperature process has been demonstrated using a gaseous source of fluorine such as IF_2, ClF_2, and BrF_2 [15,16]. This process is said to produce a semi-ionic form of CF with a higher discharge voltage. Also, mesophase pitch microfibers have been proposed as the carbon source produced by direct reaction with fluorine in a narrow temperature range (420°C–450°C). These have very high-rate capability, but the degree of fluoridation must be controlled to leave at least 10% of the carbon unreacted in the core of the carbon microfibers for good performance [17].

The development of implanted medical defibrillators required a high-rate, long-life battery system. In defibrillators, the CFx is used in combination with silver vanadium oxide (SVO) cathode materials [17]. A binary mixture of CFx and SVO are combined to form the cathode, giving the best features of SVO and CFx. Compared to CFx, the SVO has superior pulse current capability, but lower energy storage capability. The cell reactions are given in Equations 10.7 and 10.8.

$$xLi + CFx = C + xLiCFx \tag{10.7}$$

$$Li + Ag_2V_4O_{11} = Li_7Ag_2V_4O_{11} \tag{10.8}$$

The electrode capacities are adjusted so that the CFx cathode has about 10% less capacity than the SVO. During an operation of the defibrillator, the SVO supplies the pulse requirement and the CFx recharges the SVO during low demand periods. As a result, a drop in cell voltage indicates an end to the life of the battery. Physicians can use the change in cell voltage to determine when to replace the device. The principal use of Li–CFx is in miniature cells for use in memory protection, watches and cameras.

10.3.3 Lithium Manganese Dioxide Cells

The second lithium primary cell is the Li–MnO_2 [18]. For use in lithium cells, the high purity EMD must be heat treated at 350°C–375°C to activate and modify the crystal structure, as well as remove any water. Carbon black is used as a conductive diluent in the cathode. The cell electrolyte is propylene carbonate and 1,2-dimethoxyethane with $LiAsF_6$. The cell reaction is

$$Li + MnO_2 = LiMnO_2 \tag{10.9}$$

Recently, it was reported that high-rate performance of Li–MnO_2 cells could be improved through application of a partially graphitized carbon black such as PUREBLACK™ Carbon in the cathode formulation [19]. Here, a petroleum oil-derived synthetic carbon is heat treated at temperatures in excess of 2000°C to graphitize the outside of each particle, while the core of particles remaining amorphous.

The Li–MnO_2 cells are used for memory protection and to power smoke detectors, watches, and cameras. Although the manganese system has lower energy storage, it is less expensive and is preferred in commerce.

10.3.4 Specialty Cells

Both the lithium sulfur dioxide (Li–SO_2) and lithium thionyl chloride (Li–$SOCl_2$) cells may be classified as liquid cathode systems. In these systems, SO_2 and $SOCl_2$ function as solvents for the electrolyte, and as the active materials at the cathode to provide voltage and ampere capacity. As liquids, these solvents permeate the porous carbon cathode material. Lithium metal serves as the anode, and a polymer-bonded porous carbon is the cathode current collector in both systems. Both cells use a Teflon-bonded acetylene black cathode structure with metallic lithium as the anode. The Li–SO_2 is used in a spirally wound, jelly-roll construction to increase the surface area and improve performance. The principal application is to power military equipment, such as radio transmitters and night vision binoculars.

The Li–SOCl$_2$ cell in a molded porous carbon bobbin construction is also used in meter reading applications. The cell has excellent low drain characteristics, operates over a wide temperature range and gives long life.

The Li–SOCl$_2$ is produced in the spirally wound, jelly-roll configuration to power high rate applications, such as instrumentation for down-hole oil well logging to determine the characteristics of oil bearing strata. It is also produced in a low surface area, bobbin construction for low drain applications, such as automatic reading of home gas meters.

10.4 RECHARGEABLE BATTERIES

10.4.1 LEAD ACID BATTERIES

Lead acid batteries constitute the largest segment in the battery market. The largest application is for batteries used in starting, lighting, and ignition (SLI) for automobiles. Two other large market segments are stationary energy storage and motive power. Stationary energy storage includes emergency power supplies for computer and telephone applications to provide "ride-through" during utility outages. Continuous power also is essential for manufacturing operations, such as gasoline refineries, where power outages can create hazardous conditions. The motive power segment includes forklifts and golf cars, as well as fishing boats, in places where gasoline engines are banned. All cells use a polypropylene or polycarbonate polymer case to contain the unit cells.

As a lead acid battery ages, it has a tendency to develop "sulfation," in which particles of lead sulfate accumulate and grow in the negative plate. Essentially, all lead acid batteries use soot or lamp black as a component of the negative electrode. These small particles of conductive soot serve as nuclei for the deposition of small lead particles, restoring the capacity and performance of the negative lead electrode [20]. The term "expander" is used to describe the conversion of low surface area solid lead sulfate particles to soluble lead ions that nucleate and deposit onto the soot particle. This results in a high surface area lead metal particle in the active negative electrode. The resulting gray appearance of the negative plate helps distinguish it from positive plates in manufacturing operations. The carbon also increases cold cranking performance of the batteries. Lately, there has been an increasing tendency to use graphite (flake and expanded graphite) instead of carbon blacks and lamp blacks in the negative plates. The 2007 lead acid market is estimated to be in excess of $30 billion.

10.4.2 LITHIUM-ION BATTERIES

The search for rechargeable batteries with higher energy storage capability led to the development of the lithium-ion battery system. The era of the lithium-ion battery began with its introduction by Sony in 1991 [21]. Previous attempts had been made to develop a rechargeable lithium metal battery. However, significant safety concerns resulted from dendrite and mossy lithium metal deposits. Lithium could reversibly intercalate into graphite/carbon and could prevent the deposition of lithium metal. Sony's first cell used a hard carbon produced by thermal decomposition of polyfurfuryl alcohol resin as the anode and the cathode was lithium cobalt oxide, Equation 10.10.

$$\text{Li C}_6 + \text{LiCoO}_2 = \text{Li}_x\text{C}_6 + \text{Li}_{1-x}\text{CoO}_2 \tag{10.10}$$

Sony dubbed the new cell, "lithium-ion," as only lithium-ions and not lithium metal are involved in the electrode reactions. The lithiated carbon had a voltage of about 0.05 V vs. lithium metal and avoided the safety issues of mossy and dendritic lithium metal deposits. The lithium-ion rechargeable battery system has replaced the heavier, bulkier, Ni–Cd and Ni–MH cells in most applications, and has enabled the development of a wide variety of portable electronic devices such as cellular telephones, notebook computers, and digital cameras. Several good sourcebooks exist for the technology associated with the lithium-ion system [22–24].

The lithium-ion battery market has experienced rapid growth due to the development of portable electronic devices such as mobile phones and notebook computers. It has undergone continuous improvement since its introduction, as depicted in Figure 10.14. In 2006, the lithium-ion market was about 3 billion units. Today, lithium-ion batteries are poised to enter a new era of high performance and market growth. Figure 10.14 shows the segmentation into two markets. One segment continues the trend for increasing performance for portable electronic applications such as notebook computers, cellular phones, and camcorders. The other segment emphasizes steady performance with the development of new low cost materials to serve in power tools and hybrid electric vehicles. This segment will continue to replace Ni–Cd and Ni–MH systems, and will eventually challenge the lead acid battery for stationary energy storage and mobile power. The lithium-ion market was about $6 billion in 2007.

Carbon/graphite is the active material of choice for the negative electrode of lithium-ion batteries. These batteries have the ability to intercalate lithium into the structure with a small amount of expansion, depending on the type of graphite. The internal characteristics of the graphite structure play an important role in the use of graphite as an intercalation anode material. The lithium initially enters into the graphite structure in a random manner. As the base plane is inert, the lithium intercalation reaction in Equation 10.10 takes place at the electrochemically active edge plane of the graphite crystallites. The lithium atoms diffuse into the carbon matrix, with each lithium atom associated with a six-member ring. A slight expansion of the distance between the graphene layers occurs as the lithium enters the lattice. As the number of lithium atoms increases, the internal graphite structure forms into stages, as the various graphene layers become saturated with lithium. Dahn and others have identified five stages in the intercalation process [25–28]. The diffusion constant for lithium in graphite varies from 10^{-5} to 10^{-11} cm^2/s depending on the amount of lithium in the lattice [29].

When the anode is first charged, it slowly approaches the lithium potential and begins to react with the electrolyte to form a film on the surface of the electrode. This film is composed of products resulting from the reduction reactions of the anode with the electrolyte. This film is called the solid electrolyte interphase (SEI) layer [30]. Proper formation of the SEI layer is essential to good performance [31–34]. A low surface area is desirable for all anode materials to minimize the first charge related to the formation of SEI layer. Since the lithium in the cell comes from the lithium in the active cathode materials, any loss by formation of the SEI layer lowers the cell capacity. As a result, preferred anode materials are those with a low Brunauer, Emmett, and Teller (BET) surface area

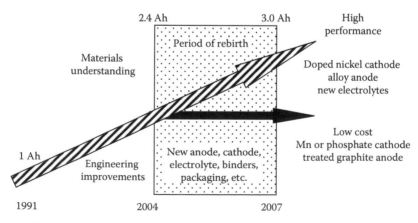

FIGURE 10.14 Depiction of the time line for the development of the lithium-ion battery system.

in the range of $1\,m^2/g$. Any contact resistance, between particles of the active graphite, will reduce performance. Conductive diluents are chosen that do not intercalate lithium.

The pace of successful growth of the lithium-ion battery chemistry, correspondingly, resulted in a steadily increasing demand for the lithium-ion grade carbons. Thus, for reference, the actual worldwide sales volumes of graphitic carbon powders into this market amounted to approximately 3000 t in 2001, propelled up to 6000 t in 2003, and well exceeded 11,000 t in 2004. Due to fierce competition among carbon manufacturers and also the lithium-ion battery becoming a commodity, the prices for anode-grade carbons declined from the levels of $50 per kg to approximate levels of $20 per kg within the same time frame. It is estimated that as of the end of 2004, the market for anode materials for lithium-ion batteries amounted to more than US$ 220 million per year [35].

10.4.2.1 Carbon Materials for Li-Ion Batteries

Graphite is found in nature, but can be produced artificially by pyrolysis of carbonaceous materials, followed by a high temperature treatment in the range of up to 3000°C. Two types of artificial carbon can be distinguished: (1) soft carbons have an ordered array of small graphite crystallites that can be graphitized by treatment at 3000°C, (2) hard carbons have a disordered stacking of the graphite crystallites that are difficult, if not impossible, to graphitize. These materials have fissures and slit pores that provide an apparent high capacity on first charge, but display a sloping discharge curve (which is a shortcoming in most segments of the lithium-ion battery market).

Typical anode-active materials for lithium-ion cells include (1) natural graphite, (2) graphitized mesocarbons or microbeads, formed by graphitization of mesophase pitch materials, (3) hard carbons formed by the pyrolysis of polymeric materials, and (4) natural or artificial graphite materials coated with a hard or a soft carbon surface layer. The mesophase pitch fibers and "balls," or spheres, have been popular, but are being phased out in favor of natural graphite due to cost considerations. Even if given a surface treatment, natural graphite materials are less expensive. At the present time, anode materials for rechargeable lithium-ion batteries are not standardized. Purity is a key issue in the performance of the anode materials. Any impurity (including trace moisture) can create side reactions, loss of charge, and deterioration of capacity on cycling.

Conductive diluent is also required in both the anode and cathode for good cycling performance. Many of the preferred anode materials have relatively low conductivity and surface morphologies that prevent good electrical contact to the active material. The addition of a small amount (0.5%–1%) of conductive additive in the anodes improves contact and cycle life. Concentration of conductive diluents in the cathodes is typically higher and may reach up to 10 wt% in certain applications. A graphite conductive diluent with a particle size of less than 5 μm is preferred mostly for high voltage (such as Mn-based cathode systems). A nonmanganese-based cathode will, typically, use a mixture of acetylene-type carbon black with graphite or an acetylene-type carbon black alone. The size, shape, and specific surface area of the conductive diluent particles affects the rheology of the coating slurry, and may impact the composition and coating parameters required for good electrode performance.

A number of advanced batteries now use a vapor-grown carbon fiber (VGCF), which is costly, and very highly conductive nanomaterial. The advantage of this material comes with the enhancement of conductivity of cathode-active matrixes of lithium-ion batteries. The conductivity of a typical cathode matrix with 5 wt% of acetylene-type carbon black may be similar to that of the matrix containing as little as 0.5 wt% of VGCF—a feature, which does not really offset the cost of the cathode, but gives a lot of leverage to the manufacturers as far as boosting either a high rate capability, or capacity of the cathodes.

The intercalation process on the anode side takes place in stages as more and more lithium enters the crystal lattice. A typical electrolyte in lithium-ion cells contains ethylene carbonate and a mixture of aliphatic carbonates such as methyl carbonate, and ethyl methyl carbonate, along with 1 M $LiPF_6$ salt. The propylene carbonate containing electrolyte, used in primary lithium cells, could not be used in lithium-ion cells until recently, as it caused exfoliation of the carbon surface. Latest advancements in the development of the anode materials, mainly spherodized natural graphites with an amorphous surface coating, resulted in the appearance of carbon materials that can be successfully used in the partially PC-based mixed solvent electrolytes [36].

Lithium-ion cells are assembled in the discharged state. The cells are activated by charging. Two processes occur in the first charge. First is the intercalation of lithium into the graphite crystal lattice, and the second is the formation of a protective SEI layer on the surface of the anode. The cathode provides the lithium for intercalation, so the concentration of the cell electrolyte remains constant. This SEI layer is composed of the reaction/reductive polymerization products of lithium with the cell electrolyte. The energy spent in formation of the SEI layer represents a loss in anode capacity on discharging. The SEI layer protects the graphite surface from reaction with the electrolyte while providing a path for Li^+ to enter and leave the anode structure. Additives, such as vinylene carbonate, can be added to the electrolyte to speed SEI formation, and reduce the amount of charge needed to form a stable SEI.

Cathode-active materials such as $LiCoO_2$, $LiMn_2O_4$, $LiCo_{0.15}Ni_{0.8}Al_{0.05}O_2$, and $LiMn_{1/3}Ni_{1/3}Co_{1/3}O_2$ are poor conductors and require the addition of a conductive carbon diluent to the cathode composition to improve their performance. The amount of conductive material added follows a similar determination as described in the introduction. Common additives include finely sized expanded and synthetic graphite, carbon blacks, such as the Super P and ENSARCO materials, acetylene black (Denka and Shawinigan), Ketchen black and VGCF. In addition, a highly conductive form of partially graphitized carbon black, PUREBLACK Carbon can be used. The conductive matrix should contact the surface of all the active particles and provide a path for electron flow from the current collector to the active materials. The effect of the conductive diluent on the coating operation, in preparation of the electrodes for cell assembly, must be recognized to optimize performance.

10.5   ULTRACAPACITORS (ELECTROCHEMICAL DOUBLE-LAYER CAPACITORS)

Although there were earlier studies of the capacitance of high surface area carbon electrodes, the first commercial electrochemical double-layer capacitors (EDLC) was a carbon–carbon device that emerged from the work at SOHIO [37,38]. The main initial application was memory protection in appliance control systems. The market was about \$700–800 million in 1996. Emerging applications include stationary energy storage, and automotive applications, especially as a power source in hybrid and electric vehicles. The key characteristics are extremely long cycle life coupled with the ability to absorb and deliver high bursts of power. The capacitance, C, of a parallel plate capacitor is given by Equation 10.11 where A is the area of the plates, d is the distance between the parallel plates, and ε is the dielectric constant of the medium between the plates. Equation 10.11 applies to EDLCs, just as it does for a parallel plate capacitor.

$$C = A\ \varepsilon/4d \tag{10.11}$$

Carbons used as electrode materials in ultracapacitors have a high surface area of $1000\,m^2$ or more, a controlled pore structure and are stable in aqueous or nonaqueous electrolytes. Since these carbons lack good conductivity, up to 10 wt% of conductive diluents may be added to electrode composition. Carbons for ultracapacitors are corrosion resistant, low cost, and good high temperature stability.

In studying the capacitance of the basal plane of highly oriented pyrolytic graphite (HOPG) in aqueous solutions, Randin and Yeager [39,40] showed that the distribution of charge carriers in the basal plane of HOPG provides a semiconducting space charge region on the carbon side of the interface. Therefore, some of the applied potential extends into the carbon, and a space-charge capacitance develops. Gerischer, using density of states calculations, showed that the capacitance and conductivity of the electrode is directly related to the density of the electronic states in the conductor, which increases on either side of the zero point of charge (ZPC) [41]. Therefore, when the electrode departs from the ZPC, the formation of the double layer effectively adds capacitance to the surface, and enhances the conductivity of the carbon. Both are limited by the number of charge carriers within the space charge region of the carbon. This implies that during the activation process, if the pore walls of the carbon become thinner and approach the thickness of the space charge region, the space charge regions overlap and generate less capacitance. To explain their results, Hahn and Baertschi used this phenomenon to account for the fact that increased capacitance and electrode conductivity followed each other [42]. As a corollary, if activation produces a carbon with thin pore walls, the space charge regions in the electrode and electrolyte overlap, reducing capacitance and electrode conductivity. The result is a higher surface area, but a lower capacitance.

A relationship exists between pore size, ion size, and specific capacitance. Surrounding each ion in an electrolyte is a solvation sheath of the solvent molecules. The solvation sheath controls the distance of closest approach of the ion to the surface of the conductor. Therefore, this sheath also controls the capacitance. The pores of the carbon need to be larger than the diameter of the ion and its sheath to accommodate the solvated ion. However, if the carbon has a pore size smaller than the solvated ion, ions may still enter the pore during charging, if they deform or lose their solvation sheaths. This decreases the distance, d, in Equation 10.11, and results in a significantly higher capacitance than would be expected.

Dash and coworkers [43–45] used a templating process to produce porous carbons with controlled pore size tunable in the range of 0.5–3 nm, with an accuracy of 0.05 nm. The carbons had surface areas of up to 2000 m^2/g. The capacitance was shown to increase substantially when the pore size decreased below 1 nm, resulting in a volumetric capacitance in the range of 55–80 F/cm^3. As the ions squeezed into the smaller pores, the solvation sheath became distorted, allowing the ions to approach closer to the conducting surface, resulting in a higher capacitance value. As anticipated by the work of Randin and Yeager, interpreted by Gerischer, this phenomenon would be possible if the wall thickness of the pores is controlled and remains greater than the space charge region.

10.6   CLOSING REMARKS

The future remains bright for the use of carbon materials in batteries. In the past several years, several new carbon materials have appeared: mesophase pitch fibers, expanded graphite, and carbon nanotubes. New electrolyte additives for lithium-ion cells permit the use of low-cost, PC-based electrolytes with surface-coated natural graphite anodes. Carbon nanotubes are attractive new materials, and should be available in large quantity in the near future. They should find application in symmetric as well as asymmetric ultracapacitor constructions. They have a high base-plane to edge-plane ratio. The wave of the future might be an ultracapacitor application in which an electronically conductive polymer is deposited on the surface of carbon nanotubes. Lithium-ion batteries may see another breakthrough when stable, long cycle life graphite–Si composites become available.

ACKNOWLEDGMENT

Author gratefully thanks Prof. E. Frackowiak and Dr. I. Barsukov for their assistance in the preparation of this chapter.

REFERENCES

1. Sarangapani, S., Akridge, J., and Schuum, B., Eds., 1984, *The Electrochemistry of Carbon*, The Electrochemical Society, Pennington, NJ, PV84–5.
2. Kinoshita, K., 1988, *Carbon Electrochemical and Physicochemical Properties*, John Wiley & Sons, New York.
3. Takamura, T. and Brodd, R., 2006, in *New Carbon Based Materials for Electrochemical Energy Storage Systems*, Eds., Barsukov, I., Johnson, C., Doninger, J., and Barsukov, V., NATO Science Series, II Mathematics, Physics and Chemistry, Vol. 229, 157, Springer, Dordrecht.
4. Besenhard, J., Ed., *Handbook of Battery Materials*, 1999, Wiley-VCH, Weinheim, Germany.

5. Barsukov, I., 2004, *Ind. Mineral.*, **8**:79.
6. Brodd, R., 1966, *Electrochim. Acta*, **11**:1107.
7. Brodd, R. and Kozawa, A., 1976, in *Experimental Methods of Electrochemistry*, Vol. 3, Ed., Salkind, A., John Wiley & Sons, New York, 199.
8. Newman, J. and Thomas-Alyea, K., 2004, *Electrochemical Systems*, 3rd Edn., Wiley-Interscience, New York.
9. Barsukov, I., Doninger, J., Zaleski, P., and Derwin, D., 2000, *New Technol. Med.*, **2**:106.
10. Fukuda, M. and Iijima, I., 1975, in *Power Sources 5*, Ed., Collins, D., Academic Press, New York, 713.
11. Watanabe, N. and Fukuda, M., 1970, U.S. Patent 3,536,532.
12. Watanabe, N. and Fukuda, M., 1972, U.S. Patent 3,700,502.
13. Toyoguchi, Y., Iijima, H., and Fukuda, M., 1981, U.S. Patent 4,271,242.
14. Ngasubramanian, G. and Rodriguez, M., 2007, *J. Power Sources*, **170**:179.
15. Hamwi, A., Daoud, M., and Coussins, J.C., 1989, *J. Power Sources*, **27**:81.
16. Hany, P., Yazami, R., and Hamwi, A., 1997, *J. Power Sources*, **68**:708.
17. Ikeda, H., Saito, T., and Tamura, H., 1975, in *Proceedings of the Manganese dioxide Symposium*, Vol. 1, Eds., Kozawa, A. and Brodd, R., 394, Cleveland, OH, IC Common Sample Office.
18. Gan, H., Rubino, R., and Takeuchi, E., 2005, *J. Power Sources*, **146**:101.
19. Barsukov, I., Gallego, M., and Doninger, J., 2006, *J. Power Sources*, **153**:288.
20. Pandolfo, A.G. and Hollenkamp, A., 2006, *J. Power Sources*, **157**:3.
21. Nagura, T. and Tozawa, T., 1991, *Prog. Batteries Solar Cells*, **10**:218.
22. Wakihawa, M. and Yamamoto, O., 1998, *Lithium Ion Batteries*, Wiley VCH, Weinheim, Germany.
23. van Schalkwijk, W. and Scrosati, B., 2002, *Advances in Lithium-Ion Batteries*, Kluwer Academic/Plenum Publishers, New York.
24. Winter, M. and Besenhard, J., 1999, in *Handbook of Battery Materials*, Ed. J.O. Besenhard, Wiley-VCH, New York, 383.
25. Dahn, J., Zheng, T., Liu, Y., and Xeu, J., 1995, *Science*, **270**:590.
26. Dahn, J., 1991, *Phys. Rev.*, **B44**:9170.
27. Ohzuku, T., Iwakoshi, Y., and Sawai, K., *J. Electrochem. Soc.*, 1993, **140**:2490.
28. Jiang, Z., Alamair, M., and Abraham, K., 1995, *J. Electrochem. Soc.*, **142**:333.
29. Funabiki, A., Inaba, M., Ogumi, Z., Yuasa, S., and Mizutani, Y., 1998, *J. Electrochem. Soc.*, **145**:172.
30. Peled, E., 1979, in *Handbook of Battery Materials*, Ed. J.O. Besenhard, Wiley-VCH, Weinheim, Germany, 1999, 419.
31. Ein-Eli, Y., Markovsky, B., Aurbach, D., Carmeli, Y., Yamin, H., and Luski, S., 1994, *Electrochim. Acta*, **39**:2559.
32. Aurbach, D., Markovsky, B., Schechter, A., and Ein-Eli, E., 1996, *J. Electrochem. Soc.*, **143**:3809.
33. Mori, S., Asahina, H., Suzuki, H., Yonei, A., and Yokoto, K., 1997, *J. Power Sources*, **68**:59.
34. Suzuki, J., Omae, O., Sekine, K., and Takamura, T., 2002, *Solid State Ionics*, **152–153**:111.
35. Barsukov, I., in *New Carbon Based Materials for Electrochemical Energy Storage Systems*, Eds., Barsukov, I., Johnson, C., Doninger, J., and Barsukov, V., NATO Science Series, II Mathematics, Physics and Chemistry, Vol. 229, 153, Springer, Dordrecht.
36. Liu, J., Vissers, D., Amine, K., Barsukov, I., and Doninger, J., 2006, in *New Carbon Based Materials for Electrochemical Energy Storage Systems*, Eds., Barsukov, I., Johnson, C., Doninger, J., and Barsukov, V., NATO Science Series, II Mathematics, Physics and Chemistry, Vol. 229, 283–292, Springer, Dordrecht.
37. Rightmire, R.A., 1966, U.S. Patent 3,288,641.
38. Boos, D.L., 1970, U.S. Patent, 3,536,963.
39. Randin, J.-P. and Yeager, E., 1972, *Electroanal. Chem. Interfacial Electrochem.*, **36**:257.
40. Randin, J.-P. and Yeager, E., 1975, *Electroanal. Chem. Interfacial Electrochem.*, **58**:313.
41. Gerischer, H., 1985, *J. Phys. Chem.*, **89**:4249.
42. Hahn, U., and Baertschi, U., Barbieri, O., Sauter, J-C., Kotz, R and Galley, R., 2004, *Electrochim. Solid-State Lett.*, **7**:A33.
43. Chmiola, J., Yushin, G., Dash, R., and Gogotsi, Y., 2006, *J. Power Sources*, **158**:765.
44. Chmiola, J., Yushin, G., Gogotsi, Y., Portet, C., Simon, P., and Taberna, P., 2006, *Science*, **313**:1760.
45. Dash, R., Yushin, G., and Gogotsi, Y., 2005, *Micropor. Mesopor. Mat.*, **86**:50.

11 Industrial Production of Double-Layer Capacitors

Roland Gallay and Hamid Gualous

CONTENTS

11.1   INTRODUCTION

Thanks to the recent technological developments, the double-layer capacitor (DLC) market is growing fast. Two main application categories may be distinguished to understand the DLC market structure: power delivery applications and energy storage applications.

In the case of the market segment for power delivery, DLCs fill the gap existing between batteries and electrolytic capacitors. In comparison with DLCs, batteries have approximately 10 times more energy density and 10 times less power density. On the other hand, electrolytic capacitors have approximately 10 times less energy density and 10 times more power density. In addition to the fact that the DLCs can provide more power than batteries, they may also be deeply cycled in voltage several millions of times. Moreover, they do not need any maintenance to fulfill their function without failure over a longer lifetime. The major applications for power DLCs are expected in the automotive market.

The DLC commercialization actually started in the 1970s in Japan with the introduction of the NEC Corporation (NEC) Supercapacitor based on an aqueous electrolyte. The market for high-power DLC components and modules has effectively started to grow since the year 2000 with the launching of industrial applications such as the

- Power supply of the pitch system of windmill blades
- Energy recovery and the power burst supply in metro substations

For the market segment of energy storage, the main application is the memory backup with components in the range of 1 F capacitance and 5 V operating voltage. This market started in the early 1990s and is reaching soon the billion dollars size. The advantage over batteries is the maintenance-free operation capability. Here, the priorities are set on high-energy density and small self-discharge and not on the power density.

The developments of the design and production processes of commercial DLC are driven by the optimization of the materials and production costs, by the processes and component reliability and by the required technology performances [1].

DLC specifications are based on the

- Application requirements defined by the customer or by the market
- Environmental, safety, and security conditions
- Specific requirements of country legislations, transport regulations, etc.

The "commercial" DLC must also fulfill its functional task in a wide range of conditions. For example, specifications define the superposition of the following working conditions: a temperature domain defined between $-40°C$ and $70°C$, combined with a voltage cycling between 1.35 and 2.7 V, 50 G mechanical shocks, and a current intensity distribution between -0.2 and $+0.2$ A/F.

The main technical goals for the development of DLCs, based on the requirements stated above, are reaching the highest power density, the highest energy density, the highest voltage, and the highest reliability. In other words, it does mean that the parameters to improve are

- Cell operating voltage
- Active material capacitance density
- Equivalent series resistance (ESR) of the active material and of the construction, topology

If the operating voltage of the individual cell is increased, the number of cells connected in series to reach the application voltage may be reduced. The weight and volume of the DLCs are inversely proportional to the square of the operating voltage when the comparison is performed at constant amount of power and energy. The capacitance performance is addressed by improving the surface density of the carbon accessible areas (target value in the range of $1700\,m^2/g$), and the electrolyte ion size and mobility. The series resistance may be decreased by mixing the active material, generally an activated carbon, with conducting particles, generally graphitic carbons. For the same purpose, it is important to improve the adhesion of the active layer on the current collector and to reduce the collector width. This last measure leads to a typical "flat" shape component where the cell height has been minimized to reduce the electronic path in the conductors.

11.2   DOUBLE-LAYER CAPACITOR DESIGN KEY PARAMETERS

The first development step for a DLC manufacturer is to make a choice between the different technologies and the different design types available: For example, aqueous or organic electrolyte, prismatic stack or winding. The choice is mainly driven by the required performance, the manufacturability and the cost. In this paragraph, a survey of the main functions required from the different component parts is described. A focus on the commercial technology based on activated carbon and organic electrolyte is treated.

From a conceptual point of view, the DLC cell may be considered as an assembly of three main parts:

- Active part consisting of the electrodes and the separator. The electrode being made itself with the electroactive material deposited on a current collector
- Electrolyte
- Packaging

The DLC active part is made in most of the case of two identical electrodes. There is a spacer between the electrodes, called separator, whose function is to provide the electronic insulation between the electrodes, while ensuring the ionic conduction by allowing the ions to move through thanks to its porosity. The active part is impregnated with an electrolyte made of a solvent containing a disassociated salt and the unit's seated in a hermetic package [2].

11.2.1   ELECTRODE FUNCTION

The electrode is the key component of the DLC which determines its capacitance and partially determines its series resistance and self-discharge characteristics. Different technology options are available for the choice of the electrode:

- Carbon [3–5]
- Metal oxide [6,7]
- Solid polymer [8,9]

For carbon electrodes, there are two contributors to the DLC capacitance: the double-layer capacitance and the pseudocapacitance [10]. The latter is attributed to functional groups mainly located on carbon edges on the surface. The double-layer capacitance value is proportional to the accessible carbon surface. Today the stability over time under electrical and thermal stresses of the different capacitance contributions is not well established. It seems that the CO generating functional groups contrarily to CO_2 generating ones [11–14] have no capacitance contribution during thermoprogrammed desorption. To maximize the double-layer capacitance, engineers develop high-surface carbon with a morphology which provides good accessibility for the ions [15–17].

To optimize the device volumetric capacitance density, once the DLC geometric parameters such as the cell size, the electrode thickness, and width have been fixed, the development efforts must be concentrated on the research of the carbon performance. Typical commercial carbons [18] have a capacitance density in the range of $50\,F/cm^3$. Their capacitance specific density is in the range of $100\,F/g$. Among the best-performing carbons available, there are those derived from metal carbide (carbide derived carbon [CDC]) [19,20]. They may reach a capacitance density of $130–140\,F/g$. At that point, to avoid confusion, it is worth mentioning the difference between carbon or electrode capacitance and DLC capacitance. The later is exactly four times smaller because of the series connection of two electrodes whose volume is half of the total electrode volume.

The basic problem with activated carbon is that, intrinsically, it is a poor electrical conductor. Moreover, the use of small particles instead of a bulk crystal adds a contribution to the contact resistance. A binder must be mixed with the powder to stick the carbon particles together. The choice of binder material type and amount is influenced by the carbon surface properties.

In general, this "low" conductivity weakness is circumvented by coating a thin active carbon layer on an aluminum current collector which provides a much better conductivity by minimizing the charge path length in the carbon. The electrode performance is sensitive to the adhesion of the carbon layer in contact with the current collector surface [21]. It is important to ensure a good quality of this contact over time, even in presence of a solvent at elevated temperature, in order to maintain a low series resistance of the device.

The activated carbon conductivity may also be improved by adding some small amount of a more conductive carbon [22]. There are carbon categories, some of which are more conductive than others, for example, carbon black, graphite, or glassy carbon are generally better conductors than activated carbon. The origin and the proportion of the different products making the electrode, as high-surface resistive carbon, low-surface conductive carbon, and binder, are manufacturer's recipes disclosed in patents or kept secret.

Many investigations have been performed on carbon nanotube (CNT) use in DLC electrodes [23–25]. Today it is well established that CNTs offer a poor surface accessibility for the ions with a resulting low-capacitance density. Ongoing studies show promise in the use of CNTs in small proportion as an additional material to enhance the electronic conductivity and the mechanical properties of the electrode. The CNTs are also used as a support for high-capacitive polymeric redox material in order to increase their weak conductivity [26].

The electrode properties are not only determined by the carbon properties, but also determined by some geometric factors. They are the results of a compromise between the required energy and power densities. To get high power, the ionic and the electronic paths must be minimized in the system. The electrode thickness is made as small as possible to reduce the path length for the ions. The electrode width is made as short as possible to reduce the path length for the electrons. The direct consequence of these two measures is that, in a given volume, the number of parallel layers which are stacked or rolled is increased. It follows that the cross section for the current is increased while the distance to be crossed is decreased. These geometric consideration leads to low series resistance and to high power capabilities.

The carbon purity is one of the key properties, which must be maintained during all the production process as it will influence the potential window in which the DLC is stable during its use. The presence of impurities and possible oxidation–reduction reactions result in a pressure build up in the DLC cell. Excessive pressure may lead to the opening of the cell with the consequence of electrolyte leakage. The presence of impurities will also accelerate the self-discharge and the aging of the component. A corresponding capacitance decrease and a series resistance increase are also observed.

There are activated carbons of high purity available on the market. The general rule is that synthetic-based activated carbons have much lower impurity content than natural-based activated carbons. They are nevertheless not widely used because their prices are 4–10 times much more expensive.

11.2.2 Separator Function

The separator has two functions to fulfill in DLC. It must provide the

- Electronic insulation between the electrodes of opposite polarization
- Ionic conduction from one electrode to the other

The separator series resistance is proportional to the separator thickness which is in the range of 10–100 µm and inversely proportional to its area. The conductivity through the separator is proportional to its porosity (40%–70%) and is given by Equation 11.1.

$$\sigma = \sigma_0 p^\alpha \quad \text{with } 1.5 \leq \alpha \leq 2 \tag{11.1}$$

where
σ_0 is the electrolyte conductance
p is the porosity
α is a power factor [27]

Most of the separators available on the market are cellulose-based material or porous polypropylene. They are produced in the form of self-standing films. Some other special materials, as ceramic and polytetrafluorethylene (PTFE), have been investigated to overcome the limitations brought by high temperature in classical materials. The film thickness should be as small as possible. Its minimum size is nevertheless limited by the risk of electrical shortening failure due to the penetration of free carbon particles [28,29]. These later may create a contact between the two electrodes. The thickness limitation is also due to the reduced mechanical strength, which may lead to a tearing of the film during the winding process.

11.2.3 Electrolyte Function

The electrolyte provides the ions, which transport the electric charges between the electrodes of opposite polarities. The electrolyte choice leads to strategic decisions in the DLC design [30]. The three electrolyte families available are

- Ionic liquid or molten salt (may be mixed also with a solvent) [31–33]
- Salts in aqueous solvents
- Salts in organic solvents

The ionic liquid electrolytes are more viscous than the liquid electrolytes made with a solvent. The ionic conductivity in ionic liquid is strongly reduced by the limited mobility of the ions. The consequence is a series resistance which is at least 10 times larger. The capacitance is also reduced, typically to half of that of an electrolyte made of a salt dissolved in an organic solvent [14]. A first cause is the larger size of ionic liquid ions. They are unable to gain access to the full carbon area. A second cause is the higher viscosity which increases the migration time required by the ions. The use of ionic liquids may be interesting at very high temperature where the series resistance is acceptable and where the other solvent-based electrolytes would need to operate beyond the solvent boiling temperature. Even if the ionic liquid decomposition voltage is expected to be equal to about 2.5 V at anodic potential and −4 V at cathodic potential, it tends to decompose and polymerize after several hundred charge/discharge cycles at cell voltage around 2 V [34]. This behavior is probably caused by the other materials of the system which reduce the potential window width. To bring ionic liquids into commercial usage, it is still necessary to study the window potential width at higher temperature, where they could be useful and where the activation energy is smaller. The main limits for introducing ionic liquids in the industrial DLCs are their price and their conditioning. The price is at least 10 times more expensive than organic electrolyte. Maintaining the required high purity in the system during the production requires complex processes.

The choice between aqueous and organic electrolyte [35] depends on four parameters: the series resistance, the capacitance, the manufacturability, and the potential window size in which the system is electrochemically stable. These parameters have often opposite influence on the DLC performances.

For given current frequency and intensity measurement conditions, the carbon capacitance density in the aqueous electrolytes is about twice that in an organic solvent while the series resistance is at least 10 times smaller [14].

The aqueous electrolytes have the advantage of being able to be manipulated in atmospheric condition. The production process is simplified in the sense that the capacitive elements do not require any drying before the electrolyte impregnation. No protection against moisture contamination (glove box or dry room) must be undertaken during the impregnation. The main difficulties come from the sealing of the housing system when using acidic or basic liquids and in the cost of the current collector. Inexpensive aluminum does not tolerate these conditions.

These electrolytes are unfortunately strictly limited by the maximum operating value of 1.23 VDC, which corresponds to the decomposition potential of the water molecules to hydrogen and oxygen gas. Modern organic electrolytes may be operated up to voltages of 3 VDC at room temperature during thousands of hours, as it is the case for the acetonitrile-based BCAP0350 DLC [36]. To take advantage of the organic electrolyte's large potential window, it is necessary to eliminate the presence of water from the system. This is a serious drawback of these types of electrolyte.

The final choice is mainly driven by the solvent window potential size which, for the organic electrolyte, leads to a typical improvement of the energy and power density by five.

Within the class of organic electrolyte the discussion is focused today on the use or not of acetonitrile (AN, http://en.wikipedia.org/wiki/Acetonitrile). This solvent provides 10 times more ionic conductivity than propylene carbonate (PC) in the low-temperature range of the specification [37]. In 2004, AN has been declassified in the EU from toxic to harmful. The change is currently under inspection in the State. ISO14000 environmental standard do not prohibit any product as it is sometime mentioned. It requires a control and an improvement of the environmental and security situation.

The ionic conductivity is provided by the addition of a salt in the solvent. The most used salt is the tetraethyl ammonium tetrafluoroborate $(C_2H_5)_4NBF_4$ because of its small size. Each molecule disassociates into a positive cation $[(C_2H_5)_4N]^+$ and a negative anion $[BF_4]^-$. The electrolyte requirement is to provide the highest ionic conductivity in the full-specified temperature range. In this domain, it must neither freeze nor vaporize, the salt must not precipitate and the potential window must remain large enough to allow deep cycling. The salt concentration must be high enough to avoid ion depletion at "high" voltage [38]. A typical concentration for $(C_2H_5)_4NBF_4$ in AN is 1 M.

11.2.4 PACKAGING FUNCTION

The system composed by the electrodes and the separator, may be provided in the form of a spiral winding (see Figure 11.1) [39], also called jelly roll, or in the form of a stack [40]. In the radial construction, the two terminals for the two different polarities may be disposed on the same packaging face. In the axial construction, they are disposed on opposite faces. To obtain a DLC with a low series resistance, the connections between the electrodes and the DLC terminals must have a low resistance.

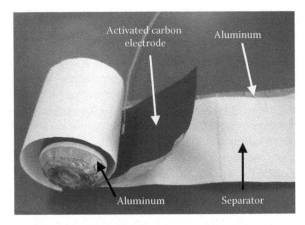

FIGURE 11.1 DLC active element: Spiral winding comprising carbon electrodes, aluminum collectors, and separators.

Different methods are used:

1. Aluminum "tabs" are inserted between superimposed active layers. The tabs are bands of aluminum, which have about the same thickness as the current collector and a width of about 3–10 mm. The tabs protrude either on each side of the winding in the axial construction or on the same side in the radial construction. The resistance is improved when they are welded to the current collector because it ensures a stable connection [41]. On the terminal side, they are screwed, riveted, or preferably, welded to the base of the terminal.
2. Aluminum collectors of each electrode protrude on each side of the jelly roll and permit the contact of a conductor on their surface [42,43]. This construction is made with electrodes which have a longitudinal side not covered with the active material. Both electrodes are shifted on each side in order to superimposed the area covered with the active material.

For both designs, the important point is to contact each collector layer. It is a major requirement in DLC manufacturing and it is the main difference when compared with battery construction which does not provide such low series resistance.

The packaging functions also keep the electrochemical system free of oxygen and water vapor, keep the electrolyte inside the cell, and provide two isolated conductive polarities for current transport. The electrochemical system must be protected from oxygen to avoid oxidation transformations which often produce gas. The choice of the container material type is driven by the electrochemical stability, the electrical conductivity, and the mechanical strength. The container must be able to sustain the internal pressure generated by any possible electrochemical decomposition. Generally, a weak mechanical point is designed on the container to allow a soft safe cell opening in the case of abusive applied conditions like overcharge. More sophisticated devices, as resealable valve or membrane with selective permeation capability, are used in some products, rather than a rupture location. The challenge for a DLC designer is to keep the container weight as low as possible so as not to affect the component performance. The problem is especially acute for small cells. The packaging factor, in the case of a BCAP0350 capacitor is 0.9; in other words, the container weight is only 10% of the total weight of the cell. The mechanical construction must also take into account that the cells must sustain mechanical shocks and vibrations defined by the different application specifications. These requirements are especially important in the transportation domain and for the windmill generators.

11.3 DOUBLE-LAYER CAPACITOR PROPERTIES, PERFORMANCES, AND CHARACTERIZATION

From an electrical point of view, the DLC is a complex system. It is built with the materials developed for both the battery domain (carbon,…) and the electrolytic capacitor domain (current collector,…). It has nevertheless capacitor like properties. Besides to its mechanical function, the aluminum collector contributes only to the electronic conduction improvement of the system. The separator function provides both the electrolyte ionic conduction and the electronic insulation between the two electrodes. In the carbon electrode, both the electronic conduction inside the carbon particle and the ionic conduction in the electrolyte around the carbon particles contribute to the electrical charge transport [44].

The characterization of DLCs is sensitive to the measurement method used. The IEC standard organization has issued the IEC 62391 series of directives [45–47] to settle a base for the testing conditions. Capacitor conditioning before the measurement is an important parameter which may lead to great divergences in results. DLCs need hours to reach an equilibrium state because some areas are very far or are difficult of access for the ions. It does mean, for example, that after a long polarization time, the measured capacitance and ESR values will be different to those found in the absence of a previous polarization. Moreover, this measurement effect will be mixed with the capacitance fading caused by the aging. Consequently, it is important to always precisely define the experimental conditions of the measurements.

11.3.1 Capacitance and Series Resistance Characterization in the Time Domain

There are different methods used for determining the capacitance and the series resistance in the time domain. The charging/discharging conditions may be either a constant current, a constant power, or a constant load. The first solution (see Figure 11.2) is the most widely used in the laboratories because it is the easiest and the cheapest. The measurement sequence starts at time t_0 by switching on a given current which is kept constant during the charge. The instantaneous voltage step observed, between U_0 and U_1, is caused by the ohmic current flow in the capacitor series resistance. The current is switched off at time t_2 when the capacitor has reached the nominal voltage, U_n. The voltage drops instantaneously from U_n to U_3 because of the disappearance of the current. When the capacitor is left floating between time t_3 and t_4, the voltage drops from U_3 to U_4 due to self-discharge and charge redistribution. The constant discharging current is switched on at time t_4 and switched off at time t_7 when the voltage has reached half of the nominal value $U_n/2$ or even lower values to respect particular specifications. The value of the current intensity is defined in the IEC 62391 standards and depends on the type of DLC (power, backup, etc,…).

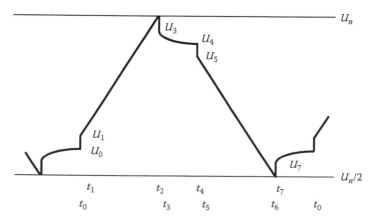

FIGURE 11.2 Charging and discharging voltage profile of a DLC when the current is constant. The profile is almost linear. When the current is switched on or off, a voltage step due to the current in the series resistance is observed.

The mean capacitance, C_n, is determined from the time necessary to discharge the capacitor between two voltage limits with a constant current, I, with the relation (Equation 11.2).

$$C = I \frac{\Delta t}{\Delta U} \tag{11.2}$$

The IEC 62391 standard, which has been written to be adapted to high series resistance DLC measurement, defines that the mean value should be calculated between 80% and 40% of the nominal voltage value.

The series resistance is calculated from the voltage drop which occurs at the nominal voltage between t_2 and t_4 during the current interruption (Equation 11.3). The first part of the drop between t_2 and t_3 is really caused by the series resistance. The second part between t_3 and t_4 is driven by charge recombination and self-discharge. If t_4 was equal to t_3, the drop would reflect exactly the value of the series resistance.

$$R_s = \frac{\Delta U_{23}}{I} \tag{11.3}$$

Of course, this value is difficult to obtain experimentally because of the band path width limitation of the measurement apparatus. The IEC 62391 standard defines the method used to measure the series resistance. The self-discharge due either to the leakage current or to the charge redistribution adds a contribution to the voltage drop corresponding to R_p, the parallel resistance in the model.

11.3.2 CAPACITANCE AND SERIES RESISTANCE CHARACTERIZATION IN THE FREQUENCY DOMAIN

Impedance spectroscopy has been extensively used to characterize carbon, electrode, and capacitor properties [27,48,49]. From the frequency impedance spectrum shape, it is possible to understand the physical origins of the observed characteristics, especially the different factors contributing to the series resistance.

The capacitance behavior is strongly dependent on the contributing surface position in the electrode [50]. If the considered surface lays deep into carbon micropores, the access time for the ions to this area will be longer and more "resistive" than for external carbon surfaces. In other words, it means that the deep internal areas are relatively inaccessible for short-electric impulses or for high-frequency currents. These deep surfaces contribute to the capacitance with a high RC time constant. As a result, the capacitance amplitude drops with increasing signal frequency.

The DLC series resistance drops also with increasing frequency. At high frequency, the deep carbon surfaces are no more accessible because the ions in the electrolyte have not the time to reach them. As a consequence, the ions will neither contribute to the current transport nor contribute to the ohmic losses. The de Levie transmission-line model [51] is an equivalent electrical model for the DLC which gives a simple explanation for the observed frequency behavior. The parallel string at the end of the transmission line, corresponding to the deep carbon micropore surfaces, has a high resistance which limits the current intensity through these paths. Finally as a rule of thumb, by comparing the datasheet values published by the DLC manufacturers, it is possible to state that the DC series resistance is approximately equal to twice the high-frequency series resistance.

11.3.3 CAPACITANCE AND ESR AS A FUNCTION OF THE VOLTAGE

In some DLC, especially in the case of organic electrolyte and "natural" carbon, the capacitance increases with the applied voltage. In the literature, the causes of this behavior are explained with different theories:

- Reduction of the solvent layer thickness caused by the increasing coulombic forces on the ions when the electric field in the double layer is increasing.
- Increase of the solvent dielectric constant caused by the compression of the solvent layer, if the concept of dielectric constant has still a meaning with such thin layers. The "Microscopic" Maxwell equations formalism would be more appropriate. ("Macroscopic" Maxwell's equations are applied to "macroscopic averages" of the fields, which vary wildly on a microscopic scale closed to individual atoms. It is only in this averaged sense that one can define quantities such as the permittivity, and permeability of a material, as well as the polarization and induction field).
- Increase of the electronic state density in the carbon pore walls with the voltage. Hahn et al. [52] have measured double-layer capacitance and electronic conductance of an activated carbon electrode in an aprotic electrolyte solution, $1 \, mol/dm^3$ $(C_2H_5)_4NBF_4$ in acetonitrile. Both quantities show a similar dependency on the electrode potential with distinct minima near the potential of zero charge. This correlation suggests that the capacitance, like the conductance, is governed substantially by the electronic properties of the solid, rather than by the ionic properties of the solution in the interface of the double layer.
- Salitra et al. [53] attributed this finding to the potential dependency of the ion penetration in nanopores, which was assumed to be minimal at the point of zero charge (pzc).

Figures 11.3 and 11.4 show the frequency spectra of 2600 F, 0.5 mΩ BCAP0010 DLC capacitance and series resistance for three different polarization voltages. It is interesting to observe the low-capacitance value when there is no voltage polarization. This phenomenon should be studied as a function of the electrode thickness. If ionic depletion was the cause, a thick electrode should display a more pronounce effect.

11.3.4 CAPACITANCE AND ESR AS A FUNCTION OF THE TEMPERATURE

In the high-frequency range ($f > 10 \, Hz$), the series resistance variation with temperature can be neglected. In the low-frequency range, the ESR increases when the temperature decreases [54]. This is caused by the electrolyte ionic resistance R_T which is strongly influenced by the temperature. Above 0°C R_T varies slowly with the temperature. Below 0°C, the temperature dependency is more

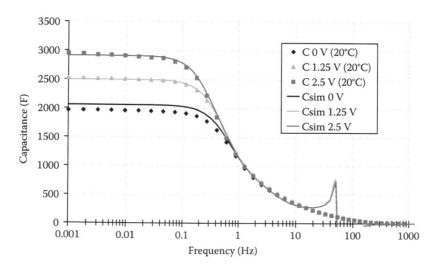

FIGURE 11.3 Maxwell Technologies BCAP0010 capacitance as a function of the frequency for three different values of the polarization voltage. The data have been obtained with an impedance spectrometer with a DC bias voltage option.

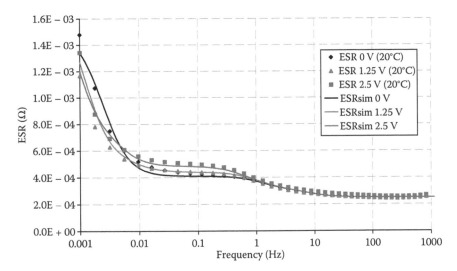

FIGURE 11.4 BCAP0010 series resistance as a function of the frequency for three different values of the polarization voltage in the same conditions as for Figure 11.3.

important, especially for the series resistance. It is caused by the rapid increase in electrolyte viscosity in the low-temperatures range [55].

In the case of $1 \, mol/dm^3$ $(C_2H_5)_4NBF_4$ in acetonitrile a relationship between R_T and the temperature has been established from experimental results. It is given by Equation 11.4:

$$R_T = R_{20} \frac{(1 + \exp(-k_T(T - T_{20})))}{2}$$ (11.4)

where
R_{20} is the resistance at 20°C
T is the surrounding temperature
k_T is the temperature coefficient $k_T = 0.025°C^{-1}$

At ultralow frequencies ($f < 0.1 \, Hz$), the capacitance is almost constant according to the temperature. It means that the ions penetrate deep in the electrode pores regardless of the temperature because they always have the time to reach the total surface. The capacitance is influenced by the temperature mainly in the intermediate frequency range between 0.1 and 10 Hz. This corresponds to the actual domain of the DLC working condition. In this case the ions have not the time to reach all the surface.

Increasing the temperature will have the main effect to reduce the electrolyte viscosity and to improve the accessibility of the surface for the ions. The ions, thanks to their higher mobility in the warm solvent, will be able to reach deeper carbon area in a shorter time. The increased accessible surface area results in a reduced DLC series resistance and in an increased capacitance with the temperature. Figures 11.5 and 11.6 show the measured and simulated frequency spectra of the BCAP0010 DLC capacitance and series resistance for different temperatures.

11.3.5 Self-Discharge and Leakage Current

The self-discharge is an important parameter for applications in which the DLCs are not connected to an electric network and therefore need to maintain their state of charge. In those applications, the device is supposed to be able to deliver power with a performance not deteriorated by the rest time.

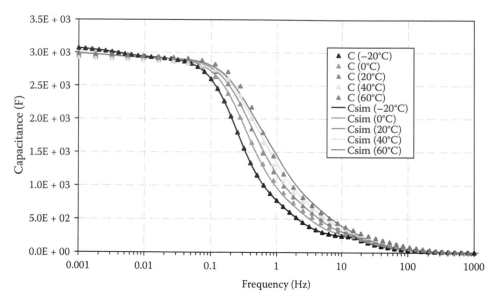

FIGURE 11.5 Maxwell Technologies BCAP0010 capacitance as a function of the frequency for five different temperatures.

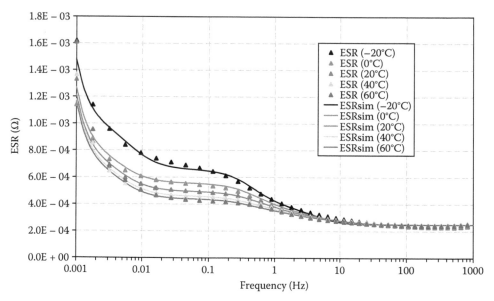

FIGURE 11.6 BCAP0010 series resistance as a function of the frequency for five different temperatures in the same conditions as for Figure 11.5.

A typical example is the starting of a car engine after a week spent at rest in an airport parking lot. In this case, it is necessary that the storage device maintains its voltage as high as possible because the available power and the stored energy decline with the square of this voltage.

The voltage drop of a charged capacitor in floating mode as a function of the time may be caused by different discharging mechanisms, which are the

- leakage current
- charge redistribution

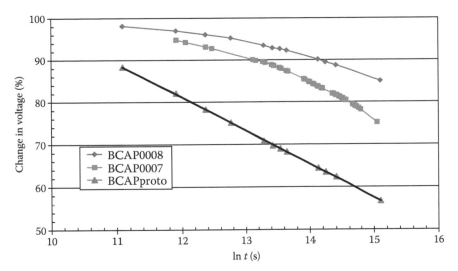

FIGURE 11.7 DLC self-discharge controlled by oxidation–reduction reactions shows a linear voltage fading when plotted in a time logarithmic scale. This is the case for the particular prototype BCAPproto.

The charge redistribution corresponds to the displacement of charges from an easy accessible area of the electrode with a short access time, to a more difficult one, with a longer access time. The charge redistribution, in contrast to the self-discharge, may lead to a voltage increase measured at the capacitor terminal. After a fast capacitor discharge, the voltage may increase because of the arrival of charges previously stored in "slow" areas. The leakage current may be due to unwanted oxidation–reduction reactions, ionic charge diffusion or/and electronic partial discharge through the separator. The self-discharge rate is determined, either by measuring the current necessary to keep a constant voltage (in the microamperes range) or by recording the floating capacitor voltage as a function of time.

The self-discharge diagnostic can be performed according to the voltage time dependency. If the voltage drops with an unwanted logarithmic law, $U(t)$ vs. $\log t$ (see Figure 11.7), the mechanisms are controlled by Faradic self-discharge from overcharge or from impurities oxidation–reduction reactions. If the voltage drops following a square root function of the time, $U(t)$ vs. $t^{1/2}$ (see Figure 11.8), the mechanisms are controlled by diffusion processes.

These self-discharge behaviors have been demonstrated with different capacitor samples. BCAP0007 and BCAP0008 are Maxwell commercial products. They show only a diffusion-driven self-discharge mechanism. BCAPproto which is a prototype with a high impurity content undergoes oxidation–reduction reactions. The two different plots show the respective linear drop in their respective representation.

The DLC self-discharge performance is the result of a compromise with its power capability. The manufacturers could use a thicker separator to improve the voltage retention but this operation would increase in the same time the series ionic resistance.

11.3.6 STATISTICAL DISPERSION OF CAPACITANCE, ESR, AND PARALLEL RESISTANCE VALUES

A DLC module performance is directly affected by the variation amplitude in both the cell capacitance and parallel resistance values. In steady-state voltage applications, the voltage repartition on the cells in the series connection is driven by the parallel resistances. Cells which undergo a greater leakage current discharge faster. In transient voltage applications, the voltage repartition is driven by the cell capacitances. Because the current in the series connected DLC is the same, a small capacitance cell is charged or discharged much faster than a larger one.

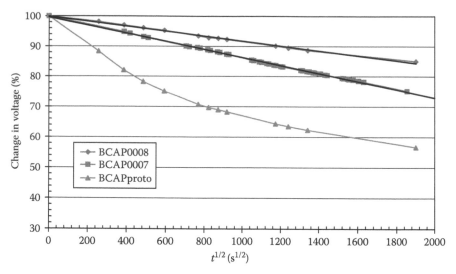

FIGURE 11.8 DLC self-discharge controlled by diffusion shows a linear voltage fading when plotted in a time square root scale. This is the case for standard commercial DLC.

TABLE 11.1

Capacitance, Series Resistance, and Self-Discharge BCAP0350 Production Dispersion

Parameter	Value	σ	Minimum Value	Maximum Value
C (F)	384	3.5	373.5	394.5
ESR at 0.6 VDC (mΩ)	2.27	0.07	2.06	2.48
Self-discharge (VDC)	0.074	0.009	0.047	0.101

The parameters which affect the spread of both capacitance and parallel resistance values are material and process variability, temperature gradient in the module, aging state, etc.

The particular production case of the DLC-type BCAP0350E250 from Maxwell Technologies has been analyzed. Table 11.1 shows the data of a production sample of 2800 cells. The capacitance data are collected with a 10 A charge between 0 and 0.6 VDC. The series resistance is measured during the current interruption at 0.6 VDC. The cell datasheet gives a nominal capacitance of 350 F (± 20% or 280/420 F) and an ESR of 3.2 mΩ (± 25% or 2.4/4.0 mΩ).

The 3.5 F measured standard deviation affecting the cell capacitance is small. It represents less than 1% of the cell nominal capacitance and means that 99% of the production is included in a capacitance range of ± 3%. The spread on the parallel resistance measured indirectly with the self-discharge after 24 h at room temperature, starting from a voltage of 0.6 VDC (for production speed reason), is more important. It is in the range of 12%. This routine measurement is nevertheless not significant because the leakage current stabilizes a lot with temperature and polarization time. After 72 h of polarization, the leakage current is 100 times lesser than in the beginning.

To obtain a module with low cell voltage variations, the use of external circuitry is required. These are the so-called "balancing circuit" and are connected across each cell in the stack. Their main role is to reduce the voltage spread and to get the cell voltages equally distributed in the stack. Several balancing systems have been described in the literature. They can be classified in two groups:

- Passive systems are those which bypass a portion of the current flow, across all the cells, through resistors or diodes connected in parallel with the cells
- Active systems control the individual cell voltages and react to maintain the cells in a safe-controlled situation

Most of the active balancing devices, specifically developed for the DLC, turn out to be a subgroup of the passive systems. The cell voltages are monitored by a survey equipment. When one of these cells reaches or overcomes a preset threshold level, a bypass is activated by an interrupter (transistor). This opens a path for the current which may flow in a dissipative resistor [56,57]. This method limits the self-discharge amplitude to a minimum value because the system is working only when necessary.

More sophisticated systems are able to remove charges from cells which have a higher voltage than the others and to bring them to cells which have a lower voltage in the series connection [58]. Among the main converter topologies, the bidirectional nondissipative buck-boost converter [59,60] and the flyback converter [61,62] are the most studied. Active systems appear to be sophisticated circuits and thus may be expensive in relation to the cost of the cell they are intended to protect.

For example, with a 60% spread of the initial leakage resistance, the use of a passive balancing circuit, with resistor 15 times smaller than the averaged leakage resistance, will result in a voltage spread of 2.4%. This value can be used to estimate the number of cells required for a stack for a given voltage of 450 VDC. Only 185 cells are required with the balancing resistor instead of 288 without. Increasing the improvement factor implies the use of lower resistance values. On one hand there is a drawback: the self-discharge rate increases proportionally to the ESR reduction, and on the other hand there is an advantage: the transient time constant is drastically reduced so the balancing system becomes faster. The designer will have to make a compromise. The use of active balancer circuits allows the use of higher current values, thus improving the general performances of the system. The main benefit of such circuits is the ability to switch the system in a standby mode where almost no equalizing current flows. In this case, the self-discharge rate is improved several orders of magnitude.

It has also been shown that a module in a vehicle is able to work for years without any on-board balancing. The principle is to equalize the maximum voltage of the DLCs in the module from time-to-time with an external device [63,64]. The cells must have a relatively uniform self-discharge performance for this method to be successful. It is nevertheless necessary to have an on-board surveillance system to avoid problems caused by unexpected failures.

11.4   DOUBLE-LAYER CAPACITOR ELECTRICAL EQUIVALENT MODEL

The DLC is a physical component which has not only a requested capacitance, but also an unavoidable parasitic inductance due to its geometry, a series resistance caused by the electronic and ionic conductor ohmic resistances and a parallel resistance due to the leakage current between the electrodes.

The ESR is a combination of the series resistance, R_s, and the parallel resistance, R_p, (see Figure 11.9). It is responsible for the electric losses which generate the internal heating.

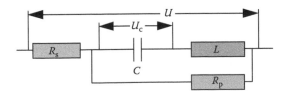

FIGURE 11.9   Basic capacitor electrical equivalent circuit comprising a capacitance, a series inductance, a series resistance, and a parallel resistance. This simple model can fit a DLC behavior in first approximation for a given frequency.

To work with high power, it is necessary to have a low ESR to limit the ohmic losses. The parallel resistance, R_p, has an effect visible only at ultralow frequency (below the millihertz range). It is responsible for the capacitor self-discharge. Its value must be as high as possible to limit the leakage current. The time constant τ of the self-discharge is equal to $\tau = R_pC$.

The transmission-line basic model used to describe the frequency behavior of both the capacitance and the series resistance has been originally proposed by de Levie [51,65]. This theory does not take into account the voltage and temperature dependencies of the capacitance and series resistance. A simple model which considers an additional linear dependency of the capacitance on the voltage has been proposed by Zubieta et al. [66]. Similar models have been used also by Dougal et al. [67] and Belhachemi et al. [68]. The capacitance is composed by a constant part C_0 and a linear voltage dependent one $C_v = K_vU$, where K_v is a coefficient which depends on the technology. The total capacitance at the voltage U is given by $C = C_0 + C_v$.

The relation between the current and the voltage must be derived from the relation between the current and the charge (Equation 11.5) which remains always true.

$$i = \frac{dQ}{dt} \qquad (11.5)$$

Substituting the Q expression as a function of U and C and considering the indirect dependency of C with the time, it is easy to show [69] that the current is given by Equation 11.6.

$$i(t) = (C_0 + 2K_vU)\frac{dU}{dt}, \qquad (11.6)$$

By analogy with the classical relation, one may define a differential DLC capacitance (Equation 11.7) as

$$C_{\text{diff}} = C_0 + 2K_vU \qquad (11.7)$$

The stored energy is equal to the time integral of the power that leads to the relation (Equation 11.8):

$$E(U) = \left(C_0 + \frac{4}{3}K|U|\right) * \frac{U^2}{2} \qquad (11.8)$$

In conclusion, the current and the energy for a given voltage are larger than they were expected based on the classical expressions for a constant capacitance.

The capacitance and the series resistance have values which are not constant over the frequency spectrum. The performances may be determined with an impedance spectrum analyzer [70]. To take into account the voltage, the temperature, and the frequency dependencies, a simple equivalent electrical circuit has been developed (Figure 11.10). It is a combination of de Levie frequency model and Zubieta voltage model with the addition of a function to consider the temperature dependency.

The equivalent scheme is composed of a fixed capacitance C_0 connected in parallel with a variable capacitance C_v. C_v increases linearly with the voltage. The series resistance and capacitance voltage dependencies are active only in the low-frequency domain which may be taken into account with the simple R_vC_v circuit. To get the frequency dependency, the resistance R_T and the capacitance C_R have been introduced in circuit 2 (Figure 11.10). Their behavior is the one of a low-pass filter with a cut-off frequency $\tau = R_TC_R$. "Circuit 1" has been added to take into account the thermal dependency. Circuit 3 shown in Figure 11.10 is required to complete the modeling regarding the leakage current and the charge redistribution. It includes two RC parallel branches with two different time constants. It also includes a parallel R_F resistance which gives the longtime leakage current.

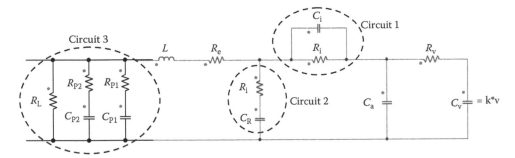

FIGURE 11.10 DLC equivalent circuit for capacitance and series resistance dependencies as a function of frequency, voltage, and temperature.

Basically the available capacitance is maximum at low frequency. This may be explained with the longer time available for the ions in the electrolyte to reach the surfaces which are located deep in the carbon pores. At higher frequency, only the superficial carbon surface is accessible for the ions. The capacitance is consequently much smaller.

The series resistance is composed of an electronic and an ionic part. The electronic contribution comes from the ohmic resistance in the conductor and in the carbon particles. The ionic contribution comes from the ions mobility in the electrolyte.

11.5 METHODOLOGY FOR DLC MODULE SIZING

The optimal sizing of the electrochemical storage system needs some iterative calculations because the actual available energy for the application load depends on the current intensity. The first step is to evaluate the application energy requirement with a small current and to calculate the corresponding storage device volume. Then the losses are calculated based on the power requirement. To compensate for this energy loss, an additional volume is introduced to the storage device. During each sizing step of the DLC, special care is dedicated to possible temperature increase. In order to maintain the mechanical stresses inside the component in an acceptable range, a temperature elevation threshold must not be exceeded.

11.5.1 Energy Requirement Consideration

First, it is necessary to evaluate the energy required by the load which must be supplied. Once these values are known, the efficiency must be evaluated and the storage system resized with a supply correction factor. The performances must be available during the component lifetime, even when it reaches its end of life (e.g., when the capacitance has dropped by 20% or the series resistance has increased by 100%). This introduces an aging sizing factor to compensate for the capacitance drop. To get an ideal sizing of the device, it is important to have a good knowledge of the application voltage and temperature time distribution during the component life. High temperatures and high voltages outside the potential window stability range accelerate the DLC aging mechanisms.

From the user view, a reasonable voltage range of working for the DLCs is comprised between the nominal voltage and half of this value. In this voltage, range ¾ of the energy content is available. To recover, the energy stored at lower voltage would be detrimental to the efficiency.

11.5.2 Efficiency Consideration

A point to consider in the calculation of the storage device sizing is of course the efficiency of the electric circuit which connects the storage system with the load or the supply. The losses in this circuitry must be compensated by an increased storage capacity. The choice of the converter technology is a key factor for getting a high efficiency from the system.

11.5.3 Voltage Requirement Consideration

This part is mainly driven by the power-electronic requirements. Basically the tendency is to maximize the working voltage to reduce the conductor weights and to increase the electronic conversion efficiency. As the voltage on the DLC module varies according to the state of charge, there is a need to integrate an inverter in the system in almost all the applications.

The voltage repartition on the cells must be equalized to maintain the safety of operation. This function has an energy consumption which must be considered in the storage sizing.

Each cell must not overcome the determined operating voltage which allows the system to reach the application lifetime requirement. To fix the voltage, which may be applied on each cell, it is necessary to consider the derating law presented before in this chapter. Moreover, it is often necessary to consider the voltage distribution of the different application stresses defined in the customer specification to optimize the storage system size.

11.5.4 Power Requirement Consideration

Most of the power applications may be classified in three categories as a function of their behavior during the electrical discharge:

*   Constant power: Voltage decay acceleration and current increase [71]
*   Constant current: Quasiconstant decay of the voltage
*   Constant load: Exponential decay of the voltage and of the current

In the case of a constant power discharge, the sizing of the conductors must be done considering the low-voltage situation. In the case of a discharge in a constant load, the power and the current are maximum at the beginning of the discharge, when the potential is still high. Generally, the final limitation is given by the heat dissipated in the DLC. As it will be shown, the higher the current, the higher will be the losses inside the component.

The capacitor voltage, U_t, can be measured only across the capacitance and the series resistance taken together. At time zero, there is an instantaneous voltage drop due to the current in the series resistance, $R_s\ i(0)$. The discharging process across a load resistance, R, has then a time constant (Equation 11.9).

$$\tau = (R + R_S)C \tag{11.9}$$

The available power is time dependent. The initial discharge instantaneous power, P_0, is given by Equation 11.10.

$$P_0 = \frac{U_0^2}{R}\left(\frac{R}{R+R_S}\right)^2 \tag{11.10}$$

The mean power available during the period τ is given by Equation 11.11.

$$P_\tau = \frac{1}{2}P_0 \tag{11.11}$$

A part of the energy content is burnt in the DLC internal series resistance instead of being used in the load. This amount depends on the current intensity and is given by the "Ragone plot" which is a relation between the available energy density and the power density at the nominal voltage. The higher the current is, the more energy is dissipated inside the capacitor.

At matched impedance, half of the energy is lost inside the capacitor in its internal resistance. This occurrence is represented by the position of the last point on the right in the Ragone plot. From

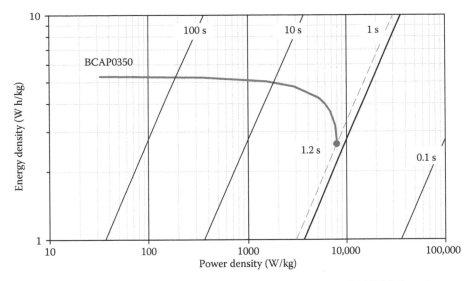

FIGURE 11.11 Ragone diagram based on the Maxwell Technologies BCAP0350 DLC performances. The chart shows that for high power, a great part of the energy is dissipated in the component and is no more available to supply the system. At matched impedance, when the load resistance equals the series resistance, exactly half of the stored energy is dissipated internally in the DLC.

the 5.2 W h/kg total energy stored in the capacitor, only 2.6 W h/kg are available for supplying the load request. The corresponding circuit charging/discharging time constant is equal to $\tau = 2\,R_S C$, 1.2 s in the case of our example (see Figure 11.11).

11.5.5 THERMAL AGING CONSIDERATION

The last point to consider in the sizing of the storage system is the lifetime reduction caused by the temperature elevation induced by the power losses. The severity of the problem is proportional to the storage system volume optimization. It is often necessary to increase the storage system size to reduce the temperature elevation in an acceptable range. In any case, it is important to consider the hot spot of the storage system, usually in its center, for determining its lifetime. If a resizing is operated at this point, it is necessary to reconsider the previous points where the DLC efficiency is calculated.

11.6 DOUBLE-LAYER CAPACITOR THERMAL PROPERTIES

In a majority of applications, DLCs are charged and discharged at high current rate. This operating mode produces a big amount of heat inside the DLC cell which limits its performances. In the particular case of DLCs made with an activated carbon and an organic electrolyte, the manufacturers specify an operating temperature range limited between −40°C and 65°C. The temperature has a reversible effect on DLC properties: the ESR is decreasing, while the capacitance and the self-discharge are increasing with the increasing temperature. In contrast, the temperature has also an irreversible effect on the DLC: the lifetime expectancy is reduced with increasing temperature. It is therefore a goal to remove or at least minimize the heat sources. This is achieved by selecting high-purity activated carbon, low-resistance electronic conductors, and high-conductive electrolyte.

It is known that the DLC heat management problems depend on the power solicitation and on the thermal environment of the device [72]. Therefore, an optimum heat loss can be achieved by maximizing the surface-to-volume ratio. This condition is in opposition with the needs of high energy density and high current levels. The resulting component heating is due to the balance between the Joule losses heat generated by the electrical current flowing through the ESR and the component thermal dissipation capability.

The studied BCAP0350 DLC has a D-cell battery shape factor which is defined in the standard with a 33 mm outside diameter and a 61.5 mm length. The total external surface is about 80 cm². The production of losses inside the DLC is assumed to be uniform in the volume. In the case of a 30 A charge/discharge current the dissipated power is equal to 2.88 W. The measurements have been performed at room temperature $T_a = 20°C$ which was constant during the experiment. The DLC is only cooled with a slowly moving airflow due to the natural convection.

The temperature distribution is determined by solving the well-known heat equation

$$\nabla^2 T + \frac{P}{\lambda} = \frac{\rho C_p}{\lambda} \frac{\partial T}{\partial t} \tag{11.12}$$

where
    T is the temperature
    t is the time
    ρ is the mass density
    λ is the thermal conductivity
    C_p is the heat specific capacity
    P is the local power volumetric density [73]

In the described cylindrical topology (see Figure 11.12), the modeling of the DLC internal construction has given a heat transfer coefficient of 12 W/m²/K. The radial thermal conductivity is equal to $\lambda = 0.5$ W/m/K and is much smaller than the axial thermal conductivity which is equal to 210 W/m/K.

The succession of conductive and insulating layers in the radial direction constitutes a thermal barrier. The initial temperature of the DLC is assumed to be uniform and equal to 20°C. When the charging current is switched on the DLC temperature increases exponentially over time (see Figure 11.13).

In this model, the thermal resistance of each layer is assumed to be equivalent to the thermal resistance of a long cylinder of inside radius r_i, outside radius r_o, and length L. For each layer, the length is considered as much larger than the radius. The thermal resistance may be written as

$$R_{th} = \frac{\ln(r_o/r_i)}{2\pi\lambda L} \tag{11.13}$$

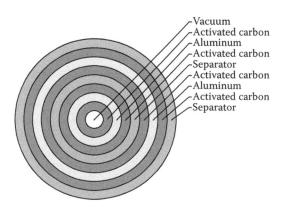

FIGURE 11.12   Cross section of a simplified DLC construction used for thermal modeling.

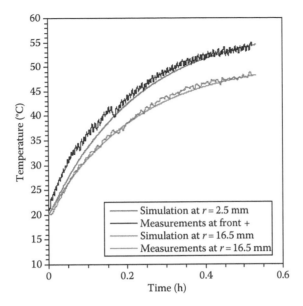

FIGURE 11.13 Measured and simulated BCAP0350 internal temperatures closed to the center and closed to the can. The charge and discharge currents were constant and equal to 30 A.

The thermal resistance concept may be used for multiple cylindrical layers. For four layers (Figure 11.12), the equivalent thermal resistance along the radial direction is given by

$$R_{th} = \frac{1}{2\pi\lambda_c L}\ln\left(\frac{r_i + e_c}{r_i}\right) + \frac{1}{2\pi\lambda_s L}\ln\left(\frac{r_i + 2e_c + e_a + e_s}{r_i + 2e_c + e_a}\right) + \frac{l}{2\pi\lambda_c L}\ln\left(\frac{r_i + 2e_c + e_a}{r_i + e_c + e_a}\right)$$
$$+ \frac{1}{2\pi\lambda_a L}\ln\left(\frac{r_i + e_c + e_a}{r_i + e_c}\right) \tag{11.14}$$

where
   e_c is the activated carbon thickness
   e_a is the aluminum thickness
   e_s is the separator thickness
   λ_c is the activated carbon conductivity
   λ_a is the aluminum thermal conductivity
   λ_s is the separator thermal conductivity

It is assumed that $e_a \ll e_c$ and $e_s \ll e_c$. With these conditions, the equivalent thermal resistance is approximatively equal to the thermal resistance of the activated carbon. Therefore, the equivalent thermal conductivity along the radial direction is considered as equal to the activated carbon conductivity ($\lambda_r \sim \lambda_c$). Along the axial direction, the thermal conductivity, λ_z, is assumed to be the same as the aluminum conductivity. This condition is deduced from the electrical analog used to represent the heat flow inside the DLC by the parallel thermal resistances as follows:

$$\frac{1}{R_{th}} = \sum_{n=1}^{N} \frac{1}{R_{th,n}} \tag{11.15}$$

where
   $R_{th,n}$ is the thermal resistance of the layer number "n"
   N is the total number of the layers (carbon, aluminum, and separator)

TABLE 11.2
BCAP0350 Thermal Resistances

I (A)	R_{th} (K/W)
10	10.670
15	10.734
20	10.755
25	10.764
30	10.665
35	10.775
40	10.777

The experimental and the predicted values (Figure 11.13) are in good agreement. The small difference between the measurement results and the predicted values for $r = 2.5$ mm is a consequence of the assumption that the measured temperature at the DLC positive pole is equal to the temperature inside the DLC.

Using the developed model, the thermal resistance of the BCAP0350 F is calculated for different DLC charge and discharge current values.

Table 11.2 gives the BCAP0350 DLC thermal resistances when the device is charged and discharged, with 10, 15, 20, 25, 30, 35, and 40 A. These results show that there is a small difference between the measured values: the difference is less than 1%.

In most of the industrial applications, several DLCs are connected either in series or in series/parallel. They are generally subjected to very high currents. Consequently, the heat produced by Joule effect must be dissipated with cooling systems like fans or air distribution channels. The choice of the cooling system depends on the level of the heat transfer coefficient and the maximum allowed operating temperature. The chosen cooling system should be sufficient to keep the DLC temperature at a tolerable temperature level which leads to a longer lifetime.

11.7 DOUBLE-LAYER CAPACITOR LIFETIME EXPECTANCY

A "failure" is defined as the lack of ability of a component to perform its intended function as designed. A failure may be the result of one or many faults. "Reliability" is the ability of an entity to perform under "normal" conditions. "Safety" is the ability of an entity to perform under "abnormal" conditions and "security" is the ability of an entity to perform in the presence of "malevolent" environment.

"Early failures" and "wear out failures" are reflected in the curve known as the "bathtub" curve (see Figure 11.14).

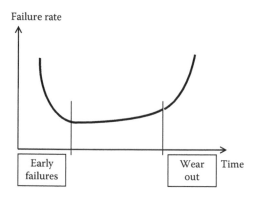

FIGURE 11.14 Failure bathtub curve statistic. The early failure drop is generally canceled with design, type, and routine tests.

Design tests during the conception, type tests before the launching of the product, and routine tests after the production are performed by the manufacturers. They are intended to eliminate the component early failures and to provide a high reliability during the given lifetime. The lifetime expectancy estimation presented below concerns the statistical failures of the components considering only the wear out failures.

The Survivor function, $F(t)$, is the number of elements of the statistical sample which have not failed or lost their function at time t and are still working. In the case of a Weibull statistic model, the survivor function is given by the relation:

$$F(t) = \exp^{-(\lambda_o t)^p} \qquad (11.16)$$

where
   $1/\lambda_o$ is the characteristic time corresponding to the loss of 63.3% of the initial quantity of samples
   p is a factor which governs the "slope" of the failure process

The failure rate, $\lambda(t)$, is not the number of failures observed in a unit of time; but the ratio of the failures per unit of time over the number of survivors.

$$\lambda(t) = \lambda_o p(\lambda_o t)^{p-1} \qquad (11.17)$$

The failure rate is given in failure in time (FIT) which is the number of failures occurring during 10^9 h of working of one object. $\lambda(t)$ and λ_o must not be confused. The latter is a constant (independent of time, but dependent on the temperature and voltage) which corresponds to the inverse of the time necessary for 63% of the sample to fail. $\lambda(t)$ is the inverse of the mean time between failure (MTBF). In the wear out region $\lambda(t)$ is increasing with the time. The manufacturer's specifications give the maximum value of $\lambda(t)$ within the announced lifetime: 50 FIT and 10 years of lifetime expectancy.

The DLC failure modes may be, for example:

1. *Cell container opening due to an internal overpressure* [74]. The voltage and the temperature generate a gas pressure inside the cell which increases with the working time. When the pressure reaches a determined limit, a mechanical fuse, generally a groove on the can wall or a pressure relief, may open softly, avoiding an explosion of the device.
2. *More than 20% loss of capacitance.* The accessible carbon surface and the ions availability are reduced during the electrochemical cycling.
3. *More than 100% increase of the ESR.* The electrode adhesion on the collector is weakening with time and temperature. The ion availability is reduced.

In these examples, the given numerical values (20% for the capacitance and 100% for the series resistance) are subject to be adapted for other particular applications, for which requirements may be different.

To present the observed statistic of the cell opening, a special "Weibull" scale (see Figure 11.15) is used to verify a straight line in the case of Weibull failure process.

The steeper the slope is with large value of p, the more the process is under control. In that case almost all the failures occur during a short-time period just before the given lifetime expectancy limit. The measurements presented in Figure 11.15 have been performed on 14 DLCs pieces BCAP0140 (140 F, 7 mΩ) at 70°C, with an applied DC voltage of 2.85 V. In this example, the failure of one piece represents a loss of 7.1% of the tested samples. The first point in the bottom of the curve is given by the opening of the first cell after 250 h.

To get an estimation of the time required to reach 20% of electrode capacitance loss, the coefficients λ_o and p must be determined for all the operating temperatures and for all the operating

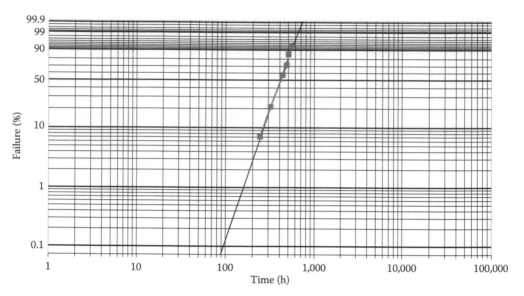

FIGURE 11.15 DLC Weibull overpressure failure statistic during a 2.85 VDC voltage solicitation at 70°C. Power factor $p = 4.5$ and mean lifetime $1/\lambda_o = 440$ h.

voltages. They are measured for a set of discrete values, and then, the coefficients λ_o and p for the other temperatures and voltages are calculated using the following "derating" relationship:

$$\frac{t_1}{t_2} = \left(\frac{V_2}{V_1}\right)^n \exp\left[\frac{E_a}{k}\left(\frac{1}{T_{1_{abs}}} - \frac{1}{T_{2_{abs}}}\right)\right] \tag{11.18}$$

where
  t_1 is the lifetime at the temperature T_1 and voltage V_1
  t_2 is the lifetime at the temperature T_2 and voltage V_2
  E_a is the activation energy determined by the experimental data
  k is the Boltzmann constant
  n is a constant determined experimentally

An extensive study has been performed to demonstrate a general approach to assess electrochemical capacitor reliability as a function of operating conditions on commercial capacitor cells [75,76]. For the temperature dependency an Arrhenius law is used, whereas for the voltage dependency an inverse power law is used. Some electronic apparatus concepts are already available to estimate *in situ* the DLC residual life by monitoring the temperature and voltage constraints of the application [77]. DLC capacitance lifetime expectancies are displayed in Figure 11.16 as a function of the temperature for different values of the applied DC voltage.

It is noteworthy that when the voltage is close to the electrochemical decomposition voltage there is an acceleration of the degradation phenomena, especially in the higher temperature domain.

The challenge in most industrial applications is to size the storage system as small as possible, both for cost-saving and available volume reasons. To optimize the dimensions of the components and setting the reserve to a minimum value, all the available field data, which allow fitting precisely the actual application requirements, must be considered. Practically the duration, temperature, and voltage solicitation stresses must be individually estimated. An equivalent stress weight is calculated for each contribution, based on the "derating" Equation 11.18 given above. The lifetime expectation is evaluated by summing all the estimated contributions.

It is also interesting to determine not only the time necessary to get a capacitance drop of 20%, but also the curve shape of capacitance drop and series resistance increase as a function of the time.

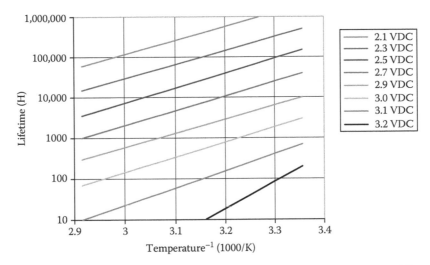

FIGURE 11.16 DLC capacitance lifetime expectation as a function of the temperature for different voltages. The DLC is considered as failed when its capacitance drops below 80% of its nominal value.

The experimental work (see Figure 11.17) shows that three phases may be distinguished on the capacitance drop curve.

A first phase which lasts between 1 and 12 h, depending on the temperature and the voltage, is in fact a measurement effect. It is caused by the charging of the electrode areas which have no counter electrode in their neighborhood. The ions need a long time to reach these distant areas. This leads to a discharge of the charges initially located on the paired electrode toward the unpaired electrode areas.

The second capacitance fading process is exponential. In the voltage and temperature conditions of Figure 11.17, the time constant is in the range of 2000 h. This part of the process may be attributed to a potential window shift combined with the aging due to the construction asymmetry. The third one occurs after an approximate drop of the capacitance of 15%–20%. It appears as linear, but may also be exponential with a long time constant. This is not possible to find out in the limited experimental time. The attribution of the second and third processes to a physical phenomenon is not well established. An attempt to correlate the exponential decay with leakage current measurement has been presented in the literature [54].

The increase of the ESR is linear with the time. The values measured during the aging experiment at 50°C and 2.5 V are aligned on a line whose slope is equal to 7.5×10^{-4} [%/h].

The physical origin of the aging is not well established. It is attributed to different phenomena as the oxidation of the carbon surface, the closing of the pores access, or/and to the ionic depletion in the electrode [78]. When a DLC is opened, after an aging period under large stress, the oxidation of the separator may be observed. A brown coloration appears on the surface, especially on the side exposed to the positive electrode. The electrolyte undergoes irreversible transformations which are accentuated with voltage and temperature. The electrochemical decomposition of the electrolyte generates a gas overpressure in the DLC package (e.g., generation of H_2 in the case of acetonitrile and CO_2 in the case of polypropylene carbonate). Charging and discharging creates mechanical stresses in the electrode. It has been shown that the application of a voltage induces a reversible expansion of the electrode [79,80]. This mechanical motion, especially in the case of ionic insertion in the electrode, is known to be one of the origins of aging in the battery domain.

The measurements done to perform an aging characterization are sensitive to their polarization history. In particular, the first measurement, used as the reference initial value, is obtained starting from a nonpolarized state. In contrast, all the other subsequent values are obtained starting the measurement from a nominal voltage polarization state.

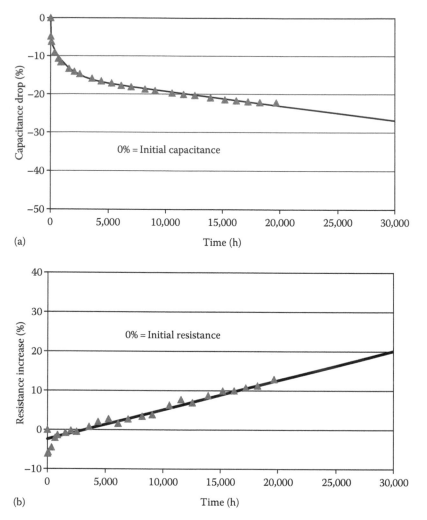

FIGURE 11.17 BCAP0350 lifetime experiment. The capacitance and series resistance are recorded as a function of the duration of a constant voltage polarization of 2.5 VDC and a constant temperature of 50°C.

11.8 APPLICATIONS OF DOUBLE-LAYER CAPACITORS

There are basically two main application categories for which the DLCs are interesting:

1. High-power applications for which batteries have a too high series resistance.
2. Maintenance-free applications which cannot be provided by the batteries.

Due to their high power density, the DLCs are destined for existing and new applications in automotive engineering, railway traction, telecommunication systems, industrial, medical, and consumer electronics. The following examples give an overview of the different possibilities.

11.8.1 AUTOMOTIVE DOMAIN

In the automotive domain, the DLC applications are classified in four categories:

- On-board electrical systems: Electromagnetic valve control, catalysts preheating, brake actuators, and steering
- Microhybrid: Integrated starter-generator, electrohydraulic, or mechanical braking

- Mild hybrid: Energy storage for the traction assistance
- Strong hybrid: Energy storage for the traction

A brief overview of application examples is presented in the following paragraphs.

11.8.1.1   Combustion Engine Starting

Enhanced starting of automobile engines is an attractive application for DLCs [81,82]. Today the energy required to crank an internal combustion engine (ICE) is stored in batteries, either Pb or Ni–Cd. Because of their high internal resistance, which limits the initial peak current, they have to be oversized. The fast battery discharging and the cold environmental temperature strongly affect battery performance. The principal advantage of the DLC compared to the battery in this application is that it can support several tens of thousands of charge/discharge cycles with very high currents. The battery cyclability is limited to only a few hundreds of cycles with currents much weaker than for DLCs. In addition, for the starting of a thermal engine, the battery is discharged with high currents. This obliges the manufacturers to install expensive high-power batteries.

Two demonstration cases of an ICE starting have been done with DLCs module with a diesel engine assembled on a test bench. In this application, an additional low-power battery is sufficient because it is only used for the initial charge of the DLCs which may be performed with low current. Once the engine is launched, the alternator of the vehicle charges the DLCs. It thus results in a reduction of the size of the battery and a greater longevity of operation.

The first example concerns the start of a standard vehicle where the battery is removed and replaced by a DLC module. This later is composed by six DLC cells of 2600 F connected in series. The module has the following characteristics:

- Total capacitance: 433 F
- ESR: 3.6 mΩ
- Maximal voltage: 15 V
- Weight: 3.15 kg
- Volume: 3 dm^3

Figure 11.18 shows the experimental results obtained when a standard ICE is started using a DLC module. In the beginning of the operation, the DLC module is charged at a voltage of 11 VDC. When the ICE starts, the current demand rises to 220 A, after which, the current of the vehicle alternator is used to charge the DLCs module until 13 VDC. The ICE is started three times in this example.

The second example shows a diesel engine assembled on a test bench in the laboratory. The characteristics of the engine used in the experiment are as follows: Peugeot 604 turbo diesel engine of 105 kW, resistive torque to the starting of 32 daN m, and minimal speed of launching of 130 rpm. In this configuration, the DLC module consists of a four-series connection of 2600 F cells and has a nominal voltage of 10 VDC. This module can start the engine without any problem for more than nine times. Figure 11.19 shows the evolution of both the current and voltage of the module. The voltage drop after each starting varies between 0.2 and 0.75 VDC. In this example, the module is oversized for the cranking requirements defined in standards.

11.8.1.2   42 V Electrical Power Net

The development of innovative automotive systems is determined by the demand for comfort improvement, reduced fuel consumption, reduced environmental pollution, and increased efficiency. A result is the substitution of mechanic by electric systems (such as power steering, electromagnetic valve control, electric water pump, electromechanical braking, electric air conditioning, catalyst preheating, etc.), as well as the introduction of new drive train functions like start–stop and recuperative braking. These functions show a total power demand in the range of 8–20 kW [83].

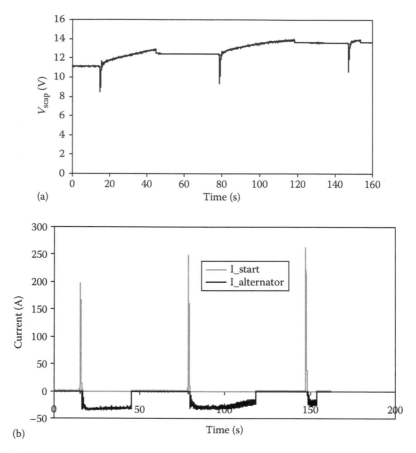

FIGURE 11.18 Current and voltage of a six pieces 2600 F DLC module for several successive crankings of an ICE engine. The module is recharged by the alternator between the crankings.

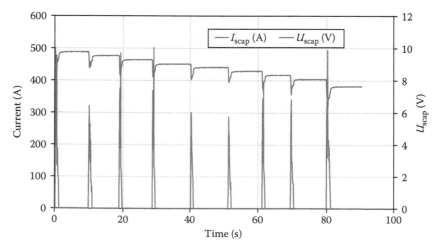

FIGURE 11.19 Voltage and current as a function of the time during cranking of a car ICE with a module consisting of four pieces of 3500 F DLC connected in series.

Future 42 V electric subsystems will be greatly enhanced using DLCs. Their long life and high cycle life are ideal for the variable power load required of new electrical subsystems. Localized load leveling of pulse loads will reduce the need to run high-current wires for long distances in the vehicle.

11.8.1.3 Integrated Starter Generators

The short high-current requirement of engine starting, especially in cold weather, is an excellent application for DLCs. Start–stop motoring is familiar in urban traffic. Currently, the kinetic energy during braking is converted to heat and expelled into the environment. In view of increasing energy costs evolution as well as the need to reduce environmental pollution, it is imperative to recover the braking energy in order to reuse it during acceleration. This method of energy saving can also be usefully applied to vehicles with ICEs, especially in combination with improved alternators used as braking generators, the so-called integrated starter generators (ISG). Such crankshaft starter generators will need up to 10 kW of power in the near future. Conventional lead-acid batteries cannot deliver the required energy in the seconds range because of the slow chemical processes. DLCs are designed to store the energy generated within a very short time and release the energy with high efficiency thanks to the very low internal resistance. The ISG will allow fuel consumption reductions, mainly in urban traffic where start–stop function is frequent, of up to 25%. In near future, the ISGs, placed in the drive train between the engine and the gearbox, will furnish enough energy to power the so-called small hybrids. Such vehicles will recuperate the braking energy during 30 s and reuse it during acceleration of 10 s. Thanks to the ISG a faster engine starting will be done in about 300 ms.

11.8.1.4 Mild and Strong Hybrid Vehicles

These environmentally friendly drive systems are based on the combination of an ICE with an electric power train. The DLCs store the converted kinetic energy from braking and release it later to accelerate the vehicle. In addition, the DLC can supply the energy requirements of auxiliary electrical power equipments. The duration and magnitude of typical acceleration and braking events determine the size of the DLCs bank. The DLCs can also be a device useful to improve the lifetime of a storage system. They present a high number of charge/discharge cycles capability, withstand wide temperature ranges and require little maintenance.

11.8.1.5 Fuel Cell Vehicles

In the future, the energy obtained from fuel combustion in a thermal engine could be replaced by electricity produced by a fuel cell and used to drive an electric engine. The promise of fuel cell technology has had a recent resurgence due to new advancements—not just in fuel cells, but also thanks to DLCs availability. Indeed, high-power energy storage is required in all types of fuel cell applications and DLCs are ideally suited to provide it. These improvements open up opportunities for the development of new power train and subsystem architectures—using both DLCs and fuel cells—which can improve performance, efficiency, and cleanliness in electric and hybrid vehicle technology [84–86].

In collaboration with the Paul Scherrer Institute, the Volkswagen group and other partners have built a fuel cell vehicle incorporating BOOSTCAP DLCs [87]. The fuel cell, which acts as a primary power source, is sized for the continuous load requirement. The DLC bank, which acts as the secondary power source, is sized for peak load leveling events such as vehicle starting, acceleration, and braking. These short duration events are experienced many thousands of times throughout the life of the vehicle and require relatively little energy but substantial power.

11.8.1.6 Hybrid Buses

Hybrid buses used for city transit applications are unique in that batteries are poorly suited for them, due primarily to thermal management issues and secondarily to cycle life considerations [88]. In California, several hundreds of hybrid buses are already on duty. In 2002, General Electric has produced a hybrid diesel-electric bus able to deliver 110 kW for 10 s with a DC link voltage of 240 V. In 2003, ISE started the production of a series of hybrid gasoline-electric buses (720 V, 200 kW peak). A total of 27 buses in 2004, 100 buses in 2005, and >150 buses in 2006 have been produced. In Europe, Vossloh Kiepe has developed a gasoline-electric bus (720 V, 140 kW) which will be introduced in the Milan city bus network in 2008. Scania has presented its 198 kW ethanol hybrid bus in Helsinki in 2007. The energy storage has been made with four modules of DLCs of 125 V. The available energy is greater than 400 W h. This technology offers at least 25% fuel saving and reduces the CO_2 emissions by up to 90%. The buses are an interesting application for DLCs because of the elevated number of stops the vehicle has to do in commercial operation.

11.8.2 TRACTION APPLICATIONS

Several projects are now running in the field of traction applications. For example, tram supply without catenary (overhead lines), voltage drop compensation for weak distribution network, or big diesel locomotive engine cranking.

11.8.2.1 Energy Storage for Traction

The installation of energy storage systems for the vehicle traction management presents a great interest. Today the engineers have to face two major problems: the increase of the mean vehicle power requirement and the increase of the number of vehicles circulating on the network. The distribution of energy storage systems inside or close to the vehicles allows a reduction of the peak power request to the network. It is possible to supply more vehicles without infrastructure investment because the electrical conductors do not need to be oversized [89]. Besides this cost saving the energy storage module provides a reduction of the financial penalties in the electricity bill based on the maximum peak power demand.

The storage systems bring also a global energy saving. The energy stored during the vehicle braking is delivered for its acceleration after the stop at the station. It has been shown that with today electricity cost, the storage system has a return on investment of 5 years only taking into account the energy savings. In the future, this performance will be improved because of the expected energy cost increase.

Manufacturers are proposing either fixed or on-board systems. Both solutions fulfill the same function. In the case of the fixed solution, the DLC are placed in an electric network substation and the main advantages are

*   They do not occupy space inside the vehicle, leaving more room for the customers.
*   There is no added weight on the vehicle to accelerate.

In the case of the on-board solution, to preserve customer space, the storage system is placed either on the roof or under the deck of the vehicle. The on-board solution has the main advantage to offer 500 m of catenary-free operation, in historic city blocks or for maintenance purpose for example.

September 2003 Bombardier started the testing of the "MITRAC" energy saver system (see Figure 11.20) in a Light Rail Vehicle (LVR) of the German city of Mannheim. The module is built with 576 DLC which have a capacitance of 2700 F and a nominal voltage of 2.7 VDC; 192 pieces are connected in series and three in parallel. The total energy content of the 42 F module is equal to 1.5 kWh. It is mounted on the roof of the LRV in a container which has a size of $1.9 \times 0.95 \times 4.5$ m and

FIGURE 11.20 Bombardier Mitrac on-board energy storage system for railway applications. The module is built with 576 DLC of 2700 F.

has a weight of 450 kg. The energy cost-savings are estimated at €30,000 per year for each vehicle thanks to the 30% energy consumption reduction.

The Siemens "SITRAS SES" energy storage system (see Figure 11.21) has a capacitance of 64 F and a nominal voltage of 750 VDC. It is composed of 1344 BCAP0010 DLC which has a capacitance of 2600 F and a nominal voltage of 2.5 VDC. With the cabinet, the inverter and the control power-electronic, the system has a volume of 2.8 × 2.9 × 2.8 m and has a weight of 5.5 ton. It is able to supply a maximum power of 1 MW in an area of 3 km. The module of DLCs allows a 50 kW reduction of the mean power consumption.

11.8.2.2   Starter Application in the Traction

A BOOSTCAP module has been used to crank a 1520 kW (2400 HP) Pielstick diesel motor which equips the BB67000 locomotives belonging to the SCNF, the French railway national company. Today the storage device consists of a 72 V Ni–Cd battery of 140 A h with a weight of 500 kg. The test has been performed in the SNCF maintenance site of "Quatres-Mares" in the neighborhood of Rouen.

The bank was made of 32 series connections of six parallel DLC rated at 800 F, 2.5 VDC and a series resistance of less than 2 mΩ. The complete bank presented a capacitance of 150 F, a series resistance of 10 mΩ, and a nominal voltage of 80 V. To fulfill all the requirements, the DLC module is connected in parallel with a battery of about a third of the size of the original one. The 60 kg module has provided a peak current of 2060 A; it does mean about 350 A in each DLC. The peak power delivered into the starter, limited by the load resistance, was 107 kW, which corresponds to a

FIGURE 11.21 Siemens "SITRAS SES" energy storage system for railway applications. The module consists of 1344 DLC pieces of 2600 F.

power density of 1.5 kW/kg. The Ni–Cd battery delivers only a peak current of 1640 A in the same conditions, which corresponds to a power density of 0.1 kW/kg. The cranking has also been performed only with DLCs in 4 s. Moreover, the test has shown that if the fuel injection is blocked, the DLCs are able to supply a cranking current longer than 15 s (the experiment is interrupted in order not to burn the coil). The full charging time of the bank of DLCs is shorter than 2 min. With a battery, the corresponding required time is longer than 1 h.

11.8.3 INDUSTRIAL APPLICATIONS

In industrial electronics, DLCs can be used in uninterruptible power supplies (UPS), elevators, pallet trucks, etc. [90]. They can be combined into large modules with integrated balancing that span outages in all power categories. In power-electronic, they are particularly suitable for backup in operation independent of the line voltage. A particular example is the power supply of wind turbine pitch system which is a market which is increasing very fast.

11.8.3.1 UPS Applications

The DLCs can handle more than 500,000 charging/discharging cycles during a minimum operating life of 10 years with almost no change in their capacitance. Thanks to DLC, maintenance cycles and test runs are completely obviated. In addition, DLCs can be completely discharged, a condition which usually is inconceivable in conventional systems. Care must be taken to ensure short charging times in case of frequent power failures particularly in the UPS sector. In this point too, DLCs are far superior to conventional batteries.

The safest and most advanced solution is the online UPS, which completely decouples the line from the load. Sensitive and highly critical loads such as complex and sensitive manufacturing systems are secured exclusively by such online UPS systems.

During voltage sags or complete interruptions of the power supply, the energy has to be delivered by local energy storage devices. It is important to note that over 98% of power outages in the low-voltage area, last less than 10 s. Thus, in contrast to batteries, DLCs are ideally suited to fulfill UPS short-time interruption requirements. The strength of DLCs lies in their ability to release energy with no delay for periods from a fraction of a second up to 10 s.

With their guaranteed low-residual current, long operating life, no maintenance or costly test runs, the ability of full discharge, and the short recharging times for frequent power failures, DLCs are an outstanding storage device for storing energy in high reliability systems.

11.8.3.2 Safety Energy Reserve for Actuators

Many DLC applications are concerned with the supply of electricity for safety systems.

A good example today is the power supply of pitch systems in wind turbines. The problem to be solved is to avoid the destruction of the wind turbine in the absence of load during a network disconnection which would lead to an acceleration of the propeller rotation. To remove the wind power, the wings must be reoriented by the pitch system. The benefit of DLCs in this application is the maintenance-free property because the interruption of the energy production has a high economical impact. Advanced wind turbines consist of three-bladed variable speed turbines. The rotor blades are adjusted and controlled via three independent electromechanical propulsion units, the pitch systems. On a pitch-controlled wind turbine, the rotor blades are slightly turned out of the wind when the power output becomes too high. Conversely, the blades are turned back to the wind whenever the wind drops again. Thus, aerodynamic efficiency and reduced loads on the drive train is assured, providing reduced maintenance and longer turbine life. To enhance the level of safety, the newest wind turbine technology uses the wind not only to produce energy but also for its own safety. The converters feature aerodynamic braking by individual pitch control. The rotor attains the full-braking effect with a 90° off-position of all blades. Even if a blade pitch unit fails, the braking process is completed safely by the other two rotor blades. To enhance the level of safety, each of the autonomous pitch systems is equipped with an emergency power pack to ensure the reliable functioning of the pitch system, for example, in the event of a total power failure or for maintenance purposes. Due to their high reliability, efficiency, and operating lifetime, DLCs have been designed into pitch systems of many wind turbine.

Pitch systems are located in the rotating rotor hub of the wind turbine (Figure 11.22). The power supply and control signals for the pitch systems are transferred by a slip ring from the nonrotating part of the nacelle. The slip ring first is connected to a unit which includes clamps for distributing power and control signals to the three individual blade drive units. Each of them consists of a switched mode power supply, a field bus, the motor converter, an emergency system, and the bank of DLCs. When the power supply is switched on, the module of DLC is charged to its nominal voltage. Typical charging time is approximately 1 min. The capacitor module has an energy content high enough to run the system for more than 30 s with nominal power. The module of DLCs is directly connected to the DC link of the motor converter.

Megawatt class turbines dominate much of the actual world market, pushing the development of multi-megawatt turbines, as the offshore market may demand such installations. The largest turbines are able to produce power up to 6 MW with a rotor diameter of up to 110 m. To ensure the functioning of the fast blade pitch system even for such large installations, bigger emergency power packs have to be integrated.

11.8.4 Consumer Market

At the beginning of the DLC market expansion, the main applications were the low power backup for clock chips and CMOS memory. This was typical "energy storage" applications.

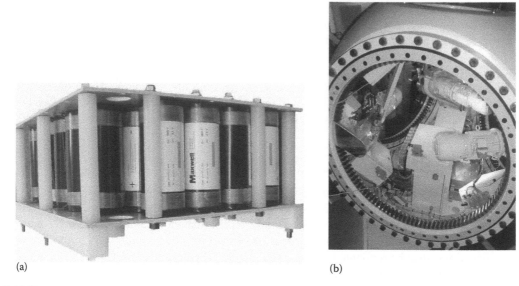

(a)   (b)

FIGURE 11.22 Rotor hub of a wind turbine with independent electromechanical pitch propulsion units and emergency power pack containing 34 large cells rated at 2700 F, a nominal capacitance of 78 F, and a nominal voltage of 76 VDC.

DLCs are nevertheless ideal in supplying peak power in electronic devices. In these applications, DLCs are used in tandem with batteries for systems that require both constant low power discharges for continuous function and a pulse of power for peak loads. In fact, DLCs have been used in various applications, ranging from portable scanners for factory bar-code reading and automated meter reading (AMR) systems to digital cameras. Typically small-sized DLCs are used in AMR systems (see Figure 11.23) that are linked through a two-way communication architecture.

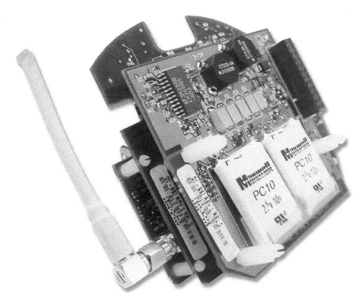

FIGURE 11.23 Digital camera and AMR system containing each 2 PC10 Maxwell DLCs for a nominal voltage of 5 VDC.

DLCs provide many benefits over traditional energy storage components. Using DLCs instead of lithium-ion or lead-acid batteries, the life expectancy of the power supply in AMRs is extended to over 10 years—representing a 100%–300% improvement over lead-acid batteries. The PC10s in each unit are also lighter and smaller, and facilitate a simpler design-in process due to the components' configuration, which allows them to be mounted flat on the board. DLCs are slightly more expensive in initial cost, but because their life is much longer, an overall cost savings of over $200 per unit is realized.

Similarly, in a digital camera application representative of a typical DLC-enhanced design, one PC10 DLC works with a battery to provide overall system power management. The DLC drives the initial power-up of the camera and drives functions involved in composing photographs, such as microprocessor, zoom, and flash. The major high-power-peak demand was observed during the microprocessor activity, i.e., writing to the disk and the LCD operation. Here, DLCs in conjunction with basic inexpensive alkaline batteries achieve the same life cycle as expensive new high-power batteries. Using the capacitor in parallel with the alkaline batteries, the overall system impedance will drop therefore allowing the battery to act as a pure energy source. Thus replaceable, low-cost off-the-shelf alkaline batteries can be employed, making the camera smaller, lighter, and truly portable.

Other current and potential portable DLC applications include two-way pagers, GSM-protocol cell phones, handheld GPS systems, PDA's, power tools, etc. [91]. As the demand for smaller portable devices increases, the flexibility, durability, and power of the DLC will help designers enhance products functionality while simultaneously decreasing their size.

11.9   CONCLUSIONS

The DLC market is expected to grow rapidly to reach the billion dollar size. Hundreds of millions of small size DLCs (1 F, 5 V) are used in the consumer electronic market for more than 20 years. Millions of larger size DLCs (350 F, 2.5 V) have already been installed in industrial applications, like wind turbine pitch systems, since 2002. Today DLCs are already used in the car industry, as for example in the breaking actuator system of the Toyota Prius hybrid car, and in the aircraft industry, as for example in the door opening system of the Airbus A380.

The DLCs provide a higher power density than the other electrochemical energy storage systems. They are often used in hybrid systems, in "parallel" with batteries or fuel cells, to improve the power performance and efficiency.

The maintenance-free aptitude of DLCs gives them an advantage in a large number of applications, especially when the accessibility is difficult or when the cost of interruptions is elevated.

The DLC manufacturer developments are driven by the cost required to meet the automotive market goals. The price pressure bounces indirectly on the carbon and electrolyte suppliers.

The present technical goals may be summarized as follows:

1. Increase the carbon accessible surface
2. Increase the carbon electronic conductivity
3. Increase the electrolyte ionic conductivity
4. Increase the potential stability window of the system

ACKNOWLEDGMENT

The authors would like to thank the EU commission, which through the financing of the "HYHEELS" project has supported this work.

REFERENCES

1. Conway BE. *Electrochemical Supercapacitors—Scientific Fundamentals and Technological Applications.* New York: Kluwer Academic; 1999.
2. Yoshida A, Imoto K, Yoneda H, Nishino A. An electric double-layer capacitor with high capacitance and low resistance. *IEEE Transactions on Components, Hybrids, and Manufacturing Technology* 1992;15(1):133–138.
3. Kinoshita K. *Carbon: Electrochemical and Physicochemical Properties.* New York: Wiley; 1987.
4. Frackowiak E, Béguin F. Carbon materials for the electrochemical storage of energy in capacitors. *Carbon* 2001;39:937–950.
5. Pandolfo AG, Hollenkamp AF. Carbon properties and their role in supercapacitors. *Journal of Power Sources* 2007;157:11–27.
6. Cottineau T, Toupin M, Delahaye T, Brousse T, Bélanger D. Nanostructured transition metal oxides for aqueous hybrid electrochemical supercapacitors. *Applied Physics* 2006;A82:599–606.
7. Staiti P, Lufrano F. A study of the electrochemical behaviour of electrodes in operating solid-state supercapacitors. *Electrochimica Acta* 2006;53:710–719.
8. Khomenko V, Raymundo-Pinero E, Béguin F, Frackowiak E. High-voltage asymmetric supercapacitors operating in aqueous electrolyte. *Applied Physics* 2006;A82:567–573.
9. Snook GA, Peng C, Chen GZ, Fray DJ. Achieving high electrode specific capacitance with materials of low mass specific capacitance: Potentiostatically grown thick micro-nanoporous PEDOT film. *Electrochemistry Communications* 2007;9:83–88.
10. Conway BE, Birss V, Wojtowicz J. The role and utilization of pseudocapacitance for energy storage by supercapacitors. *Journal of Power Sources* 1997;66:1–14.
11. Azaïs P. Recherche des causes du vieillissement de supercondensateurs à électrolyte organique à base de carbones actives. Thèse Université d'Orléans; 2003.
12. Bleda-Martinez MJ, Macia-Agullo JA, Lozano-Castello D, Morallon E, Cazorla-Amoros D, Linares A. Role of surface chemistry on electric double layer capacitance of carbon materials. *Carbon* 2005; 43:2677–2684.
13. Centeno TA, Stoeckli F. The role of textural characteristics and oxygen-containing surface groups in the supercapacitor performances of activated carbons. *Electrochimica Acta* 2006;52:560–566.
14. Centeno TA, Hahn M, Fernández JA, Kötz R, Stoeckli F. Correlation between capacitances of porous carbons in acidic and aprotic EDLC electrolytes. *Electrochemistry Communications* 2007;9:1242–1246.
15. Koresh J, Soffer A. Double layer capacitance and charging rate of ultramicroporous carbon electrodes. *Journal of the Electrochemical Society* 1977;124(9):1379–1385.
16. Ania CO, Pernak J, Stefaniak F, Raymundo-Pinero E, Béguin F. Solvent-free ionic liquids as in situ probes for assessing the effect of ion size on the performance of electrical double layer capacitors. *Carbon* 2006;44:3113–3148.
17. Eliad L, Pollak E, Levy N, Salitra G, Soffer A, Aurbach D. Assessing optimal pore-to-ion size relations in the design of porous poly(vinylidene chloride) carbons for EDL capacitors. *Applied Physics A* 2006;82:607–613.
18. For example, supercapacitor capacitance density in Kuraray product information for BP20 activated carbon for supercapacitor. http://www.kuraraychemical.com/Products/SC/capacitor.htm
19. Arulepp M, Leis J, Lätt M, Miller F, Rumma K, Lust E, Burke AF. The advanced carbide-derived carbon based supercapacitor. *Journal of Power Sources* 2006;162:1460–1466.
20. Chmiola J, Yushin G, Gogotsi Y, Portet C, Simon P, Taberna PL. Anomalous increase in carbon capacitance at pore sizes less than 1 nanometer. *Science* 2006;313:1760–1763.
21. Kazuya H, Takeshi M, Manabu S, Takeshi K, Manabu T. Asahi glass. Electric double layer capacitor having an electrode bonded to a current collector via a carbon type conductive adhesive layer. US patent /6072692.
22. Michael MS, Prabaharan SRS. High voltage electrochemical double layer capacitors using conductive carbons as additives. *Journal of Power Sources* 2004;136:250–256.
23. Che G, Lakshmi BB, Fisher ER, Martin CR. Carbon nanotubule membranes for electrochemical energy storage and production. *Nature* 1998;393:346–349.
24. Emmenegger C, Mauron P, Sudan P, Wenger P, Hermann V, Gallay R, Züttel A. Investigation of electrochemical double-layer (ECDL) capacitors electrodes based on carbon nanotubes and activated carbon materials. *Journal of Power Sources* 2003;124:321–329.
25. Viswanathan S, Tokune T. Honda Motor and University Ohio. Functionalized nanotube material for supercapacitor electrodes. WO Patent /2007/047185.

26. Frackowiak E, Khomenko V, Jurewicz K, Lota K, Béguin F. Supercapacitors based on conducting polymers/nanotubes composites. *Journal of Power Sources* 2006;153:413–418.

27. Kötz R, Carlen M. Principles and applications of electrochemical capacitors. *Electrochimica Acta* 2000;45:2483–2498.

28. Richner RP. Entwicklung neuartig gebundener Kohlenstoffmaterialien für elektrische Doppelschichtkondensatorelektroden. PhD submitted to the ETHZ, Zurich; 2001. http://e-collection. ethbib.ethz.ch/ecol-pool/diss/fulltext/eth14413.pdf

29. Wade TL. High power carbon-based supercapacitors. PhD submitted to the School of Chemistry, University of Melbourne; 2006. http://eprints.infodiv.unimelb.edu.au/archive/00002521/01/High_ Power_Carbon-Based_Supercapacitors.pdf

30. Ue M, Ida K, Mori S. Electrochemical properties of organic liquid electrolytes based on quaternary onium salts for electrical double-layer capacitors. *Journal of the Electrochemical Society* 1994; 141(11):2989–2995.

31. McEwen AB, Ngo HL, Lecompte K, Goldman JL. Electrochemical properties of imidazolium salt electrolytes for electrochemical capacitor applications. *Journal of the Electrochemical Society* 1999; 146(5):1687–1695.

32. Makoto, U. Application of ionic liquids to double-layer capacitors. In *Electrochemical Aspects of Ionic Liquids*. John Wiley & Sons, NJ; 2005, ISBN 9780471648512.

33. Balducci A, Soavi F, Mastragostino M. The use of ionic liquids as solvent-free green electrolytes for hybrid supercapacitors. *Applied Physics A* 2006;82:627–632.

34. Kurzweil P, Chwistek M, Gallay R. Capacitance determination and abusive aging studies of supercapacitors based on acetonitrile and ionic liquids. *Proceedings of the 16th International Seminar on Double Layer Capacitors*, 2006, Deerfield Beach, FL, p. 78.

35. Tanahashi I, Yoshida A, Nishino A. Comparison of the electrochemical properties of electric double-layer capacitors with an aqueous electrolyte and with a nonaqueous electrolyte. *Bulletin of the Chemical Society of Japan* 1990;63:3611–3614.

36. BCAP0350 supercapacitor from Maxwell Technologies. 350F, 3.2 mOhm. http://www.maxwell.com, 2004.

37. Liu P, Soukiazian S, Verbrugge M. Influence of temperature and electrolyte on the performance of activated-carbon supercapacitors. *Journal of Power Sources* 2006;156:712–718.

38. Zheng JP, Jow TR. The effect of salt concentration in electrolytes on the maximum energy storage for double layer capacitors. *Journal of the Electrochemical Society* 1997;144(7):2417–2420.

39. Sugalski R. General electric. Electrochemical cell having cast-in-place insulator. US patent /4320182.

40. Farahmandi C, Dispennette J, Blank E, Kolb A. Maxwell Technologies. Multi-electrode double layer capacitor having single electrolyte seal and aluminium-impregnated carbon cloth electrodes. US patent /1996/000726728.

41. Takaaki M, Yoshishige I, Noburo K. Nichicon Corporation. Aluminum electrolytic capacitor. US patent /6307733.

42. Gallay R, Guillet D, Hermann V, Schneuwly A. Electrical energy accumulating device consisting of wound strips and its manufacturing method. US patent /2002/048140.

43. Teruhisa M, Makoto F, Masafumi O, Haruhiko H, Takumi Y, Toshiyuki H. Matsushita Electrical Industrial. Capacitor. US patent /6310756.

44. Verbrugge M, Liu P. Microstructural analysis and mathematical modeling of electric double-layer supercapacitors. *Journal of the Electrochemical Society* 2005;152(5):D79–D87.

45. IEC 62391-1. Fixed electric double layer capacitors for use in electronic equipment—Part 1: Generic Specification, 1st edn. International Standard, International Electrotechnical Commission, ICS codes 31.060.10, http://webstore.iec.ch/preview/info_iec62391-1%7Bed1.0%7Den.pdf, Publication date April 10, 2006.

46. IEC 62391-2. Fixed electric double layer capacitors for use in electronic equipment—Part 2: Sectional specification-Electric double layer capacitors for power application, 1st edn. International Standard, International Electrotechnical Commission, ICS codes 31.060.10, http://webstore.iec.ch/preview/info_ iec62391-1%7Bed1.0%7Den.pdf, Publication date April 10, 2006.

47. IEC 62391-2-1. Fixed electric double layer capacitors for use in electronic equipment—Part 2-1: Blank detail specification-Electric double layer capacitors for power application–Assessment level EZ, 1st edn. International Standard, International Electrotechnical Commission, ICS codes 31.060.10, http:// webstore.iec.ch/preview/info_iec62391-1%7Bed1.0%7Den.pdf, Publication date April 10, 2006.

48. Taberna PL, Simon P, Fauvarque JF. Electrochemical characteristics and impedance spectroscopy studies of carbon–carbon supercapacitors. *Journal of the Electrochemical Society* 2003;150(3): A292–A300.

49. Kurzweil P, Fischle HJ. A new monitoring method for electrochemical aggregates by impedance spectroscopy. *Journal of Power Sources* 2004;127:331–340.

50. Rafik F, Gualous H, Gallay R, Crausaz A, Berthon A. Frequency, thermal and voltage supercapacitor characterization and modeling. *Journal of Power Sources* 2007;165:928–934.

51. de Levie R. Advances in electrochemistry and electrochemical engineering 1967;6:329–397.

52. Hahn M, Barbieri O, Campana P, Kötz R, Gallay R. Carbon based double layer capacitors with aprotic electrolyte solutions: The possible role of intercalation/insertion processes. *Applied Physics A* 2006;82:633–638.

53. Salitra G, Soffer A, Eliad L, Cohen Y, Aurbach D. Carbon electrodes for double-layer capacitors. *Journal of the Electrochemical Society* 2000;147:2486–2493.

54. Kötz R, Hahn M, Gallay R. Temperature behaviour and impedance fundamentals of supercapacitors. *Journal of Power Sources* 2006;154:550–555.

55. Gualous H, Bouquain D, Berthon A, Kauffmann JM. Experimental study of supercapacitor serial resistance and capacitance variations with temperature. *Journal of Power Sources* 2003;123:86–93.

56. Hein G. Siemens. Capacitor balancing method for a capacitor battery/bank of capacitors monitors the charging status of capacitors by applying three voltage levels available through a source of reference voltage. US patent /2004/090731.

57. Desprez P, Barrailh G, Rochard D, Rael S, Sharif F, Davat B. CIT Alcatel. Supercapacitor balancing method and system. US patent /2003/062876.

58. Kutkut NH, Divan DM. Dynamic equalization techniques for series battery stacks. *18th INTELEC* 1996; 514–521.

59. Rufer A. EPFL. Electrical energy storage system with at least two energy sources having different power characteristics. EP patent /1065775.

60. Thrap G, Gallay R, Schlunke D. Maxwell Technologies. Charge balancing circuit for double-layer capacitors. US patent /2004/263121.

61. Haerri V, Erni P, Marinkovic G, Egger S. HTA Lucerne. Current-accumulator module comprising batteries and capacitors, in particular, supercapacitors. WO patent /2002/0215363.

62. Bolz S, Gotzenberger M, Knorr R, Lugert G, Siemens AG. Device and method for equalising the charge of serially connected capacitors belonging to a double layer capacitor. WO patent /2005/074092.

63. Koetz R, Sauter JC. Conception et dev. Michelin and Paul Scherrer Institut. Detachable charge control circuit for balancing the voltage of supercapacitors connected in series. WO patent /2006/032621.

64. Koetz R, Sauter JC, Ruch P, Dietriech P, Büchi FN, Magne PA, Varenne P. Voltage balancing: Long-term experience with the 250V supercapacitor module of the hybrid fuel cell vehicle HY-LIGHT. *Journal of Power Sources* 2007;174:264–271.

65. Miller JR. Performance of mixed metal oxide pseudocapacitors: Comparison with carbon double layer capacitors, *Proceedings of the 2nd International Seminar on Double Layer Capacitors and Similar Energy Storage Devices*, Deerfield Beach, FL, 1992.

66. Zubieta L, Bonert R, Dawson F. Considerations in the design of energy storage systems using double-layer capacitors. IPEC Tokyo 2000; 1551.

67. Dougal RA, Gao L, Liu S. Ultracapacitor model with automatic order selection and capacity for dynamic system simulation. *Journal of Power Sources* 2004;126:250–257.

68. Belhachemi F, Raël S, Davat B. A physical based model of power electric double-layer supercapacitors, IEEE-IAS'00, Rome, 2000.

69. Hermann V, Schneuwly A, Gallay R. High power double-layer capacitor developments and applications, ISE2001, San Francisco, CA.

70. Kurzweil P, Fischle HJ. A new monitoring method for electrochemical aggregates by impedance spectroscopy. *Journal of Power Sources* 2004;127:331–340.

71. Verbrugge M, Liu P. Analytic solutions and experimental data for cyclic voltammetry and constant-power operation of capacitors consistent with HEV applications. *Journal of the Electrochemical Society* 2006;153(6):A1237–A1245.

72. Lajnef W, Vinassa JM, Azzopardi S, Briat O, Guédon-Gracia A, Zardini C. First step in the reliability assessment of ultracapacitors used as power source in hybrid electric vehicles. *Microelectronics Reliability* 2005;45:1746–1749.

73. El-Husseini MH, Venet P, Rojat G, Joubert C. Thermal simulation for geometric optimization of metallized polypropylene film capacitors. *IEEE Transactions on Industry Applications* 2003;38(3):713–718.

74. Hahn M, Kötz R, Gallay R, Siggel A. Pressure evolution in propylene carbonate based electrochemical double layer capacitors. *Electrochimica Acta* 2006;52:1709–1712.

75. Goltser I, Butler S, Miller JR. Reliability assessment of electrochemical capacitors: Method demonstration using 1-F commercial components, *Proceedings of the 15th International Seminar on Double Layer Capacitors and Similar Energy Storage Devices*, Deerfield Beach, FL, 2005, p. 215.

76. Miller JR, Klementov A, Butler S. Electrochemical capacitor reliability in heavy hybrid vehicles. *Proceedings of the 15th International Seminar on Double Layer Capacitors and Similar Energy Storage Devices*, Deerfield Beach, FL, 2005, p. 218.

77. Yurgil J. General Motors Corporation. Ultracapacitor useful life prediction. US patent /2006/012378.

78. Kurzweil P, Frenzel B, Gallay R. Capacitance characterization methods and ageing behavior of supercapacitors. *Proceedings of the 15th International Seminar on Double Layer Capacitors*, Deerfield Beach, FL, 2005, p. 14.

79. Hardwick L, Hahn M, Ruch P, Holzapfel M, Scheifele W, Buqa H, Krumeich F, Novák P, Kötz R. An in situ Raman study of the intercalation of supercapacitor-type electrolyte into microcrystalline graphite. *Electrochimica Acta* 2006;52:675–680.

80. Hahn M, Barbieri O, Gallay R, Kötz R. A dilatometric study of the voltage limitation of carbonaceous electrodes in aprotic EDLC type electrolytes by charge-induced strain. *Carbon* 2006;44:2523–2533.

81. Beliakov AI. *Russian Supercapacitors to Start Engines*, Battery International, April 1993, p. 102.

82. Ivanov A, Poliashov L, Radionov N. Application of ECOND's double electric layer capacitors in starting systems of internal combustion engines. *Proceedings of the 3th International Seminar on Double Layer Capacitors and Similar Energy Storage Devices*, Deerfield Beach, FL, 1993.

83. Schöttle R, Threin G. Electrical power supply systems: Present and future, VDI Berichte 2000; Nr. 1547.

84. Fuglevand W. Avista Laboratories. Fuel cell power system, method of distributing power, and method of operating a fuel cell power system. WO patent/2002/095851.

85. Raiser S. General Motors Corporation. Hybrid fuel cell system with battery capacitor energy storage system. WO patent/2006/065364.

86. Pearson M. Ballard Power Systems. Power supply and ultracapacitor based battery simulator. US patent 2004/228055.

87. Kötz R, Bärtschi M, Büchi F, Gallay R, Dietrich P. HY-POWER—a fuel cell car boosted with supercapacitors. *Proceedings of the 12th International Seminar on Double Layer Capacitors and Similar Energy Storage Devices*, Deerfield Beach, FL, 2002.

88. Viterna LA. Hybrid electric transit bus, *Proceedings of the SAE International Truck and Bus Meeting and Exposition*, Cleveland, OH, 1997; paper 973202.

89. Rufer A, Barrade P, Hotellier D. Power-electronic interface for a supercapacitor-based energy-storage substation in dc-transportation networks. *EPE Journal* 2004;14(4):43–49.

90. Varakin I, Klementov A, Litvinenko S, Starodubtsev N, Stepanov A. Application of ultracapacitors as traction energy sources. *Proceedings of the 7th International Seminar on Double Layer Capacitors and Similar Energy Storage Devices*, Deerfield Beach, FL, 1997.

91. Miller JR. Capacitor tech talk. *Battery and Energy Storage Technology (BEST) Magazine*, October 2007; 125.

12 Advanced Battery Applications of Carbons

Morinobu Endo, Yong Jung Kim, and Ki Chul Park

CONTENTS

12.1 INTRODUCTION

At the end of the last century, rechargeable lithium-ion (Li-ion) batteries could be regarded as an innovative technology in the field of portable energy-storage devices. Among their wide applications, Li-ion batteries have recently been used to supply power for electric and hybrid vehicles. They are expected to partially replace internal combustion engines and to contribute in solving the problems of air pollution and the emission of green-house gases. The applicability of this energy storage device to the innovative technological fields is strongly dependent on the constituent electrode materials. The diversity of the electrode materials imposes to examine many parameters such as the microstructure, texture, crystallinity, and morphology for the enhancement of the battery performance. This chapter reviews various carbon materials as candidates for anode materials of rechargeable Li-ion batteries and for assistant materials enhancing the conductivity in various types of batteries including the lead-acid battery, primary and secondary Li batteries. Tubular carbon materials including single-wall carbon nanotubes (SWCNTs), multiwall carbon nanotubes (MWCNTs) and mass-produced multiwalled carbon nanotubes (product name: vapor-grown carbon fiber [VGCF]) have been considered as alternative electrode materials or as additives to enhance the stability on repeated charge/discharge cycles. Furthermore, the demand and trend in the market of rechargeable Li-ion batteries are also mentioned.

Carbon materials, ranging from highly ordered graphite to disordered carbons, have been extensively investigated for their application as anode materials of batteries [1,2]. The main focus to examine the applicability of carbon materials to battery applications is laid on the enhancement of the specific capacity, the cyclic efficiency, and the cycle life of the energy storage devices. Moreover, the possibility of a new application in electric vehicles (EV) and hybrid electric vehicles (HEV), as auxiliary/main power supply system, requires superiority in both energy density (Wh/L) and power density (W/L) [3–5]. Tailoring the microstructure and morphology of carbon candidates is inevitable to obtain the best performance in practical devices. It is well demonstrated that the performance of Li-ion batteries is strongly dependent on the thermal history and morphology of carbon and graphite materials used in the electrodes [6]. Carbon and graphite materials have a wide variety of microstructures, texture, crystallinity, and morphology. Therefore, it has been considered as a crucial point to design and choose the anode material [7–9]. Two types of carbon materials, i.e., soft carbons graphitized by high-temperature treatment in the vicinity of 2800°C and hard carbons prepared at a low temperature around 1100°C [10], have been generally used as anode materials in commercial batteries. The precursor materials include cokes, polymers, and biomass materials abundant in nature. The insertion behavior of Li-ions into various kinds of carbon and graphite hosts and its mechanism are closely related to their structure and morphology. Extensive investigations by experimental and theoretical methods have been devoted to these subjects [11–16]. However, the Li-insertion mechanism and the corresponding electrochemical properties of low-temperature prepared carbons are still ambiguous. Thus, the clarification of the Li-insertion mechanism is required for further development of anode materials. The low-temperature carbons might become a very promising material for the next generation of Li-ion batteries [17]. Furthermore, they are more advantageous to reduce the production cost, because graphitization is achieved at 3000°C, which requires a high-energy cost for the mass production.

The share of Li-based batteries currently outperforms other systems, accounting for 63% of worldwide sales volume in portable batteries [18]. Since Sony Corporation has first developed and commercialized rechargeable Li-ion batteries in June 1991, they have been extended to a wide range of portable systems such as notebook computers, cellular phones, and digital video cameras, etc. [19]. Rechargeable Li-ion batteries are advantageous due to high-energy densities (120–150 Wh/kg) and high-operation voltage up to 3.6 V. In terms of environmental load by an increasing consumption of fossil fuel, more recently, Li-ion batteries have been expected to become alternative energy sources for EV/HEV. In fact, Li-ion batteries have been eagerly investigated as a promising candidate to replace the Ni/MH battery in "PRIUS," which is the first commercial hybrid automobile mass-produced and marketed in the world by Toyota Motors Corporation [20].

12.2 CLASSIFICATION OF ANODE CARBON HOST

12.2.1 Conventional Carbons

From the viewpoint of materials science as well as battery applications, carbon materials have attracted much attention due to their diverse physical and chemical properties. The diversity of material property can be attributed to the variety of crystalline lattice, morphology, and textures. The thermal history in preparing carbon materials influences the resulting structure, which is also dependent strongly on the precursor. Low-temperature carbon materials can be classified roughly into two major categories, i.e., (1) soft carbons or "graphitizable" carbons which can be converted to graphite by heat treatment up to 3000°C (generally, heat treatment above 2400°C is called graphitization) under atmospheric or lower pressure and (2) nongraphitizable carbons, or "hard carbons," which do not develop the graphite structure even at high temperatures, since the carbon layers are immobilized by cross-linkage [10].

In addition, some carbons may also include crystallites, which contain carbon layers or packages of stacked carbon layers characterized by significant misfits and misoriented angles of the stacked segments to each other (turbostratic disorder), leading to higher average layer plane spacing [21–24].

12.2.2 Nano-Carbon Hosts (Carbon Nanotubes)

Among carbon allotropes, one-dimensional (1-D) carbon nanotubes (CNTs) have attracted much attention in the past decade due to their unique properties [25,26]. Theoretically, it is possible to construct a carbon tubule by rolling up a hexagonal graphene sheet in various ways. Figure 12.1a shows a schematic illustration for the formation of a SWCNT from a rolled-up single graphene. According to the rolled-up direction of the graphene sheet, CNTs can be classified into three species. These configurations are designated as armchair, zigzag, and chiral tubes (Figure 12.1c).

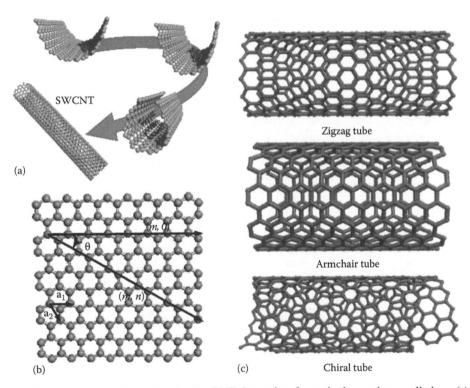

FIGURE 12.1 Conceptual illustration for SWCNT formation from single graphene rolled-up (a). Two-dimensional honeycomb lattice of graphene. Primitive lattice vectors a_1 and a_2 are depicted outlining the primitive unit cell (b). Molecular models of SWCNTs exhibiting different chiralities. According to the rolled-up direction of the graphene sheet, CNT can be classified into zigzag, armchair, and chiral conformation (c). (Reprinted from Terrones, M., *Inter Mater Rev.*, 49, 325, 2004. With permission.)

In the armchair structure, two C–C bonds on the opposite side of each hexagon are perpendicular to the tube axis, while in the zigzag arrangement, these bonds are parallel to the tube axis. All other conformations, where the C–C bonds lie at some angle to the tube axis, are known as chiral or helical structures.

Type of nanotubes, i.e., SWCNTs or MWCNTs and state of CNTs (e.g., ropes composed of nanotube arrays or entangled network) are significantly dependent on the synthesis methods and conditions. The vector that wraps around the circumference of a SWCNT lies within a 30° wedge of the in-plane graphite lattice as denoted by the solid line arrows in Figure 12.1b. In a customary notation for a SWCNT, this vector is represented by the two primitive lattice vectors, a_1 and a_2, shown in Figure 12.1b. This vector determines the direction of rolling a graphene sheet, in which a lattice point (m, n) is superimposed with an origin defined as $(0, 0)$.

12.3 BATTERY PERFORMANCE FOR CONVENTIONAL CARBON MATERIALS

12.3.1 TYPICAL CHARGE/DISCHARGE PROFILE DEPENDENT ON CRYSTALLINITY OF CARBON

Figure 12.2 shows three types of charge and discharge profiles obtained at the second cycle for representative carbon and graphite materials. The shape of the profiles and the capacity vary depending on the category of materials and the thermal history for preparation. Graphitic carbon gives a reversible capacity of 280–330 mAh/g, and Li discharge/charge shows a plateau at below 0.2 V. In the first cycle, all types of carbon materials show an irreversible capacity. After the second cycle, the irreversible capacity is reduced, and the electrode exhibits stable cyclic properties. Although the graphitic materials have some merits on discharge profile, the limited capacity of Li storage by LiC_6 stoichiometry can be said to be a main drawback.

Furthermore, the preparation of graphitic carbon materials requires high-temperature treatment, which is disadvantageous in production cost. The nongraphitic carbon obtained at around

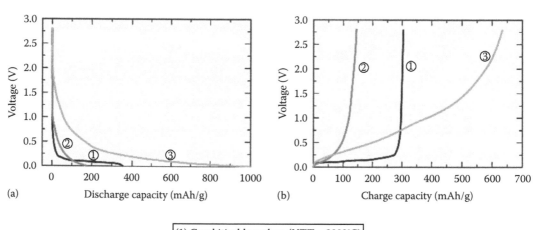

| (1) Graphitizable carbon (HTT = 3000°C) |
| (2) Graphitizable carbon (HTT = 2000°C) |
| (3) Nongraphitic carbon (HTT = 700°C) |

FIGURE 12.2 Typical voltage profiles from carbon materials with different heat-treatment temperatures and different structural variations. The profiles were obtained at the second cycle, and (a) and (b) show discharge and charge cycles by using ① graphitizable carbon heat-treated at 3000°C, ② graphitizable carbon heat-treated at 2000°C, and ③ nongraphitic carbon obtained at 700°C. (Reprinted from Endo, M., et al., *Carbon*, 38, 183, 2000. With permission.)

700°C shows a discharge capacity of 600 mAh/g. The high capacities of nongraphitic carbons can be explained by (1) the formation of Li_2 molecules between layers [11], (2) Li^+ cluster formation in nanocavities [27], and (3) hydrogen content of single graphene layers [13]. In addition, Yazami et al. [28] suggested that the excess Li capacity (more than LiC_6) originates from Li multilayers on graphite sheets. On the other hand, Zhou et al. [29] suggested that Li–C–H bonds contribute to the excess capacity. Although the mechanism causing the excess capacity is not thoroughly clarified yet, it would have deep relation to the nanostructure of carbon materials. Nongraphitic carbons show different output performance from those of graphitic carbons, i.e., slightly inclined discharge characteristics.

Figure 12.3 shows the second cycle charge capacity as a function of crystallite thickness, Lc_{002}, on various carbon-fiber and poly p-phenylene (PPP)-based-carbon electrodes [6]. Well-ordered graphite ($Lc_{002} > 20$ nm) and low crystalline materials ($Lc_{002} < 3$ nm) have larger capacities. However, for the intermediate crystallite sizes (ca. 10 nm) capacity decreases. Dahn et al. [13] reported a similar kind of dependence of the charge capacity on heat treatment temperatures and classified carbon materials suited for commercial Li-ion batteries into three groups. In the high temperature treatment (HTT)-dependence on capacity, interestingly, the disordered PPP-700 carbon (prepared by the heat treatment at 700°C) exhibits a large charge capacity of 680 mAh/g. As Lc_{002} becomes smaller than that of graphite crystal, the charge capacity decreases monotonically until the Lc_{002} value reaches about 10 nm, which is based on the Li^+ intercalation in the turbostratic carbon structures. On the other hand, a different process of doping and undoping of Li-ions may occur in $Lc_{002} < 10$ nm and is largely enhanced by decreasing the crystallite thickness. For Lc_{002} around 10 nm, both reaction processes occur incompletely, which might cause a minimum in the capacity.

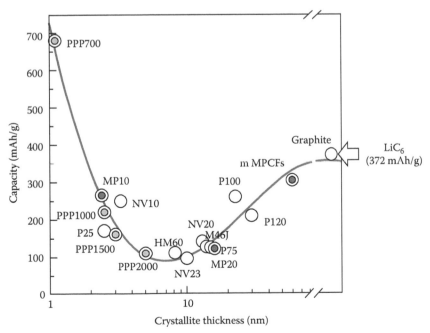

P100, P120, HM60, P25, P75, M46J; Commerical MPCFs
MP10, MP20; MPCF HTT = 1000°C, 2000°C
NV10, NV23; Vapor grown carbon fiber HTT = 1000°C, 2300°C
PPP700, PPP1000, PPP1500, PPP2000; PPP-based carbon

FIGURE 12.3 Charge capacities (C_{rev} at the second cycle) from diverse carbon materials as a function of crystallite thickness. (Reprinted from Endo, M., et al., *J. Phys. Chem. Solids*, 57, 725, 1996. With permission.)

12.3.2 NONGRAPHITIC CARBONS

The performance of Li-ion batteries depends strongly on the crystalline structure of carbon materials used for the anode electrodes [30]. Especially, Kovacic-method-based PPP carbons (see below for Kovacic method) heat-treated at the low temperature of 700°C are proved to possess a superior Li reversible capacity of 680 mAh/g, which is two times higher than that of the well-ordered graphite-based first-stage intercalation compound. This is also three times higher than the Yamamoto-method-based PPP carbons (see below for Yamamoto method), which was obtained by heat treatment at the same temperature (280 mAh/g) [31]. To enhance the cell capacities in Li-ion batteries, different types of low-temperature carbonized and nongraphitized carbons have been extensively investigated for anode materials [32], such as mesocarbon microbeads (MCMBs) [14] and phenolic resin [33]. A high excess of Li storage capacity (corresponding to LiC_3) in the range of 550–700 mAh/g has been reported [34]. However, the Li storage mechanism in such kinds of disordered carbonaceous systems is not yet understood well, and the mechanisms might be largely affected by the structure of the low HTT forms of carbons. These promising Li storage properties for anode materials of Li-ion batteries have been considered to depend on the size of the defective carbon networks, the amount of hydrogen bonded to the periphery of carbons, and the coexisting nanopores. In such defective structures, Li is considered to be stored as LiC_2, to form nanoclusters in the pores and to be absorbed on a single layer of carbons.

Figure 12.4 shows the charge/discharge profiles for PPP-based carbon, which was heat-treated from 700°C to 3000°C. Two different PPPs have been used as precursors and synthesized by the Kovacic- and Yamamoto-methods [35,36]. The synthesis routes of the two methods are described in Figure 12.5. PPP from the Kovacic method is prepared through a polymerization of benzene or a benzene-derived reagent by using an $AlCl_3$–$CuCl_2$ catalyst. The polymerization is promoted by the formation of benzene radicals. Therefore, the procedure leads to the formation of long polymer chains with some structural defects such as branching in ortho- or metapositions. The Yamamoto-method PPP is prepared from 1,4-dibromobenzene by using a Grignard reagent and a transition metal catalyst, resulting in a shorter polymer chain. In carbonization by heat treatment, the Kovacic-method-based PPP has provided higher yields, which is advantageous to reduce the production cost.

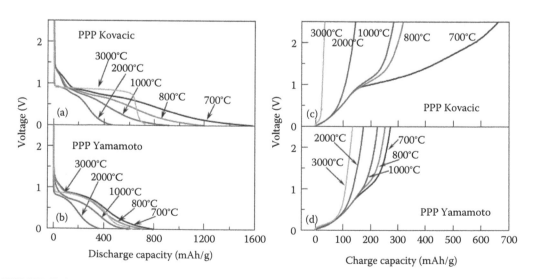

FIGURE 12.4 Discharge (a, b) and charge (c, d) curves of PPP-based carbons, which were synthesized by Kovacic-(a, c) and Yamamoto-method (b, d) as a function of HTT. ① 700°C, ② 800°C, ③ 1000°C, ④ 2000°C, and ⑤ 3000°C.

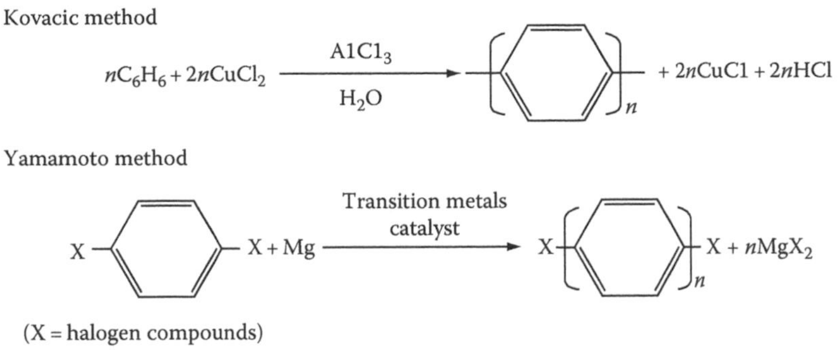

Kovacic method

$$nC_6H_6 + 2nCuCl_2 \xrightarrow[\text{H}_2\text{O}]{\text{AlCl}_3} \left[\right]_n + 2nCuCl + 2nHCl$$

Yamamoto method

$$X-\left[\right]-X + Mg \xrightarrow[\text{catalyst}]{\text{Transition metals}} X-\left[\right]_n-X + nMgX_2$$

(X = halogen compounds)

FIGURE 12.5 Chemical routes for the formation of the two different PPP on the basis of the Kovacic and Yamamoto methods. (Adapted from Endo, M., et al., *Mol. Cryst. Liq. Cryst.*, 310, 353, 1998. With permission.)

The Kovacic-method-based PPP carbons at the HTT below 2000°C show a larger Li insertion capacity than that of the Yamamoto-method-based PPP carbons. On the other hand, for the HTT above 2000°C, the Yamamoto-method-based carbons show a larger battery capacity than that of the Kovacic-method-based PPP carbons. These results might be related to the fact that the Kovacic-method-based PPP carbons at the HTT below 2000°C have a more porous texture. For anode materials in Li-ion batteries, the porous texture of the Kovacic-method-based PPP carbons could be useful for the penetration of electrolyte. Furthermore, the Kovacic-method-based PPP carbons heat-treated at 700°C show the largest insertion capacity. This might be related to the quinoid type of plate-like graphene structure and the homogeneously developed disordered carbon structure. Propylene carbonate (PC) is not likely to be proper for both types of PPP carbons heat-treated over 2000°C. During the first discharging of both samples heat-treated at 3000°C, irreversible capacity and long plateaus are observed at about 0.8 V (electrolyte decomposition).

Tascon and coworkers [37,38] have reported that polyaramid-based carbon, which is obtained from Kevlar (i.e., PPP terephthalamide), can give an additional profit from the viewpoint of manufacturing process. In this case, a stabilization step to maintain the morphology of polyaramid fibers through carbonization is not required. Figure 12.6 shows the galvanostatic charge/discharge voltage profiles of Kevlar fibers which have been treated by single- and two-step carbonization processes. Both carbons exhibit a higher charging capacity than graphite (372 mAh/g) and a charge/discharge profile similar to that usually observed in nongraphitic carbons. The results obtained for the carbon fibers prepared with an intermediate heating step are proved to be better than without this step. Particularly, their discharge capacity at near zero volt (vs. Li/Li+) is notably high. Besides, their reversible capacity is higher than that of the fibers prepared by the single-step pyrolysis process. The introduction of an intermediate isothermal step in the pyrolysis process leads to both higher carbon yields and higher amounts of micropores. The latter feature can justify the higher Li insertion capacity of the fibers prepared by the two-step pyrolysis process [39].

12.3.3 Graphitic Carbons

12.3.3.1 MPCF, MCMB, Graphites

Mesophase pitch-based carbon fibers (MPCFs) have widely been used as a filler for carbon–carbon composite due to their excellent mechanical properties such as high strength, modulus, etc. MPCFs have anisotropic structure, and it contributes to mechanical, electrical, magnetic, thermal as well as chemical properties. This anisotropy is directly related to the layered structure with strong intralayer interactions and very weak van der Waals interplanar interactions between adjacent graphene sheets. As a functional material, MPCFs with anisotropy-originating chemical and physical properties have also been utilized as anode materials for Li-ion batteries [40,41]. As it was shown in Figure 12.2,

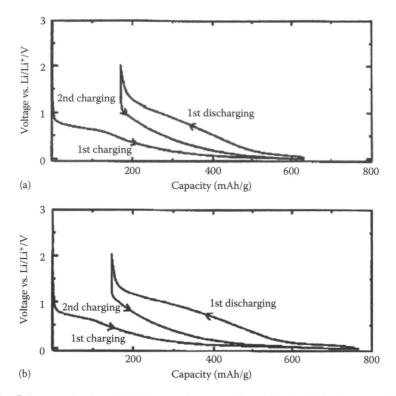

FIGURE 12.6 Galvanostatic charge/discharge voltage profiles of Kevlar-derived carbon fibers prepared (a) in a single step and (b) in two steps with an intermediate heating at 410°C. (Reprinted from Ko, K.S., et al., *Carbon*, 39, 1619, 2001. With permission.) In these curves, charge represents intercalation in the carbon host and discharge the deintercalation.

the anode performance of synthetic carbons and graphitic carbons in Li-ion batteries depends strongly on the precursor materials and the synthesis conditions. Graphitic carbons are attractive due to their good reversibility on repetitive charge/discharge cycles and the high amount of capacity close to the theoretical value of LiC_6. Figure 12.7 shows the typical charge/discharge profiles of graphitized MPCFs (Figure 12.7a and b) and artificial graphite electrodes (Figure 12.7c and d) at a low rate of 0.25 mA/cm². The charge profile for the graphitized MPCFs has one short upper potential plateau at 0.2 V and two long potential plateaus at 0.1 and 0.07 V. The reversible capacity and the coulombic efficiency of the graphitized MPCFs for the first cycle were 303 mAh/g and 94.5%. The irreversible capacity for the first cycle was 17.6 mAh/g. The reversible capacity and the efficiency during the second cycle were 305 mAh/g and 99.0%. As for the graphite, plateaus were clearly observed as indicated by arrows. The capacity and the efficiency for the first cycle were 359 mAh/g near $x = 1$ in Li_xC_6 and 91%. The irreversible capacity was 33 mAh/g. For the second cycle, the reversible capacity was 360 mAh/g and the efficiency was 97.9%. Comparing both electrodes, the maximum reversible capacity of the graphite is larger than that of the graphitized MPCF due to the higher crystallinity. However, the efficiencies for the graphitized MPCF are higher than those of the graphite. Furthermore, MPCFs have excellent cell stability on repetitive charge/discharge cycles, which maintains 86% of its initial capacity even after 400 cycles [41].

As another representative graphitic carbon, MCMBs have been intensively investigated [14,42], due to their comparable performance with natural graphite [43]. Figure 12.8 shows the charge curves (a) of MCMBs heat-treated at high temperatures ranging from 2000°C to 2800°C (200°C increment for each sample) and (b) of MCMBs treated at the HTTs of 1000°C, 2000°C, 2300°C, and 2800°C. The charge capacity increases with increasing the HTT and shows longer plateau below

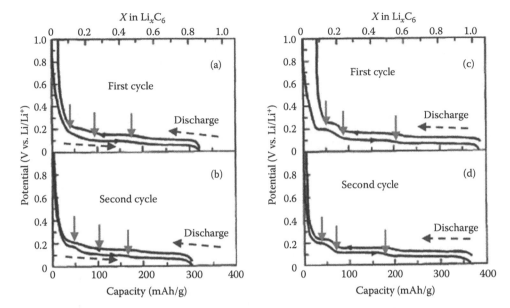

FIGURE 12.7 Charge–discharge profiles of graphitized carbon fibers (MPCF by Petoca Co.) heat-treated at 3000°C and an artificial graphite (Lonza, SFG15). (a) and (c) were obtained at the first cycle. (b) and (d) were obtained at the second cycle. (Reprinted from Takami, N., et al., *J. Electrochem. Soc.*, 142, 2564, 1995. With permission.) In these curves, charge represents intercalation in the carbon host and discharge the deintercalation.

0.25 V. Ohzuku et al. [43] have reported that the electrochemical deintercalation of Li in natural graphite proceeds below ca. 0.25 V. This suggests that the charge reaction of MCMBs below 0.25 V in EC+DEC is the deintercalation of Li from the graphitic structure with the *AB* stacking order. It is reasonable to consider that the graphitic stacking order has an effect on the Li intercalation reaction. During Li intercalation, the stacking of graphene layers along the *c*-axis becomes AA stacking. However, the change to the AA stacking is not allowed in a turbostratic structure because of the crystal defects and residual strains. In graphitic carbons, thus, it is considered that the crystallites of the graphitic stacking structure give the charge capacity only in the potential range of 0–0.25 V.

12.3.3.2 Massive Artificial Graphite (Hitachi)

Since Li-ion batteries were commercialized by Sony, their energy density has been improved by ca. 10% every year to reach 2.5 times higher value than that of the first commercial cell. The transition of battery performances is summarized in Figure 12.9.

Among the diverse carbon materials as anode candidates, Hitachi Chemical Co. Ltd. [44–46] has developed a remarkable material, i.e., massive artificial graphite (MAG), for the anode in Li-ion battery with the high capacity (362 mAh/g) comparable to the theoretical value of LiC_6 (372 mAh/g). The charge/discharge profiles of MAG are illustrated in Figure 12.10. Some plateaus are observed at 190, 95, and 65 mV in the charge and at 105, 140, and 230 mV in the discharge, which imply the formation of graphite intercalation compounds (GICs). As seen in Figure 12.11, MAGs have spherical shape and isotropic structures with high crystallinity, which can contribute to energy density. Figure 12.12 shows the capacity as a function of the discharge current density (Figure 12.12a) and the relation between the discharge capacity and the electrode density (Figure 12.12b). Compared to the other graphitic materials such as natural graphite and graphitized mesophase, MAG shows a good stability of the discharge capacity with the increase of current density and electrode density. The stability depending on current density and electrode density is attributed to pores existing

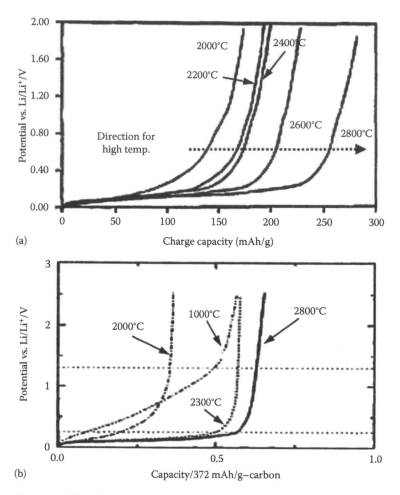

(a)

(b)

FIGURE 12.8 Charge profiles of MCMBs heat-treated at 2000°C–2800°C: potential change vs. capacity in mAh/g (a) (Adapted from Mabuchi, A., Tokumitsu, K., Fujimoto, H., and Kasuh. T., *J. Electrochem. Soc.*, 142, 1041, 1995.); potential change vs. capacity normalized by the theoretical capacity of LiC$_6$ (b) (Reproduced from Tatsumi, K., et al., *J. Electrochem. Soc.*, 142, 716, 1995. With permission.)

inside the particles. The spaces inside the particles can be expected to play a role of moderating the displacement of the graphite lattice during the Li insertion/deinsertion.

12.3.3.3 Graphite Intercalation Compounds as Positive Electrodes

Graphite can be oxidized by chemical or electrochemical means, and the resulting positive charge can be compensated by the insertion of certain anions A$^-$ between the planes of the graphite lattice. Fundamental researches on the formation of GICs with anions have suggested a potential utility of the graphite salts as a positive electrode in batteries [47,48]. Carlin et al. [49] have reported dual-intercalating molten electrolyte battery systems, which involve organic salts and the anode and cathode electrodes made from graphite. Santhanam et al. [50] have confirmed a dual-intercalation battery system using lithium perchlorate (LiClO$_4$) in PC. FDK Co. in Japan has reported that the dual-intercalation carbon system can be applied to the field of supercapacitors with high energy density [51]. The schematic illustration for the charge/discharge mechanism of the dual-carbon cell capacitor is shown in Figure 12.13. As shown in the figure, the same materials with graphitic structure are adopted in both electrodes.

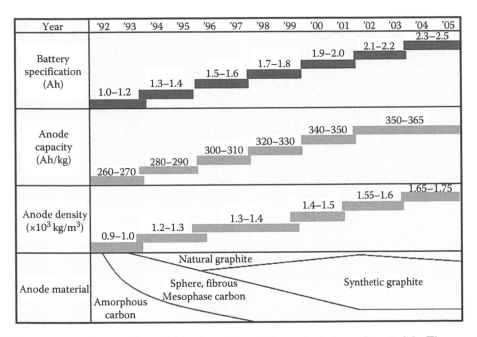

Year	'92	'93	'94	'95	'96	'97	'98	'99	'00	'01	'02	'03	'04	'05

Battery specification (Ah): 1.0–1.2, 1.3–1.4, 1.5–1.6, 1.7–1.8, 1.9–2.0, 2.1–2.2, 2.3–2.5

Anode capacity (Ah/kg): 260–270, 280–290, 300–310, 320–330, 340–350, 350–365

Anode density ($\times 10^3$ kg/m^3): 0.9–1.0, 1.2–1.3, 1.3–1.4, 1.4–1.5, 1.55–1.6, 1.65–1.75

Anode material: Amorphous carbon; Natural graphite; Sphere, fibrous Mesophase carbon; Synthetic graphite

FIGURE 12.9 Capacity transition of Li-ion batteries and change in their anode materials. The capacity of Li-ion batteries has increased by approximately 10% every year. Their anode capacities and densities have also increased. (From Ishii, Y., et al., *Hitachi Chem. Tech. Rep.*, 47, 29, 2006.)

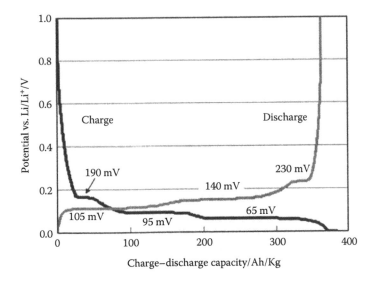

FIGURE 12.10 Charge–discharge profiles of MAG developed by Hitachi Co. In these profiles, charge indicates lithium intercalation into carbon host, and discharge indicates the deintercalation. (From Yoshito, I., *Tanso*, 225, 382, 2006. With permission.)

12.3.4 HETEROATOM-INCORPORATED CARBON ELECTRODES

Carbon materials, especially with low thermal history, usually include some impurities. Heteroatoms also can be considered as impurities. Importantly, many results have reported that heteroatoms can be introduced into the carbon lattice intentionally, which is defined as "doping" [52,53]. Doping

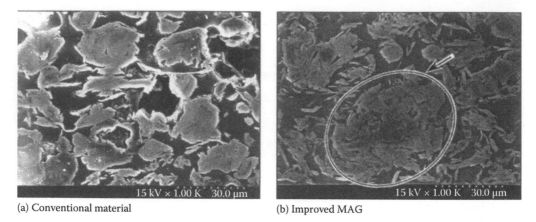

(a) Conventional material   (b) Improved MAG

FIGURE 12.11 Sectional SEM images of the MAG. The image in the left side is a conventional material (a) and the image in the right side is the improved MAG (b), which has smaller pores inside the particles. (From Ishii, Y., *Tanso*, 225, 382, 2006. With permission.)

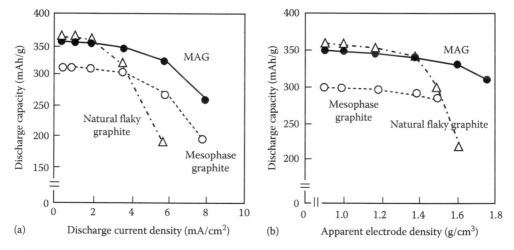

FIGURE 12.12 Comparisons of discharge capacity variation of MAG and typical graphitic carbons as a function of current density (a) and apparent electrode density (b). In these curves discharge represents the deintercalation from the carbon host. (From Endo, M., et al., *Carbon*, 39, 1287, 2001. With permission.)

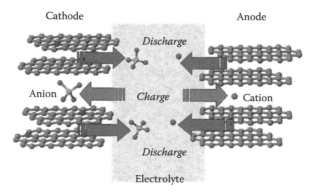

FIGURE 12.13 A schematic illustration of the charge/discharge mechanism for the dual carbon cells by FDK Co. (From http://www.fdk.co.jp/cyber-j/pi_technical08.html.)

has been used extensively in the past [54], mainly in order to change the distribution of electrons between energy levels in the carbon materials, as well as to affect the graphitization process and to modify the chemical state of the surface of the carbon materials. The doping with boron (B), nitrogen (N), and phosphorous (P) has mainly been investigated for graphite materials. Particularly, B-doped carbon materials have been experimentally and theoretically investigated not only from fundamental scientific aspects (e.g., electronic properties) but also for potential applications such as high-temperature oxidation protectors for carbon–carbon (C–C) composite and anode materials of Li-ion batteries. B-doping generates electron acceptor levels [55,56], so that the capacity has been expected to be enhanced. There have been many reports about the preparation methods of B-doped carbons by codeposition of B-containing organic molecules in chemical vapor deposition (CVD). Furthermore, a previous study has suggested the substitutional B-doping mechanism into carbon structures [57].

Figure 12.14 shows the typical voltage profiles of the second discharge and charge for the B-doped graphite cells. The samples were prepared from a mixture of the pristine material and boron carbide (B_4C) by heat treatment at 2800°C in argon atmosphere. The long plateaus below 0.2 V correspond to the reversible intercalation of Li in the graphitized pristine and the B-doped samples. It should be noted that the second discharge/charge capacities of B-doped graphitizable carbon I (B-Graphite I, Figure 12.14a) and graphitizable carbon II (B-Graphite II, Figure 12.14b) are slightly lower than

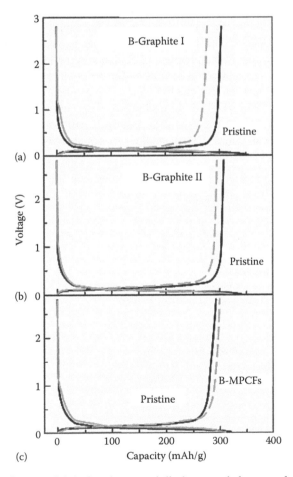

FIGURE 12.14 Change in potential during the second discharge and charge cycle of pristine graphite and B-doped samples for various types of graphite host. (From Endo, M., et al., *Carbon*, 38, 183, 2000. With permission.)

that of the pristine graphite. However, in the case of B-doped MPCFs, the second charge capacity is larger than that of the undoped pristine (Figure 12.14c). The reduced charge capacity of the B-doped samples might be related to boron atoms occupying the Li insertion active sites, such as edge-type sites in the graphite layers, which would inhibit the Li insertion process. In the discharging cycle for B-doped samples, negligible shoulder plateaus are characteristically observed at about 1.3 V, which may be caused by the induction of an electron acceptor level, so that Li insertion yields a higher voltage compared to undoped samples [57]. It is interesting that the irreversible capacity loss for some B-doped samples is lower than that of the corresponding undoped samples. These results might be related to the redistribution of the Fermi level of the B-doped samples, which is lowered by B-doping, i.e., by the introduction of an electron acceptor to the lattice.

The content of doping B atoms has been investigated by x-ray photoelectron spectrometry (XPS). Figure 12.15 shows the boron 1s (B_{1s}) peak at higher resolution. The B_{1s} peak appears in the B-doped samples, although the peak position and shape are different depending on the samples. Particularly, the B_{1s} peak of B-Graphite I is split into three peaks at 185.6, 187.7, and 189.8 eV, which were assigned to the binding energies (BE) originating from boron carbide and boron clusters substituted in the graphite plane and incorporated with N atoms, respectively. From these results, the appearance of the B_{1s} peak of B-Graphite I near 190 eV corresponds to the substitutionally incorporated B atoms into the graphite lattice, which preferentially make bonding with N atoms existing in the heat treatment atmosphere. It is also possible that the residual N atoms in the raw materials react with the boron carbide during the carbonization step and then form boron nitride and/or BC_xN compounds during the graphitization step. These phenomena should be taken into consideration for industrial process using Acheson-type furnace. Consequently, the degradation of the Li insertion capacity observed in some kinds of B-doped graphite might be highly related to the presence of B atoms in the form of boron nitride and boron carbide. Also, the unexpected opposite effects of B-doping could be related to the heterogeneous growth of the crystallites dimension, La, due to boron acting as graphitization catalyst.

In order to demonstrate topological variations by B-doping, surface analysis by scanning tunneling microscopy (STM) has been carried out [58]. Figure 12.16 shows the STM three-dimensional (3-D) surface plots with a scan range of 5 nm and their sectional analysis for pristine highly oriented

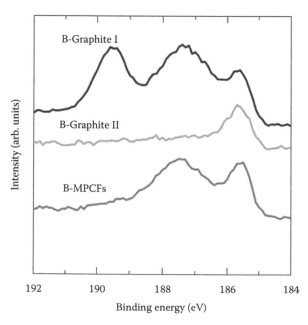

FIGURE 12.15   High resolution B_{1s} XPS spectra of two different types of boronated graphites and boronated MPCFs. (From Endo, M., et al., *Carbon*, 38, 183, 2000. With permission.)

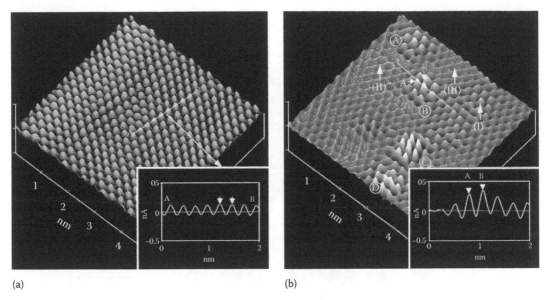

(a)   (b)

FIGURE 12.16   STM three-dimensional surface plots (a) and their section analysis for HOPG and B-doped HOPG. (From Endo, M., et al., *J. Appl. Phys.*, 90, 5670, 2001. With permission.)

pyrolytic graphite (HOPG) (Figure 12.16a) and B-doped HOPG (Figure 12.16b). The surface graphene layer of the pristine HOPG with a trigonal lattice shows a perfect superstructure with *ABAB* stacking as observed in graphite. At the B-doped HOPG graphene surface, four substituted B atoms with the highest intensity of the electron groups (from the circled A to D) are found in the image. Each bright area consists of B atoms with the highest electron density located in the center of six surrounding medium-intensity sites which correspond to carbon atoms. Also, the electron density distributions of a substituted B atom and the surrounding six carbon atoms appear in the inset of Figure 12.16b. The substituted B atom clearly shows the highest electron density in the center. The six carbon atoms closest to the B atom also show a higher electron density than the next neighbor β-site carbon atoms. This indicates that the substituted B atoms affect the electronic structure of the adjacent six β-site carbon atoms. The substitution of B atoms, which are in an electron-deficient state compared to the carbon host, produces vacancies at the top of the valence π-band, resulting in an increase in the density of states (DOS) near the Fermi level. Due to the enhancement of the DOS, the boron sites become brighter than the surrounding sites in the STM lattice image. The effect is not restricted to the boron sites themselves but extends to the surrounding carbon sites due to the delocalized nature of the boron-induced defect.

As shown in the inset of Figure 12.16a, the closest distance between the β-site carbon atoms observed in the section analysis of the STM image is measured to be 0.246 nm for the pristine HOPG. The value is similar to the a_0 spacing in the graphite lattice. On the other hand, the distance between the boron and carbon atoms located at the β-sites is 0.276 ± 0.005 nm, as shown in the inset of Figure 12.16b. The distance is slightly longer than the corresponding distance of 0.246 nm in the pristine HOPG. Turbostratic structures are found at the surface of B-doped HOPG. As can be seen in Figure 12.16b, some regions show a 3-D trigonal image (spot image) of the superstructure (Figure 12.16b-I) consistent with *ABAB* stacking. Some regions show hexagonal (Figure 12.16b-II) or linear (Figure 12.16b-III) images, indicating a turbostratic stacking [59]. Thus, the substitution of B atoms in the hexagonal network affects the stacking nature of the host material presumably because of the lattice defects and strain associated with the B substitution.

Figure 12.17 shows a schematic model for B-substituted graphite. The average distance between the boron and closest carbon (C_1) atoms at β-sites is 0.276 ± 0.005 nm, as is described above. The

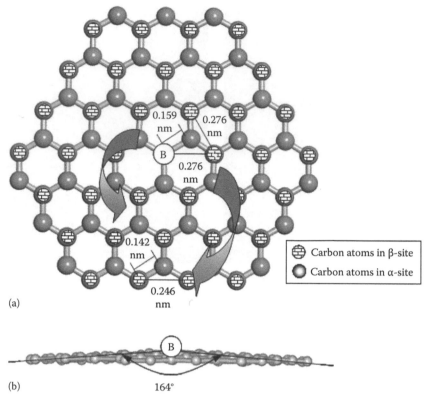

(a)

(b)   164°

FIGURE 12.17   Schematic models of the top (a) and side (b) view for a boron-substituted graphene sheet based on the measured dimensions of B–C_1 and C_1–C_1. (From Endo, M., et al., *J. Appl. Phys.*, 90, 5670, 2001. With permission.)

bond distance between the boron and adjacent carbon atoms is calculated to be 0.159 nm, and for the β-site the B–C distance is 0.276 nm. The distance of C_1–C_1 measured in the STM is also 0.276 nm. The C–C bond distance at the next β-site is evaluated to be the same as in HOPG (0.246 nm). Therefore, the substituted B atom might be located at a slightly higher position than the surrounding C atoms in the basal plane of HOPG. The substitution of boron should slightly deform the flatness of the basal plane. Hach et al. [60] reported similar results showing that the bond length between the B and C atoms is 0.154 nm for $B_2C_4H_6$ and 0.152 nm for $B_2C_{52}H_{18}$. By restricting the bond length between the B–C atoms obtained from Figure 12.16b, our group has simulated the optimized structure of a graphene sheet. As shown in Figure 12.17b, an improper torsion angle (B at the apex) is calculated to be 164°, while the original plane is almost flat with an angle of 179°. It has been reported that the electronic structure is also modified largely by B-doping, as shown by susceptibility measurement [61]. Thus, both atomically and electronically the graphite planes could be tailored by B-doping, which can contribute to modifying and controlling the properties of graphite.

12.3.5 NANO-CARBON MATERIALS

Carbon nanotubes, which can be regarded as a representative of nano-carbon architectures, have received much attention since the Iijima's report of 1991 [62], despite the fact that carbon nanotubes have been known for decades. Prior to the Iijima's report, the images of CNTs have already been shown in a paper published in 1976 [63]. The literature has reported that the pyrolysis of benzene and ferrocene at 1000°C results in tubular graphite in nanometer scale. The interesting new allotropes of carbon, CNTs, are attractive materials for Li-ion batteries. In fact, the application widely extends to cathode and anode additive materials enhancing the battery performances. In this section, the possibility of CNTs to batteries, including the practical application, is discussed.

12.3.5.1 Electrodes

12.3.5.1.1 SWCNTs and MWCNTs

Feasibility of high Li capacity in battery application has been suggested and investigated by many researchers. If all the interstitial sites (intershell van der Waals spaces, intertube channels, and inner hollows) would be accessible for Li intercalation, CNTs could achieve tremendous amount of Li storage. As was mentioned in Section 12.2.2, CNTs can be classified into two categories of SWCNTs and MWCNTs. Lithium intercalation in MWCNTs [64–78] and SWCNTs [79–89] has been actively investigated. The CNTs can be regarded as promising materials in Li-ion batteries since the large number of nanoscale sites for intercalation exceed those of the commonly used graphite electrodes. Figure 12.18 shows schematic illustration for ion-adsorption features on the CNT bundle structure. CNTs generally have bundle structure due to van der Waals force existing between their graphene layers. Intertubular vacancies (typically denoted by B) can serve as the space for Li-ion storage. This becomes a clue to discuss the charge/discharge curves different from those of graphitic carbons. Figure 12.19a shows the representative data of electrochemical intercalation in arc-discharge MWCNTs, which were used without purification [89]. 1 M solution of $LiClO_4$ in 1:1 (volume fraction) of ethylene carbonate (EC) and dimethyl carbonate (DMC) was used as an electrolyte. After a long plateau around 0.75 V, the voltage gradually dropped with the further increase of Li insertion. The voltage profile similar to that of nongraphitic carbon suggests that there is no stage transition. The total amount of Li insertion was 500 mAh/g, and 250 mAh/g was obtained by Li deinsertion. There is no drastic decrease of charge amount after the second cycle. In some reports, a reversible capacity (C_{rev}) of 100–640 mAh/g has been observed, which depends on the sample processing and annealing conditions [87–89]. In general, well-graphitized MWNTs, such as synthesized by arc-discharge methods, have a lower C_{rev} than those prepared by CVD methods. Structural studies [90,91] have shown that alkali metals can be intercalated into the intershell spaces within the individual MWNTs through defect sites. Figure 12.19b shows a typical discharge profile of SWCNTs, which indicate higher reversible and irreversible capacities than those of MWCNTs [78,79]. Ball-milling process is considered to be an effective method to

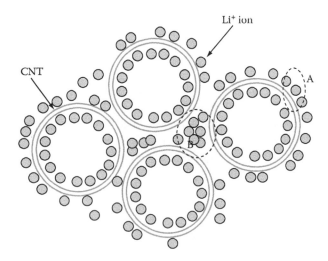

FIGURE 12.18 Schematic illustration for lithium insertion in CNT bundles. The state of lithium-inserted CNT bundles can be classified into two schemes: (1) lithium simply dispersed on the CNT surface and (2) lithium existing in the space between tubes, which might be clusters induced by Li–Li interactions.

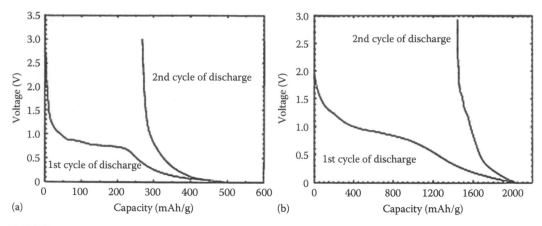

FIGURE 12.19 Voltage vs. specific lithium capacity plots obtained for pristine MWCNTs (a) and purified SWCNTs (b). Both data were collected under galvanostatic mode at 50 mA/g. (Reproduced from Gao, B., et al., *Chem. Phys. Lett.*, 307, 153, 1997. With permission.)

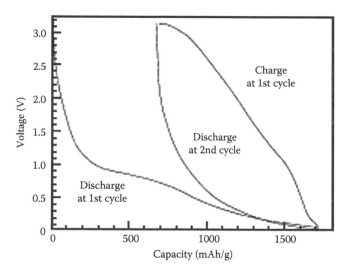

FIGURE 12.20 Charge/discharge characteristics of purified SWCNTs after ball milling. (Reproduced from Gao, B., et al., *Chem. Phys. Lett.*, 307, 153, 1997. With permission.)

improve the crucial difference of capacity between the charge and discharge of CNTs. Gao et al. [82] have reported the electrochemical properties of ball-milled SWCNTs for Li ions. Figure 12.20 shows the charge/discharge curve obtained from purified SWCNTs after ball-milling. Note that the C_{rev} has increased by 50% to reach 1000 mAh/g. No drastic deterioration of capacity has been observed by repetitive charge/discharge cycles. The mechanical process can induce disorder in the structure and cut strands of tubes. These structural and morphological variations would be at least partially related to the kinetics of the intercalation reaction, although the exact mechanism is not understood well.

12.3.5.1.2 Double-Walled Carbon Nanotube

Double-walled carbon nanotubes (DWCNTs) consist of two concentric graphene cylinders. DWCNTs, categorized different from both MWCNTs and SWCNTs, are expected to exhibit mechanical and electronic properties superior to SWCNTs. Endo et al. have suggested that

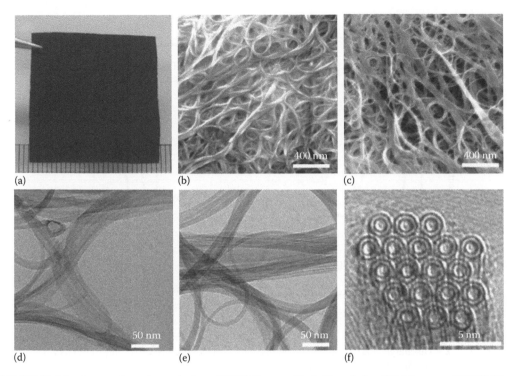

FIGURE 12.21 Image of an electrode from DWCNT paper (a); field-emission SEM images of SWCNT and DWCNT papers (b, c); low-resolution TEM images of SWCNT and DWCNT papers (d, e). It is noteworthy that both samples exist as structures with large-sized bundles; high-resolution TEM images showing a cross-section where DWCNT are stacked in a hexagonal array (f). (Reprinted from Kim, Y.A., et al., *Small*, 2, 667, 2006. With permission.)

DWCNTs have potential utility in various application fields [92–94]. DWCNTs are expected to be more advantageous than SWCNTs due to the structural stability and different pore structure originating from the features on bundling formation [93]. The pore structure characteristic of DWCNTs can offer a much better adsorption field for H_2 than SWCNTs [95]. Furthermore, a thin paper comprised of DWCNTs could be an excellent candidate of anode materials, because it offers a short path for Li-ion transfer to improve rate capability. Figure 12.21 shows a photograph of the DWCNT paper used as an electrode (Figure 12.21a), scanning electron microscopy (SEM) and transmission electron microscopy (TEM) images of a SWCNT paper (Figure 12.21b and d), SEM and TEM images of a DWCNT paper (Figure 12.21c and e), and a high-resolution TEM image of DWCNTs (Figure 12.21f). The fabrication of each nanotube paper is so easy that no binder is used. Figure 12.22 shows potential–capacity curves obtained from the SWCNT and DWCNT papers presented in Figure 12.21. The features of their potential profiles, which monotonically decreased, are very similar to nongraphitic carbons. Although both samples have very large Li-ion storage capacities (SWCNT: close to 2000 mAh/g, DWCNT: ca. 1600 mAh/g) above 1500 mAh/g, the C_{rev} values are 510 and 300 mAh/g for SWCNTs and DWCNTs, respectively. Such a big difference between charge and discharge near 1000 mAh/g is attributed to electrolyte decomposition and the formation of solid-electrolyte interphase (SEI). This is also supported by the long plateau at ca. 0.9 V observed for both samples in the first discharge.

12.3.5.1.3 Other Nanostructures

To get high power density (i.e., battery capacity at a high current rate), there have been many attempts using carbon beads, hybrid materials with CNTs and diverse oxide materials. Wang et al. [96] have

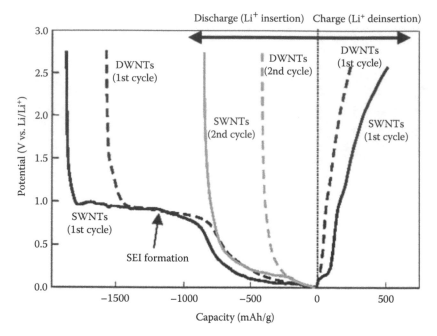

FIGURE 12.22 Discharge/charge profiles of highly pure and bundled SWCNT and DWCNT. (Reprinted from Kim, Y.A., et al., *Small*, 2, 667, 2006. With permission.)

reported that graphitized carbon nanobeads (GCNBs) are greatly available to high-rate use. The special type of nanobeads does not aggregate unlike carbon black. They explained that the short diffusion path of GCNBs and small resistance of SEI would contribute to its high-rate capability. Zhang et al. [97] has investigated Li-ion capacity by using titanate and titania nanotube and nanorods. Furthermore, Sun et al. [98] has suggested the feasibility of carbon hybrid with SnO_2 (SnO_2@C). Despite that the theoretical C_{rev} of SnO_2 is 786 mAh/g, the SnO_2@C hybrid material exhibits a high C_{rev} of 857 mAh/g at the second discharge process, which implies some synergistic effect of the SnO_2@C core-shell nanostructure. Kim et al. [99] have found that the electronic conductivity of individual SnO_2–In_2O_3 nanowires is higher by two orders of magnitude than that of pure SnO_2 nanowires, which is due to the formation of Sn-doped In_2O_3. The novel nanowires show an outstanding lithium storage capacity enough to be expected as promising Li-ion battery electrodes.

Nanostructured materials are currently of interest in Li-ion storage devices due to their high active surface area, porosity, and so on. These characteristics make it possible (1) to introduce new active reactions, (2) to increase the contact surface between the electrode and the electrolyte, (3) to decrease the path length for Li-ion transport, and (4) to improve the stability and specific capacity [100]. Moreover, composite nanostructured materials designed to develop electronic conductive paths could decrease the inner resistance of Li-ion batteries, leading to higher specific capacities even at a high charge/discharge current rate. Section 12.5 reviews the application of nanostructured materials to the electrodes of Li-ion batteries.

12.3.5.2 Additives

12.3.5.2.1 Anodes for Lithium-Ion Battery

The 1-D structure of CNTs makes it easy to form a network by the entanglement due to their relatively large aspect ratio. Furthermore, the structural integrity and its excellent electric conductivity make CNTs promising conducting materials [26]. The CNTs suffer from their high production cost originating from low production efficiency, which seriously inhibits their industrial application.

However, SDK (Showa Denko K. K.) shows one solution by accomplishing the mass production of CNTs. The CNTs mass-produced by SDK are designated as VGCF and VGCF-S [101,102]. VGCFs have been synthesized by the decomposition of hydrocarbons, such as benzene and methane, using transition metal particles as a catalyst at a growth temperature of 1000°C–1300°C [103–110]. The mass-produced but high-quality VGCFs (i.e., commercial products are graphitized) have 1-D morphology with highly preferred orientation of the graphitic basal planes parallel to the fiber axis, which gives rise to excellent mechanical properties and electrical and thermal conductivity. Therefore, VGCFs have high possibility of application to fillers of composite materials and to anode material in Li-ion batteries, including anode additives with high conductivity and surface-to-volume ratios. In fact, VGCFs have been applied to commercial Li-ion batteries. To achieve maximum battery performance, generally, electrodes used in a battery system are required to possess sufficiently high electrical and thermal conductivity, mechanical strength enough to sustain volume changes during the charge and discharge processes, and favorable penetration of the electrolyte. Therefore, in the case that VGCFs are used as electrode fillers, there are necessities not only to evaluate the basic properties of the nanotube itself but also to characterize the packed state of the nanotubes. For example, Figure 12.23a shows the carbon anode sheet made of synthetic graphite and VGCFs (shown as MWCNT in the figure), which is used in a commercial cell. As seen clearly, the VGCFs interconnect well the synthetic graphite particles to each other, contributing to the improvement of electrical conductance and the reinforcement of the electrode. VGCF-S is also expected to have favorable properties as electrode fillers in Li-ion batteries, more than VGCFs. Figure 12.23b shows a SEM image of VGCF-S. The inset in Figure 12.23b shows a TEM image of a single VGCF-S and a drawing of its computational cross-section model.

The cyclic efficiency of a synthetic graphite (HTT = 2900°C) anode with varying the proportion of VGCF-S additive is shown in Figure 12.24 [111]. As the amount of VGCF-S increases, the cyclic efficiency of the synthetic graphite anodes increases continuously. Particularly, 10 wt%-addition of VGCF-S has maintained almost 100% of cyclic efficiency up to 50 cycles. At the high addition percentage, graphite powder particles are interconnected by VGCF-S to form a continuous conductive network. Thus, the addition of VGCF-S to anode materials improves the conductivity of the anode. In addition, VGCF-S contributes to absorb and retain significant amounts of electrolyte and to provide resiliency and compressibility to the electrode structure. Therefore, the use of VGCF-S as additives enhances the performance of Li-ion batteries. Furthermore, as compared to conventional whiskers, relatively high capability of VGCF-S for Li-ion intercalation would also be beneficial for battery performance.

Although shortly mentioned above, resiliency should be emphasized as an important factor for VGCF additives to enhance cell stability. During charge/discharge, the stress is loaded to the cell by

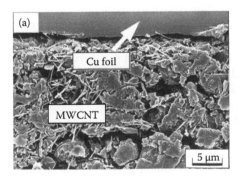
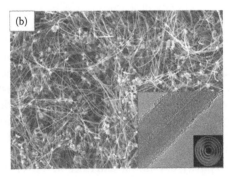

FIGURE 12.23 SEM image of a carbon anode in a commercial cell, in which the carbon sheet was made of synthetic graphite (a); SEM photograph of the additive alone (VGCF-S, which is trademark of Showa Denko K. K., SDK) (b). The inset in right side of the photo shows the TEM photograph of a single VGCF-S and the model based on computational simulation.

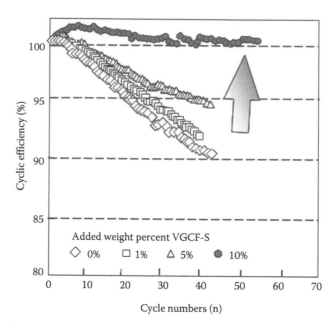

FIGURE 12.24 Cyclic efficiency of synthetic graphite (HTT = 2900°C) as a function of the added weight percent of VGCF-S. Cycle characteristic of 0.2 mA/cm² at 0–1.5 V. (From Endo, M., et al., *Carbon*, 39, 1287, 2001.)

volume change of the active material (i.e., synthetic graphite). However, VGCF additives dissipate the stress-induced displacement of the electrode by the excellent resiliency of the nanotubes, resulting in a high integrity of the electrode structure during the cycle. Therefore, the resiliency of carbon nanotubes in bulk states is greatly related to the cyclic property of anode materials in Li-ion batteries. Resiliency can be defined as the degree of restoration to the original state after constant pressure is applied to CNTs in bulk state. While the resiliency of a single CNT is largely dependent on its modulus, the resiliency property in bulk state is also affected by diameter, cross-sectional morphology, and further, the macromorphology including the size of nanotube aggregates and the degree of integrity of the entangled long nanotubes [112].

12.3.5.2.2 Cathode for Lithium-Ion Battery
Endo et al. [113] have proved that CNTs act as effective fillers for anode materials, where the electrical and resiliency properties of CNTs play an important role in extending the cycle life of Li-ion batteries, as explained in previous section. Several studies have examined the use of CNT fillers to enhance the electrical conductivity of cathode materials as alternatives to conventionally used carbon blacks [114–116]. This originates from the fact that the lithium metal oxides have low electrical conductivity as well as structural or capacity degradation during charging and discharging cycles. However, there seems to be a critical question in respect to the entire replacement of carbon black with CNTs in a cathode, because, in addition to the enhancement of conductivity, carbon black could store a significant amount of electrolyte in its primary structure. In terms of electrical, thermal, and electrolyte-adsorption properties as well as electrochemical performance, Sotowa et al. [117] have examined the advantage of adding hybrid-type fillers comprised of acetylene black and high purity, crystalline and thick MWCNTs to a LiCoO₂-based cathode by comparison with the cathodes containing either acetylene black or CNTs. Figure 12.25 shows the cycle performance of cathodes having different amounts of hybrid fillers. The capacity of the electrode containing 1 wt% of acetylene black (denoted by AB in the figure) declined to ca. 65%. In contrast, the electrode containing 1 wt% of acetylene black and 0.5 wt% of MWCNTs has shown good retention of capacity

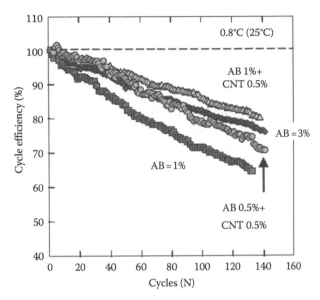

FIGURE 12.25 Cycle performance of cathodes having different amount of carbon black and hybrid fillers. (Adapted from Sotowa, C., et al., *Carbon*, 1, 911, 2008.)

(ca. 80%), which is also superior to the electrode containing 3 wt% of acetylene black. Therefore, the authors have concluded that the entire replacement of acetylene black with CNTs is not the best choice in terms of electrode performance and material cost.

12.4 APPLICATION OF CARBON MATERIALS IN OTHER TYPES OF BATTERIES

12.4.1 LEAD-ACID BATTERY

In lead-acid batteries, the addition of carbon materials with high electrical conductivity causes a significant enhancement of battery performance. The lead-acid battery, in charged state, consists of lead metal (Pb) and lead(IV) oxide (PbO_2) in an electrolyte of about 37% w/w (5.99 mol/L) sulfuric acid (H_2SO_4). In discharged state, both electrodes turn into lead(II) sulfate ($PbSO_4$), and the electrolyte loses its dissolved H_2SO_4. Lead-acid batteries are not designed for deep discharge so that they can easily be damaged by deep cycles. In this section, the effect of the addition of carbon materials to lead-acid batteries is mainly reviewed.

12.4.1.1 Conventional Carbon Additives

In lead-acid batteries, the formation of sulfate (sulfation) on the negative electrode is the most critical obstacle, which is commonly attributed to recrystallization of $PbSO_4$. The sulfation can cause deterioration on stability and safety of batteries. As a solution to this problem, the addition of diverse materials has been proposed by many researchers [118,119]. Carbon black, graphite, expanded graphite [120], and titanium oxide have been examined as additives so that life cycle is successfully improved. In laboratory-scale experiments for a valve-regulated lead-acid battery, Nakamura et al. [118] observed 20% of sulfate in the positive plate and 40% in the negative after 400 cycles. However, the addition of carbon additives to the negative paste has made the cells endure 1100 cycles. The authors suggested that the effect might be attributed to the formation of conducting bridges of carbon particles around the sulfate crystals. The suggestion would also imply that the sulfate crystals are too large for their dissolution to proceed at a sufficient rate. Thus, fluent conducting paths on negative electrodes are required for the supply of electrons closer to the crystal surface.

12.4.1.2 Carbon Nanotube Additives

As mentioned in previous section, conductive additives are inevitable for the negative electrode of lead-acid batteries. To solve the shortcoming of sulfation, conventional carbon candidates have been used as additives. On the other hand, the attractive features of CNTs, such as highly accessible area, excellent electrical conductivity, chemical stability, and mechanical strength, are expected to be useful in lead-acid batteries, especially, for the enhancement of conductivity of the electrodes. We have confirmed the addition effect of VGCF-S in lead-acid batteries. VGCF-S has been added to the active material of the positive electrode. As shown in Figure 12.26 [121], the resistivity of the electrode is lowered for 1.5% VGCF-S addition. When this sample (0.5–1 wt%) is incorporated in the negative electrode, the cycle characteristics are greatly improved in comparison with the electrode without additive (Figure 12.27). This is probably due to the ability of VGCF-S to act as a physical binder, which could suppress the mechanical disintegration and shedding of the active material in the electrode. The unusual morphology of VGCF-S, such as the concentric orientation of their graphite crystallites along the fiber cross section, would cause a high resistance to oxidation. Furthermore, the formation of a conductive network by VGCF-S would induce more effective utilization of the active material in the electrode. Therefore, the use of VGCF-S as electrode fillers improves the performance of the electrodes in lead-acid batteries compared to the electrodes using conventional graphite powders.

12.4.2 Nickel–Metal Hydride Battery

A nickel–metal hydride battery (Ni–MH), which is a type of rechargeable battery comparable to a nickel–cadmium (Ni–Cd) battery, uses a hydrogen-absorbing alloy for the negative electrode

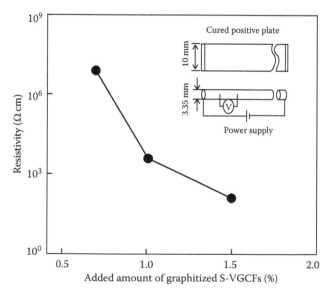

FIGURE 12.26 Variation of the resistivity in the positive plate of lead–acid batteries as a function of added weight percent of graphitized VGCF-S (width of positive plate = 10 mm, thickness of positive plate = 3.35 mm). (From Sotowa, C., et al., *Carbon*, 1, 911, 2008.)

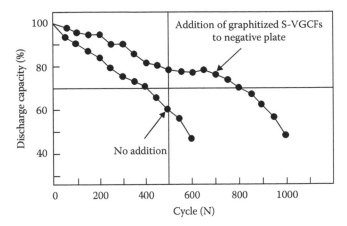

FIGURE 12.27 Cycle characteristics of seal-type lead–acid batteries (4 Ah) when graphitized VGCF-S are used as filler in the negative plate (discharge: 0.75 A to 1.70 V/cell, charge: 7.35 V (maximum 1.5 A)-6H). (From Sotowa, C., et al., *ChemSuschem*, 1, 911, 2008.)

instead of cadmium. In both batteries, the positive electrode is nickel oxyhydroxide (NiOOH). Ni–MH batteries can produce two to three times higher capacity than the equivalent size of Ni–Cd batteries. Compared to Li-ion batteries, however, the volumetric energy density of Ni–MH is lower and self-discharge is higher.

As recent issues on Ni–MH batteries, attention has been focused on the performance under high-current rate. The high-rate capability of Ni–MH batteries is strongly affected by the positive electrode. The fluent migration of electrons from the positive electrode facilitates the discharge reaction, making a high-discharge rate obtainable. However, the active material, $Ni(OH)_2$, obstructs the transfer speed of electrons, which results in the limitation of high rate capability over the whole battery configuration. In this respect, carbon materials such as graphite and carbon black are effective as conductive additives to the positive electrode in Ni–MH batteries, while intermetallic compounds can be used for the negative. Also in the field of Ni–MH batteries, the use of CNTs as conductive materials has been examined extensively. Lv et al. [122] have tried to add MWCNTs as alternative to acetylene black to the positive electrodes of Ni–MH batteries. The batteries with the CNT-added positive electrode are proved to show higher capacity and discharge voltage and better high-rate discharge performance than acetylene black-based batteries. Wu et al. [123] also have confirmed that the addition of CNTs to Ni–MH positive electrodes is effective to enhance the battery performance compared with acetylene black. The use of CNT additives to Ni–MH positive electrodes has merits of higher capacity, better cycle stability, and lower internal resistance when discharged at high current rates.

12.4.3 PRIMARY BATTERIES

In primary batteries, graphite fluoride, and graphite oxide have been investigated as cathode materials for excellent discharge performance and stability [124–126]. The structure of graphite fluoride is represented by two kinds of theoretical components, $(CF)_n$ and $(C_2F)_n$. The graphite fluorides (and oxides) have been considered to be one of diverse GICs [127]. Despite that fluorine atoms are covalently bonded with graphite, graphite fluorides can be regarded as a kind of intercalation compound. This originates from the fact that graphite fluorides have layered structures induced by insertion of fluorine atoms to the graphite interlayer spaces. Endo and coworkers [126] have already considered the mass-produced MWCNTs, i.e., VGCFs (SDK Co. Ltd., see Section 12.3.5.2.1) as novel host materials for fluorination intercalation and graphite oxide. VGCFs heat-treated at high temperature over 2900°C are very attractive to form intercalation compounds due to their highly oriented concentric graphene

TABLE 12.1

Synthetic Conditions, Stoichiometry, and Discharge Characteristics of Fluorine-Intercalated Graphite Fibers

Sample	Fluorination Temperature (°C)	Fluorination Time (h)	F/C Ratio	Component Structure	OCV (V vs. Li)	Discharge Potential (V)	Overpotential (V)	Discharge Capacity (mAh/g)	Utility (%)
A	345	46.7	0.53		3.48	2.53	−0.95	510	79
B	355	40	0.62		3.40	2.50	−0.90	550	76
C	369	40	0.68	$(C_2F)_n$	3.37	2.43	−0.94	—	—
D	456	30	0.83		3.32	2.36	−0.96	697	87
E	503	15	0.95	Mixture of	3.33	2.31	−1.02	689	81
F	513	10	0.93	$(CF)_n$ and $(C_2F)_n$	3.34	2.26	−1.08	728	87
G	614	5	0.99	$(CF)_n$	3.32	2.04	−1.28	597	69

structure. In fact, VGCFs readily make fluorine atoms intercalated at quite low reaction temperature and time. The fluorine-intercalated VGCFs can be distinguished from other intercalated materials using conventional carbon host by the difference in stoichiometry and structure.

Table 12.1 shows the reaction conditions and electrochemical characteristics of respective fluorine-intercalated VGCFs as cathodes. The stage structure of the fluorine-intercalated derivative depends on the reaction temperature. The intercalation at high temperature (614°C) corresponds to stage 1, $(CF)_n$, and the reaction at low temperature (345°C–456°C) corresponds to stage 2, $(C_2F)_n$. The intercalation at intermediate temperatures (503°C, 513°C) results in the mixed stages 1 and 2. It should be noted that the stage 2 compounds prepared at lower temperature than 400°C show higher open circuit voltage (OCV) than those prepared at higher temperature. The OCVs strongly depend on the fluorination temperature at less than 400°C, although become nearly constant beyond 400°C.

Figure 12.28 shows the discharge characteristics of the stage 2 compounds prepared by changing the reaction time at the fixed temperature of 370°C. The end point of discharge was uniformly

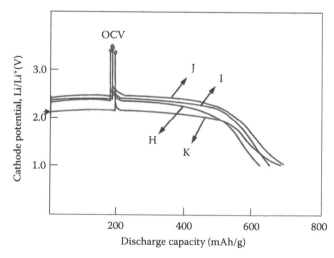

FIGURE 12.28 Effects of reaction time and heat treatment on discharge characteristics of fluorine intercalated graphite fibers. (From Touhara, H., et al., *Electrochim. Acta*, 32, 293, 1987.)

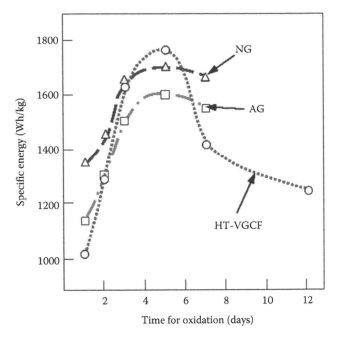

FIGURE 12.29 Variation in energy density of several graphite oxide/Li primary cells with the oxidation time of graphite. (From Touhara, H., et al., *Electrochim. Acta*, 32, 293, 1987.)

determined when the cathode reached 1 V *vs.* Li reference electrode. The reaction time of the samples H, I, and J were 50, 118, and 290 h, respectively. There was no (00l) diffraction line of residual graphite texture observed in the sample H, and the F/C ratio increased from 0.59 to 0.69 and finally 0.74. While the OCV and the discharge potential are almost the same in each case, the discharge capacity of the cathode increases with the increase of the reaction time in the order of H, I, and J. The sample K is the sample J heat-treated at 587°C for 5.5 h under fluorine atmosphere. Additional fluorination decreases the discharge potential and improves the flatness of the discharge profile.

Figure 12.29 shows the variation of the specific energy of graphite oxide/Li primary cells as a function of the oxidation time of graphite. Single crystalline natural graphite, (Nihon graphite, CSSP-B), AG (multicrystalline synthetic graphite, SDK Co. Ltd., FS), and HT-VGCF (heat treated at 3000°C) were submitted to oxidation. As seen from the figure, the maximum energy density is obtained for 5 days of oxidation in all samples. Note that the energy density of HT-VGCF reaches the highest value of 1770 Wh/kg. Therefore, VGCF oxide is considered to be the best primary battery cathode.

12.5 ALLOYS AND COMPOSITES

Modern graphite anodes have the practical capacity of 362 mAh/g [44–46], which is very close to the theoretical stoichiometry with maximum Li uptake. Thus, the possibility of improving the performance of the carbonaceous anodes is almost exhausted. On the other hand, it is unlikely that significant improvement of cathode materials can be realized in the near future. Therefore, the possibility of further improvement of Li-ion battery technology remains only in the development of alternative anodes with a capacity higher than that of graphite. Such anodes can be developed by using groups III, IV, and V light metals, particularly tin (Sn) and silicon (Si). The two elements not only offer high theoretical capacity in an appropriate potential window, but also are abundant and environmentally friendly. Si-based anodes are more attractive, because Si is 4.2 lighter than Sn (atomic weights: Si 28.086, Sn 118.690) and its alloys with maximum Li uptake have the same stoichiometry of $Li_{22}M_5$.

Therefore, composite materials containing the same amount of Si instead of Sn would have a factor of 4.2 higher overall capacity for the same final stoichiometry of Li_xM alloys.

As alternatives to Li metal, Li alloys have been investigated for a long time to solve safety problem and enhance capacity. Despite that metal alloys have substantially higher capacities than those of carbonaceous materials, a large specific volume change also occurs during Li insertion and extraction reactions. Due to the highly ionic character of Li alloys, Li_xM, they are fairly brittle. Drastic volume change during repetitive charge/discharge cycles results in rapid capacity fading [128]. In order to improve structural stability, superfine alloys, intermetallic compounds and active/inactive composite alloy materials have also been developed and investigated, which include Sn–Sb [129], Cu–Sn [130], Co–Sb [131], and Fe–Sn–C [132]. Trials to overcome the poor cycle ability of Li alloys caused by drastic volume change can be summarized as follows: (1) decreasing the particle size [133–137], (2) encapsulating and coating with carbon [137–141], (3) using thin films [142–144], (4) dispersing Si into an inactive/active matrix [145–151], and (5) formation of nanocomposite consisting of nanoparticles and/or rods with CNTs. CNTs in this case played the role of buffer for the mechanical stress caused by volume change.

12.5.1 Silicon (Si) Alloy

Si can react with Li to form an alloy with a high Li/Si ratio of 4.4 Li per Si, which yields an extremely large theoretical specific capacity of 4200 mAh/g [152]. Moreover, elemental Si shows a low voltage below 0.3 V vs. Li/Li+ [153] and high Li-ion diffusion coefficient ($\sim 7.2 \times 10^{-5} cm^2/s$ at 41.5°C) [154]. These promising characteristics have attracted considerable interest in Si-based electrode materials. Weydanz et al. [155] have investigated binary Li–Si and ternary Li–Cr–Si. The experiments on highly lithiated samples yielded charge potentials in the range of 0.3–0.65 V vs. Li/Li+ and discharge potentials of ca. 0.02–0.3 V vs. Li/Li+. The Li–Si showed reversible capacities of up to 550 mAh/g. The ternary materials showed higher reversible capacity of up to 800 mAh/g. These materials have good cycle performance, and the capacity is dependent on the initial stoichiometry with Li:Si ratios of about 1:3.5 showing the highest C_{rev}. Although the theoretical capacity of Li–Si alloys is the highest of various Li-alloys, the cycle ability is usually poor, which is attributed to large volume change during charge/discharge cycles. Wang et al. [156] reported improved cycle ability and capacity for carbon–silicon mixture prepared by extensive mechanical milling. The ball-milling process can reduce the size of the Si particles, which leads to a smaller absolute change in volume during cycles to decrease the rate of disintegration. Wilson and Dahn [157] have reported nanodispersed silicon prepared from benzene and silicon-containing precursors by CVD methods. Due to the smaller size and better dispersion of the Si centers, the material exhibits good reversibility and capacity. The C_{rev} is proved to increase by ca. 30 mAh/g per atomic percent of silicon.

Materials obtained by pyrolysis of pitch-polysilane blends have been extensively studied as carbon materials containing Si [157–161]. For some of these materials, ca. 600 mAh/g of C_{rev} for Li insertion, as well as small irreversible capacities and small hysteresis effects, were reported. It has been shown that the materials contain nanodispersion of Si–O–C and Si–O–S–C instead of nanodispersed Si particles [162–165]. Furthermore, the oxygen and sulfur contents are proved to be correlated to the irreversible capacity. There is a report about the fabrication of porous Si negative electrodes with 1-D channels, where the usefulness of the fabricated negative electrodes for rechargeable microbatteries is also suggested [166].

12.5.2 Tin (Sn) Alloy

Sn as well as Si was widely studied as an anodic material alternative to carbon for Li-ion batteries due to its much higher theoretical capacity of 991 mAh/g [167] than that of graphitic carbons. However, a pure Sn electrode suffers from poor cycle ability due to mechanical fatigue caused by volume change during Li insertion/deinsertion [168]. Similar to the case of Si-based anode materials, the improvement of cycle ability is a crucial issue for Sn-based anode materials. Extensive attention has been paid to Sn-based intermetallic compounds, i.e., Sn_xM_y (M: inactive element), such as Sn–Co [169], Sn–Sb [170], Sn–Ni [171]. These materials exhibit longer cycle ability than that of pure Sn. Besenhard and coworkers [172] inferred that the inactive element could act as a buffer for the large volume change and as a barrier against the aggregation of Sn into large grains during Li-ion insertion/deinsertion processes. However, long-term cycle will result in rapid loss of the reversible capacity and rechargeability. Although the introduction of nanosized electrodes can strongly affect the reversible capacity [173], there still remains an unsolved problem such as the aggregation of Sn during cycling. To overcome this problem, Ke et al. [174] have reported an easy process to prepare 3-D macroporous Sn–Ni alloys. The porous structure contributes to the enhancement of the cycle performance to achieve the C_{rev} of 536 mAh/g at the 75th cycles. As another trial, the improvement of cycle ability by encapsulation of carbon materials has also been suggested by Ke et al. [175]. The resulting Cu_6Sn_5-encapsulated microsphere anode shows much better cycle ability than that of Cu_6Sn_5 alloy powders.

Among diverse alloys, amorphous Sn composite oxide (ATCO) reported by Fuji photo film has caused a great deal of renewed interests in Li alloys as an alternative for use in the negative electrode of Li-ion batteries [176,177]. The ATCO provides a gravimetric capacity of >600 mAh/g for reversible Li adsorption and release, which corresponds to more than 2200 mAh/cm^3 in terms of reversible capacity per unit volume, that is, about twice that of the carbon materials.

12.5.3 OTHER ALLOYS

In addition to Si and Sn, many other alloys can be easily prepared and applied to an electrode for Li-ion secondary batteries. With some exception (Ti, Ni, Mo, Nb, and Be [178]), Li alloys are formed at ambient temperature by polarizing the metals to sufficiently negative potential in Li-ion containing electrolytes. Among matrix metals, aluminum (Al), Pb, indium (In), bismuth (Bi), antimony (Sb), silver (Ag), and some binary alloys have been investigated [179–204]. Generally, the matrix metals undergo major structural changes while alloying with Li. Also in the case of Al, alloying with Li occurs with the formation of several successive phases as α-phase, β-phase, and several Li-rich phases. Crosnier et al. [205] reported that Bi was used for its relatively high potential of Li–Bi alloy formation in the range of 0.8–0.6 V, which prevents other components within the electrode.

12.6 COMPOSITES

Another trial to improve the cycle ability of Li-ion batteries lies in the use of composite materials. Ahn et al. [206] have reported the usefulness of the nanocomposite of Sn nanodots embedded in a Si electrode. The addition of Sn nanodots to Si particles is proved to remarkably improve the cycle performance of Li-ion batteries. Kang et al. [207] have prepared Si–Cu/carbon composite by using a mechanical ball-milling process, where MCMBs were employed to provide fluent electron paths. The composite shows a remarkable reversible capacity of 925 mAh/g and outstanding cycle properties. Hosono et al. [208] have proposed nano/micro hierarchical Fe_2O_3/Ni micrometer wires. Nanocrystalline and mesoporous Fe_2O_3 films were formed on nickel mesh and wires via pyrolytic transformation of $FeO(OH)_{0.29}(NO_3)_{0.22}\cdot0.6H_2O$ film. The resulting composites show high specific capacities at high charge/discharge current rate. Furthermore, Kim et al. [99] have prepared a multicomponent 1-D nanoheterostructure by using thermal evaporation method. The prepared SnO_2–In_2O_3 heterostructured nanowires exhibited a higher initial and reversible Li storage capacity than that of pure SnO_2 nanowires (theoretical capacity of SnO_2; 781 mAh/g), i.e., the first charge capacity of 1975 mAh/g and the C_{rev} of ca. 700 mAh/g after 10 cycles.

To get high-rate performance in Li-ion batteries, Moriguchi et al. [209] have studied the nanocomposite of TiO_2 and CNTs. SWCNT-containing mesoporous TiO_2 has shown better rate performance compared to that of the TiO_2 without CNTs.

12.7 TRENDS IN RECENT MARKET AND FUTURE PROSPECTS

Rechargeable batteries become increasingly indispensable for our life exploiting portable electronics as living necessaries [210], particularly portable telecommunication equipments as a key element in the contemporary information society. As shown in Figure 12.30, the storage capacity of Li-ion cells has continuously been advanced by the diverse technological and scientific efforts to explore carbon host materials, to improve separator performances, to make cell design more efficient, etc. [211]. Nikkei Electronics predicted in the 2004 report that the energy density (based on cylindrical cell, 18650) located around 550 Wh/L will be developed to 1000 Wh/L in 2010. The accumulation of various technical and scientific findings in battery industries will realize the prediction.

Carbonaceous electrode materials are state of the art for the negative electrode. The first commercial Li-ion secondary battery realized by Sony had used graphitizable and graphitic carbons. They were then replaced by hard carbons because of the limit in the storage capacity of Li-ions. As another movement, Hitachi Chemicals Co. has developed and commercialized MAG, which has nearly ideal capacity (362 mAh/g) of graphite materials. In addition, MAG shows excellent charge/discharge profiles and high specific electrode densities. These attractive features have made MAG occupying nearly 50% of the anode market in the world. On the other hand, the search of alternative materials to carbon anodes for high capacity has gathered momentum. The noticeable trials to enhance the specific energy, especially, gravimetric energy density (Wh/kg), include the adoption of metal alloys in carbon hosts, such as Si, Sn, and their respective alloy. Note that metallic Li (possesses 3600 mAh/g of theoretical capacity) was reconsidered by FMC Co. as an anode material with high energy density [212–214]. FMC Co. suggested that Li metal powders stabilized by impregnation into graphite carbon can suppress the Li dendrite formation. However, some special stabilization procedure for reducing the reactivity of Li metal is required for the direct application to anode material.

Despite high capacity up to 1000 mAh/g at the initial step in discharge, these metal and/or alloy materials show poor cycle stability due to irreversible ion transfer. The increase of the specific energy in the recent generation of ion transfer cells is particularly due to the use of carbon materials with higher specific capacity. And still the future trends are directed toward higher specific charge and energy densities. In this respect, attention should be devoted to alternative negative electrode materials because the charge densities of certain Li alloys exceed by far that of carbon. The Fuji film Celltech ion transfer cell, which uses an ATCO as a negative electrode material, has attracted

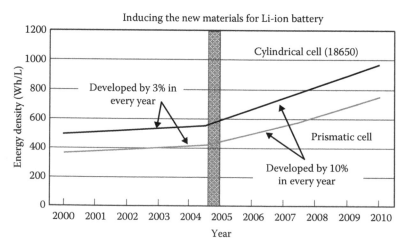

FIGURE 12.30 Enhancement in the storage capacity of rechargeable Li-ion batteries and the prospective value expected to be accomplished in 2010. (Adapted from Nikkel Electronics Cover Story, 888, 120, 2004.)

attention for a while due to the energy density superior to those of the competitors using carbon [215–220]. However, it also has a problem in reversibility on battery performance.

The power density (i.e., the ability to charge and discharge in a short time) is also a crucial point demanded in the recent application field of batteries. Although most electronics moderately require high charge/discharge rate, new applications such as regenerative braking system in HEVs, power backup and portable power tools require both high energy and power density. Kang et al. [3] reported that a novel cathode material of lithium nickel manganese oxide, Li $(Ni_{0.5}Mn_{0.5})$ O_2, can be applied as a high power electrode material by modifying its crystal structure and suggested the computational model by using ab initio method. Wang et al. [96] reported that the new type of artificial carbon, GCNBs, is applicable to a high power anode material for Li-ion batteries, which is attributed to the short diffusion path of GCNBs and the small resistance of SEI film.

In addition to the development of materials, technical improvements such as new electrode and cell designs will also contribute to the appearance of new generations of Li-ion batteries. At the present stage of development, Li-ion batteries could be the most appropriate for portable electronic devices. However, several companies have announced the appearance of EVs equipped with large Li-ion transfer batteries in this century.

12.8 CONCLUSIONS

Modern society with the drastic development of mobile device and zero emission vehicles has required further thin, light, and small energy devices with higher energy and power densities. Such demands will continue to generate many research activities for the development of new cell materials and configurations. As was mentioned in the above sections (12.3.3.2, 12.3.3.3 and 12.3.5.2), many companies have developed the anode and cathode materials with new concepts. This chapter has focused on carbon host materials for anodes in all configurations of cells. The important thing is that the effort to improve the battery performance should be made by multidisciplinary approaches such as organic and inorganic chemistry, physics, surface science, and corrosion. Through the comprehensive study we can expect significant improvements in energy density. The energy density of batteries is generally limited by the low density of host materials and their defects. Fuel cells are expected to overcome such a limited energy problem in the future. For the time being, however, Li-ion batteries will have played an important role in high-energy systems.

REFERENCES

1. Brandt K. Historical development of secondary lithium batteries. *Solid State Ionics* 1994; **69**: 173–183.
2. Delhaes P, Manceau KP, Guérard D. Physical properties of first and second stage lithium graphite intercalation compounds. *Synth Met* 1980; **2**: 277–284.
3. Kang K, Meng YS, Breger J, Grey CP, Ceder G. Electrodes with high power and high capacity for rechargeable lithium batteries. *Science* 2006; **311**: 977–980.
4. Takada K, Ohta N, Zhang L, Fukuda K, Sakaguchi I, Ma R, Osada M, Sasaki T. Interfacial modification for high-power solid-state lithium batteries. *Solid State Ionics* 2008; **179**: 1333–1337.
5. Wen Z, Huang S, Yang X, Lin B. High rate electrode materials for lithium ion batteries. *Solid State Ionics* 2008; **179**: 1800–1805.
6. Endo M, Nishimura Y, Takahashi T, Takeuchi K, Dresselhaus MS. Lithium storage behavior for various kinds of carbon anodes in Li ion secondary battery. *J Phys Chem Solids* 1996; **57**: 725–728.
7. Dresselhaus MS, Dresselhaus G, Sugihara K, Spain IL, Goldberg HA. In: Cardona M (Ed.), *Graphite Fibers and Filaments*, Springer-Verlag, New York, 1998.
8. Oberlin A. In: Thrower PA (Ed.), *Chemistry and Physics of Carbon*, vol. 22. Marcel Dekker, New York, 1989; 1–143.
9. Imanishi N, Takeda Y, Yamamoto O. (Eds.). Development of the carbon anode in lithium ion batteries. In: Wakihara M (Ed.), *Lithium Ion Batteries: Fundamentals and Performance*. Wiley-VCH, New York, 1998; 98.
10. Edwards IAS. Structure in carbons and carbon forms. In: Marsh H (Ed.), *Introduction to Carbon Science*. Butterworth-Heinemann, Portland, 1997; 1–36.

11. Sato K, Noguchi M, Demachi A, Oki N, Endo M. A mechanism of lithium storage in disordered carbons. *Science* 1994; **264**: 556–558.
12. Dahn JR. Phase diagram of Li_xC_6. *Phys Rev B* 1991; **44**: 9170–9177.
13. Dahn JR, Zheng T, Liu Y, Xue JS. Mechanisms for lithium insertion in carbonaceous materials. *Science* 1995; **270**: 590–593.
14. Mabuchi A, Tokumitsu K, Fujimoto H, Kasuh T. Charge-discharge characteristics of the mesocarbon miocrobeads heat-treated at different temperatures. *J Electrochem Soc* 1995; **142**: 1041–1046.
15. Funabiki A, Inaba M, Ogumi Z, Yuasa S, Otsuji J, Tasaka A. Impedance study on the electrochemical lithium intercalation into natural graphite Powder. *J Electrochem Soc* 1998; **145**: 172–178.
16. Flandrois S, Simon B. Carbon materials for lithium-ion rechargeable batteries. *Carbon* 1999; **37**: 165–180.
17. Noguchi M, Miyashita K, Endo M. Characteristics of heat-treated PPP(poly para phenylene) at 973K as an anode of lithium battery. *Tanso* 1992; **155**: 315–319 (in Japanese).
18. Takeshita H. Portable Li-ion, Worldwide, In: *Proc. Conf Power 2000*, San Diego, CA, 25 September 2000.
19. Nagaura T, Tozawa K. Lithium ion rechargeable battery. *Prog Batteries Solar Cells* 1990; **9**: 209–217.
20. http://en.wikipedia.org/wiki/Toyota_Prius
21. Warren BE. X-ray diffraction in random layer lattices. *Phys Rev* 1941; **59**: 693–698.
22. Franklin RE. The structure of graphitic carbons. *Acta Cryst* 1951; **4**: 253–261.
23. Pierson HO. *Handbook of Carbon, Graphite, Diamond and Fullerenes*. Noyes, Park Ridge, NJ, 1993: 43.
24. Winter M, Besenhard JO. Electrochemical intercalation of lithium ion batteries. In: Wakihara M, Yamamoto O (Eds.), *Lithium Ion Batteries: Fundamentals and Performance*. Wiley-VCH, New York, 1998; 127.
25. Terrones M, Hsu WK, Kroto HW, Walton DRM. Nanotubes: A revolution in materials science and electronics. *Top Curr Chem* 1999; **199**: 189–234.
26. Terrones M. Carbon nanotubes: synthesis and properties, electronic devices and other emerging applications, *Inter Mater Rev* 2004; **49**: 325–377.
27. Fujimoto H, Mabuchi A, Tokumitsu K, Kasuh T. Irreversible capacity of lithium secondary battery using meso-carbon micro beads as anode material. *J Power Sources* 1995; **54**: 440–443.
28. Yazami R, Munshi MZA. Novel anodes for solid state batteries. In: Munshi MZA, (Ed.), *Handbook of Solid State Batteries and Capacitors*. World Scientific, Singapore, 1995; 425–460.
29. Zhou P, Papanek P, Lee R, Fischer J. Local structure and vibrational spectroscopy of disordered carbon for Li batteries: Neutron scattering studies. *J Electrochem Soc* 1997; **144**: 1744–1750.
30. Endo M, Kim C, Nishimura K, Fujino T, Miyashita K. Recent development of carbon materials for Li ion batteries. *Carbon* 2000; **38**: 183–197.
31. Endo M, Kim C, Hiraoka T, Karaki T, Matthews MJ, Brown SDM, Dresselhaus MS. Li storage behavior in polyparaphenylene (PPP)-based disordered carbon as a negative electrode for Li ion batteries, *Mol Cryst Liq Cryst* 1998; **310**: 353–358.
32. Dahn JR, Sleigh AK, Way BMS, Weycanz WJ, Reimers NJ, Zhong Q, von. Sacken U. In: Pistoria G (Ed.), *Lithium batteries: New Materials and Perspectives*. Elsevier, North Holland, Amsterdam, 1993; 728.
33. Zheng T, Zhong Q, Dahn JR. High-capacity carbons prepared from phenolic resin for anodes of lithium-ion batteries. *J Electrochem Soc* 1995; **142**: L211–L214.
34. Matsumura Y, Wang S, Mondori J. Interactions between disordered carbon and lithium in lithium ion rechargeable batteries. *Carbon* 1995; **33**: 1457–1462.
35. Kovacic P, Kyriakis A. Polymerization of benzene to p-polyphenyl by aluminum chloride-cupric chloride. *J Am Chem Soc* 1963; **85**: 454–458.
36. Yamamoto T, Hayashi Y, Yamamoto A. A novel type of polycondensation utilizing transition metal-catalyzed C–C coupling. I. Preparation of thermostable polyphenylene type polymers. *Bull Chem Soc Jpn* 1978; **51**: 2091–2097.
37. Villar-odil S, Suarez-garcia F, Paredes JI, Martinez-Alonso A, Tascon JMD. Activated carbon materials of uniform porosity form polyaramid fibers. *Chem Mater* 2005; **17**: 5893–5908.
38. Ko KS, Park CW, Yoon SH, Oh SM. Preparation of Kevlar-derived carbon fibers and their anodic performances in Li secondary batteries. *Carbon* 2001; **39**: 1619–1625.
39. Zheng T, Xing W, Dahn JR. Carbons prepared from coals for anodes of lithium-ion cells. *Carbon* 1996; **34**: 1501–1507.
40. Kim WS, Chung KI, Lee CB, Cho JH, Sung YE, Choi YK. Studies on heat-treated MPCF anodes in Li ion batteries. *Microchem J* 2002; **72**: 185–192.

41. Takami N, Satoh A, Hara M, Ohsaki T. Rechargeable Lithium-ion cells using graphitized mesophase-pitch-based carbon fiber anodes. *J Electrochem Soc* 1995; **142**: 2564–2571.
42. Tatsumi K, Iwashita N, Sakaebe H, Shioyama H, Higuchi S, Mabuchi A, Fujimoto H. The influence of the graphitic structure on the electrochemical characteristics for the anode of secondary lithium batteries. *J Electrochem Soc* 1995; **142**: 716–720.
43. Ohzuku T, Iwakoshi Y, Sawai K. Formation of lithium-graphite intercalation compounds in nonaqueous electrolytes and their application as a negative electrode for a lithium ion (shuttlecock) cell. *J Electrochem Soc* 1993; **140**: 2490–2498.
44. Yoshito I. Carbon anode material for lithium-ion rechargeable battery. *Tanso* 2006; **225**: 382–390.
45. Ishii Y, Fujita A, Nishida T, Yamada K. High-performance anode material for lithium-ion rechargeable battery. *Hitachi Chemical Technical Report* 2001; **36**: 27–32.
46. Ishi Y, Nishida T, Suda S, Kobayashi M. Anode material for high energy density rechargeable lithium-ion battery. *Hitachi Chemical Technical Report* 2006; **47**: 29–32.
47. Rudorff W, Hofmann U, *Uber Graphitsalze*. *Z. Anorg. Allg. Chem.* 1938; **238**: 1–50.
48. Beck F, Junge H, Krohn H. Graphite intercalation compounds as positive electrodes in galvanic cells. *Electrochim Acta* 1981; **26**: 799–809.
49. Carlin RT, De Long HC, Fuller J, Trulove PC. Dual intercalating molten electrolyte batteries. *J Electrochem Soc* 1994; **141**: L73–L76.
50. Santhanam R, Noel M. Electrochemical intercalation of cationic and anionic species from a lithium perchlorate-propylene carbonate system—a rocking-chair type of dual-intercalation system. *J Power Sources* 1998; **76**: 147–152.
51. http://www.fdk.co.jp/cyber-j/pi_technical08.html
52. Lowell CE. Solid solution of boron in graphite. *J Am Ceram Soc* 1967; **50**: 142–144.
53. Kouvetakis J, Kaner RB, Sattler ML, Bartlett N. A novel graphite-like material of composition BC_3, and nitrogen–carbon graphites. *J Chem Soc Chem Commun* 1986; **24**: 1758–1759.
54. Marchand A. Electronic properties of doped carbons. In: Walker PL (Ed.), *Chemistry and Physics of Carbon*. Marcel Dekker, New York, 1971; **7**: 155–191.
55. Way BM, Way BM, Dahn JR, The effect of boron substitution in carbon on the intercalation of lithium in $Li_x(B_zC_{1-z})_6$. *J Electrochem Soc* 1994; **141**: 907–912.
56. Nakajima T, Koh K, Takashima M. Electrochemical behavior of carbon alloy C_xN prepared by CVD using a nickel catalyst. *Electrochim Acta* 1998; **43**: 883–891.
57. Nishimura Y, Yakahashi T, Tamaki T, Endo M, Dresselhaus MS. Anode performance of B-doped mesophase pitch-based carbon fibers in lithium ion secondary batteries. *Tanso* 1996; **172**: 89–94 (in Japanese).
58. Endo M, Hayashi T, Hong SH, Enoki T, Dresselhaus MS. Scanning tunneling microscope study of boron-doped highly oriented pyrolytic graphite. *J Appl Phys* 2001; **90**: 5670–5674.
59. Endo M, Oshida K, Kobori K, Takeuchi K, Takahashi K. Evidence for glide and rotation defects observed in well-ordered graphite fibers. *J Mater Res* 1995; **10**: 1461–1468.
60. Hach CT, Jones LE, Crossland C, Thrower PA. An investigation of vapor deposited boron rich carbon—a novel graphite-like material—part I: The structure of BC_x (C_6B) thin films. *Carbon* 1999; **37**: 221–230.
61. Matthews MJ, Dresselhaus MS, Dresselhaus G, Endo M, Nishimura Y, Hiraoka T, Tamaki N. Magnetic alignment of mesophase pitch-based carbon fibers. *Appl Phys Lett* 1996; **69**: 430–432.
62. Iijima S. Helical microtubules of graphitic carbon. *Nature* 1991; **354**: 56–58.
63. Oberlin A, Endo M, Koyama T. Filamentous growth of carbon through benzene decomposition, *J Cryst Growth* 1976; **32**: 335–349.
64. Nalimova VA, Sklovsky DE, Bondarenko GN, Alvergnat-Gaucher H, Bonnamy S, Béguin F. Lithium interaction with carbon nanotubes. *Synth Met* 1997; **88**: 89–93.
65. Lee JH, Kim GS, Choi YM, Park WI, Rogers JA, Paik UG. Comparison of multiwalled carbon nano-tubes and carbon black as percolative paths in aqueous-based natural graphite negative electrodes with high-rate capability for lithium-ion batteries. *J Power Sources* 2008; **184**: 308–311.
66. Frackowiak E, Gautier S, Gaucher H, Bonnamy S, Béguin F. Electrochemical storage of lithium multi-walled carbon nanotubes. *Carbon* 1999; **37**: 61–69.
67. Maurin G, Bousquet Ch, Henn F, Bernier P, Almairac R, Simon B. Electrochemical intercalation of lithium into multiwall carbon nanotubes. *Chem Phys Lett* 1999; **312**: 14–18.
68. Leroux F, Metenier K, Gautier S, Frackowiak E, Bonnamy S, Béguin F. Electrochemical insertion of lithium in catalytic multi-walled carbon nanotubes. *J Power Sources* 1999; **81–82**: 317–322.

69. Lu W, Chung DDL. Anodic performance of vapor-derived carbon filaments in lithium-ion secondary battery. *Carbon* 2001; **39**: 493–496.
70. Yang Z, Wu HQ, Simard B. Charge-discharge characteristics of raw acid-oxidized carbon nanotubes. *Electrochem Commun* 2002; **4**: 574–578.
71. Frackowiak E, Béguin F. Electrochemical storage of energy in carbon nanotubes and nanostructured carbons. *Carbon* 2002; **40**: 1775–1787.
72. Shin HC, Liu M, Sadanadan B, Rao AM. Electrochemical insertion of lithium into multi-walled carbon nanotubes prepared by catalytic decomposition. *J Power Sources* 2002; **112**: 216–221.
73. Chen WX, Lee JY, Liu Z. The nanocomposites of carbon nanotube with Sb and SnSb$_{0.5}$ as Li-ion battery anodes. *Carbon* 2003; **41**: 959–966.
74. Yoon SH, Park CW, Yang HJ, Korai Y, Mochida I, Baker RTK, Rodriguez NM. Novel carbon nanofibers of high graphitization as anodic materials for lithium ion secondary batteries. *Carbon* 2004; **42**: 21–32.
75. Wang X, Liu H, Jin Y, Chen C. Polymer-functionalized multiwalled carbon nanotubes as lithium intercalation hosts. *J Phys Chem B* 2006; **110**: 10236–10240.
76. Deng D, Lee JY. One-step synthesis of polycrystalline carbon nanofibers with periodic dome-shaped interiors and their reversible lithium ion storage properties. *Chem Mater* 2007; **19**: 4198–4204.
77. Park MS, Needham SA, Wang GX, Kang YM, Park JS, Dou SX, Liu HK. Nanostructured SnSb/carbon nanotube composites synthesized by reductive precipitation for lithium-ion batteries. *Chem Mater* 2007; **19**: 2406–2410.
78. Chen Liu Y, Minett AI, Lynam C, Wang J, Wallace GG. Flexible, aligned carbon nanotube/conducting polymer electrodes for a lithium-ion battery. *Chem Mater* 2007; **19**: 3593–3597.
79. Garau C, Frontera A, Quiñonero D, Costa A, Ballester P, Deyà PM. Ab initio investigations of lithium diffusion in single-walled carbon nanotubes. *Chem Phys* 2004; **297**: 85–91.
80. Claye AS, Fischer JE, Huffman CB, Rinzler AG, Smalley RE. Solid-state electrochemistry of the Li single wall carbon nanotube system. *J Electrochem Soc* 2000; **147**: 2845–2852.
81. Jouguelet E, Mathis C, Petit P. Controlling the electronic properties of single-wall carbon nanotubes by chemical doping. *Chem Phys Lett* 2000; **318**: 561–564.
82. Gao B, Bower C, Lorentzen JD, Fleming L, Kleinhammes A, Tang XP, McNeil LE, Wu Y, Zhou O. Enhanced saturation lithium composition in ball-milled single-walled carbon nanotubes. *Chem Phys Lett* 2000; **327**: 69–75.
83. Yang ZH, Wu HQ. The electrochemical impedance measurements of carbon nanotubes. *Chem Phys Lett* 2001; **343**: 235–240.
84. Morris RS, Dixon BG, Gennett T, Raffaelle R, Heben MJ. High-energy, rechargeable Li-ion battery based on carbon nanotube technology. *J Power Sources* 2004; **138**: 277–280.
85. Ng SH, Wang J, Guo ZP, Chen J, Wang GX, Liu HK. Single wall carbon nanotube paper as anode for lithium-ion battery. *Electrochim Acta* 2005; **51**: 23–28.
86. Udomvech A, Kerdcharoen T, Osotchan T. First principles study of Li and Li$^+$ adsorbed on carbon nanotube: Variation of tubule diameter and length. *Chem Phys Lett* 2005; **406**: 161–166.
87. Wu GT, Wang CS, Zhang XB, Yang HS, Qi ZF, Li WZ. Lithium insertion into CuO/carbon nanotubes. *J Power Sources* 1998; **75**: 175–179.
88. Béguin F, Metenier K, Pellenq R, Bonnamy S, Frackowiak E. Lithium insertion in carbon nanotubes. *Mol Cryst Liq Cryst* 2000; **340**: 547–552.
89. Gao B, Kleinhammes A, Tang XP, Bower C, Fleming L, Wu Y, Zhou O. Electrochemical intercalation of single-walled carbon nanotubes with lithium. *Chem Phys Lett* 1997; **307**: 153–157.
90. Zhou O, Fleming RM, Murphy DW, Chen CH, Haddon RC, Ramirez AP, Glarum SH. Defects in carbon nanostructures. *Science* 1994; **263**: 1744–1747.
91. Suzuki S, Tomita M. Observation of potassium-intercalated carbon nanotubes and their valence-band excitation spectra. *J Appl Phys* 1996; **79**: 3739–3743.
92. Endo M, Muramatsu H, Hayashi T, Kim YA, Terrones M, Dresselhaus MS. Buckypaper' from coaxial nanotubes. *Nature* 2005; **433**: 476.
93. Muramatsu H, Hayashi T, Kim YA, Shimamoto D, Kim YJ, Tantrakarn K, Endo M, Terrones M, Dresselhaus MS. Pore structure and oxidation stability of double-walled carbon nanotube-derived bucky paper. *Chem Phys Lett* 2005; **414**: 444–448.
94. Kim YA, Kojima M, Muramatsu H, Umemoto S, Watanabe T, Yoshida K, Sato K et al. In situ Raman study on single- and double-walled carbon nanotubes as a function of lithium insertion, *Small* 2006; **2**: 667–676.

95. Miyamoto J, Hattori Y, Noguchi D, Tanaka H, Ohba T, Utsumi S, Kanoh H et al. Efficient H_2 adsorption by nanopores of high-purity double-walled carbon nanotubes, *J Am Chem Soc* 2006; **128**: 12636–12637.

96. Wang H, Abe T, Maruyama S, Iriyama Y, Ogumi Z, Yoshikawa K. Graphitized carbon nanobeads with an onion texture as a lithium-ion battery negative electrode for high-rate use. *Adv Mater* 2005; **17**: 2857–2860.

97. Zhang H, Li GR, An LP, Yan TY, Gao XP, Zhu HY. Electrochemical lithium storage of titanate and titania nanotubes and nanorods. *J Phys Chem C* 2007; **111**: 6143–6148.

98. Sun X, Liu J, Li Y. Oxides@C core-shell nanostructures: one-pot synthesis, rational conversion, and Li storage property. *Chem Mater* 2006; **18**: 3486–3494.

99. Kim DW, Hwang IS, Kwon SJ, Kang HY, Park KS, Choi YJ, Choi KJ, Park JG. Highly conductive coaxial SnO_2–In_2O_3 heterostructured nanowires for Li ion battery electrode. *Nano Lett* 2007; **7**: 3041–3045.

100. Jiang C, Hosono E, Zhou H. Nanomaterials for lithium ion batteries. *Nano Today* 2006; **1**: 28–33.

101. http://www.sdkc.com/documents/carbon%20Nanofiber%20For%20Resins.pdf

102. http://www.sdkc.com/documents/Nanofiber%20Production.pdf

103. Koyama T, Endo M. Electrical resistivity of carbon fiber prepared from benzene. *Jpn J Appl Phys* 1974; **13**: 1175–1176.

104. Koyama T, Endo M. Structure and properties of graphitized carbon fiber. *Jpn J Appl Phys* 1974; **13**: 1933–1939.

105. Wei G, Shirai K, Fujiki K, Saitoh H, Yamauchi T, Tsubokawa N. Grafting of vinyl polymers onto VGCF surface and the electric properties of the polymer-grafted VGCF. *Carbon* 2004; **42**: 1923–1929.

106. Speck JS, Endo M, Dresselhaus MS. Structure and intercalation of thin benzene derived carbon fibers. *J Cryst Growth* 1989; **94**: 834–848.

107. Tibbetts GG. Carbon fibers produced by pyrolysis of natural gas in stainless steel tubes. *Appl Phys Lett* 1983; **42**: 666–668.

108. Benissad F, Gadelle P, Coulon M, Bonnetain L. Formation de fibres de carbone a partir du methane: I. Croissance catalytique et epaississement pyrolytique. *Carbon* 1988; **26**: 61–69.

109. Tibbetts GG, Gorkiewicz DW, Alig RL. A new reactor for growing carbon fibers from liquid- and vapor-phase hydrocarbons. *Carbon* 1993; **31**: 809–814.

110. Tatsumi K, Zaghib K, Abe H, Higuchi S, Ohsaki T, Sawada Y. A modification in the preparation process of a carbon whisker for the anode performance of lithium rechargeable batteries. *J Power Sources* 1995; **54**: 425–427.

111. Nishimura K, Kim YA, Matushita T, Hayashi T, Endo M, Dresselhaus MS. Structural characterization of boron-doped submicron vapor-grown carbon fibers and their anode performance. *J Mater Res* 2000; **15**: 1303–1313.

112. Endo M, Hayashi T, Kim YA, Tantrakarn K, Yanagisawa T, Dresselhaus MS. Evaluation of the resiliency of carbon nanotubes in the bulk state. *Carbon* 2004; **42**: 2362–2366.

113. Endo M, Kim YA, Hayashi T, Nishimura K, Matusita T, Miyashita K, Dresselhaus MS. Vapor-grown carbon tubes (VGCFs): Basic properties and their battery applications. *Carbon* 2001; **39**: 1287–1297.

114. Lin Q, Harb JN. Implementation of a thick-film composite Li-ion microcathode using carbon nanotubes as the conductive filler. *J Electrochem Soc* 2004; **151**: A1115–A1119.

115. Sheem KY, Lee YH, Lim HS. High-density positive electrodes containing carbon nanotubes for use in Li-ion cells. *J Power Sources* 2006; **158**: 1425–1430.

116. Li X, Kang F, Shen W. Multiwalled carbon nanotubes as a conducting additive in a $LiNi_{0.7}Co_{0.3}O_2$ cathode for rechargeable lithium batteries. *Carbon* 2006; **44**: 1334–1336.

117. Sotowa C, Origi G, Takeuchi M, Nishimura U, Takeuchi K, Jang IJ, Kim YJ, Hayashi T, Kim YA, Endo M, Dresselhaus MS. The reinforcing effect of combined carbon nanotubes and acetylene blacks on the cathode electrode of lithium ion batteries. *ChemSuschem* 2008; **1**: 911–915.

118. Nakamura K, Shiomi M, Takahashi K, Tsubota M. Failure modes of valve-regulated lead/acid batteries. *J Power Sources* 1996; **59**: 153–157.

119. Shiomi M, Funato T, Nakamura K, Takahashi K, Tsubota M. Effects of carbon in negative plates on cycle-life performance of valve-regulated lead/acid batteries. *J Power Sources* 1997; **64**: 147–152.

120. Valenciano J, Sanchez A, Trinidad F, Hollenkamp AF. Graphite and fiberglass additives for improving high-rate partial-state-of-charge cycle life of valve-regulated lead-acid batteries. *J Power Sources* 2006; **158**: 851–863.

121. Hojo E, Yamashita J, Kishimoto K, Nakashima H, Kasai Y. Improved valve-regulated lead–acid batteries with carbon whisker. YUASA-JIHO. *Tech Rev* 1992; **72**: 23–28.

122. Lv J, Tu JP, Zhang WK, Wu JB, Wu HM, Zhang B. Effects of carbon nanotubes on the high-rate discharge properties of nickel/metal hydride batteries. *J Power Sources* 2004; **132**: 282–287.
123. Wu JB, Tu JP, Yu Z, Zhang XB. Electrochemical investigation of carbon nanotubes as additives in positive electrodes of Ni/MH batteries. *J Electrochem Soc* 2006; **153**: A1847–A1851.
124. Endo M, Shikata M, Touhara H, Kadono K. Li battery with fluorinated vapor-grown carbon fibers as cathode electrode. *Trans IEE of Japan* 1985; **105-A**: 27–35.
125. Endo M, Nakamura J, Touhara H, Morimoto S. Li primary cell with fibrous graphitic oxide as cathode electrode. *Trans IEE Japan* 1988; **108-A**: 82–88.
126. Touhara H, Fujimoto H, Kadono K, Watanabe N, Endo M. Electrochemical characteristics of fluorine intercalated graphite fiber-lithium cells. *Electrochim Acta* 1987; **32**: 293–298.
127. Nobuatsu W, Rika H, Tsuyoshi N, Hidekazu T, Kazuo U. Solvents effects on electrochemical characteristics of graphite fluoride-lithium batteries, *Electrochim Acta* 1982; **27**: 1615–1619.
128. Winter M, Besenhard JO. Electrochemical lithiation of tin and tin-based intermetallics and composites. *Electrochim Acta* 1999; **45**: 31–50.
129. Yang J, Takeda Y, Imanish N, Yamamoto O. Ultrafine Sn and $SnSb_{0.14}$ powders for lithium storage matrices in lithium ion batteries. *J Electrochem Soc* 1999; **146**: 4009–4013.
130. Kepler KD, Vaughey JT, Thackeray MM. $LixCu_6Sn_5$ ($0<x<13$): An intermetallic insertion electrode for rechargeable lithium batteries. *Electrochem Solid-State Lett* 1999; **2**: 307–309.
131. Alcantara R, Fernandez-Madrigal FJ, Lavela P, Tirado JL, Jumas JC, Olivier-Fourcade J. Electrochemical reaction of lithium with the $CoSb_3$ Skutterudite. *J Mater Chem* 1999; **9**: 2517–2521.
132. Mao O, Dunlap RA, Dahn JR. Mechanically alloyed Sn-Fe (-C) powders as anode materials for Li-ion batteries. I. The Sn_2Fe-C system. *J Electrochem Soc* 1999; **146**: 405–413.
133. Gao B, Sinha S, Fleming L, Zhou O. Alloy formation in nanostructured silicon. *Adv Mater* 2001; **13**: 816–819.
134. Green M, Fielder E, Scrosati B, Wachtler M, Moreno JS. Structured silicon anodes for lithium battery applications. *Electrochem Solid-State Lett* 2003; **6**: A75–A79.
135. Graetz J, Ahn CC, Yazami R, Fultz B. Highly reversible lithium storage in nanostructured silicon. *Electrochem Solid-State Lett* 2003; **6**: A194–A197.
136. Li H, Huang X, Chen L, Wu Z, Liang Y. A high capacity nano-Si composite anode material for lithium rechargeable batteries. *Electrochem Solid-State Lett* 1999; **2**: 547–549.
137. Zhou GW, Li H, Sun HP, Yu DP, Wang YQ, Huang XJ, Chen LQ, Zhang Z. Controlled Li doping of Si nanowires by electrochemical insertion method. *Appl Phys Lett* 1999; **75**: 2447–2449.
138. Holzapfel M, Buqa H, Scheifele W, Novak P, Petrat FM. A new type of nano-sized silicon/carbon composite electrode for reversible lithium insertion. *Chem Commun* 2005; **12**: 1566–1568.
139. Holzapfel M, Buqa H, Krumeich F, Novak P, Petrat FM, Veit C. Chemical vapor deposited silicon/graphite compound materials as negative electrode for lithium-ion batteries. *Electrochem Solid-State Lett* 2005; **8**: A516–A520.
140. Chen L, Wang KE, Xie X, Xie J. Enhancing electrochemical performance of silicon film anode by vinylene carbonate electrolyte additive. *Electrochem Solid-State Lett* 2006; **9**: A512–A515.
141. Baranchugov V, Markevich E, Pollak E, Salitra G, Aurbach D. Amorphous silicon thin films as a high capacity anodes for Li-ion batteries in ionic liquid electrolytes. *Electrochem Commun* 2007; **9**: 796–800.
142. Zhang XN, Huang PX, Li GR, Yan TY, Pan GL, Gao XP. Si-AB_5 composites as anode materials for lithium ion batteries. *Electrochem Commun* 2007; **9**: 713–717.
143. Yang J, Wang BF, Wang K, Liu Y, Xie JY, Wen ZS. Si/C composites for high capacity lithium storage materials. *Electrochem Solid-State Lett* 2003; **6**: A154–A156.
144. Liu WR, Wu NL, Shieh DT, Wu HC, Yang MH, Korepp C, Besenhard JO, Winter M. Synthesis and characterization of nanoporous NiSi-Si composite anode for lithium-ion batteries. *J Electrochem Soc* 2007; **154**: A97–A102.
145. Umeno T, Fukuda K, Wang H, Dimov N, Iwao T, Yoshio M. Novel anode material for lithium-ion batteries: Carbon-coated silicon prepared by thermal vapor decomposition. *Chem Lett* 2001; **30**: 1186–1187.
146. Ng SH, Wang J, Wexler D, Konstantinov K, Guo ZP, Liu HK. Highly reversible lithium storage in spheroidal carbon-coated silicon nanocomposites as anode for lithium-ion batteries. *Angew Chem* 2006; **118**: 7050–7053.
147. Chew SY, Guo ZP, Wang JZ, Chen J, Munore P, Ng SH, Zhao L, Liu HK. Novel nano-silicon/polypyrrole composites for lithium storage. *Electrochem Commun* 2007; **9**: 941–946.

148. Liu Y, Hanai K, Yang J, Imanishi N, Hirano A, Takeda Y. Silicon/carbon composites as anode materials for Li-ion batteries. *Electrochem Solid-State Lett* 2004; **7**: A369–A372.
149. Uono H, Kim BC, Fuse T, Ue M, Yamaki JI. Optimized structure of silicon/carbon/graphite composites as an anode material for Li-ion batteries. *J Electrochem Soc* 2006; **153**: A1708–A1713.
150. Shi DQ, Tu JP, Yuan YF, Wu HM, Li Y, Zhao XB. Preparation and electrochemical properties of mesoporous Si/ZrO$_2$ nanocomposite film as anode material for lithium ion battery. *Electrochem Commun* 2006; **8**: 1610–1614.
151. Saint J, Morcrette M, Larcher D, Laffont L, Beattie S, Peres JP, Talaga D, Couzi M, Tarascon JM. Towards a fundamental understanding of the improved electrochemical performances of silicon-carbon composites. *Adv Funct Mater* 2007; **9**: 1765–1774.
152. Boukamp BA, Lesh GC, Huggins RA. All-solid lithium electrodes with mixed-conductor matrix. *J Electrochem Soc* 1981; **128**: 725–729.
153. Anani A, Huggins RA. Multinary alloy electrodes for solid state batteries; I. A phase diagram approach for the selection and storage properties determination of candidate electrode materials. *J Power Sources* 1992; **38**: 351–362.
154. Wen CJ, Huggins RA. Chemical diffusion in intermediate phase in the lithium-silicon system. *J Solid State Chem* 1981; **37**: 271–278.
155. Weydanz WJ, Wohlfahrt-Mehrens M, Huggins RA. A room temperature study of the binary lithium-silicon and the ternary lithium-chromium-silicon system for use in rechargeable lithium batteries. *J Power Sources* 1999; **81–82**: 237–242.
156. Wang CS, Wu GT, Zhang XB, Qi ZF, Li WZ. Lithium insertion in carbon-silicon composite materials produced by mechanical milling. *J Electrochem Soc* 1998; **145**: 2751–2758.
157. Wilson AM, Dahn JR. Lithium insertion in carbon containing nanodispersed silicon. *J Electrochem Soc* 1995; **142**: 326–332.
158. Xing W, Wilson AM, Zank G, Dahn JR. Pyrolysed pitch-polysilane blends for use as anode materials in lithium ion batteries. *Solid State Ionics* 1997; **93**: 239–244.
159. Wilson AM, Zang G, Eguchi K, Xing W, Dahn JR. Pyrolysed silicon-containing polymers as high capacity anodes for lithium-ion batteries. *J Power Sources* 1997; **68**: 195–200.
160. Wilson AM, Xing W, Zank G, Yates B, Dahn JR. Pyrolysed pitch-polysilane blends for use as anode materials in lithium ion batteries II: The effect of oxygen. *Solid States Ionics* 1997; **100**: 259–266.
161. Larcher D, Mudalige C, George AE, Porter V, Gharghouri M, Dahn JR. Si-containing disordered carbons prepared by pyrolysis of pitch/polysilane blends: effect of oxygen and sulfur. *Solid State Ionics* 1999; **122**: 71–83.
162. Wilson AM, Way BM, Dahn JR, van Buuren T. Nanodispersed silicon in pregraphitic carbons. *J Appl Phys* 1995; **77**: 2363–2369.
163. Yang J, Winter M, Besenhard JO. Small particle size multiphase Li-alloy anodes for lithium-ionbatteries. *Soid State Ionics* 1996; **90**: 281–287
164. Wilson AM, Reimers JN, Fuller EW, Dahn JR. Lithium insertion in pyrolyzed siloxane polymers. *Solid State Ionics* 1994; **74**: 249–254.
165. Xing W, Wilson AM, Eguchi K, Zank G, Dahn JR. Pyrolyzed polysiloxanes for use as anode materials in lithium-ion batteries. *J Electrochem Soc* 1997; **144**: 2410–2416.
166. Shin HC, Corno JA, Gole JL, Liu M. Porous silicon negative electrodes for rechargeable lithium batteries. *J Power Sources* 2005; **139**: 314–320.
167. Courtney IA, Dahn JR. Electrochemical and in situ X-ray diffraction studies of the reaction of lithium with Tin oxide composites. *J Electrochem Soc* 1997; **144**: 2045–2052.
168. Courtney IA, Dahn JR. Key factors controlling the reversibility of the reaction of lithium with SnO$_2$ and Sn$_2$BPO$_6$ Glass. *J Electrochem Soc* 1997; **144**: 2943–2948.
169. Tamura N, Fujimoto M, Kamino M, Fujitani S. Mechanical stability of Sn–Co alloy anodes for lithium secondary batteries. *Electrochim Acta* 2004; **49**: 1949–1956.
170. Wachtler M, Winter M, Besenhard JO. Anodic materials for rechargeable Li-batteries. *J Power Sources* 2002; **105**: 151–160.
171. Mukaibo H, Sumi T, Yokoshima T, Momma T, Osaka T. Electrodeposited Sn-Ni alloy film as a high capacity anode material for lithium-ion secondary batteries. *Electrochem Solid-State Lett* 2003; **6**: A218–A220.
172. Winter M, Besenhard JO, Spahr ME, Novák P. Insertion electrode materials for rechargeable lithium batteries. *Adv Mater* 1998; **10**: 725–763.

173. Besenhard JO, Yang J, Winter M. Will advanced lithium-alloy anodes have a chance in lithium-ion batteries? *J Power Sources* 1997; **68**: 87–90.
174. Ke FS, Huang L, Jiang H, Wei H, Yang F, Sun S. Fabrication and properties of three-dimensional macroporous Sn-Ni alloy electrodes of high preferential (110) orientation for lithium ion batteries. *Electrochem Commun* 2007; **9**: 228–232.
175. Ke W, He X, Wang L, Ren J, Jiang C, Wan C. Preparation of Cu_6Sn_5-encapsulated carbon microsphere anode materials for Li-ion batteries by carbothermal reduction of oxides. *J Electrochem Soc* 2006; **153**: A1859–A1862.
176. Idota Y, Kubota T, Matsufuji A, Maekawa Y, Miyasaka T. Tin-based amorphous oxide: A high-capacity lithium-ion-storage material. *Science* 1997; **276**: 1395–1397.
177. Huggins RA. Lithium alloy negative electrodes. *J Power Sources* 1999; **81–82**: 13–19.
178. Feng J, Hennig RG, Ashcroft NW, Hoffmann R. Emergent reduction of electronic state dimensionality in dense ordered Li-Be alloys. *Nature* 2008; **451**: 445–448.
179. Dey AN. Electrochemical alloying of lithium in organic electrolyte. *J Electrochem Soc* 1971; **118**: 1547–1549.
180. Huggins RA. Materials science principle related to alloys of potential use in rechargeable lithium cells. *J Power Sources* 1989; **26**: 109–120.
181. Wang J, Raistrick ID, Huggins RA. Behavior of some binary lithium alloys as negative electrodes in organic solvent-based electrolyte. *J Electrochem Soc* 1986; **133**: 457–460.
182. Besenhard JO, Fritz HP. Reversibles elektrochemisches legieren von metallen der V. hauptgruppe in organischen Li^+-Lösungen. *Electrochim Acta* 1975; **20**: 513–517.
183. Anani A, Crouch-Baker S, Huggins RA. Kinetic and thermodynamic parameters of several binary lithium alloy negative electrode materials at ambient temperature. *J Electrochem Soc* 1987; **134**: 3098–3102.
184. Sazhin SV, Gorodyskii AV, Khimchenko YM. Lithium rechargeability on different substrates. *J Power Sources* 1994; **47**: 57–62.
185. Shembel EM, Maksyuta IM, Neduzhko LI, Belosokhov AI, Naumenko AF, Rozhkov VV. Influence of the composition of lithium-based alloys, non-aqueous electrolytes and cycling conditions on the anode properties. *J Power Sources* 1995; **54**: 416–420.
186. Huggins RA. Polyphase alloys as rechargeable electrodes in advanced battery systems. *J Power Sources* 1988; **22**: 341–350.
187. Weppner W, Huggins RA. Determination of the kinetic parameters of mixed-conducting electrodes and application to the system Li_3Sb. *J Electrochem Soc* 1977; **124**: 1569–1578.
188. Morita M, Matsuda Y. Effects of alloying substrates on the characteristics of the lithium negative electrode. *J Power Sources* 1989; **26**: 573–578.
189. Nimon ES, Churikov AV. Electrochemical behaviour of Li–Sn, Li–Cd and Li–Sn–Cd alloys in propylene carbonate solution. *Electrochim Acta* 1996; **41**: 1455–1464.
190. Churikov AV, Nimon ES, Lvov AL. Impedance of Li–Sn, Li–Cd and Li–Sn–Cd alloys in propylene carbonate solution. *Electrochim Acta* 1997; **42**: 179–189.
191. Rao BML, Francis RW, Christopher HA. Lithium-aluminum electrode. *J Electrochem Soc* 1997; **124**: 1490–1492.
192. Besenhard JO, Cycling behaviour and corrosion of Li–Al electrodes in organic electrolytes. *J Electroanal Chem* 1978; **94**: 77–81.
193. Garreau M, Thevenin J, Fekir M. On the processes responsible for the degradation of the aluminum-lithium electrode used as anode material in lithium aprotic electrolyte batteries. *J Power Sources* 1983; **9**: 235–238.
194. Moshtev RA, Zlatilova P, Puresheva B, Manev V, Kozawa A. Cycling performance of the Li_xAl anode prepared by the compression method. *J Power Sources* 1994; **51**: 409–423.
195. Morita M, Okada Y, Matsuda Y. Lithium cycling efficiency on the aluminum substrate in blends sulfone-ether systems. *J Electrochem Soc* 1987; **134**: 2665–2669.
196. Biallozor S, Lieder M. Study on the electrochemical formation of Al–Li alloys in γ-butyrolactone electrolyte. *J Electrochem Soc* 1993; **140**: 2537–2540.
197. Epelboin I, Froment M, Garreau M, Thevenin J, Warin D. Behavior of secondary lithium and aluminum-lithium electrodes in propylene carbonate. *J Electrochem Soc* 1980; **127**: 2100–2104.
198. Jow TR, Liang CC. Lithium–aluminum electrodes at ambient temperatures. *J Electrochem Soc* 1982; **129**: 1429–1434.
199. Wen CJ, Boukamp BA, Huggins RA, Weppner W. Thermodynamic and mass transport properties of LiAl. *J Electrochem Soc* 1979; **126**: 2258–2266.

200. Frazer EJ. Electrochemical formation of lithium–aluminum alloys in propylene carbonate electrolytes. *J Electroanal Chem* 1981; **121**: 329–339.
201. Zlatilova P, Balkanov I, Geronov Y. Thin foil lithium-aluminum electrode. The effect of thermal treatment on its electrochemical behaviour in nonaqueous media. *J Power Sources* 1988; **24**: 71–79.
202. Baranski AS, Fawcett WR. The formation of lithium–aluminum alloys at an aluminum electrode in propylene carbonate. *J Electrochem Soc* 1982; **129**: 901–907.
203. Geronov Y, Zlatilova P, Moshtev RV. The secondary lithium–aluminum electrode at room temperature: I. Cycling in LiClO$_4$–propylene carbonate solutions. *J Power Sources* 1984; **12**: 145–153.
204. Geronov Y, Zlatilova P, Staikov G. The secondary lithium–aluminum electrode at room temperature: II. Kinetics of the electrochemical formation of the lithium–aluminum alloy. *J Power Sources* 1984; **12**: 155–165.
205. Crosnier O, Brousse T, Devaux X, Fragnaud P, Schleich DM. New anode systems for lithium ion cells. *J Power Sources* 2001; **94**: 169–174.
206. Ahn HJ, Kim YS, Park KW, Seong TY. Use of Sn–Si nanocomposite electrodes for Li rechargeable batteries. *Chem Commun* 2005; **43**: 43–45.
207. Kang YM, Park MS, Lee JY, Liu HK. Si-Cu/carbon composites with a core-shell structure for Li-ion secondary battery. *Carbon* 2007; **45**: 1928–1933.
208. Hosono E, Fujihara S, Honma I, Ichihara M, Zhou H. Fabrication of nano/micro hierarchical Fe$_2$O$_3$/Ni micrometer-wire structure and characteristics for high rate Li rechargeable battery. *J Electrochem Soc* 2006; **153**: A1273–A1278.
209. Moriguchi I, Hidaka R, Yamada H, Kudo T, Murakami H, Nakashima N. A mesoporous nanocomposite of TiO$_2$ and carbon nanotubes as a high-rate Li-intercalation electrode material. *Adv Mater* 2006; **18**: 69–73.
210. Scrosati B. Challenge of portable power. *Nature* 1995; **373**: 557–558.
211. Nikkei Electronics Cover Story, 2004; **888**: 120 (issued in Dec.).
212. Dover BT, Kamienski CW, Morrison RC, Currin RT. Process for producing alkali metal dispersions. US Patent, 5567474, 1996.
213. Dover BT, Kamienski CW, Morrison RC, Currin RT, Schwindeman JA. Alkali metal dispersions. US Patent, 5776369, 1998.
214. Dover BT, Kamienski CW, Morrison RC, Currin RT, Schwindeman JA. Organoalkali compounds and their preparation. US Patent, 5976403, 1999.
215. Pokhodenko VD, Koshechko VG, Krylov VA. New electrolytes and polymer cathode materials for lithium batteries. *J Power Sources* 1993; **45**: 1–5.
216. Sharma N, Shaju KM, Subba RGV, Chowdari BVR. Sol–gel derived nano-crystalline CaSnO$_3$ as high capacity anode material for Li-ion batteries. *Electrochem Commun* 2002; **4**: 947–952.
217. Kuribayashi I, Yokoyama M, Yamashita M. Battery characteristics with various carbonaceous materials. *J Power Sources* 1995; **54**: 1–5.
218. Leroux F, Koene BE, Lazar LF. Electrochemical lithium intercalation into a polyaniline/V$_2$O$_5$ nanocomposite. *J Electrochem Soc* 1996; **143**: L181–L183.
219. Kuwabata S, Kishimoto A, Tanaka T, Yoneyama H. Electrochemical synthesis of composite films of manganese dioxide and polypyrrole and their properties as an active material in lithium secondary batteries. *J Electrochem Soc* 1994; **141**: 10–15.
220. Gemeay AH, Nishiya H, Kuwabata S, Yoneyama H. Chemical preparation of manganese dioxide/polypyrrole composites and their use as cathode active materials for rechargeable lithium batteries. *J Electrochem Soc* 1995; **142**: 4190–4195.

Index

Milton Keynes UK
Ingram Content Group UK Ltd.
UKHW052025071024
449327UK00027B/2432